Principles of Computational Modelling in Neuroscience

Second Edition

Taking a step-by-step approach to modelling neurons and neural circuitry, this textbook teaches students how to use computational techniques to understand the nervous system at all levels, using case studies throughout to illustrate fundamental principles. Starting with a simple model of a neuron, the authors gradually introduce neuronal morphology, synapses, ion channels and intracellular signalling. This fully updated new edition contains additional examples and case studies on specific modelling techniques, suggestions on different ways to use this book and new chapters covering plasticity, modelling extracellular influences on brain circuits, modelling experimental measurement processes and choosing appropriate model structures and their parameters. The online resources offer exercises and simulation code that recreate many of the book's figures, allowing students to practise as they learn. Requiring an elementary background in neuroscience and high-school mathematics, this is an ideal resource for a course on computational neuroscience.

David Sterratt is Lecturer and Deputy Director of Learning and Teaching in the Institute for Adaptive and Neural Computation, School of Informatics, at the University of Edinburgh. He developed material for this book whilst teaching computational neuroscience to informatics, neuroscience and neuroinformatics master's students. He has developed and maintains several scientific software packages.

Bruce Graham is Emeritus Professor in Computing Science in the Faculty of Natural Sciences at the University of Stirling. He has been a researcher in computational neuroscience for more than 30 years and has served as a board member of the Organisation of Computational Neurosciences.

Andrew Gillies is Chief Technology Officer of Grid Software at GE Vernova. He has been actively involved in computational neuroscience research and his simulation model of the subthalamic nucleus projection neuron is recognised as a standard. He has taught neuroscience modelling at master's and PhD level.

Gaute Einevoll is Professor of Physics at the Norwegian University of Life Sciences and the University of Oslo, working on modelling of nerve cells, networks of nerve cells, brain tissue and brain signals, and on the development of neuroinformatics software tools, including LFPy.

David Willshaw is Emeritus Professor of Computational Neurobiology in the Institute for Adaptive and Neural Computation at the University of Edinburgh, where he led their innovative doctoral training programme in neuroinformatics and computational neuroscience. With over 40 years' research experience, he has received several awards, including, most recently, the Braitenberg Award in Computational Neuroscience.

'This new edition builds superbly on its predecessor. Expository excellence and beautifully clear figures remain, whilst extra material has been added throughout. New chapters cover critical topics such as modelling the way that neural signals are measured, and the details of model optimisation and selection. Its impressive combination of depth and breadth makes the text perfect source material for a wide variety of courses.'

Peter Dayan, Managing Director, Max-Planck Institute for Biological Cybernetics, Tuebingen

'*Principles of Computational Modelling in Neuroscience* has long been my choice for my students from various backgrounds, but now it is even better! With the addition of comprehensive coverage for modelling extracellular activity, neural plasticity and experimental stimulation and measurements, this text offers everything that my students need for a foundation in computational neuroscience. The new chapter on "Model Selection and Optimisation" makes a difficult topic easy to understand. *Principles of Computational Modelling in Neuroscience* remains the best choice for a text that is rigorous but accessible to students from a variety of backgrounds.'

Sharon Crook, Arizona State University

'I am thrilled to endorse the second edition of *Principles of Computational Modelling in Neuroscience*. This comprehensive and well-written text is an engaging resource on the fundamentals of biophysical modelling and computational neuroscience. The range of topics and clarity of exposition provide a spectacular display of the power of computational techniques to provide insight into the mysteries of central nervous system function across its diverse hierarchical levels. This second edition contains many additional examples and new chapters on synaptic plasticity and learning, modelling of experimental measurements, as well as model selection and optimisation. The authors have a remarkable ability to explain complex concepts in a clear and accessible manner, making this book an ideal choice for both students and researchers. The gradually increasing sophistication of each chapter makes it highly suited as a textbook for an introductory undergraduate course focusing on computational aspects of cell and systems neuroscience.'

Greg Conradi Smith, William & Mary, Virginia, and author of *Cellular Biophysics and Modeling* (Cambridge, 2019)

'A valuable resource for both students and practitioners of computational modelling in neuroscience. The authors provide a step-by-step guide to modelling, from simple neurons to complex brain tissue. Each model is carefully explained and illustrated, put into context and critiqued. A definitive text for anyone wanting to develop their own models of neural circuitry!'

Rosemary Fricker, Keele University Medical School, and Wolfson College, University of Cambridge

'This textbook is useful for engineers, mathematicians, biologists and computer scientists. It provides a deep understanding of neuron computational modelling. The approach used, from ion channels to the new plasticity chapter, provides the student with a full picture of neural network behaviour. Furthermore, this new edition includes some cases of studies where the concepts are applied to real experimental data.'

Fernando Perez-Peña, University of Cadiz

'We have been using the book by Sterratt et al. for many years on our courses in theoretical neuroscience, to teach the basics of biophysics. The book does an excellent job of providing the basics in a highly understandable way for beginners in the field, with lots of practical examples and additional background material for the implementation of models. The new edition of the book is even more useful since it integrates a lot of new material: nonlinear dynamics and bifurcations, neural field models, the modelling of measurement methods, and optimisation. In addition, programming examples have been added, making the book even more useful for teaching.'

Martin Giese, University of Tuebingen

'This book is an invaluable resource for computational neuroscientists, particularly for students starting their doctoral research. The comprehensive coverage, clear prose and extensive examples make it easy for a newcomer with only basic background knowledge to get to grips with a wide variety of complex computational modelling topics as a preparation for starting their own investigations.'

Abigail Morrison, RWTH Aachen University

Principles of Computational Modelling in Neuroscience

Second Edition

David Sterratt
University of Edinburgh

Bruce Graham
University of Stirling

Andrew Gillies
GE Vernova

Gaute Einevoll
Norwegian University of Life Sciences (NMBU)

David Willshaw
University of Edinburgh

CAMBRIDGE
UNIVERSITY PRESS

CAMBRIDGE
UNIVERSITY PRESS

Shaftesbury Road, Cambridge CB2 8EA, United Kingdom

One Liberty Plaza, 20th Floor, New York, NY 10006, USA

477 Williamstown Road, Port Melbourne, VIC 3207, Australia

314–321, 3rd Floor, Plot 3, Splendor Forum, Jasola District Centre,
New Delhi – 110025, India

103 Penang Road, #05–06/07, Visioncrest Commercial, Singapore 238467

Cambridge University Press is part of Cambridge University Press & Assessment,
a department of the University of Cambridge.

We share the University's mission to contribute to society through the pursuit of
education, learning and research at the highest international levels of excellence.

www.cambridge.org
Information on this title: www.cambridge.org/highereducation/isbn/
9781108483148

DOI: 10.1017/9781108672955

First published 2011
Second edition published 2024

A catalogue record for this publication is available from the British Library

Library of Congress Cataloging-in-Publication Data
Names: Sterratt, David, 1973– author.
Title: Principles of computational modelling in neuroscience / David
Sterratt, University of Edinburgh, Bruce Graham, University of Stirling,
Andrew Gillies, Psymetrix Limited, Gaute Einevoll, Norwegian University
of Life Sciences (NMBU), David Willshaw, University of Edinburgh.
Description: Second edition. | Cambridge, United Kingdom ; New York, NY :
Cambridge University Press, [2024] | Includes bibliographical
references and index.
Identifiers: LCCN 2023014949 | ISBN 9781108483148 (hardback) |
ISBN 9781108672955 (ebook)
Subjects: LCSH: Computational neuroscience.
Classification: LCC QP357.5 .P75 2024 | DDC 612.8/233–dc23/eng/20230712
LC record available at https://lccn.loc.gov/2023014949

ISBN 978-1-108-48314-8 Hardback
ISBN 978-1-108-71642-0 Paperback

Additional resources for this publication at www.cambridge.org/Sterratt2e

Contents

Preface

To understand the nervous system of even the simplest of animals requires an understanding of the nervous system at many different levels, over a wide range of both spatial and temporal scales. We need to know at least the properties of the neuron itself and of its specialist structures such as synapses, and how neurons become connected together and what the properties of networks of neurons are.

The complexity of nervous systems makes it very difficult to theorise cogently about how such systems are put together and how they function. To aid our thought processes, we can represent our theory as a computational model, in the form of a set of mathematical equations. The variables of the equations represent specific neurobiological quantities, such as the rate at which impulses are propagated along an axon or the frequency of opening of a specific type of ion channel. The equations themselves represent how these quantities interact according to the theory being expressed in the model. Solving these equations by analytical or simulation techniques enables us to show the behaviour of the model under the given circumstances and thus addresses the questions that the theory was designed to answer. Models of this type can be used as explanatory or predictive tools.

This field of research is known by a number of largely synonymous names, principally, **theoretical neuroscience**, or **computational neurobiology**. Most attempts to analyse computational models of the nervous system involve using the powerful computers now available to find numerical solutions to the complex sets of equations needed to construct an appropriate model.

To develop a computational model in neuroscience, the researcher has to decide how to construct and apply a model that will link the neurobiological reality with a more abstract formulation that is analytical or computationally tractable. Guided by the neurobiology, decisions have to be taken about the level at which the model should be constructed, the nature and properties of the elements in the model and their number, and the ways in which these elements interact. Having done all this, the performance of the model has to be assessed in the context of the scientific question being addressed.

This book describes how to construct computational models of this type. It arose out of our experiences in teaching master's level courses to students with backgrounds from the physical, mathematical and computer sciences, as well as the biological sciences. In addition, we have given short computational modelling courses to neurobiologists and to people trained in the quantitative sciences, at all levels from postgraduate to faculty members. Our students wanted to know the principles involved in designing computational models of the nervous system and its components, to enable them to develop their own models. They also wanted to know the mathematical basis in as far as it describes neurobiological processes. They wanted to have more than the basic recipes for running the simulation programs which now exist for modelling the nervous system at the various different levels.

This book is intended for anyone interested in how to design and use computational models of the nervous system. It is aimed at people with a bachelor's degree and is suitable for study at master's or PhD level or for people at all levels of research in academic, clinical and commercial settings. We have assumed a knowledge of basic concepts such as neurons, axons and synapses. The mathematics given in the book is necessary to understand the concepts introduced in mathematical terms. Therefore, we have assumed some knowledge of mathematics, principally of functions such as logarithms and exponentials, of vectors and of the techniques of differentiation and integration. Some of the later chapters contain more specialised mathematics which have been put in boxes and smaller technical points are given in the margins. For non-specialists, we have given verbal descriptions of the mathematical concepts we use.

Many of the models we discuss exist as open-source simulation packages. In many cases, the original code is available.

Our intention is that several different types of people will be attracted to read this book, which will include:

Experimental or clinical neuroscientists. We hope that they will become interested in the computational approach to neuroscience.

Teachers of computational neuroscience. This book can be used as the basis of a hands-on course.

Interested students from the physical sciences. We hope that the book will motivate graduate students, postdoctoral researchers or faculty members in other fields of the physical, mathematical or information sciences to enter the field of computational neuroscience.

Changes Made in the Second Edition

In this substantially expanded edition, we have introduced new material to the topics already discussed to reflect contemporary work published since the first edition. Further examples and case studies clearly demonstrate specific modelling techniques. Exercises and computer simulation code are available electronically. Whilst the overall structure of the book remains unchanged, we have added four new chapters. To emphasise the importance of modelling plasticity in neural networks, Chapter 11 treats in detail the modelling of plasticity for learning and memory and homeostatic plasticity in adult systems. The discussion of the important topic of how to choose an appropriate model structure and to set its parameter values has been expanded and is the subject of Chapter 14. Finally, two chapters introduce topics which we believe should belong in a computational neuroscience textbook: modelling extracellular influences on brain circuits (Chapter 10) and modelling the experimental measurement process (Chapter 13) so that model output can be matched more closely to experimental data.

Acknowledgements

There are many people who have inspired and helped us throughout the writing of this book. We are particularly grateful for the critical comments and suggestions from Fiona Williams, Jeff Wickens, Gordon Arbuthnott, Mark van Rossum, Matt Nolan, Matthias Hennig, Irina Erchova, Stephen Eglen and Ewa Henderson. Many thanks to David Boas, Haavard Grunnvaag, Espen Hagen, Geir Halnes, Risto Ilmoniemi, Torbjørn V. Ness, Klas H. Pettersen, Jurgis Pods, Jan-Eirik Welle Skaar, Andreas Solbrå and Marte Julie Sætra, who helped with text and figures in the second edition. We are indebted to Meg Waraczynski, Peter Hebden, Richard Faville and Geoff Goodhill, who alerted us to errors in the first edition, which we have rectified in the second edition. We are very grateful to our publishers at Cambridge University Press, particularly Gavin Swanson, with whom we discussed the initial project, and then to Martin Griffiths, Katrina Halliday, Megan Keirnan, Susie Francis, Emily Watton and Rachel Norridge who have been responsible successively for the two editions. Finally, we appreciate the great help, support and forbearance of our family members.

Abbreviations

ABC	approximate Bayesian computation
AC	alternating current
ADP	adenosine diphosphate
AHP	afterhyperpolarisation
AI	artificial intelligence; asynchronous irregular
AIC	Akaike Information Criterion
AMPA	α-amino-3-hydroxy-5-methyl-4-isoxalone propionic acid
AMPAR	AMPA receptor
ANN	artificial neural network
AP	action potential
AsI	asynchronous irregular
ATP	adenosine triphosphate
BAC	back-propagating action potential-activated calcium
BAPTA	bis(aminophenoxy)ethanetetraacetic acid
BCM	Bienenstock–Cooper–Munro
BNGL	BioNetGen Language
BSL	Bayesian synthetic likelihood
CaMKII	calcium/calmodulin-dependent kinase II
cAMP	cyclic adenosine monophosphate
CCD	charge-coupled device
cDNA	cloned DNA
cGMP	cyclic guanosine monophosphate
CHO	Chinese hamster ovary
CICR	calcium-induced calcium release
CME	chemical master equation
CNG	cyclic-nucleotide-gated channel family
CNS	central nervous system
CPG	central pattern generator
cryo-EM	cryo-electron microscopy
CSD	current-source density
CSF	cerebrospinal fluid
CV	coefficient of variation
DAG	diacylglycerol
DBS	deep brain stimulation
DC	direct current
DCM	Dual Constraint Model
dIN	descending interneuron
DNA	deoxyribonucleic acid
DOG	difference-of-Gaussians
EA	evolutionary algorithm
EBA	excess buffer approximation
ECoG	electrocorticography
ECS	extracellular space
eDOG	extended difference-of-Gaussians
EEG	electroencephalography
EGTA	ethylene glycol tetraacetic acid
EP	extracellular potential

EPP	endplate potential
EPSC	excitatory postsynaptic current
ER	endoplasmic reticulum
ES	evolution strategies
fAHP	fast afterhyperpolarisation
FEM	finite element modelling
fMRI	functional magnetic resonance imaging
GA	genetic algorithm
GABA	γ-aminobutyric acid
GHK	Goldman–Hodgkin–Katz
GIF	generalised integrate-and-fire model
GPi	globus pallidus internal segment
GR	geometric ratio
HCN	hyperpolarisation-activated cyclic-nucleotide-gated channel family
HD-MEA	high-density microelectrode array
HEK	human embryonic kidney
HFS	high-frequency stimulation
HH	Hodgkin–Huxley
HVA	high-voltage-activated
ICA	independent components analysis
ICS	intracellular space
iCSD	inverse CSD
IF	integrate-and-fire model neuron
ING	interneuronal network gamma
IP_3	inositol 1,4,5-triphosphate
IPSC	inhibitory postsynaptic current
ISI	interspike interval histogram
ISSA	inhomogeneous SSA
IUPHAR	International Union of Pharmacology
kCSD	kernel CSD
KDE	kernel density estimation
KNP	Kirchhoff–Nernst–Planck
LFP	local field potential
LFS	low-frequency stimulation
LGN	lateral geniculate nucleus
LPA	laminar population analysis
LTD	long-term depression
LTP	long-term potentiation
LVA	low voltage-activated
mAHP	medium afterhyperpolarisation
MAP	microtubule-associated protein; maximum a posteriori
MC	Monte Carlo
MCMC	Markov-Chain Monte Carlo
MD	molecular dynamics
MEA	microelectrode array
MEG	magnetoencephalography
MEPP	miniature endplate potential
mGluR	metabotropic glutamate receptor
MLE	maximum likelihood estimation
MoI	Method of Images

MOO	multi-objective optimisation
MOOSE	Multiscale-Object-Oriented Simulation Environment
MPTP	1-methyl-4-phenyl-1,2,3,6-tetrahydropyridine
MRI	magnetic resonance imaging
mRNA	messenger RNA
MSE	mean squared error
MSO	multi-stage optimisation; medial superior olive
MST	minimum spanning tree
NADPH-d	nicotinamide adenine dinucleotide phosphate-diaphorase
NMDA	N-methyl-D-aspartate
nNOS	neuronal nitric oxide synthase
NO	nitric oxide
ODE	ordinary differential equation
O-LM	oriens-lacunosum moleculare
OPM	optically pumped magnetometer
PC	pyramidal cell
PCA	principal components analysis
PDE	partial differential equation
PET	positron emission tomography
PING	pyramidal–interneuronal network gamma
PIP_2	phosphatidylinositol 4,5-bisphosphate
PIPP	pairwise interaction point process
PKA	protein kinase A
PKC	protein kinase C
PLC	phospholipase C
PMCA	plasma membrane Ca^{2+}-ATPase
PNP	Poisson–Nernst–Planck
PRP	plasticity-related protein
PSC	postsynaptic current
PSD	postsynaptic density
RBA	rapid buffer approximation
RDME	reaction–diffusion master equation
RGC	retinal ganglion cell
RNA	ribonucleic acid
RRVP	readily releasable vesicle pool
RyR	ryanodine receptor
sAHP	slow afterhyperpolarisation
SBML	Systems Biology Markup Language
SDE	stochastic differential equation
SERCA	sarcoplasmic reticulum Ca^{2+}-ATPase
SI	synchronous irregular
SQUID	superconducting quantum interference device
SR	synchronous regular
SRM	spike response model
SSA	Stochastic Simulation Algorithm
STC	synaptic tagging and capture
STDP	spike timing-dependent plasticity
STN	subthalamic nucleus
STP	short-term plasticity
TEA	tetraethylammonium
TES	transcranial electrical stimulation

TMS	transcranial magnetic stimulation
TPC	two-pore channel family
TRP	transient receptor potential channel family
TTX	tetrodotoxin
VGCC	voltage-gated calcium channel
VSD	voltage-sensing domain
VSDI	voltage-sensitive dye imaging

CHAPTER 1

Introduction

1.1 What Is This Book About?

This book is about how to construct and use computational models of specific parts of the nervous system, such as a neuron, a part of a neuron or a network of neurons, as well as their measurable signals. It is designed to be read by people from a wide range of backgrounds from the neurobiological, physical and computational sciences. The word 'model' can mean different things in different disciplines, and even researchers in the same field may disagree on the nuances of its meaning. For example, to biologists, this term can mean 'animal model'; in particle physics, the 'standard model' is a step towards a complete theory of fundamental particles and interactions. We therefore attempt to clarify what we mean by modelling and computational models in the context of neuroscience. Before giving a brief chapter-by-chapter overview of the book, we discuss what might be called the philosophy of modelling: general issues in computational modelling that recur throughout the book.

1.1.1 Theories and Mathematical Models

In our attempts to understand the natural world, we all come up with theories. Theories are possible explanations for how the phenomena under investigation arise, and from theories we can derive predictions about the results of new experiments. If the experimental results disagree with the predictions, the theory can be rejected, and if the results agree, the theory is validated – for the time being. Typically, the theory will contain assumptions which are about the properties of elements or mechanisms which have not yet been quantified, or even observed. In this case, a full test of the theory will also involve trying to find out if the assumptions are correct.

Mendel's Laws of Inheritance form a good example of a theory formulated on the basis of the interactions of elements whose existence was not known at the time. These elements are now known as genes.

In the first instance, a theory is described in words, or perhaps with a diagram. To derive predictions from the theory, we can deploy verbal reasoning and further diagrams. Verbal reasoning and diagrams are crucial tools for theorising. However, as the following example from ecology demonstrates, it can be risky to rely on them alone.

Suppose we want to understand how populations of a species in an ecosystem grow or decline through time. We might assume that the larger the population, the more likely it will grow and therefore the faster it will increase in size. From this theory, we can derive the prediction, as did Malthus (1798), that the population will grow infinitely large, which is incorrect. The reasoning from theory to prediction is correct, but the prediction is wrong and so logic dictates that the theory is wrong. Clearly, in the real world, the resources consumed by members of the species are only replenished at a finite rate. We could add to the theory the stipulation that for large populations, the rate of growth slows down, being limited by finite resources. From this, we can make the reasonable prediction that the population will stabilise at a certain level at which there is zero growth.

We might go on to think about what would happen if there are two species, one of which is a predator and one of which is the prey. Our theory might now state that: (1) the prey population grows in proportion to its size but declines as the predator population grows and eats it; and (2) the predator population grows in proportion to its size and the amount of the prey, but declines in the absence of the prey. From this theory, we would predict that the prey population grows initially. As the prey population grows, the predator population can grow faster. As the predator population grows, this limits the rate at which the prey population can grow. At some point, an equilibrium is reached when both predator and prey sizes are in balance.

We might then wonder whether there is a second possible prediction from the theory. Perhaps the predator population grows so quickly that it is able to make the prey population extinct. Therefore, once the prey has gone, the predator is also doomed to extinction. Now we are faced with the problem that there is one theory, but two possible conclusions; the theory is logically inconsistent.

The problem has arisen for two reasons. Firstly, the theory was not clearly specified to start with. Exactly how does the rate of increase of the predator population depend on its size and the size of the prey population? How fast is the decline of the predator population? Secondly, the theory is now too complex for qualitative verbal reasoning to be able to turn it into a prediction.

The solution to this problem is to specify the theory more precisely, in the language of mathematics. In the equations corresponding to the theory, the relationships between predator and prey are made precisely and unambiguously. The equations can then be solved to produce a prediction. We call a theory that has been specified by sets of equations a **mathematical model**.

It so happens that all three of our verbal theories about population growth have been formalised in mathematical models, as shown in Box 1.1. Each model can be represented as one or more differential equations. To predict the time evolution of a quantity under particular circumstances, the equations of the model need to be solved. In the relatively simple cases of unlimited growth, and limited growth of one species, it is possible to solve these equations analytically to give equations for the solutions. These are shown in Figure 1.1a and Figure 1.1b, and validate the conclusions we came to verbally.

Box 1.1 | Mathematical models

Mathematical models of population growth are classic examples of describing how particular variables in the system under investigation change over space and time according to the given theory.

According to the Malthusian, or exponential, growth model (Malthus, 1798), a population of size $P(t)$ grows in direct proportion to this size. This is expressed by an **ordinary differential equation** that describes the rate of change of P:

$$dP/dt = P/\tau,$$

where the proportionality constant is expressed in terms of the time constant τ, which determines how quickly the population grows. Integration of this equation with respect to time shows that at time t, a population with initial size P_0 will have size $P(t)$, given as:

$$P(t) = P_0 \exp(t/\tau).$$

This model is unrealistic as it predicts unlimited growth (Figure 1.1a). A more complex model, commonly used in ecology that does not have this defect (Verhulst, 1845), is one where the population growth rate dP/dt depends on the Verhulst, or **logistic function** of the population P:

$$dP/dt = P(1 - P/K)/\tau.$$

Here K is the maximum allowable size of the population. The solution to this equation (Figure 1.1b) is:

$$P(t) = \frac{K P_0 \exp(t/\tau)}{K + P_0(\exp(t/\tau) - 1)}.$$

A more complicated situation is where there are two types of species and one is a predator of the other. For a prey population with size $N(t)$ and a predator population with size $P(t)$, it is assumed that: (1) the prey population grows in a Malthusian fashion and declines in proportion to the rate at which predator and prey meet (assumed to be the product of the two population sizes NP); and (2) conversely, there is an increase in predator size in proportion to NP and an exponential decline in the absence of prey. This gives the following mathematical model:

$$dN/dt = N(a - bP), \qquad dP/dt = P(cN - d).$$

The parameters a, b, c and d are constants. As shown in Figure 1.1c, these equations have periodic solutions in time, depending on the values of these parameters. The two population sizes are out of phase with each other, with large prey populations co-occurring with small predator populations, and vice versa. In this model, proposed independently by Lotka (1925) and by Volterra (1926), predation is the only factor that limits growth of the prey population, but the equations can be modified to incorporate other factors. These types of models are used widely in the mathematical modelling of competitive systems found in, for example, ecology and epidemiology.

As can be seen in these three examples, even the simplest models contain parameters whose values are required if the model is to be understood; the number of these parameters can be large and the problem of how to specify their values has to be addressed.

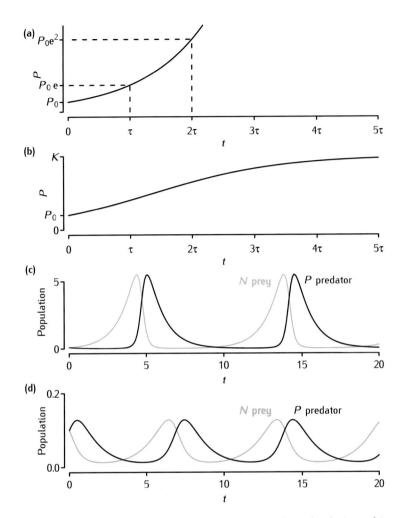

Fig 1.1 Behaviour of the mathematical models described in Box 1.1. **(a)** Malthusian, or exponential, growth: with increasing time t, the population size P grows increasingly rapidly and without bounds. **(b)** Logistic growth: the population increases with time, up to a maximum value of K. **(c)** Behaviour of the Lotka–Volterra model of predator–prey interactions, with parameters $a = b = c = d = 1$. The prey population is shown by the blue line, and the predator population by the black line. Since the predator population is dependent on the supply of prey, the predator population size always lags behind the prey size, in a repeating fashion. **(d)** Behaviour of the Lotka–Volterra model with a second set of parameters: $a = 1$, $b = 20$, $c = 20$ and $d = 1$.

In the case of the predator and prey model, analytical solution of its differential equations is not possible and so the equations have to be solved using numerical integration. In the past, this would have been carried out laboriously by hand and brain, but nowadays, the computer is used. The resulting sizes of predator and prey populations over time are shown in Figure 1.1c and Figure 1.1d. It turns out that neither of our guesses was correct. Instead of both species surviving in equilibrium or going extinct, the predator and prey populations oscillate over time. At the start of each cycle, the prey population grows. After a lag, the predator population starts to grow, due to the abundance of prey. This causes a sharp decrease in prey, which almost causes its extinction, but not quite. Thereafter, the predator population declines and the cycle repeats. In fact, this behaviour is observed approximately in some systems of predator and prey in ecosystems (Edelstein-Keshet, 1988).

In the restatement of the model's behaviour in words, it might now seem obvious that oscillations would be predicted by the model. However, the step of putting the theory into equations was required in order to reach this understanding. We might disagree with the assumptions encoded in the mathematical model. However, this type of disagreement is better than the inconsistencies between predictions from a verbal theory.

A mathematical model may have an analytical solution that allows exact calculation of quantities, or may require a numerical solution that approximates the true, unobtainable values. Often this requires computer simulation – hence a **computational model**.

The process of modelling described in this book almost always ends with calculating the numerical values of quantities such as neuronal membrane potentials. The elements of the model can represent elements of the nervous system, such as nerve cells or synapses, and therefore, the model can be regarded as a direct representation of the neurobiology. In this sense, it can be thought of as a **mechanistic model**. Where a **mathematical model** is used as a convenient way of summarising the data with elements which do not represent the neurobiology directly, it is a **descriptive model**.

1.1.2 Why Do Computational Modelling?

As the predator–prey model shows, a well-constructed and useful model is one that can be used to increase our understanding of the phenomena under investigation and to predict reliably the behaviour of the system under the given circumstances.

A very successful use of computational modelling which became well known during writing the second edition of this book is in predicting the spread of the **SARS-CoV-2 virus**. An influential model has been the SIR model (Bertozzi et al., 2020) where there are three variables: the number of susceptible individuals (S); the number of infected individuals (I); and the number of resistant individuals (R). Differential equations similar to those in Box 1.1 show the relations between these three numbers. For example, the rate of change of the number of susceptible individuals S is assumed to be proportional to the product of S and I. The values of the proportionality constants are found by fitting the models to historical data. The basic model is then extended by introducing more variables to differentiate effects due to, for example, location or age. In the extreme case, the properties of individuals can be modelled.

In neuroscience, the classic use of a computational model is Hodgkin and Huxley's model of propagation of a nerve impulse (action potential) along an axon (Chapter 3).

Whilst ultimately a theory will be validated or rejected by experiment, computational modelling is now regarded widely as an essential part of the neuroscientist's toolbox. The reasons for this are:

(1) Modelling is used as an aid to reasoning. Often the consequences derived from hypotheses involving a large number of interacting elements forming the neural subsystem under consideration can only be found by constructing a computational model. Also, experiments often only provide indirect measurements of the quantities of interest, and models are used to infer the behaviour of the interesting variables. An example of this is given in Box 1.2.

(2) Modelling removes ambiguity from theories. Verbal theories can mean different things to different people, but formalising them in a mathematical model removes that ambiguity. Use of a mathematical model ensures that the assumptions of the model are explicit and logically consistent. The predictions of what behaviour results from a fully specified

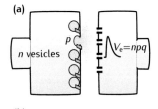

(a)

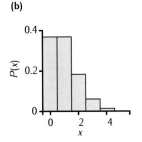

(b)

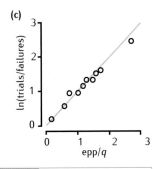

(c)

Fig 1.2 **(a)** Quantal hypothesis of synaptic transmission. **(b)** Example **Poisson distribution** of the number of released quanta when $m = 1$. **(c)** Relationship between two estimates of the mean number of released quanta at a neuromuscular junction. Blue line shows where the estimates would be identical. Plotted from data in Table 1 of Del Castillo and Katz (1954a), following their Figure 6.

Box 1.2 | Reasoning with models

An example in neuroscience where mathematical models have been key to reasoning about a system is chemical synaptic transmission. Though more direct experiments are becoming possible, much of what we know about the mechanisms underpinning synaptic transmission must be inferred from recordings of the postsynaptic response. Statistical models of neurotransmitter release are a vital tool.

The **quantal hypothesis** was put forward by Del Castillo and Katz (1954a) as an aid to understanding how release of acetylcholine at the neuromuscular synapse results in an **endplate potential** (EPP) in the muscle. In the absence of presynaptic activity, spontaneous **miniature endplate potentials (MEPPs)** of relatively uniform size were recorded in frog muscles. The working hypothesis was that the EPPs evoked by a presynaptic action potential were made up by the sum of many MEPPs, each of which contributed a discrete amount, or 'quantum', to the overall response. The proposed underlying model is that each quantum is released with a mean probability p contributing an amount q, the quantal amplitude, to the evoked EPP. The mean number of quanta m in an evoked EPP and its mean amplitude V_e are given by:

$$m = np, \qquad V_e = npq,$$

where n quanta of acetylcholine are available to be released (Figure 1.2a).

To test their hypothesis, Del Castillo and Katz (1954a) reduced synaptic transmission by lowering calcium and raising magnesium in their experimental preparation, allowing them to evoke and record small EPPs, putatively made up of only a few quanta. Given that n is large and p is very small, the number released on a trial-by-trial basis should follow a Poisson distribution (Box 8.5), which gives a relationship between the probability of a certain number x of quanta being released on a given trial and the mean number m of quanta per EPP. This leads to two different ways of obtaining a value for m from the experimental data. Firstly, m is the mean amplitude of the evoked EPPs divided by the mean amplitude of recorded MEPPs. Secondly, the low release probability leads to many failures of release. The probability of no release can be calculated from the Poisson distribution, from which another estimate of m, as ln((number of trials)/(number of failures)), can be obtained. If the model is correct, these two ways of determining m should agree with each other.

Plots of the experimental data confirmed that this was the case (Figure 1.2c), lending strong support to the quantal hypothesis. Such **quantal analysis** is still a major tool in analysing synaptic responses, particularly for identifying the pre- and postsynaptic loci of biophysical changes underpinning short- and long-term synaptic plasticity (Ran et al., 2009; Redman, 1990). More complex and dynamic models are explored in Chapter 7.

mathematical model are unambiguous and can be checked by solving again the equations representing the model.

(3) The models that have been developed for many neurobiological systems, particularly at the cellular level, have reached such a degree of

sophistication that they are accepted as being adequate representations of the neurobiology. Detailed compartmental models of neurons are one example (Chapter 5).

(4) Advances in computer technology mean that the number of interacting elements, such as neurons, that can be simulated is very large and representative of the system being modelled.

(5) In principle, testing hypotheses by computational modelling could supplement experiments in some cases. Though experiments are vital in developing a model and setting initial parameter values, it might be possible to use modelling to extend the effective range of experimentation.

Building a computational model of a neural system is not a simple task. Major problems are: deciding what type of model to use; at what level to model; what aspects of the system to model; and how to deal with parameters that have not been, or cannot be, measured experimentally. At each stage of this book, we try to provide possible answers to these questions as a guide to the modelling process. Often, there is no single correct answer, but it is a matter of skilled and informed judgement.

1.1.3 Levels of Analysis

To understand the nervous system requires analysis at many different levels (Figure 1.3), from molecules to behaviour, and computational models exist at all levels. The nature of the scientific question that drives the modelling work will largely determine the level at which the model is to be constructed. For example, to model how ion channels open and close requires a model in which ion channels and their dynamics are represented; to model how information is stored in the cerebellar cortex through changes in synaptic strengths requires a model of the cerebellar circuitry involving interactions between nerve cells through modifiable synapses.

1.1.4 Levels of Detail

Models that are constructed at the same level of analysis may be constructed to different levels of detail. For example, some models of the propagation of electrical activity along the axon assume that the electrical impulse can be represented as a square pulse train; in some others, the form of the impulse is modelled more precisely as the voltage waveform generated by the opening and closing of sodium and potassium channels. The level of detail adopted also depends on the question being asked. An investigation into how the relative timing of the synaptic impulses arriving along different axons affects the excitability of a target neuron may only require knowledge of the impulse arrival times, and not the actual impulse waveform.

Whatever the level of detail represented in a given model, there is always a more detailed model that can be constructed, and so ultimately how detailed the model should be is a matter of judgement. The modeller is faced perpetually with the choice between a more detailed model with a large number of parameter values that have to be assigned by experiment or by other means, and a less detailed but more tractable model with few undetermined parameters. The choice of what level of detail is appropriate for the model is also a question of practical necessity when running the model on the com-

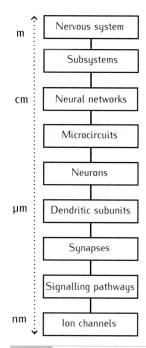

Fig 1.3 To understand the nervous system requires an understanding at many different levels, at spatial scales ranging from metres to nanometres or smaller. At each of these levels, there are detailed computational models for how the elements at that level function and interact, be they, for example, neurons, networks of neurons, synapses or molecules involved in signalling pathways.

In deciding how much detail to include in a model, we could take guidance from Albert Einstein, who is reported as saying, 'Make everything as simple as possible, but not simpler.'

puter; the more details there are in the model, the more computationally expensive the model is. More complicated models also require more effort, and lines of computer code, to construct.

As with experimental results, it should be possible to reproduce computational results from a model. The ultimate test of reproducibility is to read the description of a model in a scientific paper, and then redo the calculations, possibly by writing a new version of the computer code, to produce the same results. A weaker test is to download the original computer code of the model, and check that the code is correct, that is, that it does what is described of it in the paper. The difficulty of both tests of reproducibility increases with the complexity of the model. Complicating the model needs to be justified as much as simplifying it, because it can sometimes come at the cost of understandability.

1.1.5 Parameters

A key aspect of computational modelling is in determining values for model parameters. Often these will be estimates at best, or even complete guesses. Using the model to show how sensitive a solution is to the varying parameter values is a crucial use of the model.

Returning to the predator–prey model, Figure 1.1c shows the behaviour of only one of an infinitely large range of models described by the final equation in Box 1.1. This equation contains four parameters: a, b, c and d. A parameter is a constant in a mathematical model which takes a particular value when producing a numerical solution of the equations, and which can be adjusted between solutions. We might argue that this model only produced oscillations because of the set of parameter values used, and try to find a different set of parameter values that gives steady state behaviour. In Figure 1.1d, the behaviour of the model with a different set of parameter values is shown; there are still oscillations in the predator and prey populations, though they are at a different frequency.

In order to determine whether there are parameter values for which there are no oscillations, we could search the **parameter space**, which in this case is made up of all possible values of a, b, c and d in combination. Even if we restrict the search and vary each parameter between, say, 0.1 and 10 in steps of 0.1, to search all possible combinations of values would require 100^4 (100 million) numerical solutions to the equations. This is clearly a formidable task, even with the aid of computers.

In the case of this particularly simple model, the dynamical systems theory (Section 8.2) can be applied. This analysis shows that there are oscillations for *all* parameter settings.

However, often the models we devise in neuroscience are considerably more complex than this one, and mathematical analysis is of less help. Furthermore, the equations in a mathematical model often contain a large number of parameters. Whilst some of the values can be specified (e.g. from experimental data), usually not all parameter values are known. In some cases, additional experiments can be run to determine some values, but many parameters will remain **free parameters** (i.e. not known in advance).

How to determine the values of free parameters is a general modelling issue and not exclusive to neuroscience. An essential part of the modeller's toolkit is a set of techniques that enable free parameter values to be estimated. We devote Chapter 14 to all aspects of model selection and finding the best parameter values.

1.2 Overview of the Book

To illustrate the principles of computational modelling described in each chapter, we describe a series of computational models, constructed at different levels of analysis and detail. The level of analysis considered ranges from ion channels to networks of neurons, grouped around models of the nerve cell. In some chapters, general techniques are covered and a list of these is given in Table 1.1. We now give brief descriptions of the chapters.

Chapter 2, The Basis of Electrical Activity in the Neuron, describes how the concepts of membrane biophysics are used to model electrical activity in the neuron. We show how a simple model of neuronal firing can be represented as a simple electrical circuit.

Chapter 3, The Hodgkin–Huxley Model of the Action Potential, describes in detail this landmark model for the generation of the nerve impulse which incorporates the effects on membrane potential of the voltage-gated sodium and potassium ion channels. This model is widely heralded as the first successful example of combining experimental and computational studies in neuroscience.

Chapter 4, Models of Active Ion Channels, examines the consequences of introducing into a model of the neuron the many types of active ion channel known, in addition to the sodium and potassium voltage-gated ion channels studied in Chapter 3.

Chapter 5, Modelling Neurons over Space and Time, shows how to model the complex dendritic and axonal morphology of a neuron in order to capture the spatial and temporal dimensions of neuronal activity. This involves using the compartmental modelling approach.

Chapter 6, Intracellular Mechanisms, shows ionic channel dynamics are influenced heavily by intracellular signalling. We show how the various important effects due to calcium can be modelled. We also look at models for other signalling pathways involving more complex enzymatic reactions.

Chapter 7, The Synapse, examines several models of chemical synapses, ranging from electrical circuit-based schemes to complex stochastic models, including vesicle recycling and release. Different types of excitatory and inhibitory chemical synapses are considered, as well as models of electrical synapses (gap junctions).

Chapter 8, Simplified Models of the Neuron, examines the issues surrounding the construction of models of single neurons that are simpler than those described already. Several models with different degrees of simplification are described that are particularly useful for incorporating in networks since they are computationally more efficient and, in some cases, can be analysed mathematically.

Chapter 9, Networks of Neurons, describes the simplifications that are necessary in order to construct a network of neurons. Fundamental questions have to be asked, such as what each unit making up the network represents, what their functionality is and how units are interconnected.

In **Chapter 10, Brain Tissue**, we discuss how to model the extracellular influences on neurons, particularly those from the non-neural glial cells, which do not generate action potentials. We describe modelling the effects of the diffusion of charged and uncharged particles in the extracellular space.

Chapter 11, Plasticity, discusses modelling the various types of plasticity known to exist in the nervous system. The emphasis is on synaptic plasticity involved in learning and memory, and homeostatic plasticity which enables the nervous system to maintain stability in the face of fluctuations in its environment.

Chapter 12, Development of the Nervous System, illustrates the approaches used in modelling neural development. We discuss modelling of the development of individual nerve cells and of nerve connections. Models for development often contain fundamental assumptions that are as yet untested, which we illustrate by presenting different models for the same phenomenon.

Chapter 13, Modelling Measurements and Stimulation. In order to compare a model with experiments, it is essential to have a good model of the measurement process. In this chapter, we describe the modelling of different types of electrical signal, including the action potentials, local field potentials and signals recorded by electroencephalography. Modelling magnetic and optical signals is also discussed.

Chapter 14, Model Selection and Optimisation. It is essential to choose the appropriate form of the model for the particular phenomena being modelled. This chapter discusses methods for model selection, the choice of parameters and the optimisation of parameter values.

Chapter 15, Farewell, summarises our views on the current state of computational neuroscience and its future as a tool within neuroscience research.

Table 1.1	General topics	
Topic	Section	Page
Numerical methods	2.6.2, 5.7	33, 138
Maximum likelihood estimation	5.5.1, 14.7.1	132, 461
Kernel density estimation	5.5.1	134
Monte Carlo simulation methods	6.8.1, 14.6.2	178, 457
Dynamical systems theory	8.2	217
Poisson processes	8.5.3	237
Event-based simulation	9.3	259
Statistics of time-varying signals	9.3.1	262
Optimisation algorithms	14.3	439
Parameter sensitivity analysis	14.6	454
Uncertainty analysis	14.6.2	455
Bayesian inference for parameter estimation	14.7.1	459

1.2.1 Pathways through the Book

Whereas we hope that the entire book will be read by all, we appreciate that certain chapters will be appropriate for different types of readers. We have therefore identified several different pathways through the book.

- For those who are looking for a one-semester introductory course in computational neuroscience, Chapter 2 through to Chapter 9 would be suitable.
- To complement this introductory course:
  - Chapter 10 and Chapter 13 extend the classical computational neuroscience curriculum.
  - Chapter 12 is concerned with developmental aspects.
  - Chapter 14 has a strong mathematical perspective.
- Those interested in single cell modelling should read Chapter 2 through to Chapter 6.
- Those interested in the Hodgkin–Huxley model should read Chapter 2 and Chapter 3.

CHAPTER 2

The Basis of Electrical Activity in the Neuron

This chapter introduces the physical principles underlying models of the electrical activity of neurons. Starting with the neuronal cell membrane, we explore how its permeability to different ions and the maintenance by ionic pumps of concentration gradients across the membrane underpin the resting membrane potential. We show how these properties can be represented by an equivalent electrical circuit, which allows us to compute the response of the membrane potential over time to input current. We conclude by describing the integrate-and-fire neuron model, which is based on the equivalent electrical circuit.

The first reports of intracellular recordings from cells date from the mid-1930s (Hoyle, 1983). Sharp electrodes, as shown in Figure 2.1, are made of glass and puncture the cell membrane.

A nerve cell, or neuron, can be studied at many levels of analysis, but much of the computational modelling work in neuroscience is at the level of the electrical properties of neurons. Neurons have a tree-like structure with branches called **dendrites** and **axons**, which are used to receive and send signals from and to other neurons, sensory cells and muscle. In neurons, as in other cells, a measurement of the voltage across the membrane using an intracellular electrode (Figure 2.1) shows that there is an electrical potential difference across the cell membrane, called the **membrane potential**. In neurons, the membrane potential is used to transmit and integrate signals, sometimes over large distances. The **resting membrane potential** is typically around $-65\,\mathrm{mV}$, meaning that the potential inside the cell is more negative than that outside.

For the purpose of understanding its electrical activity, a neuron can be represented as an electrical circuit. The first part of this chapter (Sections 2.1–2.5) explains why this is so in terms of basic physical concepts such as diffusion, electric fields and capacitance. Some material in this chapter does not relate directly to computational models of neurons, but it informs decisions about which processes need to be modelled and how they are modelled. For example, changes in the concentrations of ions sometimes alter the electrical and signalling properties of the cell significantly, but sometimes they are so small that they can be ignored.

The second part of this chapter (Section 2.6) explores basic properties of the electrical circuit model neurons described in the first part, in particular

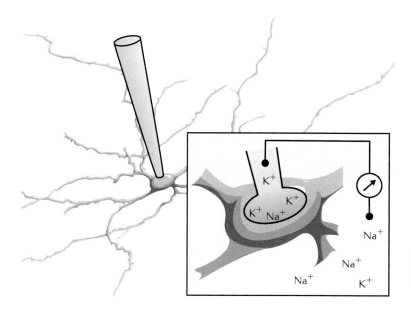

Fig 2.1 Differences in the intracellular and extracellular ion compositions and their separation by the cell membrane are the starting point for understanding the electrical properties of the neuron. The inset shows that for a typical neuron in the central nervous system (CNS), the concentration of sodium ions is greater outside the cell than inside it, and the concentration of potassium ions is greater inside the cell than outside it. Inserting an electrode into the cell allows the membrane potential to be measured.

the response of the membrane potential over time to current input. We illustrate the principle of how the response is computed by solving the ordinary differential equation describing the time evolution of the membrane potential. The response is characterised by the membrane time constant and the input impedance, which are related to the membrane properties.

The neuron model described so far is purely passive: any current inputs to it decay over time. As will be covered in Chapters 3 and 4, many real neurons are active: current inputs are amplified, which can lead to neurons 'firing', that is, producing action potentials or 'spikes'. In the third part (Section 2.7), we add a firing mechanism to the circuit model, giving rise to the integrate-and-fire neuron, the oldest model of neural activity.

The models in this chapter embody a number of fundamental concepts, can be used to understand aspects of the electrical activity of neurons and provide the foundations that will be built on in later chapters. However, more sophisticated biophysical models should be employed in some circumstances, as we discuss in Section 2.8.

2.1 | The Neuronal Membrane

The electrical properties which underlie the membrane potential arise from the separation of intracellular and extracellular space by a cell membrane. The intracellular medium, the **cytoplasm**, and the extracellular medium contain differing concentrations of various ions. Some key inorganic ions in nerve cells are positively charged **cations**, including sodium (Na^+), potassium (K^+), calcium (Ca^{2+}) and magnesium (Mg^{2+}), and negatively charged **anions** such as chloride (Cl^-). Within the cell, the charge carried by anions and cations is usually almost balanced, and the same is true of the extracellular space. Typically, there is a greater concentration of extracellular sodium than intracellular sodium, and conversely for potassium (Figure 2.1).

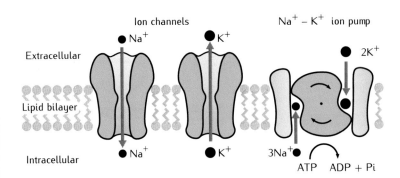

Fig 2.2 Constituents of the membrane. Three types of component play important electrical roles in a neuron's membrane. The lipid bilayer forms a virtually impermeable barrier to ions, though it is permeable to water and some other uncharged molecules. The ion channel is a protein or cluster of proteins that form a pore through the membrane, allowing certain ions to pass through. The ionic pump, or ion exchanger, pumps or exchanges certain ion types across the membrane. This example shows the Na^+–K^+ pump which exchanges three Na^+ ions from inside with two K^+ ions from outside, using energy from the hydrolysis of ATP into ADP and a phosphate ion (Pi).

The key components of the membrane are shown in Figure 2.2. The bulk of the membrane is composed of a 5 nm-thick **lipid bilayer**. It is made up of two layers of lipids, which have their hydrophilic ends pointing outwards and their hydrophobic ends pointing inwards, which makes it virtually impermeable to water molecules and ions. This impermeability allows positive ions to build up on one side of the membrane, balanced by negative ions on the other. This leads to an electrical field across the membrane, similar to that found between the plates of an ideal electrical capacitor (Table 2.1).

Ion channels are pores in the lipid bilayer, made of proteins, which can allow certain ions to flow through the membrane. As will be described in Chapters 3 and 4, a large body of biophysical work has shown that many types of ion channels, referred to as **active channels**, can exist in open states, where it is possible for ions to pass through the channel, and closed states, in which ions cannot permeate through the channel. Whether an active channel is in an open or closed state may depend on the membrane potential, ionic concentrations or the presence of bound ligands, such as neurotransmitters. Ion channels that are sensitive to neurotransmitters such as glutamate or acetyl choline are classified as **receptors**. They form one half of a chemical synapse, which transduces signals from one neuron to another.

In contrast to active channels, **passive channels** do not change their permeability in response to changes in the membrane potential. The response of some channels to the membrane potential is so mild as to be virtually passive.

Both passive channels and active channels in the open state exhibit **selective permeability** to different types of ion. Channels are often labelled by the ion to which they are most permeable. For example, potassium channels primarily allow potassium ions to pass through. There are many types of ion channels, each of which has a different permeability to each type of ion.

In this chapter, we consider how to model the flow of ions through passive channels. The opening and closing of active channels is a separate topic, which is covered in detail in Chapters 3 and 4; the concepts presented in this chapter are fundamental to describing the flow of ions through active channels in the open state. It will be shown how the combination of selective permeability of ion channels and ionic concentration gradients leads to the membrane having properties that can be approximated by ideal resistors and batteries (Table 2.1). This approximation and a fuller account of the electrical properties arising from the permeable and impermeable aspects of the membrane are explored in Sections 2.3–2.5.

Table 2.1 | Review of electrical circuit components

Component	Symbols and units	Function
Battery	E (volts, V)	Pumps charge around a circuit
Current source	I (amps, A)	Provides a specified current (which may vary with time)
Resistor	R (ohms, Ω)	Resists the flow of current in a circuit
Capacitor	C (farad, F)	Stores charge. Current flows *onto* (not through) a capacitor
Voltage recording	V (volts, V)	Records voltage
Current recording	I (amps, A)	Records current

For each component, the circuit symbol, the mathematical symbol, the Système International (SI) unit and the abbreviated form of the SI unit are shown.

Ionic pumps are membrane-spanning protein structures that actively pump specific ions and molecules in and out of the cell. Particles moving freely in a region of space always move so that their concentration is uniform throughout the space. Thus, on the high concentration side of the membrane, ions tend to flow to the side with low concentration, thus diminishing the concentration gradient. Pumps counteract this by pumping ions against the concentration gradient. Each type of pump moves a different combination of ions. The sodium–potassium exchanger pushes K^+ into the cell and Na^+ out of the cell. For every two K^+ ions pumped into the cell, three Na^+ ions are pumped out. This requires energy, which is provided by the hydrolysis of one molecule of adenosine triphosphate (ATP), producing adenosine diphosphate (ADP) and a phosphate ion (Figure 2.2).

Because the sodium–potassium exchanger causes a net loss of charge in the neuron, it is said to be **electrogenic**. An example of a pump which is not electrogenic is the sodium–hydrogen exchanger, which pumps one H^+ ion out of the cell against its concentration gradient for every Na^+ ion it pumps in. In this pump, Na^+ flows down its concentration gradient, supplying the energy required to extrude the H^+ ion; there is no hydrolysis of ATP. Other pumps, such as the sodium–calcium exchanger, are also driven by the Na^+ concentration gradient (Blaustein and Hodgkin, 1969). These pumps consume ATP indirectly as they increase the intracellular Na^+ concentration, giving the sodium–potassium exchanger more work to do.

Since ions are constantly leaking through ion channels, ionic pumps have to work constantly to maintain the concentration. It is possible to create an 'energy budget of the brain' to estimate how much energy the brain uses to maintain the resting membrane potential (Attwell and Laughlin, 2001; Howarth et al., 2012). Other behaviours, such as producing action potentials and synaptic vesicle release, also use energy.

(a)

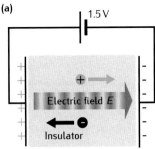

(b) Potential energy w.r.t. $x=0$

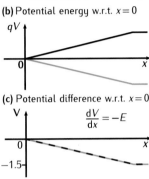

(c) Potential difference w.r.t. $x=0$

$$\frac{dV}{dx} = -E$$

Fig 2.3 Electric fields, electric potential and the forces acting on charged ions. **(a)** The positive and negative charges on the plates of a capacitor create an electric field. In this electric field, a positively charged ion (blue circle) would experience a force (blue arrow) in the same direction as the electric field. A negative charge (black circle) would experience a force (black arrow) in the opposite direction to the electric field. **(b)** The potential energy of the positive (blue) and negative (black) ions would gain or lose at points along the x axis, compared to their potential energy at the left-hand side of the field ($x = 0$). **(c)** The potential difference V between the left-hand side of the field and points along the x axis, which is the same for both ions. The gradient of the potential difference is the negative of the electric field E.

In this chapter, ionic pumps are not considered explicitly; rather we assume steady concentration gradients of each ion type. The effects of ionic pumps are considered in more detail in Chapter 6.

2.2 | Physical Basis of Ion Movement in Neurons

The basis of electrical activity in neurons is movement of ions within the cytoplasm and through ion channels in the cell membrane. Before proceeding to fully fledged models of electrical activity, it is important to understand the physical principles which govern the movement of ions through channels and within **neurites**, the term we use for parts of axons or dendrites.

Firstly, the electric force on ions is introduced. We then look at how to describe the diffusion of ions in solution from regions of high to low concentration in the absence of an electric field. This is a first step to understanding movement of ions through channels. We go on to look at electrical drift, caused by electric fields acting on ions which are concentrated uniformly within a region. This can be used to model the movement of ions longitudinally through the cytoplasm. When there are both electric fields and non-uniform ion concentrations, the movement of ions is described by a combination of electrical drift and diffusion, termed **electrodiffusion**. This is the final step required to understand the passage of ions through channels. Finally, the relationship between the movement of ions and electrical current is described.

2.2.1 The Electric Force on Ions

As ions are electrically charged, they exert forces on, and experience forces from, other ions. We use the concept of the **electric field** to describe these forces. The electric field at any point in space is defined as the force experienced by an object with a unit of positive charge. Electric fields are created by distributions of positive and negative charges in space. For example, in a parallel plate capacitor (Figure 2.3a), two flat metal plates are arranged so they are facing each other, separated by an electrical insulator. One of the plates is connected to the positive terminal of a battery, and the other to the negative terminal. The battery attracts electrons (which are negatively charged) into its positive terminal and pushes them out through its negative terminal. The plate connected to the negative terminal therefore has an excess of negative charge on it, and the plate connected to the positive terminal has an excess of positive charge. The separation of charges sets up an electric field between the plates of the capacitor.

Suppose we could place ions in the insulator. A positively charged ion in the electric field would experience a force acting in the direction of the electric field; a negatively charged ion would experience a force acting in exactly the opposite direction to the electric field (Figure 2.3a). This is consistent with the law that unlike charges attract and like charges repel. The force acting on an ion is proportional to the ion's charge q, measured in **coulombs** (symbol C), and the strength of the electric field E, measured in volts per metre.

We often use electric potential to represent electric fields. At any point in an electric field, a charge has an electrical potential energy (qV)

(Figure 2.3b), which is the energy it would gain or lose by being moved from the left-hand plate of the capacitor ($x = 0$) to the point x. The difference in the potential energy per unit charge between any two points in the field is called the **potential difference** (Figure 2.3c), denoted V and measured in **volts** (symbol V).

Because of the relationship between electric field and potential, there is also a potential difference across the charged capacitor. The potential difference is equal to the **electromotive force** of the battery. For example, a battery with an electromotive force of 1.5 V creates a potential difference of 1.5 V between the plates of the capacitor.

The strength of the electric field set up through the separation of ions between the plates of the capacitor is proportional to the magnitude of the excess charge q on the plates. As the potential difference is proportional to the electric field, this means that the charge is proportional to the potential difference. The constant of proportionality is called the **capacitance** and is measured in **farads** (symbol F). It is usually denoted by C and indicates how much charge can be stored on a particular capacitor for a given potential difference across it:

$$q = CV. \tag{2.1}$$

Capacitance depends on the electrical properties of the insulator and on the size and distance between the plates.

2.2.2 Diffusion

Individual freely moving particles, such as dissociated ions, suspended in a liquid or gas appear to move randomly, a phenomenon known as **Brownian motion**. However, in the behaviour of large groups of particles, statistical regularities can be observed. **Diffusion** is the net movement of particles from regions in which they are highly concentrated to regions in which they have low concentration. For example, when ink drips into a glass of water, initially a region of highly concentrated ink will form, but over time this will spread out until the water is uniformly coloured. As shown by Einstein (1905), diffusion, a phenomenon exhibited by groups of particles, arises from the random movement of individual particles. The rate of diffusion depends on characteristics of the diffusing particle and the medium in which it is diffusing. It also depends on temperature; the higher the temperature, the more vigorous the Brownian motion and the faster the diffusion.

In the ink example, molecules diffuse in three dimensions, and the concentration of the molecule in a small region changes with time until the final steady state of uniform concentration is reached. In Chapters 6 and 10, we will consider diffusion in three dimensions, which is necessary for some types of modelling. However, in this chapter, we need to understand how molecules diffuse from one side of the membrane to the other through channels. The channels are barely wider than the diffusing molecules, and so can be thought of as being one-dimensional.

The concentration of an arbitrary molecule or ion X is denoted $[X]$. When $[X]$ is different on the two sides of the membrane, molecules will diffuse through the channels down the concentration gradient, from the side with higher concentration to the side with lower concentration (Figure 2.4). **Flux** is the amount of X that flows through a cross-section of unit area

A singly-charged positive ion, such as sodium (Na^+) or potassium (K^+), has a charge exactly equal to the **elementary charge**, which is the magnitude of the charge on an electron, 1.602×10^{-19} C.

The capacitance of an ideal parallel plate capacitor is proportional to the area a of the plates and inversely proportional to the distance d between the plates:

$$C = \frac{\epsilon a}{d},$$

where ϵ is the permittivity of the insulator, a measure of how hard it is to form an electric field in the material, with units of Fm^{-1}.

Concentration is typically measured in moles per unit volume. One mole contains Avogadro's number (approximately 6.02×10^{23}) atoms or molecules. **Molarity** denotes the number of moles of a given substance per litre of solution (the units are $mol\,L^{-1}$, often shortened to M).

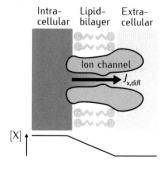

Fig 2.4 Fick's first law in the context of an ion channel spanning a neuronal membrane.

Box 2.1 | Voltage and current conventions in cells

By convention, the membrane potential, the potential difference across a cell membrane, is defined as the potential inside the cell minus the potential outside the cell. The convention for current flowing through or onto the membrane is that it is *positive* when positive charge flows *outwards*, and *negative* when positive charge flows *inwards*.

According to these conventions, when the inside of the cell is more positively charged than the outside, the membrane potential is positive. Positive charges in the cell will be repelled by the other positive charges in the cell, and will therefore have a propensity to move out of the cell. Any movement of positive charge out of the cell is regarded as a positive current. It follows that a positive membrane potential tends to lead to a positive current flowing across the membrane. Thus, the voltage and current conventions fit with the notion that current flows from higher to lower voltages.

The older convention of defining the membrane potential as the potential outside minus the potential inside is not used in this book.

per unit time. Typical units for flux are $mol\,cm^{-2}\,s^{-1}$, and its sign depends on the direction in which the molecules are flowing. To fit in with the convention for current (Box 2.1), we define the flux as positive when the flow of molecules is out of the cell, and negative when the flow is inward. Fick (1855) provided an empirical description relating the molar flux $J_{X,diff}$, arising from the diffusion of a molecule X, to its **concentration gradient** $d[X]/dx$ (here in one dimension):

$$J_{X,diff} = -D_X \frac{d[X]}{dx},$$ (2.2)

where D_X is defined as the **diffusion coefficient** of molecule X. The diffusion coefficient has units of $cm^2\,s^{-1}$. This equation captures the notion that larger concentration gradients lead to larger fluxes. The negative sign indicates that the flux is in the opposite direction to that in which the concentration gradient increases; that is, molecules flow from high to low concentrations (Figure 2.4).

2.2.3 Electrical Drift

Although they experience a force due to being in an electric field, ions on the surface of a membrane are not free to move across the insulator which separates them. In contrast, ions in the cytoplasm and within channels are able to move. Our starting point for thinking about how electric fields affect ion mobility is to consider a narrow cylindrical tube in which there is a solution containing positively and negatively charged ions such as K^+ and Cl^-. The concentration of both ions in the tube is assumed to be uniform, so there is no concentration gradient to drive diffusion of ions along the tube. Apart from lacking intracellular structures, such as microtubules, the endoplasmic reticulum and mitochondria, this tube is analogous to a section of neurite.

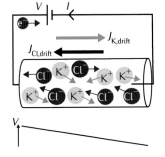

Fig 2.5 Electrical drift. The cylinder represents a section of neurite containing positively charged potassium ions and negatively charged chloride ions. Under the influence of a potential difference between the ends, the potassium ions tend to drift towards the positive terminal, and the chloride ions towards the negative terminal. In the wire, the current is transported by electrons.

Now suppose that electrodes connected to a battery are placed in the ends of the tube to give one end of the tube a higher electrical potential than the other, as shown in Figure 2.5. The K^+ ions will experience an electrical force pushing them down the potential gradient, and the Cl^- ions, because of their negative charge, will experience an electrical force in the opposite direction. If there were no other molecules present, both types of ion would accelerate up or down the neurite. But the presence of other molecules causes frequent collisions with the K^+ and Cl^- ions, preventing them from accelerating. The result is that both K^+ and Cl^- molecules travel at an average speed (**drift velocity**) that depends on the strength of the field. Assuming there is no concentration gradient of potassium or chloride, the flux is:

$$J_{X,\text{drift}} = -\frac{D_X F}{RT} z_X [X] \frac{dV}{dx}, \tag{2.3}$$

where z_X is the ion's signed **valency** (the charge of the ion measured as a multiple of the elementary charge). The other constants are: R, the gas constant; T, the temperature in kelvins; and F, Faraday's constant, which is the charge per mole of monovalent ions.

2.2.4 Electrodiffusion

Diffusion describes the movement of ions due to a concentration gradient alone, and electrical drift describes the movement of ions in response to a potential gradient alone. To complete the picture, we consider **electrodiffusion**, in which both voltage and concentration gradients are present, as is usually the case in ion channels. The total flux J_X of an ion X is the sum of the diffusion and drift fluxes from Equations 2.2 and 2.3:

$$J_X = J_{X,\text{diff}} + J_{X,\text{drift}} = -D_X \left(\frac{d[X]}{dx} + \frac{z_X F}{RT} [X] \frac{dV}{dx} \right). \tag{2.4}$$

This equation, developed by Nernst (1888) and Planck (1890), is called the **Nernst–Planck equation** and is a general description of how charged ions move in solution in electric fields. It is used to derive the expected relationships between the membrane potential and ionic current flowing through channels (Section 2.4).

2.2.5 Flux and Current Density

So far, movement of ions has been quantified using flux, the number of moles of an ion flowing through a cross-section of unit area. However, often we are interested in the flow of the charge carried by molecules rather than the flow of the molecules themselves. The amount of positive charge flowing per unit of time past a point in a conductor, such as an ion channel or neurite, is called the **current** and is measured in amperes (denoted A). The **current density** is the amount of charge flowing per unit of time per unit of cross-sectional area. In this book, we denote current density with the symbol I, with typical units $\mu A\ cm^{-2}$.

The current density I_X due to a particular ion X is proportional to the molar flux of that ion and the charge that it carries. We can express this as:

$$I_X = F z_X J_X, \tag{2.5}$$

z_X is +2 for calcium ions, +1 for potassium ions and −1 for chloride ions.

$R = 8.3145\ \text{J}\,\text{K}^{-1}\text{mol}^{-1}$
$F = 9.6485 \times 10^4\ \text{C}\,\text{mol}^{-1}$

The universal convention is to use the symbol R to denote both the gas constant and electrical resistance. However, what R is referring to is usually obvious from the context: when R refers to the universal gas constant, it is very often next to temperature T.

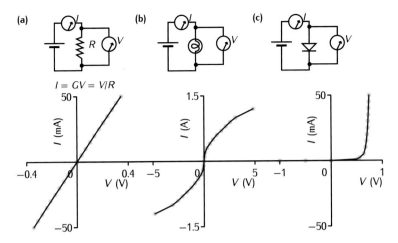

Fig 2.6 *I–V* characteristics of a number of electrical devices. In a typical high–school experiment to determine the *I–V* characteristics of various components, the voltage *V* across the component is varied, and the current *I* flowing through the component is measured using an ammeter. *I* is then plotted against *V*. **(a)** The *I–V* characteristic of a 1 m length of wire, which shows that in the wire, the current is proportional to the voltage. Thus, the wire obeys Ohm's law in the range measured. The constant of proportionality is the conductance *G*, which is measured in siemens. The inverse of the conductance is resistance *R*, measured in ohms. Ohm's law is thus *I* = *V*/*R*. **(b)** The *I–V* characteristic of a filament light bulb. The current is not proportional to the voltage, in part due to the temperature effects of the bulb being hotter when more current flows through it. **(c)** The *I–V* characteristic of a silicon diode. The magnitude of the current is much greater when the voltage is positive than when it is negative. As it is easier for the current to flow in one direction than in the other, the diode exhibits rectification. Data from unpublished A-level physics practical, undertaken by Sterratt, 1989.

where F is Faraday's constant and z_X is the ion's signed valency. As with the flux of an ion, the sign of the current depends on the direction in which the charged particles are flowing. As defined earlier, the flux of molecules or ions through channels is positive when they are flowing out of the cell. Thus, the current due to positively charged ions, such as Na^+ and K^+, will be positive when they are flowing out of the cell, and negative when they flow into the cell, since z_X is positive for these ions (Box 2.1). However, for negatively charged ions, such as Cl^-, when their flux is positive, the current they carry is negative, and vice versa. A negative ion flowing into the cell has the same effect on the net charge balance as a positive ion flowing out of it.

The total current density flowing in a neurite or through a channel is the sum of the contributions from the individual ions. For example, the total ion flow due to sodium, potassium and chloride ions is:

$$I_i = I_{Na} + I_K + I_{Cl} = F z_{Na} J_{Na} + F z_K J_K + F z_{Cl} J_{Cl}$$
$$= F J_{Na} + F J_K - F J_{Cl}. \tag{2.6}$$

The last line follows because the valencies of sodium, potassium and chloride ions are +1, +1 and −1, respectively.

2.2.6 *I–V* Characteristics

Returning to the case of electrodiffusion along a neurite (Section 2.2.4), Equations 2.3 and 2.6 show that the current flowing along the neurite, referred to as the **axial current**, should be proportional to the voltage between the ends of the neurite. Thus, the axial current is expected to obey **Ohm's law** (Figure 2.6a), which states that, at a fixed temperature, the current *I* flowing through a conductor is proportional to the potential difference *V* between the ends of the conductor. The constant of proportionality *G* is the **conductance** of the conductor in question, and its reciprocal *R* is known as the **resistance**. In electronics, an ideal resistor obeys Ohm's law, so we can use the symbol for a resistor to represent the electrical properties along a section of the neurite.

It is worth emphasising that Ohm's law does not apply to all conductors. Conductors that obey Ohm's law are called **ohmic**, whereas those that do not are non-ohmic. Determining whether an electrical component is ohmic

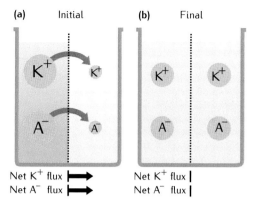

Net K^+ flux ➡
Net A^- flux ➡

Net K^+ flux |
Net A^- flux |

or not can be done by applying a range of known potential differences across it and measuring the current flowing through it in each case. The resulting plot of current versus potential is known as an *I–V* **characteristic**. The *I–V* characteristic of a component that obeys Ohm's law is a straight line passing through the origin, as demonstrated by the *I–V* characteristic of a wire shown in Figure 2.6a. The *I–V* characteristic of a filament light bulb, shown in Figure 2.6b, demonstrates that in some components, the current is not proportional to the voltage, with the resistance going up as the voltage increases. The filament may, in fact, be an ohmic conductor, but this could be masked in this experiment by the increase in the filament's temperature as the amount of current flowing through it increases.

An example of a truly non-ohmic electrical component is the diode, where, in the range tested, current can flow in one direction only (Figure 2.6c). This is an example of **rectification**, the property of allowing current to flow more freely in one direction than in another.

Whilst the flow of current along a neurite is approximately ohmic, the flow of ions through channels in the membrane is not. The reason for this difference is that there is a diffusive flow of ions across the membrane due to the concentration difference, as well as an electrical drift due to the potential difference. We explore this in more detail in Section 2.4.

2.3 The Resting Membrane Potential: The Nernst Equation

The ion channels which span the lipid bilayer confer upon the neuronal cell membrane the property of permeability to multiple types of ion. The first step towards understanding the origin of the resting membrane potential is to consider diffusion and electrical drift of ions through the membrane in a sequence of thought experiments.

The initial setup of the first thought experiment, shown in Figure 2.7a, is a container divided into two compartments by a membrane. The left-hand half represents the inside of a cell, and the right-hand half the outside. Into the left (intracellular) half, we place a high concentration of a potassium solution, consisting of equal numbers of potassium ions K^+ and anions A^-. Into the right (extracellular) half, we place a low concentration of the same

Fig 2.7 Setup of a thought experiment to explore the effects of diffusion across the membrane. In this experiment, a container is divided by a membrane that is permeable to both K^+, a cation, and A^-, an anion. The grey arrows indicate the diffusion flux of both types of ion. **(a)** Initially, the concentrations of both K^+ and A^- are greater than their concentrations on the right-hand side. Both molecules start to diffuse through the membrane down their concentration gradients, to the right. **(b)** Eventually the system reaches an equilibrium, in which the concentrations are the same on each side.

Fig 2.8 The emergence of a voltage across a semipermeable membrane. The grey arrows indicate the net diffusion flux of the potassium ions, and the blue arrows the flow due to the induced electric field.
(a) Initially, K^+ ions begin to move down their concentration gradient (from the more concentrated left side to the right side with a lower concentration). The anions A^- cannot cross the membrane. **(b)** This movement creates an electrical potential across the membrane. **(c)** The potential creates an electric field that opposes the movement of ions down their concentration gradient, so there is no net movement of ions; the system attains equilibrium.

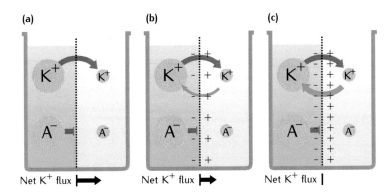

We might ask why the excess charges sit on the surface of the membrane (actually in a nanometre-thick layer), rather than being distributed throughout the solution, and why the solution itself is neutral. As described in Section 10.3.2 and Box 10.6 in Chapter 10, it happens by a process called **charge relaxation**, resulting in a Debye layer of charge around the membrane.

solution. If the membrane is permeable to both types of ions, both populations of ions will diffuse from the half with a high concentration to the half with a low concentration. This will continue until both halves have the same concentration, as seen in Figure 2.7b. This diffusion is driven by the concentration gradient; as we have seen, where there is a concentration gradient, particles or ions move down the gradient.

In the second thought experiment, we suppose that the membrane is permeable only to K^+ ions, and *not* to the anions (Figure 2.8a). In this situation, only K^+ ions can diffuse down their concentration gradient (from left to right in this figure). Once this begins to happen, it creates an excess of positively charged ions on the right-hand surface of the membrane and an excess of negatively charged anions on the left-hand surface. As when the plates of a capacitor are charged, this creates an electric field, and hence a potential difference across the membrane (Figure 2.8b).

The electric field influences the potassium ions, causing an electrical drift of the ions back across the membrane opposite to their direction of diffusion (from right to left in the figure). The potential difference across the membrane grows until it provides an electric field driving a net electrical drift that is equal and opposite to the net flux resulting from diffusion. Potassium ions will flow across the membrane either by diffusion in one direction or by electrical drift in the other direction, until there is no net movement of ions. The system is then at equilibrium, with equal numbers of positive ions flowing rightwards due to diffusion and leftwards due to the electrical drift. At equilibrium, we can measure a stable potential difference across the membrane (Figure 2.8c). This potential difference, called the **equilibrium potential** for that ion, depends on the concentrations on either side of the membrane. Larger concentration gradients lead to larger diffusion fluxes (Fick's first law, Equation 2.2).

Nernst (1888) formulated the **Nernst equation** to calculate the equilibrium potential resulting from **permeability** to a single ion:

$$E_X = \frac{RT}{z_X F} \ln \frac{[X]_{out}}{[X]_{in}}, \tag{2.7}$$

where X is the membrane-permeable ion and $[X]_{in}$ and $[X]_{out}$ are the intracellular and extracellular concentrations of X, and E_X is the equilibrium potential, also called the **Nernst potential**, for that ion. As shown in Box 2.2, the Nernst equation can be derived from the Nernst–Planck equation.

Box 2.2 | Derivation of the Nernst equation

The Nernst equation is derived by assuming diffusion in one dimension along a line (x-coordinate). For there to be no flow of current, the flux is zero throughout, so from Equation 2.4, the Nernst–Planck equation, it follows that:

$$\frac{1}{[X]}\frac{d[X]}{dx} = -\frac{z_X F}{RT}\frac{dV}{dx}.$$

Integrating along the line, that is, across the membrane, we obtain:

$$-\int_{E_m}^{0} dV = \int_{[X]_{in}}^{[X]_{out}} \frac{RT}{z_X F[X]} d[X].$$

Evaluating the integrals gives:

$$E_m = \frac{RT}{z_X F}\ln\frac{[X]_{out}}{[X]_{in}},$$

which is the Nernst equation, Equation 2.7.

As an example, consider the equilibrium potential for K^+. Suppose the intracellular and extracellular concentrations are similar to those of the squid giant axon (400 mM and 20 mM, respectively) and the recording temperature is 6.3°C (279.45 K). Substituting these values into the Nernst equation:

The squid giant axon is an accessible preparation used by Hodgkin and Huxley to develop the first model of the action potential (Chapter 3).

$$E_K = \frac{RT}{z_K F}\ln\frac{[K^+]_{out}}{[K^+]_{in}} = \frac{(8.3145)(279.45)}{(+1)(9.6485 \times 10^4)}\ln\frac{20}{400} = -72.1\,\text{mV}. \qquad (2.8)$$

Table 2.2 shows the intracellular and extracellular concentrations of various important ions in the squid giant axon and the equilibrium potentials calculated for them at a temperature of 6.3°C.

Since Na^+ ions are positively charged, and their concentration is greater outside than inside, the sodium equilibrium potential is positive. On the other hand, K^+ ions have a greater concentration inside than outside and so have a negative equilibrium potential. Like Na^+, Cl^- ions are more concentrated outside than inside, but because they are negatively charged, their equilibrium potential is negative.

This thought experiment demonstrates that the lipid bilayer forming the cell membrane acts as a capacitor, with the surfaces of the thin insulating membrane being the plates of the capacitor. The specific **membrane capacitance** of various types of neurons is often treated as a 'biological constant' of 0.9 μF cm^{-2} which is often rounded up to 1 μF cm^{-2}. This value is roughly at the centre of the range of values measured in the squid giant axon and various types of rodent neurons (Gentet et al., 2000). However, this 'biological constant' is not constant across species and neuron types; human neocortical cells have been found to have a membrane capacitance of around 0.5 μF cm^{-2}, about half the accepted value (Eyal et al., 2016).

So far, we have neglected the fact that in the final resting state of our second thought experiment, the concentration of K^+ ions on either side will differ from the initial concentration, as some ions have passed through the membrane. We might ask if this change in concentration is significant in neurons. We can use the definition of capacitance, $q = CV$ (Equation 2.1), to compute the number of ions required to charge the membrane to its resting

Table 2.2 | The concentrations of various ions in the squid giant axon and outside the axon, in the animal's blood (Hodgkin, 1964) and in a typical mammalian cell (Johnston and Wu, 1995). Equilibrium potentials are derived from these values using the Nernst equation, assuming a temperature of 6.3°C for the squid giant axon or 37°C for mammalian cells. The amount of free intracellular calcium is shown (Baker et al., 1971). There is actually a much greater total concentration of intracellular calcium (0.4 mM) than shown, but the vast bulk of it is bound to other molecules. The table does not show the concentrations of membrane-impermeable ions, which balance the concentrations of the ions shown to ensure the intracellular and extracellular media are neutral.

	Ion	K^+	Na^+	Cl^-	Ca^{2+}
Squid giant axon at 6.3°C	Concentration inside (mM)	400	50	40	10^{-4}
	Concentration outside (mM)	20	440	560	10
	Equilibrium potential (mV)	−72	52	−64	139
Typical mammalian cell at 37°C	Concentration inside (mM)	140	5–15	4	10^{-4}
	Concentration outside (mM)	5	145	110	2.5–5
	Equilibrium potential (mV)	−89	61–90	−89	135–145

A dendritic spine is a very short branch found on some types of neuron.

potential. This computation, carried out in Box 2.3, shows that in large-diameter neurites, the total number of ions required to charge the membrane is usually a tiny fraction of the total number of potassium or sodium ions in the cytoplasm, and therefore changes the concentration by a very small amount. Therefore, when computing the resting potential, the intracellular and extracellular concentrations are often treated as constants.

However, in small neurites, such as the spines found on dendrites of many neurons, the number of ions required to change the membrane potential by a few millivolts can change the intracellular concentration of the ion significantly. This is particularly true of calcium ions, which have a very low free intracellular concentration. Here the intracellular ionic concentrations cannot be treated as constants and have to be modelled explicitly. Another reason for modelling intracellular Ca^{2+} concentration is its critical role in intracellular signalling pathways. Modelling intracellular ionic concentrations and signalling pathways will be dealt with in Chapter 6. When there is repeated production of action potentials, it may be necessary to model both the intracellular and extracellular concentrations of higher-concentration ions such as sodium and potassium ions, using one of the methods described in Chapter 10.

What is the physiological significance of equilibrium potentials? In squid, the resting membrane potential is −65 mV, approximately the same as the potassium and chloride equilibrium potentials. Although originally it was thought that the resting membrane potential might be due to potassium, precise intracellular recordings of the resting membrane potential show that the two potentials differ. This suggests that other ions also contribute towards the resting membrane potential. In order to predict the resting membrane potential, a membrane permeable to more than one type of ion must be considered.

Box 2.3 | How many ions charge the membrane?

We consider a cylindrical section of the squid giant axon, 500 μm in diameter and 1 μm long, at a resting potential of −70 mV. Its surface area is 500π μm^2, and so its total capacitance is $500\pi \times 10^{-8}$ μF (1 μF cm^{-2} is the same as 10^{-8} μF μm^{-2}). As charge is the product of voltage and capacitance (Equation 2.1), the charge on the membrane is therefore $500\pi \times 10^{-8} \times 70 \times 10^{-3}$ μC. Dividing by Faraday's constant gives the number of moles of monovalent ions that charge the membrane: 1.139×10^{-17}. The volume of the axonal section is $\pi(500/2)^2$ μm^3, which is the same as $\pi(500/2)^2 \times 10^{-15}$ litres. Therefore, if the concentration of potassium ions in the volume is 400 mM (Table 2.2), the number of moles of potassium is $\pi(500/2)^2 \times 10^{-15} \times 400 \times 10^{-3} = 7.85 \times 10^{-11}$. Thus, there are roughly 6.9×10^6 times as many ions in the cytoplasm than on the membrane, and so in this case, the potassium ions charging and discharging the membrane have a negligible effect on the concentration of ions in the cytoplasm.

In contrast, the change in calcium concentration in the head of a dendritic spine required to change the membrane potential by 10 mV cannot be neglected. We leave it as an exercise for the reader to calculate: (i) the number of calcium ions at rest in a spine head modelled as a cylinder with a diameter of 0.4 μm and a length of 0.2 μm, assuming a resting calcium concentration of 70 nM; and (ii) the number of calcium ions required to increase the membrane potential by 10 mV.

2.4 Membrane Ionic Currents Not at Equilibrium: The Goldman–Hodgkin–Katz Equations

To understand the situation when a membrane is permeable to more than one type of ion, we continue our thought experiment using a container divided by a semipermeable membrane (Figure 2.9a). The solutions on either side of the membrane now contain two types of membrane-permeable ions, K$^+$ and Na$^+$, as well as membrane-impermeable anions, which are omitted from the diagram for clarity. Initially, there is a high concentration of K$^+$ and a very low concentration of Na$^+$ on the left, similar to the situation inside a typical neuron. On the right (outside), there are low concentrations of K$^+$ and Na$^+$ (Figure 2.9a).

In this example, the concentrations have been arranged so the concentration difference of K$^+$ is greater than the concentration difference of Na$^+$. Thus, according to Fick's first law, the flux of K$^+$ flowing from left to right down the K$^+$ concentration gradient is bigger than the flux of Na$^+$ from right to left flowing down its concentration gradient. This causes a net movement of positive charge from left to right, and positive charge builds up on the right-hand side of the membrane (Figure 2.9b). In turn, this creates an electric field which causes electrical drift of both Na$^+$ and K$^+$ to the left. This reduces the net K$^+$ flux to the right and increases the net Na$^+$ flux to the left. Eventually, the membrane potential grows enough to make the

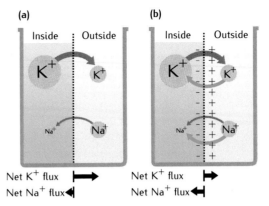

Fig 2.9 The diffusion and electrical drift of two ions with different concentration ratios on either side of a semipermeable membrane. The arrows have the same significance as in Figure 2.8, and anions have been omitted from the diagram for clarity. **(a)** Initially, K^+ ions diffuse from the left to right side, and Na^+ ions from the right to left side, as each ion flows down its concentration gradient.
(b) This results in an electrical potential across the membrane. The potential arises similarly to Figure 2.8. However, here it is influenced both by ions and by their diffusion. At equilibrium, there is still a flow of ions across the membrane; the electrical effect of the movement of one sodium ion (right to left) is neutralised by the effect of the movement of one potassium ion (left to right).

K^+ flux and Na^+ flux equal in magnitude, but opposite in direction. When the net flow of charge is zero, the charge on either side of the membrane is constant, so the membrane potential is steady.

While there is no net flow of charge across the membrane in this state, there is net flow of Na^+ and K^+, and over time this would cause the concentration gradients to run down. As it is the concentration differences that are responsible for the potential difference across the membrane, the membrane potential would reduce to zero. In living cells, ionic pumps counteract this effect. In this chapter, pumps are modelled implicitly by assuming that they maintain the concentrations through time. It is also possible to model the effect of pumps on intracellular ion concentrations explicitly (Section 6.3), and the interaction of pumps and extracellular concentrations can also be modelled (Chapter 10).

From the thought experiment, we can deduce that the resting membrane potential should lie between the sodium and potassium equilibrium potentials calculated using Equation 2.7, the Nernst equation, from their intracellular and extracellular concentrations. Because there is not enough positive charge on the right to prevent the flow of K^+ from left to right, the resting potential must be greater than the potassium equilibrium potential. Likewise, because there is not enough positive charge on the left to prevent the flow of sodium from right to left, the resting potential must be less than the sodium equilibrium potential.

To make a quantitative prediction of the resting membrane potential, we make use of the theory of current flow through the membrane devised by Goldman (1943) and Hodgkin and Katz (1949). By making a number of assumptions, they were able to derive a formula, referred to as the **Goldman–Hodgkin–Katz (GHK) current equation**, which predicts the current I_X mediated by a single ionic species X flowing across a membrane when the membrane potential is V. The GHK current equation and the assumptions from which it was derived are shown in Box 2.4, and the corresponding I–V curves are shown in Figure 2.10.

There are a number of properties worth noting from these curves:

(1) No current flows when the voltage is equal to the equilibrium potential for the ion. This is because at this potential, current flow due to electrical drift and that due to diffusion are equal and opposite. For the concentra-

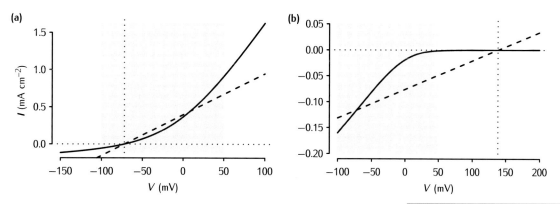

tions of ions shown in Table 2.2, the equilibrium potential of potassium is −72 mV and the equilibrium potential of calcium is +139 mV.

(2) The current changes direction (reverses) at the equilibrium potential. The current is negative (positive charge inwards) when the membrane voltage is below the equilibrium potential, and positive above it. For this reason, the equilibrium potential of an ion is also known as its **reversal potential**.

(3) The individual ions do not obey Ohm's law since the current is not proportional to the voltage.

(4) A consequence of this is that the *I–V* characteristics display rectification, defined in Section 2.2.6. The potassium characteristic favours outward currents and is described as **outward rectifying** (Figure 2.10a). The calcium characteristic favours inward currents and is described as **inward rectifying** (Figure 2.10b). The rectification effect for calcium is particularly pronounced. The GHK current equation shows that, for positive ions such as calcium, when the extracellular concentration is greater than the intracellular concentration, the *I–V* characteristic is inward rectifying; when the converse is true, it is outward rectifying.

Fig 2.10 The *I–V* characteristics for (a) K^+ and (b) Ca^{2+} ions. The solid lines show the *I–V* relationship given by the GHK current equation. The vertical dotted lines show the voltage at which no current flows (i.e. the equilibrium potential). The dashed lines show a linear approximation to the GHK *I–V* characteristic that also yields no current at the equilibrium potential. The shaded regions illustrate the range of voltage within which a neuron usually operates. The concentration values are taken from the squid axon (Hodgkin, 1964; Baker et al., 1971), with current densities calculated at 6.3°C.

We can now calculate the *I–V* characteristic of a membrane permeable to more than one ion type. Assuming that ions flow through the membrane independently, the total current flowing across the membrane is the sum of the ionic currents (Equation 2.6) predicted by the GHK current equations. We can therefore calculate the total current flowing across the membrane for a given value of the membrane potential. The resulting characteristic is broadly similar to the characteristics for the individual ions, in that the current is negative at low potentials and then increases as the membrane potential is raised.

We recall that the reversal potential is defined as the membrane potential at which the current reverses direction. By setting the sum of the ionic currents to zero and solving this equation for voltage, we obtain the **Goldman–Hodgkin–Katz voltage equation** for the reversal potential when there is more than one type of ion. For a membrane permeable to Na^+, K^+ and Cl^-, it reads:

$$E_m = \frac{RT}{F} \ln \frac{P_K[K^+]_{out} + P_{Na}[Na^+]_{out} + P_{Cl}[Cl^-]_{in}}{P_K[K^+]_{in} + P_{Na}[Na^+]_{in} + P_{Cl}[Cl^-]_{out}}, \qquad (2.9)$$

Box 2.4 | The GHK equations

Goldman (1943) and Hodgkin and Katz (1949) developed a formalism for describing the currents through, and voltages across, semipermeable membranes. This formalism models the diffusion of ions through a uniformly permeable membrane, pre-dating the notion of channels or pores through the membrane. It is assumed that ions cross the membrane independently (the **independence principle**) and that the electric field within the membrane is constant. The flux or movement of ions within the membrane is governed by the internal concentration gradient and the electric field arising from the potential difference, calculated by the Nernst–Planck equation.

From these assumptions, the Goldman–Hodgkin–Katz current equation can be derived (Johnston and Wu, 1995):

$$I_X = P_X z_X F \frac{z_X F V}{RT} \left(\frac{[X]_{in} - [X]_{out}\, e^{-z_X FV/RT}}{1 - e^{-z_X FV/RT}} \right).$$

This equation predicts the net flow I_X per unit area of membrane, measured in $\mu A\, cm^{-2}$ of an arbitrary ion type X with valency z_X. P_X is the **permeability** of the membrane to ion X, with units of $cm\, s^{-1}$. It characterises the ability of an ion X to diffuse through the membrane and is defined by the empirical relationship between molar flux J and the concentration difference across the membrane:

$$J_X = -P_X([X]_{in} - [X]_{out}).$$

In the GHK model of the membrane, permeability is proportional to the diffusion coefficient D_X, defined in Fick's first law (Equation 2.2). Hille (2001) discusses the relationship in more detail.

The GHK equation pre-dates the notion of membrane channels and treats the membrane as homogeneous. In active membranes, we can interpret the diffusion coefficient D_X as variable – an increase in the number of open channels in the membrane will increase the membrane permeability. Because of the assumption of a constant electric field in the membrane, the GHK equations are sometimes referred to as the **constant-field equations**.

where P_K, P_{Na} and P_{Cl} are the membrane permeabilities to K^+, Na^+ and Cl^-, respectively (membrane permeability is described in Box 2.4). The pattern of this equation is followed for other sets of monovalent ions, with the numerator containing the *external* concentrations of the *positively* charged ions and the *internal* concentrations of the *negatively* charged ions. The equation shows that the reversal potential for more than one ion type lies between the equilibrium potentials of the individual ions.

As the permeabilities occur in the numerator and the denominator, it is sufficient to know only relative permeabilities to compute the voltage at equilibrium. The relative permeabilities of the membrane of the squid giant axon to K^+, Na^+ and Cl^- ions are 1.0, 0.03 and 0.1, respectively. With these values, and the concentrations from Table 2.2, the resting membrane potential of the squid giant axon predicted by the GHK voltage equation is $-60\,mV$ at $6.3°C$.

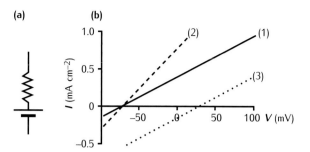

(a) **(b)**

Fig 2.11 Interpretation of the approximation of the GHK current equation. **(a)** The approximation can be viewed as a resistor, or conductance, in series with a battery. **(b)** The graph shows three different *I–V* characteristics from this circuit given different conductances and battery voltages.
(1) $g_X = 5.5\,\text{mS cm}^{-2}$, $E_X = -72\,\text{mV}$; this line is the same as the K^+ approximation in Figure 2.10a;
(2) $g_X = 11.0\,\text{mS cm}^{-2}$, $E_X = -72\,\text{mV}$;
(3) $g_X = 5.5\,\text{mS cm}^{-2}$, $E_X = 28\,\text{mV}$.

Equation 2.9, the GHK voltage equation, looks similar to the Nernst equation. Indeed, it reduces to the equivalent Nernst equation when the permeability of two of the three ions is zero. However, this equation also demonstrates that the membrane potential with two ion types is not the sum of the individual equilibrium potentials.

2.4.1 An Electrical Circuit Approximation of the GHK Current Equation

It is often sufficient to use a simpler equation in place of the GHK current equation. In the potassium characteristic shown in Figure 2.10a, the straight line that gives zero current at the equilibrium potential ($-72\,\text{mV}$) is a close approximation of the *I–V* characteristic for membrane potentials between about $-100\,\text{mV}$ and $50\,\text{mV}$, the voltage range within which cells normally operate. The equation describing this line is:

$$I_X = g_X(V - E_X), \tag{2.10}$$

where X is the ion of interest, E_X its equilibrium potential and g_X is the gradient of the line with the units of conductance per unit area, often mS cm^{-2}. The term in brackets $(V - E_X)$ is called the **driving force**. When the membrane potential is at the equilibrium potential for X, the driving force is zero.

In some cases, such as for calcium in Figure 2.10b, the GHK *I–V* characteristic rectifies too much for a linear approximation to be valid.

Making this linear approximation is similar to assuming Ohm's law, $I = GV$, where conductance G is a constant. Since the straight line does not necessarily pass through the origin, the correspondence is not exact and this form of linear *I–V* relation is called **quasi-ohmic**. There is still a useful interpretation of this approximation in terms of electrical components. The *I–V* characteristic is the same as for a battery with **electromotive force** equal to the equilibrium potential in series with a resistor of resistance $1/g_X$ (Figure 2.11).

2.5 The Capacitive Current

We now have equations that describe how the net flow of current through the different types of channels depends on the membrane potential V. In order to complete the description of the system, we need to know how the current affects the voltage. The total injected current is an extensive quantity, whereas the injected current density is an intensive quantity (Box 2.5).

Box 2.5 | Intensive and extensive quantities

A summary of key passive quantities and their typical units is given in Table 2.3. It is usual to quote the parameters of the membrane as **intensive quantities**. An **intensive quantity** is a physical quantity whose value does not depend on the amount or dimensions of the property being measured. An example of an intensive quantity is the specific membrane capacitance, the capacitance per unit area of membrane. In contrast, an **extensive quantity** does depend on the dimensions of the property being measured, for example the total capacitance of a cell's membrane.

To avoid adding extra symbols, we use intensive quantities in our electrical circuits and equations. Supposing that the area of our patch of membrane is a, its membrane resistance is proportional to the specific membrane resistance divided by the area: R_m/a. Since conductance is the inverse of resistance, the membrane conductance of the patch is proportional to the area: $g_m a$; its membrane capacitance is proportional to the specific membrane capacitance: $C_m a$. Similarly, the current is the product of a and the current density, which has units $\mu A\ cm^{-2}$. For example, the absolute capacitive current in area a is $I_c a$.

Table 2.3 | Passive quantities

Quantity	Description	Typical units
R_m	Specific membrane resistance	$\Omega\ cm^2$
C_m	Specific membrane capacitance	$\mu F\ cm^{-2}$
V	Membrane potential	mV
E_m	Leakage reversal potential due to different ions	mV
I_m	Membrane current density	$\mu A\ cm^{-2}$
i_{el}	Injected current	nA
I_c	Capacitive current	nA
I_i	Ionic current	nA

The units of R_m can often seem counter-intuitive. It can sometimes be more convenient to consider its inverse quantity, the specific membrane conductance, with units of $S\ cm^{-2}$.

Ionic current flowing through the membrane charges or discharges the membrane capacitance; that is, changes the amount of charge q on the membrane. We define the **capacitive current** as the rate of change of charge on the membrane $I_c = dq/dt$. Because the lipid bilayer acts as a capacitor, the charge on the membrane is proportional to the voltage across it (Equation 2.1). By differentiating Equation 2.1 for the charge stored on a capacitor with respect to time, we obtain a differential equation that links V and I_c:

$$I_c = \frac{dq}{dt} = C\frac{dV}{dt}. \tag{2.11}$$

This shows that the rate of change of the membrane potential is proportional to the capacitive current. The change in voltage over time, during the charging or discharging of the membrane, is inversely proportional to the capacitance – it takes longer to charge up a bigger capacitor.

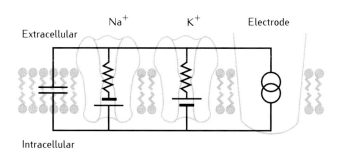

Na$^+$ K$^+$ Electrode

Extracellular

Intracellular

Fig 2.12 The equivalent electrical circuit of a patch of membrane.

2.6 The Equivalent Electrical Circuit of a Patch of Membrane

We have seen how we can represent the permeable and impermeable properties of the membrane as electrical components. Figure 2.12 shows how these components fit together to form an **equivalent electrical circuit** of a small patch of membrane. It comprises the membrane capacitance in parallel with one resistor and battery in series for each type of ion channel; in this example, there are sodium and potassium channels. We have also inserted an electrode into the membrane, which is acting as a current source. We imagine that it is in **current clamp mode**, in which we control the amount of current it delivers. Here we assume that it is delivering a constant amount of current i_{el}, which in electrophysiological applications is usually measured in nanoamps (nA). We suppose that the area of our patch of membrane is a, so the injected current density is i_{el}/a.

By **Kirchhoff's current law**, the injected current density i_{el}/a is equal to the sum of the capacitive current I_c and, in the example shown in Figure 2.12, the sodium and potassium currents:

$$i_{el}/a = I_c + I_{Na} + I_K. \tag{2.12}$$

We can now substitute the capacitive current (Equation 2.11) and the ionic currents (Equation 2.10) to get an equation for how the rate of change of voltage depends on the voltage and the level of current injection:

$$C_m \frac{dV}{dt} = -g_{Na}(V - E_{Na}) - g_K(V - E_K) + i_{el}/a. \tag{2.13}$$

This is a first-order **ordinary differential equation** (ODE) for the membrane potential V. It specifies how, at every instant in time, the rate of change of the membrane potential depends on the membrane potential itself and the current injected. Given an initial membrane potential and time course of injected current pulse, it determines the time course of the membrane potential.

In Equation 2.13, the membrane capacitance C_m is a constant. In this chapter, we consider passive ion channels, with constant conductances g_{Na} and g_K, and we assume constant reversal potentials E_{Na} and E_K. With these assumptions, the right-hand side of the equation is linear in V, which means we can solve the equation to determine the time course of voltage over time, as will be shown in Section 2.6.2, after we have made further simplification.

Kirchhoff's current law is based on the principle of conservation of electrical charge. It states that at any point in an electrical circuit, the sum of currents flowing towards that point is equal to the sum of currents flowing away from that point.

(a)   extracellular

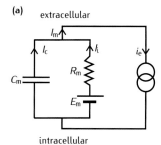

intracellular

(b)

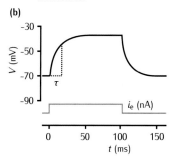

Fig 2.13 (a) The electrical circuit representing a passive patch of membrane. (b) The behaviour of the membrane potential in an RC circuit in response to an injected current pulse, shown below. Here $\tau = RC = 10$ ms.

In general, the conductances g_{Na} and g_K are not constant since ion channels can be active, changing their permeability in response to changes in the membrane potential (Chapters 3 and 4), and the reversal potentials E_{Na} and E_K may also change over time if we model changes to intracellular and extracellular ionic concentrations (Chapters 6 and 10). However, considering passive membranes is a step towards understanding the behaviour of active membranes. In addition, for small deviations of the membrane potential from the resting potential, active channels can be treated as passive channels.

2.6.1 Simplification of the Equivalent Electrical Circuit

We can simplify the electrical circuit representing a patch of passive membrane, such as the circuit shown in Figure 2.12, by lumping together the potassium and sodium currents into one effective **ionic current**. Figure 2.13a shows this simplified circuit. In place of the two resistor/battery pairs in Figure 2.12, there is one pair with a resistor, which we call the specific membrane resistance R_m, measured in $\Omega\,cm^2$, and a membrane battery with an electromotive force E_m, and the ionic current flowing through the resistor and battery is given by the quasi-ohmic relation in Equation 2.10:

$$I_i = \frac{V - E_m}{R_m}.$$ (2.14)

We can derive the values of E_m and R_m from the conductances and reversal potentials of the individual ions using **Thévenin's theorem**. For channels X, Y and Z combined, the equivalent electromotive force and membrane resistance are:

$$E_m = \frac{g_X E_X + g_Y E_Y + g_Z E_Z}{g_X + g_Y + g_Z}$$ (2.15)

$$\frac{1}{R_m} = g_m = g_X + g_Y + g_Z.$$

Note that Equation 2.15 is the ohmic equivalent of the GHK voltage equation (Equation 2.9).

Besides the capacitive current I_c and the ionic current I_i, Figure 2.13 also shows the **membrane current** I_m, defined as the sum of the currents flowing onto or through the membrane, that is, the capacitive current and the ionic current I_i, the sum of I_{Na}, I_K, etc.:

$$I_m = I_c + I_i.$$ (2.16)

Here I_m is just equal to the electrode current density i_{el}/a; were there no electrode, I_m would be zero.

The circuit in Figure 2.12a would be a reasonable description of an idealised 'electrically small' cell. However, in the many cells that are 'electrically large', for example a mammalian cortical pyramidal cell whose dendrites can be millimetres long, there will be contributions to the membrane current from other parts of the cell. In the compartmental modelling method (Chapter 5), we extend the RC circuit to model electrically large cells.

Thévenin's theorem states that any combination of voltage sources and resistances across two terminals can be replaced by a single voltage source and a single series resistor. The voltage is the open circuit voltage E at the terminals and the resistance is E divided by the current with the terminals short-circuited.

As will be described in Chapter 5, we can model neurites in electrically large cells as a chain of connected RC circuits, and a non-zero I_m arises from axial currents flowing from neighbouring circuits. The chain of connected RC circuits forms a compartmental model. We will also clarify what is meant by 'electrically small' and 'electrically large'.

Box 2.6 | Solving differential equations numerically

There are several methods for solving differential equations numerically. Common to all methods is the idea of splitting time into small chunks of length Δt. The smaller Δt is, the more accurate the solution will be, but the longer it will take to generate a solution. To illustrate the principle of numerical integration, we describe the simplest method, called **forward Euler**.

Take, for example, this differential equation linking the rate of change of voltage (here measured in millivolts) with respect to time (here measured in milliseconds):

$$\frac{dV}{dt} = -5\,V, \tag{a}$$

with the starting condition that at $t = 0$ ms, $V(0) = 1$ mV.

In forward Euler, we extrapolate from time t to time $t + \Delta t$ by using the rate of change at time t:

$$V(t + \Delta t) = V(t) + \frac{dV}{dt}\Delta t. \tag{b}$$

Substituting Equation (b) into Equation (a), the change in voltage over time Δt is $\Delta V = dV/dt = -5V\Delta t$. We will assume we have chosen $\Delta t = 0.1$ ms. Dropping units from now on, at $t = 0$, $V(0) = 1$, so the voltage at $t = \Delta t = 0.1$ is $V(0.1) = V(0) + \Delta V = 1 - 5 \times 1 \times 0.1 = 0.5$. We can repeat the calculation to get the voltage at $t = 0.2$: $V(0.2) = V(0.1) + \Delta V = 0.5 - 5 \times 0.5 \times 0.1 = 0.25$.

Figure 2.14 shows the numerical solution resulting from repeating the process, where the circles show the times at which V has been computed using the procedure, that is, $\Delta t, 2\Delta t \ldots$. The exact solution ($V(t) = \exp(-5t)$) is shown and is quite far from the numerical solution. We could make the solution more exact by choosing a smaller Δt – but this would require more calculations, and therefore more computation time.

While forward Euler illustrates the principles of splitting up time, it is not a recommended method of solving differential equations. Section 5.7.1 will demonstrate more practical methods.

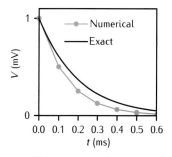

Fig 2.14 Numerical integration (cyan) and exact integration (black) of $dV/dt = -5\,V$, starting at $V(0) = 1$ mV.

2.6.2 The RC Circuit

The simplified circuit shown in Figure 2.13a is well known in electronics, where it is called an RC circuit, since its main elements are a resistor R and a capacitor C. In order to find out how the membrane potential changes when current is injected into the circuit, we replace the contributions from potassium and sodium in Equation 2.13 with the simplified expression for the ionic current in Equation 2.14 to give the simpler differential equation:

$$C_m\frac{dV}{dt} = \frac{E_m - V}{R_m} + \frac{i_{el}}{a}. \tag{2.17}$$

Solving the differential equation is the process of using this equation to calculate how the membrane potential varies over time. In general, differential equations can be solved using appropriate numerical methods, which are programmed into neural simulation computer software, such as *NEURON*

NEURON is a well-known and widely used open-source neural simulator. It computes numerical solutions to the differential equations describing the spatiotemporal variation in the neuron membrane potential. This simulator can be applied to a single neuron or a network of interconnected neurons. We have used *NEURON* to produce the plots in this and other chapters. Other simulators are available.

Solving an equation analytically means that an expression for how the membrane potential (in this case) depends on position and time can be derived as a function of the various parameters of the system. The alternative is to solve the equation numerically.

(Carnevale and Hines, 2006). It is therefore not strictly necessary to know the numerical methods in depth.

However, it is useful to have a basic understanding of the principle of numerical integration, as outlined in Box 2.6. Figure 2.13b shows the result of solving the equation numerically when the injected current is a square pulse of magnitude i_{el} and duration t_e. On the rising edge of the pulse, the membrane potential starts to rise steeply. This rise away from the resting potential is referred to as **depolarisation**, because the amount of positive and negative charge on the membrane is reducing. As the pulse continues, the rise in voltage becomes less steep and the voltage gets closer and closer to a limiting value. On the falling edge of the pulse, the membrane potential starts to fall quite steeply. The rate of fall decreases as the membrane potential gets close to its original value. As the charge on the membrane is building back up to resting levels, this phase is called **repolarisation**. By injecting negative current, it is possible to reduce the membrane potential below its resting level, which is referred to as **hyperpolarisation**.

Generally, it is difficult, and often not possible, to solve differential equations analytically. However, Equation 2.17 is sufficiently simple to allow an analytical solution. We assume that the membrane is initially at rest, so that $V = E_m$ at time $t = 0$. We then integrate Equation 2.17 to predict the response of the membrane potential during the current pulse, giving:

$$V = E_m + \frac{R_m i_{el}}{a}\left(1 - \exp\left(-\frac{t}{R_m C_m}\right)\right). \tag{2.18}$$

This is an inverted decaying exponential that approaches the steady state value $E_m + R_m i_{el}/a$ as time t grows large (Figure 2.13b, around $t = 50$ ms). Defining V_0 as the value the membrane potential has reached at the end of the current pulse at $t = t_e$, the response of the membrane after the end of the pulse is given by:

$$V = E_m + (V_0 - E_m)\exp\left(-\frac{t - t_e}{R_m C_m}\right), \tag{2.19}$$

which is a decaying exponential (Figure 2.13b, from $t = 100$ ms).

In both rising and falling responses, the denominator inside the exponential is the product of the membrane resistance and membrane capacitance $R_m C_m$. This factor has the units of time, and it characterises the length of time taken for the membrane potential to get to $1/e$ (about one-third) of the way from the final value. For this reason, the product $R_m C_m$ is defined as the **membrane time constant** τ_m. It is a measure of how long the membrane 're-members' its original value. Typical values of τ_m for neurons range between 1 and 20 ms. It is possible to measure the membrane time constant for use in a model RC-type circuit. The assumptions that are made when doing this and the effects of measurement accuracy are discussed in Chapter 5.

Another important quantity that characterises the response of neurons to injected current is the **input resistance**, defined as the change in the steady state membrane potential divided by the injected current causing it (Koch, 1999). To determine the input resistance of any cell in which current is injected, the resting membrane potential is first measured. Next, a small amount of current i_{el} is injected, and the membrane potential is allowed to reach a steady state V_∞. The input resistance is then given by:

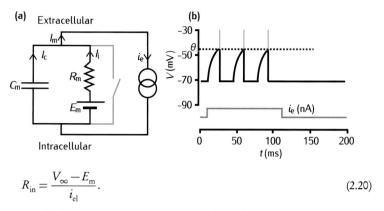

$$R_{in} = \frac{V_\infty - E_m}{i_{el}}. \tag{2.20}$$

Fig 2.15 The IF model. **(a)** The circuit diagram of the model. This is based on an RC circuit (Figure 2.13). When the membrane potential reaches a threshold voltage θ, the neuron is considered to have fired a spike and the switch, in blue on the circuit diagram, closes. This short-circuits the membrane resistance, bringing the membrane potential back to the resting membrane potential E_m. **(b)** Response of the IF circuit to superthreshold current injection. The membrane potential approaches a fixed value exponentially, but before doing so, it hits the threshold, producing a spike (represented by the blue line). The membrane potential is then reset to the resting membrane potential, and after a refractory period, the switch opens, allowing the membrane potential to rise again.

For a single RC circuit representation of a cell, the input resistance can be calculated from the properties of the cell. From Equation 2.17, by setting $dV/dt = 0$, the steady state membrane potential can be shown to be $V_\infty = E_m + (R_m/a)i_{el}$. By substituting this value of V_∞ into Equation 2.20, the input resistance is found to be $R_{in} = R_m/a$. This is a quasi-ohmic current–voltage relation where the constant of proportionality is the input resistance, given by R_m/a.

The input resistance measures the response to a steady state input. A more general concept is the **input impedance**, which measures the amplitude and phase lag of the membrane potential in response to a sinusoidal injection current of a particular frequency. The input impedance of the RC circuit can be computed, which shows that the RC circuit acts as a low-pass filter, reducing the amplitude of high-frequency components of the input signal. The topic of input impedance and the frequency response of neurons is covered in depth by Koch (1999).

2.7 Integrate–and–Fire Neurons

A fundamental feature of many neurons is the **action potential**, or 'spike', the rapid increase and decrease in membrane potential that occurs when the neuron receives sufficient input from synapses or electrodes. The RC circuit (Figure 2.13) is the basis for the Hodgkin–Huxley model (Hodgkin and Huxley, 1952d), which explains the mechanism by which action potentials are generated, and is described in Chapter 3.

The same RC circuit is also the basis for a simpler model of the neuron, the **integrate-and-fire neuron**. It captures the notion of the membrane being charged by currents flowing into the cell and, upon the membrane potential exceeding a threshold, firing an action potential and discharging. However, it does not explain the mechanism of action potential generation. It is the oldest neuron model, often attributed to Lapicque (1907), though only analysed later (Hill, 1936). The name 'integrate-and-fire neuron' was first used by Knight (1972). It can be used in theoretical analysis (e.g. Section 8.5.1) and in network models (e.g. Section 9.3).

The IF model is depicted in the circuit diagram in Figure 2.15. Spike generation and reset of the membrane potential are represented by a switch, which is closed when the membrane potential reaches a specified threshold

The term 'spike' describes the appearance of action potentials when recorded using extracellular electrodes, as described in Chapters 10 and 13.

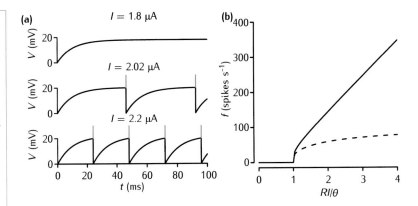

Fig 2.16 Response of integrate-and-fire neurons to current injection. The parameters are $R_{in} = 10\,k\Omega$, $C_m = 1\,\mu F$, $\theta = 20\,mV$ and $E_m = 0\,mV$, and the membrane time constant is $\tau_m = R_m C_m = 10\,ms$. **(a)** Time course of the membrane potential in response to differing levels of current injection. The level of current injection is indicated above each trace. A current of $1.8\,\mu A$ causes the membrane potential to increase but does not take it to the firing threshold of 20 mV. For a level of current injection of $2.02\,\mu A$, which is just above the threshold, the neuron spikes when the membrane potential reaches the threshold of 20 mV. The membrane potential is then reset to rest instantly and the membrane starts charging again. For a higher level of current injection of $2.2\,\mu A$, the same process is repeated more frequently. **(b)** Integrate-and-fire neuron f–I curves. The solid line shows the f–I curve for an integrate-and-fire neuron without an absolute refractory period. The dashed line shows the f–I curve for an integrate-and-fire neuron with an absolute refractory period of 10 ms.

level. It then short-circuits the membrane resistance, bringing the membrane potential back to rest. After a refractory period, the switch opens, allowing the membrane to charge again. This is illustrated in the response of the membrane potential to current injection shown in Figure 2.16a.

When the voltage is below this threshold, its value is determined by the equation for an RC circuit (Equation 2.17), written in terms of the membrane time constant τ_m and the input resistance R_{in}:

$$\tau_m \frac{dV}{dt} = -V + E_m + R_{in} I. \tag{2.21}$$

Here I is the total current flowing into the cell, which could come from an electrode or synapses. When the membrane potential V reaches the threshold, denoted by θ, the neuron fires a spike and the membrane potential V is reset to E_m.

Figure 2.16a shows the behaviour of an integrate-and-fire neuron when steady currents of various magnitudes are injected. Solving for V in Equation 2.21, with small currents, the membrane potential increases from zero and follows an exponential time course, saturating at $R_{in} I$:

$$V = E_m + R_{in} I (1 - \exp(-t/\tau_m)). \tag{2.22}$$

However, if $R_{in} I$ is bigger than the threshold θ, the voltage will cross the threshold at some point in time. The greater the current, the sooner this will happen. The membrane potential then resets to zero and the process repeats.

When there is above-threshold constant input, the integrate-and-fire neuron fires at a constant frequency. The dependence of the firing frequency f on the current I in a basic integrate-and-fire model is shown in the solid f–I curve in Figure 2.16b (Stein, 1965; Knight, 1972). There is no firing ($f = 0$) when the current is below threshold, and the firing frequency increases as the current increases.

This f–I curve is derived by calculating, for a given input current I starting at time $t = 0$, the time T_s at which a spike occurs. The spike occurs when the membrane potential is equal to θ. By substituting $V = \theta$ and $t = T_s$ in Equation 2.22 and rearranging the equation, the time to the spike T_s is:

$$T_s = -\tau_m \ln\left(1 - \frac{\theta}{R_{in} I}\right). \tag{2.23}$$

When $R_\mathrm{m}I$ exceeds θ (a necessary condition for action potentials), the argument of the logarithm is between zero and one. This makes the logarithm negative which, combined with the negative sign in front of τ_m in Equation 2.23, makes the time to the spike positive, as it should be. As the current increases, the $\theta/R_\mathrm{in}I$ term gets smaller, and the argument of the logarithm approaches one, making the magnitude of the logarithm smaller. Thus, the time to spike is shorter for greater input currents, as expected.

The interval between consecutive spikes during constant current injection is the sum of the time to the spike and the absolute refractory period, $T_\mathrm{s} + \tau_\mathrm{r}$. The frequency f is the reciprocal of this interval:

$$f(I) = \frac{1}{\tau_\mathrm{r} + T_\mathrm{s}} = \frac{1}{\tau_\mathrm{r} - \tau_\mathrm{m} \ln(1 - \theta/(R_\mathrm{in}I))}. \qquad (2.24)$$

The solid curve in Figure 2.16b depicts the f–I curve when there is no refractory period ($\tau_\mathrm{r} = 0$). Above a threshold level of current, the firing frequency increases with the current, with no upper bound. The dashed curve in Figure 2.16b shows an f–I curve of an integrate-and-fire neuron with an absolute refractory period of 10 ms. The firing frequency increases more gradually with the current, approaching a maximum rate of 100 Hz.

Chapter 8 contains examples of more sophisticated IF neurons that can mimic the spiking behaviour of various types of neurons. The simple mathematical form of IF neurons allows them to be analysed mathematically and simulated quickly. They are thus often used in large network simulations. Chapter 9 illustrates how they can be used to gain insights into the variability of neuronal firing and networks of reciprocally connected excitatory and inhibitory neurons.

In integrate-and-fire models, the leak battery is often omitted from the circuit. The only effect this has is to make the resting membrane potential 0 mV rather than E_m; this does not affect the dynamics of the membrane potential.

2.8 | Accuracy of Modelling

This chapter has introduced some of the biophysics describing the electrical properties of the membrane, which is needed to build models of the dynamics of electrical activity in neurons. However, it is worth noting the existence of more sophisticated biophysical models than the ones presented here.

Both the approximations expressed by the GHK current equations and quasi-ohmic electrical circuit approximation are used in models. However, neither should be considered a perfect representation of currents through the membrane. The GHK equations were originally used to describe ion permeability through a uniform membrane, whereas today they are used primarily to describe the movement of ions through channels. Assumptions on which the equations are based, such as the independence of movement of ions through the membrane (the independence principle; Box 2.4) and of constant electric fields, are generally not valid within the restricted space of a single channel. It is therefore not surprising that experiments reveal that the flux through channels saturates at large ionic concentrations, rather than increasing without limit as the GHK equations would predict (Hille, 2001).

The passage of ions through ion channels can be modelled in more detail than assumed for the GHK and quasi-ohmic descriptions (Hille, 2001), including using molecular dynamics (Box 4.2 in Chapter 4), but these more detailed descriptions are not typically used in computational models of the electrical activity of neurons. How can we justify using a more inaccurate description when more accurate ones exist? In answer, modelling itself is the process of making approximations or simplifications in order to understand particular aspects of the system under investigation. What simplifications or approximations are appropriate depends on the question that the model is designed to address. For certain questions, the level of abstraction offered by the quasi-ohmic approximation has proved extremely valuable, as we see in Chapter 3. Similarly, the GHK equation is used in many modelling and theoretical approaches to membrane permeability.

When choosing which of these approximations is most appropriate, there are a number of issues to consider. Most ion types do not have a strongly rectifying I–V characteristic in the region of typical membrane potentials, and so the quasi-ohmic approximation is useful. However, if the I–V characteristic is very strongly rectifying (as in the example of calcium), the GHK current equation should give a better fit. Even with fairly weak rectification, the GHK current equation can fit the data better than the quasi-ohmic approximation (Sah et al., 1988).

We might want to model how changes in intracellular concentration affect the I–V characteristic, particularly in the case of calcium, since its intracellular concentration is so low that relatively small influxes can change its concentration by an order of magnitude. The GHK current equations are commonly used to compute the current as a function of changing intracellular and extracellular concentrations and the channel conductance. Moreover, we may need to consider modelling imperfect (and more realistic) ion-selective channels which have permeabilities to more than one ion. All ion-selective channels allow some level of permeability to certain other ions, and so the GHK voltage equation can be used to calculate the reversal potential of these channels.

In the IF model introduced in Section 2.7, all the details mentioned above are cheerfully ignored – and this is perfectly fine if the aim is to understand how noise affects neuronal firing (Chapter 8) or behaviour of a network (Chapter 9). Although the simple integrate-and-fire neuron model cannot reproduce the firing patterns of real neurons, extensions to the model allow it to describe a wide range of neurons (Chapter 8).

2.9 | Summary

This chapter has outlined the primary electrical properties of neurons that provide a basis for the development of neuronal models. The physical properties of certain cell components, such as lipid membranes, intracellular and extracellular solutions and passive membrane channels, are drawn together to build an electrical circuit model of the neurite. This RC circuit model is an approximation of the passive electrical properties and is based on assumptions such as linear I–V characteristics for ions traversing the membrane, that is, passive membrane channels acting as electrical resistors. The Goldman–Hodgkin–Katz theory of current flow through the membrane provides an alternative model that demonstrates that the linear assumptions made in the electrical model are inappropriate for ions such as Ca^{2+}. Models of multiple channel types will generally involve combinations of these approaches (Chapter 4). We have also used the RC circuit as the basis for the simple integrate-and-fire model of neurons, which will be developed and used in Chapters 8 and 9.

Modelling the membrane potential along a length of a neurite can be achieved by connecting together individual electrical circuits, or compartments, as described in Chapter 5. This compartmental modelling is the fundamental approach used for simulating the electrical properties over complex neuronal morphologies.

CHAPTER 3

The Hodgkin–Huxley Model
of the Action Potential

This chapter presents the first quantitative model of active membrane properties, the **Hodgkin–Huxley model**. This was used to calculate the form of the action potentials in the squid giant axon. Our step-by-step account of the construction of the model shows how Hodgkin and Huxley used the **voltage clamp** to produce the experimental data required to construct mathematical descriptions of how the sodium, potassium and leak currents depend on the membrane potential. Simulations of the model produce action potentials similar to experimentally recorded ones and account for the threshold and refractory effects observed experimentally. While subsequent experiments have uncovered limitations in the Hodgkin–Huxley model descriptions of the currents carried by different ions, the **Hodgkin–Huxley formalism** is a useful and popular technique for modelling channel types.

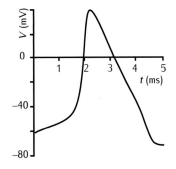

Fig 3.1 Simulated action potential in the squid giant axon at 6.3°C.

3.1 | The Action Potential

In the previous chapter, we described the basis of the membrane resting potential. We now explain a widespread feature of signalling in the nervous system: the **action potential**.

Intracellular recordings (Figure 3.1) demonstrate that action potentials are characterised by a sharp increase in the membrane potential (**depolarisation** of the membrane) followed by a somewhat less sharp decrease towards the resting potential (**repolarisation**). This may be followed by an afterhyperpolarisation phase in which the membrane potential falls below the resting potential before recovering gradually to the resting potential. The main difference between the propagation of action potentials and passive propagation of signals is that action potentials are **regenerative**, so their magnitude does not decay during propagation.

Hodgkin and Huxley (partly in collaboration with Katz) were the first to describe the active mechanisms quantitatively (Hodgkin et al., 1952; Hodgkin and Huxley, 1952a, b, c, d). Their work proceeded in three main stages:

(1) They recorded intracellularly from the squid giant axon. They used a voltage clamp amplifier in space clamp configuration (Box 3.1) to look at how current flow depends on voltage. By changing the extracellular concentration of sodium, they were able to infer how much of the current was carried by sodium ions and how much by other ions, principally potassium.

(2) They fitted these results to a mathematical model. Part of the model is the theoretically motivated framework developed in Chapter 2. Another part is based on the idea of ion-selective, voltage-dependent gates controlled by multiple gating particles. The remainder of the model is determined by fitting curves to experimental data. The model is expressed in terms of a set of equations which are called collectively the **Hodgkin–Huxley model**, or **HH model** for short.

(3) They solved the equations defining the model to describe the behaviour of the membrane potential under various conditions. This involved solving the equations numerically. The simulated action potentials were very similar to the recorded ones. The threshold, propagation speed and refractory properties of the simulated action potentials also matched those of the recorded action potentials.

Their work earned them a Nobel Prize in 1963, shared with Eccles for his work on synaptic transmission.

Hodgkin and Huxley were not able to deduce the molecular mechanisms underlying the active properties of the membrane, which was what they had set out to do (Box 3.3). Nevertheless, their ideas were the starting point for the biophysical understanding of the structures now known as ion channels, the basics of which are outlined in Chapter 4. Hille (2001) provides a comprehensive treatment of the structure and function of ion channels.

The HH model characterises two types of active channel present in the squid giant axon, namely a sodium channel and a potassium channel belonging to the family of potassium delayed rectifier channels. Work since 1952 in preparations from many species has uncovered numerous other types of active channel. Despite the age and limited scope of the HH model, a whole chapter of this book is devoted to it as a good deal of Hodgkin and Huxley's methodology is still used today:

(1) Voltage clamp experiments are carried out to determine the kinetics of a particular type of channel, though now the methods of recording and isolating currents flowing through particular channel types are more advanced.

(2) A model of a channel type is constructed by fitting equations, often of the same mathematical form, to the recordings. Modern methods of fitting equation parameters to data are covered later on, in Chapter 14.

(3) Models of axons, dendrites or entire neurons are constructed by incorporating models of individual channel types in the compartmental models that will be introduced in Chapter 5. Once the equations for the models are solved, albeit using fast computers rather than by hand, action potentials and other behaviours of the membrane potential can be simulated.

The authors discussed whether we should highlight Nobel Prizes. One view is that the work is so great that it does not need reference to prizes to back it up. The opposing, and prevailing, view is that the information is of interest and emphasises the importance of this foundational material.

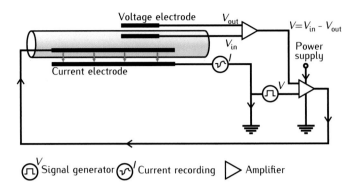

Fig 3.2 The voltage clamp. See Box 3.1 for details.

Box 3.1 | The voltage clamp

The next great experimental advance after intracellular recording (Chapter 2) was the voltage clamp, developed by Cole and Marmont in the 1940s at the University of Chicago (Marmont, 1949; Cole, 1968). Hodgkin, who was already working on a similar idea, learnt about the technique from Cole in 1947. The basic idea is to clamp the membrane potential to a steady value or to a time-varying profile, determined by the experimenter (Figure 3.2). As with a current clamp (Chapter 2), an electrode is used to inject current i_{el} into the cell. At the same time, a voltage electrode records the membrane potential. The apparatus adjusts the injected current continually so that it is just enough to counteract deviations of the recorded membrane potential from the desired voltage value. This ensures that the membrane potential remains at the desired steady value or follows the required time-varying profile.

Hodgkin and Huxley used a **space clamp** configuration, where the electrodes are long, thin wires that short-circuit the electrical resistance of the cytoplasm and the extracellular space. This ensures that the potential is uniform over a large region of membrane and that therefore there is no axial current in the region. There is thus no contribution to the membrane current (I_m; Equation 2.16) from the axial current. In this configuration, the membrane current is identical to the electrode current, so the membrane current can be measured exactly as the amount of electrode current to be supplied to keep the membrane at the desired value.

To understand the utility of the voltage clamp, we recall that the membrane current I_m comprises a capacitive and an ionic current (Equation 3.1). When the voltage clamp is used to set the membrane potential to a constant value, no capacitive current flows as the rate of change in membrane potential, dV/dt, is zero. The voltage clamp current is then equal to the ionic current. Therefore, measuring the voltage clamp current means that the ionic current is being measured directly.

In this chapter, we focus on the second (modelling) and third (simulation) parts of the procedure. In Section 3.2, we begin with a step-by-step description of how Hodgkin and Huxley used the voltage clamp data, physical intuition and curve-fitting to produce their mathematical model. In Section 3.3,

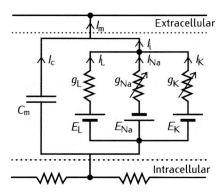

Fig 3.3 The Hodgkin–Huxley equivalent electrical circuit.

we look at simulations of action potentials in a space-clamped axon, and compare these with the experimental recordings. The concepts required to model non-space-clamped axons are first introduced in Chapter 5, but in Section 3.4 we preview simulations of non-space-clamped axons, which give rise to propagating action potentials. In Section 3.5, we consider how Hodgkin and Huxley corrected for temperature. Finally, in Section 3.6, we consider the applicability and accuracy of the HH model.

3.2 The Development of the Model

The starting point of the HH model is the equivalent electrical circuit of a compartment shown in Figure 3.3. There are three types of ionic current in the circuit: a **sodium current** I_{Na}; a **potassium current** I_K, and a current that Hodgkin and Huxley dubbed the **leak current** I_L, which is mostly made up of chloride ions. The key difference between this circuit and the one presented in Chapter 2 is that the sodium and potassium conductances can vary, as indicated by the arrow through their resistors. Their conductance changes with the voltage across them, so we refer to them as **active** rather than **passive** elements (Section 2.1).

The equation that corresponds to the equivalent electrical circuit is:

$$I_m = I_c + I_i = C_m \frac{dV}{dt} + I_i, \tag{3.1}$$

where the **membrane current** I_m and the **capacitive current** I_c are as defined in Chapter 2. The total **ionic current** I_i is the sum of sodium, potassium and leak currents:

$$I_i = I_{Na} + I_K + I_L. \tag{3.2}$$

The magnitude of each type of ionic current is calculated from the product of the ion's driving force and the membrane conductance for that ion:

$$I_{Na} = g_{Na}(V - E_{Na}), \tag{3.3}$$
$$I_K = g_K(V - E_K), \tag{3.4}$$
$$I_L = \overline{g}_L(V - E_L), \tag{3.5}$$

where the sodium, potassium and leak conductances are g_{Na}, g_K and $\overline{g}_L$, respectively, and E_{Na}, E_K and E_L are the corresponding equilibrium potentials.

As defined in Section 2.4.1, the **driving force** of an ion is the difference between the membrane potential and the equilibrium potential of that ion. Hence, the sodium driving force is $V - E_{Na}$.

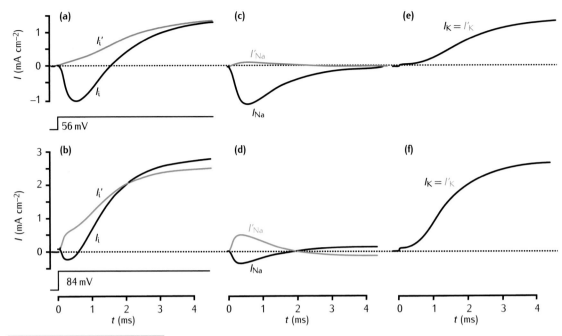

Fig 3.4 The sodium current separated from the other currents using the ion substitution method (Box 3.2). **(a)** The ionic current in seawater (I_i) and in choline water (I'_i) in response to a voltage clamp of 56 mV (seawater) or 60 mV (choline water) above the resting potential. **(b)** The same traces as **(a)**, but in response to a voltage clamp of 84 mV (seawater) and 88 mV (choline water) above the resting potential. **(c,d)** The sodium currents in seawater (I_{Na}) and in choline water (I'_{Na}) inferred from pairs of ionic currents in **(a)** and **(b)**. **(e,f)** The potassium current in seawater (I_K) and in choline water (I'_K) inferred from the pairs of ionic currents in **(a)** and **(b)**, as described in the text. These two currents are, in fact, identical. The recording temperature was 8.5°C. Adapted from Hodgkin and Huxley (1952a), with permission from John Wiley & Sons Ltd.

The bar on the leakage conductance $\overline{g}_L$ indicates that it is a constant, in contrast with the sodium and potassium conductances which depend on the recent history of the membrane potential.

3.2.1 The Potassium Current

Hodgkin and Huxley measured the potassium conductance for a number of voltage clamp holding potentials. After first isolating the potassium current (Box 3.2 and Figure 3.4), they calculated the conductance using Equation 3.4. The form of the curves at each holding potential is similar to the example of the response to a holding potential of 25 mV above rest, shown in

---

Box 3.2 | The ion substitution method

In order to fit the parameters of their model, Hodgkin and Huxley needed to isolate the current carried by each type of ion. To do this they used the **ion substitution method**. They lowered the extracellular sodium concentration by replacing a proportion of the sodium ions in the standard extracellular solution (seawater) with impermeant choline ions, to give a solution called choline water. As can be calculated by evaluating the Nernst equation (Equation 2.7) with the extracellular concentrations set to the normal or reduced values, the sodium equilibrium potential in choline water is lower than in seawater. Thus, at a given voltage clamp level, the currents carried by sodium ions in seawater and choline water differ. On the assumption that the independence principle holds (Box 2.4), the potassium and other ionic currents remain the same. Therefore, the difference between currents recorded in sodium water and choline water can be used to infer the sodium current (Figure 3.4). Having isolated the sodium current and calculated the leak current by other methods, the potassium current can be deduced by subtracting the sodium and leak currents from the total current.

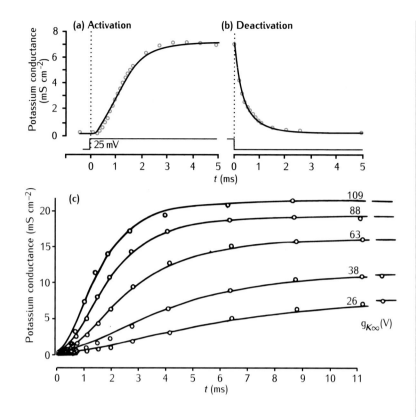

Fig 3.5 Time course of the potassium conductance in a voltage clamp with: **(a)** a step from the resting potential to 25 mV above resting potential; and **(b)** return to the resting potential. The open circles represent data points derived from the experiment. The solid lines are fits to the data (see text). **(c)** Time course of potassium conductance in response to voltage clamp steps to varying holding potentials; the voltage of the holding potential relative to rest is shown on each curve. Note that the activation of the conductance in response to a holding potential of 26 mV is slower than the activation in response to almost the same holding potential in **(a)**. This is due to a difference in recording temperatures: 21°C in **(a)** and **(b)**, compared to 6°C in **(c)**. Adapted from Hodgkin and Huxley (1952d), with permission from John Wiley & Sons Ltd.

Figure 3.5a. Upon depolarisation, the conductance rises to a constant value. This rise in conductance is referred to as **activation**. The conductance stays at this peak value until the voltage is stepped back down to rest, where the conductance then decays exponentially (Figure 3.5b). The fall in conductance is called **deactivation**.

The family of conductance activation curves (Figure 3.5c) show that there are two features of the curve that depend on the level of the voltage clamp holding potential:

(1) The value that the conductance reaches over time, $g_{K\infty}$, increases as the holding potential is increased. It approaches a maximum at high holding potentials. This implied that there was a **maximum potassium conductance** per unit area of membrane, which Hodgkin and Huxley denoted $\overline{g}_K$ and were able to estimate.

(2) The speed at which the limiting conductance is approached becomes faster at higher depolarising holding potentials.

The conductance curves show that the limiting conductance and the rate at which this limit is approached depend on the membrane voltage. Hodgkin and Huxley considered a number of models for describing this voltage dependence (Box 3.3). They settled on the idea of the membrane containing a number of **gates** which can be either closed to the passage of all ions or open to the passage of potassium ions. Each gate is controlled by a number of independent **gating particles**, each of which can be in either an open or a

Box 3.3 | Gating particles

Hodgkin and Huxley's goal had been to deduce the molecular mechanisms underlying the permeability changes evident in their experimental data. Reflecting on this later, Hodgkin (1976) wrote:

> although we had obtained much new information the overall conclusion was basically a disappointment…As soon as we began to think about molecular mechanisms it became clear that the electrical data would by itself yield only very general information about the class of system likely to be involved. So we settled for the more pedestrian aim of finding a simple set of mathematical equations which might plausibly represent the movement of electrically charged gating particles.

Their initial hypothesis was that sodium ions were carried across the membrane by negatively charged carrier particles or dipoles. At rest, these would be held by electrostatic forces. Consequently, they would not carry sodium ions in this state and, on depolarisation, they could carry sodium into the membrane. However, Hodgkin and Huxley's data pointed to a voltage-dependent gate. They settled on deriving a set of equations that would represent the theoretical movement of charged gating particles acting independently in a voltage-dependent manner.

In the contemporary view, the idea of gating particles can be taken to imply the notion of gated channels, but the hypothesis of ion pores or channels was not established at that time. Thus, though Hodgkin and Huxley proposed charged gating particles, it is perhaps tenuous to suggest that they predicted the structure of gated channels. Nevertheless, there is a correspondence between the choice of the fourth power for potassium conductance and the four subunits of the tetrameric potassium channel (Section 4.1).

---

closed position. For potassium ions to flow through a gate, *all* of the gating particles in the gate have to be in the open position.

The movement of gating particles between their closed and open positions is controlled by the membrane potential. The **gating variable** n is the probability of a single potassium gating particle being in the open state. As the gating particles are assumed to act independently of each other, the probability of the entire gate being open is equal to n^x, where x is the number of gating particles in the gate. Although, as described in Chapter 4, gating particles do not act independently, this assumption serves reasonably well in the case of potassium conductance in the squid giant axon. When there are large numbers of particles present, the large numbers ensure the proportion of particles being in the open position is very close to the probability n of an individual channel being in the open position, and the expected proportion of gates open is also the same as the probability of an individual gate being open, n^x.

The conductance of the membrane is given by the maximum conductance multiplied by the probability of a gate being open. For example, if each gate is controlled by four gating particles, as Hodgkin and Huxley's experiments suggested, the relationship between the potassium conductance g_K and the gating particle open probability n is:

The intuition behind the formula n^x can be understood by analogy to a rolling dice. The probability of rolling one six is 1/6. The probability of rolling two sixes is $1/6 \times 1/6 = (1/6)^2 = 1/36$. The probability of rolling x sixes is $(1/6)^x$. If we replace 1/6 by the probability n, we get n^x.

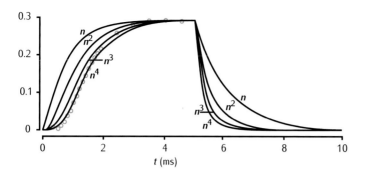

Fig 3.6 A family of curves showing the time course of n raised to various powers. From top to bottom, curves with n raised to the power 1, 2, 3 and 4 are shown. The parameters are as in Figure 3.5: $\tau_n(V_0) = 1.1$ ms, $\tau_n(V_1) = 0.75$ ms, $g_{K\infty}(V_0) = 0.09$ mS cm^{-2} and $g_{K\infty}(V_1) = 7.06$ mS cm^{-2}. To compare the curves, the time course of n raised to the powers 2, 3 and 4 have initial and final values of n, given by $(g_{K\infty}/\overline{g}_K)^{1/2}$, $(g_{K\infty}/\overline{g}_K)^{1/3}$ and $(g_{K\infty}/\overline{g}_K)^{1/4}$. The circular data points shown are the same as in Figure 3.5. Adapted from Hodgkin and Huxley (1952d), with permission from John Wiley & Sons Ltd.

$$g_K = \overline{g}_K n^4. \tag{3.6}$$

If each potassium gate were dependent solely on a single theoretical gating particle, the conductance would be $\overline{g}_K n$.

The transition of a gating particle between its closed (C) and open (O) positions can be expressed as a reversible chemical reaction:

$$C \underset{\beta_n}{\overset{\alpha_n}{\rightleftharpoons}} O. \tag{3.7}$$

The fraction of gating particles that are in the O state is n, and the fraction in the C state is $1-n$. The variables α_n and β_n are **rate coefficients** which depend on the membrane potential; sometimes they are written $\alpha_n(V)$ and $\beta_n(V)$ to highlight their dependence on voltage. Just as rate laws govern the evolution of concentrations in chemical reactions, there is a **rate law** or **first-order kinetic equation** corresponding to Equation 3.7, which specifies how the gating variable n changes over time:

$$\frac{dn}{dt} = \alpha_n(1-n) - \beta_n n. \tag{3.8}$$

The time course of the response of the gating variable n to a step change in membrane potential to a particular voltage V_1 can be determined by integrating Equation 3.8. A solution for the response of n to a voltage step is shown in Figure 3.6, along with the time courses of n raised to various powers. The curve for n looks roughly like the conductance curve shown in Figure 3.5. The main difference is that the theoretical time course of n is not S-shaped like the experimental curve; it has no initial inflection. As Figure 3.6 shows, when the time course of n in response to a positive voltage step is squared, cubed or raised to the power four, the resulting rising curve does have an inflection. The decaying part of the curve retains its decaying exponential shape. Hodgkin and Huxley found that raising n to the power four could give a better fit than cubing or squaring, suggesting that each gate contains four gating particles.

The general form of the time course for $n(t)$ in response to a voltage step is:

$$n(t) = n_\infty(V_1) - (n_\infty(V_1) - n_0)\exp(-t/\tau_n(V_1)), \tag{3.9}$$

where n_0 is the value of n at the start of the step, defined to be at time zero; the variables $n_\infty(V)$ and $\tau_n(V)$ are related to the rate coefficients $\alpha_n(V)$ and $\beta_n(V)$ by:

$$n_\infty = \frac{\alpha_n}{\alpha_n + \beta_n} \quad \text{and} \quad \tau_n = \frac{1}{\alpha_n + \beta_n}, \tag{3.10}$$

where n_∞ is the limiting probability of a gating particle being open if the membrane potential is steady as t approaches infinity, and τ_n is a time constant. When the membrane potential is clamped to V_1, the rate coefficients will immediately move to new values $\alpha_n(V_1)$ and $\beta_n(V_1)$. This means that, with the membrane potential set at V_1, over time n will approach the limiting value $n_\infty(V_1)$ at a rate determined by $\tau_n(V_1)$. The variables n_∞ and τ_n allow Equation 3.8 to be rewritten as:

$$\frac{\mathrm{d}n}{\mathrm{d}t} = \frac{n_\infty - n}{\tau_n}. \tag{3.11}$$

The final step in modelling the potassium current is to determine how the rate coefficients α_n and β_n in the kinetic equation of n (Equation 3.8) depend on the membrane potential. In using experimental data to determine these parameters, it is convenient to use the alternative quantities n_∞ and τ_n (Equation 3.10). The value of n_∞ at a specific voltage V may be determined experimentally by recording the maximum conductance attained at that voltage step, called $g_{K\infty}(V)$. Using Equation 3.6, the value of n_∞ at voltage V is then given by:

$$n_\infty(V) = \left(\frac{g_{K\infty}(V)}{\overline{g}_K} \right)^{\frac{1}{4}}. \tag{3.12}$$

The value for τ_n at a particular membrane potential is obtained by adjusting it to give the best match predicted time course of n given in Equation 3.9 and the data (Figure 3.5).

This process provides values for n_∞ and τ_n at various voltages. Hodgkin and Huxley converted them to the values for α_n and β_n using the inverse formulae to Equation 3.10:

$$\alpha_n = \frac{n_\infty}{\tau_n} \quad \text{and} \quad \beta_n = \frac{1 - n_\infty}{\tau_n}. \tag{3.13}$$

These experimental data points are shown in Figure 3.7, along with plots of the final fitted functions for α_n and β_n; see also Figure 3.11 for the equivalent

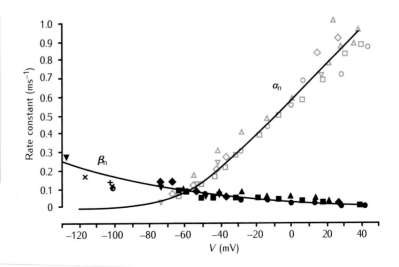

Fig 3.7 Potassium rate coefficients α_n and β_n as a function of membrane potential. Blue symbols refer to measurements of α_n, and black symbols to β_n. The shapes of the symbols identify the axon in which the value was recorded. Adapted from Hodgkin and Huxley (1952d), with permission from John Wiley & Sons Ltd.

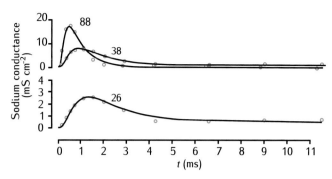

Fig 3.8 Time course of the sodium conductance in a voltage clamp with a step change in voltage from the resting potential to 26 mV, 38 mV and 88 mV above the resting potential (indicated by traces). The open circles represent data points derived from the experiment. The solid lines are fits to the data (see text). Adapted from Hodgkin and Huxley (1952d), with permission from John Wiley & Sons Ltd.

n_∞ and τ_n plots. The equations for the functions $\alpha_n(V)$ and $\beta_n(V)$ are given in the summary of the entire set of equations describing the potassium ionic current through the membrane:

$$I_K = \overline{g}_K n^4 (V - E_K),$$
$$\frac{dn}{dt} = \alpha_n(1-n) - \beta_n n,$$
$$\alpha_n = 0.01 \frac{V + 55}{1 - \exp(-(V+55)/10)},$$
$$\beta_n = 0.125 \exp(-(V+65)/80).$$

(3.14)

Note that the numerical constants 55, 10, 65 and 80 are in the units of millivolts, so V is also measured in millivolts. Alternatively – and more correctly from a physicist's viewpoint – we could add 'mV' after the constants, but this will lead to more cluttered equations. We will follow the convention set by Hodgkin and Huxley of assuming that V is measured in millivolts. The rate constants α_n and β_n are in units of ms^{-1}.

3.2.2 The Sodium Ionic Current

In a similar manner to the procedure used for potassium conductance, Hodgkin and Huxley isolated the sodium current and calculated the sodium conductance curves over a range of voltage clamp steps. The time course of the sodium conductance is illustrated in Figure 3.8. The most notable difference from the potassium conductance is that the sodium conductance reaches a peak and then decays back to rest, even while the clamped voltage remains in a sustained depolarising step. This reduction in conductance is termed **inactivation**, in contrast to deactivation (Section 3.2.1) when the reduction in conductance is due to termination of a voltage step. The time course of the conductance during inactivation differs from the time course during deactivation, and this suggested that two distinct processes can act to reduce the conductance.

The inactivation of the sodium conductance meant that Hodgkin and Huxley could not use the description they used for potassium, where there was just one gating variable n. In order to quantify the inactivation process, Hodgkin and Huxley applied a range of voltage clamp experiments and protocols (Box 3.4, and Figures 3.9 and 3.10). They introduced a gating type variable, called h, to represent the level of inactivation. It could be in either the state of 'not inactivated' or the state of 'inactivated'. The rate of transition between these states is voltage-dependent and governed by a first-order kinetic equation similar to n:

Box 3.4 | Fitting inactivation kinetics

In order to quantify inactivation, Hodgkin and Huxley applied a voltage clamp protocol using two pulses. The first pulse was a long (30 ms) **conditioning pulse**. This was set to a range of different voltages, and its purpose was to give the sodium conductance enough time to inactivate fully at that holding potential. The second pulse was a **test pulse**, which was set to the same value each time. Figure 3.9a shows that the response to the conditioning pulse was similar to the response to a prolonged pulse: the sodium conductance rises to a peak with a height that increases with membrane depolarisation and then decays. The response to the test pulse is similar, but the height of the test pulse depends on the level of the conditioning pulse. The higher the conditioning pulse, the smaller the current amplitude at the test pulse. At a conditioning pulse depolarisation of -41 mV, there is virtually no response to the test pulse. Conversely, when the membrane is hyperpolarised to -116 mV, the amplitude of the current at the pulse reaches a limiting value. This allowed Hodgkin and Huxley to isolate the amount of inactivated conductance at different voltages. By performing many of these experiments over a range of conditioning voltages, they were able to fit the data to produce the voltage-dependent inactivation function h_∞ (Figure 3.9b).

To measure the time constant τ_h of inactivation, a different form of the two-pulse experiment was used (Figure 3.10a). A short depolarising pulse is followed by an interval in which the membrane is clamped to a **recovery potential** and then by a depolarising pulse identical to the first. The peak sodium conductance in both test pulses is measured. The ratio of the two gives a measure of how much the sodium conductance has recovered from inactivation during the time the membrane has been held at the recovery potential. Plotting the ratio against the time of the recovery pulse gives the exponential curve shown in Figure 3.10b, from which the time constant of recovery from inactivation τ_h can be obtained at that particular recovery potential. Over a range of recovery potentials, the voltage dependence of τ_h can be assessed.

$$\frac{dh}{dt} = \alpha_h(1-h) - \beta_h h. \tag{3.15}$$

As with the n gating particle, the voltage-dependent rate coefficients α_h and β_h can be re-expressed in terms of a limiting value h_∞ and a time constant τ_h. Hodgkin and Huxley's experiments suggested that sodium conductance was proportional to the inactivation variable h.

Hodgkin and Huxley completed their model of sodium conductance by introducing another gating particle which, like n, may be viewed as the proportion of theoretical gating particles that are in an open state, determining sodium conductance activation. They called this sodium activation particle m. As with n and h, the time course of m was governed by a first-order kinetic equation with voltage-dependent forward and backward rates α_m and β_m:

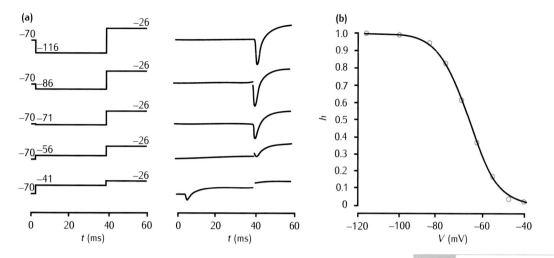

$$\frac{\mathrm{d}m}{\mathrm{d}t} = \alpha_{\mathrm{m}}(1-m) - \beta_{\mathrm{m}}m. \tag{3.16}$$

As with potassium (Figure 3.6), the activation curve of the sodium conductance is inflected. The inflection was modelled satisfactorily by using three independent m gating particles, making the sodium conductance:

$$g_{\mathrm{Na}} = \overline{g}_{\mathrm{Na}}m^3 h. \tag{3.17}$$

This enabled a good fit to be made to experimental recordings by adjusting m_∞ and τ_{m} for different holding potentials and $\overline{g}_{\mathrm{Na}}$ for all holding potentials. As with the gating variable n, Hodgkin and Huxley converted the limiting values and time constants of the m and h variables into rate coefficients (α_{m}, β_{m} and α_{h}, β_{h}) and plotted each as a function of voltage. They then found a fit to each rate coefficient that matched their experimental data. The final model of the sodium current is given by the following set of equations:

$$I_{\mathrm{Na}} = \overline{g}_{\mathrm{Na}}m^3\, h(V - E_{\mathrm{Na}}),$$

$$\frac{\mathrm{d}m}{\mathrm{d}t} = \alpha_{\mathrm{m}}(1-m) - \beta_{\mathrm{m}}m, \qquad \frac{\mathrm{d}h}{\mathrm{d}t} = \alpha_{\mathrm{h}}(1-h) - \beta_{\mathrm{h}}h,$$

$$\alpha_{\mathrm{m}} = 0.1\,\frac{V+40}{1-\exp(-(V+40)/10)}, \quad \alpha_{\mathrm{h}} = 0.07\,\exp(-(V+65)/20), \tag{3.18}$$

$$\beta_{\mathrm{m}} = 4\,\exp(-(V+65)/18), \qquad \beta_{\mathrm{h}} = \frac{1}{\exp(-(V+35)/10)+1},$$

where, as in Equation 3.14, V refers to the numerical value of the membrane potential in millivolts, and the rate constants α_{m}, α_{h}, β_{m} and β_{h} are in units of ms^{-1}.

Fig 3.9 **(a)** Two-pulse protocol used to calculate the influence of membrane potential on the inactivation of sodium in the squid giant axon. From a rest holding potential of -70 mV, the membrane is shifted to a test potential and then stepped to a fixed potential (-26 mV). The sodium current recorded in response to the final step (right) is influenced by the level of inactivation resulting from the test potential. **(b)** The level of inactivation as a function of the test potential (recorded current relative to the maximum current). Adapted from Hodgkin and Huxley (1952c), with permission from John Wiley & Sons Ltd.

3.2.3 The Leak Current

Hodgkin and Huxley's evidence suggested that while potassium is a major part of the non-sodium ionic current, other ions besides sodium might carry current across the membrane. At the potassium equilibrium potential, they found that some non-sodium current still flows. This current could not be due to potassium ions since the driving force $V - E_{\mathrm{K}}$ was zero. Hodgkin and

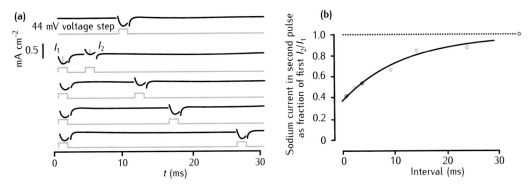

Fig 3.10 **(a)** The membrane currents associated with two square waves applied in succession to the squid giant axon (square wave amplitude 44 mV and duration 1.8 ms, blue traces). The interval between the two pulses is varied, allowing the recovery from inactivation to be plotted as a function of time for different intervals. **(b)** The ratio I_2/I_1 of the amplitude of the current in the second pulse I_2 to the amplitude of the current in the first pulse I_1 is plotted against the interval between the pulses. Adapted from Hodgkin and Huxley (1952c), with permission from John Wiley & Sons Ltd.

Huxley proposed that it was due to a mixture of ions, and they dubbed it the leak current I_L. They assumed this was a resting background conductance that was not dependent on voltage. Using a quasi-ohmic current–voltage relationship, they derived E_L and $\overline{g}_L$ from their experimental results. Both the leakage conductance and equilibrium potential are due largely to the permeability of the membrane to chloride ions. The leak current is modelled by:

$$I_L = \overline{g}_L(V - E_L). \tag{3.19}$$

Although the leak conductance $\overline{g}_L$ in the HH circuit and the membrane resistance R_m in the passive circuit (Chapter 2) appear similar, they have different meanings. In the HH model, the resting membrane potential differs from the electromotive force of the leak battery and the resting membrane resistance is not equal to the inverse of the leak conductance. Instead, the resting membrane potential and the resting membrane resistance are determined by the sodium, potassium and leak resting conductances.

3.2.4 The Complete Model

In the final paper of the series, Hodgkin and Huxley (1952d) inserted their expressions for the three ionic currents (Equations 3.3–3.5) into the membrane equation (Equation 3.1) to give a description of how the membrane potential in a small region of the squid giant axon changes over time:

$$C_m \frac{dV}{dt} = -\overline{g}_L(V - E_L) - \overline{g}_{Na} m^3 h(V - E_{Na}) - \overline{g}_K n^4(V - E_K) + I_m, \tag{3.20}$$

where the membrane current I_m is, for the space clamp conditions, set to zero. When the space clamp is not present, the membrane current is supplied by the net contribution of the axial current from neighbouring regions of the axon, which is referred to as the **local circuit current**. As will be described in Chapter 5, the local circuit currents depend on the variation of the membrane potential along the axon. When Equation 3.20 with $I_m = 0$ is put together with the differential equations for the gating variables n, m and h and the expressions for the rate coefficients (Equations 3.14 and 3.18), the resulting set of four coupled differential equations forms the HH model under space clamp conditions, which is summarised in Box 3.5.

Box 3.5 Summary of the Hodgkin–Huxley model under space clamp conditions

Under space clamp conditions, i.e. no axial current:

$$C_m \frac{dV}{dt} = -\overline{g}_L(V - E_L) - \overline{g}_{Na}m^3 h(V - E_{Na}) - \overline{g}_K n^4(V - E_K).$$

Sodium activation and inactivation gating variables (V refers to the membrane potential in millivolts, and the rate constants α_m, α_h, β_m and β_h are in units of ms^{-1}):

$$\frac{dm}{dt} = \alpha_m(1 - m) - \beta_m m, \qquad \frac{dh}{dt} = \alpha_h(1 - h) - \beta_h h,$$

$$\alpha_m = 0.1 \frac{V + 40}{1 - \exp(-(V + 40)/10)}, \qquad \alpha_h = 0.07 \exp(-(V + 65)/20),$$

$$\beta_m = 4 \exp(-(V + 65)/18), \qquad \beta_h = \frac{1}{\exp(-(V + 35)/10) + 1}.$$

Potassium activation gating variable (V refers to the membrane potential in millivolts, and the rate constants α_n and β_n are in units of ms^{-1}):

$$\frac{dn}{dt} = \alpha_n(1 - n) - \beta_n n,$$

$$\alpha_n = 0.01 \frac{V + 55}{1 - \exp(-(V + 55)/10)},$$

$$\beta_n = 0.125 \exp(-(V + 65)/80).$$

Parameter values (from Hodgkin and Huxley, 1952d):

$$
\begin{aligned}
C_m &= 1.0\,\mu\text{F cm}^{-2} \\
E_{Na} &= 50\,\text{mV} & \overline{g}_{Na} &= 120\,\text{mS cm}^{-2} \\
E_K &= -77\,\text{mV} & \overline{g}_K &= 36\,\text{mS cm}^{-2} \\
E_L &= -54.4\,\text{mV} & \overline{g}_L &= 0.3\,\text{mS cm}^{-2}
\end{aligned}
$$

See Figure 3.11 for plots of the voltage dependence of the gating particle rate coefficients.

3.3 Simulating Action Potentials

In order to predict how the membrane potential changes over time, the complete system of coupled non-linear differential equations comprising the HH model (Box 3.5) has to be solved. Hodgkin and Huxley used numerical integration methods (Section 5.7) more sophisticated than the forward Euler method described in Box 2.6. It took them three weeks' work on a hand-operated calculator. Nowadays, it takes a matter of milliseconds for fast computers to solve the many coupled differential equations in a compartmental formulation of the HH model.

In this section, we look at the action potentials that these equations predict. This will lead us to comparisons with experimental recordings and a brief review of the insights that this model provided. It is worth noting that the recordings in this section were all made at 6.3°C, and the equations and simulations all apply to this temperature. Hodgkin and Huxley discovered

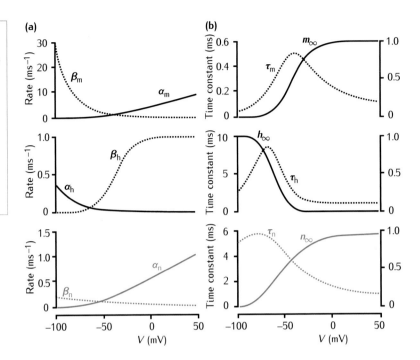

Fig 3.11 Voltage dependence of rate coefficients and limiting values and time constants for the Hodgkin-Huxley gating variables. **(a)** Graphs of forward rate variables α_m, α_h and α_n (solid lines) and backward rate variables β_m, β_h and β_n (dashed lines) for the m, h and n gating particles. **(b)** The equivalent graphs for m_∞, h_∞ and n_∞ (solid lines) and τ_m, τ_h and τ_n (dotted lines).

that temperature has a strong influence on the rate coefficients of the gating variables, but were able to correct for this, as will be discussed in Section 3.5.

In one set of experiments under space clamp (but not voltage clamp) conditions, Hodgkin and Huxley depolarised the membrane potential to varying levels by charging the membrane quickly with a brief current clamp pulse. Small depolarisations led to the membrane potential decaying back to its resting value, but when the membrane was depolarised above a threshold of around 10 mV above the resting potential, action potentials were initiated (Figure 3.12). Hodgkin and Huxley referred to these action potentials induced under space clamp conditions as **membrane action potentials**.

To simulate the different depolarisations in experiments, they integrated the equations of their space-clamped model with different initial conditions for the membrane potential. Because the current pulse that caused the initial depolarisation was short, it was safe to assume that initially n, m and h were at their resting levels.

The numerical solutions were remarkably similar to the experimental results (Figure 3.12). Just as in the experimental recordings, super-threshold depolarisations led to action potentials and sub-threshold ones did not, though the threshold depolarisation was about 6 mV above rest instead of 10 mV. The time courses of the observed and calculated action potentials were very similar, although the peaks of the calculated action potentials were too sharp and there was a kink in the falling part of the action potential curve.

Besides reproducing the action potential, the HH model offers insights into the mechanisms underlying it, which experiments alone were not able to do. Figure 3.13 shows how the sodium and potassium conductances and the gating variables change during a membrane action potential. At the start of the recording, the membrane has been depolarised to above the threshold.

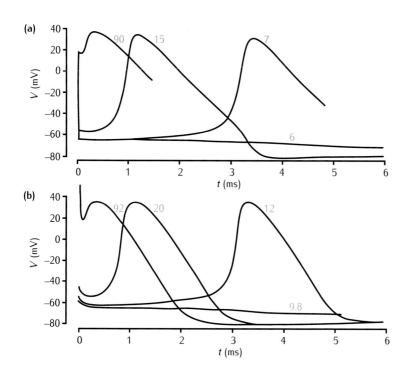

Fig 3.12 Simulated and experimental membrane action potentials. (a) Solutions of the Hodgkin-Huxley equations for isopotential membrane for initial depolarisations of 90, 15, 7 and 6 mV above the resting potential at 6.3°C. (b) Experimental recordings with a similar set of initial depolarisations at 6.3°C. Adapted from Hodgkin and Huxley (1952d), with permission from John Wiley & Sons Ltd.

This causes activation of the sodium current, as reflected in the increase in m and g_{Na}. Recall that the dependence of m on the membrane potential is roughly sigmoidal (Figure 3.11). As the membrane potential reaches the sharply rising part of this sigmoid curve, the g_{Na} activation increases greatly. As the sodium reversal potential is much higher than the resting potential, the voltage increases further, causing the sodium conductance to increase still further. This snowball effect produces a sharp rise in the membrane potential.

The slower potassium conductance g_K, the n gating variable, starts to activate soon after the sharp depolarisation of the membrane. The potassium conductance allows current to flow out of the neuron because of the low potassium reversal potential. The outward current flow starts to repolarise the cell, taking the membrane potential back down towards rest. It is the delay in its activation and repolarising action that leads to this type of potassium current being referred to as the **delayed rectifier** current. Note that this usage of the term 'rectifier' has a different meaning to rectification as a property of I-V characteristics, introduced in Section 2.2.6.

The repolarisation of the membrane is also assisted by the inactivating sodium variable h, which decreases as the membrane depolarises, causing the inactivation of g_{Na} and reduction of the sodium current flow into the cell. The membrane potential quickly swoops back down to its resting level, overshooting somewhat to hyperpolarise the neuron. This causes the rapid deactivation of the sodium current (m reduces) and its **deinactivation**, whereby the inactivation is released (h increases). In this phase, the potassium conductance also deactivates. Eventually all the state variables return to their resting states and the membrane potential returns to its resting level.

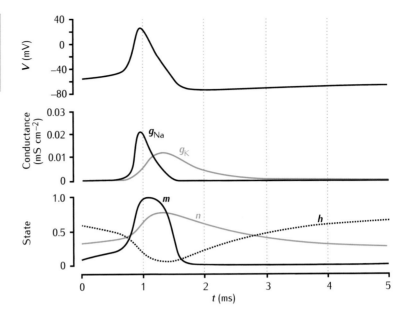

Fig 3.13 The time courses of membrane potential, conductances and gating variables during an action potential.

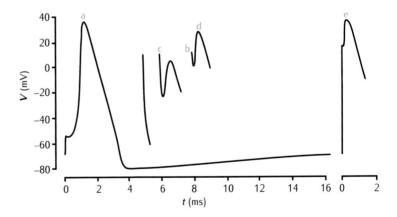

Fig 3.14 Refractory properties of the HH model. Upper curves are calculated membrane action potentials at 6.3°C. Curve **a** is the response to a fast current pulse that delivers 15 nC cm⁻². Curves **b** to **d** are the responses to a charge of 90 nC cm⁻² delivered at different times after the initial pulse. Curve **e** shows the response to a charge of 90 nC cm⁻² after the resting state has again been reached. Adapted from Hodgkin and Huxley (1952d), with permission from John Wiley & Sons Ltd.

The HH model also explains the **refractory period** of the axon. During the **absolute refractory period** after an action potential, it is impossible to generate a new action potential by injecting current. Thereafter, during the **relative refractory period**, the threshold is higher than when the membrane is at rest, and action potentials initiated in this period have a lower peak voltage. As seen in Figure 3.13, the gating variables take a long time, relative to the duration of an action potential, to recover to their resting values. It should be harder to generate an action potential during this period for two reasons. Firstly, the inactivation of the sodium conductance (low value of h) means that any increase in m due to increasing voltage will not increase the sodium conductance as much as it would when h is at its higher resting value (Figure 3.11). Secondly, the prolonged activation of the potassium conductance means that any inward sodium current has to counteract a more considerable outward potassium current than in the resting state. Hodgkin and Huxley's simulations (Figure 3.14) confirmed this view, and were in broad agreement with their experiments.

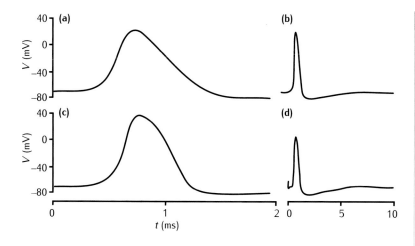

Fig 3.15 Calculated and recorded propagating action potentials. **(a)** Time course of action potential calculated from the Hodgkin-Huxley equations. The conduction velocity was $18.8\,\mathrm{m\,s^{-1}}$ and the temperature $18.5°C$. **(b)** Same action potential at a slower timescale. **(c)** Action potential recorded from the squid giant axon at $18.5°C$ on the same timescale as the simulation in **(a)**. **(d)** Action potential recorded from a different squid giant axon at $19.2°C$ at a slower timescale. Adapted from Hodgkin and Huxley (1952d), with permission from John Wiley & Sons Ltd.

3.4 | Propagating Action Potentials

Hodgkin and Huxley calculated how action potentials propagate in a length of axon not under space clamp conditions by solving an extended version of the equations in Box 3.5 that include local circuit currents. We will describe modelling spatially extended axons in Chapter 5, and here we will preview the results. The propagated action potential calculated by Hodgkin and Huxley was also remarkably similar to the experimentally recorded action potential (Figure 3.15). The value of the velocity they calculated was $18.8\,\mathrm{m\,s^{-1}}$, close to the experimental value of $21.2\,\mathrm{m\,s^{-1}}$ at $18.5°C$.

Figure 3.16 shows the capacitive, local and ionic currents flowing at different points on the membrane at a particular instant when an action potential is propagating from left to right. At the far right, local circuit currents are flowing in from the left because of the greater membrane potential there. These local circuit currents charge the membrane capacitance, leading to a rise in the membrane potential. Further to the left, the membrane is sufficiently depolarised to open sodium channels, allowing sodium ions to flow into the cell. Further left still, the sodium ionic current makes a dominant contribution to charging the membrane, leading to the opening of more sodium channels and the rapid rise in the membrane potential that characterises the initial phase of the action potential. To the left of this, the potassium conductance is activated, due to the prolonged depolarisation. Although sodium ions are flowing into the cell here, the net ionic current is outward. This outward current, along with a small local circuit contribution, discharges the membrane capacitance, leading to a decrease in the membrane potential. At the far left, in the falling part of the action potential, only potassium flows as sodium channels have inactivated. The final **afterhyperpolarisation** potential (not shown fully for reasons of space) is very small. In this part, sodium is deinactivating and potassium is deactivating. This leads to a small inward current that brings the membrane potential back up to its resting potential.

Hodgkin and Huxley had to assume that the membrane potential propagated at a constant velocity so that they could convert the partial differential equation into a second-order ordinary differential equation, giving a soluble set of equations.

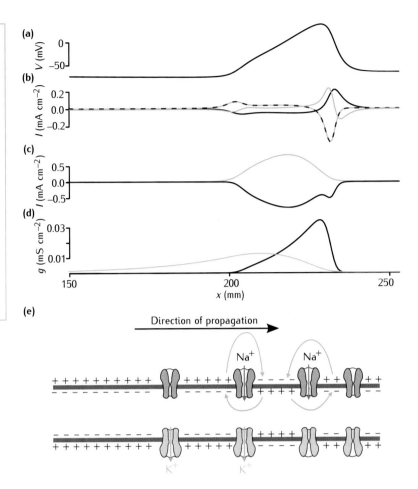

Fig 3.16 Currents flowing during a propagating action potential. **(a)** The voltage along an axon at one instant in time during a propagating action potential. **(b)** The axial current (blue line), the ionic current (dashed black–blue line) and the capacitive current (black line) at the same points. **(c)** The sodium (black) and potassium (blue) contributions to the ionic current. The leak current is not shown. **(d)** The sodium (black) and potassium (blue) conductances. **(e)** Representation of the state of ion channels, the membrane and local circuit current along the axon during a propagating action potential.

3.5 | The Effect of Temperature

Hodgkin et al. (1952) found that the temperature of the preparation affects the time course of voltage clamp recordings strongly: the rates of activation and inactivation increase with increasing temperature. In common with many biological and chemical processes, the rates increase roughly exponentially with the temperature. The Q_{10} **temperature coefficient**, a measure of the increase in rate for a 10°C temperature change, is used to quantify this temperature dependence:

$$Q_{10} = \frac{\text{rate at } T + 10°C}{\text{rate at } T}. \tag{3.21}$$

If the values of the HH voltage-dependent rate coefficients α and β at a temperature T_1 are $\alpha(V, T_1)$ and $\beta(V, T_1)$, then their values at a second temperature T_2 are:

$$\alpha(V, T_2) = \alpha(V, T_1) Q_{10}^{\frac{T_2 - T_1}{10}} \quad \text{and} \quad \beta(V, T_2) = \beta(V, T_1) Q_{10}^{\frac{T_2 - T_1}{10}}. \tag{3.22}$$

In the alternative form of the kinetic equations for the gating variables (see, for example, Equation 3.11), this adjustment due to temperature can be achieved by decreasing the time constants τ_n, τ_m and τ_h by a factor of $Q_{10}^{(T_2-T_1)/10}$ and leaving the steady state values of the gating variables n_∞, m_∞ and h_∞ unchanged.

Hodgkin et al. (1952) estimated, from recordings, a Q_{10} of about 3 for the time constants of the ionic currents. This is typical for the rate coefficients of ion channels (Hille, 2001). In fact, the principles of the transition state theory, outlined in Section 4.8.1, show that the Q_{10} itself is expected to depend on temperature: the Q_{10} at 6°C is not expected to be the same as the Q_{10} measured at 36°C. The transition state theory also allows temperature to be incorporated into the equations for the rate coefficients explicitly, rather than as a correction factor.

As well as the rate coefficients, the maximum channel conductances also increase with temperature, albeit not as strongly. If the maximum conductance for an ion type X is $\overline{g}_X(T_1)$ at temperature T_1, at temperature T_2 it will be given by:

$$\overline{g}_X(T_2) = \overline{g}_X(T_1)Q_{10}^{\frac{T_2-T_1}{10}}. \tag{3.23}$$

The Q_{10} is typically around 1.2 to 1.5 for conductances (Hodgkin et al., 1952; Rodriguez et al., 1998; Hille, 2001).

Assuming that reaction rate increases exponentially with temperature is equivalent to assuming that the effect is multiplicative. If the rate coefficient increases by a factor Q for a 1°C increase in temperature, for a 2°C increase, it is $Q \times Q$; for a 10°C increase, it is $Q_{10} \equiv Q^{10}$ and for an increase from T_1 to T_2, it is $Q^{(T_2-T_1)}$ or $Q_{10}^{(T_2-T_1)/10}$.

3.6 Applicability and Accuracy of the Hodgkin–Huxley Model

The set of equations that make up the HH model (Box 3.5) were constructed to explain the generation and propagation of action potentials specifically in the squid giant axon. How relevant is the HH model to other preparations? While the parameters and equations for the rate coefficients present in the HH model are particular to the squid giant axon, the general idea of gates comprising independent gating particles is used widely to describe other types of channel. In this section, we explore the model assumptions and highlight the constraints imposed by the formalism of independent gating particles. Although the assumptions and approximations mean that the model is not a completely accurate representation of what is known about the biology, the approximations are not so gross as to destroy the explanatory power of the model.

3.6.1 Partial Channel Selectivity

Implicit in the HH model is the notion that channels are selective for only one type of ion. In fact, all ion channels are somewhat permeable to ions other than the dominant permeant ion (Section 2.1). Voltage-gated sodium channels in the squid giant axon are about 8% as permeable to potassium as they are to sodium, and potassium channels are typically around 1% as permeable to sodium as they are to potassium (Hille, 2001).

3.6.2 The Independence Principle

As it is assumed that each type of current does not depend on the concentrations of other types of ion, these equations imply that the independence principle holds (Box 2.4). Hodgkin and Huxley (1952a) verified, to the limit of the resolving power of their experiments, that the independence principle holds for the sodium current. However, improved experimental techniques have revealed that this principle of independence does not hold exactly in general (Section 2.8).

3.6.3 The Linear Instantaneous *I–V* Characteristic

One of the key elements of the HH model is that all the ionic currents that flow through open gates have a linear, quasi-ohmic dependence on the membrane potential (Equations 3.3–3.5), for example:

$$I_{Na} = g_{Na}(V - E_{Na}). \tag{3.3}$$

As described in Chapter 2, this relation is an approximation of the non-linear Goldman–Hodgkin–Katz current equation, which itself is derived theoretically from assumptions such as there being a constant electric field in the membrane.

Hodgkin and Huxley (1952b) did not take these assumptions for granted, and carried out experiments to check the validity of Equation 3.3, and the corresponding equation for potassium. Testing this relation appears to be a matter of measuring an *I–V* characteristic, but in fact it is more complicated, since, as seen earlier in the chapter, the conductance g_{Na} changes over time, and the desired measurements are values of current and voltage at a *fixed* value of the conductance. It was not possible for Hodgkin and Huxley to fix the conductance, but they made use of their observation that it is the *rate of change* of an ionic conductance that depends directly on voltage, not the ionic conductance itself. Therefore, in a voltage clamp experiment, if the voltage is changed quickly, the conductance has little chance to change, and the values of current and voltage just before and after the voltage step can be used to acquire two pairs of current and voltage measurements. If this procedure is repeated with the same starting voltage level and a range of second voltages, an *I–V* characteristic can be obtained.

As explained in more detail in Box 3.6, Hodgkin and Huxley found that the quasi-ohmic *I–V* characteristics given in Equations 3.3–3.5 were appropriate for the squid giant axon. They referred to this type of *I–V* characteristic as the **instantaneous *I–V* characteristic**, since the conductance is given no time to change between the voltage steps. In contrast, if the voltage clamp current is allowed time to reach a steady state after setting the voltage clamp holding potential, the *I–V* characteristic measured is called the **steady state *I–V* characteristic**. In contrast to the instantaneous *I–V* characteristic, this is non-linear in the squid giant axon. With the advent of single channel recording (Chapter 4), it is possible to measure the *I–V* characteristic of an open channel directly in the open and closed states, as, for example, do Schrempf et al. (1995).

A potentially more accurate way to model the *I–V* characteristics would be to use the GHK current equation (Box 2.4). For example, the sodium current would be given by:

$$I_{Na}(t) = P_{Na}(t) \frac{F^2 V(t)}{RT} \left(\frac{[Na^+]_{in} - [Na^+]_{out} e^{-FV(t)/RT}}{1 - e^{-FV(t)/RT}} \right), \tag{3.24}$$

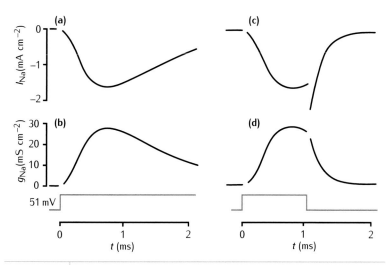

Fig 3.17 Sodium current and sodium conductance under two different voltage clamp conditions. (a) The sodium current measured in response to a voltage clamp step at $t = 0$ of 51 mV above the resting potential. (b) The conductance calculated from the current using Equation 3.3. (c) The current measured in response to a voltage step of the same amplitude, but which only lasted for 1.1 ms before returning to the resting potential. The current is discontinuous at the end of the voltage step. The gap in the record is due to the capacitive surge being filtered out. (d) The conductance calculated from recording (c) and the value of the membrane potential. Although there is still a gap in the curve due to the capacitive surge, the conductance appears to be continuous at the point where the current was discontinuous. Adapted from Hodgkin and Huxley (1952b), with permission from John Wiley & Sons Ltd.

Box 3.6 | Verifying the quasi-ohmic *I–V* characteristic

To verify that the instantaneous *I–V* characteristics of the sodium and potassium currents were quasi-ohmic, Hodgkin and Huxley (1952a) made a series of recordings using a two-step voltage clamp protocol (Figure 3.17). In contrast to the one-step protocol encountered so far (included for comparison in Figure 3.17a, b) in every recording from the two-step protocol (Figure 3.17c, d), the first step was of the same duration, and depolarised the membrane to the same level. This caused sodium and potassium channels to open. The second step was to a different voltage in each experiment in the series. The ion substitution method allowed the sodium and potassium currents to be separated.

Figure 3.17c shows one such recording of the sodium current. At the end of the step, the current increases discontinuously and then decays to zero. There is a small gap due to the capacitive surge. The current just after the discontinuous leap (I_2) depends on the voltage of the second step (V_2). When I_2 was plotted against V_2, a linear relationship passing through the sodium equilibrium potential E_{Na} was seen. The gradient of the straight line was the conductance at the time of the start of the second voltage step. The linear relationship justified the calculation of the conductance from the current and driving force according to Equation 3.3. Figure 3.17d shows the conductance so calculated. In contrast to the current, it is continuous at the end of the voltage step, apart from the gap due to the capacitive surge.

where $P_{Na}(t)$ is the permeability to sodium at time t. This equation could be rearranged to determine the permeability over time from voltage clamp recordings, and then a gating particle model for the permeability (for example, of the form $P_{Na} = \bar{P}_{Na} m^3 h$) could be derived. Sometimes it is desirable to use this form of the model, particularly where the *I–V* characteristic is non-linear and better fitted by the GHK equation. This is particularly the case for ions whose concentration differences across the membrane are large, such as in the case of calcium (Figure 2.10b).

3.6.4 The Independence of Gating Particles

Alternative interpretations and fits of the voltage clamp data have been proposed. For example, Hoyt (1963, 1968) suggested that activation and inactivation are coupled. This was later confirmed through experiments that removed the inactivation in the squid giant axon using the enzyme pronase (Bezanilla and Armstrong, 1977). Subsequent isolation of the inactivation time course revealed a lag in its onset that did not conform to the independent particle hypothesis. Consequently, more accurate models of sodium activation and inactivation require a more complex set of coupled equations (Goldman and Schauf, 1972). Unrestricted kinetic schemes, described in Section 4.5.3, provide a way to model dependencies such as this.

3.7 Summary

In their model, Hodgkin and Huxley introduced active elements into the passive membrane equation. These active currents are specified by theoretical membrane-bound gated channels, or gates, each gate comprising a number of independent gating particles. While the HH formalism does not relate directly to the physical structure of channels, it does provide a framework within which to describe experimental data. In particular, kinetic reaction equations allow the system to be fitted to voltage-dependent characteristics of the active membrane currents through the voltage dependence of the kinetic rate coefficients. Putative functions for the kinetic rate coefficients are fitted to experimental voltage clamp data. The resulting quantitative model not only replicates the voltage clamp experiments to which it is tuned, but also reproduces the main features of the action potential.

This chapter has described a specific system, the squid giant axon under space clamp conditions. The next two chapters will generalise the model to describe electrical activity in almost any type of neuron. Chapter 4 describes how types of voltage-gated channel beyond the squid sodium and potassium channels can be modelled using the HH formalism, or by more complex models. The full explanation of how to model propagating action potentials (Figure 3.16) is given in Chapter 5, which describes how to model electrical activity in axons and dendrites with any morphology. Also introduced in Chapter 4 are channels dependent on ionic concentrations, and Chapter 7 covers ligand-gated channels, as found in chemical synapses.

CHAPTER 4

Models of Active Ion Channels

There are many types of active ion channel beyond the squid giant axon sodium and potassium voltage-gated ion channels studied in Chapter 3, including channels gated by ligands such as calcium. The aim of this chapter is to present methods for modelling the kinetics of any voltage-gated or ligand-gated ion channel. The formulation used by Hodgkin and Huxley of independent gating particles can be extended to describe many types of ion channel. This formulation is the foundation for **thermodynamic models**, which provide functional forms for the rate coefficients derived from basic physical principles. To improve on the fits to data offered by models with independent gating particles, the more flexible **Markov models** are introduced. When and how to interpret kinetic schemes probabilistically to model the stochastic behaviour of single ion channels will be considered. Experimental techniques for characterising channels are outlined and an overview of the biophysics of channels relevant to modelling channels is given.

Since the Hodgkin–Huxley (HH) model was published, over 100 types of ion channel have been identified. Each type of channel has a distinct response to the membrane potential, intracellular ligands, such as calcium, and extracellular ligands, such as neurotransmitters. The membrane of a single neuron may contain a dozen or more different types, with the density of each type depending on its location in the membrane. The distribution of ion channels over a neuron affects many aspects of neuronal function, including the shape of the action potential, the duration of the refractory period, how synaptic inputs are integrated and the influx of calcium into the cell. When channels are malformed due to genetic mutations, diseases such as epilepsy, chronic pain, migraine and deafness can result (Jentsch, 2000; Catterall et al., 2008).

Models of neurons containing multiple channel types are an invaluable aid to understanding how combinations of ion channels can affect the time course of the membrane potential. For example, later on in this chapter, it will be shown that the addition of just one type of channel to a HH model neuron can have profound effects on its firing properties.

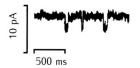

Fig 4.1 Single channel recording of the current passing through an acetylcholine-activated channel recorded from frog muscle in the presence of acetylcholine (Neher and Sakmann, 1976). Though there is some noise, the current can be seen to flip between two different levels. Reprinted by permission from Macmillan Publishers Ltd: *Nature* 260, 779–802, © 1976.

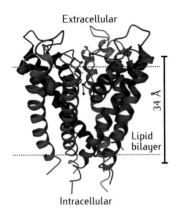

Extracellular

34 Å

Lipid
bilayer

Intracellular

Fig 4.2 Structure of a potassium channel (the KcsA channel) from the soil bacterium *Streptomyces lividans*, determined by X-ray crystallography by Doyle et al. (1998). The *kcsA* gene was identified, expressed and characterised electrophysiologically by Schrempf et al. (1995). The KcsA channel is blocked by caesium ions, giving rise to its name. The colours indicate the four subunits of the channel. Image from the Research Collaboratory for Structural Bioinformatics Protein Data Bank (RCSB PDB, rcsb.org) of PDB ID 1BL8 (Doyle et al., 1998).

In order to build neuron models with multiple channel types, models of individual constituent channel types must be constructed. In this chapter, it will be shown how the independent gating formalism used in the HH model (Chapter 3) can be extended to model different types of voltage- and ligand-gated channels.

Independent gating particle models are sufficiently accurate to explore many questions about neuronal electrical activity, but even with optimal parameter tuning, there are discrepancies between their behaviour and the behaviour of certain types of channel. An improved fit to the data can be achieved by using Markov models, which are not constrained by the idea of independent gating particles and consider the state of the entire channel, rather than constituent gating particles.

A further modification to channel models is required in order to explain voltage clamp recordings from single channels made using the patch clamp technique, pioneered by Neher and Sakmann (1976). Rather than being smoothly varying, single channel currents switch randomly between zero and a fixed amplitude (Figure 4.1). This chapter will show how probabilistic Markov models can be used to understand single channel data and will introduce a method for simulating a single channel stochastically. It will also consider under what circumstances this type of simulation is necessary.

Whilst these models are phenomenological in nature, modelling studies also can be informed by the concepts of ion channel structure and function. In this chapter, the theory of chemical reaction rates is applied to channel gating to produce thermodynamic models of channels, which naturally incorporate temperature and voltage dependence.

We will also touch on the now considerable understanding of the structure and function of ion channels at the level of the movement of molecules within channel proteins. The use of X-ray crystallography (Doyle et al., 1998) and, more recently, cryo-electron microscopy (cryo-EM) (Lau et al., 2018; Catterall et al., 2020) has enabled the derivation of the 3D structure of ion channels (Figure 4.2). Based on this structural information, molecular dynamics (MD) simulations (Roux et al., 2004; DeMarco et al., 2019; Flood et al., 2019) enable the modelling of ion channels at the level of all of the molecules involved, including those making up the ion channel proteins as well as the movement of charged ions through the ion channel pore.

Much of this understanding is more detailed than needed and, for MD, too computationally intensive, for modelling the electrical behaviour of neurons over timescales of milliseconds to seconds or longer. However, incorporating physical theory into channel models is desirable as it is likely to make them more accurate. A fundamental understanding of ion channel structure and function is also important when interpreting the data on which channel models are based.

4.1 Ion Channel Structure and Function

A vast range of biochemical, biophysical and molecular biological techniques have contributed to the huge gain in knowledge of channel structure and function since Hodgkin and Huxley's seminal work. This section provides

---

Box 4.1 | The IUPHAR scheme for naming channels

The International Union of Pharmacology (IUPHAR) has formalised a naming system for ion channel proteins in the voltage-gated-like superfamily based on both structural protein motifs and primary functional characteristics (Yu et al., 2005).

Under this scheme, channels are organised and named based on prominent functional characteristics and structural relationships. Where there is a principal permeating ion, the name begins with the chemical symbol of the ion. This is followed by the principal physiological regulator or classifier, often written as a subscript. For example, if voltage is the principal regulator of the channel, the subscript is 'v', as in Na_v or Ca_v. Where calcium concentration is the principal channel regulator, the subscript is 'Ca', as in K_{Ca}. Examples of other structural classifications are the potassium channel families K_{ir} and K_{2P}. Two numbers separated by a dot follow the subscript, the first representing the gene subfamily and the second the specific channel isoform (e.g. $K_v3.1$).

Where there is no principal permeating ion in a channel family, the family can be identified by the gating regulator or classifier alone. Examples are the cyclic-nucleotide-gated channel family (CNG), the hyperpolarisation-activated cyclic-nucleotide-gated channel family (HCN), the transient receptor potential channel family (TRP) and the two-pore channel family (TPC).

---

a very brief outline of our current understanding; comprehensive accounts can be found in Hille (2001), Tombola et al. (2006) and Catterall et al. (2020).

Each ion channel is constructed from one or more protein subunits. Those that form the pore within the membrane are called **principal subunits**. There may be one principal subunit (e.g. in Na^+ and Ca^{2+} channels), or more than one (e.g. in voltage-gated K^+ channels). Ion channels with more than one principal subunit are called **multimers**; if the subunits are all identical, they are **homomers**, and if not, they are **heteromers**. Multimeric channel structures were presaged by Hodgkin and Huxley's idea (Chapter 3) of channels as gates containing multiple gating particles. However, there is not generally a direct correspondence between the model gating particles and channel subunits. As the number of known channel proteins has increased, systematic naming schemes for them, such as the IUPHAR naming scheme (Box 4.1), have been developed.

An ion channel may also have **auxiliary subunits** attached to the principal subunits or to other auxiliary subunits. The auxiliary subunits may be in the membrane or the cytoplasm and can modulate, or even change drastically, the function of the primary subunits. For example, when the $K_v1.5$ type of the principal subunit is expressed in oocytes alone (Section 4.3.3) or with the $K_v\beta2$ auxiliary subunit, the resulting channels are non-inactivating. However, when expressed with the $K_v\beta1$ subunit, the channels are inactivating (Heinemann et al., 1996).

The secondary structure of one of the four principal subunits of a voltage-gated potassium channel is shown in Figure 4.3a. The polypeptide chain is arranged into segments organised into α-helices and connecting

In the HH model, there are four independent gating particles for the potassium conductance. This happens to correspond to the four subunits of the tetrameric delayed rectifier potassium channel. However, this type of correspondence is not true of all independent gating particle models. For example, in the HH model of the sodium channel, there are three activating particles, but the sodium channel has one principal subunit that contains four sets of voltage-sensing domains and pore-forming domains.

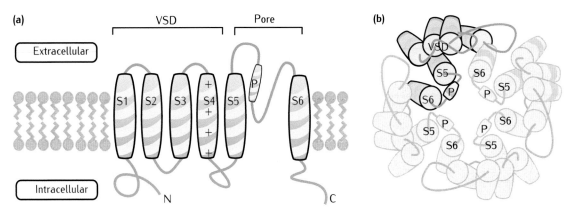

Fig 4.3 **(a)** Secondary structure of one subunit of a voltage-gated potassium channel (based on Tombola et al., 2006). A number of segments of the polypeptide chain are arranged to form α-helices, which are connected together by loops. Six of these α-helices span the membrane and are labelled S1–S6, and there is also a P α-helix in the loop that joins the S5 and S6 transmembrane segments. The S4 segment contains a number of charged residues and, along with segments S1–S3, makes up the voltage-sensing domain (VSD) of the subunit. Segments S5, S6 and P make up the pore region. Both the N-terminal and C-terminal of the polypeptide project into the cytoplasm. **(b)** The tertiary structure of the principal subunits in a closed conformation of the channel, viewed from above. The four subunits (one of which is highlighted) form a central pore, lined by the S5, S6 and P segments of each subunit.

loops. Figure 4.3b shows the 3D arrangement of the four principal subunits in a closed conformation of the channel protein. The S5 and S6 segments form the lining of the pore, and the S1–S4 segments form a voltage-sensing domain (VSD).

Since the determination of the 3D structure of the weakly voltage-gated KcsA potassium channel shown in Figure 4.2 (Doyle et al., 1998), the 3D structure of an increasing number of ion channels has been elucidated, using X-ray crystallography and cryo-EM (Lau et al., 2018; Catterall et al., 2020). This includes voltage-gated sodium and calcium channels (Catterall et al., 2020), which show a similar subunit structure to potassium channels (Figure 4.3), as well as ligand-gated channels, such as $GABA_A$ and G protein-coupled receptors (García-Nafría and Tate, 2020). This structural information, in combination with MD simulations (Box 4.2), is producing insights into the structural basis for voltage-dependent activation and inactivation, channel conductance and ion selectivity (Lau et al., 2018; Catterall et al., 2020). It is also enabling determination of the interaction of channel proteins with the surrounding lipid membrane and the identification of targets for therapeutic drugs (Duncan et al., 2020).

As seen in Chapter 3, Hodgkin and Huxley proposed a voltage-sensing mechanism consisting of the movement of charged particles within the membrane. Such a mechanism has been broadly confirmed in the S4 segment of voltage-gated potassium, sodium and calcium channel VSDs (Tombola et al., 2006; Catterall et al., 2020). A number of positively charged residuals within the S4 segment (called **gating charges**; typically positively charged arginine) experience the electric force due to the membrane potential. The resultant movements of the gating charges lead to other segments in the channel moving. A number of models for gating charge movements have been proposed (Tombola et al., 2006; DeMarco et al., 2019), with recent data from modified bacterial sodium channels lending strong support for the **sliding helix model** (Wisedchaisri et al., 2019; Catterall et al., 2020). In this model, the S4 segment is drawn intracellularly and rotates, thus moving the gating charges through the transmembrane electric field (Figure 4.4). This shift leads to a change in the pore configuration through an interaction of the S4–S5 link with the *neighbouring* subunit's S6 segment. It is hypothesised that a hinge in the S6 helical segment forms the basis of the gate, allowing ions to flow in

Box 4.2 | Molecular dynamics modelling

Ion channel activity can be studied at an atomic level using **molecular dynamics (MD)** modelling and simulations. In essence, MD uses 3D structural information on ion channels, obtained from experimental methods such as X-ray crystallography and cryo-EM, or predicted by computational deep learning methods such as AlphaFold (Jumper et al., 2021) and RoseTTAFold (Baek et al., 2021), to develop an all-atom model for simulation (Figure 4.5). During the simulation, trajectories of atomic movement are generated by solving Newton's equation of motion, $F = ma$, for potential functions representing the microscopic forces between atoms. Flood et al. (2019) and Roux et al. (2004) give detailed introductions to the complex methods involved in MD. DeMarco et al. (2019) give an overview of the strengths and limitations of MD for modelling ion channels.

MD is extremely computationally intensive: a typical model incorporating a single channel embedded in a small patch of membrane and surrounded by a volume of intracellular and extracellular fluid will involve hundreds of thousands of atoms, with vast numbers of interactions to be accounted for at each time step (DeMarco et al., 2019; Flood et al., 2019). Accurate calculation of the equations of motion requires time steps on the order of femtoseconds (10^{-15} s), resulting in one billion iterations of these calculations to simulate 1 μs of time. Simulations in the 100 ns to 1 μs range are certainly feasible with sufficient computing power, but key dynamics of ion channels, such as activation and inactivation, operate on the millisecond to second range at least. Computational techniques are being developed to allow MD simulations over these time frames (DeMarco et al., 2019; Flood et al., 2019).

However, MD simulations are providing considerable insight into ion permeation and selectivity mechanisms (Köpfer et al., 2014; Kutzner et al., 2016), protein conformations underpinning channel activation (Starek et al., 2017) and channel modulation due to lipid interactions (Kasimova et al., 2014; Hedger and Sansom, 2016), leading to identification of possible druggable targets (Hedger and Sansom, 2016; Duncan et al., 2020). Mimicking of electrophysiology is confined to determination of steady-state I–V curves with some level of accuracy, due to model formulation and computational constraints (Köpfer et al., 2014; Kutzner et al., 2016; Flood et al., 2019).

A number of software packages are available for MD, such as *NAMD* (Phillips et al., 2020) and *GROMACS* (Abraham et al., 2015). Workflows are defined for setting up MD models, running simulations and then analysing and visualising the results (Wilson et al., 2021). MD simulations are suitable for running on GPU and supercomputer hardware with many thousands of compute cores. Pre-aligned, embedded and solvated models for many membrane proteins, including ion channels, can be found in the *MemProtMD* database (Newport et al., 2018).

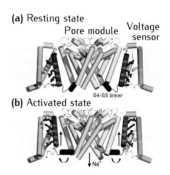

(a) Resting state

(b) Activated state

Fig 4.4 **(a)** Resting and **(b)** activated (open) configurations of a voltage-gated sodium channel. From Wisedchaisri *et al.* (2019). Reprinted with permission from Elsevier Inc.: *Cell* 178, 993–1003, © 2019.

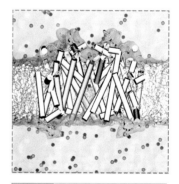

Fig 4.5 Example ion channel all-atom MD simulation. White-grey shows membrane spanned by channel protein. Blue-green surface is water. Purple balls are potassium ions; green balls are chloride ions. From DeMarco *et al.* 2019. Reprinted with permission from John Wiley and Sons: *The Journal of Physiology* 597, 679–698, © 2019.

the open state and limiting flow in the closed state. Evidence for this comes both from mutations in the S6 helical segment and data from X-ray crystallography and cryo-EM (Tombola et al., 2006; Lau et al., 2018; Catterall et al., 2020).

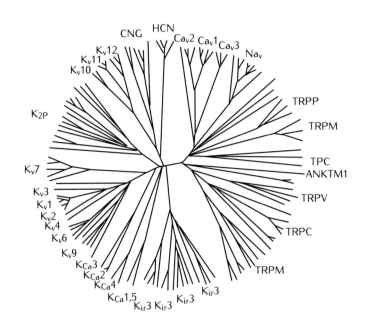

Fig 4.6 The superfamily of mammalian voltage-gated-like ion channels, organised as a phylogenetic tree. Each leaf in the tree represents the amino acid sequences of the pore regions of the channels. Each node (branching point) represents a sequence and the lengths of the lines show the number of changes in the sequence required to transform the sequence from node to node or from node to leaf. There are 143 members of the superfamily, organised into a number of family groups: Na_v, Ca_v, K_v, K_{Ca} CNG/HCN, K_{2P}/K_{ir}, TRP and TPC. Although the $K_{Ca}2$ and $K_{Ca}3$ subfamilies are purely ligand-gated, because of the structural similarity of the pore, they still belong to the voltage-gated superfamily. Adapted from Yu *et al.* (2005), with permission from the American Society for Pharmacology and Experimental Therapeutics.

4.2 Ion Channel Nomenclature

In the early 1980s, the genes of ion channels began to be sequenced. With the completion of the human genome, genes for over 140 channels have been discovered. Each channel gene is named according to a scheme specific to the organism in which it occurs. The gene prefix gives some information about the type of channel to which it refers; for example, genes beginning with KCN are for potassium channels. This **gene nomenclature** is set by organism-specific committees; for example, the Human Genome Organisation Gene Nomenclature Committee is responsible for the naming of human genes.

Family trees of ion channels can be made by grouping the amino acid sequences corresponding to each gene (Figure 4.6). Within each family, channels with the same principal permeant ion are grouped together, and there are subgroups of channels which are voltage-gated or ligand-gated. The phylogenetic tree also reflects the similarities in the channel structures such as the number of transmembrane segments.

This structure is reflected in the system of naming cation channels adopted by the IUPHAR, sometimes referred to as the **clone nomenclature** (Hille, 2001). The scheme incorporates the ion selectivity and the principal activator, as well as the degree of sequence similarity (Box 4.1). Because of its functional relevance, the IUPHAR scheme is in common use among biophysicists and increasingly by electrophysiologists. The gene nomenclature does not reflect as much of the function and structure as the IUPHAR scheme, particularly in the case of potassium channels (Gutman et al., 2005). Nevertheless, even the IUPHAR scheme has quirks, such as including in the calcium-activated potassium family $K_{Ca}5.1$, a type of potassium channel whose gating is dependent on intracellular Na^+ and Cl^-, but not on Ca^{2+}.

Table 4.1 Families of channel genes and their corresponding proteins

Human gene prefix	IUPHAR protein prefix	Ion selectivity	Activators ↑ or inactivators ↓
SCN	Na_v	Na^+	$V\uparrow$
CACN	Ca_v	Ca^{2+}	$V\uparrow$
KCN	K_v	K^+	$V\uparrow$
KCNA	K_v1	K^+	$V\uparrow$
KCNB	K_v2	K^+	$V\uparrow$
KCNC	K_v3	K^+	$V\uparrow$
KCND	K_v4	K^+	$V\uparrow$
KCNF	K_v5	K^+	$V\uparrow$
KCNG	K_v6	K^+	$V\uparrow$
KCNH	$K_v10,11,12$	K^+	$V\uparrow$
KCNQ	K_v7	K^+	$V\uparrow$
KCNS	K_v9	K^+	$V\uparrow$
KCNV	K_v8	K^+	$V\uparrow$
KCNMA	$K_{Ca}1$	K^+	Ca^{2+}, $V\uparrow$
KCNN	$K_{Ca}2$	K^+	Ca^{2+} ↑
KCNJ	K_{ir}	K^+	G-proteins, $V\uparrow$
KCNK	K_{2P}	K^+	Leak, various modulators
HCN	HCN	K^+, Na^+	$V\downarrow$
CNG	CNG	Ca^{2+}, K^+, Na^+	cAMP, cGMP
TRP	TRP	Ca^{2+}, Na^+	Heat, second messengers
CLCN	–	Cl^-	$V\downarrow$, pH
CLCA	–	Cl^-	Ca^{2+}

Data from IUPHAR *Compendium of Voltage-Gated Ion Channels* (Catterall et al., 2005a, b; Gutman et al., 2005; Wei et al., 2005; Kubo et al., 2005; Goldstein et al., 2005; Hofmann et al., 2005; Clapham et al., 2005; Jentsch et al., 2005) and *The Concise Guide to Pharmacology: Ion channels* (Alexander et al., 2019).

A number of gene families of ion channels, excluding channels activated by extracellular ligands such as synaptic channels, are shown in Table 4.1. Apart from the chloride channels, all the families belong to the superfamily of voltage-gated-like channels, which has at least 143 members (Yu et al., 2005). The superfamily is called voltage-gated-*like* because whilst most of its channels are voltage-gated, some are gated by intracellular ligands, second messengers or stimuli such as heat. For example, CNG channels, activated by the cyclic nucleotides cAMP (cyclic adenosine monophosphate) and cGMP (cyclic guanosine monophosphate), are expressed in rod and cone photoreceptor neurons and the cilia of olfactory neurons (Hofmann et al., 2005). Similarly, members of the TRP family are involved in heat or chemical sensing.

Heterologous expression of the cloned DNA of a channel (Section 4.3.3) allows the physiological characteristics of channel proteins to be determined. Currents are recorded under voltage clamp conditions and can be matched to the existing gamut of currents which have been measured using other techniques such as channel blocking, and given ad hoc names such as the

Table 4.2 | Summary of important currents, their corresponding channel types and sample parameters

Current	Channel proteins	Other names	Activation $V_{1/2}$ (mV)	k (mV)	τ (ms)	Inactivation $V_{1/2}$ (mV)	k (mV)	τ (ms)	Note
I_{Na}	Na_v1.1–1.3,1.6		−30	6	0.2	−67	−7	5	a
I_{NaP}	Na_v1.1–1.3,1.6		−52	5	0.2	−49	−10	1500	b
I_{CaL}	Ca_v1.1–1.4	HVA_l	9	6	0.5	2	−4	400	c
I_{CaN}	Ca_v2.2	HVA_m	−15	9	0.5	−13	−19	100	d
I_{CaR}	Ca_v2.3	HVA_m	3	8	0.5	−39	−9	20	e
I_{CaT}	Ca_v3.1–3.3	LVA	−32	7	0.5	−70	−7	10	f
I_{PO}	K_v3.1	Fast rectifier	−5	9	10	—	—	—	g
I_{DR}	K_v2.2, K_v3.2…	Delayed rectifier	−5	14	2	−68	−27	90	h
I_A	K_v1.4,3.4,4.1,4.2…		−1	15	0.2	−56	−8	5	i
I_M	K_v7.1–7.5	Muscarinic	−45	4	8	—	—	—	j
I_D	K_v1.1–1.2		−63	9	1	−87	−8	500	k
I_h	HCN1–4	Hyperpolarisation–activated	−75	−6	1000	—	—	—	l
I_C	K_{Ca}1.1	BK, maxi-K(Ca), fAHP	V & Ca^{2+}-dep			—	—	—	m
I_{AHP}	K_{Ca}2.1–2.3	SK1–3, mAHP	0.7 µM		40	—	—	—	n
I_{sAHP}	K_{Ca}3.1	Slow AHP	0.08 µM		200	—	—	—	o

Parameters are for voltage-dependent activation and inactivation in one model of a channel; other models have significantly different parameters. $K_{0.5}$ (Section 4.6) is given for calcium-dependent K^+ channels. Time constants are adjusted to 37°C using Q_{10}.

Notes:

[a] Rat hippocampal CA1 pyramidal cells (Magee and Johnston, 1995; Hoffman et al., 1997).

[b] Rat entorhinal cortex layer II cells (Magistretti and Alonso, 1999). The same subtypes can underlie both I_{Na} and I_{NaP} due to modulation by G-proteins; auxiliary subunits can slow down the inactivation of some Na_v subtypes (Köhling, 2002).

[c] Activation kinetics from rat CA1 cells in vitro (Jaffe et al., 1994; Magee and Johnston, 1995). Inactivation kinetics from cultured chick dorsal root ganglion neurons (Fox et al., 1987). Calcium-dependent inactivation can also be modelled (Gillies and Willshaw, 2006).

[d] Rat neocortical pyramidal neurons (Brown et al., 1993).

[e] CA1 cells in rat hippocampus (Jaffe et al., 1994; Magee and Johnston, 1995).

[f] CA1 cells in rat hippocampus (Jaffe et al., 1994; Magee and Johnston, 1995).

[g] Gillies and Willshaw (2006), based on rat subthalamic nucleus (Wigmore and Lacey, 2000).

[h] Guinea pig hippocampal CA1 cells (Sah et al., 1988).

[i] Rat hippocampal CA1 pyramidal cells; underlying channel protein probably K_v4.2 (Hoffman et al., 1997). Expression of different auxiliary subunits can convert DR currents into A-type currents (Heinemann et al., 1996).

[j] Guinea pig hippocampal CA1 pyramidal cells (Halliwell and Adams, 1982); modelled by Borg-Graham (1989). See Jentsch (2000) for identification of channel proteins.

[k] Rat hippocampal CA1 pyramidal neurons (Storm, 1988); modelled by Borg-Graham (1999). See Metz et al. (2007) for identification of channel proteins.

[l] Guinea pig thalamic relay cells in vitro; $E_h = -43$ mV (Huguenard and McCormick, 1992).

[m] Rat muscle (Moczydlowski and Latorre, 1983) and guinea pig hippocampal CA3 cells (Brown and Griffith, 1983). Inactivation sometimes modelled (Borg-Graham, 1999; Section 4.6).

[n] Rat K_{Ca}2.2 expressed in *Xenopus* oocytes (Hirschberg et al., 1998). Model in Section 4.6.

[o] Borg-Graham (1999) based on various data. Channel proteins underlying I_{sAHP} are uncertain (Section 4.6).

A-type current I_A or the T-type current I_{CaT}. This **ad hoc nomenclature** for channels is sometimes still used when the combination of genes expressed in a preparation is not known. However, as use of the IUPHAR system by neurophysiologists and modellers becomes more prevalent (see, for example, Maurice et al., 2004; Ranjan et al., 2019), terms such as 'Ca$_v$3.1-like' (instead of I_{CaT}) are also now used when the presence of a particular channel protein is not known. The currents that are associated with some well-known voltage- and ligand-gated channel proteins are shown in Table 4.2. The table includes parameters that characterise the current activation and inactivation of voltage-dependent channels: the half-activation voltage $V_{1/2}$ and the inverse slope k of a sigmoid activation or inactivation curve $(1/(1-\exp(-(V-V_{1/2})/k)))$, and an approximate time constant τ corrected for Q_{10} (Section 3.2.4 and Section 3.5).

The presence of the gene that encodes a particular channel protein, as opposed to current, can be determined using knockout studies (Stocker, 2004). One channel gene can give rise to many different channel proteins, due to alternate splicing of RNA and RNA editing (Hille, 2001). Also, these channel types refer only to the sequences of the principal subunits; as covered in Section 4.1, the coexpression of auxiliary subunits modifies the behaviour of the principal subunits, sometimes dramatically. There is thus even more channel diversity than the plethora of gene sequences suggests.

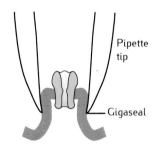

Pipette tip

Gigaseal

Fig 4.7 The patch clamp technique. A hollow glass pipette with a tip of less than 5 μm in diameter is placed on the surface of a cell and suction is applied so that a very tight gigaseal forms between the pipette and the membrane surrounding a single channel. If desired, the pipette can then be withdrawn from the cell membrane, ripping the small patch of membrane away from the cell, to form an 'inside-out' patch.

4.3 | Experimental Techniques

4.3.1 Single Channel Recordings

A major advance in understanding channels came with the development of the **patch clamp technique** (Neher and Sakmann, 1976), where a fine glass pipette is pressed against the side of a cell (Figure 4.7). The pipette rim forms a high-resistance seal around a small patch of the membrane, so that most of the current flowing through the patch of membrane has to flow through the pipette and then into the electrode contained within the pipette. Seals with resistances of the order of giga-ohms can be made, and are called **gigaseals** (Hamill et al., 1981). This allows for low-noise recordings and makes it possible to record from very small patches of membrane.

In the low-noise recordings from some patches, the current switches back and forth between zero and up to a few picoamperes (Figure 4.8). This is interpreted as being caused by the opening and closing of a single channel. The times of opening and closing are apparently random, though order can be seen in the statistics extracted from the recordings. For example, repeating the same voltage clamp experiment a number of times leads to an **ensemble** of recordings, which can be aligned in time and then averaged (Figure 4.8). This average reflects the probability of a channel being open at any time. The **macroscopic** currents (e.g. those recorded by Hodgkin and Huxley) appear smooth as they are an ensemble average over a large population of **microscopic** currents due to stochastic channel events.

Neher and Sakmann received a Nobel Prize for their work in 1991. The methods of forming high-resistance seals between the glass pipette and the membrane proved fundamental to resolving single-channel currents from background noise.

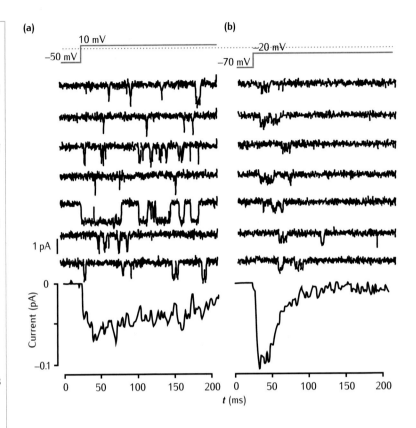

Fig 4.8 Single channel recordings of two types of voltage-gated calcium channel in guinea pig ventricular cells (Nilius et al., 1985), with 110 mM Ba^{2+} in the pipette. (a) The single channel currents seen on seven occasions in response to a voltage step to a test potential of 10 mV (top). The bottom trace shows the average of 294 such responses. It resembles the HH sodium current. This channel, named **L-type** for its long-lasting current by Nilius et al. (1985), belongs to the Ca_v1 family of mammalian calcium channels. (b) Single channel and ensemble responses to a voltage step to the lower potential of −20 mV. The ensemble-averaged current inactivates more quickly. This channel, named **T-type** because of its transient nature by Nilius et al. (1985), belongs to the Ca_v3 family of mammalian channels. Reprinted by permission from Macmillan Publishers Ltd: *Nature* 316, 443–446, © 1985.

The puffer fish is a delicacy in Japan. Sushi chefs are specially trained to prepare the fish so as to remove the organs which contain most of the toxin.

4.3.2 Channel Isolation by Blockers

Hodgkin and Huxley (1952d) succeeded in isolating three currents with distinct kinetics (the sodium, potassium delayed rectifier and leak currents). Since then, many more currents with distinct kinetics have been isolated. The process was expedited by the discovery of **channel blockers**, pharmacological compounds which prevent certain types of current flowing. First to be discovered was tetrodotoxin (TTX), which is isolated from the Japanese puffer fish and blocks Na^+ channels involved in generating action potentials (Narahashi et al., 1964). Likewise, in other early discoveries, tetraethylammonium (TEA) was found to block some types of K^+ channel (Hagiwara and Saito, 1959) and bungarotoxin, derived from snake venom, blocks nicotinic acetylcholine receptors (Chang, 1999). Subsequently, many more compounds affecting ion channel behaviour have been discovered (Hille, 2001). Comprehensive lists of blockers for various channel subtypes can be found online in the *Concise Guide to Pharmacology* (Alexander et al., 2021) and its ion channels chapter (Alexander et al., 2019).

4.3.3 Channel Isolation by mRNA Transfection

Since the 1980s, the molecular biological method of **heterologous expression** has been used to isolate channels. Heterologous expression occurs when the cloned DNA (cDNA) or messenger RNA (mRNA) of a protein is expressed in a cell which does not normally express that protein (Hille, 2001). Sumikawa et al. (1981) were the first to apply the method to ion channels, by transfecting oocytes of the amphibian *Xenopus laevis* with the mRNA of acetylcholine receptors and demonstrating the presence of acetylcholine channels in the oocyte by the binding of bungarotoxin. Subsequent voltage clamp experiments demonstrated the expression of functional acetylcholine channel proteins in the membrane (Mishina et al., 1984). This approach has been extended to mammalian cell lines such as CHO (Chinese hamster ovary), HEK (human embryonic kidney) and COS (monkey kidney) (Hille, 2001). Thus, channel properties can be explored in isolation from the original cell.

Whilst heterologous expression gives a clean preparation, there are also potential pitfalls, as the function of ion channels can be modulated significantly by a host of factors that might not be present in the cell line, such as intracellular ligands or auxiliary subunits (Hille, 2001).

4.3.4 Gating Current

The movement of the gating charges as the channel protein changes conformation leads to an electric current called the **gating current**, often referred to as I_g (Hille, 2001). Gating currents tend to be much smaller than the ionic currents flowing through the membrane. In order to measure the gating current, the ionic current is reduced, either by replacing permeant ions with impermeant ones or by using channel blockers, though the channel blocker itself may interfere with the gating mechanism. Other methods have to be employed to eliminate leak and capacitive currents. Figure 4.9 shows recordings by Armstrong and Bezanilla (1973) of the gating current and the sodium ionic current in response to a voltage step. The gating current is outward since the positively charged residues on the membrane protein are moving outwards. It also peaks before the ionic current peaks.

Gating currents are a useful tool for the development of kinetic models of channel activation. The measurement of the gating current confirmed the idea of charged gating particles predicted by the HH model. However, gating current measurements in the squid giant axon have shown that the HH model is not correct at the finer level of detail (Section 4.5.3).

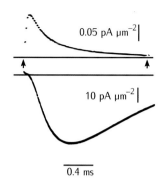

Fig 4.9 Gating current (above) and ionic current (below) in squid giant axon sodium channels, measured by Armstrong and Bezanilla (1973). Reprinted by permission from Macmillan Publishers Ltd: *Nature* 242, 459–461, © 1973.

4.4 | Gating Particle Models of Ion Channels

Building models that reproduce the electrophysiological responses of a neuron requires the inclusion of computationally efficient models for the average ion channel conductance over a patch of membrane, possibly for many different ion channel types. The HH formalism, covered in detail in Chapter 3, is now widely used to model a range of ion channel types (Podlaski et al., 2017; Ranjan et al., 2019). These models rely on the concept of independent gating particles that control the permeability of a channel to ions. We consider how to develop such models from scratch. However, in practice, it may be possible to choose from the many existing published models (Box 4.3).

4.4.1 Fitting the Hodgkin–Huxley Formalism to Data

The HH formalism for a channel comprises:

(1) an instantaneous *I–V* characteristic, e.g. quasi-ohmic or Goldman–Hodgkin–Katz (GHK) current equation;

> **Box 4.3 | Choosing an ion channel model**
>
> Published ion channel models, largely based on the HH formalism, now number in the thousands (Podlaski et al., 2017). Developing a biophysically based model of a neuron will likely involve making direct use of, or modifying, published models for at least some of the ion channels required. The long-standing, publicly available neural modelling database *ModelDB* (Hines et al., 2004) is a good source of ion channel models, with code mostly as implemented for the *NEURON* simulator.
>
> Choosing the best models for your purposes, however, is difficult. Podlaski et al. (2017) have attempted to categorise and compare ion channel models on the basis of the currents they provide, such as A-type potassium (Box 4.4). Often models for the same type of current can vary significantly in their dynamics and will have been developed using different underlying data sets from experiments in many neurons and animals. To help with finding an ion channel model that most closely matches your requirements, Podlaski et al. (2017) have built the *ICGenealogy* web-based tool for finding and comparing models across a wide range of criteria.
>
> If you do wish to build a model from scratch, or retune an existing model, then you require access to appropriate data. To this end, the *Channelpedia* online database (Ranjan et al., 2019) maintains a repository of experimental data for the kinetics of ion channels, with full metadata as to their provenance. This repository has been used to create a kinetic map of voltage-gated potassium channels (Ranjan et al., 2019).

(2) one or more gating variables (such as m and h) and the powers to which those gating variables are raised;

(3) expressions for the forward and backward rate coefficients for these variables as a function of voltage.

The data required for all the quantities are voltage clamp recordings using various protocols of holding potential of the current passing through the channel type in question. This requires that the channel be isolated by some method, such as the ion substitution method (Box 3.2), channel blockers (Section 4.3.2) or the expression in oocytes (Section 4.3.3). The data required for each are now discussed.

Linear *I–V* Characteristic

For greatest accuracy, the instantaneous *I–V* characteristic should be measured. Even the GHK current equation might not be able to capture some features of the characteristic. Also, the reversal potential may differ significantly from the equilibrium potential of the dominant permeant ion if there are other ions to which the channel is significantly permeable. In practice, the quasi-ohmic approximation is often used with a measured reversal potential as the equilibrium potential. When the intracellular and extracellular concentration differences are great, such as in the case of calcium, the GHK current equation may be used.

Gating Variables

If the channel displays no inactivation, only one gating variable is required, but if there is inactivation, an extra variable will be needed. The gating variable is raised to the power of the number of activation particles needed to capture the inflection in conductance activation, which then determines the voltage-dependent rate coefficient functions α_n, β_n of Equation 3.7.

Coefficients for Each Gating Variable

The voltage dependence of the forward and backward reaction coefficients α and β for each gating particle needs to be determined. The basis for this is the data from voltage clamp experiments with different holding potentials.

These can be obtained using the types of methods described in Chapter 3 to determine plots of steady state activation and inactivation and time constants against voltage. With modern parameter estimation techniques (Chapter 14), it is sometimes possible to short-circuit these methods. Instead, the parameters of a model can be adjusted to make the behaviour of the model as similar as possible to recordings under voltage clamp conditions.

The steady state variables, for instance n_∞ and τ_n in the case of potassium, need not be converted into rate coefficients such as α_n and β_n, since the kinetics of the gating variable can be specified using n_∞ and τ_n (Equation 3.11). This approach is taken, for example, by Connor et al. (1977) in their model of the A-type potassium current (Box 4.4). Hodgkin and Huxley fit smooth functions to their data points, but some modellers (Connor and Stevens, 1971c) connect their data points with straight lines in order to make a piecewise linear approximation of the underlying function.

If functions are to be fitted, the question arises of what form they should take. The functions used by Hodgkin and Huxley (1952d) took three different forms, each of which corresponds to a model of how the gating particles moved in the membrane (Section 4.8.3). From the point of view of modelling the behaviour of the membrane potential at a particular temperature, it does not really matter which two quantities are fitted to data or what functional forms are used, as long as they describe the data well. However, from the point of view of understanding the biophysics of channels, more physically principled fitting functions (Section 4.8) are better than arbitrary functions. Such functions can include temperature dependence, rather than having to adjust for temperature using the Q_{10}.

4.4.2 Example: A-type Potassium Current

To give a further example of the use of the HH formalism, we consider a model that seeks to reproduce the potassium ion channel conductance that gives the A-type current. The potassium A-type current, often denoted I_A, has distinct kinetics from the potassium delayed rectifier current, denoted I_{DR} or I_K or $I_{K,DR}$, originally discovered by Hodgkin and Huxley. Connor and Stevens (1971a, c) isolated the current by using ion substitution and by the differences in current flow during different voltage clamp protocols in the somata of cells of marine gastropods. The A-type current has also been characterised in mammalian hippocampal CA1 and CA3 pyramidal cells using TTX to block sodium channels (Figure 4.10).

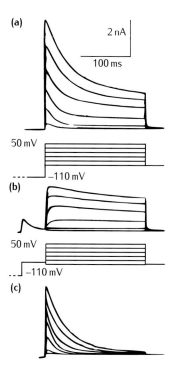

(a)
2 nA
100 ms
50 mV
−110 mV

(b)
50 mV
−110 mV

(c)

Fig 4.10 Recordings of two types of potassium channel revealed by different voltage clamp protocols in hippocampal CA1 cells. **(a)** Family of voltage clamp current recordings from CA1 cells subjected to the voltage step protocol described underneath. **(b)** The voltage was clamped as in **(a)**, except that there was a delay of 50 ms before the step to the depolarising voltage. **(c)** Subtraction of the trace in **(b)** from the trace in **(a)** reveals a transient outward current known as the A-type current. Adapted from Klee *et al.* (1995), with permission from The American Physiological Society.

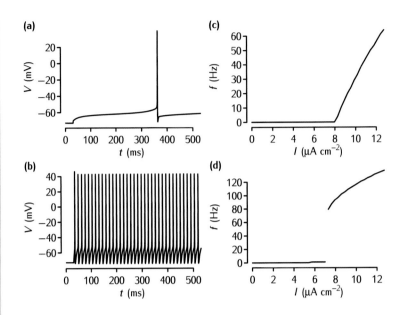

Fig 4.11 The effect of the A-type current on neuronal firing in simulations. **(a)** The time course of the membrane potential in the model described in Box 4.4 in response to current injection of 8.21 μA cm^{-2} that is just superthreshold. The delay from the start of the current injection to the neuron firing is over 300 ms. **(b)** The response of the model to a just suprathreshold input (7.83 μA cm^{-2}) when the A-type current is removed. The spiking rate is much faster. In order to maintain a resting potential similar to the neuron with the A-type current, the leak equilibrium potential E_L is set to −72.8 mV in this simulation. **(c)** A plot of firing rate f versus input current I in the model with the A-type conductance shown in **(a)**. Note the gradual increase of the firing rate just above the threshold, the defining characteristic of Type I neurons. **(d)** The f–I plot of the neuron with no A-type conductance shown in **(b)**. Note the abrupt increase in firing rate, the defining characteristic of Type II neurons.

In contrast to the delayed rectifier current, the A-type current is inactivating and has a lower activation threshold. It has been modelled using independent gating particles by a number of authors (Connor and Stevens, 1971b; Connor et al., 1977; Hoffman et al., 1997; Poirazi et al., 2003). In the model of Connor et al. (1977), the A-type current in the crustacean *Cancer magister* (Box 4.4) depends on three independent activating gating particles and one inactivating particle. In contrast, Connor and Stevens (1971b) found that raising the activating gating variable to the fourth power, rather than the third power, gave the best fit to the A-type current they recorded from the somata of marine gastropods.

The significance of the A-type current is illustrated clearly in simulations of two neurons, one containing sodium and potassium conductances modelled using the HH equations, and the other also containing an A-type conductance (Figure 4.11a, b). In the model containing the A-type conductance, the action potentials are delayed compared to the action potentials in the pure HH model. This is because the A-type potassium channel is open as the membrane potential increases towards the spiking threshold, slowing the rise of the membrane potential. However, eventually the A-type current inactivates, reducing the pull on the membrane potential towards the potassium equilibrium potential and allowing the cell to fire.

Another important difference caused by the insertion of the A-type channels is apparent from plots of firing frequency versus sustained current injection (Figure 4.11c, d). Both types of model exhibit a threshold level of current below which the neuron is quiescent. In the model with the A-type conductance, the firing frequency just above the threshold is very close to zero and increases gradually. By contrast, in the HH model, as soon as the threshold is crossed, the model starts firing at a rate much greater than zero. Hodgkin (1948) had noticed the two different types of f–I curve in axons from the crustacean *Carcinus maenas*. According to his classification, neurons which produce the continuous curve (Figure 4.11c) are **Type I**

Box 4.4 | Model of potassium A-type current

In their model of the conductances in axons from the crab *Cancer magister*, Connor et al. (1977) added an A-type potassium conductance g_A with an associated equilibrium potential E_A to a current equation which includes modified versions of the HH sodium and potassium delayed rectifier conductances:

$$C_m \frac{dV}{dt} = -g_{Na}(V - E_{Na}) - g_K(V - E_K) - g_A(V - E_A) - g_L(V - E_L).$$

The A-type conductance was derived from the experiments of Connor et al. (1977) at 18°C and was modelled using independent gating particles: three activating particles a and an inactivating particle b. The kinetic equations (Section 3.2.1) written in terms of the steady state activation curves $a_\infty(V)$ and $b_\infty(V)$ and the relaxation time constants $\tau_a(V)$ and $\tau_b(V)$ are:

$$I_A = g_A(V - E_A), \qquad g_A = \overline{g}_A a^3 b,$$

$$a_\infty = \left(\frac{0.0761 \exp\left(\frac{V+99.22}{31.84}\right)}{1 + \exp\left(\frac{V+6.17}{28.93}\right)} \right)^{\frac{1}{3}}, \qquad \tau_a = 0.3632 + \frac{1.158}{1 + \exp\left(\frac{V+60.96}{20.12}\right)},$$

$$b_\infty = \frac{1}{\left(1 + \exp\left(\frac{V+58.3}{14.54}\right)\right)^4}, \qquad \tau_b = 1.24 + \frac{2.678}{1 + \exp\left(\frac{V-55}{16.027}\right)}.$$

The HH sodium and potassium delayed rectifier conductances are modified by shifting the steady state equilibrium curves of m, h and n, multiplying the rate coefficients by a Q_{10}-derived factor of 3.8 to adjust the temperature from 6.3°C to 18°C and slowing down the n variable by a factor of 2:

$$g_{Na} = \overline{g}_{Na} m^3 h, \qquad\qquad g_K = \overline{g}_K n^4,$$

$$\alpha_m = 3.8 \frac{-0.1(V + 34.7)}{\exp(-(V + 34.7)/10) - 1}, \qquad \beta_m = 3.8 \times 4 \exp(-(V + 59.7)/18),$$

$$\alpha_h = 3.8 \times 0.07 \exp(-(V + 53)/20), \qquad \beta_h = 3.8 \frac{1}{\exp(-(V + 23)/10) + 1},$$

$$\alpha_n = \frac{3.8}{2} \frac{-0.01(V + 50.7)}{\exp(-(V + 50.7)/10) - 1}, \qquad \beta_n = \frac{3.8}{2} 0.125 \exp(-(V + 60.7)/80),$$

where V refers to the numerical value of the membrane potential in millivolts, and the rate constants α_m, β_m, α_h, β_h, α_n and β_n are in units of ms^{-1}. The remaining parameters of the model are:

$$C_m = 1 \, \mu\text{F cm}^{-2}, \quad E_{Na} = 50 \, \text{mV}, \quad \overline{g}_{Na} = 120.0 \, \text{mS cm}^{-2},$$
$$E_K = -77 \, \text{mV}, \quad \overline{g}_K = 20.0 \, \text{mS cm}^{-2},$$
$$E_A = -80 \, \text{mV}, \quad \overline{g}_A = 47.7 \, \text{mS cm}^{-2},$$
$$E_L = -22 \, \text{mV}, \quad g_L = 0.3 \, \text{mS cm}^{-2}.$$

In all the equations described here, the voltage is 5 mV lower than the values of Connor et al. (1977), to match the parameters used in Chapter 3.

This approach of adapting a model from a different organism contrasts with that taken in the earlier model of *Anisodoris* (Connor and Stevens, 1971b), where many of the values for the steady state variables are piecewise linear fits to recorded data. The two models give similar results.

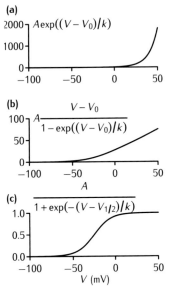

(a) $A\exp((V-V_0)/k)$

(b) $A\dfrac{V-V_0}{1-\exp((V-V_0)/k)}$

(c) $\dfrac{A}{1+\exp(-(V-V_{1/2})/k)}$

V (mV)

Fig 4.12 In the HH model (Box 3.5), the voltage-dependent rate coefficients are described by three different types of equation. **(a)** The rate coefficients β_m, α_h and β_n are described by exponential functions of the voltage V. A, V_0 and k are constants. This fits the empirical rate coefficients α_m and α_n at low membrane potentials, but overestimates the rates at higher membrane potentials. **(b)** Linear-exponential functions produce lower rate coefficients at higher voltages and fit the data well. However, it gives rate coefficients that are too high for the gating variable h at high voltages, where β_h saturates. **(c)** β_h can be described by a rate coefficient with a sigmoid function where $V_{1/2}$ is the half activation voltage and where k is the inverse slope.

neurons and those which produce the curve with the abrupt change at threshold (Figure 4.11d) are **Type II** neurons. In Chapter 8, reduced models of neurons will be introduced to gain understanding of the features of the models that give rise to Type I and Type II firing patterns.

4.4.3 Thermodynamic Models

In Box 4.4, there are effectively five different forms of function to fit the dependence on voltage of the rate coefficients $\alpha_m(V)$, $\beta_m(V)$ and so on: three for the HH sodium and potassium channels (Figure 4.12), and two for the A-type potassium channel. All these forms satisfy the critical requirement of fitting the data well. However, it is desirable to base the form of these functions as much as possible on the biophysical theory of channels, since fitting to the most principled function is likely to minimise errors due to fitting.

In thermodynamic models (Borg-Graham, 1989; Destexhe and Huguenard, 2000), the rate coefficients are given by functions derived from the **transition state theory** (or energy barrier model) of chemical reactions, to be discussed in Section 4.8. For a gating particle represented by a gating variable x, the steady state activation is given by a sigmoid curve:

$$x_\infty = \frac{1}{1+\exp(-(V-V_{1/2})/k)}, \tag{4.1}$$

where $V_{1/2}$ is the half-activation voltage and k is the inverse slope, as shown in Figure 4.13. The corresponding time constant is:

$$\tau_x = \frac{1}{\alpha'(V)+\beta'(V)} + \tau_0, \tag{4.2}$$

where τ_0 is a rate-limiting factor and the expressions for α'_x and β'_x are exponentials that depend on $V_{1/2}$ and k, a maximum rate parameter K and a parameter δ, which controls the skew of the τ curve:

$$\alpha'_x(V) = K\exp\left(\frac{\delta(V-V_{1/2})}{k}\right),$$
$$\beta'_x(V) = K\exp\left(\frac{-(1-\delta)(V-V_{1/2})}{k}\right). \tag{4.3}$$

Figure 4.13 shows plots of the time constant as a function of voltage.

The term τ_0 is in fact an addition to the basic transition state theory account. However, if it is set to zero, the effective time constant $\tau_x = \alpha_x/(\alpha_x+\beta_x)$ can go to zero. In practice, transitions tend not to happen this quickly (Patlak, 1991), and it is evident from the equations for τ_x that the rate-limiting factor τ_0 leads to a minimum time constant.

The steady state value x_∞ and time constant τ_x can be converted into equations for the rate coefficients α_x and β_x (Equation 3.10), giving:

$$\alpha_x(V) = \frac{\alpha'_x(V)}{\tau_0(\alpha'_x(V)+\beta'_x(V))+1}$$
$$\beta_x(V) = \frac{\beta'_x(V)}{\tau_0(\alpha'_x(V)+\beta'_x(V))+1}. \tag{4.4}$$

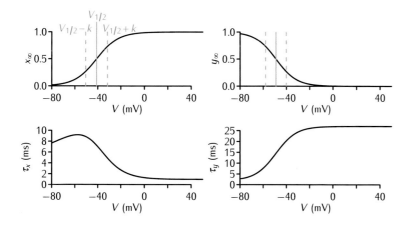

Fig 4.13 Plots of the steady states x_∞ and y_∞ of the activation and inactivation and of the corresponding time constants τ_x and τ_y of a thermodynamic model of the A-type current due to Borg-Graham (1999). The curves for the steady states and time constants are derived from Equations 4.1 and 4.2, respectively. In the steady state curves, the blue solid vertical line indicates the half-activation voltage $V_{1/2}$, and the two dashed blue flanking lines indicate $V_{1/2} - k$ and $V_{1/2} + k$. The parameters for the activation (x) curves are: $V_{1/2} = -41$ mV, $k = 9.54$ mV, $K = 8 \times 10^2$ ms^{-1}, $\delta = 0.85$, $\tau_0 = 1$ ms. The parameters for the inactivation (y) curves are $V_{1/2} = -49$ mV, $k = -8.90$ mV, $K = 4 \times 10^2$ ms^{-1}, $\delta = 1$, $\tau_0 = 2$ ms. The sign of k determines whether the curve has a positive or negative slope.

Calcium Channels

Similar principles apply to modelling calcium channels, such as the T- and L-types, voltage clamp recordings of which are shown in Figure 4.8. The only difference is that because of the inward rectifying nature of the I–V relationship for calcium due to low intracellular concentration of calcium, the GHK current equation (Box 2.4) is often used in modelling calcium channels. For the non-inactivating L-type channel, the permeability can be modelled with two activating particles m, and the inactivating T-type channel can be modelled with two activating particles m and one inactivating particle h (Borg-Graham, 1999). In both cases, $1/K$ is small compared to τ_0, so the time constants τ_m and τ_h are effectively independent of voltage. Table 4.2 shows typical values of $V_{1/2}$, k and τ_0 for these channels. There is evidence that the L-type channels require calcium to inactivate. This could be modelled using an extra calcium-dependent inactivation variable (Section 4.6).

Other Types of Voltage-Gated Channels

There are many more types of current, some of which are listed in Table 4.2. They can be characterised broadly according to whether they are activated by voltage or calcium or both, in the case of I_C, and whether they display fast or slow activation and inactivation.

The half activation voltage and the slope of these curves varies between currents. The values of these quantities listed in Table 4.2 are only indicative as there can be substantial variations between different preparations, for example, variations in temperature (Section 3.5), expression of auxiliary subunits (Section 4.1) or which modulators are present inside and outside the cell. Table 4.2 also lists the principal channel proteins which are proposed to underlie each type of current. In some cases, the same protein appears to be responsible for different currents; for example, Na$_v$1.1 appears to underlie I_{Na} and the persistent sodium current I_{NaP} (Köhling, 2002). This may be possible because of the differences in auxiliary subunit expression.

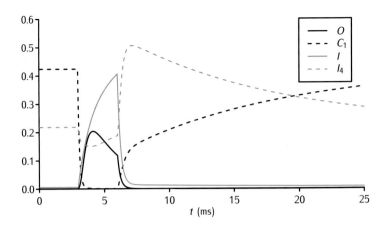

Fig 4.14 Time evolution of a selection of the state variables in the Vandenberg and Bezanilla (1991) model of the sodium channel in response to a 3 ms-long voltage pulse from -65 mV to -20 mV, and back to -65 mV. See Scheme 4.11 for the significance of the state variables shown. In the interests of clarity, only the evolution of the state variables I, O, I_4 and C_1 are shown.

4.5 Markov Models of Ion Channels

In the channel models covered so far, the gating variables, such as n, m and h in the HH model, represent the probability of one of a number of gating particles being in an open position; the probability of the entire gate (or channel) being open is the product of these variables raised to a power, indicating that the gating particles act independently. This section introduces **Markov models** of ion channels in which the probability of the entire ion channel being in one of a number of possible states is represented, and one or more of these states may correspond to the ion channel being open. This allows data to be fitted more accurately, though at the expense of having a greater number of parameters to fit. Ideally, each state would correspond to one channel protein conformation, but in practice, even complex Markov models are approximations of the actual dynamics of the channel.

Markov models are fundamentally probabilistic models in which the state changes are random. This makes them an appropriate framework in which to model the microscopic currents due to the opening and closing of single ion channels or ensembles of a few ion channels. However, when large numbers of channels are present, the recorded currents are smooth because the fluctuations of individual channels are averaged out, and it is approximately correct to interpret Markov models deterministically. In this section, the deterministic interpretation of Markov models will be introduced. The techniques required for the stochastic interpretation of Markov models are introduced in Section 4.7.

4.5.1 Kinetic Schemes

The term 'kinetic scheme' is often used interchangeably with 'Markov model', especially when interpreting Markov models deterministically (Cannon and D'Alessandro, 2006).

In a Markov model, the states and possible transitions between them are represented by a **kinetic scheme**, for example:

$$C_1 \underset{k_{-1}}{\overset{k_1}{\rightleftharpoons}} C_2 \underset{k_{-2}}{\overset{k_2}{\rightleftharpoons}} C_3 \underset{k_{-3}}{\overset{k_3}{\rightleftharpoons}} C_4 \underset{k_{-4}}{\overset{k_4}{\rightleftharpoons}} O. \tag{4.5}$$

The symbols C_1, C_2, C_3 and C_4 represent four different closed states of the channel, and there is one open state O. Each pair of reversible reaction arrows represents a possible transition between states of a single channel, with the rate coefficients $k_1, k_{-1}, k_2, \ldots$, which may depend on voltage or ligand concentration, specifying the speeds of the transitions. When the scheme is interpreted deterministically, the fractions of channels C_1, C_2, C_3, C_4 and O in each state are the relevant variables. Their dynamics are described by the following set of coupled ODEs:

$$\frac{dC_1}{dt} = k_{-1}C_2 - k_1C_1,$$
$$\frac{dC_2}{dt} = k_1C_1 + k_{-2}C_3 - (k_{-1} + k_2)C_2,$$
$$\frac{dC_3}{dt} = k_2C_2 + k_{-3}C_4 - (k_{-2} + k_3)C_3, \qquad (4.6)$$
$$\frac{dC_4}{dt} = k_3C_3 + k_{-4}O - (k_{-3} + k_4)C_4,$$
$$O = 1 - (C_1 + C_2 + C_3 + C_4).$$

An example of the time evolution of a selection of the gating variables in a complex scheme for a sodium channel (Vandenberg and Bezanilla, 1991; Section 4.5.3) is shown in Figure 4.14.

Any kinetic scheme can be interpreted deterministically and converted to a set of coupled ODEs (Stevens, 1978) by following the pattern exemplified above. The number of equations required is always one fewer than the number of states. This is because the fraction of channels in one state can be obtained by subtracting the fractions of all of the other channels from 1, as in the equation for O in Equation 4.6.

Some channels exhibit multiple open states (Hille, 2001). This can be modelled by associating each state j of a channel with the conductance γ_j of a single channel in that state. If the **state variable** representing the fraction of channels in a state j in a patch containing N single channels is denoted by x_j, then the expected number of channels in each state X_j is equal to Nx_j. The total conductance of the patch is then:

$$g(t) = \sum_{j=1}^{N} \gamma_j X_j(t). \qquad (4.7)$$

In the deterministic, macroscopic interpretation of the kinetic scheme, the variables X_j are continuous. This equation can also be used for stochastic, microscopic simulations, where the number of channels X_j is a discrete variable that represents the actual (not expected) number of channels in each state.

4.5.2 Independent Gating Models as Kinetic Schemes

By linking the rate coefficients appropriately, Scheme 4.5 can be made to be equivalent to an independent gating particle model with four gating particles, such as the HH model for the potassium delayed rectifier (see sidebox). With this set of linked parameters, C_1 is the state in which all four gating particles are in their closed positions. A transition to C_2, where exactly one of the four gating particles is in the open position, occurs when *any* one of the four particles moves to the open position. There are thus four routes out of C_1, so the forward rate coefficient is four times the rate coefficient of the

Scheme 4.5, with four closed states, can be made to be equivalent to an independent gating particle scheme with four gating particles by setting the rate coefficients as follows:

$$k_1 = 4\alpha_n \quad k_{-1} = \beta_n$$
$$k_2 = 3\alpha_n \quad k_{-2} = 2\beta_n$$
$$k_3 = 2\alpha_n \quad k_{-3} = 3\beta_n$$
$$k_4 = \alpha_n \quad k_{-4} = 4\beta_n.$$

The coefficients α_n and β_n are forward and backward rate coefficients, such as the ones defined in Chapter 3 (Equation 3.14).

reaction of a single gating particle, i.e. $4\alpha_n$. A transition back from state C_2 to state C_1 requires that the one gating particle in the open position moves to its closed position, governed by the rate coefficient β_n; a transition to C_3, where two of the four gating particles are open, occurs when any one of the three closed gating particles are open. The forward rate coefficient from C_2 to C_3 is therefore $3\alpha_n$. In state C_4, only one of the gating particles is closed. Its transition to the open position, with rate coefficient α_n, leads to the open state O.

The HH sodium channel can be represented by an eight-state diagram:

$$C_1 \underset{\beta_m}{\overset{3\alpha_m}{\rightleftharpoons}} C_2 \underset{2\beta_m}{\overset{2\alpha_m}{\rightleftharpoons}} C_3 \underset{3\beta_m}{\overset{\alpha_m}{\rightleftharpoons}} O$$
$$\beta_h \Big\| \alpha_h \qquad \beta_h \Big\| \alpha_h \qquad \beta_h \Big\| \alpha_h \qquad \beta_h \Big\| \alpha_h \qquad (4.8)$$
$$I_1 \underset{\beta_m}{\overset{3\alpha_m}{\rightleftharpoons}} I_2 \underset{2\beta_m}{\overset{2\alpha_m}{\rightleftharpoons}} I_3 \underset{3\beta_m}{\overset{\alpha_m}{\rightleftharpoons}} I_4$$

where I_1–I_4 are inactivated states. In all the inactivation states, the inactivating particle is in the closed position, and there are zero, one, two or three activation particles in the open position. This independent gating model can serve as a starting point for unrestricted kinetic schemes. To model the data accurately, changes can be made to some of the rate coefficients.

4.5.3 Unrestricted Kinetic Schemes

By not restricting the parameters of kinetic schemes to correspond to those in independent gating particle models, a better fit to data can be obtained. For example, studies of the squid giant axon sodium channel after Hodgkin and Huxley revealed significant inaccuracies in their model (Patlak, 1991):

(1) The measured deactivation time constant is slower than in the HH model. In the model, the time constant of the tail current following a voltage step is $\tau_m/3$ (Chapter 3), but experiments show that the time constant should be τ_m.

(2) The gating current is not as predicted by the HH model. The model predicts the gating current is proportional to the rate of movement of gating particles dm/dt. Thus, at the start of a voltage clamp step, the gating current should rise instantaneously, and then decay with an exponential time course. However, the measured time course of the gating current displays a continuous rising time course (Figure 4.9).

(3) When inactivation is removed by the enzyme pronase in the cytoplasm, the mean open dwell times recorded from single channels (Section 4.7) are six times longer than would be expected from the value predicted from the HH value of $1/\beta_m$ (Bezanilla and Armstrong, 1977).

(4) There is a delay before inactivation sets in, which suggests that inactivation of the channel depends on activation.

In order to account for these discrepancies, Markov kinetic schemes which are not equivalent to independent gating particles have been devised. In contrast to the HH scheme for the sodium channel (Scheme 4.8), the inactivated state or states can only be reached from the open state or one of the closed states. For example, in the scheme of Vandenberg and Bezanilla (1991), based on the gating current, single channel recordings and macroscopic ionic recordings, there are five closed states, three inactivated states and one open state:

$$C_1 \underset{k_{-1}}{\overset{k_1}{\rightleftharpoons}} C_2 \underset{k_{-1}}{\overset{k_1}{\rightleftharpoons}} C_3 \underset{k_{-1}}{\overset{k_1}{\rightleftharpoons}} C_4 \underset{k_{-2}}{\overset{k_2}{\rightleftharpoons}} C_5 \underset{k_{-3}}{\overset{k_3}{\rightleftharpoons}} O$$

$$k_{-5} \Big\Uparrow k_5 \qquad\qquad k_{-4} \Big\Uparrow k_4 \qquad (4.9)$$

$$I_4 \underset{k_{-2}}{\overset{k_2}{\rightleftharpoons}} I_5 \underset{k_{-3}}{\overset{k_3}{\rightleftharpoons}} I$$

The deactivation kinetics of the HH model and the Vandenberg and Bezanilla (1991) model are shown in Figure 4.15. The membrane is clamped to a depolarising voltage long enough to activate sodium channels, but not long enough to inactivate them much. Then the membrane is clamped back down to a testing potential. The tail currents that result illustrate the slower deactivation kinetics of the Vandenberg and Bezanilla (1991) model, in line with the experimental observations.

4.5.4 Fitting Kinetic Schemes to Data

Creating a Markov model of an ion channel involves both determination of the structure of the kinetic scheme as well as obtaining values for all of the transition rate parameters. Deciding on the kinetic scheme is a process of **model selection**. Finding values for the rate parameters is the process of **parameter estimation**. We cover both model selection and parameter estimation in more general terms and with an outline of technical details in Chapter 14. For Markov models, both processes are likely to be underdetermined by the available experimental data, leading to the likelihood that multiple models may fit the data equally well. Nonetheless, there is considerable ongoing research into the best approaches to take.

The quality and richness of the available experimental data are key to the development of good models (Cannon and D'Alessandro, 2006). The data need to include coverage of the magnitude and time course of the different states of an ion channel, including activation, closing and inactivation, if possible. States may be sensitive to multiple factors, such as membrane voltage and ligand concentrations (Section 4.6). Transition rates between states may be difficult to discriminate if their time constants are similar. Clerx et al. (2019) studied the quality of four different kinds of data for fitting a voltage-sensitive ion channel model: (1) the HH approach of fitting directly to time constants, steady-state magnitudes and I–V curves; (2) fitting these same values derived from model simulations to the experimental data; (3) fitting simulated current traces to the experimental traces over different protocols and their full time course; and (4) fitting a single current trace obtained by a dynamic voltage clamp experiment involving short and rapidly changing voltage transients. Of these, approaches (3) and (4) had the best discriminative power. The use of dynamic clamps (Figure 14.12) can result in the richest data in the shortest experimental time, though such data are not widely available.

Dynamic clamps can also be used for real-time model fitting during an experiment (Milescu et al., 2008). This can be done by firstly pharmacologically

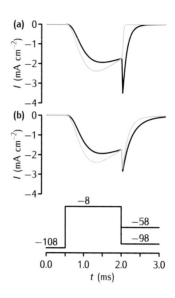

Fig 4.15 Tail currents in the Vandenberg and Bezanilla (1991) model (VB, black traces) and the Hodgkin and Huxley (1952d) model (HH, blue traces) of the sodium current in the squid giant axon in response to the voltage clamp protocol shown at the base, where the final testing potential could be **(a)** −98 mV or **(b)** −58 mV. During the testing potential, the tail currents show a noticeably slower rate of decay in the VB model than in the HH model. To assist comparison, the conductance in the VB model was scaled so as to make the current just before the tail current the same as in the HH model. Also, a quasi-ohmic I–V characteristic with the same E_{Na} as the HH model was used rather than the GHK I–V characteristic used in the VB model. This does not affect the time constants of the currents, since the voltage during each step is constant.

blocking the ion channel of interest in the experiment and then using a dynamic current clamp driven by the output of the putative ion channel model as a replacement for the blocked channel current during the experiment. Model parameter values are then adjusted online until the experimental membrane is able to reproduce responses recorded before the ion channel was blocked.

To have sufficient data, it is often necessary to aggregate results from multiple patch clamp experiments, perhaps conducted over multiple membrane patches from different cells. Variability between these different recordings can result both from differences in the exact recording conditions as well as from possible differences in ion channel characteristics between cells. Lei et al. (2020) have shown that by modelling the experimental artefacts in patch clamp recordings and then removing them from the recorded traces, the resulting identified ion channel models are typically simpler and more consistent in parameter values across multiple recordings.

Model selection involves searching the space of likely kinetic scheme structures to find models that best fit the available data during the parameter estimation process. Search methods such as genetic algorithms provide a wide coverage of the search space, assessing multiple candidate solutions at each iteration (Menon et al., 2009). Both scheme structure and parameter values are mutated as the search progresses. An alternative, that may be more effective for ranking competing models according to their complexity, is to do a complete search through a specified set of plausible structures (Mangold et al., 2021).

Attention must be paid to finding the simplest model structure that fits the data well with tightly estimated parameter values. Increasingly complex schemes will reduce the error between model output and the experimental data being used to estimate the parameter values, but at the cost of overfitting the data. It is essential to withhold a validation data set from the estimation step to enable assessment of the predictive ability of an identified model. Overly complex models will likely do no better, or even worse, than simpler models on validation data sets. A complexity cut-off can then be applied when validation scores stop dropping (Menon et al., 2009; Mangold et al., 2021).

Another problem that arises, particularly as model complexity increases, is the identifiability of model parameter values (Mangold et al., 2021). If the experimental data are strongly dependent on a particular parameter, then their estimated value will fall within tight bounds on all parameter estimation trials. However, it may become apparent that the value of some parameters impacts very little on the quality of model fitting. Thus these parameters are not identifiable from the available data and the model structure should be changed to remove them. Issues of parameter uncertainty and sensitivity are explored further in Chapter 14.

4.5.5 Independent Gating Particles versus Unrestricted Kinetic Schemes

Biophysicists have characterised the voltage-dependent behaviour of a number of ion channels precisely using Markov kinetic schemes (for example, Vandenberg and Bezanilla, 1991; Rodriguez et al., 1998; Irvine et al., 1999).

Although we refer to unrestricted schemes here, thermodynamics does impose some restrictions on the rate coefficients (Box 4.10).

Despite this, the approach formulated over 50 years ago by Hodgkin and Huxley of modelling channel kinetics with independent gating particles is still widespread in the computational neuroscience community and even some biophysicists (Patlak, 1991) regard them as the gold standard of channel kinetics. This is because independent gating particle models require less data to fit than kinetic schemes due to their relatively small number of parameters. Furthermore, independent gating particle models are sufficient to explain a large number of electrophysiological data, such as the origin of the action potential and spike adaptation due to potassium channels mediating A-type currents.

Nevertheless, Markov kinetic schemes can describe the data more accurately than independent gating models, and could be especially important in understanding particular phenomena such as repetitive firing (Bean, 2007). If a Markov kinetic scheme has been derived for the channel in question under comparable experimental conditions, most probably it will be a better choice than using the HH formalism or a thermodynamic model. This applies especially in simulations where the time course needs to have sub-millisecond precision. However, care must be taken that the experimental conditions under which the channel recording was performed are a sufficiently close match to the cell being modelled. Much work on characterisation of ion channels is carried out using heterologous expression in transfected cell lines (Section 4.3.3), in which the biochemical environment is likely to be different from the cell under study. This may have a strong effect on some channel parameters, for example, half activation voltages $V_{1/2}$. Ideally, characteristics of the channel, such as the activation and inactivation curves, should be measured in the cell to be modelled. Examples of studies where experiment and modelling using kinetic schemes have been combined include Maurice et al. (2004) and Magistretti et al. (2006).

Another approach is to have two versions of the model, one with an independent gating model and one with a kinetic scheme model of the same channel. Comparing the behaviour of both models will give an indication of the functional importance of the ion channel description.

If a Markov kinetic scheme model is not available, is it worth doing the extra experiments needed to create it compared with doing the experiments required to constrain the parameters of independent gating particle models and thermodynamic models? There is no straightforward answer to this question, but before expending effort to make one part of a neuronal model very accurate, the accuracy (or inaccuracy) of the entire model should be considered. For example, it may be more important to put effort into determining the distribution of channel conductances over an entire cell rather than getting the kinetics of one channel as accurate as possible.

4.6 Modelling Ligand–Gated Channels

In contrast to voltage-gated channels, some channels are activated by intracellular or extracellular ligands. Examples of extracellular ligands are the neurotransmitters acetylcholine, glutamate and glycine. Channels which have receptors on the extracellular membrane surface for these ligands perform the postsynaptic function of fast chemical synapses, which are discussed in Chapter 7. Among the intracellular ligands are second messengers such as cyclic nucleotides and calcium ions. The focus in this section will be on calcium-dependent potassium channels, but similar techniques can be applied to other intracellular ligands, for example, modulation of I_h by cyclic AMP (Wang et al., 2002).

Calcium-activated potassium channels are responsible for the afterhyperpolarisation that can occur after a burst of spikes. Calcium channels activated during the burst allow calcium into the cell, which in turn activates the K_{Ca} channels (Stocker, 2004). There are three main Ca^{2+}-dependent potassium currents (for a review, see Sah and Faber, 2002):

(1) I_C, also known as $I_{K(C)}$. This current depends on voltage and calcium concentration, and is due to $K_{Ca}1.1$ channel proteins. It is responsible for the fast afterhyperpolarisation (fAHP).
(2) I_{AHP}, also known as $I_{K(AHP)}$. This current depends only on calcium concentration and is due to channels in the $K_{Ca}2$ subfamily. It is responsible for the medium afterhyperpolarisation (mAHP).
(3) I_{sAHP}, also known as $I_{K\text{-slow}}$ and, confusingly, I_{AHP}. This also only depends on calcium concentration. It is currently not known what underlies this current, but mechanisms are likely to be heterogeneous across the nervous system (Andrade et al., 2012). Recent experiments implicate $K_{Ca}3.1$ channels in the cortex (Turner et al., 2016; Roshchin et al., 2020) and K_v7 and $K_{ir}6$ in the dentate gyrus (Laker et al., 2021). Modulation of potassium channels by PIP2 is also a possibility (Andrade et al., 2012). I_{sAHP} is responsible for slow afterhyperpolarisations (sAHP), which lead to spike train adaptation.

A prerequisite for incorporating ligand-gated ion channels into a model neuron is knowledge of the concentration of the ligand in the submembrane region. Intracellular ion concentrations do not feature explicitly in the equivalent electrical circuit models of the membrane considered in previous chapters, but models which include the influx, efflux, accumulation and diffusion of ligands such as Ca^{2+} will be described in detail in Chapter 6. For now, it is assumed that the time course of the Ca^{2+} concentration in the vicinity of a Ca^{2+}-gated ion channel is known.

4.6.1 Calcium–Activated Potassium Channels

In channels in the $K_{Ca}2$ subfamily, the open probability depends principally on intracellular calcium concentration (Figure 4.16). This dependence can be fitted using the **Hill equation**:

$$\text{Prob (channel open)} \propto \frac{[Ca^{2+}]^n}{K_{0.5}^n + [Ca^{2+}]^n}, \tag{4.10}$$

where $K_{0.5}$ is the half maximal effective concentration of intracellular Ca^{2+}, sometimes referred to as EC_{50}, which is the concentration at which the open probability is $1/2$, and n is the **Hill coefficient**.

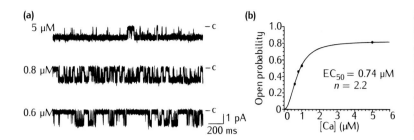

This can be incorporated into a model by introducing an activation variable whose steady state reflects the open probability. For example, Gillies and Willshaw (2006) use experimental data (Hirschberg et al., 1998; Figure 4.16) to model the small conductance calcium-dependent potassium channel $K_{Ca}2.1$, which underlies I_{AHP}. The model is defined as:

$$g_{sKCa} = \overline{g}_{sKCa}w, \quad \frac{dw}{dt} = -\frac{w - w_\infty}{\tau_w}, \quad w_\infty = 0.81\frac{[Ca^{2+}]^n}{K_{0.5}^n + [Ca^{2+}]^n},$$

$$K_{0.5} = 0.74\,\mu M, \quad n = 2.2, \quad \tau_w = 40\,ms, \tag{4.11}$$

where $K_{0.5}$ is the EC_{50}. A similar approach can be taken for the AHP current I_{sAHP} (Yamanda et al., 1998; Borg-Graham, 1999).

4.6.2 Calcium- and Voltage-Activated Potassium Channels

The gating of potassium channels in the $K_{Ca}1$ family, also known as I_C, $I_{K(C)}$, Slo1, BK or maxi K^+, depends on the intracellular calcium concentration and membrane potential (Moczydlowski and Latorre, 1983; Cui et al., 1997). This has been modelled using a number of kinetic schemes, most of which have only open and closed states (Moczydlowski and Latorre, 1983; Cui et al., 1997), though inactivating states are sometimes included on physiological grounds (Borg-Graham, 1999).

A model that has been used in a number of studies (e.g. Jaffe et al., 1994; Migliore et al., 1995) is the one derived by Moczydlowski and Latorre (1983) from their single channel recordings of $K_{Ca}1$ channels from rat muscle incorporated into planar lipid bilayers. The channel is described by a four-state kinetic scheme with two open states and two closed states:

$$C \underset{k_{-1}(V)}{\overset{k_1(V)[Ca^{2+}]}{\rightleftharpoons}} C \cdot Ca^{2+} \underset{\beta}{\overset{\alpha}{\rightleftharpoons}} O \cdot Ca^{2+} \underset{k_{-4}(V)}{\overset{k_4(V)[Ca^{2+}]}{\rightleftharpoons}} O \cdot Ca_2^{2+}. \tag{4.12}$$

The closed state can move to a closed state with one bound Ca^{2+} ion. From this state, it can go to an open state, and thence to an open state with two bound Ca^{2+} ions. Whilst the transition between the open and closed states is independent of voltage, the calcium binding steps are voltage-dependent. Under the assumption that the calcium binding steps are much faster than the step between the closed and open states, the expression for the fraction of channels in either of the open states reduces to a first-order ODE (Box 4.5).

The Hill equation was originally devised to describe the binding of oxygen to haemoglobin and is an approximate mathematical description of the fraction of binding sites on a receptor that are occupied by a ligand. The Hill coefficient is a measure of the cooperativity of the binding. It is not equivalent to the number of binding sites, providing only a minimum estimate (Weiss, 1997). See also the Michaelis–Menten equation for enzymatic reactions in Section 6.1.2.

Box 4.5 | The Moczydlowski and Latorre (1983) Model

Since it is assumed that the calcium binding transitions are fast, compared to the change in conformation in Scheme 4.12, the fractions of channels in the two closed states are at equilibrium, as are the fractions of channels in the two open states. Thus, the ratios between the two types of closed states and the two types of open states are:

$$\frac{[C \cdot Ca^{2+}]}{[C]} = \frac{k_1[Ca^{2+}]}{k_{-1}} \quad \text{and} \quad \frac{[O \cdot Ca_2^{2+}]}{[O \cdot Ca^{2+}]} = \frac{k_4[Ca^{2+}]}{k_{-4}}. \tag{a}$$

We define the total fraction of channels in the closed state C and the open state O as:

$$C = [C] + [C \cdot Ca^{2+}] \quad \text{and} \quad O = [O \cdot Ca^{2+}] + [O \cdot Ca_2^{2+}].$$

Along with Equation (a), this allows us to state the relationship between the fraction of channels in the closed and open states, and the fractions in the states on either side of the conformational transition:

$$C = \left(1 + \frac{k_{-1}}{k_1[Ca^{2+}]}\right)[C \cdot Ca^{2+}] \quad \text{and} \quad O = \left(1 + \frac{k_4[Ca^{2+}]}{k_{-4}}\right)[O \cdot Ca^{2+}]. \tag{b}$$

The first-order ODE that describes the dynamics of the aggregated open and closed states is:

$$\frac{dO}{dt} = \alpha[C \cdot Ca^{2+}] - \beta[O \cdot Ca^{2+}].$$

By substituting in the expressions in Equations (b) for $[C \cdot Ca^{2+}]$ and $[O \cdot Ca^{2+}]$, this ODE can be written as:

$$\frac{dO}{dt} = aC - bO = a(1 - O) - bO,$$

where

$$a = \alpha \left(1 + \frac{k_{-1}}{k_1[Ca^{2+}]}\right)^{-1} \quad \text{and} \quad b = \beta \left(1 + \frac{k_4[Ca^{2+}]}{k_{-4}}\right)^{-1}.$$

To complete the model, the ratios of the forward and backward rate constants are required. Moczydlowski and Latorre (1983) found that the ratios could be described by:

$$\frac{k_{-1}(V)}{k_1(V)} = K_1 \exp(-2\delta_1 FV/RT) \quad \text{and} \quad \frac{k_{-4}(V)}{k_4(V)} = K_4 \exp(-2\delta_4 FV/RT),$$

where F is Faraday's constant, R is the molar gas constant, K_1 and K_4 are constants with the units of concentration, and δ_1 and δ_4 are parameters related to the movement of gating charge (Section 4.8). The model parameters varied from channel to channel. For one channel, these were $K_1 = 0.18\,\text{mM}$, $\delta_1 = 0.84$, $K_4 = 0.011\,\text{mM}$, $\delta_4 = 1.0$, $\alpha = 480\,\text{s}^{-1}$ and $\beta = 280\,\text{s}^{-1}$.

4.7 Modelling Single Channel Data

Markov models of channels were introduced in Section 4.5, but only their deterministic interpretation was considered. In this section, the underlying probabilistic basis of Markov models is outlined. This provides tools for analysing single channel data and for simulating Markov models.

4.7.1 Markov Models for Single Channels

Markov models all obey the **Markov property**: the probability of a state transition depends only on the state the channel is in and the probabilities of transitions leading from that state, not on the previous history of transitions. This can be illustrated by considering a very simple kinetic scheme in which the channel can be in an open state (O) or a closed state (C):

$$C \underset{\beta}{\overset{\alpha}{\rightleftharpoons}} O, \tag{4.13}$$

where α and β are transition probabilities which can depend on voltage or ligand concentration, analogous to the rate coefficients in the deterministic interpretation. With the channel in the closed state, in an infinitesimally small length of time Δt, it has a probability of $\alpha \Delta t$ of moving to the open state; if it is in the open state, it has a probability of $\beta \Delta t$ of moving back to the closed state. This scheme can be simulated exactly using the algorithm to be described in Section 4.7.2 to produce conductance changes such as those shown in Figure 4.17a. Each simulation run of the scheme produces a sequence of random switches between the C and O states.

A key statistic of single channels is the distribution of times for which they dwell in open or closed conductance states. Histograms of the channel open and closed times can be plotted, as shown in Figures 4.17b and 4.17c for the most basic two-state scheme (Scheme 4.13). By considering the time steps Δt to be infinitesimally small, the transition from one state to another acts as a Poisson process, in which the inter-event intervals are distributed exponentially:

$$\begin{aligned} \text{Prob (in closed state for time } t) &= \alpha \exp(-\alpha t) \\ \text{Prob (in open state for time } t) &= \beta \exp(-\beta t). \end{aligned} \tag{4.14}$$

The mean closed time is $1/\alpha$, meaning that the higher the forward reaction rate α, the shorter the time during which the channel will stay in the closed state. Similarly, the mean open time is $1/\beta$.

Open and closed time histograms extracted from experimental recordings of single channel currents do not tend to have this simple exponential structure. For example, the closed time distribution of calcium channels in bovine chromaffin cells (Figure 4.18) is fitted more closely by a double exponential than by a single exponential (Fenwick et al., 1982). As each exponential has a characteristic time constant, this indicates that there are at least three timescales in the system. Since each transition is associated with a time constant, this means that a kinetic scheme with at least three states is required to model the data. The data shown in Figure 4.18 can be modelled

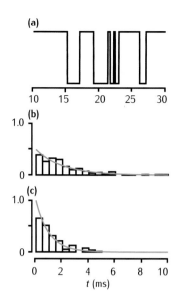

Fig 4.17 Features of a two-state kinetic scheme. **(a)** Simulated sample of the time course of channel conductance of the kinetic scheme described in Scheme 4.13. The parameters are $\alpha = 1\,\text{ms}^{-1}$ and $\beta = 0.5\,\text{ms}^{-1}$. **(b)** Histogram of the open times in a simulation and the theoretical prediction of $\beta e^{-\beta t}$ from Equation 4.14. **(c)** Histogram of the simulated closed times and the theoretical prediction $\alpha e^{-\alpha t}$.

The number of events generated by a Poisson process in a given interval is distributed according to a Poisson distribution (Box 1.2).

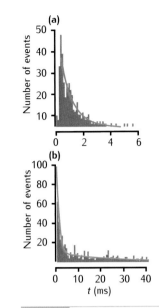

Fig 4.18 Distributions of
(a) open and **(b)** closed times of
single Ca^{2+} channels from
bovine chromaffin cells recorded
by Fenwick et al. (1982). The
distribution of open times is
fitted by a single exponential
with a time constant of 0.81 ms,
but the distribution of closed
times is fitted by two
exponentials with time constants
of 1.05 ms and 25.5 ms. Adapted
from Fenwick et al. (1982), with
permission from John Wiley &
Sons Ltd.

by a three-state scheme with two closed states (C_1 and C_2) and one open state (O):

$$C_1 \underset{\beta_1}{\overset{\alpha_1}{\rightleftharpoons}} C_2 \underset{\beta_2}{\overset{\alpha_2}{\rightleftharpoons}} O.$$ (4.15)

In this case, it is possible to determine the four transition probabilities from the time constant of the open time distribution, the fast and slow time constants of the closed time distribution and the ratio of the fast and slow components of the closed time distribution.

4.7.2 Simulating a Single Channel

Simulating the opening and closing of a single channel over time requires the generation of random numbers to determine the transitions between channel states and their dwell times. This is known as a stochastic, or Monte Carlo, simulation. A patch of membrane will contain a population of ion channels of a particular type – the larger the area of the patch and the larger the channel density, the greater number of ion channels expected in the population. The simulation of each ion channel individually, as considered here, is less efficient than the population-based method of the Stochastic Simulation Algorithm, which is presented in detail in Section 6.8, for modelling general kinetic schemes.

As an example of simulating a single channel, consider the five-state potassium channel scheme (Scheme 4.5) with the rate coefficients given by the HH model. In this kinetic description, transitions between states depend only on the membrane potential. The simulation method described in this section is efficient but only works when the membrane potential is steady, as it is under voltage clamp conditions.

In order to simulate an individual potassium channel using this scheme, an initial state for the channel must be chosen. It is usual to select a state consistent with the system having been at rest initially. The probability that the channel is in each of the five states can then be calculated. For example, the probability of being in state C_1, given initial voltage V_0, is:

$$P_{C_1} = (1 - n_\infty(V_0))^4 = \left(1 - \frac{\alpha_n(V_0)}{\alpha_n(V_0) + \beta_n(V_0)}\right)^4.$$ (4.16)

C_4 is the state in which all four of the particles are closed. As described in Chapter 3, the probability of a particle being in the closed state is given by $1 - n_\infty(V_0)$. Consequently, the probability that all four of the particles are in the closed state is $(1 - n_\infty(V_0))^4$. The probability of being in state C_2, where exactly one of the four particles is in the open state, with the remainder closed, is:

$$P_{C_2} = \binom{4}{3}(1 - n_\infty(V_0))^3 n_\infty(V_0),$$ (4.17)

where $\binom{4}{3}$ is the number of possible combinations in which any three of the four particles can be in the closed state. Similarly, the probabilities of the particle being in states C_3, C_4 and O, respectively, given the initial voltage V_0, are given by:

$$P_{C_3} = \binom{4}{2}(1 - n_\infty(V_0))^2 n_\infty(V_0)^2,$$

$$P_{C_4} = \binom{4}{1}(1 - n_\infty(V_0)) n_\infty(V_0)^3,$$ (4.18)

$$P_O = n_\infty(V_0)^4.$$

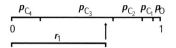

Fig 4.19 The initial probabilities of each state of the kinetic model can be used to divide a unit line. The example uses the five-state Hodgkin and Huxley channel model given in Scheme 4.5, with initial voltage $V_0 = -60\,\text{mV}$.

The initial state for the simulation is selected stochastically by assigning to each probability a section of a line running from 0 to 1 (Figure 4.19), drawing a random number r_1 between 0 and 1 from a uniform distribution and selecting the initial state according to where r_1 lies on the line.

The next step in the simulation is to determine how long the system resides in a state, given the membrane potential V. Suppose the channel is in state C_1. As state transitions act as a Poisson process (Section 4.7.1), the probability that the system remains in state C_1 for duration τ is given by:

$$P_{C_1}(\tau) = 4\alpha_n \exp(-4\alpha_n \tau),$$ (4.19)

where $4\alpha_n$ is the rate at which state C_1 makes the transition to state C_2 at voltage V (the V dependency has been omitted from α_n for clarity). By converting this distribution into a cumulative distribution, another random number r_2, between 0 and 1, can be used to calculate the duration:

$$\tau = -\frac{\ln(r_2)}{4\alpha_n}.$$ (4.20)

Similar probabilities for the other states can be calculated:

$$P_{C_2}(\tau) = (3\alpha_n + \beta_n)\exp(-(3\alpha_n + \beta_n)\tau),$$

$$P_{C_3}(\tau) = (2\alpha_n + 2\beta_n)\exp(-(2\alpha_n + 2\beta_n)\tau),$$

$$P_{C_4}(\tau) = (\alpha_n + 3\beta_n)\exp(-(\alpha_n + 3\beta_n)\tau),$$ (4.21)

$$P_O(\tau) = 4\beta_n \exp(-4\beta_n \tau),$$

and the random duration calculated by replacing $4\alpha_n$ in Equation 4.20 with the appropriate rate.

Finally, once the system has resided in this state for the calculated duration, its next state must be chosen. In this example, in states C_1 and O, there is no choice to be made and transitions from those states go to one place only. For the intermediate states, transition probabilities are used to select stochastically the next state. The probabilities of the transitions from state C_2 to C_1 or C_3 are given by:

$$P_{C_2,C_1} = \frac{\beta_n}{\beta_n + 3\alpha_n},$$ (4.22)

$$P_{C_2,C_3} = \frac{3\alpha_n}{\beta_n + 3\alpha_n}.$$

Clearly, the sum of transition probabilities away from any state adds to 1. To choose the new state stochastically, we can use the technique for selecting the initial state illustrated in Figure 4.19.

The **Monte Carlo simulation** approach is used for systems that contain a random element that precludes the direct calculation of desired quantities. Values for random variables, such as the dwell time of an ion channel state, are obtained by sampling from their probability distributions. The system simulation is run many times with different outcomes, due to the random sampling. A quantitative average outcome may be calculated from the aggregation of many simulations.

$\binom{n}{k}$ is the number of combinations, each of size k, that can be drawn from an unordered set of n elements. This is given by:

$$\binom{n}{k} = \frac{n!}{k!(n-k)!}.$$

4.7.3 Ensemble versus Stochastic Simulation

As seen in Figure 4.8, when the microscopic single channel records are aligned in time and averaged, the result looks like data from macroscopic patches of membrane. A similar effect appears when multiple channels in

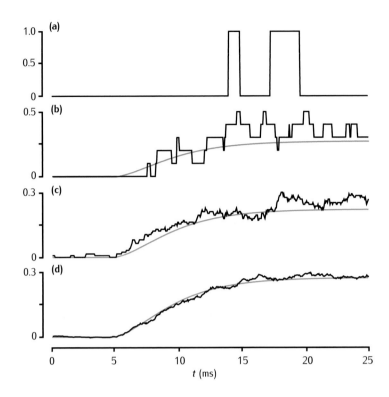

Fig 4.20 Simulations of varying numbers of the stochastic potassium channel described in Scheme 4.5 during a voltage step from −60 mV to −30 mV at $t = 5$ ms. The conductance of a single channel is normalised to $1/N$, where N is the number of channels in the simulation. **(a)** Simulation of a single channel. **(b)** 10 channels. **(c)** 100 channels. **(d)** 1000 channels. The smooth blue line in **(b)**–**(d)** is the plot of n^4, the Hodgkin and Huxley potassium gating variable, for the same voltage step.

parallel are simulated (Figure 4.20): the more channels contribute to the current, the smoother the current. As the number of channels increases, the current tends towards what would be predicted by interpreting the kinetic scheme as the deterministic evolution of the fraction of channels in each state.

An important modelling question is: when is it appropriate to model multiple single channels stochastically, and when to model the average properties of an ensemble of channels? The answer depends strongly on the size of the membrane patch in which the channels are situated. For a given density of channels and probability of channels being open, the expected specific membrane conductance does not depend on the area of the membrane patch, but the size of fluctuations in the specific membrane conductance is inversely proportional to the square root of the area of the patch (Box 4.6). It is sometimes possible to predict the fluctuations in voltage arising from fluctuations in conductance, though this is not straightforward, as the fluctuations in the membrane potential depend on a number of factors, including the membrane time constant, channel kinetics and the difference between the membrane potential and the reversal potential of the current of interest (Chow and White, 1996). However, the larger fluctuations in conductance present in patches with smaller areas will tend to lead to larger fluctuations in voltage (Koch, 1999). If the fluctuations in voltage are small compared to the maximum slope of the activation curves of all channels in the system, the noise due to stochastic opening and closing of channels is unlikely to play a large part in neuronal function. Nevertheless, even small fluctuations could be important if the cell is driven to just below its firing threshold, as small fluctuations in voltage could cause the membrane potential to exceed this

Box 4.6 | Magnitude of specific conductance fluctuations

Consider a membrane patch of area a, in which there is a density ρ (number of channels per unit area) of a particular channel type with a single channel conductance of γ. At a particular point in time, the probability of an individual channel being open is m.

The number of channels in the patch is $N = \rho a$. The number of channels expected to be open is Nm. Therefore the expected specific membrane conductance due to this type of channel is:

$$\frac{Nm\gamma}{a} = \rho m\gamma.$$

Thus, the expected specific conductance is independent of the area of the patch. Assuming that channels open and close independently, the standard deviation in the number of channels open is, according to binomial statistics, $\sqrt{Nm(1-m)}$. Therefore, the standard deviation in the specific conductance is:

$$\frac{\sqrt{Nm(1-m)}\gamma}{a} = \frac{\sqrt{\rho m(1-m)}\gamma}{\sqrt{a}}.$$

Thus, the fluctuation of the specific membrane conductance is inversely proportional to the square root of the area of the patch.

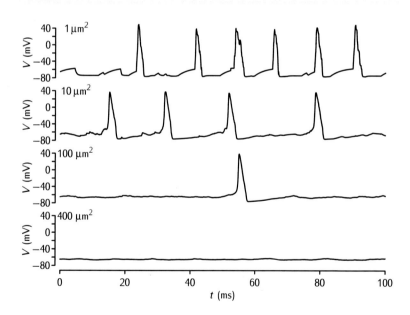

Fig 4.21 Simulations of patches of membrane of various sizes with stochastic Na^+ and K^+ channels. The density of Na^+ and K^+ channels is 60 channels per μm^2 and 18 channels per μm^2, respectively, and the single channel conductance of both types of channel $\gamma = 2\,pS$. The Markov kinetic scheme versions of the HH potassium and sodium channel models were used (Schemes 4.5 and 4.8). The leak conductance and capacitance are modelled as in the standard HH model. No current is applied; the action potentials are due to the random opening of a number of channels taking the membrane potential to above threshold.

threshold and fire an action potential (Strassberg and DeFelice, 1993; Chow and White, 1996).

Simulations of a stochastic version of the HH model in small patches of membrane, based on those of Chow and White (1996) and displayed in Figure 4.21, show that in squid giant axon patches of area less than $1\,\mu m^2$, the opening of single channels has an appreciable effect on the membrane potential. In the simulations, no current is applied, yet when the area of the membrane is small enough, random fluctuations in the number of sodium and potassium channels open are sufficient to push the membrane potential above threshold. The action potentials generated are jagged compared to the smooth action potentials produced in the deterministic HH model. As the area of the membrane increases, and the total membrane capacitance of the patch becomes large compared to the conductance of a single channel, the membrane potential behaves more and more like it would do with deterministic channels.

With increasing amounts of computing power, simulating multiple stochastic channels is more feasible than it used to be, boosting the arguments in favour of stochastic simulation. Some simulator packages make it fairly straightforward to simulate the same kinetic scheme either deterministically or stochastically. Detailed neuron models likely involve some subcellular compartments with a surface area on the order of only a few square micrometres at most, such as thin dendritic branches and dendritic spines, containing 100s rather than 1000s of channels. For such models, it may be worth running the initial simulations deterministically, and then running a number of simulations with stochastic dynamics in order to assess how important the effect of stochastic channel opening and closing is in the system. Cannon et al. (2010) demonstrated that stochastic ion channel openings can significantly affect the occurrence of dendritic and somatic spikes in a model of a hippocampal CA1 pyramidal cell.

4.7.4 Fitting Kinetic Schemes to Single Channel Data

Fitting a stochastic Markov model to single channel data requires different techniques to the error-minimising optimisation needed for deterministic models with macroscopic channel data (Section 4.5.4). Now both the model output and experimental data are characterised by the probability distributions of open and closed times. Consequently, probabilistic fitting techniques are needed to match the model and experimental distributions.

The most common method is to find model parameters (transition rate constants) that maximise the likelihood that the experimental data are the result of a process corresponding to the model (Colquhoun et al., 1996; Qin et al., 1996, 1997; Colquhoun et al., 2003; Sivilotti and Colquhoun, 2016); see Box 5.8 for a summary of maximum likelihood estimation (MLE). Techniques for MLE of ion channel models include methods for correcting for missed open/closed events due to the finite time sampling resolution in the data (Sivilotti and Colquhoun, 2016) and for coping with unknown numbers of channels (more than a single channel) contributing to the recorded data (Milescu et al., 2005). Software such as HJCFIT (Colquhoun et al., 1996; Sivilotti and Colquhoun, 2016) and QuB (Qin et al., 1996; Nicolai and Sachs, 2013) have been designed specifically for fitting stochastic ion channel models to data using MLE.

An alternative and more informative approach, but that is computationally more expensive, is to use Bayesian inference with Markov–Chain Monte Carlo (see Box 14.6) techniques (Siekmann et al., 2012; Epstein et al., 2016). Bayesian inference seeks to find model parameter values that correspond to the model output being most probable, given the experimental data and prior information about the likely distribution of parameter values. The advantage over MLE, which only gives a point estimate of parameter values, is that Bayesian inference results in an estimate of the probability distribution of parameter values. This provides valuable information about parameter identifiability (see Section 4.5.4): a very broad distribution indicates that there is considerable uncertainty in the particular parameter's value, given the available data. An introductory treatment of Bayesian inference in neural modelling is given in Section 14.7 and Hines (2015).

4.8 The Transition State Theory Approach to Rate Coefficients

This section describes the **transition state theory** and shows how it can be applied to derive equations for the voltage-dependence of rate coefficients, commonly used in Markov models built by biophysicists. The section will also show how the thermodynamic models of ion channels introduced in Section 4.4.3 can be derived from the transition state theory.

4.8.1 Transition State Theory

The transition state theory describes how the rate of a chemical reaction depends on temperature. The key concept, put forward by Arrhenius (1889), is that during the conversion of reactants into products, there is an intermediate step in which the reactants form an activated complex. Forming the activated complex requires work to be done, and the energy for this work comes from thermal fluctuations. Once the activated complex is formed, it is converted into the products, releasing some energy. There is thus an **energy barrier** between the initial and final states of the system. Arrhenius postulated that the reaction rate coefficient k is given by:

$$k \propto \exp\left(-\frac{E_a}{RT}\right),\tag{4.23}$$

where E_a is the **activation energy** required to surmount the energy barrier.

Box 4.7 | Q_{10} and the Arrhenius equation

Equation 4.23 can be used to derive the **Arrhenius equation** (Arrhenius, 1889), which describes the ratio of rate coefficients $k(T_1)$ and $k(T_2)$ at temperatures T_1 and T_2, respectively:

$$\log\left(\frac{k(T_2)}{k(T_1)}\right) = \frac{E_a}{R}\left(\frac{1}{T_1} - \frac{1}{T_2}\right).$$

As described in Section 3.5, the Q_{10} measured at temperature T is defined as the ratio $k(T+10)/k(T)$. By substituting $T_1 = T$ and $T_2 = T + 10$ into the Arrhenius equation, a relationship for the Q_{10} in terms of the activation energy E_a can be derived:

$$\log(Q_{10}) = \frac{E_a}{R}\frac{10}{T(T+10)}.$$

This equation can be used to estimate the activation energy from the Q_{10}. It also shows that, assuming that the activation energy is independent of temperature, the Q_{10} depends on temperature. However, this dependence is insignificant for the ranges considered in mammalian biology. For example, the Q_{10} of a reaction calculated from rates at 5°C and 15°C (278 K and 288 K) is expected to differ from the Q_{10} calculated at 27°C and 37°C (300 K and 310 K) by a factor of 1.000017.

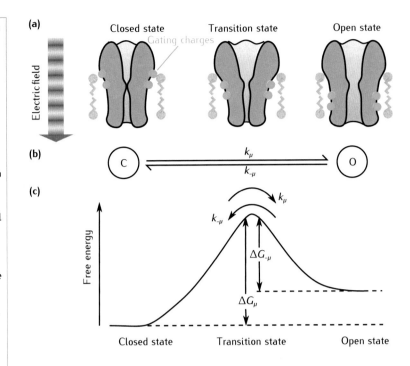

Fig 4.22 The application of the transition state theory to determining rate constants. **(a)** Representation of a hypothetical channel protein with two stable states: a closed state and an open state. Whilst changing conformation between these states, the channel passes through a **transition state**. In each state, the gating charges (in blue) have a different position in the electric field. **(b)** Representation of the channel as a Markov scheme. The transition state does not feature in the scheme. The forward reaction, indexed by μ, has a rate constant k_μ, and the backward reaction, indexed by $-\mu$, has a rate constant $k_{-\mu}$. **(c)** The curve shows the free energy as a function of the progress of the channel protein between the closed and the open states. In the forward reaction, the size of the free energy barrier that the reaction or channel has to surmount is ΔG_μ. In the backward reaction, the free energy barrier is $\Delta G_{-\mu}$. Both ΔG_μ and $\Delta G_{-\mu}$ depend on the voltage, as described in the text.

Temperature T is always measured in kelvins in these equations. The units of the activation energy E_a are $J\,mol^{-1}$ and the gas constant $R = 8.314\,J\,K^{-1}mol^{-1}$.

Boltzmann's constant $k_B = 1.3807 \times 10^{-23}\,J\,K^{-1}$.
Planck's constant $h = 6.6261 \times 10^{-34}\,J\,s$.

The activation energy depends on the reaction in question. According to Equation 4.23, the larger the activation energy, the slower the reaction. The rate coefficient also depends on temperature: the higher the temperature, the faster the reaction; the larger the activation energy, the stronger the dependence on temperature. This control of temperature dependence suggests a link between the activation energy and the Q_{10}, which is expanded on in Box 4.7.

The Arrhenius formula can be applied to a transition between channel states by: (1) associating the reactants with the initial state of the channel; (2) associating the activated complex with a transitional, high-energy conformation of the ion channel; and (3) associating the product with the final conformation of the channel (Figure 4.22). As in the chemical reaction, each state has a certain level of energy.

To describe the full range of temperature dependence of chemical reactions and channel transitions, an extension to the concept of Arrhenius activation energy using the concepts of Gibbs free energy and entropy from chemical thermodynamics (Box 4.8) is necessary. This is embodied in the **Eyring equation** (Eyring, 1935; Pollak and Talkner, 2005), here presented for a reaction indexed by μ:

$$k_\mu = \frac{k_B T}{h} \exp\left(-\frac{\Delta G_\mu}{RT}\right) = \frac{k_B T}{h} \exp\left(\frac{\Delta S_\mu}{R}\right) \exp\left(-\frac{\Delta H_\mu}{RT}\right), \tag{4.24}$$

where k_B is Boltzmann's constant and h is Planck's constant; ΔG_μ, ΔH_μ and ΔS_μ are, respectively, the differences in Gibbs free energy, potential energy and entropy between the base and transition states.

Box 4.8 | Gibbs free energy

In thermodynamics, the ability of a system in a particular state to do work depends not only on its potential energy, but also on how disordered that state is. The degree of disorder associated with the state is quantified by **entropy** S. Entropy reduces the ability of the state to donate energy to do work. This is expressed in the definition of **Gibbs free energy** G (also known as Gibbs energy), which is a measure of how much energy there is available to do useful work:

$$G = H - TS = U + pV - TS,$$

where H is the **heat energy** (or just **enthalpy** in the context of chemical reactions), U is the potential energy of the state and p and V are pressure and volume, respectively. The enthalpy is the sum of the potential energy of the state plus the product of pressure and volume: $H = U + pV$. Since the channel transitions occur at constant pressure and volume, changes in enthalpy and potential energy between states are equal: $\Delta H = \Delta U$.

To apply the concept of Gibbs free energy to channel transitions, the activation energy E_a is replaced by the difference in Gibbs free energy ΔG_μ between the starting state and the transition state, the activated complex, of the μth reaction, which in turn depends on the potential energy and entropy differences:

$$\Delta G_\mu = \Delta H_\mu - T\Delta S_\mu.$$

Along with the $k_B T/h$ dependence of the constant of proportionality, this leads to the Eyring equation (Equation 4.24).

If the transition state is more ordered than the base state (i.e. ΔS_μ is negative), this can allow for a negative potential energy difference ΔH_μ, and hence for the reaction rate to decrease with increasing temperature, giving a Q_{10} of less than one, as measured in some chemical reactions (Eyring, 1935).

4.8.2 Voltage-Dependent Transition State Theory

Although Equation 4.24 describes how the rate coefficient depends on temperature, it does not depend on the membrane potential explicitly. However, the membrane potential is in the equation implicitly, because besides depending on the conformation (Figure 4.22a) of the ion channel in each state, the potential energy difference ΔH_μ depends on the membrane potential (Borg-Graham, 1989). This is because the gating charges move when the channel protein changes conformation (Section 4.1), and movement of charges in an electric field requires work or releases energy, the amount of which depends on the electrical potential difference between the initial and final positions.

The movement of all of the gating charges of the channel proteins from state 1 through the transition state to state 2 can be reduced to movement of an **equivalent gating charge** z_μ from one side to the other of the membrane, with the transition state occurring at a fractional distance δ_μ through the membrane (Figure 4.23 and Box 4.9). If the potential energy difference between state 1 and the transition state when the membrane potential is zero

Fig 4.23 The gating charges in a channel. **(a)** Each channel, represented by the box, contains a number of charged regions, each of which contains a number of units of charge. There are three charges, with valencies z_1, z_2 and z_3, represented here by the circles. In state 1, the charges are at the left-hand positions. During a gating event, they move to the transitional positions, and then to the right-hand positions. **(b)** The multiple charges can be considered as an equivalent gating charge that moves from the inside to the outside of the membrane. The potential energy of these particles in their resting positions is the same as a charge with valency z_μ at the inside of the membrane. The difference in energy between state 1 and the transitional state is the same as when the equivalent charge has moved a fractional distance δ_μ through the membrane, that is, $-z_\mu F \delta_\mu V$. In the reverse direction, the gating particle has to move a fractional distance $\delta_{-\mu} = 1 - \delta_\mu$ through the membrane. The energy of the gating charge in the field therefore contributes $z_\mu(1 - \delta_\mu)V$ to the enthalpy.

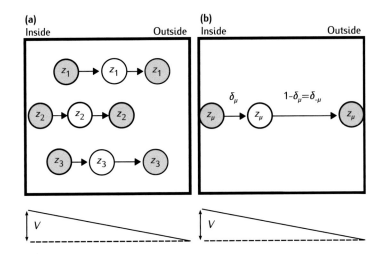

is given by $\Delta H_\mu^{(0)}$, the potential energy difference when the membrane potential is V is given by:

$$\Delta H_\mu(V) = \Delta H_\mu^{(0)} - \delta_\mu z_\mu F V, \tag{4.25}$$

where z_μ is the effective valency of the equivalent gating charge and δ_μ is a number between 0 and 1 representing the distance travelled by the equivalent charge when the channel is in the activated state as a fraction of the total distance between the start and finish states (Figure 4.23). Since the channel protein can move back to its original conformation, taking the gating charge back to its original position, the potential energy change of the reverse reaction, labelled $-\mu$, is:

$$\Delta H_{-\mu}(V) = \Delta H_{-\mu}^{(0)} + (1 - \delta_\mu) z_\mu F V. \tag{4.26}$$

Substituting these expressions into Equation 4.24 gives a physically principled expression for the rate coefficient that depends on membrane potential and temperature:

$$k_\mu = \frac{k_B T}{h} \exp\left(\frac{\Delta S_\mu}{R}\right) \exp\left(-\frac{\Delta H_\mu^{(0)} - \delta_\mu z_\mu F V}{RT}\right), \tag{4.27}$$

$$k_{-\mu} = \frac{k_B T}{h} \exp\left(\frac{\Delta S_{-\mu}}{R}\right) \exp\left(-\frac{\Delta H_{-\mu}^{(0)} + (1 - \delta_\mu) z_\mu F V}{RT}\right). \tag{4.28}$$

Thus, the forward and backward rates of each transition can be described by six parameters: ΔS_μ, $\Delta S_{-\mu}$, ΔH_μ, $\Delta H_{-\mu}$, z_μ and δ_μ.

In order to determine the potential energy and entropy components, experiments at different temperatures must be undertaken. A number of researchers have done this and produced Markov kinetic schemes where the enthalpy and entropy of every transition is known (Rodriguez et al., 1998; Irvine et al., 1999).

Box 4.9 | Equivalent gating charge

Suppose there are a number of gating charges, indexed by j, on the protein that moves in reaction μ. The total energy required to move from state 1 to the transition state is:

$$-FV \sum_j z_{\mu j} \delta_{\mu j},$$

where $\delta_{\mu j}$ is the fractional distance through the membrane that each charge travels. In the reverse direction, the total energy needed to move the gating charges from their positions in state 2 to their transition state positions is:

$$FV \sum_j z_{\mu j} \delta_{-\mu j}.$$

The equivalent gating charge z_μ is defined as:

$$z_\mu = \sum_j z_{\mu j} \delta_{\mu j} + \sum_j z_{\mu j} \delta_{-\mu j}.$$

In state 1, this equivalent gating charge is on the inside edge, and in state 2, it is on the outside. The fractional distance that an equivalent charge would move through the membrane to get from state 1 (inside) to the transition state is:

$$\delta_\mu = \frac{\sum_j z_{\mu j} \delta_{\mu j}}{z_\mu},$$

and the fractional distance from the position of the equivalent charge in state 2 to its position in the transition state is:

$$\delta_{-\mu} = 1 - \delta_\mu.$$

4.8.3 The Thermodynamic Formalism

When rate coefficients derived from the transition state theory are used in independent gating models, the result is the thermodynamic formalism described in Section 4.4.3. This can be shown by considering an ensemble interpretation of the two-state system:

$$C \underset{k_{-1}}{\overset{k_1}{\rightleftharpoons}} O. \tag{4.29}$$

By using the expressions for k_1 and k_{-1} from Equations 4.27 and 4.28, the sigmoidal dependence of O_∞ on V as in Equation 4.1 is obtained:

$$O_\infty = \frac{1}{1 + \exp(-(V - V_{1/2})/k)}, \tag{4.30}$$

where the inverse slope k depends on the molar gas constant R, the temperature T, the Faraday constant F and the effective valency of the gating charges z:

$$k = \frac{RT}{zF}, \tag{4.31}$$

and where the half-activation voltage $V_{1/2}$ also depends on temperature:

$$V_{1/2} = \frac{\Delta H_1^{(0)} - \Delta H_{-1}^{(0)}}{zF} - \frac{\Delta S_1^{(0)} - \Delta S_{-1}^{(0)}}{zF} T. \tag{4.32}$$

According to the Boltzmann distribution, at equilibrium, the fraction of particles in a state i is proportional to $\exp(-G_i/kT)$ where G_i is the free energy of the ith state. Equation 4.1 can also be derived from the Boltzmann distribution.

The constant of proportionality K in Equations 4.3 for $\alpha'_x(V)$ and $\beta'_x(V)$ is related to the thermodynamic parameters according to:

$$K = \frac{k_B T}{h} \exp\left(\frac{\delta_1 \Delta S_{-1}^0 + \delta_{-1} \Delta S_1^0}{R}\right) \exp\left(-\frac{\delta_1 \Delta H_{-1}^0 + \delta_{-1} \Delta H_1^0}{RT}\right), \tag{4.33}$$

where $\delta_{-1} = 1 - \delta_1$.

This corresponds to the thermodynamic models presented in Section 4.4.3 when the rate-limiting factor τ_0 is zero. There is an exponential temperature dependence, which can be approximated by the Q_{10} (Box 4.7), built into this in the third factor. This is a theoretical basis for the Q_{10} factor applied to the gating variable time constants in thermodynamic models.

This derivation does not include the rate-limiting factor τ_0. When this factor is non-zero, the exponential temperature dependence is no longer predicted to hold exactly. A more principled means of incorporating rate limiting is to use a model of the gating particle with more states. For example, the rate-limiting linear exponential form for the rate coefficients used by Hodgkin and Huxley (Figure 4.12) can be obtained from a multi-well model (Tsien and Noble, 1969; Hille, 2001):

The multi-well model (Scheme 4.34) is also a representation of the constant-field Nernst–Planck equation for electrodiffusion (Section 2.2.4) and corresponds to the idea that the gating charges permeate the membrane in the same way in which ions move through channels.

$$C_1 \underset{k_{-1}}{\overset{k_1}{\rightleftarrows}} C_2 \underset{k_{-2}}{\overset{k_2}{\rightleftarrows}} C_3 \ldots C_{n-1} \underset{k_{-(n-1)}}{\overset{k_{n-1}}{\rightleftarrows}} C_n \underset{k_{-n}}{\overset{k_n}{\rightleftarrows}} O, \tag{4.34}$$

where the transitions between the initial closed state and the open state are voltage-independent.

This approach can also be used to obtain other forms of rate coefficient, such as the sigmoidal form used by Hodgkin and Huxley (Figure 4.12), which can be obtained from the kinetic scheme:

$$C_1 \underset{k_{-1}(V)}{\overset{k_1(V)}{\rightleftarrows}} C_2 \overset{k_2}{\longrightarrow} O, \tag{4.35}$$

where all the rate coefficients use the thermodynamic formalism, but with k_2 being a voltage-independent coefficient. In principle, more accurate temperature dependencies could be generated by finding the thermodynamic parameters of each reaction rather than by applying a Q_{10} correction to the rate generated by the entire scheme.

4.8.4 Higher-Order Models

The potential energy differences between the various states may also be affected by processes such as deformation of the electric field within the membrane that depend on higher powers of V (Hill and Chen, 1972; Stevens, 1978; Destexhe and Huguenard, 2000). These can be incorporated by adding terms to the expression for the potential energy:

$$\Delta H_{-\mu}(V) = \Delta H_{-\mu}^{(0)} - (1 - \delta_\mu) z_\mu F V + b_\mu V^2 + c_\mu V^3, \tag{4.36}$$

Box 4.10 | Microscopic reversibility

One important constraint on kinetic schemes which contain loops (Scheme 4.9, for instance) is that of **microscopic reversibility**, also known as **detailed balance** (Hille, 2001). The principle is that the sum of the energy differences in the loop must be zero. For example, in the hypothetical scheme:

$$C \underset{k_{-1}}{\overset{k_1}{\rightleftharpoons}} O$$

$$k_4 \Big\updownarrow k_{-4} \quad k_{-2} \Big\updownarrow k_2$$

$$I \underset{k_3}{\overset{k_{-3}}{\rightleftharpoons}} I_2$$

the enthalpy and the entropy differences around the circuit must all sum to zero:

$$\Delta H_1 - \Delta H_{-1} + \Delta H_2 - \Delta H_{-2} + \Delta H_3 - \Delta H_{-3} + \Delta H_4 - \Delta H_{-4} = 0,$$
$$\Delta S_1 - \Delta S_{-1} + \Delta S_2 - \Delta S_{-2} + \Delta S_3 - \Delta S_{-3} + \Delta S_4 - \Delta S_{-4} = 0.$$

From this, it is possible to conclude that the product of the reaction coefficients going round the loop one way must equal the product in the other direction:

$$k_1 k_2 k_3 k_4 = k_{-1} k_{-2} k_{-3} k_{-4}.$$

With an enthalpy that depends linearly on the membrane potential (Equations 4.25 and 4.26), microscopic reversibility implies that the sum of effective gating charges around the loop is zero:

$$z_1 + z_2 + z_3 + z_4 = 0.$$

where b_μ and c_μ are per-reaction constants. Using non-linear terms of voltage in the exponent (Equation 4.36) can improve the fit of thermodynamic models to data (Destexhe and Huguenard, 2000).

4.9 | Summary

This chapter has concentrated on the theory of modelling ion channels. It has been shown how ion channel models of varying levels of complexity can be used to describe voltage- and ligand-gated ion channel types. Three broad formalisms have been used:

(1) The HH formalism, which has independent gating particles, and no constraints on the voltage dependence of the rate coefficients (Chapter 3 and Section 4.4).
(2) The thermodynamic formalism (Section 4.4.3), which has independent gating particles, but where the rate coefficients are constrained by the transition state theory (Section 4.8).
(3) Markov kinetic schemes with rate coefficients constrained by the transition state theory (Sections 4.5 and 4.7).

Ideally, in order to construct a detailed model of a neuron, the computational neuroscientist should follow Hodgkin and Huxley by characterising the behaviour of each type of channel in the neuron at the temperature of interest and producing, at least, an independent gating model of the channel. In practice, this does not tend to happen because the effort involved is prohibitive compared to the rewards. When faced with building a neuron model containing a dozen channel types, rather than running experiments, typically the computational neuroscientist searches the literature for data from which to construct a required ion channel model. The data are not necessarily from the neuron type, brain area or even species in question, and quite probably have been recorded at a temperature that differs from the model temperature. More likely, the scientist will search for existing computational models (see Box 4.3), but these also will have the same provisos as experimental data.

As with many aspects of neuronal modelling, whether developing a model from scratch, or in trying to use or adapt a published model, there is a range of options of varying levels of detail available to the modeller. How to decide which type of ion channel model to develop can be hard. The fundamental choices to make are:

- Are there enough experimental data to model at the desired level of detail?
- Is the HH or a thermodynamic gating particle formalism good enough, or is a Markov kinetic scheme required?
- Will a deterministic model suffice or does the model need to be stochastic?

Whilst building detailed channel models can be tempting, they should be considered in the context of the entire cell model and the question that is being addressed. Most complex neuron models (see Chapter 5) make use of deterministic gating particle ion channel models for computational simplicity. A deterministic Markov model may be required to capture the fine details of an ion channel's dynamics and (in)activation properties. To investigate the effects of the noise due to random opening and closing of ion channels on a neuron's electrical activity, then stochastic ion channel models will be needed.

This chapter has discussed the modelling of ligand-gated channels (Section 4.6). Calcium is a key intracellular ligand for some ion channels, in particular calcium-dependent potassium channels. How to model the calcium signals that activate these channels is covered in Chapter 6. Modelling of synapses, the postsynaptic side of which are channels activated by extracellular ligands, will be covered in Chapter 7.

CHAPTER 5

Modelling Neurons over Space and Time

The membrane potential of a neuron can vary widely across the spatial extent of a neuron. The membrane may have spatially heterogeneous passive and active properties due to distinct distributions of ion channels. Synaptic inputs arrive at specific dendritic locations and propagate to the cell body. Action potentials propagate along the axon. To capture these factors, we need neuron models that include the spatial, as well as temporal, dimensions. The most common and flexible approach is known as compartmental modelling, in which the spatial extent of a neuron is approximated by a series of small compartments that are all assumed to be isopotential. In some restricted cases of simple neuron geometry, analytical solutions for the membrane potential at any point along a neurite can be obtained through the use of cable theory. We describe both modelling approaches here, with emphasis on compartmental modelling. Two case studies demonstrate the power of compartmental modelling: (1) action potential propagation along axons; and (2) synaptic signal integration in pyramidal cell dendrites. A brief introduction is given to the numerical methods used to solve compartmental models.

In Chapter 2, we considered the electrical properties of a patch of membrane or small neuron to be homogeneous across the spatial extent of the membrane. This is appropriate when considering an area of membrane over which the membrane potential is effectively constant, or isopotential. However, most neurons have a significant spatial extent of membrane that cannot be considered isopotential throughout. At any point in time, the membrane potential may be significantly different across the neuron, resulting in axial currents flowing along the neurites. For example, during the propagation of action potentials (APs) (Section 3.4), only a part of the axon is experiencing the AP at any given time. Similarly, dendrites cannot generally be treated as isopotential, with spatially distributed synaptic potentials and active membrane processes.

Fortunately, it is straightforward to incorporate the electrical circuit model of a patch of membrane into a model of spatially extended neurites.

Fig 5.1 A diagram of a compartmental neuron model. (a) The cell morphology is represented by (b) a set of connected cylinders. An electrical circuit consisting of (c) interconnected RC circuits is then built from the geometrical properties of the cylinders, together with the membrane properties of the cell.

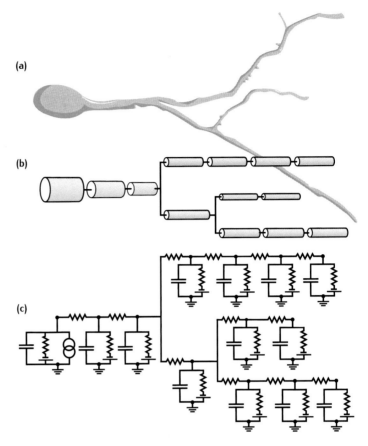

(a)

(b)

(c)

This is achieved by creating an electrical circuit model in which the circuits representing assumed isopotential patches of membrane, commonly known as **compartments**, are connected by axial resistances to form the complete neuron model circuit. The relationship between a real neuron, its approximated compartmental morphology and the overall equivalent electrical circuit, or **compartmental model**, is illustrated in Figure 5.1. In Section 5.1, we outline the compartmental modelling approach, which enables the numerical modelling of the membrane potential across arbitrary neurite morphology, with both passive and active membrane properties. We then illustrate its use through a case study of the speed and reliability of AP propagation along uniform and branched axons (Section 5.2).

Compartmental modelling is built upon the approximation of dividing the neuron into discrete isopotential compartments. If the neurite morphology we wish to model is relatively simple, such as a single, unbranched section of uniform diameter, it is sometimes possible to calculate the membrane potential over time and continuous space analytically using the cable equation. We introduce the cable equation in Section 5.3 and demonstrate the types of results that can be obtained with it.

If we do want to model the complex spatial extent of a real neuron, for example, to study the integration of spatially segregated synaptic inputs across the dendrites, then compartmental modelling is necessary. In Section 5.4, we discuss the issues and approaches when using detailed neuronal morphologies as the basis for our neuron model. In Section 5.6, we present a case study of using a detailed compartmental model of a neocortical pyramidal cell to investigate the forward and backward propagation of signals in the neuron's dendrites.

As additional ion channels and/or compartments are added to a compartmental model, it can quickly become very complex. It is not unusual for such models to be described mathematically by hundreds, if not thousands, of coupled differential equations. This is not a problem for modern-day computers to solve numerically. We conclude this chapter in Section 5.7 with an introduction to the numerical integration techniques used to solve compartmental models.

How to select and constrain the many parameters in such systems, how to construct useful and informative simulations and how to analyse vast quantities of measurable variables are all issues to be tackled. Our case studies introduce the types of measurements that can be made and how they can be visualised. The issue of estimating values for the many parameters is considered in Chapter 14.

5.1 Constructing a Compartmental Model

The basic approach in compartmental modelling is to split up the neurite into cylindrical compartments (Figure 5.2). Each compartment has a length l and a diameter d, making its surface area $a = \pi d l$. Within each compartment, current can flow onto the membrane capacitance or through the membrane resistance. This is described by the RC circuit for a patch of membrane (Section 2.6). Additionally, current can flow longitudinally through the cytoplasm and the extracellular media. This is modelled by axial resistances that link the compartments. How this approach relates to the spatially continuous cable equation (Section 5.3) is examined in Box 5.2.

Since it is usually assumed that the intracellular resistance is much greater than the extracellular resistance, it may be acceptable to consider the extracellular component of this resistance to be effectively zero (implying that the main longitudinal contribution is intracellular resistivity). Modelling the extracellular medium as electrical ground, it acts in an isopotential manner (as shown in Figure 5.2). For many research questions, such as modelling intracellular potentials, this assumption is valid. In Chapter 10, we consider

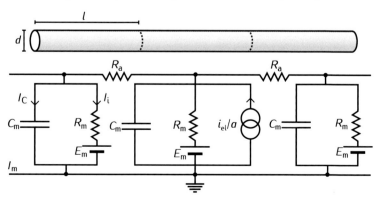

Fig 5.2 A length of passive membrane described by a compartmental model.

Table 5.1 | Passive quantities for spatially extended neurites

Quantity	Description	Typical units	Relationships
d	Diameter of neurite	μm	
l	Length of compartment	μm	
R_a	Specific axial resistance (resistivity)	$\Omega\,cm$	
r_m	Membrane resistance per inverse unit length	$\Omega\,cm$	$r_m = \frac{R_m}{\pi d}$
c_m	Membrane capacitance per unit length	$\mu F\,cm^{-1}$	$c_m = C_m \pi d$
r_a	Axial resistance per unit length	$\Omega\,cm^{-1}$	$r_a = \frac{4R_a}{\pi d^2}$

The quantities r_m, r_a and c_m are useful alternatives to their specific counterparts R_m, R_a and C_m. They express key electrical properties of a neurite of specific diameter and can clarify the equations representing a neurite of arbitrary length. See Table 2.3 for other quantities.

when regions of the extracellular space may not be isopotential and how to model electrodiffusion in this space.

We assume here a circuit as given in Figure 5.2, with the extracellular medium modelled as ground. The axial resistance of a compartment is proportional to its length l and inversely proportional to the cylinder's cross-sectional area $\pi d^2/4$. The **axial resistivity**, also known as the specific **axial resistance** R_a, has units $\Omega\,cm$ and gives the resistivity properties of the intracellular medium. The axial resistance of the cylindrical compartment is then $4R_a l/\pi d^2$. Compartments with longer lengths have larger axial resistance, and those with larger cross-sectional areas have reduced resistances. This and other passive quantities needed in compartmental modelling are listed in Table 5.1.

Assuming each compartment is isopotential, we can describe the electrical circuit representing the neurite with one equation per compartment. We number the compartments in sequence using the subscript j. For example, V_j denotes the membrane potential in the jth compartment and $i_{el,j}$ is the current injected into the jth compartment. Following the procedure used for a single compartment (Section 2.6), we can use Kirchhoff's current law, the quasi-ohmic relation (Equation 2.10) and the equation for the capacitive current (Equation 2.11) to derive our circuit equations. The main difference from the treatment for a single compartment is that, rather than the compartment being isolated, the membrane current $I_{m,j}a$ is now able to spread both leftwards and rightwards within the cytoplasm, that is, the membrane current is equal to the sum of the leftward and rightward axial currents, each given by Ohm's law:

$$I_{m,j}a = \frac{V_{j+1} - V_j}{4R_a l/\pi d^2} + \frac{V_{j-1} - V_j}{4R_a l/\pi d^2}. \tag{5.1}$$

In this case, we are assuming all compartments have the same cylindrical dimensions. Substituting for this membrane current into Equation 2.16, we obtain:

$$I_{c,j}a + I_{i,j}a = I_{m,j}a + i_{el,j}$$

$$I_{c,j}a + I_{i,j}a = \frac{V_{j+1} - V_j}{4R_a l/\pi d^2} + \frac{V_{j-1} - V_j}{4R_a l/\pi d^2} + i_{el,j}. \tag{5.2}$$

This leads to an equation that is similar to Equation 2.17 for a patch of membrane, but now has two extra terms, describing the current flowing along the axial resistances into the two neighbouring compartments $j-1$ and $j+1$:

$$\pi d l C_m \frac{dV_j}{dt} = \frac{E_m - V_j}{R_m / \pi d l} + \frac{V_{j+1} - V_j}{4 R_a l / \pi d^2} + \frac{V_{j-1} - V_j}{4 R_a l / \pi d^2} + i_{el,j}. \tag{5.3}$$

We have used the surface area of the cylinder $\pi d l$ as the area a. Dividing through by this area gives a somewhat less complicated-looking equation:

$$C_m \frac{dV_j}{dt} = \frac{E_m - V_j}{R_m} + \frac{d}{4 R_a} \left(\frac{V_{j+1} - V_j}{l^2} + \frac{V_{j-1} - V_j}{l^2} \right) + \frac{i_{el,j}}{\pi d l}. \tag{5.4}$$

This is the fundamental equation of a compartmental model and there is one such equation for each compartment in a model. It can be solved numerically (Section 5.7) to allow calculation of the membrane potential over time and space. This is commonly referred to as simulating the neuronal response, and a number of computer software packages to enable the construction and carrying out of such simulations using this numerical approach are available (sidebox).

The *NEURON* simulator implements appropriate numerical techniques for compartmental models. The model developer uses scripting languages to specify the model structure and simulation procedures, without having to worry about the low-level numerical details. Note that *NEURON* refers to isopotential compartments as **segments**, which subdivide unbranched **sections** of the neurite.

5.1.1 Boundary Conditions

The equations above assume that each compartment j has two neighbouring compartments $j-1$ and $j+1$, but this is not true in the compartments corresponding to the ends of neurites. Special treatment is needed for these compartments, which depends on the condition of the end of the neurite being modelled.

The simplest case is that of a **killed end**, in which the end of the neurite has been cut. This can arise in some preparations such as dissociated cells, and it means that the intracellular and extracellular media are directly connected at the end of the neurite. Thus, the membrane potential at the end of the neurite is equal to the extracellular potential. To model this, in the equation for the membrane potential, V_0 in the first compartment is set to zero, as illustrated in Figure 5.3a. This allows Equation 5.4 to be used. Setting $V_0 = 0$ is called a **boundary condition** as it specifies the behaviour of the system at one of its edges. Explicitly specifying the value of a quantity at the boundary is called a **Dirichlet** boundary condition.

If the end of the neurite is intact, a different boundary condition is required. Here, because the membrane surface area at the tip of the neurite is very small, its resistance is very high. In this **sealed end** boundary condition, illustrated in electric circuit form in Figure 5.3b, we assume that the resistance is so high that a negligible amount of current flows out through the end. Since the axial current is proportional to the gradient of the membrane potential along the neurite, zero current flowing through the end implies that the gradient of the membrane potential at the end is zero. This boundary condition is modelled by setting a notional compartment $V_{-1} = V_1$, giving a modified version of Equation 5.4 for compartment 0:

$$C_m \frac{dV_0}{dt} = \frac{E_m - V_0}{R_m} + \frac{d}{4 R_a} \left(\frac{2(V_1 - V_0)}{l^2} \right) + \frac{i_{el,j}}{\pi d l}. \tag{5.5}$$

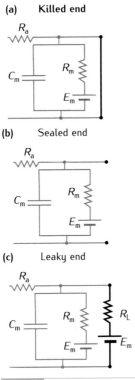

(a) Killed end

(b) Sealed end

(c) Leaky end

Fig 5.3 Circuit illustration of three types of cable terminal conditions.

Specifying the spatial derivative of a quantity at the boundary is called a **Neumann** boundary condition. This is the usual boundary condition assumed for the end of every neurite when modelling a complete, healthy neuron.

It can also be assumed that there is a **leaky end**; in other words, that the resistance at the end of the cable has a finite absolute value R_L (Figure 5.3c). In this case, the boundary condition is derived by equating the axial current, which depends on the spatial gradient of the membrane potential, to the current flowing through the end $(V - E_m)/R_L$.

5.1.2 Example: Ball-and-Stick Neuron

To give a first example of a compartmental model, Figure 5.4 shows the membrane potential response over time at multiple spatial points in a simple ball-and-stick neuron configuration, consisting of a single-compartment soma attached to a long dendrite modelled as an unbranched, multicompartment cylinder. In this simulation, the dendrite is driven by an injected conductance waveform that mimics the response to an excitatory synaptic input (Figure 5.4a) at a point 80% along the dendrite, away from the soma. The resultant synaptic current (Figure 5.4b) propagates passively to the soma, where it is still strong enough to depolarise the cell sufficiently to cause an action potential (AP), which then propagates back into the dendrite (Figure 5.4c).

We consider how to model synaptic responses in detail in Chapter 7. In the example here, the synaptic conductance waveform is modelled as the response to a dual exponential conductance change, given by:

$$g_{syn}(t) = \overline{g}_{syn} \frac{\tau_1 \tau_2}{\tau_1 - \tau_2} \left(\exp\left(-\frac{t - t_s}{\tau_1}\right) - \exp\left(-\frac{t - t_s}{\tau_2}\right) \right), \quad (5.6)$$

where t_s (set to 2 ms in the illustrated simulation) is the time of arrival of a presynaptic AP; the time course is governed by two time constants $\tau_1 = 1$ ms and $\tau_2 = 1.1$ ms; and $\overline{g}_{syn}$ is the maximum possible conductance. The resulting synaptic current is given by:

$$I_{syn}(t) = g_{syn}(t)(V(t) - E_{syn}), \quad (5.7)$$

where $E_{syn} = 0$ mV is the synaptic current reversal potential. This synapse generates a depolarising inward current (Figure 5.4b) and hence is excitatory.

In the simulation, $\overline{g}_{syn}$ is chosen so that the resulting postsynaptic current magnitude at the synaptic site is just large enough that once it has propagated to the soma, it still triggers an AP (Figure 5.4c). Note that the back-propagating AP (Figure 5.4c, blue line) causes a transient decrease in the magnitude of the synaptic current (Figure 5.4b), as the membrane potential at the site of the synapse approaches the synaptic reversal potential.

5.1.3 Variations in Neurite Properties

Membrane and morphological properties are likely to vary along the length of a neurite. These can include changes in membrane capacitance and resistance, ion channel densities and axial resistivity and diameter. In the compartmental approach, changes in membrane properties are handled easily as all cross-membrane currents are calculated on a per-compartment basis.

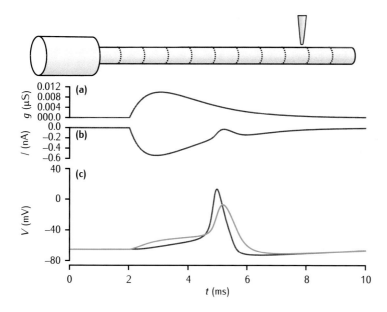

Fig 5.4 The time-varying membrane potential measured at different points along a ball-and-stick model neuron. As illustrated in the schematic (not to scale), the soma is a cylinder, 40 μm long and 40 μm in diameter, modelled as a single compartment. The dendritic cable is a cylinder, 200 μm long and 4 μm in diameter, divided into 11 equal compartments. The passive membrane parameters are $R_a = 150\,\Omega\,cm$, $R_m = 30\,k\Omega\,cm^2$ and $C_m = 1\,\mu F\,cm^{-2}$. The soma contains the Hodgkin–Huxley sodium and potassium ion channel models for action potential generation, with temperature adjusted to 18 °C. **(a)** Synaptic conductance at the site of synapse, 80% along the dendrite, away from the soma (electrode in schematic). **(b)** Synaptic current at the site of synapse. **(c)** Somatic (black line) and dendritic (at site of synapse; blue line) membrane potentials.

This necessitates specifying capacitance, membrane resistance and ion channel densities uniquely for each compartment.

However, changes in diameter and axial resistivity affect intracellular current flow between adjacent compartments. This requires averaging values for these parameters between compartments, slightly complicating the resulting voltage equation. For variations in diameter, when calculating the voltage in compartment j, we calculate the cross-sectional area between, say, compartments j and $j + 1$ using the average diameter: $\pi((d_{j+1} + d_j)/2)^2/4$. To get the term coupling this compartment with compartment $j + 1$, this area is divided by the surface area of compartment j multiplied by its length, resulting in:

$$\frac{(d_{j+1} + d_j)^2}{16d_j R_a l^2}\left(V_{j+1} - V_j\right).$$

A full treatment of how to deal with such variations is given in Mascagni and Sherman (1998).

Incorporating Branch Points

Dendrites and axons are not just single long cables, but form complex branching structures. When representing a branching structure by a compartmental model, at a branch point, each compartment will have more than two neighbours. Fortunately, in real neurites (dendrites and axons), branches seem to be exclusively bifurcations, that is, the parent branch splits into two daughter branches. Thus each compartment at the branch point will have three, rather than two, neighbours. So Equation 5.4 gains an extra term that accounts for the membrane potential in a third neighbouring compartment. This complicates the overall system of differential equations that must be solved to calculate the membrane potential in each compartment. Well-designed indexing of branches in a complex neurite can help with the numerical solution and various schemes have been proposed: this is beyond the scope of this book, but see Mascagni and Sherman (1998) for further details.

5.1.4 Adding Active Channels

So far we have only given the equations for passive compartmental models. However, with the Hodgkin–Huxley model of the AP, we have already seen how active ion channel currents can be combined with the equations for passive response of a single compartment. This approach simply generalises when we have multiple compartments, such as in the ball-and-stick model neuron considered above, with the active ion channel currents considered separately for each compartment. So the basic compartmental equation is now:

$$C_{\mathrm{m}} \frac{\mathrm{d}V_j}{\mathrm{d}t} = -\sum_k I_{\mathrm{i},k,j} + \frac{d}{4R_{\mathrm{a}}} \frac{V_{j+1} - V_j}{l^2} + \frac{d}{4R_{\mathrm{a}}} \frac{V_{j-1} - V_j}{l^2} + \frac{i_{\mathrm{el},j}}{\pi d l}, \quad (5.8)$$

where the right-hand side now includes the sum of all ionic currents in compartment j, $I_{\mathrm{i},k,j}$.

Ion Channel Distributions

For spatially extensive neurons, the location of different types of ion channels has a critical effect on neuronal function (Gillies and Willshaw, 2006). It is possible to some extent to get quantitative data on ion channel densities experimentally. Immunolabelling with fluorescent proteins or gold particles, with light or electron microscopic examination of the labelled tissue, can give estimates of specific ion channel locations and densities (Kerti et al., 2012). Electrophysiological recordings using patch clamping with channel blocking at different locations in the dendrites can reveal the magnitudes of specific ion currents, from which channel densities can be inferred (Hoffman et al., 1997). In practice, such distribution information is not likely known with any certainty and the densities of channels in each compartment are free parameters in the model. In models with many compartments, this can lead to an unmanageably large number of parameters: at least one conductance density parameter per channel type per compartment. Thus one of the immediate consequences of adding active channels is the explosion in the number of parameters in the model and a significant increase in the number of degrees of freedom (Rall, 1990). Parameter estimation procedures to determine ion channel densities are considered in detail in Chapter 14.

One approach to reducing the number of degrees of freedom is to parameterise channel densities throughout a dendrite in terms of distribution functions (Migliore and Shepherd, 2002). The simplest of these would be a uniform distribution where a single maximum conductance density parameter is used in all compartments of the tree (Figure 5.5a). Linear distributions, where the conductance density changes with a linear gradient as a function of distance from the soma, is an example of a slightly more complex distribution function (Figure 5.5b). Knowledge of the particular cell and channel composition is crucial to guide the simplifications. For example, separate distributions may be required within different neuron dendritic processes (Figure 5.5c), such as oblique branches, as opposed to the apical trunk in hippocampal pyramidal cells (Poirazi et al., 2003).

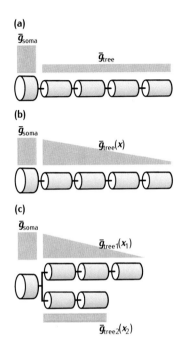

(a) $\bar{g}_{\mathrm{soma}}$ $\bar{g}_{\mathrm{tree}}$

(b) $\bar{g}_{\mathrm{soma}}$ $\bar{g}_{\mathrm{tree}}(x)$

(c) $\bar{g}_{\mathrm{soma}}$ $\bar{g}_{\mathrm{tree1}}(x_1)$ $\bar{g}_{\mathrm{tree2}}(x_2)$

Fig 5.5 Different channel distribution functions. **(a)** Uniform distribution in the tree with an independent soma channel distribution. **(b)** Linear distribution in the tree specified as a function of distance x from the soma. **(c)** Two trees with different distribution functions.

5.2 Case Study: Action Potential Propagation along an Axon

To give a flavour of the utility of compartmental modelling, here we use compartmental models of axons to explore the fundamental issues of the speed and reliability of AP propagation along spatially extensive and branched axons. Such issues are not easily explored experimentally as they require the measurement of membrane potential at multiple points along an axon. With a compartmental model, we can both measure and visualise the membrane potential along the entire length of the axon (Figure 3.16). Furthermore, with the model, we have full control over the axon morphology and the electrophysiological proporties of its membrane, allowing us to investigate the impact of these properties on AP propagation. A key issue to setting up an accurate compartmental model is the choice of the number of compartments to use: approaches to making a sensible choice are discussed in Box 5.1.

5.2.1 Speed of Action Potential Propagation

In our first model, we measure and visualise the AP travelling along a long axon of uniform diameter. This is illustrated in Figures 5.6 and 5.7 for the model of the squid giant axon, as formulated by Hodgkin and Huxley (Chapter 3), with adjustments for temperature of the maximum conductances and time constants. This axon contains a uniform density of fast sodium and delayed-rectifier potassium channels along its length.

The results shown in Figure 5.6 illustrate that the speed of AP propagation is affected by the axon diameter and temperature. At 18.5 °C, the propagation speed measured from the simulation is 18.8 m s^{-1} (Figure 5.6a). This reduces to 13.2 m s^{-1} if the axon diameter is halved (Figure 5.6b). Temperature also has a significant effect on the propagation speed. As detailed in Section 3.5, temperature strongly affects the kinetics of ion channels, and lowering the temperature of the axon slows the opening and closing of the

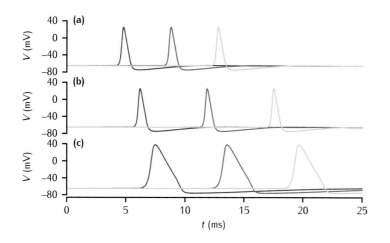

Fig 5.6 Propagation over time of the AP at 25% (left), 50% (middle) and 75% (right) along a spatially uniform squid axon with a length of 300 mm and a diameter of 476 μm. Hodgkin and Huxley sodium and potassium ion channels are distributed throughout the membrane with uniform density. The model axon is divided into 3000 compartments of equal length. The AP is initiated in the first compartment by a brief current injection at 0.5 ms. **(a)** T=18.5 °C; **(b)** T=18.5 °C, diameter reduced by half; **(c)** T=6.3 °C.

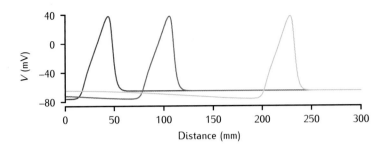

Fig 5.7 Propagation over space of the AP along a uniform axon, as specified in Figure 5.6. The model axon is divided into 3000 compartments of equal length. T=18.5 °C. Membrane potential over space at times 5 ms (left), 10 ms (middle) and 20 ms (right) after AP initiation at the start of the axon (distance 0).

sodium and potassium channels. This broadens the AP and slows its propagation to $12.3 \, \mathrm{m \, s^{-1}}$ at 6.3 °C (Figure 5.6c).

Our simulation also allows us to visualise the AP along the spatial extent of the axon at particular points in time (Figure 5.7). The depolarisation of the AP affects nearly 50 mm of axon at any one time, with the subsequent hyperpolarisation (which contributes to the refractory period of AP generation) covering over 100 mm of axon behind the depolarising wave.

5.2.2 Myelination

The continuous conduction of APs, as illustrated above, is actually quite slow as it requires ion channels throughout the axon membrane, with AP generation and propagation speed being limited by the ion channel dynamics. As a consequence, our nervous systems use a different system for long axons where signal transmission needs to be very fast to avoid significant delays in, for example, responding to sensory input or generating motor output.

Such axons are covered in an insulating **myelin sheath**, as a result of specialised glial cells wrapping their membranes around the axon (Figure 5.9). The myelin sheath provides high trans-sheath resistance and low capacitance, due to the large number of myelin layers that make up the sheath. The myelin does not provide a uniform covering, as there are regular gaps called **nodes of Ranvier**, and active sodium and potassium ion channels are largely restricted to these gaps. To either side of each node are paranodal regions in which the myelin sheath is in close juxtaposition to the axon, and the axon begins to thicken to its full internodal diameter. The remaining internodal axon typically is of larger diameter than the nodes and maintains a small periaxonal space between the axon membrane and the surrounding myelin sheath which provides a high resistance to external current flow.

This arrangement corresponds to an equivalent electrical circuit (Figure 5.9b) in which most axon (non-nodal) compartments are purely passive but also are connected in series to a high-resistance passive myelin compartment. The nodal compartment is the familiar active circuit, as in the continuous conductance model demonstrated above. This results in **saltatory conduction** of APs, where the AP is regenerated at each node and propagates passively between nodes, with little loss because of the high resistivity provided by the added myelin.

To demonstrate AP propagation along a myelinated axon, we use the particular model of McIntyre et al. (2002). The active nodal membrane contains fast sodium, slow potassium and leak currents, and also a persistent

Box 5.1 | Choosing the number of compartments

A fundamental choice in setting up a compartmental model is to decide how many compartments should be used to subdivide the spatial extent of the neurites, as illustrated in Figure 5.8.

The model only generates voltage values at discrete points (corresponding to each compartment) along continuous neurites such as axons. A voltage value is assumed constant along the length of a compartment (isopotential assumption). This affects the accuracy with which our model can calculate voltages changing dynamically over space. The results in Figure 5.6 are for a 3000-compartment model, so each compartment covers 100 μm of axon. Decreasing this to 300 compartments (equivalent to 1000 μm) gives indistinguishable results. However, increasing the number of compartments to 30 000 (or 10 μm) gives a slightly slower propagation speed of $18.55 \, \text{m s}^{-1}$ at 18.5 °C.

The higher the number of compartments, the more accurately the model will capture variations in potential along the length of a neurite. But this comes at a computational cost and so usually we want to choose the smallest number of compartments that still gives reasonable accuracy. Two rules of thumb are commonly used to help with this choice.

Firstly, consider a constant point source of current in a long passive neurite: it can be shown that the membrane potential decays exponentially with distance from this point, with the rate of decay given by the passive length constant $\lambda \equiv \sqrt{R_m d/4 R_a}$ (Section 5.3.1). If we choose the number of compartments so that each compartment size is around 10% of this length constant, then the assumption of isopotential compartments is a reasonable approximation to the actual small voltage decay along one compartment's length, and our model will give an accurate voltage decay along the neurite.

However, many neuronal signals, such as action potentials, and synaptic potentials, are transient and strongly affected by membrane capacitance. So an alternative is to consider a frequency-dependent length constant (Box 5.3) of the form (Carnevale and Hines, 2006): $\lambda_f \equiv 1/2\sqrt{d/\pi f R_a C_m}$, where the frequency $f \gg 0$ is chosen to capture the time course of rapidly changing signals in the neuron, with a frequency on the order of 50–100 Hz being reasonable. Thus, another rule of thumb is to choose the compartment size to be 10% of λ_f. This quantity will be smaller than if using the passive length constant λ, thus leading to a larger number of compartments.

For the squid axon (Figure 5.7) $R_m \approx 3333 \, \Omega \, \text{cm}^2$ (approximately, as this is derived purely from the passive leak conductance for simplicity's sake), $R_a = 35.4 \, \Omega \, \text{cm}$, $C_m = 1 \, \mu\text{F cm}^{-2}$ and $d = 476 \, \mu\text{m}$. Taking frequency $f = 100 \, \text{Hz}$, these quantities give a passive length constant $\lambda = 10\,585 \, \mu\text{m}$, and a frequency-dependent constant $\lambda_f = 10\,344 \, \mu\text{m}$. So in this case, both length constants are very similar and of the order of $10\,000 \, \mu\text{m}$ (10 mm). Therefore, around 300 compartments, each with a size of 1000 μm, will give us reasonable spatial accuracy, and this is borne out by the simulations.

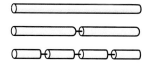

Fig 5.8 An unbranched length of neurite (axon or dendrite) can be split into different numbers of compartments.

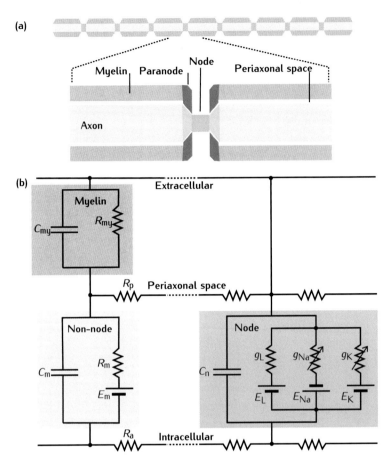

Fig 5.9 **(a)** Schematic of a myelinated axon, with details of a nodal region. **(b)** The equivalent electrical circuit.

sodium current. The channel kinetics are tuned to match the characteristics of mammalian nerve fibres at 36 °C, and so are somewhat different from the squid ion channels of the Hodgkin–Huxley model (see McIntyre et al. (2002) for details). The node is significantly thinner than the surrounding myelinated axon. The paranodal regions are simply modelled as single compartments with the same diameter as a node, but with a surrounding myelin sheath with a small periaxonal space. Multiple paranodal compartments of increasing diameter could be used to more accurately model the transition between node and internodal axon.

AP propagation along this model axon is shown in Figure 5.10. APs are (re)generated at each node. Membrane potential changes drop rapidly in the internodal region (Figure 5.10b) and there is very little membrane current at internode segments, with almost all current flowing axially. AP propagation speed is now 100 m s^{-1}, which is much faster than along a length of unmyelinated axon of the same diameter (around 1 m s^{-1}).

5.2.3 Reliability of Propagation in Branching Axons

Compartmental modelling allows the construction of axon models with arbitrarily complex branching structures, as exhibited by real axons. Such models enable the investigation of the impact of branching on the speed and reliability of AP propagation along the full extent of a complex axon.

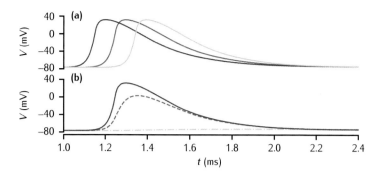

Fig 5.10 Propagation over time of the AP along a myelinated axon of length 28 000 μm modelled using 221 compartments consisting of: 21 nodes, each 1 μm in length, 4.7 μm in diameter and 1400 μm apart; 40 paranodes of length 3 μm and diameter 4.7 μm; 40 internodal segments of length 1394 μm (divided into 8 compartments) and diameter 10.4 μm; the myelin sheath at paranodes and internodal segments give a total fibre diameter of 14 μm; the periaxonal space is 0.002 μm at a paranode and 0.004 μm at an internodal region. (a) Example nodal membrane potentials from node numbers 5, 10 and 15 (from left to right). (b) Membrane potentials approaching node 10: node 10 itself (top, solid); paranode (middle, dashed); and preceding internodal compartment (bottom, dot-dashed).

Goldstein and Rall (1974) undertook the first detailed study of the effects of specific branching geometries on AP propagation. To illustrate this we consider a simple axon with a single bifurcation where the parent axon splits into two daughter branches (Figure 5.11). Depending on the diameters of the daughter branches, relative to the parent branch, this bifurcation, or branch point, may change the electrical load on AP generation at and near the branch point. Fundamental to this is the geometric ratio GR between the parent and daughter branches (Goldstein and Rall, 1974), given by:

$$\text{GR} \equiv \frac{d_L^{3/2} + d_R^{3/2}}{d_P^{3/2}}. \tag{5.9}$$

If GR = 1 and the membrane properties are uniform in all branches, then the branch point does not exert any change in electrical load and an AP will pass through the branch point unimpeded (Figure 5.11: top plots). However, if GR > 1, then the branches impose an increased electrical load at the branch point, resulting in the AP slowing down and being reduced in amplitude as it approaches the branch point (Figure 5.11: middle plots). Once the AP is established in the daughter branches, it returns to full amplitude and now propagates more quickly due to the larger axon diameter, compared with the parent. In contrast, if GR < 1, then the branches impose a decreased electrical load at the branch point, resulting in the AP accelerating and being increased in amplitude as it approaches the branch point (Figure 5.11: bottom plots). Past the branch point, the AP again returns to the initial amplitude, but now it propagates more slowly, due to the smaller axon diameter. Goldstein and Rall (1974) established that the AP speed is constant in units of the passive length constant ($\lambda \equiv \sqrt{R_m d / 4 R_a}$; Equation 5.12), so an increase in diameter results in an increase in absolute speed. In Section 5.4.4, a value of GR = 1 forms the basis of being able to reduce a branching geometry to an equivalent single cylinder.

5.3 | The Cable Equation

We saw in Section 2.6 that it is possible to solve analytically the equation representing a single membrane compartment. This gave us an equation from which it is easy to see the basis for the time course of the membrane potential, and the important concept of the membrane time constant. Previously

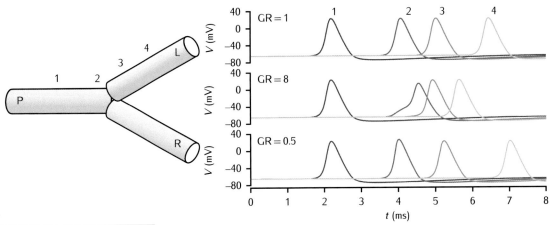

Fig 5.11 Axon potential propagation along a branched axon. The axon is 4 mm long in total divided between the parent (P; 2 mm) and two daughters (L, R; each 2 mm). Left: Schematic diagram of the axon and the location of four recording points: 1 – halfway along the parent axon; 2 – at the end of the parent; 3 – 20% along the left daughter; 4 – halfway along the left daughter. Right: APs at the four recording points (left-to-right). The parent axon has a diameter of 1 μm in each example; top plots: daughter diameters are 0.63 μm (GR = 1); middle plots: daughter diameters are 2.52 μm (GR = 8); bottom plots: daughter diameters are 0.3969 μm (GR = 0.5).

in this chapter, extra compartments were added to allow spatially extended neurites to be described, but this has come at the expense of being able to solve the equations analytically. Although modern computers can numerically integrate the equations of the compartmental model at very high spatial resolutions by using many compartments, looking at analytical solutions can give a deeper understanding of the behaviour of the system.

We now introduce the **cable equation**, which allows the spatiotemporal evolution of the membrane potential to be solved analytically. As shown in more detail in Box 5.2, the cable equation is derived from the equations of a compartmental model (Equation 5.4) by effectively splitting a neurite into an infinite number of infinitesimally small compartments. This gives a partial differential equation (PDE) with the form:

$$C_m \frac{\partial V}{\partial t} = \frac{E_m - V}{R_m} + \frac{d}{4R_a} \frac{\partial^2 V}{\partial x^2} + \frac{i_{el}}{\pi d}. \tag{5.10}$$

In the cable equation, the membrane potential $V(x, t)$ is a function of distance x along a continuous cable, and time t. $i_{el}(x, t)$ is the current injected per unit length at position x. It is very similar to the equation for a single compartment (Equation 2.17), except that the derivative dV/dt has been replaced by the partial derivative $\partial V/\partial t$ and there is an extra term $d/4R_a \partial^2 V/\partial x^2$. The extra term is the net density of current flowing along the length of the cable into point x.

5.3.1 Steady-State Behaviour of the Membrane Potential

The simplest situation to examine is the steady-state case, in which a constant current is injected into the cable; this situation often arises in experiments. In the steady state, when the system has settled and the voltage no longer changes through time, the derivative $\partial V/\partial t$ in Equation 5.10 is zero. This equation then turns into a second-order ordinary differential equation (ODE), which is considerably easier to solve.

Semi-Infinite Cable

We start by considering a semi-infinite cable (Figure 5.12). It has one sealed end from which it extends an infinite distance, and a current with an absolute

Box 5.2 | Derivation of the cable equation

To derive the cable equation from the discrete equations for the compartmental model (Equation 5.4), we set the compartment length l to the small quantity δx. A compartment indexed by j is at a position $x = j\delta x$ along the cable, and therefore, the membrane potentials in compartments $j - 1$, j and $j + 1$ can be written:

$$V_j = V(x, t) \quad V_{j-1} = V(x - \delta x, t) \quad V_{j+1} = V(x + \delta x, t).$$

Also, we define the current injected per unit length as $i_{el}(x, t) = i_{el,j}/\delta x$. This allows Equation 5.4 to be rewritten as:

$$C_m \frac{\partial V(x, t)}{\partial t} = \frac{E_m - V(x, t)}{R_m}$$
$$+ \frac{d}{4R_a} \left[\frac{1}{\delta x} \left(\frac{V(x + \delta x, t) - V(x, t)}{\delta x} - \frac{V(x, t) - V(x - \delta x, t)}{\delta x} \right) \right] + \frac{i_{el}(x, t)}{\pi d}.$$

$$(a)$$

The derivative of V with respect to t is now a partial derivative to signify that the membrane potential is a function of more than one variable.

The length δx of each compartment can be made arbitrarily small, so that eventually there is an infinite number of infinitesimally short compartments. In the limit as δx goes to 0, the term in square brackets in Equation (a) above becomes the same as the definition of the second partial derivative of distance:

$$\frac{\partial^2 V(x, t)}{\partial x^2} = \lim_{\delta x \to 0} \frac{1}{\delta x} \left(\frac{V(x + \delta x, t) - V(x, t)}{\delta x} - \frac{V(x, t) - V(x - \delta x, t)}{\delta x} \right). \quad (b)$$

Substituting this definition into Equation (a) leads to Equation 5.10, the cable equation.

In the case of discrete cables, the sealed end boundary condition gives:

$$\frac{d}{4R_a} \frac{V_1 - V_0}{\delta x^2} = \frac{i_{el,1}}{\pi d \delta x} + \frac{E_m - V_1}{\pi d \delta x R_L}.$$

In the limit of $\delta x \to 0$, at the $x = 0$ end of the cable, this is:

$$-\frac{d}{4R_a} \frac{\partial V}{\partial x} = \frac{i_{el}(0, t)}{\pi d} + \frac{E_m - V(0, t)}{\pi d R_L}.$$

At the $x = l$ end of the cable, this is:

$$\frac{d}{4R_a} \frac{\partial V}{\partial x} = \frac{i_{el}(l, t)}{\pi d} + \frac{E_m - V(l, t)}{\pi d R_L},$$

assuming a sealed end means that the axial current at the sealed end is zero and therefore that the gradient of the voltage at the end is also zero.

value of i_{el} is injected into the cable at the sealed end. Although this is unrealistic, it gives us a feel for how voltage changes over large distances from a single injection site. The analytical solution to Equation 5.10, along with the sealed end boundary conditions (Box 5.2), shows that, in agreement with the numerical solution of the discrete cable equation shown in Figure 5.12a,

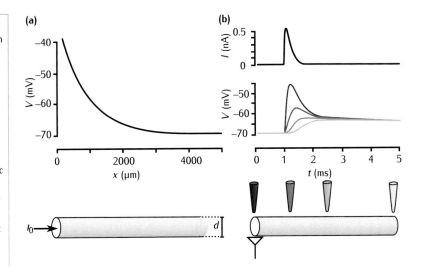

Fig 5.12 **(a)** Steady-state membrane potential as a function of distance in a semi-infinite cable in response to current injection at one end ($x = 0\,\mu m$). The parameters are $d = 1\,\mu m$, $R_a = 35.4\,\Omega\,cm$ and $R_m = 10\,k\Omega\,cm^2$, which, from Equation 5.12, gives the length constant $\lambda = 840\,\mu m$. **(b)** Top: a simulated excitatory postsynaptic current (EPSC). Below: the membrane potential measured at different points along a cable in response to the EPSC evoked at the left-hand end of the cable. The colour of each trace corresponds to the locations of the electrodes. The neurite is 500 μm long, divided into 50 compartments for discrete simulation; other parameters are the same as the semi-infinite cable shown in **(a)**.

the steady-state membrane potential is a decaying exponential function of distance along the neurite:

$$V(x) = E_m + R_\infty i_{el} e^{-x/\lambda}. \tag{5.11}$$

The quantity λ is called the **length constant** of the cable, and R_∞ is the input resistance (defined in Section 2.6.2) of a semi-infinite cable.

The value of λ determines the shape of the exponential voltage decay along the length of the cable. It is determined by the specific membrane resistance, the axial resistivity and the diameter of the cable:

$$\lambda \equiv \sqrt{\frac{R_m d}{4R_a}} = \sqrt{\frac{r_m}{r_a}}. \tag{5.12}$$

This equation shows that the smaller the membrane resistance is relative to the axial resistance, the smaller the length constant will be. The leakier the membrane is (smaller r_m), the more current is lost earlier in its journey along the neurite. Just as the membrane time constant sets the temporal scale of a neurite, so the length constant sets the spatial scale. This is very informative for choosing the number of compartments to use to maintain spatial accuracy in the discrete compartmental model (Box 5.1). However, many neuronal signals are transient and will decay faster with distance than a steady-state signal. It is possible also to derive a length constant that is a function of the frequency of a sinusoidal alternating current (AC) source (Box 5.3).

The input resistance of a semi-infinite cable R_∞ is determined by the specific membrane resistance, the axial resistivity and the diameter:

$$R_\infty = \frac{R_m}{\pi d \lambda} = \sqrt{\frac{4R_m R_a}{\pi^2 d^3}} = \sqrt{r_m r_a}. \tag{5.13}$$

This tells us that we should expect the input resistance of thinner neurites to be higher than that of thicker ones. Thus a given injection current will have a greater effect on the membrane potential of a thinner neurite than on

In Equation 5.12, we have introduced two diameter-specific constants, r_m and r_a, defined in Table 5.1. These are convenient quantities that express the key passive electrical properties of a specific cable of arbitrary length. They are often used to simplify the equations representing a neurite of specific diameter.

Box 5.3 | Frequency-dependent length constant

The cable equation is generally solved assuming constant, unchanging current sources. However, it is also possible to solve it for time-varying alternating current (AC) sources. In this situation, the membrane capacitative current now plays a significant role. For the semi-infinite cable, this yields an AC frequency-dependent expression for a new length constant λ_f that captures the decay over space of the amplitude of the resulting voltage changes (Johnston and Wu, 1995):

$$\lambda_f = \lambda \sqrt{\frac{2}{1 + \sqrt{1 + (2\pi f \tau_m)^2}}}, \tag{a}$$

where f is the frequency of the driving current, $\tau_m = R_m C_m$ is the membrane time constant and λ is the steady-state (direct current, DC) length constant. When f is zero (DC), then $\lambda_f = \lambda$, and then $\lambda_f < \lambda$ for $f > 0$, that is, the frequency-dependent length constant is always less than, or equal to, the steady-state length constant.

In neuronal membranes, τ_m is on the order of tens of milliseconds and f is tens of Hertz (neuronal oscillations and synaptic activity). Assuming $2\pi f \tau_m \gg 1$, then a reasonable approximation to λ_f is (Carnevale and Hines, 2006):

$$\lambda_f \approx \lambda \sqrt{\frac{1}{\pi f \tau_m}} = \frac{1}{2} \sqrt{\frac{d}{\pi f R_a C_m}}, \tag{b}$$

where the final expression comes from substituting in for $\lambda \equiv \sqrt{R_m d / 4 R_a}$. Now it is clear that λ_f is essentially independent of the specific membrane resistance R_m, but does depend on the capacitance C_m.

---

a thicker one. As we will see, this general idea also applies with time-varying input and in branching dendrites.

Finite Cable

The situation of a cable of finite length is more complicated than the infinite cable, as the boundary conditions of the cable at the far end (sealed, killed or leaky) come into play. It is possible to solve the cable equation analytically with a constant current injection applied to a finite cable. This will give an expression for the membrane potential as a function of distance along the cable that also depends on the injection current i_{el} and the type of end condition of the cable. The end condition is represented by a resistance R_L at the end of the cable. For a sealed end, the end resistance is considered to be so large that R_L is effectively infinite. For leaky end conditions, R_L is assumed to be finite. A killed end is a short circuit ($R_L = 0$) where the intracellular and extracellular media meet and there is zero potential difference at the end of the axon. The analytical solution to the finite cable equation in these cases is given in Box 5.4.

Examples of how different end conditions alter the change in voltage over the length of the axon are plotted in Figure 5.13. The solid black line shows the membrane potential in a semi-infinite cable, and serves as a reference. The two solid grey lines show the membrane potential in two ca-

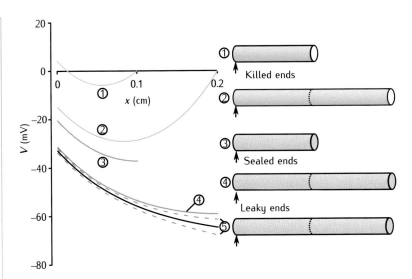

Fig 5.13 The membrane potential as a function of distance for different lengths of cable with different end terminal resistances when a current of 0.5 nA is injected at one end (shown by arrows). The passive cable properties are: $R_m = 6000\,\Omega\,cm^2$, $R_a = 35.4\,\Omega\,cm$ and $d = 2.5\,\mu m$. From Equation 5.12, the length constant $\lambda = 1029\,\mu m$. The black line shows the membrane potential in a semi-infinite cable. The solid grey lines (3, 4) refer to cables with sealed ends, one of length 1000 μm (approximately one length constant) and one of length 2000 μm (approximately two length constants). The solid blue lines (1, 2) refer to cables with one killed end. The dashed grey lines (5) refer to cables with leaky ends; see text for details.

bles with sealed ends, but of different lengths, one of length $l = 1000\,\mu m$ and the other $l = 2000\,\mu m$. Given that the displacement of the membrane potential from its resting value of $-70\,mV$ is proportional to the input resistance, Figure 5.13 shows that the shorter cable has a higher input resistance than both the longer one and the semi-infinite cable. This makes sense since the shorter cable offers fewer paths to the extracellular medium than the longer one. The membrane potential of the longer cable is quite close to that of the semi-infinite cable. As the cable gets longer, the difference between the two will become negligible. Note that the gradient of the membrane potential at the end of a sealed end cable is zero. Since the gradient of the curve is proportional to the current flowing along the axon, a zero gradient means that there is no axial current flowing at the end of the cable, which has an infinitely large resistance.

The two blue lines show what happens under a killed end condition. The membrane potential at the far end of the cable is equal to the extracellular membrane potential (0 mV). This is because the circuit is effectively 'short-circuited' by a zero end resistance.

The two dashed grey lines show what happens when there is a leaky end to the cable. Here, there is a finite resistance R_L at the cable end. The upper dashed line shows the situation when R_L is greater than R_∞, and the lower line shows R_L being less than R_∞. The importance of this situation will become apparent in Section 5.4.4 where we consider simplifying branched dendrites.

5.3.2 Time-Dependent Behaviour of the Membrane Potential

So far we have ignored time in our study of the cable equation. It is possible to solve the cable equation to give mathematical expressions for the time course of the membrane potential at different points along a passive cable in response to pulses of current or continuous input. At any point along the dendrite, the time course of the membrane potential will be given by:

$$V(x,t) = E_m + C_0(x)e^{-t/\tau_0} + C_1(x)e^{-t/\tau_1} + C_2(x)e^{-t/\tau_2} + \dots, \tag{5.14}$$

Box 5.4 | Solutions to the cable equation

It is often useful to express the length along the neurite or cable in relation to the length constant. We denote this normalised length as X, defined as $X = x/\lambda$. The quantity X is dimensionless, leading to clearer formulae. For example, the steady-state membrane potential along a semi-infinite cable (compare with Equation 5.11) becomes:

$$V(X) = E_m + R_\infty i_{el} e^{-X}.$$

For clarity, we look at the finite cable solutions for the sealed end and leaky end boundary conditions. We do not present the killed end case here.

Given a resistance R_L at the end of a leaky cable and an injection current i_{el}, the membrane potential as a function of length X is given by:

$$V(X) = E_m + R_\infty i_{el} \frac{R_L/R_\infty \cosh(L - X) + \sinh(L - X)}{R_L/R_\infty \sinh L + \cosh L}, \tag{a}$$

where R_∞ is the input resistance of a semi-infinite cable with the same diameter, membrane resistance and cytoplasmic resistivity (Equation 5.13) and $L = l/\lambda$ is the electrotonic length of the cable where l is the anatomical length and λ is the length constant. The hyperbolic functions sinh and cosh are the hyperbolic sine and hyperbolic cosine, defined as:

$$\sinh x = \frac{e^x - e^{-x}}{2}, \qquad \cosh x = \frac{e^x + e^{-x}}{2}.$$

According to the definition of input resistance (Equation b), the input resistance of the leaky cable is:

$$R_{in} = \frac{V(0) - E_m}{i_{el}} = R_\infty \frac{R_L/R_\infty \cosh L + \sinh L}{R_L/R_\infty \sinh L + \cosh L}.$$

In the case of a sealed end, where $R_L = \infty$, the membrane potential as a function of length in Equation (a) simplifies to:

$$V(X) = E_m + R_\infty i_{el} \frac{\cosh(L - X)}{\sinh L}, \tag{b}$$

and the input resistance simplifies to:

$$R_{in} = R_\infty \frac{\cosh L}{\sinh L} = R_\infty \coth L,$$

where the function coth is the hyperbolic cotangent, defined as:

$$\coth x = \frac{\cosh x}{\sinh x} = \frac{e^x + e^{-x}}{e^x - e^{-x}}.$$

where the coefficients $C_n(x)$ depend on the distance along the cable, τ_0 is the membrane time constant and τ_1, τ_2, and so on, are time constants with successively smaller values (Rall, 1969). A method for determining multiple time constants experimentally is described in Section 14.4.2.

Figure 5.12b shows the simulation of the membrane potential at different points along a cable following synaptic input at one end. After about 2 ms, in this simulation, the membrane potential at all points has equalised and the membrane potential decays exponentially to its resting value. The time

Box 5.5 | Eccles, Rall and the charging time constant of motor neurons

A dispute between Eccles and Rall – described in detail in Segev et al. (1995) – over how to interpret the charging curves of motor neurons demonstrates the importance of time-dependent solutions to the cable equation. Recall that when a current is injected into a small passive neuron, the membrane potential responds by shifting to a new steady state value (Chapter 2). The time course of the approach to the new potential varies exponentially with the membrane time constant. Coombs et al. (1956) injected current into motor neurons and recorded the membrane potential as a function of time (thick black line in Figure 5.14). These data could be fitted by an exponential function with a time constant of 2.5 ms (Figure 5.14, dashed curve). Under the implicit assumption that a spatially extended motor neuron has the equivalent electrical behaviour to a neuron composed of a soma only, Coombs et al. concluded that the membrane time constant was 2.5 ms.

Rall showed that this method of analysing the data gives an answer for the membrane time constant that is too small by a factor of two (Rall, 1957). In Figure 5.14, the blue line shows Rall's solution of the full time-dependent cable equation for a 'ball-and-stick' model of the motor neuron, a soma with a single dendrite attached to it, in which the membrane time constant is 5 ms. This solution can be seen to be very similar to the charging curve of a lone soma with a membrane time constant of 2.5 ms. For comparison, the charging curve of a lone soma with a membrane time constant of 5 ms is shown in black.

The Eccles group was effectively using the lone soma model to analyse data from a soma and dendrites. They therefore had to fit the experimental data (dashed line) with a curve with a shorter time constant instead of fitting the curve generated from the ball-and-stick model with a longer time constant (black line); this procedure therefore gave the wrong result.

The expression for the charging curve of the ball-and-stick model is $V/V_0 = \frac{1}{6}(1 - \exp(-t/\tau)) + \frac{5}{6} \, \text{erf} \, \sqrt{t/\tau}$, where the function 'erf' is the error function, defined below. The factors $\frac{1}{6}$ and $\frac{5}{6}$ derive from Rall's assumption that in the steady state, one-sixth of the current injected flows out through the soma and the remaining five-sixths through the dendrites.

The error function $\text{erf} \, x$ is the area under the Gaussian $\frac{2}{\sqrt{\pi}} \exp(u^2)$ between 0 and x:

$$\text{erf} \, x = \frac{2}{\sqrt{\pi}} \int_0^x \exp(u^2) \mathrm{d}u.$$

constant of this final decay is the membrane time constant τ_0, as this is the longest time constant ($\tau_0 > \tau_1$, etc.). The contributions of the faster time constants τ_1, τ_2, etc., become smaller as t becomes large.

The solutions of the time-dependent cable equation are not just of descriptive value, but have also been decisive in resolving interpretations of data (Box 5.5).

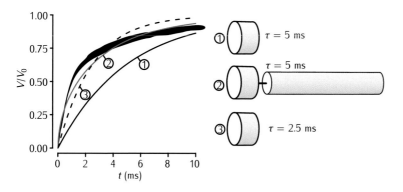

Fig 5.14 Membrane charging curves in three models. The thick black line shows the original data of Coombs et al. (1956). The solid black curve (1) shows the membrane charging curve of a lone soma neuron, comprising a cell body only, with a membrane time constant of 5 ms. The blue curve (2) shows the charging curve of a 'ball-and-stick' neuron in which the membrane time constant is 5 ms. The dashed black curve (3) shows the charging curve of a lone soma neuron which has a membrane time constant of 2.5 ms. The charging curve of the ball-and-stick neuron is similar to the curve of the lone soma neuron with a membrane time constant lower by a factor of two. All membrane potentials are shown relative to the resting potential, and as a fraction of the steady-state displacement of the membrane potential from rest V_0.

5.4 Models Based on Complex Neuronal Morphologies

Now we return to the issue of modelling complex neuronal morphologies in which active membrane properties may vary widely over space. This requires the use of the compartmental modelling approach with numerical simulations, as analytical solutions are not generally possible. Here we consider some of the issues with trying to obtain and use highly realistic morphologies from actual neurons as the basis for our compartmental models.

5.4.1 Obtaining Actual Neuron Morphologies

Compartmental models are often based on morphological reconstructions of real cells and work to produce such reconstructions has a long and extensive history (Halavi et al., 2012). Morphological data are obtained via microscopic imaging of a labelled neuron. Visualising neurons stained with dye dates back to the nineteenth century (Ramón y Cajal, 1911). Nowadays, fluorescent dyes are introduced into a neuron by direct injection or via transgenic approaches. A variety of light microscopic methods, such as confocal or two-photon laser-scanning microscopy, can be used to then produce a 3D image stack containing the processes (soma, dendrites, axon) of the labelled neuron. The challenge then is to extract a set of 3D points, with associated diameters, that accurately capture in some detail the morphology of the neuron. Computer-aided microscope packages, such as *Neurolucida*, enable this to be done (Halavi et al., 2012). It is a laborious and error-prone process to do manually, with errors arising particularly for segment diameters in thin, tapering dendrites (Ascoli et al., 2001; Ascoli, 2002). Significant differences as reported from different laboratories can exist between the characteristics of digitised cells of the same type (Scorcioni et al., 2004; Szilágyi and De Schutter, 2004). There are ongoing efforts to produce computer algorithms to automate this process (Meijering, 2010; Peng et al., 2015; Radojević and Meijering, 2019).

Despite the difficulties in obtaining neuronal reconstructions, there are many thousands of reconstructed cells available online in databases such as *Neuromorpho.org* and the *Allen Brain Map* that are suitable for use in compartmental models (Cannon et al., 1998; Ascoli et al., 2001; Ascoli, 2006; Halavi et al., 2008). The case study in Section 5.6 gives an example of a compartmental model based on a full morphological reconstruction of a neocortical layer 5 pyramidal cell. However, if there are insufficient reconstructed cells available for your purposes, then it is possible to use computational algorithms to construct model cells with statistically appropriate morphologies. We outline such algorithms in Section 5.5.

5.4.2 Mapping Morphology to Simple Geometric Objects

No matter how complex the morphology of a target cell may be, the corresponding compartmental model will use simple geometric objects, such as spheres, ellipsoids and cylinders, to represent the anatomical structures observed.

A cell body is usually represented by either a sphere or a cylinder, and modelled as a single RC circuit. The soma surface area a_s is calculated from the geometry and is used to calculate the electrical properties of the circuit, for example, the soma membrane capacitance is given by $C_m a_s$. Note that if the soma is represented by a cylinder, the membrane surface area is generally only calculated as the cylindrical surface, with the two end faces not included, as membrane currents are only deemed to flow across this surface.

Axonal and dendritic processes are typically represented as collections of connected cylinders. Again, surface areas for membrane currents are calculated as the cylindrical surface only. Axial currents flow across the end faces. Diameters of processes can vary greatly along their length, particularly in dendrites. A rule for deciding when unbranched dendrites should be represented by more than one cylinder with different diameters has to be devised. For example, the point at which the real diameter changes by a preset amount (e.g. 0.1 μm) along the dendrite can be chosen as a suitable criterion (Figure 5.15).

There may not necessarily be a one-to-one correspondence between the representation of morphology with simple geometric shapes and the final electrical circuit. A single long dendrite may be represented adequately by a single cylinder, but to model the voltage variations along the dendrite, it should be represented by multiple compartments. Choosing the number and size of compartments was considered in Box 5.1. In addition, there may be more morphological information available from the real neuron than is required to specify the relationships between geometric elements. Three-dimensional spatial information specifying the relative positions and orientations of each element may also have been recorded in the reconstruction procedure. Although this may not be required in models that do not represent spatial aspects of a cell's environment, this information can be useful in certain situations, for example, in modelling the input to cells with processes in different cortical layers; critically, it is needed to compute extracellular potentials (Section 13.2).

Simulation packages designed specifically for building compartmental neuron models, such as *NEURON*, often provide cylinders as the only geometrical object used to represent different morphologies. This means, for example, that a spherical soma or synaptic bouton must be translated to a cylinder with an equivalent surface area. If the soma or bouton is represented by a single RC circuit (i.e. a single electrical compartment), representing it as a cylinder makes no electrical difference. *NEURON* also uses representations

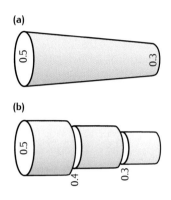

(a)

0.5 0.3

(b)

0.5 0.4 0.3

Fig 5.15 **(a)** An unbranched tapered dendrite can be split into individual cylinders with **(b)** progressively smaller diameters (units in μm).

that reflect the division between morphology and electrical compartments. This facility makes it straightforward to change the number of compartments electrically representing a single cylindrical branch, without changing the geometric representation of the morphology. The spatial accuracy is then conveniently abstracted from the actual representation of the morphology.

5.4.3 Incorporating Spines

For many neurons, dendritic spines, which are short branches protruding a few micrometres from a main dendrite (schematic in Figure 5.16), are a conspicuous part of the morphology. Their functional role is the subject of a large body of experimental and modelling work (Shepherd, 1996). To include all possible spines in a detailed compartmental model of such a cell is computationally very expensive and is likely to involve many assumptions about the size, shape and location of the spines.

If the aim of the model is to explore local dendritic interactions between individual synapses, where the voltage transient or calcium influx at particular synapses is to be tracked, then it is crucial to include explicitly at least the spines on which the synapses occur. As shown in Figure 5.16, the voltage transient in an active spine head has a much greater amplitude than the subsequent transient in the attached dendrite. However, transmission from the dendrite into a spine is not attenuated, and nearby spines will see a very similar transient to that in the adjacent dendrite. In Section 6.5.4 and Box 6.4, we show that longitudinal diffusion of calcium from a spine head usually has a much shorter length constant than the membrane voltage. Thus concentration transients are also highly local to the spine head.

Spines and small processes can have a significant impact on the overall surface area of the cell. For example, spines can account for half of the surface area of a typical pyramidal cell (Larkman et al., 1991). It is possible to include their general effects on the electrical properties of the cell using data about their size, number and distribution. This can be done by adjusting the membrane surface area appropriately and keeping the ratio of length-to-diameter-squared to a constant value (Figure 5.17). The axial resistance of a cylindrical compartment of length l and diameter d is:

$$r_a l = \frac{4R_a l}{\pi d^2}. \tag{5.15}$$

Preserving the ratio of the length-to-diameter-squared ensures that the axial resistance remains constant (Stratford et al., 1989). Alternatively, the values of specific electrical properties R_m and C_m can be adjusted (Holmes and Rall, 1992c) to account for the missing membrane. If preserving cell geometry is required, for visualisation, formation of network connections in 3D space or calculating extracellular potentials (Chapter 13), for example, then this method is to be preferred.

If the main aim of the model is to explore the somatic voltage response to synaptic input, then adjusting dendritic lengths and diameters to include the spine membrane, or adjusting cellular membrane resistance and capacitance, may provide a reasonable and computationally efficient approximation to the contribution of spines. If such an adjustment is not made, then the quantitative contribution of a synaptic excitatory postsynaptic potential (EPSP) will be overestimated at the soma (Figure 5.18).

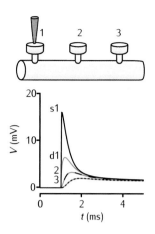

Fig 5.16 Voltage response to a single synaptic excitatory postsynaptic potential (EPSP) at a spine on a dendrite. There are three spines, 100 μm apart, on a 1000-μm long dendrite of 2-μm diameter. EPSP occurs at spine 1. Black lines show the voltage at each spine head. Blue lines are the voltage response in the dendrite at the base of each spine. Dendrite modelled with 1006 compartments (1000 for the main dendrite plus two for each spine); passive membrane throughout with $R_m = 28$ kΩ cm^2, $R_a = 150$ Ω cm, $C_m = 1$ μF cm^{-2}; spine head: diameter 0.5 μm, length 0.5 μm; spine neck: diameter 0.2 μm, length 0.5 μm. For a similar circuit, see Woolf et al. (1991).

Spines are typically modelled by one or two cylindrical compartments, modelling the spine shaft and head (Figure 5.17a). If only a small number of spines are modelled, then the dendritic surface area covered by the spine shaft is not usually subtracted from area calculations. If there are many spines on a dendritic compartment, then this might not be a good approximation.

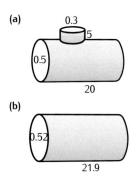

(a)

0.3

5

0.5

20

(b)

0.52

21.9

Fig 5.17 **(a)** A dendritic cylinder with an attached spine can be represented as **(b)** a single cylinder with an enlarged surface area. The axial resistance is identical in the two dendritic cylinders, as the ratio of length-to-diameter squared in the cylinders is kept the same.

R_m and C_m can be modified in a specific compartment to absorb the additional spine surface area a_{spine} whilst leaving the compartment dimensions l and d unchanged. Compartment-specific values are, respectively, given as:

$$R_m \frac{\pi d l}{\pi d l + a_{spine}},$$

$$C_m \frac{\pi d l + a_{spine}}{\pi d l}.$$

5.4.4 Simplifying the Morphology

Although neural simulators, such as *NEURON*, can simulate large multi-compartmental models efficiently, there are situations in which simpler models are desirable, for example, if we want to run large numbers of simulations to explore the effects of changes in specific parameter values. Different approaches to simplifying neuron models and their consequences are discussed in Chapter 8.

Here, we focus on one principled approach which allows the creation of compartmental models with reduced complexity and a number of compartments that have certain passive electrical properties that are identical to more morphologically complex models. It is based on the theory that the passive membrane properties (resistance and capacitance) of branched dendrites can be equivalent to the passive properties of a single, unbranched cylinder. The limitations of this approach are also noted, including when it can legitimately be applied and what distortions it may introduce when simulating different scenarios such as spatially distributed synaptic input. Rall (1964) showed that passive dendrites are equivalent electrically to a single cylinder, provided that they obey:

(1) Specific membrane resistance (R_m) and specific axial resistance (R_a) are the same in all branches.
(2) All terminal branches end with the same boundary conditions (e.g. a sealed end).
(3) The end of each terminal branch is the same total electrotonic distance from the origin at the base of the tree.
(4) The relationship between each parent branch diameter (d_P) and its two child branch diameters (d_L and d_R) obeys the '3/2' diameter rule:

$$d_P^{3/2} = d_L^{3/2} + d_R^{3/2}. \tag{5.16}$$

Before using this equivalent cylinder simplification, it is important to assess if these conditions apply to the morphological structures being examined. Applying this simplification to neurons whose morphologies or electrical properties do not meet the required assumptions can lead to erroneous results (Holmes and Rall, 1992a). At first glance, conditions **(3)** and **(4)** above may appear to be too restrictive. Some dendritic trees conform to the 3/2 diameter rule, such as the cat motor neuron (Rall, 1977) and cat lateral geniculate nucleus relay cells (Bloomfield et al., 1987). Other trees, such as apical trunks of cortical pyramidal neurons, violate the 3/2 assumption (Hillman, 1979). Note that these conditions do not imply that the tree must be symmetrical. For example, it is possible to have two branches from a parent with different lengths, provided the total electrotonic distance from the origin at the base of the tree to the ends of each terminal branch is the same. This can be the case when the longer branch has a larger diameter.

Using a model that has been simplified by using equivalent cylinders can significantly reduce the number of parameters and allow a clearer understanding of its responses. This was done for the cat motor neuron

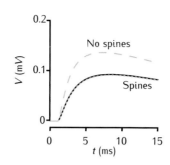

Fig 5.18 Somatic voltage
response to a single synaptic
EPSP on a spine at a distance
of 500 μm in the apical dendrites
of a CA1 pyramidal cell model
(blue dot on cell image). The
voltage amplitude is reduced
(black line) when 20 000 inactive
spines are evenly distributed in
the dendrites, compared to when
no spines are included (blue
dashed line). The somatic voltage
response is corrected if R_m is
decreased and C_m is increased in
proportion with the additional
membrane area due to spines
(blue dotted line). The CA1
pyramidal cell was obtained from
ModelDB with accession number
55035, as used in Migliore et al.
(2005). The model has 455
compartments (with no spines);
passive membrane throughout
with $R_m = 28\,k\Omega\,cm^2$,
$R_a = 150\,\Omega\,cm$, $C_m = 1\,\mu F\,cm^{-2}$;
spine head: diameter 0.5 μm,
length 0.5 μm; spine neck:
diameter 0.2 μm, length 0.5 μm;
20 000 spines increase the total
membrane area by 1.4 and
reduce the somatic voltage
response by around 30%. For a
similar comparison, see Holmes
and Rall (1992b).

(Rall, 1977). Furthermore, the simplification is still useful with the addition of active membrane properties.

To demonstrate how a tree satisfying the above conditions can act electrically as a single cylinder, we now investigate the small symmetrical tree in Figure 5.19a. It is assumed that the morphology is completely specified, except for the diameter of the parent cylinder d_p. We first assume all the conditions mentioned above are met, including that the diameter of the cylinders satisfies the 3/2 rule. The voltage along the length of the entire tree in response to a constant injected current at the left-hand end of the tree (central line, Figure 5.19b) decreases smoothly as if it were from a single cylinder – there is no abrupt change in the gradient at the tree branch point. Compare this with the situation where the diameter of the parent cylinder is set to a value which is half of what would satisfy the 3/2 rule ($d_p = 0.4\,\mu m$). Now the voltage along the tree shows an abrupt change in gradient at the branch point (lower line in Figure 5.19b). Similarly, if the diameter of the parent cylinder is greater than the value required for the 3/2 rule ($d_p = 1.6\,\mu m$), there is still an abrupt change in the gradient of the voltage plot (upper line, Figure 5.19b). Only where all the conditions are met does the voltage change smoothly, consistent with a single equivalent cylinder.

The equivalent cylinder can be constructed with the same diameter and length constant as the parent, and the length given by the total electrotonic length of the tree. It can also be shown that the surface area of the tree and the surface area of such an equivalent cylinder are the same. Box 5.6 provides a demonstration, using the cable equation, of how these four conditions allow the construction of an equivalent cylinder.

Further limitations to using this simplification arise when modelling inputs over the dendritic tree. Representing a branching dendritic tree by a single cylinder limits the spatial distribution of inputs that can be modelled. Input at a single location on the equivalent cylinder corresponds to the sum of individual inputs, each with an identical voltage time course, simultaneously arriving at all electrotonically equivalent dendritic locations on the corresponding tree. For a fuller discussion of the class of trees that have mathematically equivalent cylinders and the use of this simplification, see Rall et al. (1992). For a different examination of the electrotonics of dendrites, Box 5.7 introduces the **morphoelectrotonic transform**.

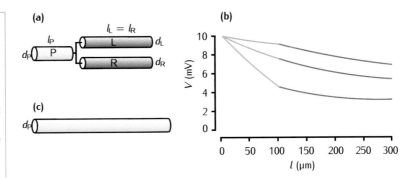

Fig 5.19 **(a)** A simple example of a branching tree. **(b)** Curves showing the voltage at any point along the tree illustrated in **(a)**, in response to a constant injected current at the left-hand end of the tree, for three different diameters of the parent cylinder. The centre line shows the voltage along the tree for $d_P = 0.8\,\mu m$, which satisfies the 3/2 rule (blue indicates the voltage along the parent cylinder, and grey along the left-hand daughter). The upper and lower curves illustrate the voltage along the tree when the parent diameter does not satisfy the 3/2 rule ($d_P = 1.6\,\mu m$ and $d_P = 0.4\,\mu m$, respectively). $R_m = 6000\,\Omega\,cm^2$, $R_a = 150\,\Omega\,cm$ and $E_L = 0\,mV$. **(c)** An equivalent cylinder for the tree in **(a)**, with diameter $d_P = 0.8\,\mu m$ and length $l = 352\,\mu m$. l is calculated from the electrotonic length of the tree, using $L = l/\lambda$.

5.5 Generation of Realistic Model Cell Morphologies

Compartmental modelling studies may require populations of morphologically distinct neurons of a particular type, for example, in creating neural network models or for determining whether differences in morphology can account for the variations in cell properties across a population such as input resistance (Winslow et al., 1999). If insufficient examples are available from staining real neurons, then new examples with the same morphological characteristics can be generated using a computational algorithm. This approach is being used to construct large-scale neural network models with realistic variation in cell properties such as for the circuitry in a neocortical column (Markram et al., 2015).

5.5.1 Algorithms Based on Shape Parameters

Neurite construction algorithms (van Pelt and Uylings, 1999) require a specification of parameters that appropriately define the structural characteristics of neurites, as required for a compartmental model. Experimental distributions of these parameter values must be obtainable. Hillman (1979) defined seven fundamental parameters, collected from particular types of neurite (dendrite or axon). These parameters are illustrated in Figure 5.21. Construction algorithms create neurites of statistically similar morphologies by appropriate sampling from the experimental distributions (Hillman, 1979; Burke et al., 1992; Ascoli, 2002; Burke and Marks, 2002; Donohue et al., 2002). The outcome is the specification of the diameter and length of each unbranched segment of neurite and whether each segment terminates or ends in a branch point leading to a bifurcation.

Histograms of segment diameters and lengths are collected from experimental tracings of real neurites. These histograms are then fitted with univariate (single variable) and multivariate (multiple variable) distributions; for example, the bivariate diameter and length distribution describe the probability that a neurite segment will have a particular diameter and length; similarly, the trivariate distribution of parent and two daughter diameters at branch points describes the probability that the parent and its daughter segments will have particular individual diameters. This can be done by using **maximum likelihood estimation** (Box 5.8) to fit parametric probability

Box 5.6 Equivalent cylinder for complex trees

To demonstrate the existence of an equivalent cylinder for an appropriate branching tree, we consider the small symmetrical tree shown in Figure 5.19a and assume that it meets all four conditions for cylinder equivalence (Section 5.4.4). Following Box 5.4, we use the dimensionless quantity L to measure cable branch lengths: L is the branch length l in terms of the length constant λ, $L = l/\lambda$. The resistance at the end of the parent cable, R_{LP}, can be calculated from the input resistances of the children, R_{inL} and R_{inR}:

$$\frac{1}{R_{LP}} = \frac{1}{R_{inL}} + \frac{1}{R_{inR}}. \tag{a}$$

The child branch input resistances with sealed ends are:

$$R_{inL} = R_{\infty L} \coth L_L, \qquad R_{inR} = R_{\infty R} \coth L_R, \tag{b}$$

where $R_{\infty L}$ and $R_{\infty R}$ are the input resistances of semi-infinite cables with the same electrical properties as the tree and with diameters d_L and d_R, respectively (Box 5.4). The end of each terminal branch has the same total electrotonic length from the origin at the base of the tree (condition 3). In our example tree, $L_P + L_L = L_P + L_R$, and we define $L_D = L_L = L_R$.

Using $1/R_{\infty} = \sqrt{\pi^2/4 R_m R_a}\, d^{3/2}$, derived from Equation 5.13, and applying the 3/2 rule (condition 4), results in a relationship between the semi-infinite input resistances of the parent and child branches:

$$\frac{1}{R_{\infty P}} = \frac{1}{R_{\infty L}} + \frac{1}{R_{\infty R}}. \tag{c}$$

Substituting for R_{inL} and R_{inR} from Equations (b) into Equation (a):

$$\frac{1}{R_{LP}} = \frac{1}{\coth L_D}\left(\frac{1}{R_{\infty L}} + \frac{1}{R_{\infty R}}\right),$$

then, using Equation (c), the resistance at the end of the parent cylinder is:

$$R_{LP} = R_{\infty P} \coth L_D.$$

Therefore, the resistance at the end of the parent cylinder is equivalent to the input resistance of a single cylinder with the same diameter as the parent and length $l = L_D \lambda_P$.

Thus the entire tree can be described by a single cylinder of diameter d_P, length constant λ_P and length $l = (L_P + L_D)\lambda_P$.

distributions, such as uniform, Gaussian or gamma, to the data (Ascoli, 2002). In general, common parametric distributions may not provide a good fit. In this case, it is more appropriate not to make any assumptions about the shape of the underlying distribution and produce a non-parametric model. One approach to non-parametric modelling is **kernel density estimation** (Box 5.9).

The construction algorithms proceed by making assumptions about the dependencies between parameters, on the basis of the experimental distributions. A usual starting point is to use the diameter of a segment to constrain the choice of length and probability of termination or bifurcation.

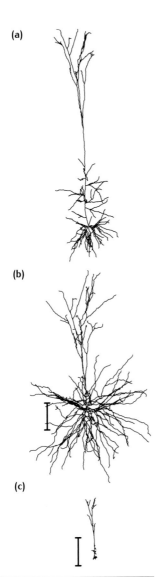

(a)

(b)

(c)

Fig 5.20 Morphoelectrotonic
transforms of the layer 5
pyramidal cell model. **(a)** Cell
morphology.
(b) Morphoelectrotonic transform
for DC signals travelling to the
soma from all dendritic tips.
(c) Transform for DC signals
travelling from the soma. These
transforms were produced using
the *NEURON* simulator's
ImpShape tool.

Box 5.7 | The morphoelectrotonic transform

The morphoelectrotonic transform (Zador et al., 1995) is a way of visualising
the functional length of neurites in a cell, as opposed to the anatomical
length. A simple measure is to consider the log-attenuation of a voltage
signal from its site of origin to a target site, given by:

$$L_m = \ln\left(\frac{V_{in}}{V_{targ}}\right). \tag{a}$$

The input voltage V_{in} can be a DC level or the amplitude of a sinusoidal
signal. The resultant length, $L_m \geq 0$, increases with increasing attenuation of
the signal at the target location V_{targ}, irrespective of the anatomical distance
between the two points; it is zero if there is no attenuation.

We illustrate this transform in Figure 5.20 for the layer 5 pyramidal
cell of the case study in Section 5.6. The cell is quite compact for DC
signals originating at the soma and propagating out into the dendrites
(Figure 5.20c). Much of the attenuation is in the main apical trunks. A
dynamic signal (not shown) is attenuated more than a DC signal. On the
other hand, signals originating at dendritic tips are significantly attenuated
on their journey to the soma (Figure 5.20b), so the cell is far from compact
for distal synaptic inputs. Strikingly, the functional lengths of the basal
dendrites are similar to the apical lengths, even though the apical tips
are anatomically further distant. So the thick apical trunks provide some
equalisation of functional lengths across the cell. Again, dynamic signals
suffer greater attenuation (not shown).

An example algorithm is given in Figure 5.22. This works by sampling an
initial segment diameter. Then for each iteration, the segment is either increased in length by some fixed amount or terminated or branched to form
two daughter segments (Lindsay et al., 2007). The probabilities for continuation, termination or branching all depend on the current length and diameter, and the forms of these probability functions have to be supplied to the
algorithm. A taper rate may be specified, such that the diameter changes
(usually decreases) with length (Burke et al., 1992; Burke and Marks, 2002;
Donohue et al., 2002). When a bifurcation takes place, the daughter diameters are selected from distributions that relate parent to daughter diameters,
and daughter diameters to each other.

The final outcome is the topological tree structure of the neurite, which
can be illustrated as a **dendrogram** (Figure 5.24). This structure contains
no information about how the cell appears in 3D space, but is sufficient to
enable the specification of a compartmental model of the neuron.

The ability of an algorithm to reproduce real neurite topology can be
tested by comparing the distributions of shape parameters between real and
model neurites. Of particular interest are those parameters that emerge from
the reconstruction process, such as the number of terminals, the path lengths
to those terminals and the total length of the neurite. Simple algorithms may
reproduce some, but not all, of the features of the neurites being modelled
(Donohue et al., 2002; Donohue and Ascoli, 2005). It is common for extra

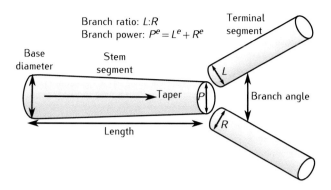

Fig 5.21 Hillman's fundamental parameters of neurite morphology. This is based on the reasonable assumption that all branches are bifurcations, in which a parent branch has two daughter branches, referred to as the child and the sibling. P: parent branch diameter; L: left child branch diameter; R: right child branch diameter; e: branch power relationship.

probability functions to be used to increase the match to data, such as one which specifies a reduction in branching probability with path distance from the cell body (Ascoli, 2002).

Alternative algorithms to the diameter-dependent example given here consider branching to be a function of distance, branch order (number of branch points between a segment and the cell body) and the expected number of terminal segments for the dendrite (Kliemann, 1987; Carriquiry et al., 1991; Uemura et al., 1995; Winslow et al., 1999; Burke and Marks, 2002; Samsonovich and Ascoli, 2005a, b). A variety of temporal and spatial effects are likely to influence neurite outgrowth, including interaction with signalling molecules and other neurites in the external environment. Such factors may underpin these different dependencies but are not modelled explicitly.

Embedding in 3D Space

A dendrogram only specifies the topology (branching structure) of a neurite, and not how it is shaped within 3D space. This information may be required, for example, to determine the expected connectivity between neurons and incoming axons when constructing a network model (Markram et al., 2015) and to compute extracellular potentials (Chapter 13). Constructing dendritic orientation in 3D space has been the subject of some modelling work (Ascoli, 2002; Burke and Marks, 2002). Realistic branch angles can be obtained by assuming a principle of volume minimisation at a bifurcation point (Tamori, 1993; Cherniak et al., 1999, 2002). Orientation of dendritic branches can be described by a **tropism** rule in which branches have a strong tendency to grow straight and away from the cell body (Samsonovich and Ascoli, 2003).

5.5.2 Minimum Spanning Tree Algorithm

A different algorithmic approach has been taken by Cuntz et al. (2010): given a distribution in 3D space of putative topological points of a neurite, known as carrier points, they use a greedy extended **minimum spanning tree (MST) algorithm** to construct the neurite by adding carrier points as tree nodes, in a step-by-step manner (Figure 5.25a). Such nodes may be continuation points of elongated branches, branch points or terminal points. At each step, the carrier point that is closest to the current tree according to a cost function is added as a new tree node, connecting to the existing node to which it is closest. The key factor is the cost function, which

A spanning tree connects a set of points to form an undirected graph with no cycles (loops). A minimum spanning tree connects the points in such a way as to minimise the total cost, where each edge in the graph (line connecting two points) has a cost associated with it. For dendrites, the cost is the wiring length of the tree.

Box 5.8 | Maximum likelihood estimation (MLE)

Suppose we have a sample value x for a random variable X. Then we define the **likelihood** of the sample, $L \equiv L(x)$, as the probability $p(X = x)$ if X is discrete, or the probability density $f(X = x)$ if X is continuous. To take the Gaussian distribution as an example, the probability density is a function defined by two parameters, the mean μ and variance σ^2. MLE seeks to find values of μ and σ^2 that maximise the likelihood of sample x. If we have n independent samples, $x_1, \ldots, x_n$, then the likelihood of these samples is the joint probability density $L = f(x_1, x_2, \ldots, x_n) = f(x_1)f(x_2)\ldots f(x_n)$ where:

$$f(x_i) = \frac{1}{\sqrt{2\pi\sigma^2}} \exp\left(-\frac{(x_i - \mu)^2}{2\sigma^2}\right). \tag{a}$$

The likelihood is maximised with respect to the parameters when:

$$\frac{dL}{d\mu} = \frac{dL}{d\sigma^2} = 0. \tag{b}$$

Given the functional form for $f(x)$, this is solved easily, yielding sample estimates for the mean and variance:

$$\hat{\mu} = \frac{1}{n}\sum_{i=1}^{n} x_i, \quad \hat{\sigma}^2 = \frac{1}{n}\sum_{i=1}^{n}(x_i - \hat{\mu})^2. \tag{c}$$

This procedure can be carried out for any parameterised distribution function, though finding the solution may be more or less difficult than for the Gaussian distribution. Note that minimising the root-mean–squared-error between model output and experimental data is equivalent to carrying out MLE if we can assume that the experimental measurements are subject to independent Gaussian noise with a constant variance (Press et al., 1987).

The minimum spanning tree idea was inspired by Ramón y Cajal's specification of fundamental anatomical principles of nerve cell organization (Ramón y Cajal, 1911; Cuntz et al., 2010), including conserving material costs and conduction times in neurites. Material cost corresponds to the wiring cost of a dendritic tree, with conduction times being a function of path lengths in the tree.

is the weighted sum of the wiring cost of connecting to the carrier point (distance to the nearest existing node) plus the path length from the root node to the carrier point. Thus, the algorithm consists of the following basic steps:

(1) Add a known root carrier point as the first node in the tree.
(2) Find the carrier point that is closest to the existing tree nodes according to the cost function: *Cost = (wiring cost) + bf.(path length cost)*, where the balancing factor *bf* ≤ 1 weights the relative cost of the path length.
(3) Add this carrier point as a new tree node and connect it to the existing node to which it is closest.
(4) Repeat from step 2 until all carrier points are added to the tree.

The balance between the two costs significantly affects the final tree topology, but the balancing factor *bf* can be chosen by trial and error to give topologies that match the real tree of interest (Figure 5.25b). Additional constraints, such as suppression of multi-furcations and adding spatial jitter to node locations, can be added to improve the end result.

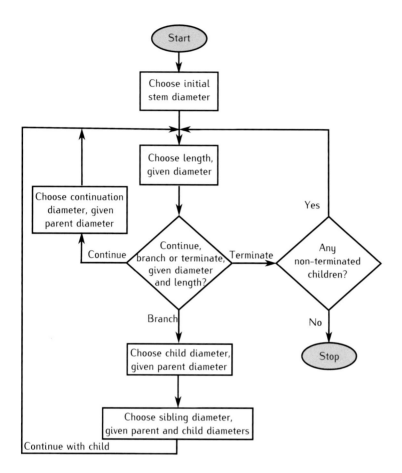

Fig 5.22 Example construction algorithm. Probability distributions for segment lengths, diameters and the likelihood of branching or termination are sampled to construct a neurite topology.

Rather than statistics of shape parameters, the algorithm relies on a reasonable specification of the distribution of carrier points in space. This is determined by collecting large numbers of topological points along the neurites of real neurons of interest and approximating their density in 3D space. Carrier points are then generated by sampling from this density distribution. For a simple planar cell, such as a stellate neuron, this could be a distance-dependent density within a prescribed hull, such as illustrated in Figure 5.25b. The volumes in 3D space of more complex cells may be approximated by combinations of spheres and cones, or more complicated shapes. This same algorithm also has proven effective in generating realistic specific cell reconstructions from image stacks of visualised neurons. In this case, the carrier points are generated directly from the cell anatomy, as revealed by microscopy, giving a distribution that is specific to the cell being reconstructed.

Unlike the reconstruction algorithms described above, the resulting tree is already embedded in 3D space, due to the initial distribution of the carrier points. Section diameters can be added through the use of a suitable distance-dependent function. This whole approach has been demonstrated to be highly effective in generating realistic morphologies for wide-ranging neuron types (Cuntz et al., 2010).

A well-documented implementation of this minimum spanning tree algorithm for neurite construction is available as the *TREES* toolbox for *Matlab* (Cuntz et al., 2010). The *T2N* framework for linking *TREES* with the *NEURON* simulator has been created to enable development of compartmental models tuned over multiple cell morphologies (Beining et al., 2017).

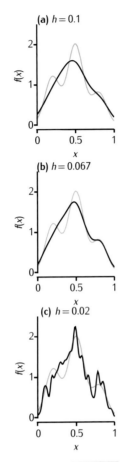

(a) $h = 0.1$

(b) $h = 0.067$

(c) $h = 0.02$

Fig 5.23 KDE example where the underlying distribution consists of a mixture of three Gaussian distributions: $\mu_1 = 0.2$, $\sigma_1 = 0.1$, $\mu_2 = 0.5$, $\sigma_2 = 0.1$, $\mu_3 = 0.8$, $\sigma_3 = 0.1$, with 30% of values drawn from the first distribution, 50% from the second and 20% from the third. KDEs using Gaussian kernels were built from 200 samples drawn from the underlying distribution. KDEs for different values of h are shown in black in each plot. Underlying distribution is shown in blue.

Box 5.9 | Kernel density estimation (KDE)

Suppose we have n measurements x_i $(i = 1, \ldots, n)$ of a random variable X. A KDE of the probability density function is:

$$\hat{f}(x) = \frac{1}{nh} \sum_{i=1}^{n} K \left(\frac{x - x_i}{h} \right), \tag{a}$$

for some **kernel function** K and bandwidth, or smoothing parameter h. K is a function that peaks at $K(0) = 1$ and decreases smoothly to 0 with increasing magnitude of the argument. So here K provides an estimate of the probability of x, given a known data point x_i. K peaks when x exactly matches the data point. How quickly K decreases with the distance of x from x_i is controlled by h. A common choice of kernel function is the Gaussian probability density function, with h specifying the standard deviation, giving:

$$\hat{f}(x) = \frac{1}{nh\sqrt{2\pi}} \sum_{i=1}^{n} \exp \left(-\frac{(x - x_i)^2}{2h^2} \right). \tag{b}$$

This KDE has only the single smoothing parameter h that needs to be specified. Figure 5.23 illustrates the effects of choosing different values for h. An appropriate choice for h is vital; if h is too large, then the kernel density model is rather smooth and may not capture well the underlying distribution of the data; conversely, if h is too small, then the model will exhibit large variations that are specific to the particular data set on which the model is based. Automatic bandwidth selection is a difficult process for which a variety of methods have been proposed. A simple rule of thumb that often works well is to assume that the data actually come from a single Gaussian distribution, and then set:

$$h = 0.9 \, S n^{-1/5}, \tag{c}$$

where S is the minimum of the estimated sample standard deviation σ and three-quarters of its interquartile range (Torben-Nielsen et al., 2008). This yields $h \approx 0.067$ for the example above.

5.6 | Case Study: Neocortical Pyramidal Cell

Compartmental modelling of a detailed neuronal morphology is typically used to explore signal integration and flow in dendrites and the resultant cell spiking output. Unlike in experiments on real cells, such a model allows the 'experimenter' full control over the cell's active membrane properties and inputs to the cell. Tuning of model parameters is done by trying to replicate particular biological experiments (Chapter 14). Once this is done, the modeller is then able to carry out simulations with the model that are not technically possible experimentally, and thus make predictions about neural signal processing. For example, perfect blocking of specific ion channels can be done by simply setting conductance values to zero. The number, location and strength of synaptic input can be completely specified and controlled.

Activity in the model cell can also be examined and visualised throughout the cell's morphology by recording voltages, currents and ion concentrations in all (or selected) model compartments during a simulation.

To illustrate the power and usefulness of compartmental modelling of a complex neuron, we show examples of model output from the Hay et al. (2011) model of a neocortical layer 5 pyramidal neuron. Further consideration of this exemplar model is given in Chapter 13 where we look at modelling extracellular potentials, and in Chapter 14 where we discuss the parameter optimisation procedures used to procure the final values of ion channel conductances and other parameters for this model.

The particular cell morphology used in these simulations can be seen in Figure 5.26: it is a thick-tufted layer 5b pyramidal cell from an adult Wistar rat (postnatal 36 days). The compartmental model consists of around 200 compartments. The model cell contains a variety of ionic currents generated by a variety of putative ion channel types distributed in the dendrites, soma and axon. These are listed in Table 5.2 using the terminology adopted in Hay et al. (2011). This differs slightly from the terminology used in Table 4.2 and illustrates the informality with which channel currents are notated. The reader is referred to Hay et al. (2011) for the complete mathematical and technical details of the channel models used here.

Experimental data are insufficient to fully constrain parameter values (Chapter 14), so assumptions have to be made. The ion channels underpinning the variety of ionic currents may be distributed differently between the following four neuronal components: axon, soma, apical dendrites and basal dendrites; and most ion channel types are assumed to be uniformly distributed within any of these components. Experiments suggest that the hyperpolarisation-activated HCN channels, underpinning I_h, increase in density in the apical dendrites with distance from the soma: this is accounted

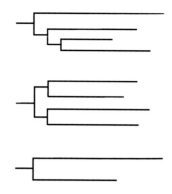

Fig 5.24 Example dendrograms produced by separate runs of a particular construction algorithm. These show the topology (connectivity structure) and diameters of neurite segments. Note that the vertical lines only illustrate connectivity and do not form part of the neurite.

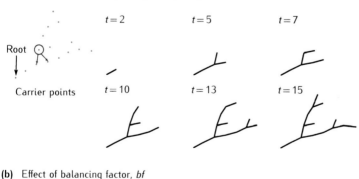

(a) Growth of minimum spanning tree (MST)

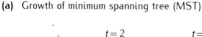

(b) Effect of balancing factor, *bf*

Fig 5.25 Examples of constructed dendrites using the *TREES* MST algorithm (Cuntz et al., 2010). **(a)** Growth of tree over successive steps *t* of the algorithm from a given set of carrier points, with balancing factor *bf* = 0.5. Open circle shows the next closest carrier point at step 5, and arrows indicate the decision at step 5 as to which tree node to connect it to, according to the cost function. **(b)** Effect of the balancing factor *bf* on the topology of constructed trees. Examples use a set of 500 carrier points randomly distributed in a 2D circular hull.

Table 5.2 | Ionic currents and their distributions in the layer 5 pyramidal cell model

Current	Description	Locations
I_{leak}	Non-specific leak	Soma, apical, basal, axon
I_{Nat}	Fast inactivating sodium	Soma, apical
I_{Nap}	Persistent sodium	Soma
I_{Kp}	Slow inactivating (persistent) potassium	Soma
I_{Kt}	Fast inactivating (transient) potassium	Soma
I_{SK}	Calcium-activated potassium	Soma, apical
$I_{Kv3.1}$	Fast, non-inactivating potassium	Soma, apical
I_{Ca_HVA}	High-voltage-activated calcium	Soma, apical
I_{Ca_LVA}	Low-voltage-activated calcium	Soma, apical
I_m	Muscarinic potassium (M)	Apical
I_h	Hyperpolarisation-activated HCN	Apical, basal

Computer code for the layer 5 pyramidal cell model by Hay et al. (2011) is freely available from *ModelDB* with model accession number 139653. The model runs in the *NEURON* simulator. Simulations with synaptic input were conducted using the version of the code developed by Shai et al. (2015), which is also available from *ModelDB* with accession number 180373.

for using an exponential (with distance) density function here. Experiments also suggest that LVA and HVA calcium channels are most strongly located in 'hot zones' in the apical dendrite, typically outwards from the main bifurcation of the apical trunk. Thus a step function with distance is used to specify their densities, with a large increase in density between 685 μm and 885 μm from the soma.

The model contains an SK-type calcium-activated potassium current (Table 4.2 and Section 4.6), and thus a model for calcium concentration within each compartment is also needed. A simple model is used that assumes a uniform calcium concentration in a submembrane shell within a compartment, with instantaneous buffering and a single decay time constant to account for calcium diffusion into the compartment interior. Diffusion of calcium between compartments is not accounted for. Thus, the calcium concentration in a compartment is given by:

$$\frac{d[Ca^{2+}]_i}{dt} = -\frac{I_{Ca}}{2\gamma F d} - \frac{[Ca^{2+}]_i - 0.0001}{\tau_{decay}}, \tag{5.17}$$

where I_{Ca} is the calcium current flowing through all calcium channels in the compartment, γ is the inverse of the calcium buffer binding ratio, F is Faraday's constant, d is the depth of the submembrane shell and τ_{decay} is the time constant of diffusion. Modelling of calcium concentrations is treated in detail in Chapter 6.

The major aim of the modelling presented in Hay et al. (2011) was to replicate the distinctive somatic and apical dendritic activity patterns recorded in real cells. In particular, a somatic AP may back-propagate into the apical dendrites and interact with depolarising input in these dendrites to generate a large, long-lasting calcium spike in the 'hot zone' of calcium channels. This calcium spike then forward-propagates to the soma, generating a burst of APs there. Thus, near-simultaneous somatic and apical input may interact non-linearly to cause a bursting response that is not generated by

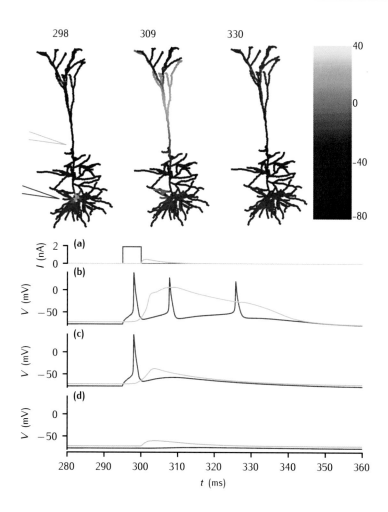

Fig 5.26 BAC firing in layer 5 pyramidal cell model due to current injection. Top: three cell images showing colour coding of membrane voltage (mV) on a 'hot' scale at time points 298 ms, 309 ms and 330 ms; arrows show stimulation points in the soma (black) and apical trunk (blue). Bottom: cell responses to current injection over time. **(a)** Injected current at the two stimulation points: soma (black) and apical tuft (blue); **(b)** membrane voltage at the two stimulation points; **(c)** voltage response to somatic current injection only; **(d)** voltage response to apical tuft current injection only.

either input alone. This is termed **BAC firing**: back-propagation-activated calcium spike firing (Larkum, 2013). The model response showing this activity pattern is illustrated in Figure 5.26. From our model, not only can we plot voltage traces over time from specific points in the cell (Figure 5.26a–d), but we can also visualise the membrane voltage across the entire cell at a point in time, using colour coding on the cell morphology (Figure 5.26, top). Movies can be generated from such visualisations made at regular, short time intervals.

The results shown in Figure 5.26 are based on replicating specific experimental results. However, a detailed compartmental model such as this can be used to conduct simulations of situations that are not possible to implement in an experiment. Shai et al. (2015) have used the Hay model to investigate BAC firing as a function of the number of concurrent synaptic inputs to basal and apical tuft dendrites. This is a much more realistic situation than using specific points of current injection. Example results of similar responses to the current injection simulations are shown in Figure 5.27.

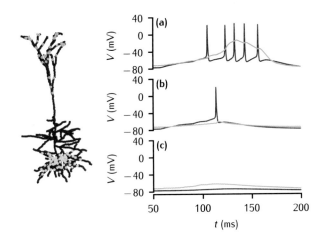

Fig 5.27 BAC firing in layer 5 pyramidal cell model due to synaptic input. Left: cell morphology, with coloured dots illustrating example sites of synapses in the basal and apical tuft dendrites. Each synapse generates a single EPSP, with response times distributed uniformly over an interval of 50 ms starting after an initial 55 ms. Right: **(a)** voltage over time due to all synaptic input recorded at the soma (black) and apical trunk (blue); **(b)** voltage response to basal input only; **(c)** voltage response to apical tuft input only.

5.7 Numerical Methods for Compartmental Models

Most of the mathematical models presented in this book involve differential equations describing the evolution in time and space of quantities such as membrane potential (this chapter) or calcium concentration (Chapter 6). The differential equations are usually too complex to allow an analytical solution that would enable the explicit calculation of a value of, say, voltage at any particular time point or spatial position. The alternative is to derive algebraic expressions that approximate the differential equations and allow the calculation of quantities at specific, predefined points in time and space. This is known as numerical integration. Methods for defining temporal and spatial grid points and formulating algebraic expressions involving these grid points from the continuous (in time and space) differential equations are known as **finite difference** and **finite element methods**. Box 5.2 shows how a compartmental model forms a finite difference approximation to the continuous cable equation.

It is not our intention to provide full details of these numerical integration methods. Instead, we will outline some of the simplest methods to illustrate how they work. This includes the Crank–Nicholson method (Crank and Nicholson, 1947), which is widely used as a basis for numerically solving the cable equation. Further details on these methods as applied to neural models can be found in Carnevale and Hines (2006) and Mascagni and Sherman (1998).

5.7.1 Ordinary Differential Equations

For compartmental modelling, we need to consider the ODE for the rate of change of membrane voltage:

$$\frac{\mathrm{d}V}{\mathrm{d}t} = f(V, t), \tag{5.18}$$

for some function f of voltage and time. A particular example is the equation that describes the response of a patch of passive membrane to an injected current:

$$C_m \frac{dV}{dt} = \frac{E_m - V}{R_m} + \frac{i_{el}}{a}. \tag{5.19}$$

As we saw in Chapter 2, if we assume at time 0 that $V = E_m$ and i_{el} is switched from 0 to a finite value at this time and then held constant, this equation has an analytical solution:

$$V = E_m + (R_m \, i_{el}/a)[1 - \exp(-t/R_m C_m)]. \tag{5.20}$$

In general, an ODE cannot be solved analytically. We now consider how numerical approximations can be derived and solved for ODEs. We compute numerical solutions to Equation 5.19 to illustrate how approximate and exact solutions can differ.

These numerical solutions are derived from algebraic equations that approximate the time derivative of the voltage. In combination with the function f, this enables the approximate calculation of the voltage at predefined time points. The **forward Euler method** (Box 2.6) estimates the time derivative at time t as the slope of the straight line passing through the points $(t, V(t))$ and $(t + \Delta t, V(t + \Delta t))$, for some small time-step Δt:

$$\frac{dV}{dt} \approx \frac{V(t + \Delta t) - V(t)}{\Delta t}. \tag{5.21}$$

This is known as a **finite difference method**, because it is estimating a quantity, the rate of change of voltage, that changes continually with time, using a measured change over a small, but finite, time interval Δt. How accurate this estimation is depends on how fast the rate of change of V is at that time. It becomes more accurate the smaller Δt is. For practical purposes in which we wish to calculate V at very many time points over a long total period of time, we want to make Δt as large as possible without sacrificing too much accuracy. Substituting this expression into our original Equation 5.18 gives:

$$\frac{V(t + \Delta t) - V(t)}{\Delta t} = f(V(t), t). \tag{5.22}$$

Rearranging, we arrive at an expression that enables us to calculate the voltage at time point $t + \Delta t$, given the value of the voltage at time t:

$$V(t + \Delta t) = V(t) + f(V(t), t)\Delta t. \tag{5.23}$$

Suppose we start at time 0 with a known voltage $V(0) \equiv V^0$. We can use this formula to calculate iteratively the voltage at future time points Δt, $2\Delta t$, $3\Delta t$ and so on. If $t = n\Delta t$ and we use the notation $V^n \equiv V(t)$ and $V^{n+1} \equiv V(t + \Delta t)$, then:

$$V^{n+1} = V^n + f(V^n, n\Delta t)\Delta t. \tag{5.24}$$

For our example of the patch of passive membrane, this approximation is:

$$V^{n+1} = V^n + \frac{\Delta t}{C_m}\left(\frac{E_m - V^n}{R_m} + \frac{i_{el}^n}{a}\right). \tag{5.25}$$

This approximation has first-order accuracy in time, because the local error between the calculated value of V and its true value is proportional to the

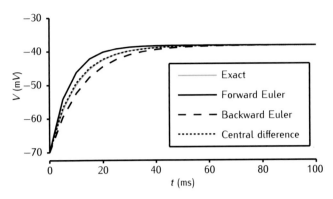

Fig 5.28 Comparison of finite difference approximations with the exact solution to current injection in passive membrane. The exact solution is plotted every 0.1 ms; the approximations use $\Delta t = 5$ ms.

size of the time-step Δt. A comparison of this method with the exact solution for the voltage response to an injected current is shown in Figure 5.28. A rather large value of 5 ms for Δt is used to illustrate that this is only an approximation. If a time-step of less than 1 ms is used, then the approximation is virtually indistinguishable from the exact solution.

Other finite difference schemes can be more accurate for a given timestep and also more stable; the error grows but remains within finite bounds as the step size is increased. The **backward Euler method** is an example of a so-called implicit method that is also first-order accurate in time, but is more stable than the forward Euler method. It results from using a past time point, rather than a future time point, in the approximation of the time derivative:

$$\frac{dV}{dt} \approx \frac{V(t) - V(t - \Delta t)}{\Delta t}. \tag{5.26}$$

The full ODE is thus approximated as:

$$\frac{V(t) - V(t - \Delta t)}{\Delta t} = f(V(t), t). \tag{5.27}$$

Shifting this to the same time points as the forward Euler method yields:

$$\frac{V(t + \Delta t) - V(t)}{\Delta t} = f(V(t + \Delta t), t + \Delta t). \tag{5.28}$$

Now both the left- and right-hand sides involve the unknown voltage at the time point $t + \Delta t$. Using the notation for iterative time points, for our example we have:

$$C_m \frac{V^{n+1} - V^n}{\Delta t} = \frac{E_m - V^{n+1}}{R_m} + \frac{i_{el}^{n+1}}{a}. \tag{5.29}$$

Fortunately, as for the forward Euler expression, this can be rearranged to give an explicit, but now slightly different, equation for V^{n+1}:

$$V^{n+1} = \left[V^n + \frac{\Delta t}{C_m} \left(\frac{E_m}{R_m} + \frac{i_{el}^{n+1}}{a} \right) \right] \bigg/ \left(1 + \frac{\Delta t}{R_m C_m} \right). \tag{5.30}$$

For the same time-step, this method tends to underestimate the rate of change in voltage in our example (Figure 5.28), whereas the forward Euler approximation overestimates it. Consequently, the backward Euler method produces an approximation that smoothly approaches, and never overshoots, the final steady-state value of V. For large step sizes, the forward Euler method may produce values of V that are greater than the steady state, leading to oscillations in V around this value. Such oscillations are not seen in the exact solution and are not desirable in a good approximation.

Note that in more complex models in which several related variables are being solved for, the backward Euler method will result in a set of equations that need to be solved simultaneously. Consequently, the backward Euler method is known as an **implicit method**. We will see this below for approximations to the cable equation. In contrast, with the forward Euler method, values at future time points of all variables can be calculated directly from values at known time points. The forward Euler method is an example of an **explicit method**.

A third method, which is both more accurate and stable than these Euler methods, is the **central difference** method, in which the time derivative is estimated from a future and a past value of the voltage:

$$\frac{dV}{dt} \approx \frac{V(t+\Delta t)-V(t-\Delta t)}{2\Delta t}. \tag{5.31}$$

By examining the equations for the Euler methods, it should be clear that this method results by taking the average of the forward and backward Euler approximations. If we use the expression for the backward Euler method involving the future voltage $V(t+\Delta t)$, then the ODE is approximated by:

$$\frac{V(t+\Delta t)-V(t)}{\Delta t} = \frac{1}{2}[f(V(t+\Delta t),t+\Delta t)+f(V(t),t)], \tag{5.32}$$

that is, we now take an average of the forward and backward Euler right-hand sides. For our example, this leads to the expression:

$$C_{\mathrm{m}}\frac{V^{n+1}-V^n}{\Delta t} = \frac{1}{2}\left(\frac{E_{\mathrm{m}}-V^n}{R_{\mathrm{m}}}+\frac{i_{\mathrm{el}}^n}{a}+\frac{E_{\mathrm{m}}-V^{n+1}}{R_{\mathrm{m}}}+\frac{i_{\mathrm{el}}^{n+1}}{a}\right). \tag{5.33}$$

This can be rearranged to give an explicit expression for V^{n+1}. This approximation is accurate, even for the rather large time-step of 5 ms (Figure 5.28). It can be shown that the central difference method is **second-order** accurate in time because the error is proportional to the square of the step size Δt.

5.7.2 Partial Differential Equations

These same methods can be used for the temporal discretisation of PDEs, but now the spatial dimension must also be discretised. Let us consider the cable equation (Section 5.3) for voltage spread along a neurite of uniform diameter d:

$$C_{\mathrm{m}}\frac{\partial V}{\partial t} = \frac{E_{\mathrm{m}}-V}{R_{\mathrm{m}}}+\frac{d}{4R_{\mathrm{a}}}\frac{\partial^2 V}{\partial x^2}+\frac{i_{\mathrm{el}}(x)}{\pi d}. \tag{5.34}$$

This involves the first derivative of V with respect to time, but the second derivative of V with respect to space. Consequently, a second-order central difference method involving values at three different spatial grid points is required to discretise the spatial dimension:

$$\frac{\partial^2 V}{\partial x^2} \approx \frac{V(x+\Delta x)-2V(x)+V(x-\Delta x)}{(\Delta x)^2}, \tag{5.35}$$

The choice of time-step depends critically on how rapidly the quantity of interest, such as membrane voltage, is changing. When simulating action potentials, a small time-step is required, on the order of 10–100 μs, to capture accurately the rapid rise and fall of the action potential. In between action potentials, however, a neuron may sit near its resting potential for a long period. During this time, the small time-step is unnecessary. Variable time-step integration methods have been developed to account for just this sort of situation. The time-step is automatically increased when a variable is only changing slowly, and decreased when rapid changes begin. These methods can drastically decrease the computation time required and are available in certain neural simulators such as *NEURON*.

for a small spatial step Δx. If we use the notation that position x is the mid-point of compartment j, $x + \Delta x$ corresponds to compartment $j + 1$, $x - \Delta x$ to $j - 1$, and the length of each compartment is $l = \Delta x$, then this is identical to the compartmental structure introduced in Section 5.1. Using this notation and the above spatial discretisation, the forward Euler numerical approximation to the full cable equation is:

$$C_{\mathrm{m}} \frac{V_j^{n+1} - V_j^n}{\Delta t} = \frac{E_{\mathrm{m}} - V_j^n}{R_{\mathrm{m}}} + \frac{d}{4R_{\mathrm{a}}} \frac{V_{j+1}^n - 2V_j^n + V_{j-1}^n}{l^2} + \frac{i_{\mathrm{el},j}^n}{\pi d l}. \qquad (5.36)$$

To make the compartmental structure clear, now we will assume that the injected current is zero in compartment j, $i_{\mathrm{el},j}^n = 0$, so that we can remove this term. We can rewrite Equation 5.36 to indicate explicitly the current flow between compartments:

$$C_{\mathrm{m}} \frac{V_j^{n+1} - V_j^n}{\Delta t} = \frac{E_{\mathrm{m}} - V_j^n}{R_{\mathrm{m}}} + \frac{c}{R_{\mathrm{a}}} \left(V_{j+1}^n - V_j^n \right) + \frac{c}{R_{\mathrm{a}}} \left(V_{j-1}^n - V_j^n \right), \quad (5.37)$$

where we define a coupling coefficient c between compartments as the cross-sectional area between compartments divided by the surface area of a compartment multiplied by the length l between compartments:

$$c \equiv \frac{\pi d^2}{4} \frac{1}{\pi d l} = \frac{d}{4 l^2}. \qquad (5.38)$$

Rearranging Equation 5.37, we arrive at an expression for the voltage V_j^{n+1} in compartment j at time-step $n + 1$ as a function of the values of the voltage at the previous time-step n in compartment j and in its two neighbours $j - 1$ and $j + 1$:

$$V_j^{n+1} = V_j^n + \frac{\Delta t}{C_{\mathrm{m}}} \left(\frac{E_{\mathrm{m}} - V_j^n}{R_{\mathrm{m}}} + \frac{c}{R_{\mathrm{a}}} \left(V_{j+1}^n - V_j^n \right) + \frac{c}{R_{\mathrm{a}}} \left(V_{j-1}^n - V_j^n \right) \right). \quad (5.39)$$

The backward Euler method uses the same spatial discretisation, but with values of V at time point $n + 1$ on the right-hand side:

$$C_{\mathrm{m}} \frac{V_j^{n+1} - V_j^n}{\Delta t} = \frac{E_{\mathrm{m}} - V_j^{n+1}}{R_{\mathrm{m}}} + \frac{c}{R_{\mathrm{a}}} \left(V_{j+1}^{n+1} - V_j^{n+1} \right) + \frac{c}{R_{\mathrm{a}}} \left(V_{j-1}^{n+1} - V_j^{n+1} \right). \quad (5.40)$$

To solve this, we rearrange it so that all unknown quantities are on the left-hand side and all known quantities are on the right:

$$-a V_{j-1}^{n+1} + b V_j^{n+1} - a V_{j-1}^{n+1} = V_j^n + \frac{\Delta t E_{\mathrm{m}}}{C_{\mathrm{m}} R_{\mathrm{m}}}, \qquad (5.41)$$

where $a = \Delta t c / C_{\mathrm{m}} R_{\mathrm{a}}$ and $b = 1 + 2a + \Delta t / C_{\mathrm{m}} R_{\mathrm{m}}$. This gives a set of equations involving values of V at the new time $n + 1$, but at different spatial points along a cable, that must be solved simultaneously. Using vector notation, this can be neatly expressed as:

$$\mathbf{A} \vec{V}^{n+1} = \vec{B}, \qquad (5.42)$$

where $\mathbf{A}$ is a tridiagonal matrix with entries $A_{j,j} = b$, $A_{j,j-1} = A_{j,j+1} = -a$ and all other entries are zero. The entries in the vector $\vec{B}$ are the right-hand side of Equation 5.41.

A more stable and accurate method for this form of PDE is the Crank–Nicholson method (Crank and Nicholson, 1947), which uses central differences for both the temporal and spatial discretisations. As with the ODE, this results from taking the right-hand side to be the average of the forward and backward Euler approximations. As for the backward Euler method, this yields a system of equations involving values for voltage at the new time point in all compartments, which must be solved simultaneously.

Another aspect of calculating a solution to the cable equation is specifying the initial value of V at each spatial point and the boundary conditions (Section 5.1.1) that specify what happens at the end of the cable, where one of the grid points, $j + 1$ or $j - 1$, will not exist. When simulating a compartmental model of a neuron, the initial values are usually the resting membrane potentials, which may differ throughout a neuron due to differences in ionic currents. Sealed end boundary conditions are assumed to apply, meaning that the spatial derivative of V is zero at the end of each neurite (Neumann boundary condition). This can be easily incorporated into the discrete equations. Suppose our first spatial grid point is 0. The central difference formula for the spatial derivative at this point is:

$$\frac{\partial V(0)}{\partial x} \approx \frac{(V_1 - V_{-1})}{2\Delta x} = 0. \tag{5.43}$$

This gives us $V_{-1} = V_1$, and we replace $V_1^n - 2V_0^n + V_{-1}^n$ with $2V_1^n - 2V_0^n$ for that compartmental equation.

When our compartmental model includes the branching structure of neurites, the system of equations that must be solved simultaneously does not result in the simple tri-diagonal structure for matrix A illustrated here. Careful labelling of branches is required to give non-zero entries in A that still allow Equation 5.42 to be solved efficiently. Details are beyond the scope of this book and the reader is referred to Mascagni and Sherman (1998) for further information.

5.8 Summary

The complex morphology of a neuron can be captured in a computational model by using the compartmental modelling approach in which the neuron is approximated as a set of isopotential compartments of membrane connected by resistors. Compartments are often represented by cylinders, with lengths and diameters derived from reconstructed morphologies, or as stereotypical approximations of real dendrites and axons.

Compartmental models can be used to conduct virtual 'experiments' that are technically challenging or even impossible to carry out in real neural tissue. Two case studies have been presented to illustrate this: (1) monitoring and visualising AP propagation along an axon; and (2) the integration of synaptic signals in dendrites as they propagate towards the cell's soma.

Model cell morphologies can be derived from the morphologies of real neurons, through suitable experimental visualisation techniques, or can be generated by computational algorithms that reconstruct the morphological characteristics of specified neuronal types.

A compartmental model of a neuron that has a complex morphology comes with a huge number of free parameters. The active and passive properties of the membrane of each compartment need to be specified. Parameter estimation techniques can be used to choose values for free parameters. These techniques select combinations of parameter values that specifically reduce an error measure reflecting the difference between model and experimentally recorded data. Such techniques are covered in detail in Chapter 14.

A significant part of the process of building multicompartmental models is using appropriate approaches to reduce the number of free parameters. This can involve simplifying the morphology or assuming easily parameterised distributions of ion channels; for example, a particular ion channel type may have the same conductance density throughout the neuron.

In Chapter 6, we move from electrophysiology to consideration of modelling intracellular ionic concentrations, particularly for calcium, and the signalling pathways in which ions, proteins and other molecules participate. Then in Chapter 7, we look in detail at how synaptic responses may be modelled.

CHAPTER 6

Intracellular Mechanisms

Intracellular molecular signalling plays a crucial role in modulating ion channel dynamics, synaptic plasticity and ultimately, the behaviour of the whole cell. In this chapter, we investigate ways of modelling intracellular signalling systems. We focus on **calcium**, as it plays an extensive role in many cell functions. Included are models of **intracellular buffering** systems, **ionic pumps** and calcium-dependent processes. This leads us to outline other **intracellular signalling pathways** involving more complex enzymatic reactions and cascades. We introduce the well-mixed approach to modelling these pathways and explore its limitations. Rule-based modelling can be used when full specification of a signalling network is infeasible. When small numbers of molecules are involved, stochastic approaches are necessary and we consider both population-based and particle-based methods for stochastic modelling. Movement of molecules through diffusion must be considered in spatially inhomogeneous systems.

So far most of the modelling we have presented is devoted to determining neuronal membrane potentials over time and space. The electrical currents underpinning the membrane potential are dependent on ionic concentrations, both intracellular and extracellular. These concentrations most often are not included explicitly in the equations used to model these currents. Instead, it is assumed that ionic concentrations do not change significantly during neuronal electrical activity so that equilibrium potentials given by the Nernst equation (Section 2.3) remain constant. To more accurately model membrane currents requires explicit calculation of ionic concentrations. This is often necessary for calcium, as intracellular concentrations of calcium can vary significantly in different neuronal compartments such as spine heads and dendritic branches.

Calcium plays a variety of roles within a neuron. We have seen that it determines the conductance of a number of types of potassium channel (Section 4.6), necessitating the inclusion of a simple model of intracellular calcium concentration in the case study of compartmental modelling of a pyramidal cell (Section 5.6). In addition, calcium plays a crucial role in a multitude of **intracellular signalling pathways** (Berridge et al., 2003). It acts as a **second messenger** in the cell, triggering processes such as (Hille, 2001) biochemical cascades that lead to the changes in receptor insertion in the membrane, which underlie synaptic plasticity; muscle contraction; secretion of neurotransmitter at nerve terminals; and gene expression. We consider models of synaptic plasticity in Chapter 11.

After an introduction to basic signalling pathways (Section 6.1), the first part of this chapter (Sections 6.2–6.6) deals mainly with modelling intracellular calcium. The techniques are also applicable to other molecular species. For further information, see one of the excellent treatments of modelling intracellular calcium (De Schutter and Smolen, 1998; Koch, 1999; Bormann et al., 2001; Smith, 2001).

The final part of the chapter (Sections 6.7–6.9) deals with modelling complex signalling pathways involving many molecular species. Both deterministic and stochastic modelling approaches are presented, with consideration of when different approaches are appropriate and the computational issues involved.

6.1 | Intracellular Signalling Pathways

The *NEURON* simulator has facilities for modelling intracellular signalling in the context of neuronal morphology and electrophysiology, as described in Sections 6.1 to 6.7. The *Multiscale Object-Oriented Simulation Environment (MOOSE)* provides facilities for building neural models that combine levels from intracellular signalling up to networks of neurons. Developed in the field of systems biology, *COPASI* (Hoops et al., 2006) is a widely used simulator for modelling general intracellular signalling networks. Further simulators, particularly for implementing stochastic models, are mentioned in Sections 6.8 and 6.9.

The cell cytoplasm is a complex milieu of molecules and organelles, which move by diffusion and active transport mechanisms, and may interact through chemical reactions. So far we have treated the opening and closing of ion channels due to voltage changes or ligand binding as simple chemical reactions described by kinetic schemes. Within the cell, molecules may react in complex ways to create new molecular products. The propensity of reactions to occur depends on the concentrations of the particular molecular species involved, which may vary with location as molecules diffuse through the intracellular space. The combination of chemical reactions and molecular diffusion results in a **reaction–diffusion system**. Such systems are particularly important in neural development, involving extracellular as well as intracellular chemical gradients (Chapter 12). Specific sequences of reactions leading from a cause (such as neurotransmitter binding to postsynaptic receptors) to an end effect (such as phosphorylation of AMPA (α-amino-3-hydroxy-5-methyl-4-isoxalone propionic acid) receptors that changes the strength of a synapse) constitute intracellular signalling pathways.

Much of the work on modelling intracellular signalling is based on the assumption of a **well-mixed system** in which the diffusion of the participant molecules is much faster than any reaction time course. We can then ignore diffusion and concentrate solely on modelling the reaction kinetics. Typical signalling pathways involve both binding and enzymatic reactions (Bhalla, 1998, 2001; Blackwell and Hellgren Kotaleski, 2002; Blackwell, 2005).

6.1.1 Binding Reactions

In the simplest **binding reaction**, molecule A binds to molecule B to form complex AB:

$$A + B \underset{k^-}{\overset{k^+}{\rightleftharpoons}} AB. \tag{6.1}$$

Molecules A and B are the substrates and AB is the product. The reaction rate depends on the concentrations of all reacting species. The forward reaction rate coefficient k^+ has units of per unit concentration per unit time; the backward rate coefficient k^- is per unit time.

If the well-mixed system is sufficiently large such that the molecular species are present in abundance, then the **law of mass action** applies and the rate equation is equivalent to a set of coupled differential equations that describe the rate of change in concentration of the different molecular species. For this binding reaction, the rate of change of species A, and equivalently of species B, is:

$$\frac{d[A]}{dt} = -k^+[A][B] + k^-[AB]. \tag{6.2}$$

The time evolution of the product $[AB]$ is the negative of this expression, given by:

$$\frac{d[AB]}{dt} = -k^-[AB] + k^+[A][B]. \tag{6.3}$$

For a closed system in which the mass does not change, given initial concentrations for the substrates and assuming no initial product, the concentrations of A and B can be calculated from the product concentration:

$$\begin{aligned}
[A]_t &= [A]_0 - [AB]_t, \\
[B]_t &= [B]_0 - [AB]_t,
\end{aligned} \tag{6.4}$$

where t denotes time, and $[A]_0$ and $[B]_0$ are the initial concentrations of A and B.

At equilibrium, when the concentrations have attained the values $[A]_\infty$, $[B]_\infty$ and $[AB]_\infty$, the relative concentration of substrates to product is expressed by the **dissociation constant**:

$$K_d = [A]_\infty [B]_\infty / [AB]_\infty = k^-/k^+, \tag{6.5}$$

which has units of concentration.

More complex binding reactions will involve more than two molecular species and may require more than a single molecule of a particular species to generate the product. Such reactions may be broken down into a sequence of simple binding reactions involving the reaction of only a single molecule of each of two species.

6.1.2 Enzymatic Reactions

Enzymatic reactions are two-step reactions in which the action of one molecule, the enzyme E, results in a substrate S being converted into a product P via a reversible reaction that produces a complex ES. E itself is not consumed. This sort of reaction was described by Michaelis and Menten (1913) as the reaction sequence:

$$E + S \underset{k^-}{\overset{k^+}{\rightleftharpoons}} ES \overset{k^c}{\longrightarrow} E + P. \tag{6.6}$$

Note that the second reaction step, leading to product P, is assumed to be irreversible, with forward rate coefficient k^c and typically with the substrate

The law of mass action assumes that molecules move by diffusion and they interact through random collisions. According to this law, the rate of action is proportional to the product of the concentrations of the reactants.

in excess, so the reaction sequence is limited by the amount of enzyme. Assuming the law of mass action applies, the production of complex ES and product P is described by the differential equations:

$$\frac{d[ES]}{dt} = k^+[E][S] - (k^- + k^c)[ES],$$ (6.7)

$$\frac{d[P]}{dt} = k^c[ES].$$

If the complex ES is in equilibrium with enzyme and substrate, then the dissociation constant is:

$$K_m = [E]_\infty[S]_\infty/[ES]_\infty = (k^- + k^c)/k^+.$$ (6.8)

If the reactions are sufficiently fast that it can be assumed that all species are effectively in equilibrium, then at all times we have:

$$k^+[E][S] - (k^- + k^c)[ES] = 0.$$ (6.9)

Substituting in $[E] = [E]_{tot} - [ES]$, where $[E]_{tot}$ is the total enzyme concentration, and rearranging, leads to:

$$[ES] = \frac{[E]_{tot}[S]}{K_m + [S]}.$$ (6.10)

Thus the flux, in units of concentration per time, of substrate S converted to product P is:

$$J_{enz} = \frac{d[P]}{dt} = V_{max}\frac{[S]}{K_m + [S]},$$ (6.11)

where $V_{max} = k^c[E]_{tot}$ is the maximum velocity in units of concentration per time, and K_m, in units of concentration, determines the half-maximal production rate (Figure 6.1). This steady-state approximation to the full enzymatic reaction sequence is often referred to as **Michaelis–Menten kinetics**, and Equation 6.11, which results from the approximation, is known as the **Michaelis–Menten function**. It is used extensively later in this chapter to model, for example, ionic pump fluxes (Section 6.3.3).

One of the features of Michaelis–Menten kinetics is that for low concentrations of S, the rate of production of P is approximately a linear function of S. This feature is not shared by more complex molecular pathways. For example, if the production of P involves the binding of a number n of identical molecules of S simultaneously, the rate of reaction is given by the Hill equation (Section 4.6.1):

$$\frac{d[P]}{dt} = V_{max}\frac{[S]^n}{K_m^n + [S]^n},$$ (6.12)

where n is the Hill coefficient.

Binding and enzymatic reactions, in combination with diffusion, are the basic building blocks for modelling intracellular signalling pathways. In Sections 6.2–6.6, they are used to develop a model for the dynamics of intracellular calcium. We then consider examples of more complex signalling pathways (Section 6.7), before addressing the problems that arise when the assumption of mass action kinetics is not reasonable (Section 6.8).

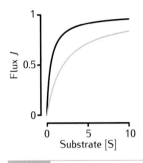

Fig 6.1 Two examples of the enzymatic flux J as a function of the substrate concentration [S] for Michaelis–Menten kinetics (Equation 6.11). Arbitrary units are used and $V_{max} = 1$. Black line: $K_m = 0.5$. Blue line: $K_m = 2$. Note that half-maximal flux occurs when $[S] = K_m$.

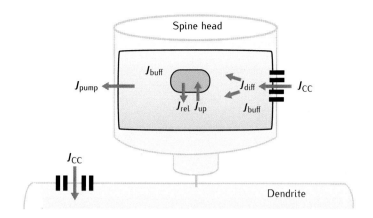

Fig 6.2 Intracellular calcium concentration is determined by a variety of fluxes, including diffusion (J_{diff}), buffering (J_{buff}), entry through voltage- and ligand-gated calcium ion channels (J_{cc}), extrusion by membrane-bound ionic pumps (J_{pump}), uptake (J_{up}) and release (J_{rel}) from intracellular stores.

6.2 Modelling Intracellular Calcium

The calcium concentration in a cellular compartment is highly dynamic and is determined by influx of calcium through voltage-gated channels, release of calcium from second messenger- and calcium-activated internal stores, diffusion into the intracellular space, buffering by mobile and fixed buffers, and extrusion by calcium pumps (Figure 6.2). The change in calcium concentration with time is determined by the sum of these various fluxes:

$$\frac{d[Ca^{2+}]}{dt} = J_{diff} - J_{buff} + J_{cc} - J_{pump} - J_{up} + J_{rel} - J_{leak}. \tag{6.13}$$

It is possible, and sometimes desirable, to model explicitly many or all of these mechanisms. Through a series of models of increasing complexity, we now explore the effects of these different components on the dynamics of intracellular calcium. We highlight when it is reasonable to make simplifications and when it is not.

6.3 Transmembrane Fluxes

Firstly, we examine the intracellular calcium dynamics that result from transmembrane fluxes into and out of a small cellular compartment such as a spine head or a short section of dendrite. Entry into the compartment is through voltage- or ligand-gated calcium channels in the cell membrane. Resting calcium levels are restored by extrusion of calcium back across the membrane by ionic pumps and by uptake into internal stores. The compartment is assumed to be well mixed, meaning that the calcium concentration is uniform throughout.

6.3.1 Ionic Calcium Currents

The flux that results from an ionic calcium current I_{Ca}, in units of current per unit area, is given by:

$$J_{cc} = -\frac{aI_{Ca}}{2Fv}, \tag{6.14}$$

Here, each flux J specifies the rate of change of the number of molecules of calcium per unit volume, that is, the rate of change of concentration (with typical units of $\mu M\,s^{-1}$). This is derived by multiplying the rate of movement of molecules across a unit surface area by the total surface area across which the movement is occurring, and dividing by the volume into which the molecules are being diluted.

J_{leak} represents a background leak flux (e.g. influx through voltage-gated calcium channels at rest) that ensures the total flux is zero at rest.

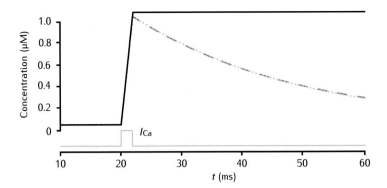

Fig 6.3 Calcium transients in a single cylindrical compartment, 1 μm in diameter and 1 μm in length (similar in size to a spine head). Initial concentration is 0.05 μM. Influx is due to a calcium current of $5\,\mu A\,cm^{-2}$, starting at 20 ms for 2 ms, across the radial surface. Black line: accumulation of calcium with no decay or extrusion. Blue line: simple decay with $\tau_{dec} = 27\,ms$. Dashed line: instantaneous pump with $V_{pump} = 4 \times 10^{-6}\,\mu M\,s^{-1}$, equivalent to $10^{-11}\,mol\,cm^{-2}\,s^{-1}$ through the surface area of this compartment, $K_{pump} = 10\,\mu M$, $J_{leak} = 0.0199 \times 10^{-6}\,\mu M\,s^{-1}$.

The expression for the flux J_{cc} arises from the need to change the amount of charge passing into the compartment per unit time into the number of calcium ions carrying that charge. Faraday's constant gives the amount of charge carried by a mole of monovalent ions: 96 490 coulombs per mole. Since calcium is divalent, we multiply this by two, then divide the current I_{Ca} by the result to get the number of moles flowing per unit area and per unit time. Multiplying this by the total surface area and dividing by the compartment volume gives the rate of change in concentration, that is, the flux.

where a is the surface area across which the current flows, v is the volume of the intracellular compartment and F is Faraday's constant. The current results in a certain number of calcium ions entering the cellular compartment per unit time. The presence of volume v in this equation turns this influx of calcium into a change in the calcium concentration in the compartment.

The rise in intracellular calcium due to a short calcium current pulse, such as might arise due to transmitter release activating N-methyl-D-aspartate (NMDA) channels at a synapse, or due to a back-propagating action potential activating voltage-gated calcium channels in the dendritic membrane, is illustrated in Figure 6.3 (black line). This short-lasting flux sees a rapid rise in the intracellular calcium concentration to a fixed level, since there are no mechanisms in this model to return the calcium to its resting level.

6.3.2 Calcium Decay

Other fluxes, such as those due to membrane-bound pumps and buffering, act to restore calcium to its resting level following such an influx. A simple model captures the fact that the calcium concentration will always return to its resting value. This model describes calcium decay by a single time constant τ_{dec} (Traub and Llinás, 1977), giving:

$$\frac{d[Ca^{2+}]}{dt} = J_{cc} - \frac{[Ca^{2+}] - [Ca^{2+}]_{res}}{\tau_{dec}}, \tag{6.15}$$

where $[Ca^{2+}]_{res}$ is the resting level. An example of such calcium decay is shown in Figure 6.3 (blue line).

This model is consistent with the calcium transients in cellular compartments imaged with fluorescent dyes. Clearance of calcium from small compartments, such as spine heads, may be quite rapid, with a time constant as small as 12 ms (Sabatini et al., 2002). More complex models include the fluxes that remove calcium from a cellular compartment such as pumps, buffering and diffusion. The sum of these fluxes may be approximated by this single exponential decay model.

6.3.3 Calcium Extrusion

Membrane-bound pumps contribute to restoring resting levels of calcium by extruding calcium ions back through the membrane, moving them against their concentration gradient. Two pumps act to remove calcium from a cell. The plasma membrane Ca^{2+}-ATPase (PMCA) pump is a high-affinity, low-capacity mechanism that can switch on rapidly following calcium entry through voltage-gated channels, and so plays a major role shaping dynamic changes in intracellular calcium. This pump is also responsible for uptake of calcium into intracellular stores (Section 6.4). The sodium–calcium exchanger acts as a low-affinity, high-capacity calcium pump that is largely responsible for maintaining resting calcium levels.

Complex pump models have been attempted, particularly of the sodium–calcium exchanger (Gabbiani et al., 1994; De Schutter and Smolen, 1998). A relatively simple approach is to treat extrusion explicitly as a chemical reaction pathway. This involves the intracellular binding and extracellular unbinding of calcium to membrane-bound pump molecules P (Migliore et al., 1995; Carnevale and Hines, 2006). Ca_i and Ca_o denote intracellular and extracellular calcium, respectively:

$$
\begin{aligned}
Ca_i + P &\underset{k_1^-}{\overset{k_1^+}{\rightleftharpoons}} CaP, \\
CaP &\underset{k_2^-}{\overset{k_2^+}{\rightleftharpoons}} Ca_o + P.
\end{aligned}
\tag{6.16}
$$

The external calcium concentration is typically much higher than intracellular calcium. Thus, relative changes in external calcium during electrical activity are very small. If we assume extracellular calcium is effectively constant, the value of k_2^- can be set to zero. Then the pump flux can be modelled as an instantaneous function of the intracellular calcium concentration, described by a modified Michaelis–Menten relationship (e.g. Jaffe et al., 1994):

$$
J_{pump} = V_{pump} \frac{[Ca^{2+}]}{K_{pump} + [Ca^{2+}]}.
\tag{6.17}
$$

Calcium extrusion has maximum velocity $V_{pump} = a k_2^+ P_m / v$ (in units of concentration per unit time; P_m is the number of pump molecules in a unit area of membrane) and is a function of the calcium concentration with half maximum flux at $K_{pump} = (k_1^- + k_2^+)/k_1^+$ (in units of concentration).

Little firm data are available about the properties of these pumps, but use of the simple Michaelis–Menten approach minimises the number of parameters that must be estimated. Pump model parameters are typically chosen to fine-tune the calcium transients in cellular compartments (Schiegg et al., 1995). Estimates of pump velocity for the PMCA pump in hippocampal pyramidal cell models range over several orders of magnitude from around 10^{-13} to 10^{-10} mol cm^{-2} s^{-1}, with half maxima at around 1 μM (Zador et al., 1990; Jaffe et al., 1994; Schiegg et al., 1995). A much higher velocity of nearly 10^{-7} mol cm^{-2} s^{-1} has been used in a Purkinje cell model (De Schutter and Smolen, 1998). The sodium–calcium exchanger will have higher values for V_{max} and K_{pump} than the PMCA pump (Schiegg et al., 1995).

An example of the effect of such a pump is given in Figure 6.3 (dashed line), which shows the response of a model with calcium influx and with removal via a pump and a constant background leak flux:

$$
\frac{d[Ca^{2+}]}{dt} = J_{cc} - J_{pump} - J_{leak}.
\tag{6.18}
$$

A high-affinity pump has a low dissociation constant K_m so that it reaches its half-maximal pump rate at a low calcium concentration. High capacity means that the maximum pump velocity V_{max} is large.

Estimates of pump velocities are usually given in units of moles per unit area per unit time; in our model, these would be multiplied by the membrane surface area and divided by the compartment volume to give velocity in units of concentration per unit time.

This model is well matched by the simple decay model with a suitable time constant; the calcium transients of the two models are indistinguishable in Figure 6.3 (blue and dashed lines).

6.4 Calcium Stores

Intracellular calcium can be sequestered into stores in such structures as the endoplasmic reticulum (ER), with release into the cytoplasm being mediated by second-messenger pathways. Stores can have large capacity, with estimates of $[Ca^{2+}]$ in stores ranging from $100\,\mu M$ to $5\,mM$. Uptake and release of calcium from these stores can result in intracellular calcium waves, or oscillations, on the timescale of seconds (Smith, 2001; Schuster et al., 2002).

Uptake and release mechanisms are modelled in a similar fashion to the transmembrane fluxes. However, rather than modelling explicitly the physical structure of stores, such as the ER, it can be assumed that they occupy a fractional volume, with an associated surface area, within the spatial compartment in which the calcium concentration is being calculated. For example, the volume may be assumed to be in the form of a long, thin cylinder, thus enabling the store membrane surface area to be calculated (De Schutter and Smolen, 1998).

6.4.1 Calcium Uptake

Uptake into the stores is via a sarcoplasmic reticulum Ca^{2+}–ATPase (SERCA) pump in the smooth ER membrane. It binds two calcium ions for each adenosine triphosphate (ATP) molecule, and so can be described by Michaelis–Menten kinetics with a Hill coefficient of 2 (Blackwell, 2005):

$$J_{up} = V_{up} \frac{[Ca^{2+}]^2}{K_{up}^2 + [Ca^{2+}]^2}. \tag{6.19}$$

Uptake is across the surface area of the ER membrane a_{ER}. A limitation of this approach is that the uptake flux does not depend on the intrastore calcium concentration, even though empty stores have a higher uptake rate (De Schutter and Smolen, 1998).

6.4.2 Calcium Release

The ER membrane contains calcium channels that are activated by calcium itself, leading to calcium-induced calcium release (CICR). The two major classes of channel correspond to membrane-bound receptors that bind ryanodine, and to receptors that are activated by inositol 1,4,5-triphosphate (IP_3). These different types of receptor tend to be localised in different parts of the ER and consequently different parts of a neuron. For example, in Purkinje cell spines, CICR is largely mediated by IP_3 receptors, whereas in hippocampal CA1 pyramidal cell spines, ryanodine receptors dominate (Berridge, 1998).

The calcium flux through the ER membrane into the cytoplasm is given by:

$$J_{rel} = V_{rel} R([Ca^{2+}])([Ca^{2+}]_{store} - [Ca^{2+}]), \tag{6.20}$$

where $R([Ca^{2+}])$ is the fraction of calcium channels in the open state, which depends on the cytoplasmic calcium concentration itself. $[Ca^{2+}]_{store}$ is the concentration of free calcium in the store.

This formulation is equivalent to the equations used to describe current flow through voltage-gated channels, where the calcium concentration takes the place of voltage. The function R has been described by the same sort of mathematical formulations as for the opening of voltage-gated channels, principally either by Markov kinetic schemes (Section 4.5) or by Hodgkin–Huxley-style gating particles (Chapter 3). The simplest approach is to use a Hill function of cytoplasmic calcium:

$$R([Ca^{2+}]) = \frac{[Ca^{2+}]^n}{K_{rel}^n + [Ca^{2+}]^n},\qquad(6.21)$$

with a suitable Hill coefficient n (Goldbeter et al., 1990). Two problems with this model are that: (1) CICR is not modulated instantaneously in concert with rapid changes in calcium (De Schutter and Smolen, 1998), so this steady state approach may not be accurate; and (2) it does not describe the IP_3 dependence of the IP_3 receptors. More complicated dynamic models for R are required.

Ryanodine Receptor Models

De Schutter and Smolen (1998) modified the Michaelis–Menten approach by assuming that the flux J_{rel} relaxes to the steady-state value given by the Michaelis–Menten formulation for R, with a fixed time constant. In this way, changes in cytoplasmic calcium due to, say, influx from outside the cell do not lead to instantaneous increases in calcium release from ryanodine-mediated stores. They introduced a threshold so that the value of J_{rel} is set to zero below a certain cytoplasmic calcium concentration. This allows the store concentration $[Ca^{2+}]_{store}$ to remain much higher than the cytoplasmic concentration $[Ca^{2+}]$, which is the likely situation.

Other models use Markov kinetic schemes to capture the calcium dependence and dynamics of ryanodine receptor opening and slow inactivation, on the timescale of seconds. Keizer and Levine (1996) model the receptors assuming four states, two open states (O_1, O_2) and two closed states (C_1, C_2):

$$C_1 \underset{k_1^-}{\overset{k_1^+}{\rightleftharpoons}} O_1 \underset{k_2^-}{\overset{k_2^+}{\rightleftharpoons}} O_2 .$$

$$\ \ k_3^- \big\Vert k_3^+$$

$$C_2\qquad\qquad\qquad(6.22)$$

The forward rate coefficients k_1^+ and k_2^+ are dependent on the cytoplasmic calcium concentration, whereas the other rates are fixed. Increasing calcium drives the receptors more strongly into the open states O_1 and O_2. The open state O_1 can proceed slowly into the closed state C_2, providing the channel inactivation. Given a large population of receptors, we can interpret the states as indicating the fraction of available receptors in a given state. Thus, the function R in our release flux Equation 6.20 is now:

$$R([Ca^{2+}]) = O_1 + O_2,\qquad(6.23)$$

which is the fraction of receptors in an open state.

Tang and Othmer (1994) use a similar four-state kinetic scheme in which the inactivation step is also calcium-dependent. Both models are phenomenological rather than directly trying to capture the states of the four subunits that make up a ryanodine receptor.

IP$_3$ Receptor Models

Considerable attention has been paid to modelling IP$_3$-induced calcium release (De Young and Keizer, 1992; De Schutter and Smolen, 1998; Kuroda et al., 2001; Schuster et al., 2002; Fraiman and Dawson, 2004; Doi et al., 2005), as it is likely to play a key role in intracellular calcium oscillations and long-term plasticity at synapses. The dynamics of the IP$_3$ receptors on the surface of a store are more complex than those of the ryanodine receptors. Firstly, in order to open, the receptors require binding of IP$_3$, as well as calcium. The receptor open probability exhibits a bell-shaped dependence on the intracellular calcium concentration and the receptors show a slow, calcium-dependent inactivation. As for the ryanodine receptor, the function R has been described by Markov kinetic schemes which capture various closed, open and inactivated states of the receptors, but also by phenomenological Hodgkin–Huxley-style equations for the open state.

A simple three-state kinetic scheme specifies that the receptors may be in a closed, open or inactivated state (Gin et al., 2006):

$$\text{C} \underset{k_1^-}{\overset{k_1^+}{\rightleftharpoons}} \text{O} \underset{k_2^-}{\overset{k_2^+}{\rightleftharpoons}} \text{I}. \tag{6.24}$$

The transition rate k_1^+ from closed to open depends on both IP$_3$ and calcium, and the rates k_1^- and k_2^+ out of the open state are calcium-dependent. Again treating the states as indicating the fraction of receptors in a given state, then the form of the function R to be used in Equation 6.20 is:

$$R(\text{IP}_3, [\text{Ca}^{2+}]) = O. \tag{6.25}$$

Doi et al. (2005) employ a more complex, seven-state kinetic model in which the receptors bind IP$_3$ and calcium sequentially to reach the open state. Binding of calcium alone leads to progressive entry into four inactivation states.

The process of rapid activation followed by inactivation is analogous to the voltage-dependent dynamics of sodium channels underlying action potential generation. Consequently, it is feasible to use Hodgkin–Huxley-style gating particles to describe these receptors. In this case, the fraction of open receptors is a function of the state of activation and inactivation particles (Li and Rinzel, 1994; De Schutter and Smolen, 1998):

$$R(\text{IP}_3, [\text{Ca}^{2+}]) = m^3 h^3, \tag{6.26}$$

where the dynamics of m and h depend on the concentrations of IP$_3$ and intracellular calcium, rather than voltage.

All these models greatly simplify the actual dynamics of both ryanodine and IP$_3$ receptors. Other molecules may influence receptor sensitivity, such as cyclic adenosine diphosphate (ADP) ribose for ryanodine receptors, and receptor opening may also be affected by intrastore calcium (Berridge, 1998; Berridge et al., 2003).

6.5 | Calcium Diffusion

So far we have considered our cellular compartment to contain a single pool of calcium in which the concentration is spatially homogeneous: a well-mixed pool. However, calcium concentrations close to a source may reach much higher values than for cellular regions far from the source. To capture such variations, it is necessary to model the transport of calcium through space by diffusion.

6.5.1 Two-Pool Model

An important requirement for many models of intracellular calcium is an accurate description of the calcium concentration immediately below the membrane. This is needed when modelling the electrical response of neuronal membrane containing calcium-activated potassium channels, which are likely to be activated only by a local calcium transient due to influx through nearby calcium channels. The simplest extension to our well-mixed model is to divide the compartment into two pools: a thin submembrane shell and the interior core (Figure 6.4). We need to add diffusion of calcium from the submembrane shell into the core to our existing models.

Due to Brownian motion of calcium molecules, there is an average drift of calcium from regions where there are many molecules to regions where there are fewer. In other words, calcium tends to flow down its concentration gradient; see Bormann et al. (2001), Koch (1999) and Section 2.2.2 for further details of this process. The resulting flux of calcium is given by Fick's first law (1855), which for a single spatial dimension x is (Equation 2.2):

$$J_{\text{diff}}^* = -a D_{\text{Ca}} \frac{d[\text{Ca}^{2+}]}{dx}, \tag{6.27}$$

where J_{diff}^* is the rate of transfer of calcium ions across cross-sectional area a and D_{Ca} is the diffusion coefficient for calcium and has units of area per time. In what follows, we consider the diffusional flux to be occurring between well-mixed pools of known volume. Dividing J_{diff}^* by the pool volume will give the rate of change in concentration, or flux, of that pool, J_{diff}.

We consider only the diffusion of calcium along a single dimension x across known cross-sectional areas. In general, diffusion can occur in three dimensions and is properly described by a partial differential equation (PDE) that can be derived from Fick's first law (Koch, 1999). An overview of diffusion PDEs in single and multiple dimensions is given in Box 6.1.

In our two-pool model, we need to calculate the flux due to diffusion between the submembrane shell and the core compartment. To do this, we make the simplest possible assumptions, resulting in a numerical scheme that approximates only crudely the underlying continuous diffusion PDE. The resulting model has limitations, which will become clear later. We assume that both pools are well mixed, so that we can talk about the concentrations in the submembrane shell, $[\text{Ca}^{2+}]_s$, and the core, $[\text{Ca}^{2+}]_c$. The two compartments have volumes v_s and v_c, respectively, and diffusion takes place across the surface area a_{sc} of the cylindrical surface that separates them. These

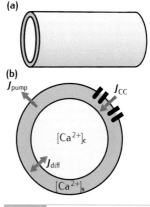

(a)

(b)

Fig 6.4 Cylindrical compartment with a thin submembrane shell and a large central core.

Box 6.1 General diffusion

Considering the flux of calcium into and out of an infinitesimally small volume over an infinitesimally small time period, from Equation 6.27, it can be shown that the rate of change in concentration over time and space is given by the PDE (Koch, 1999):

$$\frac{\partial[Ca^{2+}]}{\partial t} = D_{Ca}\frac{\partial^2[Ca^{2+}]}{\partial x^2}. \tag{a}$$

This can easily be expanded to include diffusion in two or three dimensions. In Cartesian coordinates for three spatial dimensions:

$$\frac{\partial[Ca^{2+}]}{\partial t} = D_{Ca}\left(\frac{\partial^2[Ca^{2+}]}{\partial x^2} + \frac{\partial^2[Ca^{2+}]}{\partial y^2} + \frac{\partial^2[Ca^{2+}]}{\partial z^2}\right).$$

Particularly for the compartmental modelling of neurons, it is often useful to consider this equation with alternative coordinate systems. In cylindrical coordinates, it becomes (Figure 6.8):

$$\frac{\partial[Ca^{2+}]}{\partial t} = D_{Ca}\left(\frac{\partial^2[Ca^{2+}]}{\partial r^2} + \frac{1}{r}\frac{\partial[Ca^{2+}]}{\partial r} + \frac{1}{r^2}\frac{\partial^2[Ca^{2+}]}{\partial \Theta^2} + \frac{\partial^2[Ca^{2+}]}{\partial x^2}\right),$$

for longitudinal distance x, radial distance r and axial rotation Θ.

As we are usually interested in axisymmetric radial diffusion within a compartment, or longitudinal diffusion between compartments, one dimension is sufficient for most purposes. It is only necessary to go to two or three dimensions if radial and longitudinal diffusion are being considered at the same time.

Longitudinal diffusion is handled in Cartesian coordinates by Equation (a), with x defining the longitudinal axis of the cylinder.

Radial diffusion is best handled in cylindrical coordinates. If we assume there is no concentration gradient in the Θ or x directions, the cylindrical diffusion equation reduces to:

$$\frac{\partial[Ca^{2+}]}{\partial t} = D_{Ca}\left(\frac{\partial^2[Ca^{2+}]}{\partial r^2} + \frac{1}{r}\frac{\partial[Ca^{2+}]}{\partial r}\right).$$

Diffusion in extracellular space is considered in Chapter 10.

volumes and the surface area can be calculated from the length, diameter and thickness of the submembrane compartment.

A first-order discrete version of Fick's first law gives the rate of change of calcium concentration in the cell core due to diffusion from the submembrane shell as:

$$J_{sc} = \frac{a_{sc}}{v_c}D_{Ca}\frac{[Ca^{2+}]_s - [Ca^{2+}]_c}{\Delta_{sc}}, \tag{6.28}$$

where Δ_{sc} is the distance between the midpoints of the two compartments, which is the distance over which diffusion takes place. The flux from the core to the shell is in the opposite direction and is diluted into volume v_s. This results in our two-pool model being described by two coupled ordinary differential equations (ODEs):

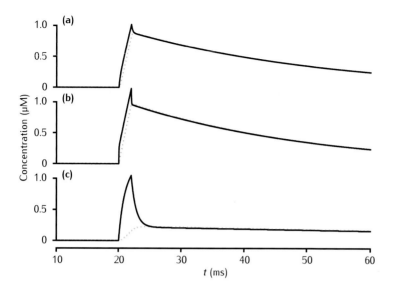

Fig 6.5 Calcium transients in the submembrane shell (black lines) and the compartment core (blue dashed lines) of a two-pool model that includes a membrane-bound pump and diffusion into the core. Initial concentration $[Ca^{2+}]_0$ is $0.05\,\mu M$ throughout. Calcium current is as in Figure 6.3. Compartment length is $1\,\mu m$. **(a)** Compartment diameter is $1\,\mu m$, and shell thickness is $0.1\,\mu m$; **(b)** Compartment diameter is $1\,\mu m$, and shell thickness is $0.01\,\mu m$; **(c)** Compartment diameter is $4\,\mu m$, and shell thickness is $0.1\,\mu m$. Diffusion: $D_{Ca} = 2.3 \times 10^{-6}\,cm^2\,s^{-1}$; pump: $V_{pump} = 10^{-11}\,mol\,cm^{-2}\,s^{-1}$, $K_{pump} = 10\,\mu M$, $J_{leak} = V_{pump}[Ca^{2+}]_0/(K_{pump} + [Ca^{2+}]_0)$.

$$\frac{d[Ca^{2+}]_s}{dt} = -D_{Ca}c_{cs}([Ca^{2+}]_s - [Ca^{2+}]_c) + J_{cc} - J_{pump} - J_{leak},$$

$$\frac{d[Ca^{2+}]_c}{dt} = D_{Ca}c_{sc}([Ca^{2+}]_s - [Ca^{2+}]_c).$$

(6.29)

The coupling coefficient c_{sc} for movement from the shell to the core is $c_{sc} = a_{sc}/v_c\Delta_{sc}$, and that from the core to the shell is $c_{cs} = a_{sc}/v_s\Delta_{sc}$. Calcium may enter the submembrane shell through voltage-gated channels with flux J_{cc}. It is extruded across the membrane, with flux J_{pump}, in addition to diffusing into the cell core.

Example calcium transients from this model are shown in Figure 6.5. For a small compartment of $1\,\mu m$ in diameter, the shell and core concentrations are always similar, with a small peak in the submembrane concentration that is not mirrored in the core (Figure 6.5a). This peak is larger when the shell is very thin ($0.01\,\mu m$; Figure 6.5b). It might be acceptable to model this compartment very approximately as containing just a single pool of calcium. However, this is not possible when the compartment has a larger diameter of $4\,\mu m$ (Figure 6.5c). In this case, calcium influx into the shell compartment is more rapid than diffusion from the shell to the core, resulting in a very large peak in the submembrane calcium concentration, which is then brought into equilibrium with the core concentration by diffusion. The core concentration never reaches levels near that of the submembrane shell as the calcium influx is diluted in the large compartmental volume.

Typically, calcium influx through voltage- or ligand-gated channels may be more rapid than the dispersal of calcium by diffusion. Hence the calcium concentration in the cellular compartment in which an influx occurs will reach a peak before diffusion, and other dispersion mechanisms, such as pumps, begin to restore the resting calcium level. This peak will be determined by the compartment volume. Thus the identical influx of calcium will cause a greater increase in concentration in a smaller compartment.

Therefore, compartment sizes must be chosen with reference to the range of calcium influence that is being modelled, for example, the average distance of calcium-activated potassium channels from the voltage-gated calcium channels through which the necessary calcium enters.

6.5.2 Three-Pool Model

In certain situations, it is necessary to subdivide the submembrane compartment further. For example, with calcium-activated potassium channels, a complication arises in that different types of such channel are apparently activated by different pools of calcium. The I_C current (Section 4.6) switches on rapidly due to calcium influx. Hence the underlying ion channels presumably are colocated with calcium channels, whereas the I_{AHP} current (Section 4.6) is much more slowly activated, probably due to these potassium channels being located further from the calcium channels. Multiple submembrane pools can be accommodated relatively simply, as proposed by Borg-Graham (1999).

In a way that does not complicate the model greatly, the submembrane shell is divided into two pools. The first pool (*local*: volume v_l) corresponds to a collection of membrane domains that are near the calcium channels, whilst the second pool (*submembrane*: volume v_s) consists of the remaining submembrane space that is further from the calcium channels (Figure 6.6). Calcium influx is directly into the local pool, with diffusion acting to move calcium into the submembrane pool. We define the volume of the local pool to correspond to some small fraction α_l ($\approx 0.001\%$) of the membrane surface area. For a thin submembrane shell, the two volumes are approximately:

$$v_l = \alpha_l a_s \Delta_s, \quad v_s = (1-\alpha_l)a_s \Delta_s, \tag{6.30}$$

where a_s is the surface area and Δ_s is the thickness of the submembrane shell.

We have only defined a volume for the local pool, but not a specific geometry. Diffusion between the local and submembrane pools must occur across an effective diffusive surface area. Without an explicit geometry for the local pool, we define the surface area between the pools as $a_{ls} = \alpha_{ls}a_s\Delta_s$, with interdigitation coefficient α_{ls} per unit length, usually set to 1.

Diffusion also takes place between the submembrane volume v_s and the cell core v_c, but not, we assume, between the local pool and the cell core. This occurs across the surface area separating the submembrane pool from the cell core, $a_{sc} = (1-\alpha_l)a_s$.

The complete model is described by the following system of ODEs:

$$\frac{d[Ca^{2+}]_l}{dt} = -D_{Ca}c_{sl}([Ca^{2+}]_l - [Ca^{2+}]_s) + J_{cc}$$

$$\frac{d[Ca^{2+}]_s}{dt} = D_{Ca}c_{ls}([Ca^{2+}]_l - [Ca^{2+}]_s)$$
$$\qquad - D_{Ca}c_{cs}([Ca^{2+}]_s - [Ca^{2+}]_c) - J_{pump} - J_{leak} \tag{6.31}$$

$$\frac{d[Ca^{2+}]_c}{dt} = D_{Ca}c_{sc}([Ca^{2+}]_s - [Ca^{2+}]_c).$$

The local pool corresponds to the surface membrane containing calcium channels, so calcium influx J_{cc} is directly into the local pool. The submembrane shell contains the remaining surface membrane in which membrane-bound pumps exist to restore resting calcium levels, J_{pump}. The diffusive

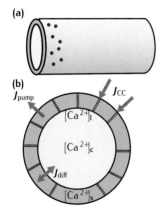

(a)

(b)

J_{pump}

J_{cc}

$[Ca^{2+}]_l$

$[Ca^{2+}]_c$

J_{diff}

$[Ca^{2+}]_s$

Fig 6.6 Three-pool model of intracellular calcium. The grey dots on the surface and thin lines across the submembrane shell represent the local pool of the membrane surrounding calcium channels.

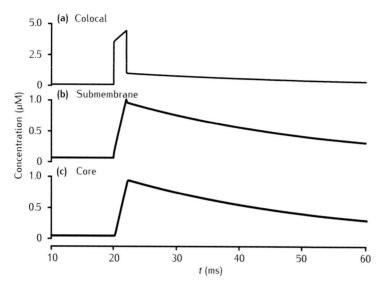

Fig 6.7 Calcium transients in (a) the potassium channel colocalised pool, (b) the larger submembrane shell, and (c) the compartment core in a three-pool model. Note that in (a), the concentration is plotted on a different scale. Compartment 1 μm in diameter with 0.1-μm thick submembrane shell; the colocalised membrane occupies 0.001% of the total membrane surface area. Calcium current, diffusion and membrane-bound pump parameters as in Figure 6.5.

coupling coefficients are again of the form $c_{ij} = a_{ij}/v_j \Delta_{ij}$, where a_{ij} is the effective area between volumes i and j, v_j is the volume into which calcium is diluted and Δ_{ij} is the distance between the source and destination volumes.

An example of the calcium transients in all three pools is shown in Figure 6.7. The transients in the submembrane (Figure 6.7b) and the core (Figure 6.7c) pools match those of the two-pool model. However, the calcium concentration in the local pool (Figure 6.7a) is very rapid and reaches a much higher level. Thus the colocalised potassium channels are driven by a very different calcium transient from that seen in an average submembrane shell.

6.5.3 Radial Diffusion

The assumption in the two- and three-pool models that the large central core of the cellular compartment is well-mixed may be reasonable for small compartments, but could be highly inaccurate for larger diameters. It may be necessary to divide the entire intracellular space into a number of thin shells, with diffusion of calcium taking place between shells, in the same way that we have so far considered diffusion from the submembrane shell into the core (Figure 6.8). This will then capture radial intracellular gradients in calcium, and also improve the accuracy of the submembrane calcium transient due to more accurate modelling of diffusion of incoming calcium away from the submembrane region.

Examples of submembrane calcium transients for different total numbers of shells, but with the same submembrane shell thickness, are shown in Figure 6.9. The remaining shells equally divide up the compartment interior. Our previous simple model containing only the submembrane shell and the cell core provides a reasonably accurate solution for both submembrane and core concentrations for a small dendritic compartment of 1 μm diameter (not shown). However, if the diameter is increased to 4 μm, at least four shells are required to model accurately the dynamics of diffusion of calcium into the interior, with a small improvement being gained by using 11 shells (Figure 6.9).

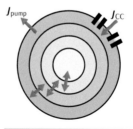

Fig 6.8 Multiple shells for modelling radial diffusion of intracellular calcium.

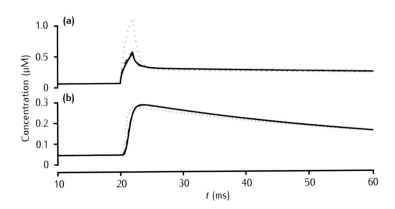

Fig 6.9 Calcium transients in the **(a)** submembrane shell and **(b)** cell core (central shell) for different total numbers of shells. Black line: 11 shells. Grey dashed line: 4 shells. Blue dotted line: 2 shells. The diameter is 4 μm and the submembrane shell is 0.1 μm thick. Other shells equally subdivide the remaining compartment radius. All other model parameters are as in Figure 6.5.

In particular, the submembrane concentration is exaggerated in the two-pool model, as in this case diffusion into the interior is slow due to the gradient being calculated over the entire distance of 2 μm from the membrane to the compartment centre. With more shells, calcium is calculated as diffusing down a steeper gradient from the membrane to the centre of the nearest interior shell, a distance of 0.48 μm with 4 shells and 0.2 μm with 11 shells.

As the submembrane concentration is often the most critical to model accurately, a reasonable number of shells is required here, even if the concentration in only the submembrane shell is used by other model components, such as calcium-activated potassium channels. A good rule of thumb is that interior shells should all have the same thickness, Δr, which is twice the thickness of the submembrane shell. The innermost core shell should have a diameter Δr (Carnevale and Hines, 2006). Our 11-shell model implements this criterion for the 4 μm-diameter compartment with a 0.1 μm-thick submembrane shell. However, this may lead to an excessive number of shells for large compartments with a thin submembrane shell. As shown here, smaller numbers of shells may still provide good solutions. It is not possible to prescribe the optimum choice of shell thickness in any given situation, and so some testing with different numbers of shells is recommended.

6.5.4 Longitudinal Diffusion

In addition to radial diffusion of calcium from the membrane to the compartment core, calcium may also diffuse longitudinally along a section of dendrite (Figure 6.10). Diffusion into a cylindrical compartment i from neighbouring compartments k and j is given by:

$$\frac{d[Ca^{2+}]_j}{dt} = D_{Ca}c_{ij}([Ca^{2+}]_i - [Ca^{2+}]_j) + D_{Ca}c_{jk}([Ca^{2+}]_k - [Ca^{2+}]_j). \quad (6.32)$$

The diffusional coupling coefficients are again of the form $c_{ij} = a_{ij}/v_j\Delta_{ij}$, where a_{ij} is the effective diffusional area between spaces i and j, v_j is the volume of the space into which calcium flows and Δ_{ij} is the distance between the source and destination volumes.

Figure 6.11 shows examples of calcium gradients along the length of a dendrite. In these examples, a 10 μm-long segment of dendrite is divided into 10 1 μm-long compartments, each of which contains four radial shells. Longitudinal diffusion is calculated along the length for each of these shells,

Fig 6.10 Longitudinal diffusion of intracellular calcium between compartments along the length of a dendrite.

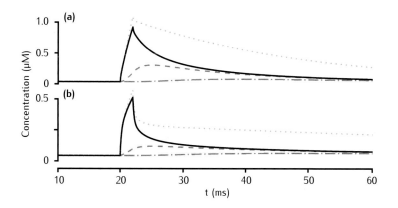

Fig 6.11 Longitudinal diffusion along a 10-μm length of dendrite. Plots show calcium transients in the 0.1 μm-thick submembrane shell at different positions along the dendrite. Solid black line: 0.5 μm; dashed: 1.5 μm; dot-dashed: 4.5 μm. **(a)** Diameter is 1 μm. **(b)** Diameter is 4 μm. Calcium influx occurs in the first 1-μm length of dendrite only. Each of 10 1 μm-long compartments contains four radial shells. Blue dotted line: calcium transient with only radial (and not longitudinal) diffusion in the first 1 μm-long compartment. All other model parameters as in Figure 6.5.

in parallel with radial diffusion between the shells. The submembrane calcium concentration drops rapidly with distance from the calcium source, which is 0.5 μm along the dendrite. For a dendrite of 1 μm diameter, peak calcium is only 34% of the amplitude in the source compartment at a distance of 1.5 μm along the dendrite, which is 1 μm from the source. For a 4 μm-diameter dendrite, the peak 1.5 μm along is only 24% of the source amplitude. However, the longitudinal diffusion clearly shapes the submembrane calcium transient in the source compartment. In Figure 6.11, compare the solid black lines with the dotted blue lines from the model without longitudinal diffusion. The peak concentration is reduced and the time course of calcium decay is significantly faster. Thus it may be necessary to include longitudinal diffusion to accurately capture the calcium transient resulting from a local calcium influx.

Under certain assumptions, it is possible to calculate a space constant for diffusion that is equivalent to the electrical space constant for voltage (Box 6.4). Typically the space constant of diffusion is much shorter than the electrical space constant, and decreases with decreasing diameter. Thus diffusion along very thin cables, such as the neck of a spine, around 0.1 μm in diameter, is often ignored in models. However, as illustrated in Figure 6.11, longitudinal diffusion will influence local calcium transients in cables of the thickness of typical dendrites and axons (of the order of 1 μm).

6.5.5 Numerical Calculation of Diffusion

The numerical solution of diffusion equations is a complex subject, the full details of which are beyond the scope of this book. A brief outline of numerical integration techniques is given in Section 5.7. Though this outline uses the membrane voltage equation as an example, this equation has the same form as the equation for the 1D diffusion of calcium. Details of the diffusional coupling coefficients that arise in various geometries of interest is given in Box 6.2. More detailed treatments of the numerical solution of the diffusion equation, including how it can be done in multiple dimensions, are to be found in De Schutter and Smolen (1998), Bormann et al. (2001) and Smith (2001).

6.5.6 Electrodiffusion and Electrogenesis

Throughout this chapter, we have been dealing with mechanisms that lead to the movement of ions, in this case calcium. Such movement generates

---

Box 6.2 | Diffusion coupling coefficients

The diffusional coupling coefficient from cellular compartment i to compartment j is:

$$c_{ij} = \frac{a_{ij}}{v_j \Delta_{ij}},$$

where a_{ij} is the surface area between compartments i and j, v_j is the volume of compartment j, and Δ_{ij} is the distance between the compartment centre points. These coefficients depend on the geometry being modelled and can take quite simple forms.

For axisymmetric radial diffusion into a cylinder of length Δx, consider two adjacent shells, both of thickness Δr, with the surface between them, the inner surface of shell i and the outer surface of cell j, having radius r. Then the surface area between the shells is $a_{ij} = 2\pi r \Delta x$. The volume of shell j is $v_j = \pi r^2 \Delta x - \pi (r - \Delta r)^2 \Delta x = \pi \Delta r (2r - \Delta r) \Delta x$. This results in the coupling coefficient:

$$c_{ij} = \frac{2r}{\Delta r^2 (2r - \Delta r)}.$$

For longitudinal diffusion along a cylinder of uniform radius r, the cross-sectional area between two compartments is always $a = \pi r^2$. If each compartment has length Δx, then the compartment volume is $v = a \Delta x$ and the coupling coefficient between any two compartments is:

$$c = \frac{a}{a \Delta x . \Delta x} = \frac{1}{\Delta x^2}.$$

---

electrical currents and, in turn, is affected by electrical fields (Chapter 2). Both of these effects have been neglected in our treatment of calcium fluxes. In the treatment of the electrical properties of a neuron, typically the actual movement of ions and resulting concentration gradients are ignored in calculating electrical current flow.

In many instances, it is reasonable to neglect the effects of potential gradients when dealing with calcium concentrations, and concentration gradients when dealing with electrical currents (De Schutter and Smolen, 1998). However, it is possible to include these effects in our models explicitly, and this is sometimes important. For example, electrical currents due to ionic pumps can be significant. In Box 6.3, we consider the electrodiffusion of ions. It should be noted that including the effects of electrodiffusion in a model adds significantly to the computational load, as well as to the model complexity. We consider electrodiffusion more generally in Chapter 10.

6.6 | Calcium Buffering

Calcium interacts with a variety of other molecules in the intracellular space. These molecules act as buffers to the free diffusion of calcium and can strongly affect the spatial and temporal characteristics of the free calcium concentration. Endogenous neuronal calcium buffers include calmodulin, calbindin and parvalbumin (Koch, 1999; Smith, 2001; Berridge et al.,

Box 6.3 | Electrodiffusion

Ions move down both their potential gradient and their concentration gradient. Changes in ionic concentrations in intracellular compartments also affect the local membrane potential. Therefore, electrical currents and ionic concentrations are intimately linked and ideally should be treated simultaneously. The effects of electrodiffusion have been treated in detail by Qian and Sejnowski (1989); see also useful discussions in De Schutter and Smolen (1998) and Koch (1999).

Consider longitudinal movement only of calcium along a cylindrical section of neurite. The calcium concentration as a function of time and space is properly derived from the Nernst–Planck equation (Koch, 1999), resulting in:

$$\frac{\partial [Ca^{2+}]}{\partial t} = D_{Ca}\frac{\partial^2 [Ca^{2+}]}{\partial x^2} + D_{Ca}\frac{z_{Ca}F}{RT}\frac{\partial}{\partial x}\left([Ca^{2+}]\frac{\partial V}{\partial x}\right) - J_{pump} - J_{leak},$$

where now the simple diffusion of calcium along the cylinder (Box 6.1) is augmented by drift down the potential gradient (the second term on the right-hand side) and by the transmembrane flux, $J_{pump} + J_{leak}$, due to voltage-gated ion channels and pumps (De Schutter and Smolen, 1998).

To solve this equation we need an expression for the membrane potential V as a function of local ionic concentrations. Each ionic species, including calcium, contributes an amount of electrical charge, depending on its concentration. The local membrane potential is determined by the total charge contributed by all ionic species X_i (Qian and Sejnowski, 1989):

$$V(x, t) = V_{rest} + \frac{rF}{2C_m}\sum_i z_i([X_i](x, t) - [X_i]_{rest})$$

for neurite cylindrical radius r, membrane capacitance C_m, ionic valence z_i and ionic concentrations $[X_i]$. For a typical intraneuronal environment, which ionic species should be included in this sum is debatable (De Schutter and Smolen, 1998). A full treatment of self-consistent schemes that include ion movement in both intracellular and extracellular spaces is given in Chapter 10.

Electrodiffusion can be implemented as a straightforward extension to standard compartmental modelling (Qian and Sejnowski, 1989), but it is computationally highly expensive. Concentration changes, particularly of calcium, can be large in small neural structures, such as spines (Box 2.3). Thus, it is important to use the GHK equation (Section 2.4) for computing membrane currents. However, given the small space constant of diffusion, implementation of the full electrodiffusion model may not be necessary (De Schutter and Smolen, 1998).

2003). Visualising intracellular calcium requires the addition of a fluorescent dye to a neuron, such as Fura-2. Such dyes bind calcium and are thus calcium buffers that also strongly affect the level of free calcium. Models of intracellular calcium based on fluorescent dye measurements of calcium transients need to account for this buffering effect. Other exogenous buffers include EGTA (ethylene glycol tetraacetic acid) and

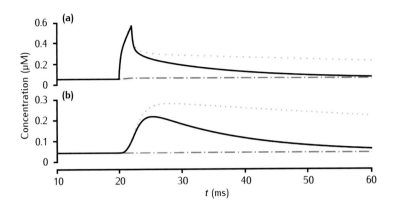

Fig 6.12 Calcium transients in **(a)** the 0.1 μm-thick submembrane shell and **(b)** the cell core of a single dendritic compartment of 4 μm diameter with four radial shells. Slow buffer with initial concentration 50 μM, forward rate $k^+ = 1.5\,\mu M^{-1}\,s^{-1}$ and backward rate $k^- = 0.3\,s^{-1}$. Blue dotted line: unbuffered calcium transient. Grey dashed line (hidden by solid line): excess buffer approximation (EBA). Grey dash-dotted line: rapid buffer approximation (RBA). Binding ratio $\kappa = 250$. All other model parameters as in Figure 6.5.

BAPTA (bis(aminophenoxy)ethanetetraacetic acid). Explicitly modelling these buffers is necessary for comparison of models with experiments in which such buffers are used.

The second-order interaction of calcium and a buffer in a single, well-mixed pool is given by the kinetic scheme:

$$Ca + B \underset{k^-}{\overset{k^+}{\rightleftharpoons}} CaB, \tag{6.33}$$

in which a single calcium ion binds with a single molecule of free buffer B to produce calcium-bound buffer CaB, with association and dissociation rate coefficients k^+ and k^-, respectively. This is the most basic buffering scheme. Common buffers, such as calmodulin, may contain multiple calcium binding sites, and so are better described by higher-order reaction schemes (Koch, 1999), but accurate estimates of all rate coefficients in such schemes are not readily available. In the following treatment, we consider only this basic second-order interaction.

Our intracellular model for calcium concentration now needs a term to account for the reaction with buffer. In addition, the model needs to account for the free [B] and bound [CaB] buffer concentrations. For a particular well-mixed cellular compartment, this leads to the system of coupled ODEs:

$$\frac{d[Ca^{2+}]}{dt} = -k^+[Ca^{2+}][B] + k^-[CaB] + J_{Ca},$$

$$\frac{d[B]}{dt} = -k^+[Ca^{2+}][B] + k^-[CaB] + J_{diffB}, \tag{6.34}$$

$$\frac{d[CaB]}{dt} = k^+[Ca^{2+}][B] - k^-[CaB] + J_{diffCaB}.$$

The calcium flux J_{Ca} lumps together all the possible sources and sinks for calcium, including diffusion between compartments. Both free and bound buffer may also diffuse with fluxes J_{diffB} and $J_{diffCaB}$, respectively. However, it is often assumed that free and bound buffer have the same diffusion coefficient D_{buf}, which is reasonable if the buffer molecules are much larger than calcium. This model can be easily extended to include multiple buffers with different characteristics. The cytoplasm may contain both fixed ($D_{buf} = 0$) and mobile ($D_{buf} \neq 0$) buffers.

The effect of buffers on calcium gradients can be drastic, depending on the buffer concentration and its binding rate. Figures 6.12 and 6.13 show the

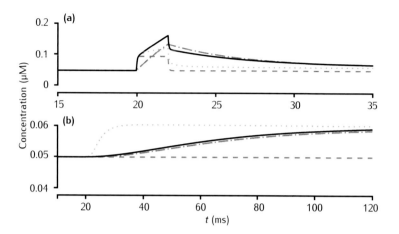

Fig 6.13 Calcium transients in (a) the 0.1 μm-thick submembrane shell and (b) the cell core of a single dendritic compartment of 4 μm diameter with four radial shells. Fast buffer with initial concentration 50 μM, forward rate $k^+ = 500\,\mu M^{-1}\,s^{-1}$ and backward rate $k^- = 1000\,s^{-1}$. Black solid line: fixed buffer. Blue dotted line: mobile buffer with diffusion rate $D_{buf} = 1\times 10^{-6}\,cm^2\,s^{-1}$. Grey dashed line: EBA. Grey dash-dotted line: RBA. Binding ratio $\kappa = 25$. All other model parameters as in Figure 6.5.

effects of both fixed and mobile buffers with slow and fast binding rates, respectively. The buffer is present in 50 μM concentration and the slow buffer has a higher affinity, with dissociation constant $K_B = k^-/k^+ = 0.2\,\mu M$, than the fast buffer, with $K_B = 2\,\mu M$.

The slow buffer reduces the calcium transient marginally in the submembrane shell and more significantly in the cell core. It also speeds up the return to resting levels, compared to the unbuffered case (Figure 6.12, blue dotted line). This situation is well captured by the excess buffer approximation, detailed in Section 6.6.1.

The fast buffer drastically reduces the peak amplitude and sharpens the submembrane calcium transient. However, now the return to equilibrium throughout the cell compartment is greatly slowed. The initial calcium influx is largely absorbed by the submembrane buffer. The slow equilibration throughout the compartment and return to resting calcium levels are dictated by the time course of unbinding of calcium from the buffer. The mobility of the buffer has little effect when the buffer is slow (not shown), but increases the apparent diffusion rate of calcium when binding is fast, resulting in a faster rise in calcium in the core compartment (Figure 6.13, blue dotted line). These fast buffer effects can be derived mathematically by considering a rapid buffer approximation, as described in Section 6.6.2.

Full modelling of buffering, which might include a number of different buffers and buffers with multiple binding sites, may involve a large number of differential equations per compartment. Some specific assumptions about buffers can result in greatly simplified models of buffering that require modifications only to the ODE for calcium and do not require explicit ODEs for buffer concentrations at all. Two important simplifications are the **excess buffer approximation** (EBA) and the **rapid buffer approximation** (RBA).

6.6.1 Excess Buffer Approximation

An approximation that may be valid for endogenous buffers and often is valid for exogenous buffers is that the buffer is in excess and cannot be saturated by incoming calcium (Smith, 2001). In the steady state, when there is no influx of calcium, then the concentrations of free calcium and unbound buffer will reach resting levels, with:

$$[B]_{res} = \frac{K_B[B]_{tot}}{K_B + [Ca^{2+}]_{res}}, \quad [CaB]_{res} = \frac{[Ca^{2+}]_{res}[B]_{tot}}{K_B + [Ca^{2+}]_{res}}, \tag{6.35}$$

where the dissociation constant $K_B = k^-/k^+$, and $[B]_{tot} = [B] + [CaB]$.

The EBA assumes that both $[B]$ and $[CaB]$ are constant, with values $[B]_{res}$ and $[CaB]_{res}$, even during calcium influx. Substituting these values into the first ODE in Equation 6.34 gives:

$$\frac{d[Ca^{2+}]}{dt} = -k^+[Ca^{2+}][B]_{res} + k^-[CaB]_{res} + J_{Ca}. \tag{6.36}$$

From Equation 6.35, it can be shown that $k^-[CaB]_{res} = k^+[Ca^{2+}]_{res}[B]_{res}$. Substituting this into Equation 6.36 gives:

$$\frac{d[Ca^{2+}]}{dt} = -k^+[B]_{res}([Ca^{2+}] - [Ca^{2+}]_{res}) + J_{Ca}. \tag{6.37}$$

This is the same as the simple calcium decay model (Equation 6.15) with $\tau_{dec} = 1/(k^+[B]_{res})$, and so this simpler model can be used.

Figure 6.12, in which the slow buffer is in excess, illustrates that this approximation can capture well the time course of the initial calcium transient as it enters a cellular compartment and binds to the buffer. However, it cannot capture the complex dynamics of calcium binding and unbinding from the buffer interacting with diffusion between cellular compartments, which is prominent with the fast buffer (Figure 6.13).

6.6.2 Rapid Buffer Approximation

A complementary approximation is that the buffering of calcium is so fast that the calcium and buffer concentrations are essentially always in equilibrium with each other, so that:

$$[B] = \frac{K_B[B]_{tot}}{K_B + [Ca^{2+}]}, \quad [CaB] = \frac{[Ca^{2+}][B]_{tot}}{K_B + [Ca^{2+}]}. \tag{6.38}$$

Calculating the differential of bound calcium with respect to free calcium gives the **calcium binding ratio** $\varkappa$:

$$\varkappa = \frac{d[CaB]}{d[Ca^{2+}]} = \frac{K_B[B]_{tot}}{(K_B + [Ca^{2+}])^2}, \tag{6.39}$$

which indicates how much calcium entering a cellular compartment becomes bound to the buffer. If calcium concentrations are much less than the dissociation constant K_B, this ratio is approximately $\varkappa \approx [B]_{tot}/K_B$.

In neuronal cells, the binding ratio $\varkappa$ is typically large, on the order of 20 or more, indicating that 95% or more of the calcium is bound to intracellular buffers. For the examples in Figures 6.12 and 6.13, for the slow buffer $\varkappa = 250$, and for the fast buffer $\varkappa = 25$.

For a fixed buffer, where $D_{buf} = 0$ and hence $J_{diffB} = 0$ and $J_{diffCaB} = 0$:

$$\frac{d[Ca^{2+}]}{dt} + \frac{d[CaB]}{dt} = J_{Ca}. \tag{6.40}$$

Using Equation 6.39, we can write:

$$\frac{d[CaB]}{dt} = \frac{d[CaB]}{d[Ca^{2+}]}\frac{d[Ca^{2+}]}{dt} = \varkappa \frac{d[Ca^{2+}]}{dt}. \tag{6.41}$$

Putting this into Equation 6.40 leads to:

$$\frac{d[Ca^{2+}]}{dt} = \frac{J_{Ca}}{1+x}.$$

(6.42)

The binding ratio x describes how rapid buffering attenuates the influence of calcium influx J_{Ca} on the free calcium concentration. The calcium influx includes the diffusion of free calcium between cellular compartments, and so the effective diffusion coefficient of free calcium is reduced to:

$$D_{eff} = \frac{D_{Ca}}{1+x}.$$

(6.43)

If the buffer can also diffuse, then it can be shown that this effective diffusion coefficient becomes approximately (Wagner and Keizer, 1994; Zador and Koch, 1994; De Schutter and Smolen, 1998; Koch, 1999; Smith, 2001):

$$D_{eff} = \frac{D_{Ca} + x D_{buf}}{1+x}.$$

(6.44)

Even if the buffer diffuses more slowly than calcium ($D_{buf} < D_{Ca}$), which is likely as the buffers are generally much larger molecules, the effective diffusion of calcium is increased by the movement of calcium bound to buffer molecules.

Figure 6.13 shows that the RBA can capture something of the fast submembrane transient and follows well the slow cell core concentration. It fails to capture the dynamics of calcium with a slow buffer, drastically overestimating the effect of the buffer (Figure 6.12).

Use of the RBA allows the definition of space and time constants for the calcium concentration in the presence of diffusion and buffering, in analogy to the electrical space and time constants. Details of these definitions are given in Box 6.4.

6.6.3 Calcium Indicator Dyes

Experimental data concerning calcium levels in cellular compartments come from fluorescence measurements using calcium indicator dyes in tissue slice preparations (Section 13.10.1). A model that seeks to match such results needs to include the binding of calcium to the indicator dye and how this affects the fluorescence of the dye. The RBA provides a simple model which can be used to extract information about peak calcium changes and endogenous buffer capacity from the imaging data.

Having matched the in vitro experimental results, the dye can be removed from the model. Now the model can be used to determine free calcium transients in response to a stimulus for the in vivo situation. This is ultimately the most interesting and explanatory use of a model, providing predictions about situations that cannot be explored experimentally.

Fluorescence measurements from calcium indicator dyes are taken at either one or two wavelengths, and at different temporal and spatial resolutions, depending on the dye and imaging equipment being used. We consider how to interpret a single wavelength fluorescence measurement, following the treatment of Maravall et al. (2000).

It is assumed that the fluorescence f is related linearly to the free calcium indicator dye concentration [F] and the calcium-bound dye concentration [CaF] via:

Box 6.4 | Space and time constants of diffusion

Longitudinal diffusion of calcium along a cylinder in the presence of a buffer is given by:

$$\frac{\partial[Ca^{2+}]}{\partial t} = D_{Ca}\frac{\partial^2[Ca^{2+}]}{\partial x^2} - k^+[Ca^{2+}][B] + k^-[CaB] + J_{Ca}.$$

If the RBA is assumed to hold, then the effect of the buffer is to alter the apparent diffusion rate of calcium, simplifying the expression for buffered diffusion of calcium to:

$$\frac{\partial[Ca^{2+}]}{\partial t} = D_{eff}\frac{\partial^2[Ca^{2+}]}{\partial x^2} + J_{Ca}.$$

Substituting in the effective diffusion rate given by Equation 6.44 and rearranging yields:

$$(1+\kappa)\frac{\partial[Ca^{2+}]}{\partial t} = (D_{Ca} + \kappa D_{buf})\frac{\partial^2[Ca^{2+}]}{\partial x^2} + J_{Ca}. \tag{a}$$

We now assume that the extra calcium flux J_{Ca} is provided by a steady calcium current, $J_{CC} = -aI_{Ca}/2Fv = -2I_{Ca}/4Fr$, and a non-saturated pump, $J_{pump} = aV_{max}[Ca^{2+}]/vK_{pump} = 2P_m[Ca^{2+}]/r$, both across the surface area of the cylinder, which has radius r. Substituting these into Equation (a) and rearranging slightly leads to an equation that is identical in form to the voltage equation (Zador and Koch, 1994; Koch, 1999):

$$\frac{r(1+\kappa)}{2}\frac{\partial[Ca^{2+}]}{\partial t} = \frac{r(D_{Ca} + \kappa D_{buf})}{2}\frac{\partial^2[Ca^{2+}]}{\partial x^2} - P_m[Ca^{2+}] - \frac{I_{Ca}}{2F}.$$

By analogy to the electrical space and time constants, the response of this system to a steady calcium current I_{Ca} in an infinite cylinder is an exponentially decaying calcium concentration with space constant:

$$\lambda_{Ca} = \sqrt{\frac{r(D_{Ca} + \kappa D_{buf})}{2P_m}},$$

and time constant:

$$\tau_{Ca} = \frac{r(1+\kappa)}{2P_m}.$$

Both space and time constants are functions of radius r. Whilst the time constant may be on the same order as the electrical time constant, depending on the diameter, the space constant is typically a thousand times smaller than the electrical space constant (De Schutter and Smolen, 1998; Koch, 1999).

$$f = S_F[F] + S_{FCa}[CaF] = S_F[F]_{tot} + (S_{FCa} - S_F)[CaF], \tag{6.45}$$

where the total dye concentration is $[F]_{tot} = [F] + [CaF]$ and the coefficients S_F and S_{FCa} specify the contribution from the dye's unbound and bound forms, respectively. The fluorescent intensity of saturated dye (i.e. maximally bound by calcium so that $[F]$ can be neglected compared to $[CaF]$) is $f_{max} = S_{FCa}[F]_{tot}$. The intensity in minimal calcium ($[CaF]$ neglected compared to $[F]$) is $f_{min} = S_F[F]_{tot}$.

Assuming that the RBA holds, following Equation 6.38, the concentration of calcium-bound dye is:

$$[CaF] = \frac{[Ca^{2+}][F]_{tot}}{K_F + [Ca^{2+}]},\tag{6.46}$$

with dissociation constant K_F. Given this approximation and the expressions for f, f_{max} and f_{min} above, the free calcium concentration can be derived as:

$$[Ca^{2+}] = K_F \frac{f - f_{min}}{f_{max} - f}.\tag{6.47}$$

In practice, it is difficult to measure f_{min} as calcium cannot be entirely removed from the experimental tissue slice preparation. Instead, this formulation is recast by introducing a new parameter, the dynamic range, $R_f = f_{max}/f_{min}$, of the indicator (Maravall et al., 2000). Substituting this into Equation 6.47 and dividing top and bottom by f_{max} gives:

$$[Ca^{2+}] = K_F \frac{f/f_{max} - 1/R_f}{1 - f/f_{max}}.\tag{6.48}$$

If R_f is large, as it is for dyes such as Fluo-3 and Fluo-4, then $1/R_f$ is much smaller than f/f_{max} and does not much influence the calculated value of $[Ca^{2+}]$. So reasonable estimates of $[Ca^{2+}]$ can be obtained by estimating f_{max} in situ and using calibrated values of R_f and K_F.

Rather than trying to estimate the free calcium concentration $[Ca^{2+}]$, more robust measurements of changes in calcium from resting conditions, $\Delta[Ca^{2+}] = [Ca^{2+}] - [Ca^{2+}]_{res}$, can be made. Now we make use of the change in fluorescence over the resting baseline, $\Delta f/f = (f - f_{res})/f_{res}$, which has the maximum value, $\Delta f_{max}/f = (f_{max} - f_{res})/f_{res}$. Using Equation 6.48 to give expressions for $[Ca^{2+}]$ and $[Ca^{2+}]_{res}$, substituting in $\Delta f/f$ and rearranging eventually leads to:

$$\Delta[Ca^{2+}] = K_F \left(1 - \frac{1}{R_f}\right) \frac{(\Delta f_{max}/f_{res} + 1)\Delta f}{(\Delta f_{max}/f_{res} - \Delta f/f_{res})\Delta f_{max}}.\tag{6.49}$$

By substituting $\Delta f_{max}/f_{res}$ into Equation 6.48 for $[Ca^{2+}]_{res}$ and rearranging, resting calcium can be estimated as:

$$[Ca^{2+}]_{res} = K_F \left[\frac{(1 - 1/R_f)}{\Delta f_{max}/f_{res}} - \frac{1}{R_f}\right].\tag{6.50}$$

Estimates of endogenous buffer capacity can also be made by measuring calcium changes resulting from a known stimulus in different concentrations of indicator dye (Maravall et al., 2000). If both the dye and the endogenous buffers are assumed to be fast, then:

$$\Delta[Ca^{2+}] = [Ca^{2+}]_{peak} - [Ca^{2+}]_{res} = \frac{\Delta[Ca^{2+}]_{tot}}{(1 + \varkappa_B + \varkappa_F)},\tag{6.51}$$

where $\varkappa_B$ ($\varkappa_F$) is the endogenous (dye) buffer capacity (calcium binding ratio). Given that $\Delta[Ca^{2+}]_{tot}$ is the same in each experiment, then the x-axis intercept of a straight-line fit to a plot of $\varkappa_F$ versus $1/\Delta[Ca^{2+}]$ will yield an estimate of $\varkappa_B$.

6.7 | Complex Intracellular Signalling Pathways

Up to this point, we have concentrated on modelling intracellular calcium, but the methods apply to modelling general intracellular signalling pathways. The deterministic modelling of well-mixed systems has been used to investigate the dynamics of a wide range of intracellular signalling cascades found in neurons (Bhalla and Iyengar, 1999; Bhalla, 2014), and particularly the temporal properties of the complex pathways underpinning long-term synaptic plasticity (Ajay and Bhalla, 2005). We consider particular signalling cascade models of synaptic plasticity in Chapter 11. In addition to mediating synaptic plasticity, chemical signalling cascades can play a variety of functional roles, such as identifying patterns of incoming spatially distributed synaptic activity that might need to be 'remembered' through synaptic plasticity (Kim et al., 2013; Bhalla, 2014). Such patterns may cover timescales from milliseconds through to hours.

A major aspect of much intracellular signalling is the quantitative determination of postsynaptic calcium transients resulting from particular synaptic stimulation protocols. Intracellular signalling pathways contribute to the time course and magnitude of calcium transients, which then drive the signalling pathways leading to changes in synaptic strength. These pathways mediate such aspects as the phosphorylation state and the number of membrane-bound neurotransmitter receptors in the postsynaptic density (PSD).

6.7.1 Example: Model of IP_3 Production

As we have seen, the calcium transient in a cellular compartment, such as a spine head, is the result of interaction between calcium fluxes through ion channels, extrusion of calcium by membrane-bound molecular pumps, binding of calcium by buffers and diffusion. An extra component is the production of IP_3 (Figure 6.14) which, together with calcium, leads to calcium release from internal stores by activation of IP_3 receptors in the ER membrane (Section 6.4.2). IP_3 is produced via a molecular cascade that begins with glutamate that is released in the synaptic cleft binding to membrane-bound metabotropic glutamate receptors (mGluRs). Activated mGluRs catalyse the activation of the Gq variety of G protein, which, in turn, activates phospholipase C (PLC). PLC then catalyses the production of IP_3 from

Fig 6.14 Signalling pathway from glutamate to IP_3 production, leading to calcium release from intracellular stores in the ER.

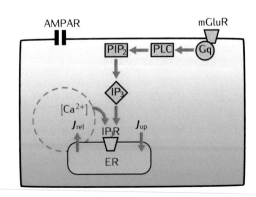

the membrane phospholipid phosphatidylinositol 4,5-bisphosphate (PIP_2) (Bhalla and Iyengar, 1999; Blackwell, 2005; Doi et al., 2005). The reaction pathways are:

$$mGluR + Glu \underset{k_1^-}{\overset{k_1^+}{\rightleftharpoons}} Glu.mGluR,$$

$$G_{\alpha\beta\gamma} + Glu.mGluR \underset{k_2^-}{\overset{k_2^+}{\rightleftharpoons}} GGlu.mGluR \xrightarrow{k_2^c} Glu.mGluR + G_{q\alpha},$$

$$PLC + G_{q\alpha} \underset{k_3^-}{\overset{k_3^+}{\rightleftharpoons}} PLC.G_{q\alpha},$$

$$PLC.G_{q\alpha} + PIP_2 \underset{k_4^-}{\overset{k_4^+}{\rightleftharpoons}} PLC.G_{q\alpha}.PIP_2 \xrightarrow{k_4^c} PLC.G_{q\alpha} + IP_3 + DAG,$$

(6.52)

where dots indicate molecular complexes. Including all complexes, this system contains 12 molecular species and requires the specification of 10 reaction rates. A complete model would also include the return to the resting state once the glutamate transient is over, by the degradation of IP_3 and inactivation of the G protein. The implementation of this model as a set of differential equations is detailed in Box 6.5.

These pathways can be combined with a model for the IP_3 receptors in the ER membrane (Section 6.4.2) to complete the path from glutamate in the synaptic cleft to calcium release from intracellular stores (Figure 6.14). Doi et al. (2005) have modelled this complete system in detail, including a seven-state model of IP_3 receptors, to investigate the magnitude and time course of calcium transients in an individual Purkinje cell spine head following

Specification and simulation of general intracellular signalling networks fall within the domain of **systems biology**. For computer implementation, signalling pathway models can be specified formally using the **Systems Biology Markup Language (SBML)**. Subsequently, computer software packages, such as *COPASI*, can read *SBML* files and simulate the specified model as a set of ODEs, or perhaps using stochastic methods (Section 6.8).

Box 6.5 | ODE model of IP_3 production

Under the assumptions that the spine head is well mixed and that all molecular species are present in abundance, the signalling system for IP_3 production can be modelled as the set of ODEs:

$$\frac{d[Glu.mGluR]}{dt} = k_1^+[mGluR][Glu] - k_1^-[Glu.mGluR]$$

$$\frac{d[GGlu.mGluR]}{dt} = k_2^+[Glu.mGluR][G_{\alpha\beta\gamma}] - (k_2^- + k_2^c)[GGlu.mGluR]$$

$$\frac{d[G_{q\alpha}]}{dt} = k_2^c[GGlu.mGluR]$$

$$\frac{d[PLC.G_{q\alpha}]}{dt} = k_3^+[PLC][G_{q\alpha}] - k_3^-[PLC.G_{q\alpha}]$$

$$\frac{d[PLC.G_{q\alpha}.PIP_2]}{dt} = k_4^+[PLC.G_{q\alpha}][PIP_2] - (k_4^- + k_4^c)[PLC.G_{q\alpha}.PIP_2]$$

$$\frac{d[IP_3]}{dt} = k_4^c[PLC.G_{q\alpha}.PIP_2].$$

One simplification would be to replace the equations relating to the enzymatic reactions with their steady-state fluxes, provided this is justified by the reaction rates (Blackwell, 2005). Note that equations for the concentrations of the remaining species should also be included, but could take the form of algebraic relationships, rather than ODEs, under the assumption of a closed system.

climbing fibre and parallel fibre activation. This single-compartment spine head model describes 21 different molecular species and contains 53 variables and 96 parameters. This relatively large-scale signalling pathway model leads to long computer simulation times. Yet it is still a small model compared to the complete set of signalling pathways in a cell. Computationally, quantitative modelling of signalling pathways in this way may be limited to around 200 different molecular species (Bhalla, 2004a). It is still impractical to model pathways at this level of detail over the full spatial extent of a neuron. Doi et al. (2005) were modelling just a single spine head, and it would be computationally prohibitive to incorporate this into a detailed compartmental model of a Purkinje cell, including a full complement of thousands of spines.

6.7.2 Parameter Estimation

Formulating models of intracellular mechanisms relies on information about protein–protein interactions, plus the likely form of chemical reactions (Bhalla, 1998, 2001). This determines the kinetic equations for the model and thus enables the resulting set of mathematical equations (mostly ODEs) to be solved. In addition, information is required about reaction rates and initial molecular concentrations. Much of this information will come from experiments in systems other than the one being modelled. In the worst case, much information will be unavailable, in particular, good estimates for reaction rates. Where values are not known, they must be assigned on the basis of reasonable heuristics or via parameter optimisation (Chapter 14).

Bhalla provides an excellent introduction to the availability and use of existing experimental data (Bhalla, 1998, 2001). Doi et al. (2005) give full details of where they obtained parameter values. They could not find values for 36 of their 96 parameters in the literature. In this situation, the model was completed using a sensible parameter estimation strategy.

In simple binding reactions (Equation 6.1), the dissociation constant $K_d = k^-/k^+$, and possibly the time constant of the reaction, may be available. If it is assumed that one of the two species A and B is present in excess, say B, then the reaction is approximately first order (depends only on the concentration of A) and the time constant of A being converted to AB is:

$$\tau = \frac{1}{k^+[\text{B}]_{\text{tot}} + k^-}. \tag{6.53}$$

If the concentration $[\text{B}]_{\text{tot}}$ is known for the experiment in which the time constant was measured, then this expression plus that for K_d allows determination of the forward and backward rate coefficients k^+ and k^-.

For enzymatic reactions (Equation 6.6), values for $K_m = (k^- + k^c)/k^+$ and $V_{max} = k^c[\text{E}]_{\text{tot}}$ may be available. If the full enzymatic reaction is to be modelled, rather than the steady-state Michaelis–Menten flux from substrate to product (Equation 6.11), then at least one rate will have to be assumed. The forward rate k^c can be determined from V_{max}, given a known enzyme concentration, but either k^+ or k^- needs to be assumed to allow determination of the other from the value for K_m. Kuroda et al. (2001) assume that k^- is 2–20 times greater than k^c, on the basis that k^- is greater than k^c in many such reactions. Bhalla typically uses a scaling of four ($k^- = 4k^c$), having discovered that many models are highly insensitive to the exact ratio of k^- to k^c (Bhalla, 1998, 2001).

Parameter Optimisation

After as many parameters as possible have been assigned values from experimental estimates, it still remains to determine values for all other parameters in a sensible fashion. Most likely this will be done by trying to minimise the difference between the model output and experimental data for the temporal concentration profiles of at least a few of the molecular species involved in the model. This may be attempted in a simple heuristic fashion (Doi et al., 2005), which is acceptable if the number of unknown parameters is reasonably small and only a qualitative fit to experimental data is required. Otherwise, mathematical optimisation techniques that will adjust parameter values to minimise the error between the model output and the data should be used (Arisi et al., 2006).

Further confidence in the model can be obtained by carrying out a thorough sensitivity analysis to determine the consequences of the choices made for parameter values. Doi et al. (2005) varied each parameter value over two orders of magnitude, whilst keeping all other values fixed, and compared the model outputs with a known experimental result that was not used during parameter estimation. This procedure indicates which particular model components (and hopefully their biochemical equivalents!) most strongly determine the system output. Further details of parameter optimisation and sensitivity analysis are given in Chapter 14.

6.7.3 Rule-Based Modelling

Computational complexity limits the size of signalling models to those covering only on the order of at most a few hundred molecular species. However, biochemical signalling networks can readily suffer from combinatorial complexity that can make full specification and then computation of all the reaction pathways simply infeasible (Stefan et al., 2014; Tapia et al., 2019). Not only may a network involve many species, but molecules of a particular species may be present in different states, such as being phosphorylated or not. As the number of conformations and phosphorylation sites on a molecule increases, the number of meaningful configurations (states) increases combinatorially.

For example, the calcium/calmodulin-dependent kinase II (CaMKII) holoenzyme, which is found in synaptic spines and involved in synaptic plasticity, comprises 12 subunits, arranged in two opposing hexamer rings (Lisman et al., 2012). In a relatively simple model of CaMKII, each subunit has two phosphorylation sites which can be phosphorylated or unphosphorylated, can be in at least two conformations (open or closed) and can be bound or unbound to calmodulin molecules. There are thus 16 possible combinations of phosphorylation, conformation and binding states in each subunit. Using the necklace function (Weisstein, 2017) to compute the number of configurations of 12 'beads' in one of 16 possible 'colours', we can estimate that there are around 10^{16} configurations of 12 subunits, each of which can be in 16 possible states. This is before we consider generating all the possible reactions between these species – Michalski and Loew (2012) estimated that it would take 290 years on a contemporary desktop computer to generate all possible reactions between the states in a model of CaMKII with 10 subunits, each with six states.

This combinatorial problem can be alleviated by model simplification, though the strategy required will depend on what is being modelled. For example, in the case of CaMKII, we might reduce the number of subunits or lump together states. A general approach to this is **rule-based modelling** (Faeder et al., 2009; Stefan et al., 2014; Chylek et al., 2015; Tapia et al., 2019), in which a set of reaction rules are specified, rather than all of the explicit reactions between the molecular species in all of their possible states. It relies on being able to make simplifying assumptions, such that for a particular reaction, only certain aspects of the states of the molecules involved actually matter. So it is possible to specify a single reaction rule with a given rate that applies for multiple molecular states (Stefan et al., 2014).

To give an example, consider protein S that can be phosphorylated and bound to another protein P at two sites (Chylek et al., 2015). In such a system, there can be up to nine different species corresponding to distinct states of phosphorylation and binding, with 12 possible reactions. Using subscripts u, p and b for unmodified, phosphorylated and bound, respectively, at each of the two sites on S, these reactions are:

$$
\begin{array}{lll}
S_{uu} \xrightarrow{r_1} S_{pu} & S_{up} \xrightarrow{r_5} S_{pp} & S_{pp} \xrightarrow{r_9} S_{pb} \\
S_{uu} \xrightarrow{r_2} S_{up} & S_{up} + P \xrightarrow{r_6} S_{ub} & S_{ub} \xrightarrow{r_{10}} S_{pb} \\
S_{pu} + P \xrightarrow{r_3} S_{bu} & S_{bu} \xrightarrow{r_7} S_{bp} & S_{bp} + P \xrightarrow{r_{11}} S_{bb} \\
S_{pu} \xrightarrow{r_4} S_{pp} & S_{pp} \xrightarrow{r_8} S_{bp} & S_{pb} + P \xrightarrow{r_{12}} S_{bb}.
\end{array}
\tag{6.54}
$$

If we assume that the rates of phosphorylation and binding at each of the two sites are the same and independent of the state of the other site, then this set of 12 reactions reduces to four reaction rules (where x designates an unspecified site state):

$$
\begin{array}{ll}
S_{ux} \xrightarrow{k_1} S_{px} & S_{px} + P \xrightarrow{k_2} S_{bx} \\
S_{xu} \xrightarrow{k_3} S_{xp} & S_{xp} + P \xrightarrow{k_4} S_{xb}.
\end{array}
\tag{6.55}
$$

Specification and Simulation of Rule-Based Models

Many computer languages have been developed to allow the specification of rule-based models and their subsequent simulation. Second-generation **rule-based modelling** languages, such as *Kappa* (Boutillier et al., 2018) or *BioNetGen Language* (*BNGL*; Faeder et al., 2009), have a well-defined, general syntax to specify binding sites and states of proteins and interactions between protein binding domains.

Simulation software, such as *BioNetGen* (Faeder et al., 2009), can expand the set of interaction rules describing a model (e.g. Equation 6.55) to generate the biological reaction network, that is, the full set of complexes and reactions needed to simulate the system (e.g. Equation 6.54; Faeder et al., 2009). These reactions then can be converted into ODEs or stochastic differential equations (SDEs), for solving numerically or simulated using a stochastic simulation method (Section 6.8).

The computation time and computer memory requirements needed to specify the full set of reactions may be prohibitive (Stefan et al., 2014), in which case, reaction network generation can proceed 'on the fly', starting with only generating the reactions involving the initial seed species and extending the network as necessary during a simulation as new species are created (Faeder et al., 2009; Stefan et al., 2014).

6.7.4 Beyond Mass Action in Well-Mixed Systems

The two basic assumptions of the models presented, namely that the cellular compartment is well mixed and that all molecular species are present in abundance, are both likely to be false for many situations of interest, including synaptic plasticity in spine heads. As already considered for calcium, gradients of diffusion can be modelled by subdividing a cellular compartment into smaller compartments between which molecules diffuse. This allows heterogeneous concentration profiles across the entire spatial system.

On the other hand, modelling a small number of molecules requires completely different modelling strategies that take into account the precise number of molecules present in a cellular compartment. Changes in the number of a molecular species will happen stochastically due to reaction with other molecules and diffusion through space. Techniques for stochastic and spatial modelling are introduced in Sections 6.8 and 6.9, respectively.

6.8 Stochastic Models

So far we have considered models that calculate the temporal and spatial evolution of the concentration of molecular species. Concentration is a continuous quantity that represents the average number of molecules of a given species per unit volume. The actual number of molecules in a volume is an integer that is greater than, or equal to, zero. Chemical reactions take place between individual molecules in a stochastic fashion, determined by their random movements in space and their subsequent interactions. The smooth changes in concentration predicted by mass action kinetics (Section 6.1.1) as two molecular species react to create a third product species masks the reality of fluctuating numbers of all three species, even at apparent equilibrium. For small numbers of molecules, these fluctuations can be large relative to the mean. This may be crucial for a bistable system that, in reality, may randomly switch from one state to another when mass action kinetics would predict that the system is stable in a particular state.

If we are modelling low concentrations of molecules in small volumes, for example, calcium in a spine head, then the number of molecules of a particular species may be very small (tens or less). The assumption of the law of mass action is then unreasonable and a different modelling approach is required which takes into account the actual number of molecules present, rather than their concentration. Individual molecules undergo reactions and we may treat the reaction kinetics as describing the probability that a particular reaction, or state transition, may take place within a given time interval, resulting in a stochastic model.

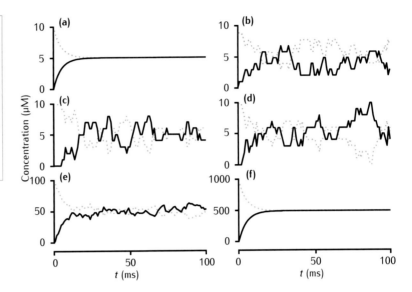

Fig 6.15 Simulations of a simple reversible reaction of S_1 (blue, dotted lines) being converted to S_2 (black, solid lines). Forward rate $k_f = 0.1$; backward rate $k_b = 0.1$. **(a)** Mass action model with $[S_1]_0 = 10 \, \mu M$; Stochastic model with initially **(b)–(d)** 10 molecules, **(e)** 100 molecules and **(f)** 1000 molecules of S_1.

Comparisons of deterministic (mass action) modelling with stochastic modelling of the simple reaction of a substrate S_1 going reversibly to a product S_2 are shown in Figure 6.15. The stochastic simulations were done using the Stochastic Simulation Algorithm (Section 6.8.1). When there are initially only 10 molecules of S_1, each simulation of the stochastic evolution of molecules of S_1 to molecules of S_2 produces large fluctuations in the amounts of S_1 and S_2, even in the final steady state. Thus the mass action model does not capture the true and continuing dynamics of this system. The fluctuations in S_1 and S_2 could have profound consequences for their ability to take part in further downstream reactions. The deterministic and stochastic models are indistinguishable when there are initially 1000 molecules of S_1 (Figure 6.15a, f).

Two approaches can be taken to create a stochastic model that accounts for the fluctuations in the amount of each molecular species over time. **Population-based methods** track the number of molecules of each species over time. Alternatively, **particle-based methods** track the existence of individual molecules. The latter are more flexible, but potentially are more expensive computationally.

6.8.1 Population-Based Methods

At the population level, the time evolution of a reacting system with multiple molecular species is described by the **chemical master equation** (CME; Gillespie, 1977) which specifies the probability distribution of the state of the system at any time, $P(X_1, \ldots, X_N; t)$, where the state of the system is given by the number X_i of molecules of each of N species S_i. Whilst generally it is possible to write down the master equation for a given system, it usually turns out to be intractable to solve analytically (Gillespie, 1977). Instead, Monte Carlo simulation techniques (introduced in Section 4.7.2) can be used to simulate how the system state changes over time. Each new simulation will produce a different time evolution of the state due to the stochastic nature of reactions occurring. We now outline these techniques.

Exact Methods

Gillespie (1977) developed a Monte Carlo scheme, called the **Stochastic Simulation Algorithm (SSA)**, that produces trajectories that are provably drawn from the distribution described by the master equation, but no explicit knowledge of the master equation is required. Such schemes are known as exact stochastic methods. The approach is to use random numbers to determine: (1) at what time in the future the next reaction occurs; and (2) which reaction happens next.

Suppose our system contains N molecular species S_i and there are M reaction pathways R_j between the species. For example, a system with five species and two reaction pathways might be:

$$
\begin{aligned}
R_1 &: S_1 + S_2 \xrightarrow{c_1} S_3, \\
R_2 &: S_1 + S_4 \xrightarrow{c_2} S_5.
\end{aligned}
\tag{6.56}
$$

To determine which reaction happens next and when, we need an expression for the reaction probability density function $P(\tau, \mu)$. Given that the number of molecules of each reaction species (system state) at time t is $(X_1, \ldots, X_N)$, $P(\tau, \mu)\Delta t$ is the probability that the next reaction will be R_μ and it will occur an infinitesimally short time after $t + \tau$.

Firstly, the probability that reaction R_μ will occur in the next infinitesimal time interval is $a_\mu \Delta t = h_\mu c_\mu \Delta t$ where h_μ is the number of distinct R_μ molecular reactant combinations in the current state (e.g. $h_1 = X_1 X_2$ where X_1 and X_2 are the current number of molecules of S_1 and S_2, respectively, in our example system). Then:

$$
P(\tau, \mu)\Delta t = P_0(\tau) a_\mu \Delta t,
\tag{6.57}
$$

where $P_0(\tau)$ is the probability that no reaction occurs in the time interval $(t, t + \tau)$. It can be deduced that this probability is exponentially distributed and is given by:

$$
P_0(\tau) = \exp\left(-\sum_{j=1}^{M} a_j \tau\right) = \exp(-a_0 \tau).
\tag{6.58}
$$

This gives us what we need to implement an algorithm for Monte Carlo simulation of state trajectories. A step-by-step approach is given in Box 6.6. The procedure is straightforward to implement in a computer simulation and produces stochastic time evolutions of the chemical system with exact reaction times. Computation time scales linearly with the number of reaction pathways M, as do deterministic, concentration-based models. This is not a problem for the simple examples we have given here, but can be if modelling a system that contains, say, proteins with multiple states due to, for example, conformational changes, with each state reacting in slightly different ways. Such systems can contain millions of reaction pathways (Firth and Bray, 2001). Computation also scales linearly with the volume of the system, as the larger the volume, the more molecules there are, and hence reactions take place more frequently. Various techniques have been employed to make the implementation of stochastic methods as efficient as possible (Gillespie, 1977, 2001; Gibson and Bruck, 2000).

The reaction constant c_j is related to, but not identical with, the reaction rate k_j in the deterministic, concentration-based description of such a system (Gillespie, 1977). If our system is spatially homogeneous, contained in a space with volume v, and is in thermal equilibrium, then $c_j \Delta t$ is the average probability that a particular combination of reactant molecules will react according to reaction pathway R_j in the next infinitesimal time interval Δt; for example, one molecule of S_1 reacts with one molecule of S_2 to form one molecule of S_3.

A number of computer software packages for simulating intracellular signalling pathways incorporate both exact and approximate population-based stochastic methods. Blackwell and Koh (2019) provide an overview of both software and techniques. Popular simulators include *COPASI* and *STEPS* (Hepburn et al., 2012). They can combine stochastic methods with deterministic ODEs to provide adaptive simulations.

Box 6.6 | Stochastic Simulation Algorithm

A Monte Carlo simulation of Gillespie's exact method proceeds in the following way:

Step 0 Initialisation. Specify the M reaction constants c_j and the N initial molecular counts X_i. Set time t to zero.

Step 1 Calculate the M reaction probabilities, given the current state of the system $a_j = h_j c_j$, where h_j depends on the molecular counts of each reaction species.

Step 2 Generate a random number r_τ from a uniform distribution on the interval $[0, 1]$. Use this number to calculate the time of the next reaction:

$$\tau = (1/a_0)\ln(1/r_\tau).$$

This is the same as drawing from the probability density function $P_\tau = a_0 \exp(-a_0\tau)$, where a_0 is the sum of the M values, a_j.

Step 3 Generate another random number r_μ from the same uniform distribution and use this to calculate which reaction R_μ happens at time τ by selecting μ to be the integer for which:

$$\frac{1}{a_0}\sum_{j=1}^{\mu-1} a_j < r_\mu \le \frac{1}{a_0}\sum_{j=1}^{\mu} a_j.$$

This is the same as drawing from the probability density function $P_\mu = a_\mu/a_0$.

Step 4 Increase time t by τ and adjust the molecular counts X_i in accordance with the occurrence of reaction R_μ. For our example system (Equation 6.56), if R_1 occurred, X_1 and X_2 would be decremented by 1 and X_3 would be incremented by 1 to adjust the counts for species S_1, S_2 and S_3, respectively.

Step 5 Return to Step 1.

Approximate Methods

An alternative approach, which is closer in principle to the numerical solution of deterministic models, is to calculate the system's time evolution on a per time-step basis. Such an approach provides only an approximation of the master equation as its accuracy depends on the size of time-step used, Δt. One advantage is that it allows the formulation of adaptive methods that switch between stochastic and deterministic solutions (Vasudeva and Bhalla, 2004). We now outline such an adaptive method.

A per time-step algorithm follows straightforwardly from the exact method (Vasudeva and Bhalla, 2004). We calculate the probability that a particular reaction R_μ takes place in a finite (but small) time interval $(t, t + \Delta t)$. This is given as 1 minus the probability that the reaction does not take place:

$$P_\mu(\Delta t) = 1 - \exp(-a_\mu \Delta t). \tag{6.59}$$

For Δt sufficiently small, a useful approximation provides:

$$P_\mu(\Delta t) \approx 1 - (1 - a_\mu \Delta t) = a_\mu \Delta t. \tag{6.60}$$

The algorithm proceeds by testing for the occurrence of each of the M reactions and adjusting the molecular counts accordingly on each time-step. The occurrence of a reaction is tested by generating a uniform random number r_μ from the unit interval and then testing if $r_\mu < P_\mu$. If the test is true, then it is assumed that a single instance of reaction R_μ has taken place in interval $(t, t + \Delta t)$.

6.8.2 Particle-Based Methods

A limitation of population-based methods is that the identity of individual molecules is not included, only the species to which they belong. Thus it is not possible to track over time the different conformational or phosphorylation states of a particular molecule. Population-based methods require maintaining separate populations for each state of each molecular species and can run into the problem of combinatorial explosion in the number of different populations to be represented.

An alternative is to explicitly model individual molecules using **particle-based methods**. Each molecule (particle) may be in a particular state, such as phosphorylated or not, and the states of individual molecules, rather than populations of molecules, are followed through the simulation (Stefan et al., 2014). For example, Michelson and Schulman (1994) simulated a CaMKII holoenzyme having a variable in each subunit that described which of five states the subunit was in. Transition probabilities between a subunit's states depended on its own state and that of its neighbouring subunit.

Particle-based methods alone do not solve the problem of how to represent the states and the transitions between states clearly and concisely (Stefan et al., 2014). Rule-based modelling (Section 6.7.3) facilitates the specification of transitions between states: 'rules' are specified in which the state of a fragment of the system is mapped to the transitions that can occur within that fragment. For example a CaMKII monomer may be phosphorylated when both it and its neighbour (the fragment) are bound to the Ca^{2+}–CaM complex (Stefan et al., 2012).

Particle-Based Simulation

Particle-based simulations can be carried out using SSA-type algorithms in which reaction probabilities are calculated on the basis of the number of molecules of each reaction species present at the current time (Danos et al., 2007; Sneddon et al., 2011; Boutillier et al., 2018). Unlike in population-based methods, once the next reaction is selected, a new step is required to select which individual molecules actually partake in the reaction. If the location of molecules within the simulated volume is not relevant, then molecules of the required species may be selected at random and then they react together according to the probability determined by the applicable reaction rate. This is the approach taken in simulators such as *KaSim* and *NFSim*.

If molecules may move through space (Section 6.9), then simulations must be carried out differently to account for movement and the effect of location on reaction probabilities. Simulations now proceed at a small, fixed time-step, with the state of each individual molecule, including its position and propensity to react, being updated (Kerr et al., 2008; Andrews, 2017). Positions of individual molecules may be adjusted at each step according to

Particle-based simulators include *KaSim* (Boutillier et al., 2018), *NFSim* (Sneddon et al., 2011), *MCELL* (Stiles and Bartol, 2001; Kerr et al., 2008) and *Smoldyn* (Andrews et al., 2010; Andrews, 2017) These simulators may use rule-based models specified with *Kappa* or *BNGL* (Section 6.7.3). An overview of available particle-based simulators is provided by Andrews (2018).

diffusive motion. They can only react if, during their movement, they collide according to a defined collision volume. Collisions between molecules thus are determined at each simulation step, with colliding molecules reacting with any applicable reaction probability. This spatial approach is taken in simulators such as *MCELL* and *Smoldyn*.

In combination with rule-based modelling, particle-based methods allow for **'network-free' simulation** in which the underlying biological reaction network is not explicitly generated (Faeder et al., 2009; Sneddon et al., 2011; Stefan et al., 2014; Tapia et al., 2019). Instead simulators create the molecular complexes that exist throughout a simulation dynamically. Now reactions are replaced by 'rules', with the probability of a rule applying being calculated from the molecular species present at the current time and the reactions specified by the rule (Danos et al., 2007; Sneddon et al., 2011). Thus only rules that can apply are selected at any update time and only the reactions embedded in a rule need to be specified explicitly when needed. Network-free methods avoid the prohibitive memory requirements needed to store all possible states in a large network (Sneddon et al., 2011) and even allow simulations with infinite numbers of potential species (Danos et al., 2007). All the particle-based simulators mentioned here can support rule-based models and 'network-free' simulation.

6.8.3 Adaptive Stochastic–Deterministic Methods

As highlighted above, the computation time for stochastic methods increases with the number of molecules in the system. However, as this number increases, the solution obtained approaches that of the deterministic concentration-based models. Neurons incorporate both large (cell bodies) and small (spine heads) volumes, and chemical signals, such as calcium concentrations, can change rapidly by an order of magnitude or more.

Efficient modelling of intracellular signalling within a neuron may be achieved by a combination of stochastic and deterministic methods, with automatic switching between methods when required. One approach is as follows (Vasudeva and Bhalla, 2004). Consider the reaction:

$$S_1 + S_2 \xrightarrow{c_1} S_3. \tag{6.61}$$

A deterministic solution can be written in terms of the number of each molecular species, where the number is treated as a continuous variable:

$$\frac{dX_1}{dt} = -c_1 X_1 X_2, \tag{6.62}$$

where X_1 (X_2) is the number of molecules of species S_1 (S_2). If a simple Euler integration scheme with time-step Δt is used to solve this equation numerically, the change in the number of molecules of S_1 in a single time-step is given by:

$$\Delta X_1 = -c_1 X_1 X_2 \Delta t = -P_{S_1}, \tag{6.63}$$

where we call P_{S_1} the propensity for species S_1 to react. This equation can be used to update deterministically the number of molecules of S_1 at each time-step.

If we divide X_1 by the volume v in which the molecules are diluted, this is then a particular implementation of a concentration-based ODE model. However, in this formulation, it is clear that when X_1 is small, say less than 10, it is a rather crude approximation to the real system. For example, does $X_1 = 1.4$ mean there are actually one or two molecules of S_1 present?

When P_{S_1} is significantly less than 1, it is equivalent to the probability that a molecule of S_1 reacts, used in the per time-step stochastic method outlined above, and could be used in this way. But P_{S_1} may also be much greater than 1, indicating that more than one molecule of S_1 may react in a single time-step. In this case, we cannot treat P_{S_1} as a probability and it is more reasonable to update X_1 deterministically according to Equation 6.63. These considerations lead naturally to an algorithm that updates X_1 either stochastically or deterministically, depending on the magnitude of P_{S_1}:

> If $P_{S_1} < P_{\text{thresh}} \ll 1$ update X_1 with probability P_{S_1},
> otherwise use the deterministic update, Equation 6.63.

This adaptive algorithm must also include a scheme for switching between continuous and discrete values for the 'number' of molecules for each species (Vasudeva and Bhalla, 2004). One scheme for going from a continuous to an integer number is to round up or down probabilistically, depending on how far the real number is from the lower and upper integers. This can eliminate bias that might be introduced by deterministic rounding to the nearest integer (Vasudeva and Bhalla, 2004).

In our example above, the situation could arise in which species S_1 is in abundance (X_1 is large), but there are only a small number of molecules of S_2. This could result in a large propensity P_{S_1} suggesting a deterministic update of the molecular numbers. However, for accuracy, X_2, the number of molecules of S_2, should really be updated stochastically. Thus, in addition to checking the propensity against a threshold value, molecular counts should also be checked against a minimum X_{min} (e.g. 100), below which they should be updated stochastically. A more complicated update is now required, as a large propensity cannot be treated simply as a probability measure.

Noting that P_{S_1} is also the propensity for a molecule of S_2 to react, if P_{S_1} is greater than 1, then it is divided into integer and fractional parts; for example, $P_{S_1} = 1.25$ is split into $P_{\text{int}} = 1$ and $P_{\text{frac}} = 0.25$. These parts are used in the update rule (Vasudeva and Bhalla, 2004):

> Choose a random number r uniformly from the interval $[0, 1]$.
> If $r < P_{\text{frac}}$ then $\Delta X_2 = -P_{\text{int}} - 1$,
> otherwise $\Delta X_2 = -P_{\text{int}}$.

For example, suppose $X_1 = 540$, $X_2 = 9$ and $P_{S_1} = 1.25$. Then X_1 would be decremented by 1.25, but 75% of the time, X_2 would be decremented by 1, and 25% of the time by 2 (but ensuring that X_2 does not go below 0).

6.9 Spatial Modelling

The modelling methods described so far assume that molecules are within a well-stirred, spatially homogeneous environment. However, the cellular environment is not homogeneous; for example, at a synapse on a dendritic spine, calcium enters through NMDA receptors on one side of the spine head. It can react with buffers on a shorter timescale than it takes to diffuse through the spine, and can exist within microdomains around the receptors briefly at high concentrations. Thus, to address some questions, it is necessary to model space explicitly.

The extension to a system in which at least some of the molecular species are subject to diffusion follows straightforwardly from the consideration of diffusible calcium and associated buffers, detailed above, when mass action kinetics can be assumed. The equations describing changes in molecular species that are mobile will include diffusive terms and constitute a reaction–diffusion system.

In situations in which deterministic calculations of mass action kinetics and diffusion are not appropriate, then movement of individual molecules, as well as their reaction, must be considered. Bulk diffusion is derived from the average movement of molecules subject to Brownian motion and thus undertaking random walks in space (Koch, 1999; Bormann et al., 2001). For population-based stochastic methods, this requires the extension of the CME (Section 6.8.1) to a **reaction–diffusion master equation** (RDME; Stundzia and Lumsden, 1996).

6.9.1 Compartmental Approaches

Movement of molecules can be approximated by restricting movement to diffusion between well-mixed pools across an intervening boundary (Bhalla, 2004b, c; Blackwell, 2006). This movement can be represented simply by another reaction pathway in which the 'reaction' is the movement of a molecule from one pool to another. This is the approach described earlier for the simple deterministic model of calcium, in which it may diffuse from a submembrane shell to the cell core. The only 'movement' then is across the barrier between two cellular compartments, and not within each compartment. This is analogous to compartmental modelling of membrane voltage (Chapter 5), in which electrical current flows through a resistor between two isopotential compartments.

Consider two cellular compartments, labelled 1 and 2, with volumes v_1 and v_2, separated by a distance Δx, with a boundary cross-sectional area a. Diffusion of molecular species A between these compartments can be described by the reaction pathway:

$$A_1 \underset{c_2}{\overset{c_1}{\rightleftharpoons}} A_2, \tag{6.64}$$

with rates $c_1 = aD_A/v_1\Delta x$ and $c_2 = aD_A/v_2\Delta x$ where D_A is the diffusion coefficient for A (Bhalla, 2004b). The adaptive algorithm described above can be employed such that the change in the number of molecules of A in compartment 1 over some small time-step Δt is given by:

$$\Delta X_1 = (-c_1 X_1 + c_2 X_2)\Delta t. \tag{6.65}$$

An alternative approach is to consider the probability that a molecule may leave a cellular compartment due to its Brownian motion. If such a molecule may only move along a single dimension x and the compartment has length Δx, then the probability that a molecule of A will leave the compartment in a short time interval Δt is (Blackwell, 2006):

$$P_A = 2D_A \frac{\Delta t}{(\Delta x)^2}. \qquad (6.66)$$

Of all the molecules of A that leave the compartment, half will move forwards and half will move backwards. The remaining molecules of A will not leave the compartment. By a single sampling of the corresponding trinomial distribution, it is possible to determine the number of molecules of A moving forwards, backwards or not moving at all (Blackwell, 2006). This approach can readily be extended to movement in two or three dimensions.

The combination of this movement across compartment boundaries with SSA simulation (Section 6.8.1) of reaction between populations of molecules within well-mixed compartments is known as 'Spatial SSA' or 'Inhomogeneous SSA' (ISSA; Blackwell and Koh, 2019).

6.9.2 Particle-Based Diffusion Approaches

A more general approach is to simulate the movement and interaction of individual molecules through 2D or 3D space as features of particle-based simulation (Section 6.8.2). Brownian motion of molecules can be specified as a probability distribution of the direction and distance a molecule may move in a fixed time-step. Such probability distributions are used in Monte Carlo simulations which adjust the position in space of diffusing molecules in a probabilistic fashion at each time-step. No discrete spatial grid is required, as each molecule may occupy any possible 3D position within an unrestricted volume.

Typically such diffusion takes place within some surrounding surface structure such as the cell membrane. Spatial models require the definition of these surfaces, defining, for example, the shape and extent of presynaptic boutons or postsynaptic spines. Surfaces may reflect, absorb or transmit colliding molecules. They may contain receptor molecules that can bind diffusing molecules, should they collide with them, for example, neurotransmitter diffusing across a synaptic cleft binding with postsynaptic receptors.

Thus the general diffusion approach not only requires the calculation of each molecule's trajectory through free space, but must also determine whether a molecule collides with a surface or another diffusing molecule and what happens as a result. This can include restriction in its movement or undergoing a chemical reaction.

Though individual molecules are likely to be very small, compared to the space in which they are moving, if there are sufficient molecules in a restricted spatial compartment, then consideration should be given to the volume occupied by the molecules and the potential impact on their diffusion rate of such molecular crowding (Andrews, 2017, 2018).

Particle-based spatial simulators, such as *MCELL* and *Smoldyn*, implement these features, allowing complex spatial geometries for membrane surfaces to be defined, such as anatomically reconstructed synaptic clefts (Stiles and Bartol, 2001; Coggan et al., 2005; Sosinsky et al., 2005). An example *MCELL* simulation of a frog neuromuscular junction is illustrated in Figure 6.16 (Dittrich et al., 2013).

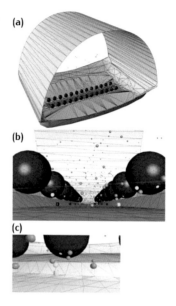

Fig 6.16 Particle-based simulation of a frog neuromuscular junction, visualised and simulated using *MCELL*. **(a)** Complete model showing a double row of synaptic vesicles (red spheres). More detail is shown in **(b–c)**: small spheres are diffusing calcium ions; cylinders are voltage-gated calcium channels (red, closed; yellow, open); dots on the underside of vesicles are calcium sensor sites (black, unbound; yellow, bound). From Dittrich et al. (2013). Reprinted with permission from Elsevier Inc.: *Biophysical Journal* 104, 2751–2763, © 2013.

6.10 | Summary

A model that recreates typical neuronal electrical behaviour may well need to include intracellular signalling pathways, particularly those involving calcium. In this chapter, we have detailed how the intracellular calcium concentration can be modelled. This involves mathematical descriptions for the variety of sources and sinks for calcium in a cellular compartment. These include voltage-gated calcium channels, membrane-bound pumps, calcium buffers and diffusion.

The techniques for modelling calcium are quite general and provide the tools for modelling other reaction–diffusion systems and intracellular signalling pathways. Examples of complex pathways of particular relevance to neuronal behaviour are discussed. A particularly important example concerns the pathways that underpin synaptic plasticity and these are considered in Chapter 11.

Consideration is given to whether it is reasonable to model cellular compartments as comprising well-mixed systems in which molecular species are present in sufficient abundance that mass action kinetics apply. In this situation, reaction schemes are modelled as systems of coupled ODEs that describe molecular concentrations. In small neuronal compartments, such as spine heads, even species such as calcium may only be present in a few tens of molecules or fewer. To model this situation requires stochastic models describing the movement and interaction of individual molecules.

CHAPTER 7

The Synapse

This chapter covers a spectrum of models for both chemical and electric-al synapses. Different levels of detail are delineated in terms of model complexity and suitability for different situations. These range from empir-ical models of voltage waveforms to more detailed kinetic schemes, and to complex stochastic models, including vesicle recycling and release. Simple static models that produce the same postsynaptic response for every presy-naptic action potential are compared with more realistic models incorp-orating short-term dynamics that produce facilitation and depression of the postsynaptic response. Different postsynaptic receptor-mediated excita-tory and inhibitory chemical synapses are described. Electrical connections formed by gap junctions are considered.

So far mostly we have considered neuronal inputs in the form of electrical stimulation via an electrode, as in an electrophysiological experiment. This is appropriate as many neuronal modelling endeavours start by trying to repro-duce the electrical activity seen in particular experiments. However, once a model is established on the basis of such experimental data, it is often desired to explore the model in more realistic settings that are not reproducible in an experiment. For example, how does the complex model neuron respond to patterns of synaptic input? How does a model network of neurons function? What sort of activity patterns can a network produce? These questions, and many others besides, require us to be able to model synaptic input.

In the example model of a ball-and-stick neuron (Section 5.1.2), we first introduced how to model the postsynaptic response to release of neurotrans-mitter at a chemical synapse in the form of a predetermined conductance waveform that generates a **postsynaptic current (PSC)**. Chemical synaptic transmission is much more complex than this, however, and can be mod-elled in a variety of ways and levels of detail (Section 7.1). We discuss chem-ical synapses in most detail as they are the principal mediators of targeted neuronal communication. Electrical synapses are discussed in Section 7.6.

A model of a chemical synapse could itself be very complex. The first step in creating a synapse model is identifying the scientific question we wish to address. This will affect the level of detail that needs to be included.

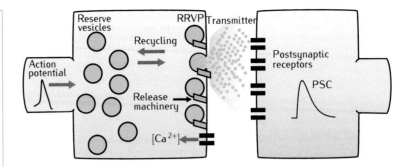

Fig 7.1 Schematic of a chemical synapse. In this example, the presynaptic terminal consists of a single active zone containing a RRVP which is replenished from a single reserve pool. A presynaptic action potential leads to calcium entry through voltage-gated calcium channels which may result in a vesicle in the RRVP fusing with the presynaptic membrane and releasing neurotransmitter into the synaptic cleft. Neurotransmitter diffuses in the cleft and binds with postsynaptic receptors which then open, inducing a PSC.

The abbreviation **IPSC**, standing for **inhibitory postsynaptic current**, is also used.

Very different models will be used if our aim is to investigate the dynamics of a neural network involving thousands of synapses compared to exploring the influence of transmitter diffusion on the time course of a miniature **excitatory postsynaptic current (EPSC)**.

In this chapter, we outline the wide range of mathematical descriptions that can be used to model both chemical and electrical synapses. We start with the simplest models that capture the essence of the postsynaptic electrical response, before including gradually increasing levels of detail (Section 7.2). We consider the dynamics of vesicle recycling and release and the binding of neurotransmitters to postsynaptic receptors, all of which contribute to facilitation and depression over time of the postsynaptic response, known as **short-term plasticity (STP)** (Sections 7.3–7.4). In Chapter 11, we look at models of long-lasting synaptic plasticity that may underpin learning and memory. Very detailed synaptic models may include pre- and postsynaptic diffusion of calcium and extracellular diffusion of neurotransmitter in the synaptic cleft (Section 7.5).

7.1 The Chemical Synapse

The **chemical synapse** is a complex signal transduction device that produces a postsynaptic response when an action potential arrives at the presynaptic terminal. A schematic of the fundamental components of a chemical synapse is shown in Figure 7.1. We describe models of chemical synapses based on the conceptual view that a synapse consists of one or more active zones that contain a presynaptic **readily releasable vesicle pool (RRVP)** which, on release, may activate a corresponding pool of postsynaptic receptors (Walmsley et al., 1998). The RRVP is replenished from a large reserve pool. The reality is likely to be more complex than this, with vesicles in the RRVP possibly consisting of a number of subpools, each in different states of readiness (Thomson, 2000b). Recycling of vesicles may also involve a number of distinguishable reserve pools (Thomson, 2000b; Rizzoli and Betz, 2005).

7.2 The Postsynaptic Response

The aim of a synapse model is to describe accurately the postsynaptic response generated by the arrival of an action potential at a presynaptic ter-

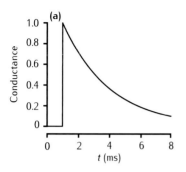
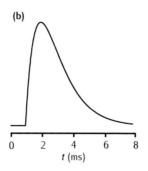
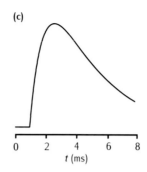

minal. We assume that the response of interest is electrical, but it could equally be chemical, such as an influx of calcium or the triggering of a second-messenger cascade. For an electrical response, the fundamental quantity to be modelled is the time course of the postsynaptic receptor conductance. This can be captured by simple phenomenological waveforms, or by more complex kinetic schemes that are analogous to the models of membrane-bound ion channels discussed in Chapter 4.

Fig 7.2 Three waveforms for synaptic conductance: **(a)** single exponential decay with $\tau = 3$ ms; **(b)** alpha function with $\tau = 1$ ms; and **(c)** dual exponential function with $\tau_1 = 3$ ms and $\tau_2 = 1$ ms. Response to a single presynaptic action potential arriving at time $= 1$ ms. All conductances are scaled to a maximum of 1 (arbitrary units).

7.2.1   Simple Conductance Waveforms

The electrical current that results from the release of a unit amount of neurotransmitter at time t_s is, for $t \geq t_s$:

$$I_{\mathrm{syn}}(t) = g_{\mathrm{syn}}(t)(V(t) - E_{\mathrm{syn}}), \tag{7.1}$$

where the effect of transmitter binding to, and opening, postsynaptic receptors is a conductance change $g_{\mathrm{syn}}(t)$ in the postsynaptic membrane. $V(t)$ is the voltage across the postsynaptic membrane, and E_{syn} is the reversal potential of the ion channels that mediate the synaptic current. For excitatory synapses, this reversal potential is typically near 0 mV, resulting in EPSCs that depolarise the membrane. The reversal potential of inhibitory synapses, however, is typically near or just below the resting membrane potential of a neuron, resulting in IPSCs hyperpolarising, depolarising or causing no change in the potential.

Simple waveforms are used to describe the time course of the synaptic conductance $g_{\mathrm{syn}}(t)$ for the time after the arrival of a presynaptic spike $t \geq t_s$. Three commonly used waveform equations are illustrated in Figure 7.2, in the following order: (a) single exponential decay; (b) alpha function (Rall, 1967); and (c) dual exponential function:

$$g_{\mathrm{syn}}(t) = \overline{g}_{\mathrm{syn}} \exp\left(-\frac{t - t_s}{\tau}\right) \tag{7.2}$$

$$g_{\mathrm{syn}}(t) = \overline{g}_{\mathrm{syn}} \frac{t - t_s}{\tau} \exp\left(-\frac{t - t_s}{\tau}\right) \tag{7.3}$$

$$g_{\mathrm{syn}}(t) = \overline{g}_{\mathrm{syn}} \frac{\tau_1 \tau_2}{\tau_1 - \tau_2}\left(\exp\left(-\frac{t - t_s}{\tau_1}\right) - \exp\left(-\frac{t - t_s}{\tau_2}\right)\right). \tag{7.4}$$

The alpha and dual exponential waveforms are more realistic representations of the conductance change at a typical synapse, and good fits of Equation 7.1 using these functions for $g_{\mathrm{syn}}(t)$ can often be obtained to recorded

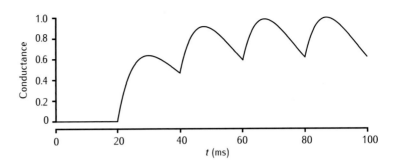

Fig 7.3 Alpha function conductance with $\tau = 10\,\text{ms}$ responding to action potentials occurring at 20, 40, 60 and 80 ms. Conductance is scaled to a maximum of 1 (arbitrary units).

synaptic currents. The dual exponential is needed when the rise and fall times must be set independently.

Response to a Train of Action Potentials

If it is required to model the synaptic response to a series of transmitter releases due to the arrival of a stream of action potentials at the presynaptic terminal, then the synaptic conductance is given by the sum of the effects of the individual waveforms resulting from each release. For example, if the alpha function is used, for the time following the arrival of the nth spike $(t > t_n)$:

$$g_{\text{syn}}(t) = \sum_{i=1}^{n} \overline{g}_{\text{syn}} \frac{t - t_i}{\tau} \exp\left(-\frac{t - t_i}{\tau}\right), \tag{7.5}$$

where the time of arrival of each spike i is t_i. An example of the response to a train of releases is shown in Figure 7.3.

A single neuron may receive thousands of inputs. Efficient numerical calculation of synaptic conductance is often crucial. In a large-scale network model, calculation of synaptic input may be the limiting factor in the speed of simulation. The three conductance waveforms considered are all solutions of the impulse response of a damped oscillator, which is given by the second-order ordinary differential equation (ODE) for the synaptic conductance:

$$\tau_1 \tau_2 \frac{d^2 g}{dt^2} + (\tau_1 + \tau_2)\frac{dg}{dt} + g = \overline{g}_{\text{syn}} x(t). \tag{7.6}$$

The function $x(t)$ represents the contribution from the stream of transmitter releases. It results in an increment in the conductance by $\overline{g}_{\text{syn}}$ if a release occurs at time t. The conductance $g(t)$ takes the single exponential form when $\tau_1 = 0$, and the alpha function form when $\tau_1 = \tau_2 = \tau$.

This ODE can be integrated using a suitable numerical integration routine to give the synaptic conductance over time (Protopapas et al., 1998) in a way that does not require storing spike times or the impulse response waveform, both of which are required for solving Equation 7.5. A method for handling Equation 7.5 directly that does not require storing spike times and is potentially faster and more accurate than numerically integrating the impulse response is proposed in Srinivasan and Chiel (1993).

Voltage Dependence of Response

These simple waveforms describe a synaptic conductance that is independent of the state of the postsynaptic cell. Certain receptor types are influenced by membrane voltage and molecular concentrations. For example, N-methyl-D-aspartate (NMDA) receptors are both voltage-sensitive and are affected by the level of extracellular magnesium (Ascher and Nowak, 1988; Jahr and Stevens, 1990a, b). The basic waveforms can be extended to capture these sorts of dependencies (Zador et al., 1990; Mel, 1993):

$$g_{NMDA}(t) = \overline{g}_{syn} \frac{\exp(-(t-t_s)/\tau_1) - \exp(-(t-t_s)/\tau_2)}{(1 + \mu[Mg^{2+}]\exp(-\gamma V))}, \tag{7.7}$$

where μ and γ set the magnesium and voltage dependencies, respectively. In this model, the magnesium concentration $[Mg^{2+}]$ is usually set at a predetermined, constant level, for example, 1 mM. The voltage V is the postsynaptic membrane potential, which will vary with time.

7.2.2 Current-Based Synaptic Responses

To make a neuronal model easier to analyse and faster to simulate with synaptic input, **current-based synapses** can be used. In these synapses, it is the time course of the synaptic current, rather than conductance, that is prescribed, for example, by a decaying exponential current:

$$I_{syn}(t) = \begin{cases} \overline{I}_{syn} \exp\left(-\frac{t-t_s}{\tau_{syn}}\right) & \text{for } t \geq t_s \\ 0 & \text{for } t < t_s, \end{cases} \tag{7.8}$$

where $\overline{I}_{syn}$ is the maximum current. There is no dependence of the current on the membrane potential; compare this with Equation 7.1, where the membrane potential V changes over time. This decaying exponential time course for $t > t_s$ can be generated by the following differential equation:

$$\tau_{syn} \frac{dI_{syn}}{dt} = -I_{syn}, \tag{7.9}$$

with I_{syn} set to $\overline{I}_{syn}$ at $t = t_s$.

For typical excitatory synapses, current-based synapses provide a reasonable approximation to conductance-based synapses. Implicit in Equation 7.8 is that $\overline{I}_{syn}$ is the product of the conductance $\overline{g}_{syn}$ and a constant driving force. If the resting membrane potential is -70 mV and, as in the case of AMPA (α-amino-3-hydroxy-5-methyl-4-isoxalone propionic acid) synapses, the synaptic reversal potential E_{syn} is around 0 mV, then the driving force $V(t) - E_{syn}$ will not vary much, in percentage terms, during an EPSC that may cause a voltage change from rest on the order of 10 mV or so.

Inhibitory synapses are not so well approximated by current-based synapses since the reversal potential of inhibitory synapses is often close to the resting potential. In a conductance-based inhibitory synapse, the current may be outward if the membrane potential is above the reversal potential, or inward if it is below. In an inhibitory current-based synapse, the current may only be outward. This can also lead to unrealistically low membrane potentials. Nevertheless, current-based inhibitory synapses can be useful models, as they do capture the notion that inhibitory synapses prevent the neuron from firing.

The effect of a current-based synapse with a decaying exponential current (Equation 7.8) on the membrane potential can be calculated. The result is a dual exponential with a decay time constant equal to the membrane time

---

Box 7.1 | Approximating short PSCs by delta functions

The decaying exponential current in Equation 7.8 can be approximated by a delta function (Section 7.3.1):

$$I_{syn}(t) = \overline{Q}_{syn}\delta(t - t_s),$$

where $\overline{Q}_{syn}$ is the total charge delivered by the synapse. To see this, note that the charge delivered by a decaying exponential synapse with finite τ_{syn} would be $\overline{Q}_{syn} = \overline{I}_{syn}\tau_{syn}$. Thus the current can be written:

$$I_{syn}(t) = \begin{cases} \overline{Q}_{syn}\left[\frac{1}{\tau_{syn}}\exp\left(-\frac{t-t_s}{\tau_{syn}}\right)\right] & \text{for } t \geq t_s \\ 0 & \text{for } t < t_s. \end{cases}$$

As τ_{syn} approaches zero, the term in square brackets becomes very close to zero, apart from at $t = t_s$ where it has the very large value $1/\tau_{syn}$. This is approximately a Dirac delta function $\delta(t - t_s)$.

The **Dirac delta function** is a useful mathematical function defined as:

$$\delta(x) = 0 \text{ for } x \neq 0, \qquad \int_{-\infty}^{\infty} \delta(x)\mathrm{d}x = 1.$$

In other words, the delta function is zero for all values of x, except at $x = 0$ where it has an infinitely large value and the area underneath this value is one unit.

---

constant τ_m and a rise time that depends on the decay time constant of the synaptic current input τ_{syn}:

$$V(t) = \begin{cases} R_m\overline{I}_{syn}\frac{\tau_{syn}}{\tau_m - \tau_{syn}}\left(\exp\left(-\frac{t-t_s}{\tau_m}\right) - \exp\left(-\frac{t-t_s}{\tau_{syn}}\right)\right) & \text{for } t \geq t_s \\ 0 & \text{for } t < t_s. \end{cases} \tag{7.10}$$

The same waveform was used above to model synaptic conductances (Figure 7.2c).

The simplest approximation to synaptic input is to make the EPSCs or IPSCs infinitesimally short and infinitely sharp by reducing the synaptic time constant τ_{syn} towards zero. In this case, the voltage response approximates a simple decaying exponential:

$$V(t) = \begin{cases} \frac{R_m\overline{Q}_{syn}}{\tau_m}\exp\left(-\frac{t-t_s}{\tau_m}\right) & \text{for } t \geq t_s \\ 0 & \text{for } t < t_s, \end{cases} \tag{7.11}$$

where $\overline{Q}_{syn} = \overline{I}_{syn}\tau_{syn}$ is the total charge contained in the very short burst of current. In order to simulate this, the quantity $R_m\overline{Q}_{syn}/\tau_m$ is simply added to the membrane potential of the model neuron at time t_s; no differential equation for I_{syn} is required. Infinitesimally short EPSCs or IPSCs can be denoted conveniently using delta functions (Box 7.1).

7.2.3 Kinetic Schemes

A significant limitation of the simple waveform description of synaptic conductance is that it does not capture the actual behaviour seen at many synapses when trains of action potentials arrive. A new release of neurotransmitter soon after a previous release should not be expected to contribute as much to the postsynaptic conductance due to saturation of postsynaptic

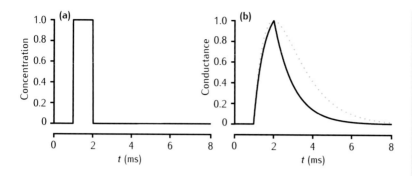

Fig 7.4 Response of the simple two-gate kinetic receptor model to a single pulse of neurotransmitter of amplitude 1 mM and duration 1 ms. Rates are $\alpha = 1\,\text{mM}^{-1}\text{ms}^{-1}$ and $\beta = 1\,\text{ms}^{-1}$. Conductance waveform scaled to an amplitude of 1 and compared with an alpha function with $\tau = 1\,\text{ms}$ (dotted line).

receptors by previously released transmitter and the fact that some receptors will already be open. Certain receptor types also exhibit **desensitisation** that prevents them (re)opening for a period after transmitter-binding, in the same way that the sodium channels underlying the action potential inactivate. To capture these phenomena successfully, kinetic – or Markov – models (Section 4.5) can be used. Here we outline this approach. More detailed treatments can be found in the work of Destexhe et al. (1994b, 1998).

Basic Model

The simplest kinetic model is a two-state scheme in which receptors can be either closed, C, or open, O, and the transition between states depends on transmitter concentration [T] in the synaptic cleft:

$$C \underset{\beta}{\overset{\alpha[T]}{\rightleftharpoons}} O, \tag{7.12}$$

where α and β are voltage-independent forward and backward rate constants. For a pool of receptors, states C and O can range from 0 to 1, and describe the fraction of receptors in the closed and open states, respectively. The synaptic conductance is:

$$g_{syn}(t) = \overline{g}_{syn}O(t). \tag{7.13}$$

A complication of this model, compared to the simple conductance waveforms discussed above, is the need to describe the time course of transmitter concentration in the synaptic cleft. One approach is to assume that each release results in an impulse of transmitter of a given amplitude T_{max} and fixed duration. This enables easy calculation of synaptic conductance with the two-state model (Box 7.2). An example response to a pulse of transmitter is shown in Figure 7.4. The response of this scheme to a train of pulses at 100 Hz is shown in Figure 7.5a. However, more complex transmitter pulses may be needed, as discussed below.

The Neurotransmitter Transient

The neurotransmitter concentration transient in the synaptic cleft following release of a vesicle is characterised typically by a fast rise time followed by a decay that may exhibit one or two time constants, due to transmitter uptake and diffusion of transmitter out of the cleft (Clements et al., 1992; Destexhe et al., 1998; Walmsley et al., 1998). This can be described by the same sort of mathematical waveform, such as the alpha function, used to

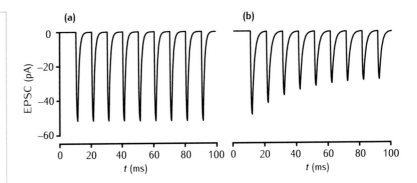

Fig 7.5 Postsynaptic current in response to 100 Hz stimulation from: **(a)** two-gate kinetic receptor model,
$\alpha = 4\,\text{mM}^{-1}\text{ms}^{-1}$, $\beta = 1\,\text{ms}^{-1}$;
(b) five-gate desensitising model,
$R_b = 13\,\text{mM}^{-1}\text{ms}^{-1}$,
$R_{u1} = 0.3\,\text{ms}^{-1}$, $R_{u2} = 200\,\text{ms}^{-1}$,
$R_d = 10\,\text{ms}^{-1}$, $R_r = 0.02\,\text{ms}^{-1}$,
$R_o = 100\,\text{ms}^{-1}$, $R_c = 1\,\text{ms}^{-1}$.
Each presynaptic action potential is assumed to result in the release of a vesicle of neurotransmitter, giving a square-wave transmitter pulse amplitude of 1 mM and a duration of 1 ms. The current is calculated as
$I_{\text{syn}}(t) = g_{\text{syn}}(t)(V(t) - E_{\text{syn}})$.
$E_{\text{syn}} = 0\,\text{mV}$ and the cell is clamped at $-65\,\text{mV}$. The value of $g_{\text{syn}}(t)$ approaches 0.8 nS on the first pulse.

model the postsynaptic conductance itself (Section 7.2.1). However, a simple square-wave pulse for the neurotransmitter transient is often a reasonable approximation for use with a kinetic model of the postsynaptic conductance (Destexhe et al., 1994a, 1998), as illustrated above.

For many synapses, or at least individual active zones, it is highly likely that, at most, a single vesicle is released per presynaptic action potential (Redman, 1990; Thomson, 2000b). This makes the use of simple phenomenological waveforms for the transmitter transient both easy and sensible. However, some synapses can exhibit multivesicular release at a single active zone (Wadiche and Jahr, 2001). The transmitter transients due to each vesicle released must then be summed to obtain the complete transient seen by the postsynaptic receptor pool. This is perhaps most easily done when a continuous function, such as the alpha function, is used for the individual transient, with the resulting calculations being the same as those required for generating the postsynaptic conductance due to a train of action potentials (Section 7.2.1).

Box 7.2 | Solving the two-state model

The basic two-state kinetic scheme (Equation 7.12) is equivalent to an ODE in which the rate of change of O is equal to the fraction converted from state C minus the fraction converted from O to C:

$$\frac{dO}{dt} = \alpha[T](1 - O) - \beta O,$$

where $O + C = 1$.

If the neurotransmitter transient is modelled as a square-wave pulse with amplitude T_{max}, then this ODE can be solved both for the duration of the transient, Equation (a), and for the period when there is no neurotransmitter, Equation (b). This leads to the numerical update scheme (Destexhe et al., 1994a, 1998):

$$O_{t+1} = O_\infty + (O_t - O_\infty) \exp\left(\frac{-\Delta t}{\tau_O}\right) \quad \text{if } [T] > 0, \tag{a}$$

$$O_{t+1} = O_t \exp(-\beta \Delta t) \quad \text{if } [T] = 0, \tag{b}$$

for time-step Δt. In the presence of neurotransmitter, the fraction of open receptors approaches $O_\infty = \alpha T_{\text{max}}/(\alpha T_{\text{max}} + \beta)$ with time constant $\tau_O = 1/(\alpha T_{\text{max}} + \beta)$.

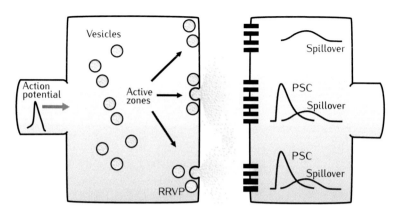

Fig 7.6 Schematic of a chemical synapse with multiple active zones. Single vesicles are releasing at the lower two active zones, resulting in spillover of neurotransmitter between zones.

A further complication is spillover of transmitter from neighbouring active zones. To compensate for this requires consideration of the spatial arrangement of active zones and the likely contribution of spillover due to diffusion (Destexhe and Sejnowski, 1995; Barbour and Häusser, 1997). This can be described by a delayed, and smaller, amplitude transmitter transient that also must be summed together with all other transients at a particular active zone. A synapse containing multiple active zones with spillover of neurotransmitter between zones is illustrated in Figure 7.6.

More Detailed Models

Postsynaptic conductances often exhibit more complex dynamics than can be captured by the simple scheme used above. EPSCs may contain fast and slow components, and may decline in amplitude on successive presynaptic pulses due to desensitisation of the postsynaptic receptors. These factors can be captured in kinetic schemes by adding further closed and open states, as well as desensitised states (Destexhe et al., 1994b, 1998).

A basic five-gate kinetic scheme that includes receptor desensitisation is:

$$C_0 \underset{R_{u1}}{\overset{R_b[T]}{\rightleftharpoons}} C_1 \underset{R_{u2}}{\overset{R_b[T]}{\rightleftharpoons}} C_2 \underset{R_c}{\overset{R_o}{\rightleftharpoons}} O, \tag{7.14}$$

$$R_r \Big\Vert R_d$$

$$D$$

where the binding of two transmitter molecules is required for opening and fully bound receptors can desensitise (state D) before opening. An example of how such a scheme may be translated into an equivalent set of ODEs is given in Section 4.5.1. The response of this scheme to 100 Hz stimulation, obtained by numerically integrating the equivalent ODEs, is shown in Figure 7.5b. The EPSC amplitude declines on successive stimuli due to receptor desensitisation.

Variations on this scheme have been used to describe AMPA, NMDA and γ-aminobutyric acid (GABA)$_A$ receptor responses (Destexhe et al., 1994b, 1998). More complex schemes include more closed, open and desensitised states to match the experimental time course of PSC responses to applied and evoked neurotransmitter pulses. Transition rates may be constant, sensitive to ligands such as neurotransmitter or neuromodulators, or voltage-sensitive.

Models of Metabotropic Receptors

The kinetic schemes discussed above are suitable for modelling the response of ionotropic receptors in which the current-carrying ion channels are directly gated by the neurotransmitter. Other receptors, such as $GABA_B$, are metabotropic receptors that gate remote ion channels through second messenger pathways. Kinetic schemes can also be used to describe this type of biochemical system. For example, the $GABA_B$ response has been modelled as (Destexhe and Sejnowski, 1995; Destexhe et al., 1998):

$$
\begin{aligned}
&R_0 + T \rightleftharpoons R \rightleftharpoons D, \\
&R + G_0 \rightleftharpoons RG \longrightarrow R + G, \\
&G \longrightarrow G_0, \\
&C + nG \rightleftharpoons O,
\end{aligned}
\tag{7.15}
$$

where the receptors enter activated R and desensitised D states when bound by transmitter T. G-protein enters an activated state G, catalysed by R, which then proceeds to open the ion channels via n independent binding sites.

7.3 | Presynaptic Neurotransmitter Release

Any postsynaptic response is dependent upon neurotransmitter being released from the presynaptic terminal. In turn, this depends on the availability of releasable vesicles of neurotransmitter and the likelihood of such a vesicle releasing due to a presynaptic action potential. A complete model of synaptic transmission needs to include terms describing the release of neurotransmitter.

The simplest model, which is commonly used when simulating neural networks, is to assume that a single vesicle releases its quantum of neurotransmitter for each action potential that arrives at a presynaptic terminal, as in Figure 7.3. In practice, this is rarely a good model for synaptic transmission at any chemical synapse. A better description is that the average release is given by np where n is the number of releasable vesicles and p is the probability that any one vesicle will release (Box 1.2). Both n and p may vary with time, resulting in either facilitation or depression of release, and hence the postsynaptic response. Such short-term synaptic plasticity operates on the timescale of milliseconds to seconds and comes in a variety of forms with distinct molecular mechanisms, largely controlled by presynaptic calcium levels (Magleby, 1987; Zucker, 1989, 1999; Thomson, 2000a, b; Zucker and Regehr, 2002).

We consider how to model n and p. We present relatively simple models that describe a single active zone with a RRVP that is replenished from a single reserve pool. Two classes of model are described, one in which an active zone contains an unlimited number of release sites and another which has a limited number of release sites (Figure 7.7).

7.3.1 Vesicle Release

Firstly we consider models for p, the probability that a vesicle will be released. There is a direct relationship between p and presynaptic calcium

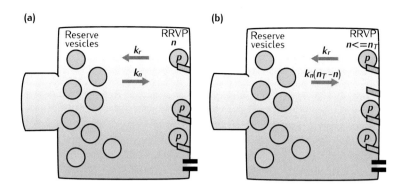

Fig 7.7 Two models of vesicle recycling and release at an active zone: **(a)** vesicle-state model in which the number of vesicles in the RRVP is limited only by vesicle recycling and release rates; **(b)** release-site model in which there is a limited number of release sites. n is the number of vesicles available for release; p is the release probability of a vesicle; k_n is the arrival rate from the reserve pool; k_r is the return rate to the reserve pool.

levels. Release probability is determined by the calcium influx through one, or very few, voltage-gated calcium channels close to a release site, with p varying according to a power (usually 3–4) of the calcium concentration (Zucker and Fogelson, 1986; Redman, 1990; Thomson, 2000b). Most release occurs in synchrony with the arrival of a presynaptic action potential, which results in a rapid influx of calcium. Residual calcium levels between action potentials can result in spontaneous asynchronous release and may enhance release on subsequent action potentials (Zucker, 1999; Thomson, 2000a, b).

Complex models can be conceived that include explicitly the voltage waveform of the action potential, the accompanying calcium current through voltage-gated calcium channels, the associated change in intracellular calcium concentration and then calcium-activated signalling pathways that eventually result in transmitter release. Unless it is the release process itself that is the subject of study, this level of detail is unnecessary. Here we use simpler approaches which describe the vesicle release probability as a direct function of a stereotypical action potential or the resultant calcium transient.

Phenomenological Model of Facilitation

At many synapses, trains of presynaptic action potentials cause the release probability to increase in response to the arrival of successive spikes, due to residual calcium priming the release mechanism (Del Castillo and Katz, 1954b; Katz and Miledi, 1968; Zucker, 1974; Thomson, 2000a, b). Release probability returns to baseline with a time constant of around 50–300 ms (Thomson, 2000a).

A simple phenomenological model that can capture the amplitude and time course of experimentally observed facilitation (Tsodyks et al., 1998) increments the probability of release p by an amount $\Delta p(1-p)$ at each action potential, with p decaying back to baseline p_0 with a fixed time constant τ_f:

$$\frac{\mathrm{d}p}{\mathrm{d}t} = -\frac{(p-p_0)}{\tau_f} + \sum_s \Delta p(1-p)\delta(t-t_s), \tag{7.16}$$

where the delta function $\delta(t-t_s)$ (Box 7.1) results in the incremental increase in p at each spike time t_s. Only the times of presynaptic action potentials t_s appear explicitly in the model. The resultant calcium influx implicitly determines p, but is not explicitly modelled.

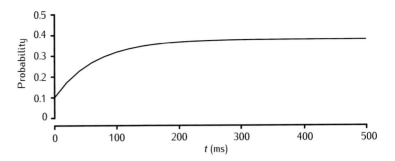

In between spikes, the solution of Equation 7.16 gives that p decays exponentially towards p_0. For an interspike interval of Δt, p will return towards p_0 by the fractional amount $1 - \exp(-\Delta t / \tau_f)$ from its value immediately following the previous action potential. This leads to an expression for the probability of release at the sth spike in terms of the probability at the $(s-1)$th spike:

$$p_s = p_{s-1}^+ + (1 - \exp(-\Delta t / \tau_f))(p_0 - p_{s-1}^+), \tag{7.17}$$

where

$$p_{s-1}^+ = p_{s-1} + \Delta p(1 - p_{s-1}), \tag{7.18}$$

is the release probability attained just following the previous action potential due to facilitation. For a stream of spikes at a constant rate, the release probability reaches a steady-state value per spike of:

$$p_\infty = \frac{p_0(1 - \exp(-\Delta t / \tau_f)) + \Delta p \exp(-\Delta t / \tau_f)}{1 - (1 - \Delta p)\exp(-\Delta t / \tau_f)}. \tag{7.19}$$

An example of facilitation in release is shown in Figure 7.8. Though p is calculated as a continuous function of time, the model is only used to determine the release probability of a vesicle at the time of arrival of each presynaptic action potential.

In this model, there is no attempt made to provide a causal mechanism for the facilitation of release. Other slightly more complex models have attempted to relate the magnitude and time course of facilitation explicitly to calcium entry (Dittman et al., 2000; Trommershäuser et al., 2003).

Kinetic Gating Scheme

A different approach to modelling p attempts to represent the interaction between calcium and the release machinery by using a kinetic scheme driven by an explicit calcium concentration to capture the time course of release, in the same way that such schemes driven by the neurotransmitter transient can be used to describe the postsynaptic conductance change (Section 7.2.3). A basic scheme (Bertram et al., 1996) specifies p as the product of the proportions of open states of a number N (typically 2–4) of different gates:

$$p(t) = O_1(t) \times O_2(t) \ldots \times O_N(t), \tag{7.20}$$

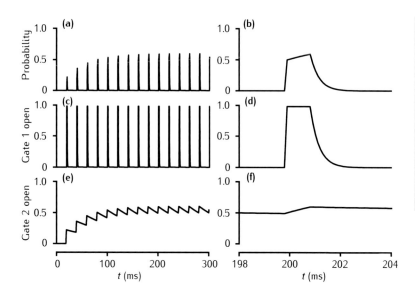

Fig 7.9 Facilitation of transmitter release in a kinetic two-gate model. **(a)**, **(c)** and **(e)** show stimulation at 50 Hz. **(b)**, **(d)** and **(f)** show release and gating transients at the tenth pulse. Fast gate:
$k_1^+ = 200\,\text{mM}^{-1}\text{ms}^{-1}$,
$k_1^- = 3\,\text{ms}^{-1}$; slow gate:
$k_2^+ = 0.25\,\text{mM}^{-1}\text{ms}^{-1}$,
$k_2^- = 0.01\,\text{ms}^{-1}$; square-wave
$[\text{Ca}^{2+}]_r$ pulse: 1 mM amplitude,
1 ms duration.

where $O_j(t)$ is the proportion of open states of gate j, which is determined by the local calcium concentration at the release site:

$$C_j \underset{k_j^-}{\overset{k_j^+[\text{Ca}^{2+}]_r}{\rightleftharpoons}} O_j. \tag{7.21}$$

This requires a model of the calcium transient $[\text{Ca}^{2+}]_r$, seen by the release site in response to the arrival of a presynaptic action potential. For many purposes, a simple stereotypical waveform, such as a brief square-wave pulse, is sufficient. This model is mathematically identical to the simple kinetic scheme for postsynaptic conductance considered earlier (Section 7.2.3). More complex models of the effect of the local calcium concentration, that include the accumulation of residual calcium, may also be used.

An example of the time course of $p(t) = O_1 O_2$ resulting from a two-gate model, with one fast gate and one slow gate, is shown in Figure 7.9. Fast gates determine the total release probability attained for each action potential. They will close completely between action potentials. Slower gates may remain partially open between action potentials and gradually summate to achieve higher open states on successive action potentials. This results in facilitation of the probability of release obtained on successive action potentials.

This model is actually specifying the average rate of vesicle release as a function of time, rather than the probability that a vesicle will release. This rate can be turned into a probability that a vesicle will release by determining the area under the curve p in the interval $[t, t+\Delta t]$, for some small time interval Δt. This gives the probability p^* that a vesicle will release during that interval. A crude, but useful, approximation is:

$$p^* = p(t)\Delta t. \tag{7.22}$$

In principle, this form of model can describe both release due to presynaptic action potentials and spontaneous release between action potentials.

However, with this gating scheme, it is usual that at least one gate will close completely between action potentials, reducing r, and hence p, to zero, as illustrated in Figure 7.9. Note that release due to an action potential is spread realistically over a short time interval (Figure 7.9b, d, f).

Other Factors

These basic models of release can be extended to include other factors that affect the release probability, such as changes in the calcium transient due to calcium channel inactivation (Forsythe et al., 1998) or activation of pre-synaptic metabotropic glutamate receptors (mGluRs). For example, mGluR activation has been modelled as a decrease by a small amount Δp_b in the baseline release probability p_b following each action potential (Billups et al., 2005):

$$\frac{dp_b}{dt} = \frac{p_0 - p_b}{\tau_b} - \sum_s \Delta p_b \cdot p_b \delta(t - t_s). \tag{7.23}$$

This baseline recovers slowly with time constant τ_b, which is on the order of several seconds, to the initial release probability p_0. Baseline release p_b is used in place of p_0 in Equation 7.16 when determining the spike-related release probability p.

7.3.2 Vesicle Availability

The second part of the presynaptic model concerns the number n of vesicles available for release at any particular time, that is, the size of the RRVP. The average size of this pool is determined by the average rate of vesicle release and the rate at which the pool is replenished from reserve pools and vesicle recycling. Depletion of this pool has long been identified as a major component of synaptic depression (Liley and North, 1953; Del Castillo and Katz, 1954b; Elmqvist and Quastel, 1965; Betz, 1970; Kusano and Landau, 1975; Thomson, 2000b).

Continuous Models

A basic model considers the RRVP V_p, to be replenished at rate k_n from a single reserve pool V_r, containing n_r vesicles. The n vesicles in the RRVP may spontaneously return to the reserve pool at rate k_r. Exocytosis (vesicle release) occurs at rate p, which typically reaches a brief maximum following arrival of a presynaptic action potential. This is described by the kinetic scheme:

$$V_r \underset{k_r}{\overset{k_n}{\rightleftharpoons}} V_p \xrightarrow{p} T. \tag{7.24}$$

This is an example of a **vesicle-state model** (Gingrich and Byrne, 1985; Heinemann et al., 1993; Weis et al., 1999), as vesicles may be either in reserve, release-ready or released, and the size of the RRVP is not limited (Figure 7.7a).

For simplicity, we assume that the number of vesicles in the RRVP and reserve pool is sufficiently large that we can treat n and n_r as real numbers, representing the average number of vesicles. This is reasonable for large synapses, such as the neuromuscular junction and giant calyceal synapses in the auditory system. It could also represent the average over many trials at a single central synapse, or the average over many synapses onto a single cell. The kinetic scheme (Equation 7.24) is then equivalent to the ODE:

$$\frac{\mathrm{d}n}{\mathrm{d}t} = k_n n_r - k_r n - n p. \tag{7.25}$$

A further simplification is to assume that the reserve pool is so large that n_r is effectively constant over the stimulation period. The model then reduces to:

$$\frac{\mathrm{d}n}{\mathrm{d}t} = k_n^* - k_r n - n p, \tag{7.26}$$

with constant replenishment rate $k_n^* = k_n n_r$.

If no action potentials arrive at the synapse, such that $p(t) = 0$ for all t, the size of the RRVP reaches a steady-state value of:

$$n_\infty = k_n^*/k_r. \tag{7.27}$$

If a vesicle release takes place with constant probability p_s on the arrival of each presynaptic action potential, the rate of vesicle release is then:

$$p(t) = p_s \delta(t - t_s), \tag{7.28}$$

for a spike arriving at time t_s, with $p(t) = 0$ for $t \neq t_s$. For action potentials arriving with a constant frequency f, the release rate averaged over time is:

$$p(t) = f p_s. \tag{7.29}$$

Substituting this value into Equation 7.26 and setting the rate of change in n to be zero, the inclusion of the synaptic stimulation rate f results in the steady-state RRVP size being decreased to:

$$n_\infty = \frac{k_n^*}{k_r + f p_s}, \tag{7.30}$$

compared to the resting size (Equation 7.27). Increasing stimulation frequency f depletes the RRVP further, since n_∞ decreases as f increases, resulting in a depression of the steady-state postsynaptic response, which is proportional to $n_\infty p$.

A typical example from this model is illustrated, in combination with the phenomenological facilitation model for release probability p (Equation 7.16), in Figure 7.10a (solid lines).

An alternative formulation of the $n p$ active zone model assumes that the number of release sites in an active zone is physically limited to a maximum size n_{max} (Figure 7.7b). This **release-site model** (Vere-Jones, 1966; Dittman and Regehr, 1998; Weis et al., 1999; Matveev and Wang, 2000) has different steady-state and dynamic characteristics from the vesicle-state model. Each release site can either contain a vesicle (state V_p) or be empty (state V_e). A site may lose its vesicle through transmitter release, or via spontaneous undocking and removal to the reserve pool. An empty site can be refilled from the reserve pool. The state of a release site is described by the kinetic scheme:

$$V_e \underset{k_r^*}{\overset{k_n}{\rightleftharpoons}} V_r, \tag{7.31}$$

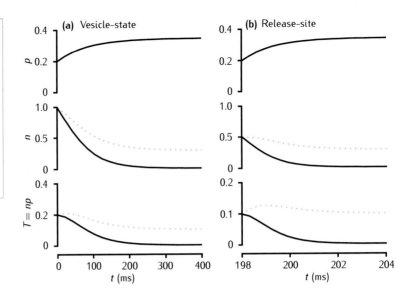

Fig 7.10 Facilitation and depression in deterministic models of short-term synaptic dynamics. Phenomenological model of facilitation: $p_0 = 0.2$; $\Delta p = 0.05$; $\tau_f = 100\,\text{ms}$. **(a)** Vesicle-state model: $k_n^* = k_r = 0.2\,\text{s}^{-1}$. **(b)** Release-site model: $n_T = 1$; $k_n = k_r = 0.2\,\text{s}^{-1}$. Solid lines: $n_s = 0$; dotted lines: $n_s = 0.1$. Synapse stimulated at 50 Hz.

where $k_r^* = p + k_r$ is the sum of the rate of vesicle release and the rate of spontaneous removal of vesicles to the reserve pool. Empty sites become filled with release-ready vesicles from an effectively infinite reserve pool at rate k_n. Given a total of n_{max} release sites, the size of the RRVP over time is described by the ODE:

$$\frac{dn}{dt} = k_n(n_{max} - n) - k_r^* n. \tag{7.32}$$

In the absence of presynaptic activity ($p(t) = 0$), the size of the RRVP reaches a steady state of:

$$n_\infty = \frac{k_n n_{max}}{k_n + k_r}, \tag{7.33}$$

which is equal to n_{max} if vesicles do not spontaneously undock, so that $k_r = 0$.

For stimulation with constant frequency f, the steady state decreases to:

$$n_\infty = \frac{k_n n_{max}}{k_n + k_r + f\,p_s}. \tag{7.34}$$

An example for this model is shown in Figure 7.10b (solid lines). Note that if n is significantly less than n_{max}, the term:

$$k_n(n_{max} - n) \approx k_n n_{max} = k_n^* \tag{7.35}$$

is constant, and this model is now equivalent to the vesicle-state model.

So far we have assumed that the refilling rate k_n of the RRVP is constant. The reality is likely to be rather more complex, with k_n being a direct or indirect function of the presynaptic calcium concentration (Gingrich and Byrne, 1985; Weis et al., 1999; Dittman et al., 2000; Trommershäuser et al., 2003). In particular, the rate of filling of the RRVP may be dependent on presynaptic activity. At the neuromuscular junction (Worden et al., 1997) and the calyx of Held (Wong et al., 2003), recorded postsynaptic responses are well fitted to the model if it is assumed that a fraction, or number, of new vesicles n_s is added to the RRVP following each presynaptic action potential.

For a regular stimulation frequency f, the release-site model is now:

$$\frac{dn}{dt} = k_n(n_{max} - n) - k_r n + (n_s(n_{max} - n) - n p_s)f. \qquad (7.36)$$

The vesicle-state model, in which there is no hard limit to the size of the RRVP, now becomes:

$$\frac{dn}{dt} = k_n^* - k_r n + (n_s - n p_s)f. \qquad (7.37)$$

The effect of n_s on the model responses is shown in Figure 7.10 (dotted lines). The activity-dependent recovery of the RRVP increases the steady-state size of the RRVP for both models. For high-frequency stimulation of the vesicle-state model, the size of the RRVP approaches the steady value:

$$n_\infty = \frac{k_n^* + f n_s}{k_r + f p_s} \approx n_s / p_s. \qquad (7.38)$$

This steady state may be less than (depletion) or greater than (facilitation) the steady-state RRVP size in the absence of activity (k_n^*/k_r), depending on the level of activity-dependent replenishment (n_s).

Many variations and extensions to these model components are possible, including one or more finite-sized reserve pools which may deplete and more complex models of vesicle recycling and refilling of the RRVP.

Stochastic Models

In the models discussed so far, we have been treating the number of pre-synaptic vesicles as a continuous variable representing an average over several spatially distributed synapses that are synchronously active, or over a number of trials at a single synapse (Tsodyks and Markram, 1997; Markram et al., 1998; Dittman et al., 2000). It is also possible to treat n as a count of the number of vesicles in the RRVP, with release of each vesicle being stochastic with probability p (Vere-Jones, 1966; Matveev and Wang, 2000; Fuhrmann et al., 2002). This latter case is computationally more expensive, but represents more realistically the result of a single trial at typical central synapses for which the number of vesicles available for release at any time is limited, and failures of release can be common. This treatment also allows determination of the variation in synaptic response on a trial-by-trial basis.

Instead of treating kinetic models for the size of the RRVP, n, as being equivalent to continuous ODEs, they can be turned into discrete stochastic models by treating n as an integer representing the exact number of vesicles currently in the RRVP. The rates of vesicle recycling k_n and k_r and release p determine the probability that a new vesicle will enter or leave the RRVP in some small time interval Δt.

Computer simulation of this type of model is carried out using Monte Carlo techniques (Section 4.7.2). An algorithm is detailed in Box 7.3. A single simulation is equivalent to a single experimental trial at a real synapse.

Box 7.3 | Simulating a stochastic synapse model

Monte Carlo simulations of a model can be carried out using a per time-step method, as outlined previously for general kinetic schemes (Section 6.8.1). The algorithm is as follows.

(1) Time is divided into sufficiently small intervals of duration Δt, so that the probability of a vesicle entering or leaving the RRVP during Δt is significantly less than 1.

(2) During each small time interval Δt, the arrival of a new vesicle from the reserve pool has probability p_a:

$$\text{Vesicle-state model:} \quad p_a = 1 - \exp(-k_n^* \Delta t) \approx k_n^* \Delta t$$

$$\text{Release-site model:} \quad p_a \approx k_n(n_{max} - n)\Delta t.$$

Arrivals are generated by testing whether a random number selected uniformly in the interval [0, 1] is less than, or equal to, p_a. If it is, n is incremented by 1.

(3) Similarly, spontaneous undocking of a vesicle from the RRVP is tested by selecting n random numbers uniformly from the interval [0, 1] and decrementing n by 1 for each number that is less than, or equal to, $k_r \Delta t$.

(4) If a presynaptic action potential occurs in a particular time interval Δt, then each of the remaining n vesicles is tested for release against release probability p_s, a release occurring if a uniform random number in the interval [0, 1] is less than, or equal to, p_s.

Fig 7.11 Stochastic vesicle-state model of short-term dynamics at a single synaptic active zone. **(a)** A single simulation run, showing the release probability p, the actual number of releasable vesicles n (initially 1) and neurotransmitter release T. **(b)** Average values of these variables, taken over 10 000 trials. Phenomenological model of facilitation: $p_0 = 0.2$; $\Delta p = 0.05$; $\tau_f = 100$ ms. Vesicle-state model: $k_n^* = k_r = 0.2\,s^{-1}$; $n_s = 0.1$. Synapse stimulated at 50 Hz.

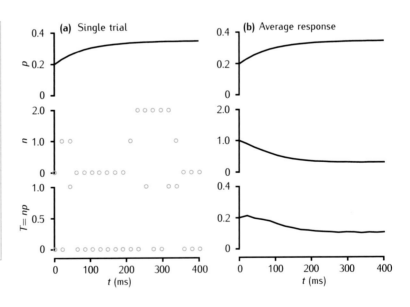

Results from multiple simulations can be averaged to give equivalent results to the deterministic ODE models. Single and average responses from a stochastic version of the vesicle-state model are shown in Figure 7.11. The average response matches the output of the deterministic model.

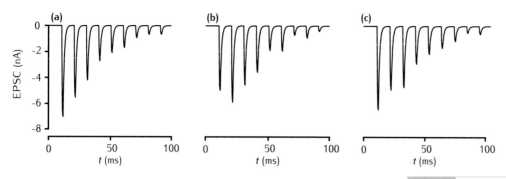

Fig 7.12 Three distinct trials of a complete synapse model with 500 independent active zones. Each active zone uses the stochastic vesicle-state model of short-term synaptic dynamics combined with the two-gate kinetic model of AMPA receptors. Phenomenological model of facilitation: $p_0 = 0.2$; $\Delta p = 0.05$; $\tau_f = 100$ ms. Vesicle-state model: $k_n^* = k_r = 0.2\,\text{s}^{-1}$; $n_s = 0$. Two-gate kinetic receptor model: $\alpha = 4\,\text{mM}^{-1}\text{ms}^{-1}$, $\beta = 1\,\text{ms}^{-1}$. $E_{syn} = 0$ mV, with postsynaptic cell clamped at -65 mV. Synapse stimulated at 100 Hz.

7.4 Complete Synaptic Models

A complete model of a synapse relates the arrival of the presynaptic action potential with a postsynaptic response. The simplest such models were described in Section 7.2, where the postsynaptic response is driven purely by the time of arrival of a presynaptic action potential. We have also considered a number of approaches to explicitly modelling the pre- and postsynaptic mechanisms involved, including presynaptic vesicle recycling and release. These different approaches may be mixed and matched to produce a complete synaptic model which contains sufficient detail for addressing the research questions under study. If the synaptic model is to be used in a large neural network involving thousands of synapses, the simplest synaptic model may be necessary for computational efficiency. More complex models will be desirable and necessary if we wish to examine in detail the short-term dynamics of a particular input pathway onto a single cell. A phenomenological model that incorporates features of short-term plasticity whilst being computationally simple enough for use in large network models is described in Box 7.4

A distinguishing feature of complete models is whether the postsynaptic mechanism is driven by: (1) the time of arrival of a presynaptic spike; (2) the average amount of transmitter released due to a presynaptic spike; or (3) the transmitter transient due to the precise number of vesicles released by a spike.

The kinetic schemes for the postsynaptic conductance are driven by an explicit neurotransmitter transient. These postsynaptic models can be coupled with the presynaptic models of the RRVP, n, and release probability p by producing a transmitter transient due to release $T = np$. If releases due to a single presynaptic spike are assumed to be synchronous, then the parameter T can be used as a scaling factor for the amplitude of a predefined transmitter waveform, assuming interspike intervals are sufficiently large that successive waveforms will not overlap. If release is asynchronous, or transmitter waveforms from successive spikes overlap in time, then the transmitter transients due to individual releases will need to be summed to give the transmitter concentration over time.

Figure 7.12 shows the output of an example model based on the giant calyx of Held (Wong et al., 2003) that combines the stochastic version of the vesicle-state model for vesicle recycling and release (Figure 7.11) with a

---

Box 7.4 | A complete phenomenological synapse model

The Tsodyks–Markram phenomenological synapse model (Tsodyks and Markram, 1997; Markram et al., 1998; Tsodyks et al., 1998) includes facilitation and depression, features of short-term plasticity (STP), in a way that is computationally efficient enough for use in neural network models (Markram et al., 2015) and also allows for some analysis of the effects of STP on neural activity.

The model assumes that a synapse has a finite amount of resource available, with at any time a fraction of this resource being in a recovering state R, an active state E or an inactive state I. The postsynaptic response is proportional to the amount of resource in the active state. On the arrival of a presynaptic action potential at time t_s, the amount of active resource is increased by an amount U_{SE}. Active resource quickly inactivates with a time constant τ_i and then slowly recovers with a time constant τ_r. This accounts for short-term depression, which may be due to depletion of presynaptic vesicles or desensitisation of postsynaptic receptors. Facilitation is accounted for by increasing U_{SE} by a fixed fraction U_{SE}^0 at each presynaptic action potential. The model equations are:

$$\frac{dR}{dt} = \frac{I}{\tau_r} - U_{SE}R\delta(t - t_s), \tag{a}$$

$$\frac{dE}{dt} = -\frac{E}{\tau_i} + U_{SE}R\delta(t - t_s), \tag{b}$$

$$I = 1 - R - E, \tag{c}$$

$$\frac{dU_{SE}}{dt} = -\frac{U_{SE}}{\tau_f} + U_{SE}^0(1 - U_{SE})\delta(t - t_s), \tag{d}$$

$$EPSC = A_{SE}E. \tag{e}$$

Since this model treats the synaptic resource as a continuous variable, it represents the synaptic activity as averaged over a large number of trials or synaptic connections. Suitable parameter values to match averaged experimental recordings from different neocortical synapses have been obtained (Tsodyks and Markram, 1997; Markram et al., 1998; Tsodyks et al., 1998).

---

simple two-gate kinetic scheme for the AMPA receptor response (Figure 7.5). This shows the summed EPSCs due to 500 independent active zones which contain, on average, a single releasable vesicle. Note the trial-to-trial variation due to stochastic release and the interplay between facilitation of release and depletion of available vesicles.

If the trial-to-trial variation in the postsynaptic response is of interest, in addition to using a stochastic model for vesicle recycling and release, variations in **quantal amplitude**, which is the variance in postsynaptic conductance on release of a single vesicle of neurotransmitter, can be included. This is done by introducing variation into the amplitude of the neurotransmitter transient due to a single vesicle, or variation in the maximum conductance that may result (Fuhrmann et al., 2002).

(a) Presynaptic calcium

(b) Neurotransmitter transient

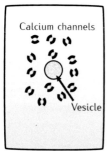

Calcium channels

Vesicle

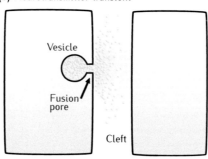

Vesicle

Fusion
pore

Cleft

Fig 7.13 **(a)** Spatial
distribution of voltage-gated
calcium channels with respect to
location of a docked vesicle
(viewed looking at the membrane
facing the synaptic cleft).
(b) Spatial and temporal profile
of neurotransmitter in the
synaptic cleft on release of a
vesicle (side-on view of pre- and
postsynaptic compartments).

7.5 Detailed Modelling of Synaptic Components

We have concentrated on models that describe synaptic input for use in either detailed compartmental models of single neurons or in neural networks. Modelling can also be used to gain greater understanding of the components of synaptic transmission, such as the relationship between presynaptic calcium concentrations and vesicle recycling and release (Zucker and Fogelson, 1986; Yamada and Zucker, 1992; Bennett et al., 2000a, b), or the spatial and temporal profile of neurotransmitter in the synaptic cleft (Destexhe and Sejnowski, 1995; Barbour and Häusser, 1997; Rao-Mirotznik et al., 1998; Smart and McCammon, 1998; Franks et al., 2002; Coggan et al., 2005; Sosinsky et al., 2005). This is illustrated in Figure 7.13.

Such studies typically require modelling molecular diffusion in either two or three dimensions. This can be done using the deterministic and stochastic approaches outlined in Chapter 6 in the context of modelling intracellular signalling pathways. Deterministic models calculate molecular concentrations in spatial compartments and the average diffusion between compartments (Zucker and Fogelson, 1986; Yamada and Zucker, 1992; Destexhe and Sejnowski, 1995; Barbour and Häusser, 1997; Rao-Mirotznik et al., 1998; Smart and McCammon, 1998). Stochastic models track the movement and reaction state of individual molecules (Bennett et al., 2000a, b; Franks et al., 2002). Increasingly, spatial finite element schemes based on 3D reconstructions of synaptic morphology are being employed (Coggan et al., 2005; Sosinsky et al., 2005). The simulation package *MCELL* is specifically designed for implementing stochastic models based on realistic morphologies (Section 6.9).

7.6 Gap Junctions – Electrical Synapses

First identified in invertebrate and vertebrate motor systems (Furshpan and Potter, 1959; Auerbach and Bennett, 1969), it is now recognised that many neurons, including those in the mammalian central nervous system (Connors and Long, 2004), may be connected by purely electrical synapses known

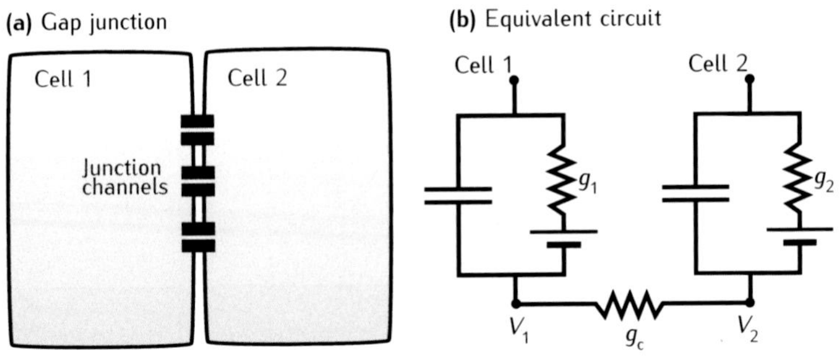

Fig 7.14 (a) Schematic of a gap junction connection between two apposed neurites (dendrites or axons), with (b) the equivalent electrical circuit.

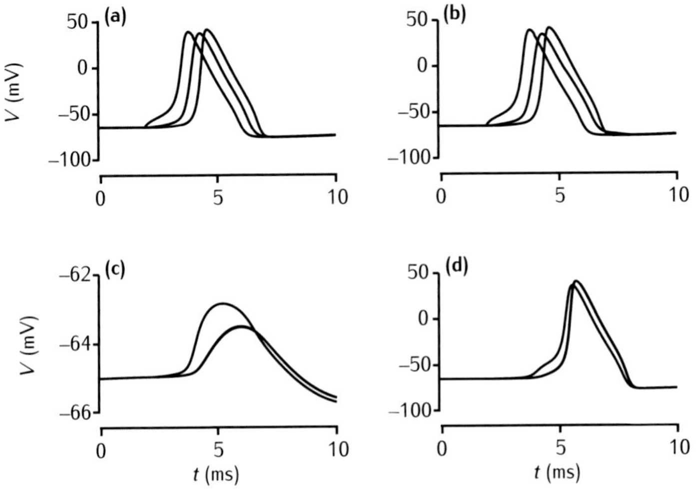

(a) Gap junction

Cell 1   Cell 2

Junction channels

(b) Equivalent circuit

Cell 1   Cell 2

g_1 g_2

V_1 g_c V_2

Fig 7.15 Simulated action potential travelling along two axons that are joined by a gap junction halfway along their length. Membrane potentials are recorded at the start, middle and end of the axons. (a, b) Axon in which an action potential is initiated by a current injection into one end. (c) Other axon to (a), with a 1 nS gap junction. (d) Other axon to (b), with a 10 nS gap junction. Axons are 100 μm long, 2 μm in diameter, with standard Hodgkin–Huxley sodium, potassium and leak channels.

as **gap junctions**. These electrical connections are typically dendrite-to-dendrite or axon-to-axon and are formed by channel proteins that span the membranes of both connected cells (Figure 7.14a). These channels are permeable to ions and other small molecules, and allow rapid, but attenuated and slowed, exchange of membrane voltage changes between cells (Bennett and Zukin, 2004). Gap junctions between cells of the same type usually are bidirectional, but junctions between cells of different types often show strong rectification, with depolarisations being transferred preferentially in one direction and hyperpolarisations in the other. This is due to the different cells contributing different protein subunits at either side of the junction (Marder, 2009; Phelan et al., 2009). Gap junction conductance can also be modulated by various G-protein-coupled receptors, leading to long-lasting changes in coupling strength as the result of neuronal activity (Bennett and Zukin, 2004), equivalent to the long-lasting changes seen at chemical synapses (Section 11.1).

A simple gap junction model assumes a particular fixed, symmetric permeability of the gap junction channels. Thus the electrical current through a gap junction is modelled as being strictly ohmic, with a coupling conductance g_c. The current flowing into each neuron is proportional to the voltage difference between the two neurons at the point of connection (Figure 7.14b):

$$I_1 = g_c(V_2 - V_1),$$
$$I_2 = g_c(V_1 - V_2). \tag{7.39}$$

An example of the effect of a gap junction between two axons is shown in Figure 7.15. The gap junction is halfway along the two axons, and an action potential is initiated at the start of one axon. If the gap junction is sufficiently strong, then the action potential in the first axon can initiate an action potential in the second axon, which then propagates in both directions along this axon.

Even this simple connection between two neurons can lead to complex effects. The response of one neuron to the voltage change in the other can be quite asymmetric between neurons, despite a symmetric coupling conductance, if the two neurons have different cellular resistances (or, equivalently, conductances). For the circuit shown in Figure 7.14b, consider current being injected into cell 1 so that it is held at voltage V_1. Kirchhoff's law for current flow stipulates that the current flowing into point V_2 must equal the current flowing out, so that:

$$(V_1 - V_2)g_c = V_2 g_2. \tag{7.40}$$

Rearranging this equation gives a coupling coefficient that describes the relative voltage seen in cell 2 for a given voltage in cell 1:

$$\frac{V_2}{V_1} = \frac{g_c}{g_2 + g_c}. \tag{7.41}$$

Clearly, V_2 is always less than V_1, and the attenuation of V_2 will be only small if the conductance of cell 2 is low (the cell has a high input resistance), relative to the coupling conductance. Similarly, if cell 2 is held at V_2, then the attenuation in V_1 is given by:

$$\frac{V_1}{V_2} = \frac{g_c}{g_1 + g_c}. \tag{7.42}$$

If $g_1 \neq g_2$, then the attenuation across the gap junction is not symmetric for voltage changes in cell 1 or cell 2.

The above derivation of coupling attenuation assumes that the cellular conductances are fixed. However, changes in membrane potential are often accompanied by, or result from, changes in membrane conductance due to the opening or closing of ion channels. This can result in a fixed, symmetric coupling also being somewhat rectifying, with hyperpolarisations being less attenuated than depolarisations, due to the decrease in cellular membrane resistance associated with depolarisations. This is a different form of rectification than that resulting from an asymmetric subunit structure of the gap junction proteins.

Computational modelling is helping to uncover the possible functionality of gap junctions between neurons. We give an example in Box 7.5.

Box 7.5 | Application – axo-axonal coupling in *Xenopus*

One advantage of electrical connections is that intercellular communication from axon to dendrite or soma becomes faster, without synaptic delays at chemical synapses. In cases where there are axo-axonic connections, the function of electrical coupling is less clear.

In *Xenopus* tadpoles, a well-characterised group of around 30 neurons with thin, unmyelinated axons originating in the brainstem contact spinal motor neurons which drive swimming. These descending interneurons (dINs) are electrically coupled. Hull et al. (2015) modelled the effects of electrical coupling, assuming that dIN cells are coupled through axo-axonal contacts. They found that axon–axon gap junctions close to the soma gave the best match to experimentally measured coupling coefficients, and axon diameter had a strong influence on coupling. Each axon had only a small number of gap junctions, meaning that only 25% of neuron pairs were coupled directly, most via one or more intermediate neurons. With the addition of active channels, action potential propagation along the thin axons was unreliable. This was improved by increasing sodium channel densities and decreasing potassium channel densities in the initial axon segment.

Whole-cell recordings showed that the dINs are the first neurons to fire on each side on each cycle of swimming (Soffe et al., 2009). When gap junction blockers are applied, swimming episodes become shorter and dIN firing becomes unreliable. These results suggest that electrical coupling is important for reliable rhythm generation and here plays a key role in generating swimming activity. However, whole-cell recordings from dINs show that they fire only once in response to step-current injection, which seems at odds with the repetitive firing seen when the whole dIN population is excited by the experimental application of excitatory synaptic agonists like glutamate (Arbas and Calabrese, 1987).

Hull et al. (2015) compared the response of a dIN to step-current injections in an electrically coupled model network with the response of a decoupled dIN. Electrically coupled dINs fired only once to all levels of injected current; with no coupling, the input resistance of individual dINs recorded at the soma increased and step-current injections caused them to fire repetitively. Their interpretation of this effect was that in a coupled network, dIN excitability is depressed by current flow through gap junctions from the stimulated neuron to its neighbours which sit at a more negative membrane potential, closer to rest. To test this hypothesis, they injected a step-current pulse simultaneously into all 30 dINs in the electrically coupled model network. The whole dIN population fired nearly synchronously and repetitively at 30 Hz during the current injection, the action potentials propagating reliably for 1500 μm along their thin axons. As the current was increased, the firing frequency increased to 80 Hz, before repetitive firing failed at higher current levels.

This modelling work is an example of how electrically coupled neurons influence neuronal activity. It offers a new interpretation of the long-standing paradox that experimentally recorded dINs only fire a single spike to a depolarising current step, whereas they fire rhythmically when depolarised. The modelling suggests that the restricted firing is a consequence of the electrical coupling, and that when the whole coupled dIN population is excited, the dINs are able to fire rhythmically as pacemakers.

7.7 | Summary

The chemical synapse is a very complex device. Consequently, a wide range of mathematical models of this type of synapse can be developed. When embarking on developing such a model, it is essential that the questions for which answers will be sought by the use of the model are clearly delineated.

In this chapter, we have presented various models for chemical synapses, ranging from the purely phenomenological to models more closely tied to the biophysics of vesicle recycling and release, and neurotransmitter gating of postsynaptic receptors. Along the way, we have highlighted when different types of model may be appropriate and useful. These models can capture STP, including facilitation and depression. Models of long-lasting changes in synaptic strength, underpinning learning and memory, are considered in Chapter 11.

Many of the modelling techniques are the same as those seen previously in Chapter 4 and Chapter 6, in particular, the use of kinetic reaction schemes and their equivalent ODEs. Stochastic algorithms have been used when the ODE approach is not reasonable, such as for the recycling and release of small numbers of vesicles at an active zone. We considered how subcomponents of the chemical synapse can also be illuminated by mathematical modelling. Reaction–diffusion systems at the ODE and stochastic levels are of importance here to investigate presynaptic calcium transients and the neurotransmitter transient in the synaptic cleft.

Finally, we explored modelling electrical gap junctions, which are simpler in operation than chemical synapses, but can result in highly asymmetric voltage changes between coupled cells.

CHAPTER 8

Simplified Models of the Neuron

In this chapter, a range of models with fewer details than those in previous chapters is considered. These simplified neuron models are particularly useful for incorporating into networks since computationally they are more efficient, and sometimes they can be analysed mathematically. **Reduced compartmental models** can be derived from large compartmental models by lumping together compartments. Additionally, the number of gating variables can be reduced whilst retaining much of the dynamical flavour of a model. These approaches make it easier to analyse the function of the model using the mathematics of dynamical systems. In the yet simpler **integrate-and-fire** model, first introduced in Chapter 2 and elaborated on in this chapter, there are no gating variables, action potentials being produced when the membrane potential crosses a threshold. At the simplest end of the spectrum, **rate-based models** communicate via firing rates rather than individual spikes.

Up until this point, the focus has been on adding details to our neuron models. In this chapter, we take the apparently paradoxical step of throwing away a lot of what is known about neurons. Given all the painstaking work that goes into building detailed models of neurons, why do this? There are at least three reasons:

(1) We wish to explain how a complicated neural model works by stripping it down to its bare essentials. This gives a simpler model in which the core mechanisms have been exposed and so are easier to understand.

(2) We wish to understand the behaviour of a network of neurons. We can use a simplified model to describe the essential function of the neurons in question. These are faster to simulate than compartmental neurons and may allow mathematical analysis of the network. Using simplified models of neurons in networks of neurons is discussed in Chapter 9.

(3) We wish to produce a summary description of the characteristics of a neural recording in order to classify the type of neuron the recording comes from.

In this chapter, the programme of simplification begins with multi-compartmental models. By reducing the number of compartments and reducing the number of state variables, models such as the two-compartment Pinsky–Rinzel model of a hippocampal pyramidal cell (Pinsky and Rinzel, 1994) can be derived. Despite having only two compartments, this model can exhibit a wide range of firing behaviour, and using it gives insights into the function of hippocampal pyramidal cells. A further reduction in the number of compartments to one and a reduction of the number of state variables to two lead to models such as the FitzHugh–Nagumo model (FitzHugh, 1961; Nagumo et al., 1962) or the Morris–Lecar model (Morris and Lecar, 1981). These models can be analysed using the mathematics of dynamical systems (Section 8.2), and the behaviour of the system can be visualised straightforwardly. Analysis using these concepts helps explain what properties of a model lead to model behaviours such as Type I or Type II firing, as defined in Section 4.4.2. However, knowledge of dynamical systems analysis is not essential for understanding the rest of this chapter.

The next level of simplification is the integrate-and-fire (IF) model (Lapicque, 1907; Hill, 1936; Knight, 1972), already introduced in Chapter 2, Section 2.7. This chapter covers variants and extensions of integrate-and-fire neurons, which allow them to reproduce a wider range of firing patterns than the basic integrate-and-fire model. These types of model are well suited to be fitted to experimental recordings from somatic current injections (Jolivet et al., 2008; Pozzorini et al., 2015). We incorporate noise intrinsic to the neuron (ion channel noise) and extrinsic to the neuron (random inputs), so that the models fire stochastically. The culmination of the sequence of integrate-and-fire models is the Generalised Integrate-and-Fire model. Chapter 9 will show how integrate-and-fire neurons can be used in models of networks of spiking neurons. In some cases, the simplicity of the integrate-and-fire neuron allows these networks to be analysed mathematically.

Finally, we remove even the spikes, to give firing-rate models, which generate and receive firing rates rather than spike times. These models are appropriate for applications in which precise firing times are not very important to the network behaviour.

As discussed by other authors (e.g. Herz et al., 2006), there are a number of factors to be taken into consideration when deciding upon what type and degree of simplification are appropriate:

The level of explanation: the elements of the model need to correspond to the physical elements being described. For example, a model incorporating models of the appropriate types of ion channels is needed in order to predict what happens to a neuron when a particular neuromodulator is released or a particular type of channel is blocked. However, if we can characterise the behaviour of the manipulated neurons, we could construct a simple model of them which would allow the behaviour of a network of such neurons to be predicted.

The data available: a complex model tends to contain more parameters. If there are not a great deal of data available, it may not be possible to constrain all these parameters; see, for example, the discussion of Markov models in Chapter 4.

The desired type of analysis: very simple models allow mathematical analysis of networks, giving insight into mechanisms of network behaviour.

Computational resources: simpler models are faster to simulate than more complex models. This is particularly a consideration when large networks of neurons are being simulated. However, as the speed of computers increases, lack of computational resources will become less of a consideration.

8.1 | Reduced Compartmental Models

This section considers two approaches to reducing the number of variables in compartmental models of neurons. The main goal of this reduction is to produce models that, while still exhibiting the types of behaviours that emerge from the full models, are simple enough to analyse.

The first approach is to reduce the number of compartments. The method due to Rall (1964) for reducing the number of compartments in a model by lumping compartments together into equivalent cylinders has already been discussed in Section 5.4.4. However, beyond this, there are few hard-and-fast principles governing *when* it is appropriate to reduce the number of compartments, and by how much. We therefore proceed by example, citing cases in which neuron models with varying numbers of compartments have been employed. In particular, we discuss Pinsky and Rinzel's (1994) two-compartmental model of a hippocampal pyramidal cell in some depth. This model has been used widely, and our discussion illustrates how a simplified model can be used to understand the behaviour of models that correspond to real neurons more exactly.

The second approach is to reduce the number of state variables. This can be achieved by changing the equations governing the membrane potential so that the number of differential equations describing the system is reduced. There are a number of models in the literature with two state variables, and we focus on the model developed by Morris and Lecar (1981).

8.1.1 Models with More than Two Compartments

Reducing the number of compartments is not always appropriate. For example, cells whose neurites are much longer than the length constant (Section 5.7.1) must be split into hundreds or thousands of compartments in order to address questions regarding the spatiotemporal behaviour of the membrane potential throughout the neuron. In their model of a Purkinje cell, De Schutter and Bower (1994a) used 1600 compartments to ensure that the passive electrotonic length of each compartment was less than 0.05 of the length constant. This level of detail allowed the response to in vitro current injection and inputs from realistic distributions of synaptic inputs to be simulated (De Schutter and Bower, 1994b). The prediction from the simulations that inhibitory inputs have more influence over spike timing than excitatory inputs was later confirmed by experiments (Jaeger and Bower, 1999). Another case in which it is necessary to model the spatial extent of the neuron is when predicting extracellular potentials rather than just the somatic membrane potential. Section 13.2 gives a more general discussion of how extracellular recordings can be interpreted.

An early example of a reduced model is that of a mitral cell in the olfactory bulb, due to Rall and Shepherd (1968). The model comprised one somatic compartment with two ionic conductances, sufficient to create an action potential, and 5–10 dendritic compartments. Simplifying assumptions about the geometry of the olfactory bulb allowed the model to predict extracellular potentials, which were compared to readings from extracellular electrodes in the olfactory bulb.

The differences between the model and experimental results allowed Rall and Shepherd to infer the existence of dendro-dendritic synapses between mitral cells and granule cells. The existence of these synapses was confirmed subsequently by electron micrograph studies (Rall et al., 1966).

Hippocampal CA1 and CA3 pyramidal cells have been modelled with varying numbers of compartments, ranging from over 1500 compartments in a model of a CA1 cell designed to investigate back-propagating action potentials (Golding et al., 2001; accession number 64167 in the modelling archive ModelDB) through models with a few hundred compartments (Poirazi et al., 2003) or 64 compartments (Traub et al., 1994), designed to investigate synaptic integration in single cells, down to models with 19 compartments (Traub et al., 1991), which were able to reproduce experimentally recorded fast bursting, and were used in simulations of oscillations in the CA3 network (Traub et al., 1992).

Even relatively small and simple compartmental models can give useful insights, as in the analysis by Agmon-Snir et al. (1998) of coincidence detection cells in the auditory brainstem. These cells have two dendritic trees emerging from the soma, and each tree receives synapses from one ear. Agmon-Snir et al. showed that a cell with the dendritic trees modelled as passive cables produced better coincidence detection than a one-compartment cell. In addition, in the model, there is an optimal length for the dendrites, which depends on the frequency of the inputs received by the cell. This mirrors the relationship between dendrite length and input frequency found in chicks (Smith and Rubel, 1979).

8.1.2 Two Compartments: the Pinsky–Rinzel Model

Pinsky and Rinzel (1994) used the ionic conductances of a 19-compartment reduced model of a CA3 cell developed by Traub et al. (1991) for their two-compartment model of the same cell type (Figure 8.1). The model, the equations for which are presented in Box 8.1, contains a soma and a dendrite compartment. The soma compartment contains sodium and potassium delayed rectifier currents (I_{Na} and I_{DR}). The dendrite contains a voltage-dependent calcium current I_{Ca} and a voltage-dependent potassium after-hyperpolarisation (AHP) current I_{AHP}, as well as a calcium-dependent potassium C current I_C. The proportion p of the membrane area occupied by the soma and the strength of the coupling conductance g_c between the soma and dendrite can be varied.

Despite the small number of compartments, the model can reproduce a variety of realistic activity patterns in response to somatic current injection

In an introduction to the papers in a retrospective collection of Rall's work (Segev et al., 1995), Shepherd explains that there was a delay in writing up the model and so the experimental work (Rall et al., 1966) was published two years before the modelling work (Rall and Shepherd, 1968).

Fig 8.1 Circuit diagram of the Pinsky–Rinzel model. Meaning of symbols: I_{Na} – sodium current; I_{Ca} – calcium current; I_{DR} – delayed rectifier potassium current; I_{AHP} – potassium after hyperpolarisation current; I_C – calcium-dependent potassium current; I_L – leak current; I_s – current injection to soma; I_{syn} – synaptic current; g_c – soma-dendrite coupling conductance; V_s – soma compartment membrane potential; V_d – dendrite compartment membrane potential.

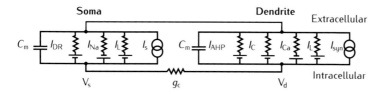

Box 8.1 | The Pinsky–Rinzel equations

The membrane equations for the two compartments in Figure 8.1 are:

$$C_m \frac{dV_s}{dt} = -\overline{g}_L(V_s - E_L) - g_{Na}(V_s - E_{Na}) - g_{DR}(V_s - E_K)$$
$$+ \frac{g_c}{p}(V_d - V_s) + \frac{I_s}{p},$$
$$C_m \frac{dV_d}{dt} = -\overline{g}_L(V_d - E_L) - g_{Ca}(V_d - E_{Ca}) - g_{AHP}(V_d - E_K) - g_C(V_d - E_K)$$
$$+ \frac{g_c}{1-p}(V_s - V_d) + \frac{I_{syn}}{1-p}.$$

$C_m = 3\,\mu F\,cm^{-2}$ is the membrane capacitance; p is the proportion of the membrane area taken up by the soma ($p = 0.5$ in the simulations shown in Figure 8.2); g_c is the coupling conductance; I_s is the somatic injection current density; and I_{syn} is the dendritic synaptic current density, set to zero in this chapter.

Voltage-dependent conductances are modelled using the Hodgkin–Huxley formalism with differential equations for the state variables h, n, s, c and q. The state variable m has a very fast time constant τ_m, so its value is well approximated by its steady-state value m_∞. To make the equations more legible, we define the following quantities: $V_1 = V + 46.9$, $V_2 = V + 19.9$, $V_3 = V + 24.9$, $V_4 = V + 40$, $V_5 = V + 20$, $V_6 = V + 8.9$, $V_7 = V + 53.5$, $V_8 = V + 50$, all measured in millivolts. With V in millivolts, the equations for the conductances and rate coefficients are:

$$g_{Na} = \overline{g}_{Na} m_\infty^2 h \qquad \alpha_m = -\frac{0.32V_1}{\exp(-V_1/4)-1} \qquad \beta_m = \frac{0.28V_2}{\exp(V_2/5)-1}$$
$$\alpha_h = 0.128\exp(\frac{-43-V}{18}) \qquad \beta_h = \frac{4}{1+\exp(-V_5/5)}$$
$$g_{DR} = \overline{g}_{DR} n \qquad \alpha_n = -\frac{0.016V_3}{\exp(-V_3/5)-1} \qquad \beta_n = 0.25\exp(\frac{-V_4}{40})$$
$$g_{Ca} = \overline{g}_{Ca} s^2 \qquad \alpha_s = \frac{1.6}{1+\exp(-0.072(V-5))} \qquad \beta_s = \frac{0.02V_6}{\exp(V_6/5)-1}$$
$$g_C = \overline{g}_C c \chi([Ca^{2+}]) \quad \alpha_c = 0.0527\left(\exp(\frac{V_8}{11} - \frac{V_7}{27})\right) \quad \text{for } V \leq -10\,mV$$
$$\alpha_c = 2\exp(-V_7/27) \qquad \text{for } V > -10\,mV$$
$$\beta_c = 2\exp(-V_7/27) - \alpha_c \qquad \text{for } V \leq -10\,mV$$
$$\beta_c = 0 \qquad \text{for } V > -10\,mV$$
$$\chi([Ca^{2+}]) = \min([Ca^{2+}]/250, 1)$$
$$g_{AHP} = \overline{g}_{AHP} q \qquad \alpha_q = \min(0.00002[Ca^{2+}], 0.01) \quad \beta_q = 0.001.$$

The calcium current $I_{Ca} = g_{Ca}(V_d - E_{Ca})$ increases the calcium concentration $[Ca^{2+}]$ in a thin submembrane shell in the dendrite, and the accumulated calcium leaks out:

$$\frac{d[Ca^{2+}]}{dt} = -0.13I_{Ca} - 0.075[Ca^{2+}].$$

Here I_{Ca} is in $\mu A\,cm^{-2}$ and t is in ms. Pinsky and Rinzel measured $[Ca^{2+}]$ in arbitrary units, since the thickness of the thin shell is not known. In simulations in this chapter, it is (arbitrarily) in millimolar. The maximum conductances in $mS\,cm^{-2}$ are:

$$\overline{g}_L = 0.1 \quad \overline{g}_{Na} = 30 \quad \overline{g}_{DR} = 15 \quad \overline{g}_{Ca} = 10 \quad \overline{g}_{AHP} = 0.8 \quad \overline{g}_C = 15.$$

The reversal potentials are:

$$E_{Na} = 60\,mV \quad E_K = -75\,mV \quad E_{Ca} = 80\,mV.$$

In the equations presented here, V has been shifted by $-60\,mV$, compared to the equations in Pinsky and Rinzel (1994), so that V refers to the actual membrane potential rather than the deflection from the resting potential.

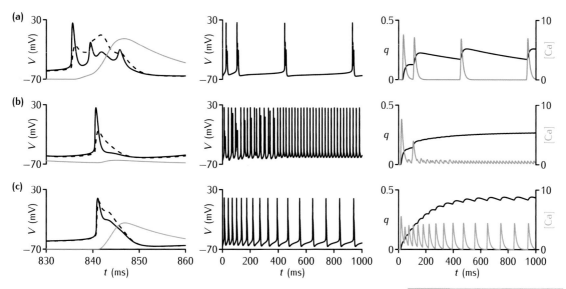

or dendritic synaptic input. Figure 8.2 shows the behaviour of the somatic and dendritic membrane potential, the dendritic calcium concentration and the I_{AHP} activation variable q for three different combinations of somatic injection current I_s and the soma–dendrite coupling conductance g_c.

In the first combination (Figure 8.2a) the coupling conductance and the somatic injection current are relatively low. During the 1 s of stimulation, the neuron appears to produce four spikes (middle), giving a firing frequency of around 4 Hz. At a higher resolution (left), each spike can actually be seen to be a **burst** of action potentials. The first spike of the burst is initiated in the soma, and then spreads to the dendrite. The soma repolarises faster than the dendrite, and the resulting voltage difference between the dendrite and the soma drives current from the dendrite into the soma, triggering another spike. This second spike allows the dendrite to remain depolarised, thus causing the calcium current to flow and leading to a prolonged depolarisation of the dendrite. This long calcium spike activates the calcium-dependent potassium current I_C, which then causes both dendrite and soma to hyperpolarise and ends the calcium spike. The calcium influx also increases the activation of I_{AHP} (right-hand plot in Figure 8.2a), which, with its slow time constant, keeps the cell hyperpolarised for hundreds of milliseconds and determines the interburst interval.

In the second parameter combination, when the level of somatic current injection is higher, conventional action potentials result (Figure 8.2b, left). This is because the more frequent action potentials lead to a higher level of activation of I_{AHP} at the start of each action potential, thus causing the dendrite to be more hyperpolarised, preventing current flowing back from the dendrite and initiating another spike in the soma. There is also much less calcium influx into the dendrite, due to it being generally more hyperpolarised. The rate at which these spikes are fired, 44 Hz, is much higher than the rate at which the bursts are fired.

In the third parameter combination, there is the same level of current injection as in the second combination, but a higher coupling conduct-

Fig 8.2 Behaviour of the Pinsky–Rinzel model (Pinsky and Rinzel, 1994) for different values of the coupling parameter g_c and the level of somatic current injection I_s. In each subfigure, the left-hand column shows the detail of the somatic membrane potential (solid line), the dendritic membrane potential (dashed line) and the calcium concentration (blue line) in a period of 30 ms around a burst of action potential. The middle column shows the behaviour of the membrane potential over 1000 ms. The right-hand column shows the behaviour of q, the I_{AHP} activation variable and the calcium concentration over the period of 1000 ms. The values of I_s in μA cm^{-2} and g_c in mS cm^{-2} in each row are: **(a)** 0.75, 2.1; **(b)** 2.5, 2.1; **(c)** 2.5, 10.5.

ance. Single spikes are produced, but with a long afterdepolarising shoulder (Figure 8.2c, left) and the firing rate of 16 Hz is between the firing rates found in the other two combinations. The larger coupling conductance makes it much easier for current to flow between the soma and the dendrite. This means that the calcium spike in the dendrite is initiated before the soma has had time to repolarise. Thus the sodium current does not have a chance to deinactivate, so there is no second somatic spike, caused by electrotonic spread from the dendritic calcium spike. The firing rate adapts as the overall level of I_{AHP} activation increases (Figure 8.2c, right).

These behaviours, and other features investigated by Pinsky and Rinzel (1994) such as f-I curves, mirror the behaviour of the more complicated Traub et al. (1991) model. The simpler formulation of the Pinsky–Rinzel model allows for a greater understanding of the essential mechanisms at work in the more complex model, and in real neurons. A slightly simpler version of the model, though still with two compartments, has been dissected using the phase plane techniques described in Section 8.2 (Booth and Rinzel, 1995).

The small size of the Pinsky–Rinzel model means that it is faster to simulate than the Traub et al. (1991) model, which is especially important when carrying out network simulations. Moreover, Pinsky and Rinzel (1994) demonstrated that a network of Pinsky–Rinzel neurons connected by excitatory synapses has similar properties, such as synchronous bursting, to a similar network of Traub neurons.

8.1.3 Single-Compartment Reduced Models

The Hodgkin–Huxley (HH) model of a patch of membrane explains the generation of action potentials well. However, because it contains one differential equation for each of the state variables V, m, n and h, it is hard to understand the interactions between the variables. Various authors have proposed sets of equations that have only two state variables – voltage and another state variable – which can recreate a number of properties of the HH model, most crucially the genesis of action potentials. Some well-known models are the FitzHugh–Nagumo model (FitzHugh, 1961; Nagumo et al., 1962), the Kepler et al. (1992) model and the Morris–Lecar model (Morris and Lecar, 1981).

The formulation of the Morris–Lecar model makes it the best of these models to use as an example of the application of the dynamical systems theory (Section 8.2). The model was developed to describe the barnacle giant muscle fibre, but it has been applied to other systems, such as lobster stomatogastric ganglion neurons (Skinner et al., 1993) and mammalian spinal sensory neurons (Prescott et al., 2008). The model contains calcium, potassium and leak conductances. Both active conductances are non-inactivating, and so they can be described by one state variable. A further reduction in state variables is made by assuming that the calcium conductance responds instantaneously to voltage. There are thus two state variables: the membrane potential V and the potassium state variable w. The parameters V_1, V_2, V_3 and V_4 determine the half-activation voltage and slope of the activation curves for the calcium and potassium conductances. Activation curves and time constants of the calcium and potassium currents for two different settings of the parameters are shown in Figure 8.3.

As described in Section 4.4.2, Hodgkin (1948) identified two types of neuron, distinguished by their f-I curves. In **Type I** neurons, the firing rate is zero below a threshold level of injected current and as the current is increased past the threshold, the firing rate starts to increase continuously from zero. In **Type II** neurons, the firing rate is also zero below the threshold current, but then jumps discontinuously as the current is increased past the threshold.

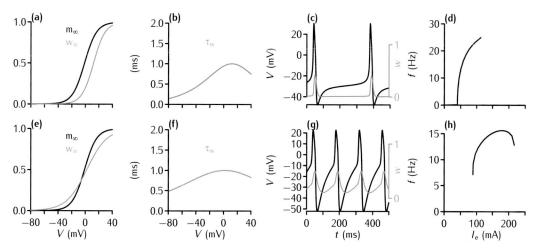

Both settings lead to trains of action potentials being produced in response to current injection (Figure 8.3c, g). The first setting of the parameters (Figure 8.3a, b) leads to Type I firing behaviour in the f–I curve (Figure 8.3d). The second set (Figure 8.3e, f) leads to Type II firing behaviour (Figure 8.3h). The full set of equations describing the Morris–Lecar model are given in Box 8.2.

Dynamical systems analysis can be used to characterise the different types of behaviour of the Morris–Lecar model, which depend on the amount of current injected and on the values of the parameters V_1–V_4 (Box 8.2). The state of the system is characterised by the values of the variables V and w. The system may be stable, with V and w settling to constant values, or unstable with both V and w varying periodically through time. The properties of this model are explored in detail in the next section, as an illustration of the application of dynamical systems theory. Rinzel and Ermentrout (1998) consider the Morris–Lecar neuron in more detail and go on to analyse reduced neuronal models with three states, which can give rise to bursting and the **aperiodic**, apparently random behaviour that is referred to as **chaos**.

8.2 | Dynamical Systems Theory

Any system of equations that describes how the state of a system changes through time is called a **dynamical system**. A simple example is the movement of a pendulum, where the state can be described by one variable, the angle of the pendulum. A more complex example is the Lotka–Volterra model (Lotka, 1925; Volterra, 1926). As discussed in Chapter 1 (Figure 1.1), this model describes how two state variables – the numbers of predators and their prey – change over time. Almost all the neuron models encountered in this book are dynamical systems. The models which incorporate active channels have at least four state variables, and multicompartmental models may have hundreds.

All dynamical systems have a number of characteristic behaviours. The state variables may reach a stationary state, such as when a pendulum is at rest. The system may behave periodically, for example, a moving pendu-

Fig 8.3 The Morris–Lecar model (Morris and Lecar, 1981) with a set of parameters giving Type I firing behaviour (a–d) and a set giving Type II behaviour (e–h). (a, e) The steady state of the Ca^{2+} activation variable m_∞ (black) and the steady state of the K^+ activation variable w_∞ (blue) as functions of voltage. (b, f) The time constant for τ_w (blue). There is no time constant for the m state variable, as it responds instantaneously to changes in voltage. (c, g) The time course of the membrane potential V and potassium activation variable w just above the firing threshold for each set of parameters. In (c), the injected current is $i_{el} = 40.3\,\mu A\,cm^{-2}$, and the action potentials are being fired at 2.95 Hz; arbitrarily low frequencies are possible with slightly less current. In (g), the injected current is $i_{el} = 89.9\,\mu A\,cm^{-2}$, and the action potentials are being fired at 9.70 Hz; significantly lower frequencies are not possible. (d, h) The f–I curves. In (d), the f–I curve is of Type I since the transition from not firing to firing is continuous, whereas the curve in (h) is Type II because of the discontinuous jump in firing frequency at the onset of firing.

Box 8.2 | Morris–Lecar model

In the formulation of Rinzel and Ermentrout (1998), the state variables are:

V, the membrane potential; and

w, the K^+ activation variable.

Other variables are:

i_{el}, the injected current;

I_i, the ionic current, comprising fast Ca^{2+}, slow K^+ and leak;

m_∞, the Ca^{2+} activation variable;

w_∞, the steady state K^+ activation;

τ_w, the K^+ activation time constant; and

ϕ, the temperature/time scaling.

$$C_m \frac{dV}{dt} = -I_i(V, w) + i_{el},$$

$$\frac{dw}{dt} = \frac{w_\infty(V) - w}{\tau_w(V)},$$

$$I_i(V, w) = \bar{g}_{Ca} m_\infty(V)(V - E_{Ca}) + \bar{g}_K w(V - E_K) + g_L(V - E_L),$$

$$m_\infty(V) = 0.5(1 + \tanh((V - V_1)/V_2)),$$

$$w_\infty(V) = 0.5(1 + \tanh((V - V_3)/V_4)),$$

$$\tau_w(V) = \phi/\cosh((V - V_3)/(2V_4)).$$

Type I parameters:

$$C_m = 20\,\mu F\,cm^{-2} \quad E_{Ca} = 120\,mV \quad \bar{g}_{Ca} = 4.0\,mS\,cm^{-2}$$
$$\phi = 0.066\,ms^{-1} \quad E_K = -84\,mV \quad \bar{g}_K = 8.0\,mS\,cm^{-2}$$
$$E_L = -60\,mV \quad g_L = 2.0\,mS\,cm^{-2}$$
$$V_1 = -1.2\,mV \quad V_2 = 18.0\,mV \quad V_3 = 12.0\,mV \quad V_4 = 17.4\,mV.$$

Type II parameters are same as Type I parameters, except: $V_3 = 2.0\,mV$, $V_4 = 30.0\,mV$, $\bar{g}_{Ca} = 4.4\,mS\,cm^{-2}$, $\phi = 0.04\,ms^{-1}$.

lum or the oscillating populations of predator and prey that can occur in the Lotka–Volterra model. There may be conditions under which aperiodic, or chaotic, behaviour occurs. **Dynamical systems theory** is the area of applied mathematics that seeks to determine under what conditions the various types of behaviour are exhibited.

This section gives a very brief introduction to concepts in dynamical systems theory, using the Morris–Lecar model of a single neuron (Box 8.2) as an example application. Although we have used a single neuron model as an example, dynamical systems theory has many other applications. A notable example is the **Wilson–Cowan oscillator**, the set of equations used by Wilson and Cowan (1972) to describe a network of coupled populations of excitatory and inhibitory neurons, to be described in Section 9.2.3.

A brief description of the mathematics of stability analysis is given in Box 8.3. For more details, see Edelstein–Keshet (1988) for the mathematics of bifurcations applied to biological problems, or Hale and Koçak (1991) who provide a more comprehensive and general treatment.

Box 8.3 | Stability analysis

The stability of an equilibrium point can be determined by considering small perturbations of the state variables from the equilibrium point. The equations of a 2D system of differential equations can be written:

$$\frac{dV}{dt} = f(V, w),$$

$$\frac{dw}{dt} = g(V, w),$$

where $f(V, w)$ and $g(V, w)$ are given non-linear functions.

At an equilibrium point (V_0, w_0), the two derivatives are equal to zero, $f(V_0, w_0) = g(V_0, w_0) = 0$. At a point $(V_0 + \Delta V, w_0 + \Delta w)$ close to the equilibrium point, the Taylor expansion of $f(V, w)$ is:

$$f(V_0 + \Delta V, w_0 + \Delta w) \approx f(V_0, w_0) + \frac{\partial f}{\partial V}\Delta V + \frac{\partial f}{\partial w}\Delta w = \frac{\partial f}{\partial V}\Delta V + \frac{\partial f}{\partial w}\Delta w,$$

where $\partial f/\partial V$ and $\partial f/\partial w$ are evaluated at (V_0, w_0). Writing a similar Taylor expansion for $g(V, w)$ allows the differential equations to be linearised:

$$\frac{d\Delta V}{dt} \approx \frac{\partial f}{\partial V}\Delta V + \frac{\partial f}{\partial w}\Delta w, \tag{a}$$

$$\frac{d\Delta w}{dt} \approx \frac{\partial g}{\partial V}\Delta V + \frac{\partial g}{\partial w}\Delta w. \tag{b}$$

The general solution of Equations (a) and (b) is:

$$\Delta V = Ae^{\lambda_1 t} + Be^{\lambda_2 t},$$

$$\Delta w = Ce^{\lambda_1 t} + De^{\lambda_2 t},$$

where A, B, C, D, λ_1 and λ_2 are constants. λ_1 and λ_2 are eigenvalues of the matrix:

$$\begin{pmatrix} \dfrac{\partial f}{\partial V} & \dfrac{\partial f}{\partial w} \\ \dfrac{\partial g}{\partial V} & \dfrac{\partial g}{\partial w} \end{pmatrix}.$$

This matrix is called the Jacobian matrix of Equations (a) and (b). The real parts, $\Re$, of the eigenvalues determine stability as follows:

$\Re(\lambda_1) < 0, \Re(\lambda_2) < 0$: stable equilibrium

$\Re(\lambda_1) > 0, \Re(\lambda_2) > 0$: unstable equilibrium

$\Re(\lambda_1) < 0, \Re(\lambda_2) > 0$ (or vice versa): saddle node.

The real parts determine the speed towards or away from the equilibrium. The imaginary parts determine the speed at which the trajectory circles around the point. At a point where there are one or two eigenvalues with zero real parts, there is a bifurcation.

8.2.1 The Phase Plane

The beauty of dynamical systems with two state variables is that they can be visualised using a 2D plot called the **phase plane** (Figure 8.4). Each axis corresponds to one of the state variables, V and w in the example described here. Each point in the phase plane represents a possible state (V, w) of the system. Figure 8.4a shows the phase (state) plane of the Morris–Lecar model with a set of parameters, V_1–V_4, that give Type II behaviour (Box 8.2, Figure 8.3),

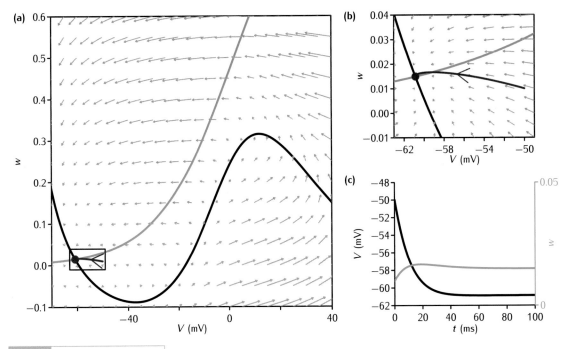

Fig 8.4 Phase plane analysis of Morris–Lecar neurons with Type II parameters and no injected current i_{el}. **(a)** The phase plane of the Morris–Lecar model. The red line with an arrow on it shows a trajectory in the phase space, which ends at the stable node, denoted by the red solid circle. The solid black line is the V-nullcline, and the light blue line is the w-nullcline. The rectangle highlights the portion of the phase plane that is shown in **(b)**. This highlights the stable equilibrium point $V = -60.85$ mV, $w = 0.01493$. **(c)** The time course of V (black line) and w (blue line) corresponding to the trajectory shown in **(a)**. Note that V and w settle to steady values of -60.9 mV and 0.0149, respectively, corresponding to their values at the equilibrium point in the phase plot.

and with the injected current parameter set to zero, $i_{el} = 0$ μA cm^{-2}. There are four types of information contained in this phase plane: a **trajectory** (the red line with the arrow on it), an **equilibrium point** (red, solid circle at the end of the trajectory), the **direction field**, indicated by arrows arranged in a grid, and two **nullclines** (the blue and black lines).

Trajectory and Equilibrium Point

The trajectory shows how the state of the system (V, w) changes over time. To visualise how it is drawn, imagine creating an animation in which each frame shows the location of the point (V, w) as a point in time. The trail left by this point is the trajectory, and the arrow on the trajectory indicates the direction in which the point moves. The trajectory shown in Figure 8.4a corresponds to the time courses of V and w obtained by numerical solution of the Morris-Lecar equations shown in Figure 8.4c. Given different initial conditions for V and w, the values of V and w move towards stable values V_0 and w_0. It can be seen from the close-up in Figure 8.4b that the trajectory ends at the pair of values (V_0, w_0). This point is called the equilibrium point, and is also referred to as a **node, steady state** or **singular point**.

Direction Field

The direction field shows the direction in which the state of the system tends to move. In the close-up in Figure 8.4b, the trajectory follows the flow suggested by the direction field. The horizontal component of each arrow plotted in the direction field is proportional to dV/dt for the values of V and w upon which the arrow is anchored, and the vertical component is proportional to dw/dt. Thus the direction of the arrow shows which direction the values of V and w will move in at that point in phase space, and the length

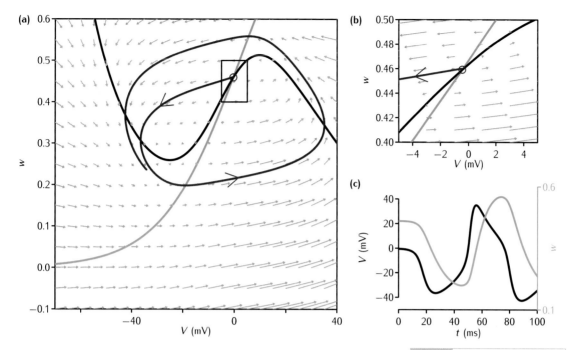

of the arrow indicates how fast it will do so. The direction field gives a feel for the behaviour of the system, but usually it is not plotted on phase planes.

Fig 8.5 Phase plane analysis of Morris–Lecar neurons with Type II parameters with injected current ($i_{el} = 150\,\mu\text{A cm}^{-2}$). Panels **(a–c)** correspond to Figure 8.4a–c. In this case, the node is unstable rather than stable. This unstable node is denoted by an unfilled circle.

Nullclines

Nullclines are points where a derivative is zero. The blue line in Figure 8.4a is the w-nullcline, showing which combinations of w and V give $\mathrm{d}w/\mathrm{d}t = 0$. The black line is the V-nullcline, where $\mathrm{d}V/\mathrm{d}t = 0$. The equilibrium point is the point of intersection of the nullclines. Here the state of the system will not change as both derivatives are zero. In the case of the Morris–Lecar neuron, the w-nullcline is the same as the steady-state activation curve for w (Figure 8.3e) since $\mathrm{d}w/\mathrm{d}t = 0$ when $w = w_{\infty}(V)$. The V-nullcline is obtained by setting the left-hand side of the equation for $\mathrm{d}V/\mathrm{d}t$ in Box 8.2 equal to zero to express w in terms of V.

8.2.2 Stability and Limit Cycles

In the enlargement of the phase plane around the equilibrium point (Figure 8.4b), the direction of the arrows suggests that, had the state of the system been started from any point close to the equilibrium point, the state of the system would be drawn inexorably towards the equilibrium point. This arrangement of arrows means that the equilibrium point is **stable**. Like a ball sitting at the bottom of a valley, if the state is moved slightly, it will return to the point of zero gradient.

Figure 8.5a shows how the phase plane changes when a constant current of $i_{el} = 150\,\mu\text{A cm}^{-2}$ is injected. The position of the w-nullcline is the same as in Figure 8.4a, but the V-nullcline has been shifted upwards and to the right, and there is a corresponding change in the direction field. There is still only one intersection between the nullclines. This intersection is an equilibrium point, but in contrast to Figure 8.4, it is **unstable**; this is denoted by an open circle rather than a solid one. The enlargement of the phase plane around the equilibrium point (Figure 8.5b) shows the arrows of the direction field pointing away from the equilibrium point. Like a ball at the top of a hill, if the state is exactly on the equilibrium point, it stays there. If the state is displaced even slightly from the equilibrium point, it will move away from the equilibrium point.

The trajectory plotted in Figure 8.5a and b shows the state moving away from the equilibrium point. Once the state has moved away from the equilibrium point, it never finds another such point. Instead, it starts to circle around the equilibrium point; this type of behaviour is called a **limit cycle**. Each loop of the state around the cycle corresponds to one period of an oscillation. In neurons, each oscillation corresponds to an action potential and the recovery period (Figure 8.5c).

It is possible to determine mathematically whether a given equilibrium point is stable or unstable (Box 8.3). Because of the non-linearity of the equations, generally it is not possible to compute the trajectory of the limit cycle analytically. However, for dynamical systems with two state variables, the **Poincaré–Bendixson theorem** (Hale and Koçak, 1991) sets out conditions under which limit cycles are guaranteed to exist.

8.2.3 Bifurcations

If the injected current is increased very gradually from zero, the V-nullcline will gradually shift its position from that in Figure 8.4 to that in Figure 8.5. There will always be an equilibrium point where the nullclines intersect, and for low values of injected current, it will be stable. However, at some threshold level of injected current, the equilibrium point will change abruptly from being stable to unstable. This abrupt change in behaviour is called a **bifurcation**. The parameter that is changed to induce the bifurcation, in this case the injected current i_{el}, is called the **bifurcation parameter**.

The Python-based PyDSTool has been used for all bifurcation analysis in this section.

Hopf bifurcations are also known as Andronov–Hopf bifurcations or Poincaré–Andronov–Hopf bifurcations.

The Hopf Bifurcation

Bifurcations come in many varieties, depending on exactly how the equilibrium point moves from being stable to unstable. In this case, when the bifurcation parameter passes through the bifurcation point, the behaviour of the system goes from being in equilibrium to being in a limit cycle with a non-zero frequency, making the f–I curve a Type II curve (Figure 8.3h). In the dynamical systems literature, this is called a **supercritical Hopf bifurcation**. According to the Poincaré–Andronov–Hopf theorem, the limit cycle that emerges at a Hopf bifurcation always has a non-zero frequency, which corresponds to the definition of Type II firing.

A **bifurcation diagram** is a summary of the types of behaviour a system exhibits. The bifurcation diagram for the Morris–Lecar neuron with the parameters used so far is given in Figure 8.6a. It was generated by solving the Morris–Lecar equations using the method of **numerical continuation** (Krauskopf et al., 2007), which is implemented in several software packages. The bifurcation parameter i_{el} is on the horizontal axis and one of the state variables, V, is on the vertical axis.

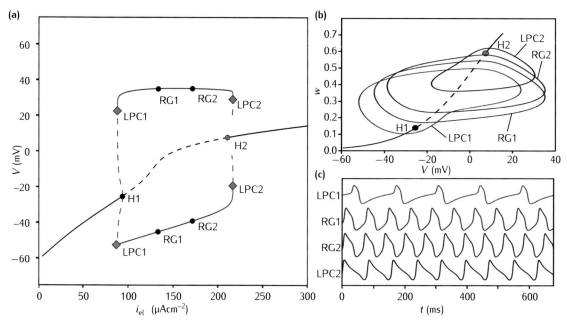

Fig 8.6 (a) Bifurcation diagram of the Morris–Lecar model with Type II parameters. The injected current i_{el} is on the x-axis and the membrane potential is plotted on the y-axis. For $i_{el} < 93.8\,\mu\text{A cm}^{-2}$ or for $i_{el} > 212.0\,\mu\text{A cm}^{-2}$, the membrane potential settles to a steady value, which is represented by the single solid line. At intermediate values, the equilibrium indicated by the dashed line is unstable, and the system enters a limit cycle (oscillation). The maximum and minimum values of the oscillation are represented by the solid lines. (b) The stable equilibrium points (black lines) and sample limit cycles plotted in phase space. (c) The time course of the membrane potential corresponding to the limit cycles plotted in the phase space shown in (b).

Starting with values of i_{el} to the left of point H1 in Figure 8.6a, the solid line shows the equilibrium value of V for a particular value of i_{el}. The point labelled H1 is a **bifurcation point**, called a supercritical Hopf bifurcation, at which the equilibrium point changes from being stable to unstable. The line continues as a dashed line to point H2, representing the value of V at the unstable equilibrium. Between H1 and H2, the only stable solution is the limit cycle, which is represented by the pair of solid lines between the bifurcation points LPC1 and LPC2. The heights of the lines show the maximum and minimum values of V during the limit cycle. At point H2, a **subcritical Hopf bifurcation**, the equilibrium point becomes stable again. The points RG1 and RG2 denote two examples of parameters between LPC1 and LPC2. Limit cycle trajectories for the points LPC1, RG1, RG2 and LPC2 are shown in the phase plane in Figure 8.6b. Figure 8.6c shows the corresponding time courses of V.

In the small range between LPC1 (at $88.2\,\mu\text{A cm}^{-2}$) and H1 (at $93.9\,\mu\text{A cm}^{-2}$), there are two stable solutions: the stable equilibrium and the limit cycle. There is also an unstable limit cycle, represented by the dashed lines between H1 and LPC1. The equilibrium solution can be reached by increasing the current slowly from below, and the stable limit cycle solution can be reached by decreasing the current slowly from above. This type of multistability is known as **hysteresis**. Similarly, between H2 ($i_{el} = 212.0\,\mu\text{A cm}^{-2}$) and LPC2 ($216.9\,\mu\text{A cm}^{-2}$), there is a stable equilibrium, a stable limit cycle and an unstable limit cycle.

The Saddle-Node Bifurcation

We now investigate the phase plane and bifurcation diagram of the Morris–Lecar neuron with the setting of the parameters $V_1 - V_4$ (Box 8.2) that gives Type I firing behaviour.

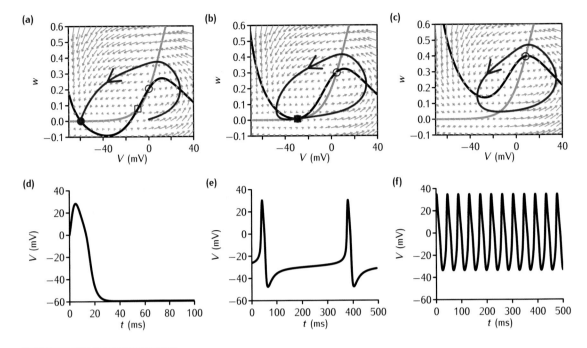

Fig 8.7 Dynamical systems analysis of Morris–Lecar neurons with Type I parameters **(a–c)** and corresponding trajectories of the membrane potential **(d–f)**. **(a)** With no applied current $(i_{el} = 0)$, the system has one stable node (filled circle) at $V = -59.5$ mV, $w = 2.70 \times 10^{-4}$, an unstable node (unfilled circle) and a saddle point (unfilled square). **(d)** When the system state is started from above threshold, one action potential results. **(b)** The saddle node bifurcation when $i_{el} = 40.4\,\mu A\,cm^{-2}$, where the stable node and the saddle point merge, but the unstable node remains. A limit cycle trajectory emerges, though its frequency is very low **(e)**. **(c)** A higher level of current injection $(i_{el} = 96.8\,\mu A\,cm^{-2})$. The saddle point and stable point have disappeared. There is one limit cycle, and its frequency is higher **(f)**.

In the phase plane shown in Figure 8.7a, in which there is no applied current, the V-nullcline is the same as the phase plane for Type II parameters with no injected current (Figure 8.4a), but the w-nullcline is steeper and shifted to the right. This has the effect of creating three intersections between the nullclines and so there are three equilibrium points.

A stability analysis shows that the left-most equilibrium point is stable and the right-most one is unstable. The equilibrium point in the middle, denoted by the open square, is a type of unstable equilibrium point called a **saddle node**. A saddle node is like the bottom of a dip in a mountain ridge: in both directions along the ridge, the gradient is upwards, but in the directions at right angles, the gradient is downwards. If the system state lies on the saddle node, small perturbations from it would cause the state to return to the stable equilibrium point. A large perturbation causes just one action potential, as shown in the trajectory in Figure 8.7a and the voltage trace in Figure 8.7d.

As more current is injected, the V-nullcline changes until the stable point and the saddle point merge (Figure 8.7b) and then disappear (Figure 8.7c). The type of bifurcation that occurs when the saddle node and an equilibrium point merge is called a **saddle node bifurcation**. At this point, the V and w variables start to oscillate. However, in contrast with the Hopf bifurcation, the oscillation is of very low frequency when i_{el} is just above its value at the saddle-node bifurcation (Figure 8.7e), and increases steadily above this value. This makes the Morris–Lecar neuron with the second set of parameters fire in a Type I fashion (Figure 8.7f).

The bifurcation diagram of the Morris–Lecar model with Type I parameters is shown in Figure 8.8. For values of i_{el} between $-9.95\,\mu A\,cm^{-2}$ (LP2)

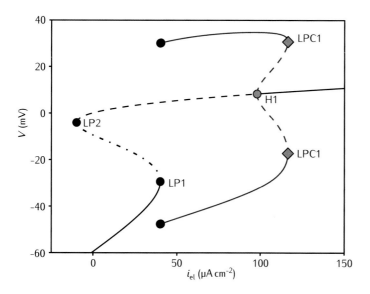

and 40.0 µA cm^{-2} (LP1), there are three fixed points. The stable fixed point is shown by the solid line, the unstable fixed point by the dashed line and the saddle node by the dash-dotted line. Above LP1 (at $i_{el} = 40.0$ µA cm^{-2}), the saddle-node bifurcation, both the stable node and the saddle node disappear, leaving the unstable node and the stable limit cycle. The bifurcation point H1 (at $i_{el} = 98.1$ µA cm^{-2}) is a subcritical Hopf bifurcation. As with the Type II neurons, the equilibrium point changes from being unstable to stable again. Between H1 and the bifurcation point LPC1 (at 116.4 µA cm^{-2}), both the equilibrium and limit cycle solutions are possible. Above LPC1, only the stable equilibrium point exists.

8.2.4 Applications of Dynamical Systems Theory

The application of dynamical systems theory to the Morris–Lecar neuron demonstrates that the types of dynamical behaviour a system exhibits can depend crucially on subtle changes in parameters. Bifurcation analysis allows us to explore a wide range of parameter settings systematically, and identify ranges of input in which the system exhibits hysteresis. Wilson (1999) provides examples of the dynamical systems theory in the context of neuroscience, including in networks.

Although conductance-based models are often reduced to 2D models to facilitate dynamical systems analysis (Izhikevich, 2007), dynamical systems theory can be applied to models with more than two dynamic variables, and it is also possible to explore more than one bifurcation parameter. For example, Verma et al. (2020) analyse the effect of the bifurcation parameters of injection current and various conductance on a HH-style model of the dorsal root ganglion cell with nine state variables. Even in two dimensions, there are many more types of bifurcations than the ones presented here. Ermentrout and Terman (2010) and Izhikevich (2007) present more types of bifurcations in the context of neural models.

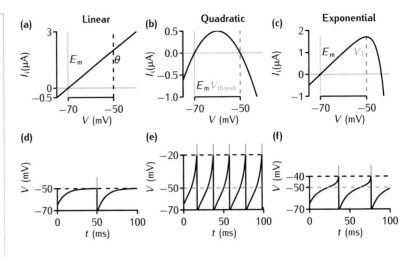

Fig 8.9 I–V curves for linear and non-linear integrate-and-fire models, and examples of firing. **(a)** I–V curve for the linear integrate-and-fire model (black) with $E_m = -70$ mV (grey line) and $R_m = 10$ kΩ; zero current shown by grey line, and firing threshold θ shown by dashed black line. **(b)** I–V curve for the quadratic integrate-and-fire model (Equation 8.1). E_m and R_m as for linear integrate-and-fire; $V_{thresh} = -50$ mV. V_{thresh} indicated by dashed blue line and θ not shown. **(c)** I–V curve for the exponential integrate-and-fire model (Equation 8.2). E_m and R_m as for linear integrate-and-fire model; $V_T = -50$ mV (dashed blue line), $\Delta_T = 3$ mV (not shown). **(d–f)** Membrane potential (black) and spikes (blue) produced by the three models in a cell with $C_m = 1$ μF (making $\tau_m = 10$ ms) and injected current $I = 0.201$ μA cm^{-2}. In the quadratic **(e)** and exponential **(f)**, the membrane potential is reset at $\theta = -20$ mV or $\theta = -40$ mV and an inflection in the membrane potential can be seen around V_{thresh} and V_T.

8.3 Extended Integrate-and-Fire Neurons

In the neuron models discussed so far in this chapter, spiking behaviour is described using two or more coupled non-linear differential equations. This allows a detailed understanding of how action potentials and more complex spiking behaviours (such as bursting; Section 8.1.2) arise. Despite the complexity inherent in the generation of action potentials, in many cases, their time course and the conditions required for their initiation can be characterised by initiating a spike when the membrane potential reaches a certain threshold – which is the integrate-and-fire neuron, introduced in Section 2.7.

One advantage of the basic integrate-and-fire neuron is its faster speed of simulation compared to a model which generates spikes using realistic conductances. However, this speed of simulation comes at the expense of a poor fit to the ionic current in a spiking neuron when the membrane potential is close to, and below, the threshold. In a basic, linear integrate-and-fire neuron, the closer the membrane potential is to the threshold, the greater the outward ionic current (Figure 8.9a). In contrast, in a model with active conductances, in the neighbourhood of the threshold, the ionic current goes from being outward to inward.

8.3.1 Quadratic Integrate-and-Fire Model

A hybrid model that models the ionic current close to the threshold better than the linear integrate-and-fire neuron is the **quadratic integrate-and-fire neuron** (Hansel and Mato, 2000; Latham et al., 2000). It replaces the linear dependence of the ionic current $I_i = (V - E_m)/R_m$ in the integrate-and-fire neuron (Equation 2.17, Figure 8.9a) with a quadratic function of V that is zero both at the resting potential E_m and a threshold V_{thresh}:

$$C_m \frac{dV}{dt} = -I_i + I = -\frac{(V - E_m)(V_{thresh} - V)}{R_m(V_{thresh} - E_m)} + I, \tag{8.1}$$

where I is the current flowing into the cell via an electrode or synapses. The I–V characteristic of ionic current I_i is plotted in Figure 8.9b, and an example of firing behaviour in Figure 8.9e. In contrast to the behaviour of the linear integrate-and-fire model with a just-superthreshold current injected (Figure 8.9d) when the membrane potential exceeds the threshold V_{thresh}, this quadratic term becomes positive, causing the neuron to depolarise still further. At a preset value of the membrane potential θ, the membrane potential is reset to the voltage V_{reset}, which can differ from the resting membrane potential.

8.3.2 Exponential Integrate-and-Fire Model

Another non-linear variation on the standard integrate-and-fire neuron is the **exponential integrate-and-fire neuron** (Fourcaud-Trocmé et al., 2003). Its form is similar to a linear integrate-and-fire neuron, except that there is an additional current that depends on the exponential of the voltage:

$$C_{\text{m}} \frac{dV}{dt} = -\left(\frac{V - E_{\text{m}}}{R_{\text{m}}} - \frac{\Delta_{\text{T}}}{R_{\text{m}}} \exp\left(\frac{V - V_{\text{T}}}{\Delta_{\text{T}}} \right) \right) + I, \qquad (8.2)$$

where I is the current flowing into the cell via an electrode or synapses, and where V_{T} is a threshold voltage and Δ_{T} is a spike slope factor that determines the sharpness of spike initiation. The I–V characteristic of the ionic current is plotted in Figure 8.9c, and an example firing behaviour in Figure 8.9f. As with the quadratic neuron, the current is zero, very close to E_{m}, rises to a maximum at $V = V_{\text{T}}$, and then falls, going below zero. If the neuron is initialised with a membrane potential below the upper crossover point, the membrane potential decays back to the resting membrane potential E_{m}. However, if the membrane potential is above the crossover point, the membrane starts to depolarise further, leading to a sharp increase in voltage, similar to the start of an action potential in the HH model. There is also a threshold θ, which is greater than V_{T}, at which the membrane potential is reset to V_{reset}.

The curve differs from the quadratic curve in that it is asymmetrical, and has a larger linear region around E_{m}. Also, the exponential dependence of the additional current matches the lower part of the sodium activation curve better than the quadratic form, suggesting that this is likely to be a more accurate simplification of neurons with fast sodium currents than the quadratic neuron. The exponential integrate-and-fire neuron behaves in a similar way to the quadratic integrate-and-fire neuron. However, differences in the behaviour of the two types of neuron are apparent during fast-spiking behaviour, as the quadratic integrate-and-fire neuron tends to take longer to produce a spike (Fourcaud-Trocmé et al., 2003).

8.4 | Integrate-and-Fire Models with Adaptation

The integrate-and-fire models described so far are able to model regularly firing cells with Type I characteristics, but cannot produce behaviours such as Type II firing, firing rate adaptation or bursting, all of which are observed in many real neurons in response to current injection. This section introduces extensions to the integrate-and-fire model that enable it to reproduce a wider repertoire of intrinsic neuronal firing patterns. In all cases, a second dynamical variable is added, which underpins the adaptation.

A further motivation for these extensions is to produce predictive models of real neurons. Creating these models requires data from real neurons, usually the membrane potential (or just spike time) response to a noisy current injected into the soma. The same noisy current is then injected into the model neuron and the model's parameters are tuned until its membrane potential time course, or spike times, resembles the real data as closely as possible. In order to quantify the goodness of fit of a model with the data, metrics have to be defined, for example, the fraction of spikes in the model that occur within 2 ms of spikes from the real neuron. The same procedure can be used when comparing two different model neurons.

Integrate-and-fire neuron models, such as the adaptive exponential integrate-and-fire (AdEx) model (Section 8.4.3), have won the Quantitative Single-Neuron Modelling Competition (Jolivet et al., 2008).

8.4.1 Firing-Rate Adaptation

In many types of neuron, the firing rate in response to a sustained current injection decreases throughout the spike train. The integrate-and-fire models described so far cannot exhibit this behaviour. However, it is possible to incorporate this behaviour into integrate-and-fire models by incorporating an extra conductance I_{adapt} that depends on the neuronal spiking (Koch, 1999; Latham et al., 2000). Whenever the neuron spikes, the adaptive conductance g_{adapt} is incremented by an amount Δg_{adapt} and otherwise it decays with a time constant of τ_{adapt}:

$$\frac{\mathrm{d}g_{\text{adapt}}}{\mathrm{d}t} = -\frac{g_{\text{adapt}}}{\tau_{\text{adapt}}} \quad \text{and} \quad I_{\text{adapt}} = g_{\text{adapt}}(V - E_{\text{m}}). \tag{8.3}$$

Figure 8.10 shows the response to a constant level of current injection of an integrate-and-fire neuron incorporating such an adapting conductance. It can be seen (Figure 8.10a) that the interspike intervals (ISIs) increase over time until, after about 10 ms, they are constant. The reason for this slowing down is the gradual build-up of the adapting conductance seen in Figure 8.10b.

A related approach is to make the threshold a variable that depends on the time since the neuron last spiked (Geisler and Goldberg, 1966). One possible function is a decaying exponential function:

$$\theta(t) = \theta_0 + \theta_1 \exp((t - t_{\text{s}})/\tau_{\text{r}}), \tag{8.4}$$

where t_{s} is the time since the neuron spiked and τ_{r} is a refractory time constant.

8.4.2 The Izhikevich Model

Model neurons that produce a wide range of realistic behaviours can be constructed by adding a recovery variable to the quadratic integrate-and-fire model. One example of this type of model is the **Izhikevich model** (Izhikevich, 2003; Izhikevich and Edelman, 2008), defined by the following equations:

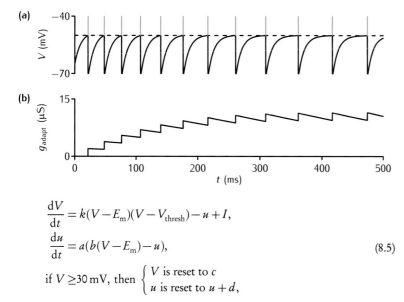

Fig 8.10 Response of an adapting integrate-and-fire neuron to a constant level of current injection. **(a)** The time course of the membrane potential (black) and the spikes resulting from crossing the threshold of $-50\,\text{mV}$ (blue). **(b)** The time course of the adaptation conductance g_{adapt}. The basic integrate-and-fire parameters are $R_{\text{m}} = 10\,\text{k}\Omega$, $C_{\text{m}} = 1\,\mu\text{F}$, $\theta = -50\,\text{mV}$ and $E_{\text{m}} = -70\,\text{mV}$, and the membrane time constant is $\tau_{\text{m}} = R_{\text{m}} C_{\text{m}} = 10\,\text{ms}$. The adaptation parameters are: $\tau_{\text{adapt}} = 300\,\text{ms}$ and $\Delta g_{\text{adapt}} = 2\,\mu\text{S}$. The level of current injection is $2.2\,\mu\text{A}$.

$$\frac{\mathrm{d}V}{\mathrm{d}t} = k(V - E_{\text{m}})(V - V_{\text{thresh}}) - u + I,$$

$$\frac{\mathrm{d}u}{\mathrm{d}t} = a(b(V - E_{\text{m}}) - u), \tag{8.5}$$

$$\text{if } V \geq 30\,\text{mV, then } \begin{cases} V \text{ is reset to } c \\ u \text{ is reset to } u + d, \end{cases}$$

where u is the recovery variable, meant to represent the difference between all inward and outward voltage-gated currents, and k, a, b, c and d are parameters. As with other integrate-and-fire models, the various terms used in the equation specifying the model are justified primarily because they reproduce firing behaviour rather than arising directly from the behaviour of ion channels. Figure 8.11 illustrates a number of the behaviours that the model can exhibit for various settings of the parameters and various current injection protocols (Izhikevich, 2004). This model is efficient to simulate as part of a network, and can be implemented using Euler integration (Section 5.7.1). For an example of a very large-scale network implemented using these neurons, see Izhikevich and Edelman (2008).

8.4.3 Adaptive Exponential Integrate–and–Fire Model

Another commonly used model that, like the Izhikevich model, can produce a wide variety of firing patterns is the **adaptive exponential integrate-and-fire (AdEx)** (Brette and Gerstner, 2005). This model combines the exponential integrate-and-fire model for spike initiation, with an extra equation to account for sub-threshold and spike-triggered adaptation. The membrane potential equation for this model is given by:

$$C_{\text{m}} \frac{\mathrm{d}V}{\mathrm{d}t} = -\left(\frac{V - E_{\text{m}}}{R_{\text{m}}} - \frac{\Delta_{\text{T}}}{R_{\text{m}}} \exp\left(\frac{V - V_{\text{T}}}{\Delta_{\text{T}}} \right) \right) + I - w, \tag{8.6}$$

which is the same as Equation 8.2, except that an **adaptation current** w is added on the right-hand side. Whenever the neuron spikes, this adaptation current is increased by an amount Δw; for the rest of the time, it evolves according to:

$$\tau_{\text{w}} \frac{\mathrm{d}w}{\mathrm{d}t} = a(V - E_{\text{m}}) - w. \tag{8.7}$$

Biophysically, this adaptation current corresponds to the small changes in current arising because of small changes in the fraction of channels open

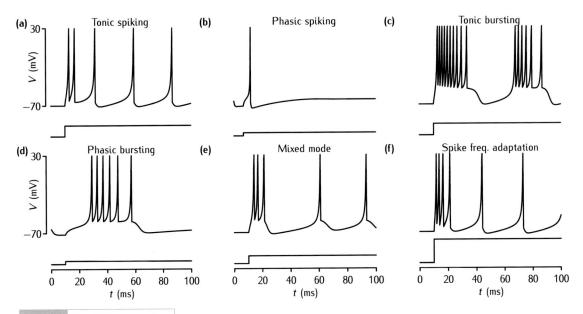

Fig 8.11 The Izhikevich model. A number of classes of waveforms can be produced using various settings of the parameters a, b, c and d. In all plots, $k = 0.04\,\text{V}\,\text{s}^{-1}$, $E_m = -70\,\text{mV}$ and $V_{\text{thresh}} = -55\,\text{mV}$. Parameters for each subplot, with units omitted for clarity: **(a)** $a = 0.02$, $b = 0.2$, $c = -65$, $d = 6$; **(b)** $a = 0.02$, $b = 0.25$, $c = -65$, $d = 6$; **(c)** $a = 0.02$, $b = 0.2$, $c = -50$, $d = 2$; **(d)** $a = 0.02$, $b = 0.25$, $c = -55$, $d = 0.05$; **(e)** $a = 0.02$, $b = 0.2$, $c = -55$, $d = 4$; **(f)** $a = 0.01$, $b = 0.2$, $c = -65$, $d = 8$. Current steps are shown below each voltage trace and are all to the same scale; the step in **(f)** has a height of $30\,\text{V}\,\text{s}^{-1}$. Figure adapted from Izhikevich (2004) and generated by code derived from that available at www.izhikevich.com.

in response to small changes in membrane potential. As with the Izhikevich model, a wide variety of firing patterns can be obtained with different choices of the parameters a and Δw (Gerstner et al., 2014). Adaptation currents can also be added to the quadratic integrate-and-fire neuron (Latham et al., 2000).

8.5 Incorporating Noise into Integrate–and–Fire Neurons

Thus far, we have considered the effect of deterministic current inputs to the basic and extended integrate-and-fire neuron models. However, unpredictable, stochastic currents contribute to the changing membrane potential in real neurons. For example, the random opening and closing of channels leads to noisy membrane currents (Chapter 4) and the firing times of neurons impinging on a neuron may also be stochastic, leading to stochastic input currents. The effect of both intrinsic noise (stochastic channels) and extrinsic noise (stochastic inputs) on the membrane potential can be modelled using **diffusive noise**. An alternative method of introducing noise is to combine deterministic integration with a noisy threshold (Gerstner et al., 2014). This form of noise, known as **escape noise**, is described in Section 8.7.

8.5.1 Intrinsic Diffusive Noise

To model diffusive noise in integrate-and-fire neurons, we separate the input current into deterministic and stochastic parts:

$$I(t) = I^{\text{det}}(t) + I^{\text{noise}}(t) \tag{8.8}$$

The deterministic part of the current $I^{\text{det}}(t)$ varies predictably over time. It might be an injection current, or a synaptic current. The stochastic part of the current $I^{\text{noise}}(t)$ is unpredictable, and might arise from channel openings and closings (Section 4.7.3), noise in the electronics controlling the current injection or the spontaneous release of vesicles at synapses (Section 7.3). Another source of noise is the random arrival of action potentials from other neurons (Section 8.5.3).

Although we cannot predict the stochastic part of the current, we assume that we do know how it is distributed at any point in time, and how strongly correlated its values at different points in time are. In what follows, we will assume that the stochastic current has a Gaussian distribution with mean of zero and standard deviation σ_I. We also assume that the value of I^{noise} at time t is completely uncorrelated with the value of I^{noise} at all other times t'. This type of noise is referred to as 'white noise', and means that the stochastic part of the current can vary rapidly. If there were correlations between the values at different times, the noisy part of the current would vary more smoothly over time.

By substituting this current into the membrane equation (Equation 2.21), we obtain a version of the membrane equation with deterministic and stochastic currents:

$$\tau_{\text{m}} \frac{\text{d}V}{\text{d}t} = -V + E_{\text{m}} + R_{\text{m}} I^{\text{det}}(t) + R_{\text{m}} I^{\text{noise}}(t), \tag{8.9}$$

where the membrane time constant is $\tau_{\text{m}} = R_{\text{m}} C_{\text{m}}$. Although it is possible to express diffusive noise in the formal mathematical framework of stochastic differential equations, for the purpose of running simulations, it is sufficient to understand the noise in the context of a simulation using forward Euler integration:

$$V(t + \Delta t) = V(t) + \frac{1}{\tau_{\text{m}}} \left(\left(E_{\text{m}} - V(t) + R_{\text{m}} I^{\text{det}}(t) \right) \Delta t + R_{\text{m}} \sigma_I \Delta W(t) \right), \tag{8.10}$$

where Δt is the time-step and $\Delta W(t)$ is a random variable drawn from a Gaussian distribution with a mean of zero and a variance of Δt. The firing and membrane potential reset are the same as in a conventional integrate-and-fire neuron.

The $\sigma_I \Delta W(t)$ term generates the noisy current $I^{\text{noise}}(t)$. It causes jumps up and down in the membrane potential. Like jumps in Brownian motion which cause the position of a molecule to diffuse over time, these jumps cause the membrane potential to diffuse over time, giving rise to the term 'diffusive noise'. The reason that the variance of ΔW, rather than its standard deviation, scales with Δt is that the variance of the sum of independent random numbers is the sum of the variances of the random numbers.

If the voltage starts at 0, after a time Δt, we would expect the mean absolute difference in voltage $|V(\Delta t)|$ to be proportional to $\sqrt{\Delta t}$, just as in Brownian motion, where the mean distance diffused is proportional to the square root of the time. The dimensions of ΔW are therefore the square root of time, and the dimensions of σ_I are current times by the square root of time.

Figure 8.12 shows an example of the evolution of the membrane potential through time and the firing pattern of a noisy integrate-and-fire neuron subjected to either constant or periodic inputs. With constant deterministic current input $I^{\text{det}}(t)$, the noisy inputs $I^{\text{noise}}(t)$ allow the neuron to fire

More generally, ΔW can be drawn from any **Wiener process**, a class of stochastic processes of which Brownian motion is one example. See Tuckwell (1988) or Gardiner (1985) for more on the theory of Wiener processes.

Others (e.g. Gerstner et al., 2014) prefer to insert a τ_{m} in Equation 8.10, meaning that σ_I has the dimensions of current, but also effectively scaling the noise with the membrane time constant. We prefer to have the slightly peculiar units for σ_I so that the effect of changing the membrane time constant on the behaviour of the neuron is clear.

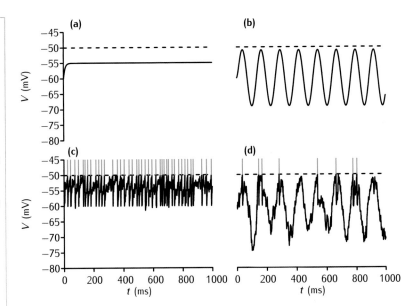

Fig 8.12 The effects of noise in integrate-and-fire neurons. The upper traces show the membrane potential of a noiseless integrate-and-fire neuron presented with either **(a)** a constant current injection or **(b)** a sinusoidally varying current, neither of which is strong enough to take the neuron to threshold. The lower traces show the membrane potential (black line) and spike times (blue strokes) produced in response to the same inputs by a noisy integrate-and-fire neuron described by Equation 8.10 with $R_m \sigma_I = 10 \, \text{mV} \, \text{ms}^{1/2}$. **(c)** In the case of constant input, the integrate-and-fire neuron fires, albeit irregularly. **(d)** In the case of the sinusoidal input, the noise is sufficient to cause the neuron to fire at some of the peaks of the oscillations. This is an example of stochastic resonance. In all simulations: $E_m = -60 \, \text{mV}$, $\tau_m = 10 \, \text{ms}$ and $\theta = -50 \, \text{mV}$. The time-step $\Delta t = 1 \, \text{ms}$.

(Figure 8.12c), even when the deterministic input would not be enough to cause a deterministic neuron with the same threshold to fire (Figure 8.12a). In contrast to the regular firing of a purely deterministic integrate-and-fire neuron with constant superthreshold input, the firing of the noisy neuron with sub-threshold constant input is irregular.

When presented with a sub-threshold periodic stimulus (Figure 8.12b), added noise allows the neuron to fire around some of the peaks of the periodic stimulus (Figure 8.12d). This phenomenon, whereby noise effectively uncovers an underlying periodic stimulus, is known as **stochastic resonance** (Benzi et al., 1981) and has been demonstrated in the crayfish mechanosensory system (Douglas et al., 1993).

Since with noise, the sub-threshold input can contribute to the firing of the neuron, the f–I curves presented in Section 2.7 are not a good description of the firing of the noisy integrate-and-fire neuron. However, it is possible to derive analytically the mean firing rate for an input current with a given mean and standard deviation (Tuckwell, 1988; Amit and Tsodyks, 1991a). The first stage is to consider an integrator with no threshold mechanism, and compute the time course of the mean depolarisation and the variance in the depolarisation. As shown in Box 8.4, for a neuron starting from the resting membrane potential E_m at $t = 0$, the time courses of the mean and variance of the membrane potential are:

$$\text{E}[V(t)] = V(0)\exp(-t/\tau_m) + \mu_V(1 - \exp(-t/\tau_m)),$$
$$\text{Var}[V(t)] = \sigma_V^2(1 - \exp(-2t/\tau_m)), \tag{8.11}$$

where the steady-state values of the mean and variance of the membrane potential are:

$$\mu_V = E_m + R_m I^{\text{det}} \quad \text{and} \quad \sigma_V^2 = \frac{R_m^2 \sigma_I^2}{2\tau_m}. \tag{8.12}$$

The variance of the membrane potential is inversely proportional to the membrane time constant. This can be understood by analogy with the vari-

Box 8.4 Mean and variance of membrane potential

With constant current input I^{det}, the theory of noiseless integrate-and-fire neurons states that the steady-state depolarisation is:

$$\mu_V = E_m + R_m I^{det}. \tag{a}$$

Using this definition and gathering terms, we can rewrite Equation 8.10:

$$V(t + \Delta t) = V(t) \left(1 - \frac{\Delta t}{\tau_m}\right) + \frac{\Delta t}{\tau_m} \left(\mu_V + \frac{R_m \sigma_I \Delta W(t)}{\Delta t}\right). \tag{b}$$

Suppose we run a simulation for time t, which we divide into n time-steps, so $\Delta t = t/n$. If we iterate Equation (b), we have:

$$V(t) = V(0) \left(1 - \frac{\Delta t}{\tau_m}\right)^n + \frac{\Delta t}{\tau_m} \sum_{i=0}^{n-1} \left(\mu_V + \frac{R_m \sigma_I \Delta W(t - (i+1)\Delta t)}{\Delta t}\right) \left(1 - \frac{\Delta t}{\tau_m}\right)^i. \tag{c}$$

From the sum of a geometric series, we can show that:

$$\frac{\Delta t}{\tau_m} \sum_{i=0}^{n-1} \left(1 - \frac{\Delta t}{\tau_m}\right)^i = 1 - \left(1 - \frac{\Delta t}{\tau_m}\right)^n, \tag{d}$$

and as $n \to \infty$, $(1 - \frac{\Delta t}{\tau_m})^n \to \exp(-n\Delta t/\tau_m) = \exp(-t/\tau_m)$, as can be shown by equating the binomial expansion to the Taylor expansion of $\exp(-t/\tau_m)$. In the limit $n \to \infty$, the summation in Equation (c) can be recast:

$$V(t) = V(0) \exp\left(-\frac{t}{\tau_m}\right) + \sum_{i=0}^{n-1} \mu_V \frac{\Delta t}{\tau_m} \exp\left(-\frac{i\Delta t}{\tau_m}\right)$$
$$+ \sum_{i=0}^{n-1} R_m \sigma_I \frac{\Delta W(t - (i+1)\Delta t)}{\tau_m} \exp\left(-\frac{i\Delta t}{\tau_m}\right). \tag{e}$$

Since the ΔW are normally distributed with mean 0, the expectation of the membrane potential over time is:

$$E[V(t)] = V(0) \exp\left(-\frac{t}{\tau_m}\right) + \sum_{i=0}^{n-1} \mu_V \frac{\Delta t}{\tau_m} \exp\left(-\frac{i\Delta t}{\tau_m}\right). \tag{f}$$

In the limit as $n \to \infty$, we can replace the summation with integration:

$$E[V(t)] = V(0) \exp\left(-\frac{t}{\tau_m}\right) + \mu_V \int_0^t \frac{1}{\tau_m} \exp\left(-\frac{t - t'}{\tau_m}\right) dt'$$
$$= V(0) \exp\left(-\frac{t}{\tau_m}\right) + \mu_V \left(1 - \exp\left(-\frac{t}{\tau_m}\right)\right). \tag{g}$$

which is the first expression in Equation 8.11. Since the variance of each noise term is Δt, the total variance of the noise terms is:

$$\text{Var}\left[\sum_{i=0}^{n-1} R_m \sigma_I \frac{\Delta W(t - (i+1)\Delta t)}{\tau_m} \exp\left(-\frac{i\Delta t}{\tau_m}\right)\right] = \sum_{i=0}^{n-1} R_m^2 \sigma_I^2 \frac{\Delta t}{\tau_m^2} \exp\left(-\frac{2i\Delta t}{\tau_m}\right). \tag{h}$$

In the limit as $n \to \infty$, we can replace the summation with integration:

$$\text{Var}[V(t)] = \frac{R_m^2 \sigma_I^2}{\tau_m^2} \int_0^t \exp\left(-\frac{2(t - t')}{\tau_m}\right) dt' = \frac{R_m^2 \sigma_I^2}{2\tau_m} \left(1 - \exp\left(-\frac{2t}{\tau_m}\right)\right), \tag{i}$$

which is the second expression in Equation 8.11.

ance of the mean in statistics – a longer membrane time constant corresponds to a larger number of samples, and therefore a smaller standard error in the mean.

We now consider firing when the membrane potential reaches a threshold θ. If the mean depolarisation μ_V is of the order of at least σ_V below the threshold, the neuron will not fire very often, and it is possible to derive an expression for the mean firing frequency in terms of μ_V and σ_V, as shown in Figure 8.13. For low levels of noise (σ_V small), the dependence of the firing frequency on the mean input is similar to the dependence on current injection (Figure 2.16b). As the level of noise increases, the firing rate curve is smoothed out so that the effective threshold of the neuron is reduced. The expression plotted in Figure 8.13 for the average firing rate in terms of μ_V and σ_V is (Tuckwell, 1988; Amit and Brunel, 1997b):

$$f(\mu_V, \sigma_V) = \left(\tau_r + \tau_m \int_{-\mu_V/\sigma_V}^{(\theta - \mu_V)/\sigma_V} \sqrt{\pi} \exp(u^2)(1 + \mathrm{erf}\, u)\,\mathrm{d}u \right)^{-1}, \qquad (8.13)$$

where $\mathrm{erf}\, u$ is the error function, defined in Box 5.5.

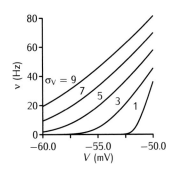

Fig 8.13 Mean firing frequency v of the diffusive noise model as a function of the mean μ_V and standard deviation σ_V (indicated in millivolts) of the membrane potential evoked by Poisson spike trains. Parameters: threshold $\theta = -50\,\mathrm{mV}$, resting membrane potential $E_m = -60\,\mathrm{mV}$, refractory period $\tau_r = 0\,\mathrm{ms}$ and membrane time constant $\tau_m = 10\,\mathrm{ms}$.

8.5.2 Populations of Noisy Integrate-and-Fire Neurons

An important application of noisy integrate-and-fire neurons is to investigate how effective groups of neurons are at transmitting time-varying or transient inputs. This is particularly motivated by the early stages of sensory systems, in which, although the firing patterns of individual cells may not be tightly coupled to the stimulus, the overall firing frequency of the population can be tightly coupled to the input.

For example, Hospedales et al. (2008) have used integrate-and-fire neurons to investigate how noise might affect the response fidelity of medial vestibular neurons, which are involved in the vestibulo-ocular reflex. They set up a model population of independently firing medial vestibular neurons, each of which fires persistently due to a constant pacemaker component of the input current $I(t)$. At the start of the simulation, each cell is initialised with a different membrane potential. In a population of cells where there is no diffusive noise, this initialisation means that each cell will be at a different point or phase in its regular cycle of firing and then being reset. If there is no additional input to the cells, they will continue firing out of phase with each other.

However, when a sinusoidal component is added to the signal, the cells tend to fire at the peak of the cycle. This leads to the firing times of the population of cells becoming synchronised over successive cycles of the sinusoidal input component (Figure 8.14a). When a population of 100 such cells is considered, the instantaneous population firing rate has sharp peaks, locked to a particular point in the cycle (Figure 8.14b). In contrast, when some diffusive noise is added to the neuron, the firing times of the neurons become desynchronised (Figure 8.14c). This leads to the population firing rate being a more accurate, albeit noisy, reproduction of the input current (Figure 8.14d).

The simplicity of integrate-and-fire neurons has allowed this type of question to be investigated analytically. Knight (1972) used elegant mathematical

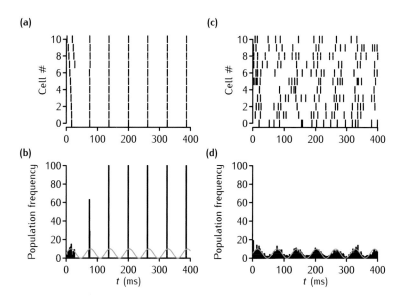

(a)

(b)

(c)

(d)

Fig 8.14 Signal transmission in deterministic and noisy integrate-and-fire neurons (Hospedales et al., 2008). In each simulation, a 16 Hz sinusoidal signal was injected into an integrate-and-fire neuron. **(a)** The spike times in 11 different deterministic neurons, whose membrane potentials at $t = 0$ range between -60 mV and -50 mV. Initially, the neurons fire asynchronously, but after two cycles of the input, the firing is synchronous. This is reflected in the firing histogram shown in **(b)**, which results from a total of 100 neurons, only 11 of which are shown in **(a)**. **(c, d)** The same simulations, but with noise added to the neurons. The spikes are now much more asynchronous, and the population firing rate is a much more faithful representation of the input current, shown as a blue line. Parameters: $\tau_m = 20$ ms, $E_m = -60$ mV, $\theta = -50$ mV, $R_m = 100$ MΩ, $I(t) = I_0 + I_1 \sin(2\pi 16 t)$, $I_0 = 0.1$ nA, $I_1 = 0.1$ nA. In **(c, d)**, $\sigma_I = 0.04$ nA ms$^{1/2}$.

methods to show that a population of leaky integrate-and-fire neurons will tend to synchronise with a suprathreshold, periodically varying input stimulus. In addition, noise in the system tends to desynchronise cells, leading to the population activity of the neurons being more informative about the underlying input stimulus. The timing precision of the first spike in response to a transient increase in input current has been investigated by van Rossum (2001), and the propagation of firing rates through layers has also been studied (van Rossum et al., 2002).

8.5.3 Extrinsic Diffusive Noise – the Stein Model

We have seen how diffusive noise can be used to model intrinsic neuronal noise sources such as ion channels. Neurons are also subject to extrinsic noise from the random firing of input neurons. An important question is therefore the relationship between the variability of inputs to neurons and the variability of their outputs. Integrate-and-fire neurons with diffusive noise have played an influential role in understanding this relationship. The **Stein model** (Figure 8.15) comprises an integrate-and-fire neuron that receives infinitesimally short input pulses from a number of excitatory and inhibitory neurons whose random firing is generated by a Poisson process (Box 8.5). The input causes the membrane potential to fluctuate, sometimes causing it to cross the threshold (Figure 8.15a).

The pattern of firing of the neuron depends strongly on the frequency and size of the excitatory and inhibitory inputs. For example, in Figure 8.15a, where the amount of excitatory and inhibitory input is finely balanced, the time course of the membrane potential and its pattern of firing times appear to be very irregular. In contrast, when there are a smaller number of excitatory inputs present, but no inhibition, the neuron can fire at the same rate, but much more regularly (Figure 8.15b).

The regularity of the spike train can be summarised in **interspike interval (ISI) histograms** (Figure 8.15c, d). Both neurons are firing at the same average rate, yet the ISI histograms look quite different. The neuron with ex-

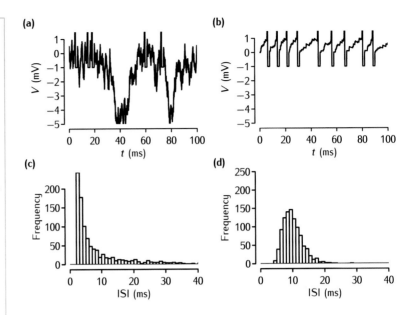

Fig 8.15 **(a)** The Stein (1965) model with balanced excitation and inhibition. The time course of the voltage over 100 ms of an integrate-and-fire neuron that receives inputs from 300 excitatory synapses and 150 inhibitory synapses. Each synapse receives Poisson spike trains at a mean frequency of 100 Hz. The threshold is set arbitrarily to 1 mV; the membrane time constant of the neuron is 10 ms, and there is a refractory period of 2 ms. Each excitatory input has a magnitude of 0.1 of the threshold, and each inhibitory input has double the strength of an excitatory input. Given the numbers of excitatory and inhibitory inputs, the expected levels of excitation and inhibition are therefore balanced. **(b)** The time course of a neuron receiving 18 excitatory synapses of the same magnitude as in **(a)**. The output firing rate of the neuron is roughly 100 Hz, about the same as the neuron in **(a)**, but the spikes appear to be more regularly spaced. **(c)** An ISI histogram of the spike times from a sample of 10 s of the firing of the neuron in **(a)**. **(d)** An ISI histogram for the neuron shown in **(b)**.

citation and inhibition—which fires more irregularly—has an ISI that appears to be exponentially distributed, with many very short intervals and a long tail of longer intervals. In contrast, the neuron that appears to fire more regularly has ISIs that appear to be distributed according to a skewed Gaussian distribution centred on 10 ms.

The ISI histograms can be summarised still further, by extracting the **coefficient of variation** (CV), defined as the standard deviation of the ISIs divided by their mean. A regularly spiking neuron would have a CV of 0, since there is no variance in the ISIs, whereas a Poisson process has a CV of 1. The CV of a neuron receiving a mix of excitation and inhibition (Figure 8.15a, c) is 1.26, whereas the more regularly firing neuron (Figure 8.15b, d) has a CV of 0.30.

A variant of the Stein model was used by Shadlen and Newsome (1994, 1995, 1998) in their debate with Softky and Koch (1993) about whether specialised cellular mechanisms are required to account for the randomness in the firing times of neocortical cells. Their model showed that a mixture of random excitatory and random inhibitory input to simple integrate-and-fire neurons with membrane time constants of around 10 ms could produce the kind of irregular spike train observed in the neocortex. Shadlen and Newsome argued that this finding implied that some elements of the much more complex compartmental model of cortical neurons of Softky and Koch (1993) were not needed to produce realistic firing patterns. In the simulations with the more complex model, only excitatory inputs had been presented to the cell, so in order to achieve the very noisy output patterns, the model had to have various conductance terms added to it to reduce the membrane time constant and therefore produce noisier output. Had Softky and Koch (1993) also used inhibitory inputs, then the extra conductance terms may not have been necessary.

The simplicity of the Stein model allows the mean firing rate to be derived mathematically as a function of the arrival rates of the inputs, the

Box 8.5 | Poisson processes

A **Poisson process** generates a sequence of discrete events randomly at times $t_1, t_2 \ldots$. Examples of sequences of discrete events that could be modelled using a Poisson process include radioactive decay of atoms, the production of action potentials, the opening and closing of channels (Section 4.7.1) and the release of synaptic vesicles (Box 1.2).

In a Poisson process, the probability of an event occurring in a short period Δt is $\lambda \Delta t$, where λ is the rate at which events are expected to occur and where Δt is small enough that $\lambda \Delta t$ is much less than 1. The probability of an event occurring is independent of the time since the last event, so the Poisson process has the Markov property (Section 4.7.1).

The expected distribution of times between events is an exponential of the form $\lambda e^{-\lambda t}$, just like the open and closed time distributions of a state of a stochastic channel (Figure 4.17b) or the interspike interval histogram of neurons that are firing in a Poisson fashion, for example, the Stein model with balanced excitation and inhibition (Figure 8.15c).

The probability $P(k)$ of k events happening in a time period of T is given by the **Poisson distribution**:

$$P(k) = \frac{e^{-\lambda T}(\lambda T)^k}{k!}.$$

The expected number of events (the mean) in the interval T is λT, and the variance in the number of events is also λT. Figure 8.16 shows examples of Poisson distributions.

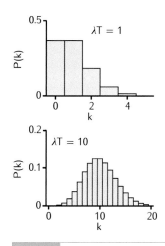

Fig 8.16 Examples of Poisson distributions, that is, the probability of k events happening in a period T at a rate λ.

threshold, the membrane time constant and the refractory period (Stein, 1965; Tuckwell, 1988; Amit and Brunel, 1997b). Suppose an integrate-and-fire neuron with membrane time constant τ_m receives spikes generated by a Poisson process at a rate ν_E through an excitatory synapse of strength w_E and Poisson-generated spikes at a rate ν_I through a synapse of strength w_I. As shown in Box 8.6, it is possible to map this setup on to the model of diffusive noise introduced in Section 8.5.1 and show that μ_V and σ_V depend on the weights, input firing rates and membrane time constant:

$$\mu_V = \tau_m(w_E \nu_E - w_I \nu_I) \quad \text{and} \quad \sigma_V^2 = \frac{\tau_m}{2}(w_E^2 \nu_E + w_I^2 \nu_I). \tag{8.14}$$

If the mean value μ_V is of the order of at least σ_V below the threshold, the neuron will not fire very often, and we can use Equation 8.13 to compute the output firing rate, as shown in Figure 8.13.

8.6 Spike–Response Model Neurons

An alternative, but related, approach to the integrate-and-fire neuron model is the **spike–response model** neuron. These model neurons lend themselves to certain types of mathematical analysis, but for simulations, integrate-and-fire neurons are more efficient (Jolivet et al., 2004). For a comprehensive treatment of spike–response model neurons, see Gerstner et al. (2014).

Box 8.6 | Analysis of the Stein model

The goal of this analysis is to compute the expected firing rate of an integrate-and-fire neuron that receives spikes generated by a Poisson process at a rate v_E through an excitatory synapse of strength w_E and Poisson-generated spikes at a rate v_I through a synapse of strength w_I. Each spike received causes the membrane potential to rise (or fall) instantaneously by w_E (or w_I), so w_E and w_I have the dimensions of voltage.

The equation for the membrane potential is:

$$\frac{dV}{dt} = -\frac{V(t)}{\tau_m} + w_E \sum_k \delta(t - t_E^k) - w_I \sum_k \delta(t - t_I^k),$$

where t_E^k are the times of spikes from excitatory neurons and t_I^k are the times of spikes from inhibitory neurons. $\delta(t)$ is the delta function defined in Section 7.3.1 and τ_m is the membrane time constant.

In a short time Δt, the expected number of excitatory inputs is $v_E \Delta t$, and the expected number of inhibitory inputs is $v_I \Delta t$. Therefore, the expected change in depolarisation is $(w_E v_E - w_I v_I)\Delta t$. We can equate this to the term $R_m I^{det}/\tau_m \Delta t$ in Equation 8.10 to find:

$$\mu_V = R_m I^{det} = \tau_m (w_E v_E - w_I v_I). \tag{a}$$

In a short time Δt, since the inputs come from a Poisson process, the expected variance in the number of excitatory and inhibitory inputs are $v_E \Delta t$ and $v_I \Delta t$, respectively. When these are weighted by the excitatory and inhibitory weights, we find a variance of $w_E^2 v_E \Delta t + w_I^2 v_I \Delta t$. The term $(R_m \sigma_I / \tau_m)\Delta W$ in Equation 8.10 has a standard deviation of $(R_m \sigma_I / \tau_m)\sqrt{\Delta t}$, so squaring this term and equating, we find:

$$(R_m \sigma_I / \tau_m)^2 = w_E^2 v_E + w_I^2 v_I. \tag{b}$$

Thus, from Equation 8.12:

$$\sigma_V^2 = \frac{R_m^2 \sigma_I^2}{2\tau_m} = \frac{\tau_m}{2}\left(w_E^2 v_E + w_I^2 v_I\right). \tag{c}$$

We have thus derived Equation 8.14 in the main text.

A key element of the spike response model is the **impulse response** of the neuron being modelled, that is, the voltage response to a very short burst of input current. It is assumed that the amount of charge injected in the impulse is small, so the membrane potential remains well below the firing threshold. In the case of an integrate-and-fire neuron, the impulse response is a decaying exponential. For most cells, it will tend to be a heavily damped oscillation. However, in the spike–response model, in principle, it can be any function that is zero for $t < 0$, so that an impulse can only affect the membrane potential *after* it has arrived. In what follows, the impulse response function will be denoted by $x(t)$. For example, the decaying exponential impulse response in a membrane with a membrane time constant τ_m is:

The **impulse response** is also known as the **Green's function**. In the context of a convolution (Equation 8.16), it is also known as the **convolution kernel**, or **kernel** for short.

$$x(t) = \exp(-t/\tau_m). \tag{8.15}$$

This is often referred to as the **impulse response kernel**.

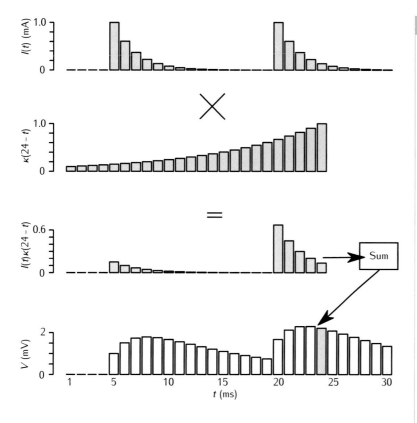

Fig 8.17 Demonstration of a spike–response model neuron. Time is split into 1 ms steps. At the top, the input current to a neuron is shown. To calculate the voltage at a particular time, for example, $t = 24$ ms, the impulse response function (second panel down) is shifted so that it is aligned with $t = 24$ ms. This shows how much influence the input current at times leading up to $t = 24$ ms has on the output voltage at time $t = 24$ ms. Each element of this is multiplied by the corresponding element of the current to produce the weighted current input shown in the third panel down. The contributions from all times are then summed to produce the voltage at $t = 24$ ms, which is highlighted by the blue bar in the fourth panel. The unshaded bars in this figure were produced by the same procedure, but with the impulse response in the second panel down shifted appropriately.

When the impulse response of a neuron is measured, the time course of the voltage is sampled at time points separated by Δt, the inverse of the sampling frequency. The times of the samples are $t_i = i\Delta t$, where the index $i = 0, 1, 2, \ldots$. The impulse response is then effectively a vector where element $\varkappa(t_i)$ corresponds to the continuous impulse response at t_i. If a neuron's impulse response is known, it can be used to predict its sub-threshold voltage response to current input from either synapses or electrodes. This is achieved by calculating the **convolution** of the time course of the input current with the neuron's impulse response:

$$V(t_i) = E_{\mathrm{m}} + \sum_{j=0}^{i} \varkappa(t_i - t_j) I(t_j). \tag{8.16}$$

This is shown graphically in Figure 8.17.

The convolution of the impulse response with the input current describes the sub-threshold response only. Spikes are modelled in a similar way to the integrate-and-fire model, by adding a threshold θ. When the membrane potential rises past this threshold, a spike is produced. The spike affects the time course of the membrane potential in two ways:

Integration of current. During a spike, the membrane conductance is high, so the memory of any input currents before the spike effectively leaks away, meaning that current inputs before the last spike have virtually no influence on the membrane potential. The conductance is still

The convolution of two discrete functions $A(t_i)$ and $B(t_i)$ is:

$$\sum_j A(t_i - t_j) B(t_j).$$

If $B(t_j)$ is zero everywhere, except for $B(0) = 1$, then the result of the convolution operation is A. The convolution of two continuous functions $A(t)$ and $B(t)$ is defined analogously:

$$\int A(t - t') B(t') \mathrm{d}t'.$$

Two-dimensional convolution is defined in Section 12.4.

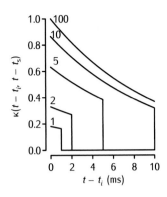

Fig 8.18 Typical spike-dependent impulse responses, or kernels, plotted at various times after the spike. The kernel is defined by $\kappa(t - t_i, t - t_s) = (1 - \exp(-(t - t_s)/\tau_s)) \exp(-(t - t_i)/\tau_m)$, where τ_m is the membrane time constant and τ_s is the refractory time constant. The kernel is zero for inputs that occur before a spike. The value of the kernel is plotted against $t - t_i$, for various periods $t - t_s$ since the spike, as indicated by the number next to each curve.

higher than at rest during the refractory period (Section 3.3), making the impulse response of the neuron to injected current dependent on the time since the spike. An input will tend to evoke a smaller voltage response right after a spike. The impulse response of the neuron must therefore depend on the time since the last spike, as well as the time since the current injection.

Spike response. Since the voltage waveform of spikes is usually highly stereotyped, the membrane potential during the spike can be modelled by adding the voltage waveform of a typical spike and any AHP or afterdepolarisation that follows it during the refractory period. This waveform is given the mathematical symbol $\eta(t_k)$ and is also called the **spike–response kernel**.

To incorporate the effect of the action potential on the integration of current, an impulse response kernel $\varkappa$ with two arguments can be used. The first argument is the time since the input current, and the second argument is the time since the spike. With the addition of the spike response kernel η, the resulting membrane potential is given by:

$$V(t_i) = E_m + \sum_{j=0}^{i} \varkappa(t_j - t_i, t_i - t_s)I(t_j) + \eta(t_i - t_s),\tag{8.17}$$

where t_s is the time of the last spike, and the impulse response kernel now depends on the time since the last spike, $t_i - t_s$. An example spike time-dependent impulse kernel is shown in Figure 8.18. The effect of this kernel is that inputs that occur before a spike are ignored altogether, and the effect of inputs that arrive shortly after a spike is less than the effect of those that arrive after a longer period has elapsed.

8.7 Generalised Integrate-and-Fire Models and Escape Noise

With new technology, robots will be able to collect large numbers of membrane potential recordings by patching large numbers of neurons automatically. To take full advantage of this opportunity, efficient tools for translating experimental data into neuron models are needed. The **generalised integrate-and-fire (GIF) model** is an extended integrate-and-fire model that is well suited to fitting to membrane potential recordings from neurons injected with a time-varying current (Pozzorini et al., 2015). The model fit to these data can predict the spiking times of the neurons in response to unseen time-varying currents accurately.

The model contains elements from both integrate-and-fire and spike–response models. The fundamental equation for the membrane potential resembles the equation for the sub-threshold dynamics of the standard integrate-and-fire neuron (Equation 2.17):

$$C_m\frac{dV}{dt} = -\frac{V - E_m}{R_m} - \sum_{t_j < t} \eta(t - t_j) + I.\tag{8.18}$$

The new element in the equation is the sum on the right-hand side, taking into account the effect of the action potential itself on the subsequent membrane dynamics. Each time an action potential is fired (at times t_j), an intrinsic current with stereotypical shape $\eta(t)$ is triggered. This spike-triggered current $\eta(t)$ is hyperpolarising when its amplitude is positive, and depolarising otherwise. As currents triggered by different spikes accumulate, the result will be spike frequency adaptation if $\eta(t)>0$, and facilitation if $\eta(t)<0$. The shape of $\eta(t)$ varies between neuron types, and the time course of $\eta(t)$ is typically not assumed beforehand, but is instead extracted from the experimental data.

Another new element is that spikes are not added deterministically every time the membrane potential reaches a fixed threshold. Instead spikes are produced stochastically according to the Poisson process (Box 8.5), with a time-varying rate $\lambda(t)$ that depends on the momentary difference between the membrane potential $V(t)$, a firing threshold $V_T(t)$ and parameters ΔV and λ_0 according to:

$$\lambda(t) = \lambda_0 \exp\left(\frac{V(t) - V_T(t)}{\Delta V}\right). \tag{8.19}$$

If the parameter ΔV is very small, the firing is deterministic: when the membrane potential is below the firing threshold ($V(t) < V_T(t)$), the rate $\lambda(t)$ is practically zero and the neuron will not fire, but as soon as the membrane potential exceeds the threshold, the rate goes very high and the neuron fires almost immediately. When ΔV is larger, the firing rate when the membrane potential is below threshold is non-zero, meaning that there is a chance that the membrane can 'escape' the threshold and produce a spike. In contrast to diffusive noise (Section 8.5), with this **escape noise**, the integration of the membrane potential is deterministic, but the firing is stochastic.

A further feature of the GIF model, akin to the spike–response model, is that the firing threshold $V_T(t)$ changes over time according to:

$$V_T(t) = V_T^* + \sum_{t_j < t} \gamma(t - t_j), \tag{8.20}$$

where V_T^* is a constant and $\gamma(t)$ describes the stereotypical time course of the change in firing threshold after the emission of a spike. Again, since the contributions from different spikes accumulate, this moving threshold constitutes an additional source of adaptation or facilitation. As with the hyperpolarising current $\eta(t)$, in the typical use of the GIF model, the functional shape of $\gamma(t)$ is not assumed beforehand but is extracted from experimental data. Similarly to other integrate-and-fire models, refractoriness of spiking is included by adding a short absolute refractory period and resetting the membrane potentials each time a spike is generated.

The structure of the model is flexible enough to enable it to predict both the spiking activity and sub-threshold dynamics of many types of neurons. The structure is also tailored for rapid parameter fitting of the model parameters to recordings of the membrane potential elicited by noisy input currents. There are three main steps to fit the model:

(1) Extract spike times from the experimental membrane potential recording.

The notation $\eta(t)$ means slightly different things in different models. In the spike-response model (SRM), $\eta(t)$ is a potential, but in the GIF model, it is a current, though in both models, it is triggered by a spike.

The parameter λ_0 is there to maintain dimensional consistency, and can be set to an arbitrary value (e.g. 1 Hz), since the parameter V_T^* (to be introduced in Equation 8.20) can compensate for it.

Fig 8.19 Forms of activation functions (*f–I* curves) for rate-based neurons. **(a)** Piecewise linear function (Equation 8.21) with threshold $\theta = 5$ and slope $k = 1$. **(b)** Sigmoid function (Equation 8.22) with $\theta = 5$ and slope $k = 1$. **(c)** Step function (Equation 8.23) with $\theta = 5$.

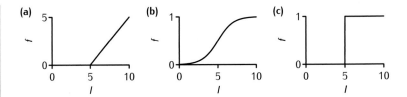

(2) Fit the sub-threshold parameters (C_m, R_m, E_m and $\eta(t)$, described using a set of basis functions with unknown parameters). Firstly the rate of change the membrane $\mathrm{d}V/\mathrm{d}t$ is extracted from the data at every time-step. Since the voltage and current are known at every time-step, Equation 8.18 is effectively a multiple regression with the rate of change of voltage as the dependent variable, the membrane potential and current as independent variables, and the passive parameters as coefficients.

(3) Simulate the sub-threshold membrane potential, and set the parameters of the voltage-dependent threshold (V_T^* and $\gamma(t)$) so as to maximise the likelihood of obtaining spikes at the time they occurred. The structure of the model means the likelihood function is concave, allowing for efficient optimisation.

It can take a matter of minutes to fit GIF models to data, making GIF models a natural choice for 'mass production' of spiking neural models from patch recordings (Pozzorini et al., 2015).

Though information is a rather general concept, the amount of information transmitted by a signal can be quantified using the **Information Theory**, invented by Shannon (1948) and used extensively in applications such as error detection and correction, cryptography, data compression and the development of phone systems and of the Internet.

8.8 | Rate-Based Models

In some neural systems, such as the fly visual system, the timing of individual spikes has been shown to encode information about stimuli (Rieke et al., 1997). Other examples include: the tendency of spikes in the auditory nerve to be locked to a particular phase of the cycle of a sinusoidal tone (Anderson et al., 1971); the representation of odours by spatiotemporal patterns of spikes in the locust antennal lobe (Laurent, 1996); and the phase of the theta rhythm at which hippocampal place cells fire, encoding how far a rat is through the place field (O'Keefe and Recce, 1993). Despite the importance of single spikes in some systems, in many systems, the average firing rate of a single neuron, determined by counting spikes in a time window or the population firing rate of a group of neurons, conveys a considerable amount of information. This observation goes back to Adrian's study of frog cutaneous receptors, in which he found that the firing frequency was proportional to the intensity of the stimulus (Adrian, 1928). It is therefore sometimes reasonable to simplify neural models further still by considering only the firing rate, and not the production of individual spikes. As described in Chapter 9, in many models, the firing rate in a population of neurons is represented by an averaged value calculated over that population.

When injected with a steady current I, an integrate-and-fire neuron fires at a steady rate f that depends on the input current. We can determine this *f–I* curve by running simulations or, in some cases, analytically. The form of the f–I curve can depend on a refractory period (Chapter 2, Equation 8.11). If there is extrinsic noise in the input (e.g. due to random firing of other synaptic inputs) or intrinsic noise in the neuron, the noise also affects the firing rate (Section 8.5.1, Figure 8.13). By contrast, in firing rate models, we specify the f–I curve via the characteristic function $f(I)$, which converts synaptic current into a firing rate directly, bypassing the integration of current and the threshold. By analogy with artificial neural networks (Box 11.2), we will call the function $f(I)$ the **activation function**. The form of the activation function $f(I)$ is not restricted to theoretical integrate-and-fire model f–I curves, though it should usually be positive; negative firing rates are not physiological.

Different forms of activation function have been considered, which define different classes of model neurons. Common examples are a piecewise linear function (Figure 8.19a):

$$f(I) = \begin{cases} 0 & \text{for } I < \theta \\ kI & \text{for } I \geq \theta, \end{cases} \tag{8.21}$$

and a sigmoid function (Figure 8.19b):

$$f(I) = \frac{\overline{f}}{1 + \exp(-k(I - \theta))}, \tag{8.22}$$

where $\overline{f}$ is the maximum firing rate, θ is a threshold and k controls the slope of the f–I curve. For a large value of k, the sigmoid curve approximates to a step function (Figure 8.19c):

$$f(I) = \begin{cases} 0 & \text{for } I < \theta \\ 1 & \text{for } I \geq \theta. \end{cases} \tag{8.23}$$

Model neurons where the activation function is a step function are sometimes referred to as **McCulloch–Pitts neurons**, in recognition of the pioneering work of McCulloch and Pitts (1943), who regarded the neurons as logical elements and showed that networks of such neurons could be constructed that would implement logical functions. Neurons with sigmoid functions form the basis of many artificial neural networks (Box 11.2).

The above characterisation of the f–I relationship holds for steady-state input to neurons, and gives a useful element for including in feedforward networks. If the input is changing over time, the assumptions underlying the derivation of the f–I curve no longer hold, and the output firing rate can depend on the history of the input as well as the history of the firing of the neuron itself. The question of characterising neurons using firing rates is often considered in the context of networks and is covered in Chapter 9.

8.9 | Summary

This chapter has described how details can be progressively removed from complex models of neurons to produce simpler models of neurons. From the large variety of models that have been devised, we have chosen to concentrate on four broad categories of neuron model.

(1) Models where the spike generation is due to a system of differential equations that include an equation for the membrane potential and equations for voltage-dependent conductances. These include compartmental models with a reduced number of compartments where also the number of state variables may have been reduced. The parameter space of models of this form with few variables can be explored thoroughly and they are amenable to insightful mathematical analysis.

(2) Models where action potentials are imposed when the membrane potential crosses a threshold. A variety of these integrate-and-fire models are examined. These models are good for mimicking the behaviour of neurons, especially for inclusion in a network. They are amenable to mathematical analysis, including their response to noise, either intrinsic to the neuron (e.g. ion channel opening and closing) or extrinsic to the neuron (e.g. randomly firing inputs). They can be extended by using extra variables to produce adaptation behaviour.

(3) Spiking models that are optimised to be fit to data from real neurons so as to be able to reproduce their input–output relationships.

(4) Models where not even action potentials are considered, but rather the firing rate of the neuron is the critical variable. Whilst many details are lost here, it is possible to construct simple networks of these types of neural models that give insights into complex computations.

The choice of model depends on the level of the explanation, the data available, whether the model should be analysable mathematically and the computational resources to hand. A fruitful approach can be to use models at different levels of complexity. Intuitions about the system being modelled can be gained from simple models, and these intuitions can then be tested in more complex models. Ideally the models should be inherently consistent, meaning that in principle the simpler models can be derived from the more complex models.

We have also introduced the tools of dynamical systems theory, which allows us to map out the possible behaviour of simplified neurons. Dynamical systems theory can be applied to elements, apart from neurons, such as populations of neurons, where each population is characterised by a firing rate (Chapter 9). One of the uses of simplified models of neurons is in the modelling of networks. Issues arising from modelling networks are discussed fully in Chapter 9.

CHAPTER 9

Networks of Neurons

When modelling networks of neurons, generally it is not possible to represent each neuron of the real system in the model. It is therefore essential to carry out appropriate simplifications for which many design questions have to be asked. These concern how each neuron should be modelled, the number of neurons in the model network and how the neurons should interact. To illustrate how these questions are addressed, networks using various types of model neuron are described. In some cases, the properties of each model neuron are represented directly in the model, and in others the averaged properties of a population of neurons. We then look at several large-scale models intended to model specific brain areas. In some of these models, the neurons are based on the neurons reconstructed from extensive anatomical and physiological measurements. The advantages and disadvantages of these different types of model are discussed.

Two severe limitations prevent the modeller from constructing a model of a neural system in which each nerve cell is represented directly by a counterpart model neuron. There are so many neurons in the neural system that having a full-scale model is computationally infeasible. In addition, usually only incomplete data are available about the structural and functional properties of the neurons, how they are arranged in space and how they interconnect.

The design issues most commonly addressed concern the numbers and types of model neurons and the topology of how they connect with each other. Another crucial issue is how the cells are situated in 3D space (Section 9.1).

The most common properties that are investigated in network models of the nervous system are the patterns of firing within the array of neurons and how such patterns are modified through specific synaptic learning rules. In this chapter, we examine the first of these properties in a variety of network models in which the neurons are modelled to differing levels of detail. Issues of plasticity are discussed in Chapter 11.

In Section 9.2, feedforward and recurrent models of rate-based neurons are introduced. We then look at rate-based models in which the number of system variables is reduced substantially by computing averaged properties

A population of neurons is a large group of neurons with similar properties which are assumed to act together to perform a specified function (Gerstner et al., 2014).

over large populations of neurons, called here neuronal units. Section 9.2.3 describes neural mass models and Section 9.2.4 describes neural field models, where the neuronal units are laid out in 3D space. We then review in detail an example of a recurrent network of integrate-and-fire neurons (Section 9.3.1), this type of model neuron being introduced in Section 2.7. We go on to discuss a model of the neocortex using integrate-and-fire neurons (Section 9.3.3). In Section 9.4, we discuss oscillations in spiking neural networks, and in Section 9.5, we conclude by giving three examples of using anatomical and physiological constraints to construct detailed models of the mammalian brain.

9.1 Network Design and Construction

In the preceding chapters, we have seen that the construction of the model of a single neuron involves a vast range of choices concerning how to model components such as cell morphology, ion channels and synaptic contacts. Each choice involves a compromise over the level of biological detail to include. How to make useful simplifications is an important part of the modelling process.

The same is true if we want to build a network of neurons. Amongst the decisions to be made are what types of neuron they should represent, what the level of detail is at which to model individual neurons and whether model neurons of the same type should have different parameter values. For a large-scale network with thousands, or hundreds of thousands, of neurons, this may require using the simplified models introduced in Chapter 8. Other issues also arise with network models. How should we handle communication between neurons? Do we need to model axons and the propagation of action potentials along them? Do we need to model short-term dynamics and stochastic neurotransmitter release at synapses? Should the synapses be modifiable?

Some neurobiological systems contain so few neurons that each real neuron can be represented by a counterpart in the model. For examples, see Abbott and Marder (1998). However, in most cases, the number of neurons will be fewer than in the real network. In the model network, does each neuron represent a single neuron or does it represent a group of neurons? How then should we scale the numbers of neurons of different classes in the network? Do we need to consider the location of neurons in space? In this section, we explore possible answers to these questions.

9.1.1 Connecting Neurons Together

Networks of neurons are formed predominantly through neurons connecting together via chemical synapses formed between axonal terminals from efferent neurons and the postsynaptic membranes of receiving neurons. The signal that passes from the efferent to the receiving neuron is the action potential. One way of modelling these connection pathways is to include in each cell model a compartmental model of its axon along which action potentials propagate. This is computationally very expensive and arguably unnecessary. Action potentials are stereotypical and the information content of signals passing from one neuron to another is carried by the arrival times of action potentials at synapses, rather than the precise voltage waveform of the action potential.

Consequently, the approach that is almost uniformly applied is to treat the signal that passes from one neuron to another to be the presence or absence of an action potential. Then the connection from one neuron to another is modelled as a **delay line** (Figure 9.1). The voltage in the soma or axon initial segment of the efferent cell is monitored continuously. If the voltage exceeds a defined threshold (e.g. 0 mV), this signals the occurrence of an action potential. The delay line then signals this occurrence to the synaptic contact on the receiving neuron at a defined time later, corresponding to the expected transmission time of the action potential along the real axon. This approach is not only vastly cheaper computationally than compartmental modelling of axons, but it is also easily implemented on parallel computers, as only spike times need to be sent between processors (Brette et al., 2007; Hines and Carnevale, 2008).

There are circumstances where it is necessary to model the detail of action potential propagation along axons. The delay line model assumes that action potential propagation is not modulated along the length of the axon. The possibility of action potential failure either at branch points or due to presynaptic inhibition is ignored. These effects have been explored using compartmental models of isolated axons (Parnas and Segev, 1979; Segev, 1990; Manor et al., 1991a, b; Graham and Redman, 1994; Walmsley et al., 1995). They could certainly be expected to influence network dynamics and thus raise the challenge of modelling action potential propagation in a network model.

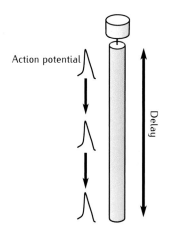

Fig 9.1 An action potential is initiated in the axon initial segment and propagates along the axon. This can be modelled as a delay line, which specifies the time taken for the action potential to travel the length of the axon. The action potential itself is not modelled.

9.1.2 Scaling Neuronal Numbers

Many networks of interest contain thousands or even millions of neurons, which often are not feasible to model. It is then necessary to model a scaled-down version of the actual network, both the numbers of neurons and the number of synapses between neurons.

Suppose our network is going to be one-tenth the size of the brain structure we are modelling. It contains three cell types: a principal excitatory neuron that makes up 80% of the cell population, and two types of inhibitory interneuron, each constituting about 10% of the population. The obvious way to scale neuronal numbers is to retain the relative proportions of cells of different types (80:10:10) in our one-tenth-sized model. Provided this results in reasonable numbers of cells of each type in the model, then this could be an appropriate choice. What may constitute a reasonable number of cells is discussed below.

The main use of the model is likely to be to study the population response of the excitatory cells. For this to be an accurate reflection of physiology, it is important that the excitatory and inhibitory synaptic inputs onto these cells represent realistic population activities. In our model network, inhibition from each population of inhibitory interneurons should be as close as possible to that experienced by a real excitatory neuron in vivo. Given that we have fewer interneurons in our model network than exist in vivo, there are two ways of achieving this:

(1) Scale up the maximum synaptic conductance of each connection from an inhibitory interneuron onto an excitatory cell by a factor of 10, in this example.

(2) Create 10 times the number of synaptic contacts from each interneuron onto each excitatory cell than exist in vivo.

Neither approach is perfect. Scaling the synaptic conductances may give an equivalent magnitude of inhibition. However, the inhibition will then be applied as a few large conductance changes at isolated points on the excitatory cell. As discussed in more detail in Section 9.3.1, the resulting voltage changes and integration with excitatory inputs will be distorted. Creating more synaptic contacts from each interneuron will result in a realistic spatial distribution of inhibitory inputs, but spikes arriving at these inputs may have unnatural correlations since groups of them are more likely to derive from the same interneuron. Unless it is possible to include physiological numbers of interneurons in the network model, one of these compromises is required. The same considerations apply to the inputs from excitatory neurons.

If it is likely that different-sized network models will be tested, it is very useful to fix the number of afferent inputs that a given cell receives from the population of cells of each type in the model. For example, the number of inhibitory inputs that each excitatory cell receives from each of two populations of inhibitory interneurons should remain fixed regardless of the actual number of each cell type in the model. When the number of cells is changed, a cell of a particular type will provide fewer or more synaptic contacts onto a target cell, but the target cell will always have the same number of synaptic inputs from the efferent cell population (Orbán et al., 2006).

Another effect of scaling the numbers of neurons is that the small populations of interneurons may be scaled to the point of having only one or a few cells representing these populations in the model. In this case, the population activity in the model of these interneurons may be a serious distortion of the activity in vivo. Real activity may involve thousands of asynchronously firing cells, with the instantaneous population activity providing a good estimate of some modulating driving force, such as slowly changing sensory input (Knight, 1972; Hospedales et al., 2008). However, a small population in the model may only provide a poor representation of the modulating input.

If this is the case, then it may be possible to scale each population of cells differently. If the excitatory cells are not strongly recurrently connected, then only a relatively small number of these cells are required in the model to allow a good study of their network activity (Orbán et al., 2006). This then allows relatively larger populations of interneurons to be modelled so that both their population activity and their inhibitory effect on the excitatory cells are much more physiological. This approach was taken by Orbán et al. (2006) in a model of the CA1 area of the hippocampus, where recurrent connections between pyramidal cells are sparse. Their network model of theta activity contained a small number (15–30) of detailed 256-compartment pyramidal cells, but with populations of up to 200 basket and 90 oriens lacunosum-moleculare cells, each cell modelled by a single compartment.

9.1.3 Positioning Neurons in Space

Real neurons have a particular location within a brain nucleus, and connectivity patterns between neurons are often distance-dependent. To capture these patterns, it may be necessary to place our model neurons in virtual space. Several cortical models allow for depth dependence (Traub et al., 2005; Potjans and Diesmann, 2014; Markram et al., 2015) and some include both depth and lateral position (Billeh et al., 2020). An early attempt is the simulation model of Marr's theory of the **cerebellar cortex** (Marr, 1969; Tyrrell and Willshaw, 1992).

For, say, a cortical column or other small part of a brain area, it may be reasonable to assume that connectivity is completely uniform (e.g. every neuron connects to every other neuron) or that there is a fixed probability that one neuron makes contact with another neuron. In this case, the precise spatial location of a neuron is not relevant and can be ignored when the task only is to compute spikes generated in the network. However, calculations of local field potentials (LFPs) as well as electroencephalography (EEG) and magnetoencephalography (MEG) signals from the same network do require information about spatial locations (Chapter 13).

In general, the cells should be laid out in a 1D, 2D or 3D arrangement that reflects the physiological layout. Typically this is done with a regular spacing between cells. When forming connections between cells, the probability that an efferent cell forms a connection onto a target cell can be a function of the distance between them. This function is often an exponential or Gaussian (Box 5.8), so that the probability of connection decreases with distance (Figure 9.2a). This reflects the basic connection arrangement in many brain nuclei. More complex connection strategies can easily be implemented. So-called **small-world network**s (Watts and Strogatz, 1998; Netoff et al., 2004; Földy et al., 2005) can be generated by first creating a network with only local connections between cells (a cell connects to a few of its nearest neighbours) and then randomly reassigning a small proportion of the connections to be much longer-range connections (Figure 9.2b).

One problem to deal with is that of edge effects, in which cells at the edge of our spatial layout receive fewer connections than interior cells. A good, biological solution might be to have a sufficiently large model network that the cells at the boundaries can be ignored. If that is not possible, then it could be assumed that the spatial arrangement wraps around, so that a cell at the end of a line is assumed to be a neighbour of the cell at the opposite end of the line (Netoff et al., 2004; Wang et al., 2004; Földy et al., 2005; Santhakumar et al., 2005), that is, the line is actually a circle (Figure 9.2).

9.1.4 Variability in Cell Properties

The vast majority of neuronal network models contain populations of cells with completely uniform properties, including morphology and membrane physiology. This does not reflect the variation seen within biological neurons and can lead to artefacts in network behaviour due to uniformity in cellular responses to synaptic input. A better approach is to introduce variation into one or more cellular properties, including membrane resistance, resting membrane potential and ion channel densities. One approach has been

Marr's theory of the cerebellar cortex as a learning machine (Marr, 1969) had great influence amongst cerebellar neuroscientists (D'Angelo, 2016). In their implementation of it in 3D, Tyrrell and Willshaw (1992) constructed a simulation model of all the circuitry associated with a single Purkinje cell. With the limited computing resources available at the time, they did this by representing each 3D layer of cells and connections in a 2D plane. To build the model, they had to guess many parameter values about the geometry as these were not available. Their simulation results agreed broadly with the analysis carried out by Marr.

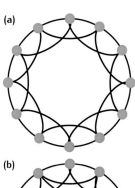

(a)

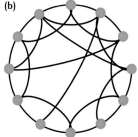

(b)

Fig 9.2 **(a)** Local connectivity in which a neuron connects only to near neighbours. **(b)** Small-world connectivity in which some of the local connections are replaced by longer-range connections.

to make a network where the large population of nerve cells is constructed from a set of stereotypes derived from real data where each cell has small randomly generated changes in structure from its parent (Section 9.5; Markram et al., 2015; Billeh et al., 2020). Alternatively, some variation can be introduced into a population of otherwise identical cell models by either starting a simulation with different initial conditions for each cell, for example, different starting membrane potentials, or providing a different background stimulus, in the form of a small depolarising or hyperpolarising current injection, to each cell (Orbán et al., 2006).

Experimental estimates of the parameters varied may be available that indicate the degree of variation in a biological population. Variations in electrophysiological responses may indicate variability in membrane ion channel densities (Aradi and Soltesz, 2002). Computational models and experiments have shown that signal integration in cells and collective network behaviour are strongly influenced by variability in individual cell characteristics (Aradi and Soltesz, 2002, 2004). Another consideration is the relative proportion of cells of different types within the network. Classification of cell types is an art form that is still evolving (Somogyi and Klausberger, 2005; Markram, 2006). Thus the number of cell populations and their relative sizes may be free variables in the network model. Simulations have shown that networks containing the same cell types, but in different proportions, can show significantly different behaviour (Földy et al., 2003, 2005).

9.2 | Networks of Rate-Based Neurons

The earliest network models, such as those developed in the pioneering studies of McCulloch and Pitts (1943) and Rosenblatt (1958), were rate-based models (Section 8.8). The threshold firing rate functions f used ensured that each neuronal unit was either an on-state ($f = 1$) or an off-state ($f = 0$), as described in Equation 8.23. Time was discretised so that the state of the network was updated in a clock-like manner. The main focus of these early studies was on how networks of neuron-like units could be used in computation, rather than trying to mimic in detail biological neural networks found in the brain. This line of inquiry later led to the development of multi-layered artificial neural networks with efficient learning rules (Rumelhart et al., 1986a), which underlies much of today's applications in artificial intelligence (Box 11.2).

Networks of such simple on–off neuronal units were also used to investigate, for example, how learning and memorisation could be achieved in brain structures such as the hippocampus (Marr, 1971). The dynamics of networks of threshold neurons will be described in the context of memory models in Chapter 11.

Another type of model, often referred to as a **neural mass model** or a **neural field model**, aims instead to describe neural dynamics in real brain tissue (Beurle, 1956; Griffith, 1963; Wilson and Cowan, 1972; Amari, 1977). Here typically the models are formulated as differential equations where the variables represent the average properties of a population of neurons. The goal is more to describe specific spatial or temporal patterns of brain activity rather than to mimic a specific brain function like learning (Sections 9.2.3 and 9.2.4).

Despite the different goals of these various network models, there are common network motifs and descriptions. Here we describe some of the simplest forms of networks involving connections between neurons designated as inputs and neurons designated as outputs. In **feedforward networks** (Section 9.2.1), the synaptic current in an output neuron is derived from the firing rates of the input neurons to which it is connected, together with the appropriate synaptic conductances; there are no loops which allow feedback. In **recurrent networks**, the connections between neurons are in both directions.

9.2.1 Feedforward and Recurrent Networks

Figure 9.3a shows a feedforward network. There is a layer of N input neurons, labelled with the subscript i, and an output neuron labelled j. Each connection from an input neuron to the output neuron has a **weight** w_{ij}, which describes the strength of the synapse from input neuron i to output neuron j. Assuming rate-based neurons (Section 8.8), the total input current flowing into the output cell is the sum of the weights multiplied by the firing rates of the corresponding input neurons. The firing rate of the output neuron is given by the activation function $f(.)$ applied to the total current I_j. If each input neuron fires at a constant rate, then so does the output neuron, and its firing rate f_j is given by:

$$I_j = \sum_i w_{ij} f_i, \quad f_j = f(I_j). \tag{9.1}$$

The weight matrix records the strengths of all the connections between neurons.

A simple example of a feedforward network model is that involved in the connections that underlie receptive fields (Figure 9.3b). Here the input neurons represent retinal ganglion cells at different locations on the **retina**. With appropriate connections, output cells can be constructed that have on-centre or off-centre characteristics, such as the cells in the lateral geniculate nucleus (LGN). An example application to the LGN is described in Section 9.2.2.

If the firing rates of the input neurons vary in time, Equation 9.1 can still be used for a feedforward network. However, for quickly changing inputs, it would not give very realistic results, since the rate of change of the output follows the rate of changes in the inputs instantaneously, whereas it would be expected that changes in the output firing rates should lag behind changes in the input firing rates. To make the model more realistic, we can adapt the differential equation used to convert spikes into decaying exponential currents (Equation 7.9) by feeding in firing rates of other neurons instead of spikes:

$$\tau_{syn} \frac{dI_j}{dt} = -I_j + \sum_i w_{ij} f_i, \quad f_j = f(I_j). \tag{9.2}$$

Although this equation is rate-based, it gives a good approximation of a network comprising integrate-and-fire neurons (Amit and Tsodyks, 1991a),

The **receptive field** of a neuron is the part of the visual field which, upon illumination, changes the neuron's firing activity. **On-centre neurons** increase activity when the centre of the visual field is illuminated, and **off-centre neurons** decrease activity.

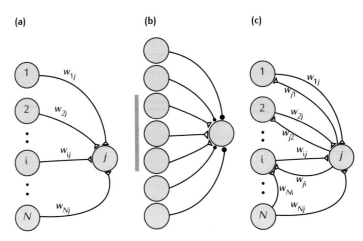

as long as the total number of spikes that would be expected to be received within a millisecond by a postsynaptic neuron is large – say of the order of tens or hundreds. If this is the case, the fluctuations in the numbers of spikes arriving within a millisecond should be low, and the function $f(I_j)$ can be approximated by the f–I curve for integrate-and-fire neurons (Equation 8.11), or by an f–I curve for integrate-and-fire neurons with diffusive noise (Section 8.5). Under different simplifying assumptions, other forms of rate-based models with slightly different equations can be derived (Dayan and Abbott, 2001; Ermentrout and Terman, 2010; Gerstner et al., 2014).

Equation 9.2 can also be used to model recurrent networks, where there may be a connection back from neuron j to neuron i (Figure 9.3c). Two examples of the types of computation that recurrent networks of rate-based neurons can perform are **autoassociative memory** (Section 11.5.3) and input integration.

9.2.2 Example: Feedforward Firing Rate Model for Response of LGN Cells to Visual Stimulation

An important application of rate models has been the description of responses of neurons in the early part of the visual system to visual stimuli. For cells in the retina and LGN, the station between the retina and cortex in the visual pathway, the receptive fields are small and circular. Furthermore, they exhibit **centre-surround antagonism**, meaning that the cells exhibit their largest firing rates when a circular stimulating light spot (for an on-centre neuron) covers the receptive field centre exactly. A light stimulus outside this centre reduces the firing activity.

Many of these cells have linear response characteristics, meaning that the neuronal response can be found by summing over contributions from light stimulation across the receptive field where each contribution is weighted by a function $g(\vec{r})$, called here the **impulse response** or **receptive field** function. In practice, this sum is done by integration:

$$v(\vec{r}) = \iint_{\vec{r}'} d^2\vec{r}'\, g(\vec{r} - \vec{r}')s(\vec{r}'), \tag{9.3}$$

Here $v(\vec{r})$ is the firing rate of a neuron at the position $\vec{r} = [x, y]$, and $s(\vec{r})$ is a function describing the light stimulation. A common choice for

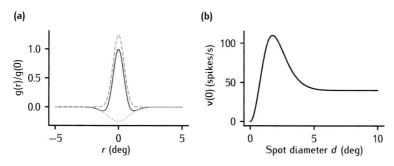

(a) Illustration of the difference-of-Gaussians (DOG) model for visual receptive fields, Equation 9.4. Gaussians (blue dashed and dotted lines) have width parameters $a = 0.6$ deg, $b = 1.2$ deg. The resulting curve (dark grey) is normalised, and only the ratio $B/A = 0.8$ affects the curve, in addition to the width parameters. **(b)** Firing rate predicted for DOG model in **(a)** for light spots of different diameters centred at the receptive field centre of the DOG model neuron, that is, $\vec{r} = 0$. This area–response curve is found by evaluating the integral in Equation 9.3. The shape of the curve is determined by a, b and the ratio B/A only, whilst the magnitude also depends on the strength (luminance) of the stimulus as well as the magnitudes of A and B; see Einevoll and Heggelund (2000) for details.

$g(\vec{r})$ capturing the centre-surround organisation of receptive fields is the **difference-of-Gaussians (DOG) model** introduced by Rodieck (1965):

$$g(\vec{r}) = \frac{A}{\pi a^2} e^{-r^2/a^2} - \frac{B}{\pi b^2} e^{-r^2/b^2}, \tag{9.4}$$

where the first and second terms correspond to the centre and surround contribution, respectively, with r measured from the centre of the receptive field. A and B (defined to be positive) are the strengths of the centre and surround, and a and b are the corresponding width parameters (Figure 9.4a).

From the formula in Equation 9.3, the response to any stimulus can be computed. A much used set of stimuli in experiments has been circular spots of increasing sizes. A characteristic feature of such **area–response curves** for cells in the retina and LGN is that, as the spot size is increased, the response will exhibit a maximum when the spot covers the receptive field centre exactly (Figure 9.4b).

Cells in the LGN receive strong feedforward inputs from the cells in the retina, and as a consequence, the receptive fields of LGN cells share the characteristic centre-surround structure of retinal cells. However, LGN cells also receive numerous feedback synaptic inputs from cells in the visual cortex, but their functional role has been unclear. In Box 9.1, it is outlined how Mobarhan et al. (2018) used a mechanistic rate model, the extended difference-of-Gaussians (eDOG) model (Einevoll and Plesser, 2012), to investigate this question.

9.2.3 Neural Mass Models

Firing rate network models for interconnected populations are referred to as neural mass models. A well-known and thoroughly analysed two-population neural mass model of recurrently connected excitatory and inhibitory populations, a key network motif in neurobiology, is known as the **Wilson–Cowan model** (Wilson and Cowan, 1972). In their simplest form, the equations for the activity in these two populations are (Figure 9.5):

$$\tau_E \frac{d\nu_E}{dt} = -\nu_E + f_E(w_{EE}\nu_E - w_{EI}\nu_I + \nu_E^{ext}),$$

$$\tau_I \frac{d\nu_I}{dt} = -\nu_I + f_I(w_{IE}\nu_E - w_{II}\nu_I + \nu_I^{ext}). \tag{9.5}$$

Here w_{xy} is the synaptic weight from population x to population y (where x and y can be E or I), ν_x^{ext} is the external input to population x, and f_E and f_I are sigmoid functions converting inputs to populations to population firing rates (Equation 8.22). The system can exhibit steady-state and

The difference-of-Gaussians model is an example of a **descriptive model** where the goal is to summarise experimental data compactly, yet accurately. The goal of such a model is thus to account mathematically for a phenomenon, not to explain it. In contrast, **mechanistic models** aim to account for the data in terms of models based on the underlying neurons and their synaptic network connections. See Chapter 1 for more discussion about different types of models.

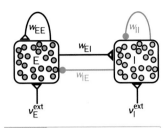

Fig 9.5 The two-population neural mass model. The excitatory (E) and inhibitory (I) populations of neurons (rectangles) each comprise a potentially large number of neurons (circles). The firing rates of all neurons in each population are averaged to give mean firing rates v_E and v_I. The averaged strengths of connections between and within the two populations are w_{EE}, w_{EI}, w_{IE} and w_{II}. There is external input to each population, v_E^{ext} and v_I^{ext}.

Box 9.1 Extended DOG model for LGN cell responses

Cells in the LGN receive strong feedforward inputs from the retina, and the receptive fields of LGN cells thus resemble those of retinal cells. Both can be well represented by the descriptive DOG model (Equation 9.4). LGN cells also receive feedback from cortical cells, and Mobarhan et al. (2018) used a mechanistic rate model, the **extended difference-of-Gaussians (eDOG) model** (Einevoll and Plesser, 2012), to investigate how this feedback modifies the receptive fields of LGN cells.

The model circuit, illustrated in Figure 9.6a, includes feedback from both cortical on-cells and off-cells. The latter increases firing when a dark spot, rather than a light spot, appears in the receptive field.

The eDOG model is not only a spatial model like the DOG model, it also predicts the time-dependent firing rate for an arbitrary space- and time-dependent visual stimulus. Whilst the eDOG model is quite complex and involves many synaptic connections, the evaluation of the neuronal response involves only the evaluation of an integral analogous to Equation 9.3: the eDOG integral is 3D rather than 2D, and the spatial DOG impulse–response function in Equation 9.4 is replaced by a spatiotemporal eDOG impulse–response function specified by the connectivity of the circuit in Figure 9.6a.

The analysis of Mobarhan et al. (2018) identified a special configuration of mixed excitatory and inhibitory cortical feedback that seems to best account for available experimental data from the literature. This configuration consisted of: (1) slow and spatially widespread inhibitory feedback, combined with (2) fast and spatially narrow excitatory feedback, in which (3) the excitatory/inhibitory on-to-on connections are accompanied, respectively, by inhibitory/excitatory off-to-off connections, that is, following a phase-reversed arrangement.

Later comprehensive experiments on mice with optogenetic control of the feedback are in accordance with this analysis (Born et al., 2021). The best agreement with experimental area–response curves were found for a version of the eDOG model with a mixed model with narrow excitatory and a broad inhibitory feedback, alternative (iii) in Figure 9.6b.

The eDOG model is available as a Python package (Mobarhan et al., 2018).

oscillating (limit cycle) behaviour, and can be analysed using the dynamical systems theory (Section 8.2).

The Wilson–Cowan equations have been used to model the response of populations of cells in the barrel cortex of rodents. Rodents use their whiskers to sense the world around then. The whiskers project topographically to the **somatosensory cortex**. Flicking an individual whisker stimulates a small cluster of cortical neurons arranged within a barrel-shaped outline. The firing of these cortical neurons depends sensitively on the temporal structure of the firing of the thalamic neurons providing their input. Rapidly varying thalamic inputs give a much larger response than slowly varying inputs (Pinto et al., 2003; Blomquist et al., 2009). This selectivity can be captured by models of the Wilson–Cowan type where the excitatory and

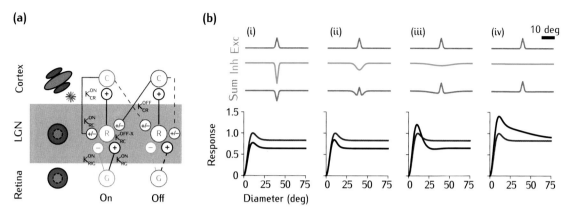

Fig 9.6 **(a)** Schematic overview of the eDOG model described in Box 9.1. Cell types are retinal ganglion cells (G), LGN relay cells (R) and cortical cells (C). Each cell type corresponds to a 2D layer (or population) of identical cells. A single cortical population is shown for each pathway (on/off), even though an arbitrary number of cortical populations is considered. Unlike the feedforward projection, the feedback is cross-symmetrical, that is, the activity of on-centre relay cells is affected by both on- and off-centre cortical cells. Off-centre LGN cells are assumed to receive the same input as the corresponding on-centre LGN cells with opposite sign. **(b)** Area–response curves for the eDOG model (bottom) with (black) and without (red) cortical feedback. Top panels show the assumed different widths of excitatory (grey) and inhibitory (blue) feedback giving different shapes of total effective feedback (dark blue). Alternative (iii), in Born et al. (2021), was found to be closest to experimental findings. Figure based on data from Born et al. (2021).

inhibitory population both receive input from the thalamus (Pinto et al., 2003). Here the time constant of the excitatory dynamics is much shorter than for the inhibitory dynamics, so that a fast and strong input can be amplified by the network before inhibition kicks in and eventually quenches the firing activity. Subsequently, network models with the same characteristics have been extracted directly from measured population firing activity by fitting models to data (Blomquist et al., 2009).

In the Wilson–Cowan model, the dynamics of the population activity is assumed to have a very simple form, identical to the sub-threshold dynamics of the simplest integrate-and-fire neuron (Section 2.7). This assumption is not founded on the true dynamics of a population of recurrently connected neurons. In particular, neural populations often respond to rapidly changing inputs with a sharp transient population firing response, a feature not captured well by the simplest Wilson–Cowan equations. For a discussion of how to construct more accurate firing rate models, see Gerstner et al. (2014, Ch. 15).

Networks of neural mass models have been used to model whole-brain neural activity, for both monkeys (Mejias et al., 2016) and humans (Sanz Leon et al., 2013). Here brain areas, or parts of brain areas, are modelled as neural mass units and connected in comprehensive networks.

Ermentrout and Terman (2010, Ch. 11) provide a thorough exposition of the properties and application of neural mass models. Byrne et al. (2020) give a perspective on such models 50 years after the birth of the Wilson–Cowan equations, including how neural mass models have been the key building block of neural field models.

9.2.4 Neural Field Models

Many phenomena observed in the brain must be described by models where space is included explicitly. One example is the propagation of neural activity waves observed in slices of brain tissue (Kim et al., 1995; Pinto et al., 2005). Another example is the existence of bumps of neural activity, which is the firing of neurons in a spatially restricted region, observed in working memory tasks in monkeys (Fuster and Alexander, 1971). To account for such phenomena, neural field models can be used where there are synaptic connections between the neuronal units positioned in 3D space, generally with specific choices of distance-dependent connection strengths to allow

Working memory is involved when, for example, you remember a phone number looked up in a directory for a few seconds before dialling it. One idea is that a neural activity 'bump' is maintained for as long as the phone number needs to be remembered and then disappears afterwards.

for self-sustaining activity bumps. Early influential examples of such models are Wilson and Cowan (1973), Nunez (1974) and Amari (1977).

An example of a neural field model used to investigate the existence and stability of bumps is the Wilson–Cowan-like model studied by Blomquist et al. (2005) (Figure 9.7):

$$\begin{aligned}\tau_\mathrm{E}\frac{\partial u_\mathrm{E}(x,t)}{\partial t} &= -u_\mathrm{E}(x,t) + \int_{-\infty}^{\infty} w_\mathrm{EE}(x-x')f_\mathrm{E}(u_\mathrm{E}(x',t))\mathrm{d}x' \\ &\quad - \int_{-\infty}^{\infty} w_\mathrm{IE}(x-x')f_\mathrm{I}(u_\mathrm{I}(x',t))\mathrm{d}x' + I_\mathrm{E}^\mathrm{ext}(x,t), \\ \tau_\mathrm{I}\frac{\partial u_\mathrm{I}(x,t)}{\partial t} &= -u_\mathrm{I}(x,t) + \int_{-\infty}^{\infty} w_\mathrm{EI}(x-x')f_\mathrm{E}(u_\mathrm{E}(x',t))\mathrm{d}x' \\ &\quad - \int_{-\infty}^{\infty} w_\mathrm{II}(x-x')f_\mathrm{I}(u_\mathrm{I}(x',t))\mathrm{d}x' + I_\mathrm{I}^\mathrm{ext}(x,t).\end{aligned} \tag{9.6}$$

Here the dynamical variable $u_{\mathrm{E/I}}$ represents the net input to neurons located at point x on an infinite line, so that the population firing rate of these neurons is given by $\nu_{\mathrm{E/I}} = f_{\mathrm{E/I}}(u_{\mathrm{E/I}})$. Otherwise, the structure of the model is very similar to the Wilson–Cowan neural mass model (Equation 9.5 in Section 9.2.3), except that the terms representing synaptic inputs are replaced with inputs from a sum (integral) of neuronal units arranged along an infinite line in the x-direction. Thus the neural field model can be regarded as a continuous version of a neural mass network model, with a function describing the spatial relationship between the neural mass units.

Blomquist et al. (2005) showed that stable bumps of activity can be formed when the spatial widths of the inhibitory kernels ($w_\mathrm{IE}, w_\mathrm{II}$) are larger than the widths of the excitatory kernels ($w_\mathrm{EE}, w_\mathrm{EI}$). The lateral extent of the bump of inhibitory activity is then larger than for excitatory activity. The physical picture here is that narrow recurrent excitation forms the local bump, whilst the wider inhibition prevents the bump from spreading out in space (Figure 9.7). However, if the inhibitory time constant τ_I becomes too large, compared to the excitatory time constant τ_E, the bump becomes unstable. Instead, there may be an oscillatory solution where the bump sizes vary in time. A later study investigated how self-sustaining bumps can be generated and annihilated by transient synaptic inputs (Yousaf et al., 2013).

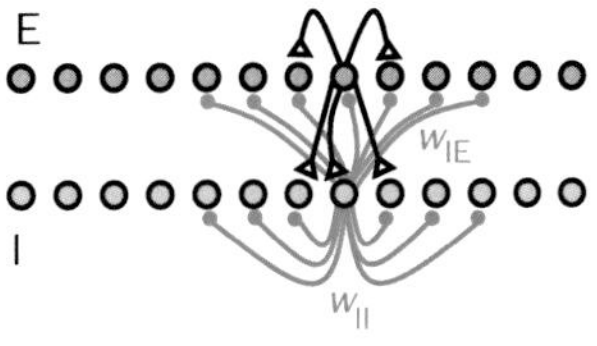

Fig 9.7 Schematic of a two-population neural field model. There is a sheet of excitatory neurons (E, black) and a sheet of inhibitory neurons (I, blue) connected to each other and to themselves. The connections are to nearby neurons in the sheets, though the inhibitory connections stretch further than the excitatory ones.

In flies, populations of neurons have been found that represent the direction in which the fly is heading by means of bump-like dynamics. With a clever combination of optical measurements and stimulation, Kim et al. (2017) demonstrated that the behaviour could be accounted for by a 1D neural field model connected in a ring-like structure. In this ring network, each position represents a specific heading direction. It was further found that a connectivity profile with narrow local excitation and broad long-range inhibition could explain successfully the experimental observations.

Neural field models have also been used to describe visual hallucinations (Ermentrout and Terman, 2010, Sec. 12.5), why neurons in the visual cortex are most responsive to stimuli oriented in certain directions (Ben-Yishai et al., 1995), and also whole-brain dynamics (Nunez, 1974; Jirsa and Haken, 1997; Liley et al., 2002). For comprehensive surveys of the use of such models, see Coombes (2005) and Coombes et al. (2014).

9.2.5 Limitations of Rate-Based Models in Networks

Whilst rate-based models can be a reasonable representation of neuronal activity and processing in a number of areas of the nervous system, and can have great explanatory power given their small number of parameters, by definition they cannot be used to investigate whether individual spikes are relevant for neural encoding, processing and the network dynamics.

One example is the response of populations of neurons receiving a common input. In Equation 9.2 for the evolution of the total synaptic current as a function of the firing rates, a sharp increase in firing rates will lead to the synaptic current increasing, after a time lag. As the output firing rate depends on the synaptic current, the increase in the output firing rate also lags behind the increase in the input firing rates.

We might assume that this finding could also be applied to a population of spiking neurons which receive common input. If the cells are quiescent before the increase in firing rates, this is indeed the case. However, if there is sufficient input to cause the neurons to be tonically firing before the increase in input, and if the firing of the neurons is asynchronous, the population firing rate can follow the input firing rate with virtually no lag (Knight, 1972; Gerstner, 2000). Gerstner et al. (2014) gives more insight into this setup.

9.3 Networks of Integrate-and-Fire Neurons

Most studies of integrate-and-fire networks have focused on general properties of spiking networks, rather than aiming to mimic a specific brain network. A key insight gained from such studies is how the very irregular and asynchronous firing observed in cortical networks can arise in recurrently connected networks of spiking neurons in which the excitatory and inhibitory inputs to each neuron are roughly equal (van Vreeswijk and Sompolinsky, 1996; Amit and Brunel, 1997a, b; Brunel, 2000). We refer to such a network as a **balanced network model**. Here one of these studies (Amit and Brunel, 1997b) will be used to exemplify the modelling of recurrent networks typically for cortical structures (Sections 9.3.1 and 9.3.2). In later work, structured networks of integrate-and-fire-neurons that model specific cortical networks have been constructed (Sections 9.3.3 and 9.5.3).

9.3.1 Example: Recurrent Network of Integrate-and-Fire Neurons

This section describes a model of a recurrent network of excitatory and inhibitory integrate-and-fire neurons presented by Amit and Brunel (1997a, b). The main goal of this study was to explore whether memories can be stored by altering the strengths of the connections between the excitatory neurons. The network was intended to represent a cortical column, the motivation being recordings from cells in the monkey temporal cortex (Miyashita, 1988). These cells fire at elevated rates for a number of seconds after a familiar visual stimulus has been presented briefly, suggesting a memory recall of the stimulus. In addition, in the absence of any stimulus, cells fire at a low background

In contrast to earlier equations for the membrane potential in this book, neither the membrane capacitance nor the membrane resistance appears in Equation 9.7. The reason for this is that the membrane resistance or capacitance scales all the weights, and so the effect of changing the resistance or the capacitance can be achieved by scaling the weights. However, the equation appears to be dimensionally incorrect; the units of current do not match the units of voltage divided by time. This can be resolved by treating the current as though it had units of $V\,s^{-1}$.

rate, and the cells in a column are thought to receive a constant background barrage of spikes from other cortical columns.

Amit and Brunel's modelling of these data proceeded in two stages. Firstly, a network of two recurrently connected populations of excitatory and inhibitory neurons was set up, similar in structure to the Wilson–Cowan neural mass model described in Section 9.2.3. The parameters were tuned to roughly reproduce the experimentally recorded firing statistics in the absence of a stimulus. To achieve this, Amit and Brunel (1997b) extended the analysis which had been applied to the Stein model (Section 8.5.3). This analysis also informed the process of scaling down the size of the realistic network to a manageable size. This part of the study is presented here, together with our own simulation results of this model. The second stage was to embed memories in the network by modifying the synaptic weights, which is presented in the chapter on plasticity (Section 11.5.5).

Network Model

The network constructed by Amit and Brunel (1997b) comprises N_E excitatory and N_I inhibitory neurons. The membrane potential V_j of each cell evolves in time according to:

$$\frac{dV_j}{dt} = -\frac{V_j}{\tau_j} + I_j^{rec} + I_j^{ext}, \tag{9.7}$$

where τ_j is the membrane time constant of the neuron, I_j^{rec} is the contribution to the current from recurrent collaterals and I_j^{ext} is the contribution from external input to the cell. When the neuron reaches the threshold θ, it emits a spike; the time of the kth spike produced by neuron j is denoted t_j^k. After firing, the neuron is reset to $V_{reset} = 10\,mV$ and there is a refractory period of $\tau_0 = 2\,ms$. The membrane time constant has the same value τ_E for all excitatory neurons, and τ_I for all inhibitory neurons.

The recurrent input to the cell comes from all the other excitatory and inhibitory cells in the network. The recurrent input received by neuron j at time t is:

$$I_j^{rec}(t) = \sum_{i \in E, k} c_{ij} w_{ij} \delta(t - \tau_{ij} - t_i^k) - \sum_{i \in I, k} c_{ij} w_{ij} \delta(t - \tau_{ij} - t_i^k). \tag{9.8}$$

The relationship between presynaptic neuron i and postsynaptic neuron j is specified by three quantities. c_{ij} is a random binary variable denoting whether a connection exists. If it does, w_{ij} is the weight of connection, and τ_{ij} is the propagation delay from i to j. The first summation is over all the spikes (numbered k) of all the neurons i in the excitatory (E) group of neurons; the second summation is over all spikes from all inhibitory (I) neurons. δ indicates a delta function, defined in Section 7.3.1. The strength of the connection w_{ij} is drawn randomly from a Gaussian distribution (Box 5.8), with a mean which depends on the classes of the two neurons it connects (w_{EE} for excitatory-to-excitatory synapses, w_{EI} for excitatory-to-inhibitory synapses and so on) and a standard deviation of Δ times the mean. The delays τ_{ij} are drawn from a uniform distribution between 0.5 ms and 1.5 ms.

The input from external columns is assumed to come from neurons firing in other cortical columns. The times of spikes $t_j^{ext,k}$ arriving at neuron j

Box 9.2 | Event-based simulation

Conventional methods for solving the time course of variables described by coupled differential equations involve splitting up time into discrete chunks of length Δt. These methods can be applied to spiking neuron models, but the precision with which the time of a spike can be specified is limited by the length of Δt.

In **event-based simulation** methods, rather than simulating every time-step of the simulation, each step in the simulation corresponds to the production or reception of a spike. For the entire network, there is a queue of event times which are expected to occur in the future. At each step, the earliest event on the list is considered. If this is a spike production event in neuron i, events tagged with the identity j of receiving neurons are added to the queue to occur at time $t + \tau_{ij}$, where τ_{ij} is the delay from i to j. If the event is a spike being received at neuron j, the time at which neuron j is next expected to fire is recomputed. For certain classes of synaptic potential in current-based neurons, this can be done with a precision that is limited only by the floating-point accuracy of the computer.

In some simulations of networks of integrate-and-fire neurons, there are significant differences between the level of synchronisation observed in event-based and time-step methods, even when the time-step is 0.01 ms, two orders of magnitude faster than the rise time of excitatory postsynaptic currents (Hansel et al., 1998).

Whilst event-based methods are more precise than time-step methods, the logic behind them and their implementation are quite intricate, and depend on the class of model neuron simulated. Event-based methods exist for integrate-and-fire models with various types of current-based synapses, and many of these methods are incorporated in simulators such as *NEURON* (Carnevale and Hines, 2006; van Elburg and van Ooyen, 2009). Event-based simulation methods for conductance-based neurons have also been developed (Brette, 2006; Rudolph and Destexhe, 2006), but again the type of conductance time courses they can simulate is limited.

---

are generated by an independent Poisson process (Box 8.5) at a rate $cN_E \nu_E^{\text{ext}}$, where ν_E^{ext} is the mean firing rate of cells in the external columns. The increase in membrane potential is set to be the same as the mean weight onto other excitatory or inhibitory neurons:

$$I_j^{\text{ext}} = \begin{cases} w_{EE} \sum_k \delta(t - t_j^{\text{ext},k}), & j \in E \\ w_{IE} \sum_k \delta(t - t_j^{\text{ext},k}), & j \in I. \end{cases} \tag{9.9}$$

The parameters of the network are set up so that ν_E^{ext} is similar to the actual mean firing rate ν_E of the excitatory neurons modelled in the cortical column. Thus the mean external input is similar to the input the column receives from local excitatory cells.

We used event-based simulation (Box 9.2) to simulate a small version of the network of Amit and Brunel (1997b), with $N_E = 6000$ excitatory neurons and $N_I = 1500$ inhibitory neurons, where the probability of a connection existing ($c_{ij} = 1$) is 0.2; Figure 9.8 shows the results. Of the three selected excitatory neurons whose membrane potentials are shown in Figure 9.8a–c, two fire irregularly at different time-averaged rates, whilst the

Fig 9.8 Our simulations of a network of recurrently connected excitatory and inhibitory neurons using Equations 9.7, 9.8 and 9.9, after Amit and Brunel (1997b). **(a–c)** The time courses of the membrane potential of three excitatory neurons. **(d)** The time course of one inhibitory neuron. **(e)** A spike raster of 60 excitatory neurons (lower traces) and 15 inhibitory neurons (upper traces, highlighted in blue). **(f)** Population firing rates of excitatory (black) and inhibitory neurons (blue). **(g)** Average autocorrelation of spikes from excitatory neurons. **(h)** Histogram of average firing rates of excitatory neurons (black) and inhibitory neurons (blue). Parameter values: $\theta = 20\,\text{mV}$, $V_{\text{reset}} = 10\,\text{mV}$, $\tau_E = 10\,\text{ms}$, $\tau_I = 5\,\text{ms}$, $w_{EE} = 0.21\,\text{V s}^{-1}$, $w_{EI} = 0.63\,\text{V s}^{-1}$, $w_{IE} = 0.35\,\text{V s}^{-1}$, $w_{II} = 1.05\,\text{V s}^{-1}$, $\Delta = 0.1$. τ_{ij} is drawn uniformly from the range $[0.5, 1.5]\,\text{ms}$ and $\nu_E^{\text{ext}} = 13\,\text{Hz}$.

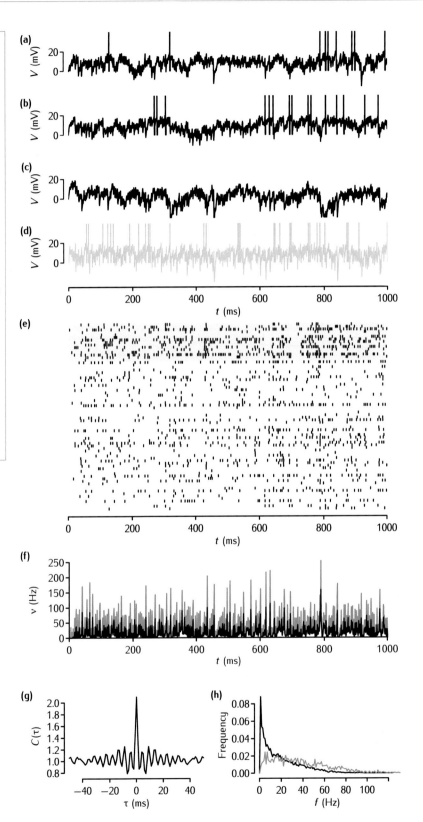

third is silent. The membrane potential and spiking activity of an inhibitory neuron (Figure 9.8d) appear similar to those of the two active excitatory neurons. Figure 9.8e shows a spike raster of 60 of the excitatory neurons, and 15 of the inhibitory neurons (highlighted). Whilst the firing times are random, there are events in which many neurons fire. These events can be seen more clearly by computing some statistics of the spike trains (Box 9.3). In the plots of the excitatory and inhibitory population firing rates (Figure 9.8f), there is an irregular oscillation in the overall activity levels. In order to quantify this, the cross-correlation of the population firing rates can be computed (Figure 9.8g). The histogram of time-averaged firing rates (Figure 9.8h) shows that the number of excitatory neurons firing at a particular frequency follows a roughly exponential distribution, whilst the distribution of inhibitory firing rates is flatter.

Insights from Analysis of the Model

So far, the behaviour of the network has been described with only one set of parameters. In order to set these parameters, Amit and Brunel (1997a, b) analysed the network under some simplifying assumptions:

(1) Each neuron receives many spikes within its integration period.
(2) The increase in membrane potential due to an incoming spike is small, compared to the threshold.
(3) The firing times of the excitatory and inhibitory neurons are independent, although the firing rates may not be.

This analysis, summarised in Box 9.4, is an extension of the analysis in Section 8.5.3.

A number of constraints on the parameters emerge:

- Inhibition has to be sufficiently strong, relative to excitation, for the stable asynchronous state to exist, but it does not have to be finely tuned – the concept of a balanced network.
- Up to a point, the faster the inhibition, the less inhibition is required.
- For a given ratio of the inhibitory and excitatory input to neurons, the weights from excitatory cells must be sufficiently large, compared to the threshold.

In fact, the simulations shown in Figure 9.8 demonstrate that the firing times are correlated, and so the assumptions of the analysis are not fulfilled strictly. However, the correlations are sufficiently small that the results of the analysis are good enough for the parameters of the network to be set so that there is a stable level of activity. More recent analysis, which uses the tools of statistical physics, can address non-stationary firing rates (Buice and Cowan, 2009).

Scaling the Amit and Brunel Model

The Amit and Brunel (1997b) model network is intended to represent a cortical column, with around 10^5 neurons. They simulated a number of networks of various sizes. Of these, the network presented here, which is the one we simulated (Figure 9.8), has only 7500 neurons, smaller by a factor

Box 9.3 | Spike statistics

In order to make sense of spike trains recorded from multiple neurons, a number of statistics can be employed. A prerequisite for computing the statistics is to bin the spike times by computing $S_i(t)$, the number of spikes produced by neuron i in the interval $[t, t + \Delta t]$. The width of a bin Δt is typically on the order of a millisecond.

The **instantaneous population firing rate** in a pool of N neurons is defined as:

$$v(t) = \frac{1}{N\Delta t} \sum_{i=1}^{N} S_i(t).$$

At the end of a simulation of duration T, the **temporally averaged firing rate** of a neuron can be computed:

$$v_i = \frac{1}{T} \sum_{k=0}^{T/\Delta t - 1} S_i(k\Delta t).$$

The **autocorrelation** is a temporally averaged quantity defined as:

$$C_i(\tau) = \frac{1}{T - \tau} \sum_{k=0}^{(T-\tau)/\Delta t - 1} S_i(k\Delta t)S_i(k\Delta t + \tau).$$

The autocorrelation indicates how well the instantaneous firing of a neuron correlates with its firing at a time τ later. For a Poisson spike train with rate v, the autocorrelation is expected to have a peak value v at τ and is expected to be $v\Delta t$ elsewhere. Autocorrelations sometimes have sidebands, which indicate periodic activity.

The **cross-correlation** is similar to the autocorrelation, except that it indicates how the instantaneous firing of one neuron i is related to the firing of another neuron j at a time τ later:

$$C_{ij}(\tau) = \frac{1}{T - \tau} \sum_{k=0}^{(T-\tau)/\Delta t - 1} S_i(k\Delta t)S_j(k\Delta t + \tau).$$

Correlated activity between neurons is revealed as a peak, which may not lie at $\tau = 0$.

Amit and Brunel (1997b) used an average cross-correlation, defined between excitatory and inhibitory neurons as:

$$\overline{C}_{EI}(\tau) = \frac{1}{T - \tau} \sum_{k=0}^{(T-\tau)/\Delta t - 1} v_E(k\Delta t)v_I(k\Delta t + \tau),$$

where $v_E(t)$ and $v_I(t)$ are the instantaneous population rates for the excitatory and inhibitory neurons.

For more detailed introductions to spike train statistics, see texts such as Dayan and Abbott (2001), Koch (1999), Gabbiani and Koch (1998) and Rieke et al. (1997).

Box 9.4 | Analysis of excitatory–inhibitory networks

Amit and Brunel's (1997a) analysis of an excitatory–inhibitory network is an extension of the analysis of the Stein model, which was outlined in Section 8.5.3. It is supposed that the network reaches a state in which each neuron fires in a random manner that can be described by Poisson statistics. The mean firing rate of the excitatory neurons v_E is a function of the mean μ_E and variance σ_E^2 of the membrane potential in excitatory neurons:

$$v_E = f(\mu_E, \sigma_E) \text{ and } v_I = f(\mu_I, \sigma_I). \tag{a}$$

Figure 8.13 shows a typical form of f. Similarly, the mean firing rate of the inhibitory neurons v_I depends on the mean μ_I and variance σ_I^2 of the membrane potential in the inhibitory neurons.

Using the same reasoning as in Section 8.5.1 (Equation 8.11) and Box 8.6, but with the excitatory and inhibitory incoming spike rates being $cN_E(v_E + v_E^{ext})$ and $cN_I v_I$, respectively, the time evolution of the means and variances of the membrane potential of the excitatory and inhibitory neurons can be written down:

$$\frac{d\mu_E}{dt} = -\frac{\mu_E}{\tau_E} + (w_{EE} c N_E (v_E + v_E^{ext}) - w_{IE} c N_I v_I),$$

$$\frac{d\sigma_E^2}{dt} = -\frac{\sigma_E^2}{\tau_E} + \frac{1}{2}(w_{EE}^2 c N_E (v_E + v_E^{ext}) + w_{IE}^2 c N_I v_I),$$

$$\frac{d\mu_I}{dt} = -\frac{\mu_I}{\tau_I} + (w_{EI} c N_E (v_E + v_E^{ext}) - w_{II} c N_I v_I), \tag{b}$$

$$\frac{d\sigma_I^2}{dt} = -\frac{\sigma_I^2}{\tau_I} + \frac{1}{2}(w_{EI}^2 c N_E (v_E + v_E^{ext}) + w_{II}^2 c N_I v_I).$$

The sets of equations (a) and (b) form a closed system of equations in the four variables μ_E, μ_I, σ_E and σ_I. A solution can be found numerically and then checked for stability (Box 8.3).

The steady state of the means and variances of the membrane potentials can be obtained by setting the left-hand side of Equations (b) to zero. The steady-state means and variances show that changing the size of the system by changing N_E and N_I changes the means and variances, and is therefore expected to change the behaviour of the network. The equations also demonstrate that this can be compensated for by scaling the connection probability c, so that cN_E and cN_I remain constant. In contrast, the mean weights cannot be used to compensate for changes in the size of the system. For example, the mean can be kept constant by scaling w_{EE} so that $N_E w_{EE}$ remains constant. However, this implies that $N_E w_{EE}^2$ changes, and so the variances change.

A contrast to this analysis of an excitatory–inhibitory network of noisy spiking neurons is that of Wilson and Cowan (1972) (Section 9.2.3).

of 13.3. Simulating this network demonstrates a number of the principles underlying the design of models that were laid out in Section 9.1.2:

(1) The fraction of excitatory and inhibitory cells in the scaled-down network is the same as in the original, 80% and 20%, respectively.

(2) The probability c of two neurons being connected has been scaled up by a factor of 4, so as to compensate partly for the smaller number of inputs.

(3) The spontaneous firing rates have been scaled up from around 1–5 Hz in vivo to around 13 Hz.

Taken together, this means that the expected number of spikes impinging on an excitatory or inhibitory neuron in one second should be roughly the same as the number of spikes arriving in one second at a neuron of the same type in the full network (Box 9.4). A large number of spikes arriving within the period of the membrane time constant is necessary for the analysis of the network (Section 9.3.1) to be valid.

To a first approximation, scaling up the connectivity c to compensate for the reduction in network size should cause no change in the network behaviour. However, the connection probability cannot be greater than 1, so the scope for increasing c is limited. For example, if the connectivity in the full network is 5%, the maximum factor by which the connectivity can be increased is 20. In contrast, scaling the synaptic weights w_{ij} would be expected to change the network behaviour, as demonstrated in Box 9.4.

In order to increase the firing rates, the function that converts the mean and variance of the depolarisation into a firing rate has to be altered, or the mean weights (w_{EE}, w_{EI}, w_{IE} and w_{II}) need to be modified. Thus scaling the firing rates does change the network behaviour. In 1997, when Amit and Brunel proposed their network, it would have been very challenging to simulate a network of the size of a cortical column (10^5 neurons) with realistic firing rates, but increased computing power has made it possible, as exemplified in Section 9.3.3.

Despite the attempt to maintain the fundamental characteristics of the network in the scaled-down version, the size of the network does have an effect on its behaviour. The larger the network, the smaller the magnitude of the oscillations in activity, as shown by the height of the peak of the cross-correlation function. Plotting the height of the cross-correlation against network size suggests that, for infinitely large networks, the cross-correlation function would be flat (Amit and Brunel, 1997b). Since this is one of the assumptions of the analysis, this suggests that the analysis would be precise in the case of an infinitely large network. With very small networks, chance correlations can cause large fluctuations, which throw the network into a persistent state of rapid firing.

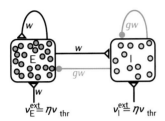

Fig 9.9 Excitatory-inhibitory network, as parameterised by Brunel (2000). The excitatory (E) and inhibitory (I) populations (rectangles) of neurons (circles) each comprise $N_E = 0.8N$ and $N_I = 0.2N$ neurons, respectively. Each neuron receives cN_E excitatory recurrent synapses and cN_I inhibitory recurrent synapses, where c is a very low fraction. All excitatory recurrent and external synapses have strength w, and all inhibitory synapses have strength gw. There is external input to each population, v_E^{ext} and v_I^{ext}. A fixed synaptic delay $\tau_{ij} = t_d = 1.5$ ms is used.

9.3.2 Rich Dynamics of Two-Population Spiking Models

The network of two populations of excitatory and inhibitory neurons studied by Amit and Brunel (1997b) in Section 9.3.1 was tuned to be in a state with very irregular firing. However, with other choices for model parameters, the network may exhibit very different dynamics. Brunel (2000) explored this rich network behaviour systematically using a combination of analytical and numerical techniques. They simplified the parameter space by assuming all excitatory synapses have a weight $w = w_{EE} = w_{EI}$ and that all inhibitory synapses are g times as strong as the excitatory ones, so $w_{IE} = w_{II} = -gw$ (Figure 9.9). The parameter η gives the external input in

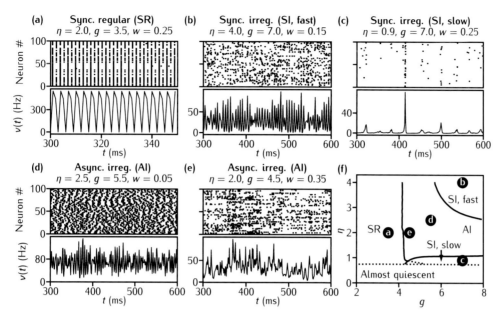

Fig 9.10 Rich dynamics in networks of excitatory and inhibitory spiking neurons. **(a)–(e)** Spike trains (top) from 100 randomly selected neurons across both excitatory and inhibitory populations, and population rate (bottom) for five combinations of the parameters, as indicated above the spike trains: η (external input), g (ratio of inhibitory-to-excitatory synaptic strengths) and w (excitatory synaptic strength). The combinations of parameters lead the firing to be synchronous regular (SR), fast or slow synchronous irregular (SI) and asynchronous irregular (AsI). **(f)** Bifurcation diagram of behaviour with different values of the two network parameters g and η, with $w = w_{EE} = w_{EI} = 0.1$. The dots labelled a–e indicate the η and g values simulated in **(a)–(e)**. Figure based on data from Skaar et al. (2020); simulation scripts available from the *OpenSourceBrain* database.

units of the 'threshold input', that is, the input needed to drive one of the neurons to fire in the absence of recurrent input. Depending on the choice of the strength of inhibition g and the amount of external input η, the network exhibits different types of behaviour, amongst them being three types illustrated in Figure 9.10a–e:

- **Synchronous regular (SR) states**, in which groups of neurons fire synchronously and regularly, causing strong regular oscillations in population activity;
- **Asynchronous irregular (AsI) states**, in which each individual neuron has an irregular firing pattern and the population of neurons fires asynchronously, leading to fluctuations in population activity without a dominant frequency; and
- **Synchronous irregular (SI) states** with oscillatory population activity, but strongly irregular individual firings.

Figure 9.10f shows a bifurcation diagram of the two network parameters g and η, showing regions in which the different types of network states occur. When the external input is low ($\eta < 0.8$), the network is almost quiescent, that is, there is very little firing. Above this input level, the irregular external input causes some firing of neurons, which leads to recurrent feedback between the excitatory and inhibitory populations. Since there are four times as many excitatory neurons as inhibitory ones, and it is assumed that they are equally likely to connect to target neurons, the inhibition strength needs to be four times larger than the excitation strength, that is, $g = 4$, for balanced excitation and inhibition. The bifurcation diagram shows that when inhibition outweighs excitation, $g > 4$, AsI firing tends to predominate, as shown in Figure 9.10d, e and in the Amit and Brunel (1997a) simulations (Section 9.3.1, Figure 9.8). The idea of balanced excitatory and inhibitory inputs to a neuron causing irregular firing can be understood in single neurons with uncorrelated external input (Section 8.5.3).

9.3.3 Potjans–Diesmann Model of a Cortical Column

The networks of integrate-and-fire neurons investigated in Amit and Brunel (1997b) (Section 9.3.1) and Brunel (2000) (Section 9.3.2) were used to investigate possible principles for how spiking networks exhibiting the irregular firing seen in experiments might learn and memorise. These model networks were idealised and not made to mimic a particular network found in the brain. Potjans and Diesmann (2014) explored whether insights from these studies could be combined with knowledge from anatomical and physiological studies to make a plausible model for a piece of the mammalian neocortex. They thus constructed a layered model corresponding to a 1-mm^2 piece of cat visual cortex comprising 80 000 integrate-and-fire neurons divided across four cortical layers (layer 2/3, layer 4, layer 5, layer 6). In total, eight populations were included, since each layer had an excitatory and an inhibitory population, similar to the two-population models of Amit and Brunel (1997b) and Brunel (2000).

A key insight from the papers of Amit and Brunel (1997b) and Brunel (2000), simultaneously discovered by van Vreeswijk and Sompolinsky (1996), was that AsI firing as seen in experiments could be achieved in a balanced network, that is, a network where on average the effects from excitatory and inhibitory inputs onto each cell cancel out (Section 8.5.3). By combining this principle with neurobiological data, the model constructed by Potjans and Diesmann (2014) also predicted background firing patterns in accordance with experiments (Figure 9.11). It also predicted stimulus-evoked responses, investigated in the model by driving the cortex model with a volley

Fig 9.11 (a) Sketch of the eight-population network consisting of 80 000 leaky integrate-and-fire neurons (Potjans and Diesmann, 2014). External input is provided by a population of thalamic (TC) neurons and unspecified other cortical cells providing a steady background input. Red/blue arrows correspond to excitatory/inhibitory connections. (b) Spontaneous ($t < 900$ ms) and stimulus-evoked spiking activity (from synchronous firing of TC neurons at time $t = 900$ ms, denoted by thin vertical line). Each dot represents the spike time of a particular neuron. Figure based on data from Hagen et al. (2016).

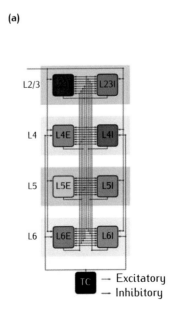

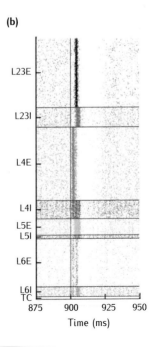

of strong synaptic inputs from the thalamus, in qualitative agreement with experiments (Figure 9.11b).

The Potjans–Diesmann model has also been converted to corresponding population firing-rate models, which cannot incorporate all details but are easier to handle both conceptually and computationally (Bos et al., 2016; Cain et al., 2016; Osborne et al., 2021). Such a multiscale approach, where the same network is modelled with two complementary models at different levels of granularity, can be powerful for gaining a better understanding (Einevoll et al., 2019).

Scripts for running simulations of the Potjans–Diesmann model are accessible from the *OpenSourceBrain* database (Gleeson et al., 2019).

9.4 Oscillatory Networks

Computational models of networks of spiking neurons have played a key role in helping our understanding of the activity properties of neurons embedded in different brain nuclei. Two major, and somewhat competing, network effects have been studied. The first is how networks of coupled excitatory and inhibitory neurons can maintain irregular firing patterns, as recorded in the neocortex. One such study has been described in the previous section. The flip side to this is that such networks have a strong tendency to exhibit more regular, rhythmic firing. Prominent **oscillations** in **local field potentials** (Section 13.4) at a range of frequencies, from a few hertz to hundreds of hertz, have been recorded in different brain areas in different behavioural states for the animal. Computational models provide considerable insight into how these rhythms may arise in different networks. Network models have been used to explore how different frequencies arise and how oscillations may be coherent over spatially distributed cell populations (Traub et al., 1999, 2004, 2005; Whittington et al., 2000; Bartos et al., 2007; Kopell et al., 2010). Details of the Traub et al. (2005) thalamocortical model are given in Section 9.5.1. Synaptic and intrinsic cell time constants are important, as is the structure of recurrent excitatory and feedback inhibitory loops. Spatial coherence may involve gap junctions that can quickly entrain the firing of groups of cells (Traub et al., 2004; Bartos et al., 2007).

Cortical oscillations in the **beta** (12–30 Hz) and **gamma** bands (30–80 Hz) are largely determined by network inhibition (Whittington et al., 2000). Gamma oscillations are recorded in many areas in the mammalian neocortex and the hippocampus. The gamma oscillation has been given central importance as a possible substrate for information coding. The basic idea is that a pattern of information is defined by neurons that fire on the same oscillation cycle (Singer, 1993; Lisman and Idiart, 1995). A network model that explores pattern recall at gamma frequency is discussed in Section 11.5.6.

Such rhythms are the natural consequence of the time course of γ-aminobutyric acid (GABA)$_A$-mediated feedback inhibition within a recurrent network of excitatory cells that also drive a population of inhibitory interneurons that feed back inhibition onto those cells. This is precisely the arrangement seen in the Amit and Brunel network model discussed above (Section 9.3.1). Irregular firing in that network arises from the delicate balancing of excitation and inhibition. If inhibition is dominant, then the network becomes an oscillator. This has been termed the pyramidal–interneuronal network gamma (**PING**) rhythm (Kopell et al., 2010). Mathematical analysis and simulations with excitatory–inhibitory (E–I) networks of simple spiking neurons have given a detailed picture of how this works (Ermentrout and Kopell, 1998; Whittington et al., 2000; Kopell et al., 2010). Gamma oscillations can also arise in purely inhibitory networks, termed **ING** for interneuronal network gamma (Whittington et al., 2000; Bartos et al., 2007; Kopell et al., 2010).

Slower oscillations, such as the **theta** (4–12 Hz) rhythm, prominent in the hippocampus, require specific neuronal properties, in addition to the network interaction of excitatory and inhibitory cells. Computational models have demonstrated the role of ionic currents, including the h-, M- and A-currents in generating theta through endowing inhibitory interneurons, such as stellate cells in the entorhinal cortex and oriens lacunosum-moleculare (O-LM)-projecting cells in the hippocampus, with membranes that resonate in this frequency range (Orbán et al., 2006; Kopell et al., 2010). Purely inhibitory networks of fast-spiking interneurons (I cells), along with O-LM interneurons (O cells) giving an I-O network, or with the addition of excitatory cells (E-I-O network), can produce theta rhythms, with gamma oscillations embedded in them (Kopell et al., 2010).

9.5 Networks of Biophysically Detailed Neuron Models

Until the 2010s, the lack of sufficient computer power prevented studies of large network studies using biophysically detailed neuron models. Whilst the standard integrate-and-fire neuron model is described mathematically by a single differential equation, typically biophysically detailed multi-compartment neurons are described by tens or hundreds of differential equations. The numerical solution of networks of thousands of interconnected neurons may easily encompass the evaluation of millions of differential equations. In addition, many parameter values must be obtained to model a single compartmental neuron, and the number multiplies when a network of such neurons are modelled.

This challenge was outlined in the pioneering study of this type by Traub et al. (2005) which provided a primer for constructing large-scale models of mammalian brain areas, giving precise detail on the cell models and network connectivity, and the model design decisions taken based on experimental data and computational feasibility. The opening two paragraphs of the paper set forth the motivations and difficulties of such a study with great clarity:

> The greatest scientific challenge, perhaps, in all of brain research is how to understand the cooperative behaviour of large numbers of neurons. Such co-operative behaviour is necessary for sensory processing and motor control, planning, and in the case of humans, at least, for thought and language. Yet

it is a truism to observe that single neurons are complicated little machines, as well as to observe that not all neurons are alike – far from it; and finally to observe that the connectional anatomy and synaptology of complex networks, in the cortex for example, have been studied long and hard, and yet are far from worked out. Any model, even of a small bit of cortex, is subject to difficulties and hazards: limited data, large numbers of parameters, criticisms that models with complexity comparable to the modelled system cannot be scientifically useful, the expense and slowness of the necessary computations, and serious uncertainties as to how a complex model can be compared with experiment and shown to be predictive.

The above difficulties and hazards are too real to be dismissed readily. In our opinion, the only way to proceed is through a state of denial that any of the difficulties need be fatal. The reader must then judge whether the results, preliminary as they must be, help our understanding.

Working in this 'state of denial', Traub et al. (2005) built a large-scale model of a single-column thalamocortical network made up of 3560 multi-compartmental neurons, as described in Section 9.5.1. Later, several large-scale models for neuronal networks in the cortex (Reimann et al., 2013; Markram et al., 2015; Neymotin et al., 2016; Billeh et al., 2020) and hippocampus (Dyhrfjeld-Johnsen et al., 2007; Bezaire et al., 2016), as well as subcortical structures (Migliore et al., 2015; Hjorth et al., 2020), have been built based on multicompartmental neurons. The execution of such large-scale models typically requires extensive software packages and access to supercomputers. Moreover, the construction of the models has typically required teams of researchers doing comprehensive investigations of the neurobiological literature, often in combination with in-house experiments tailored to provide the necessary neuron models and connectivity data. Two examples of such models, the Blue Brain model for the rat neocortex (Markram et al., 2015) and the Allen model for the mouse primary visual cortex (Billeh et al., 2020), are described in Sections 9.5.2 and 9.5.3, respectively.

The use of these large-scale models also differs from the use of simpler models with few parameters. In practice, it will be impossible to fit all the model parameters in, say, a mouse visual cortex model by comparing with experiments recorded from the visual cortex of a mouse. But this raises the question of what the goal of the modelling should be, given that there is also biological variability between the visual cortices between two mice, even if both perform, for example, the same visual perception task equally well. Should the goal be to make a model for a particular mouse or for some kind of average mouse? Should the goal be to make different instantiations of the cortex model which statistically are part of the same distributions as different mice? These are questions which must be addressed when performing such model studies.

An objection to large-scale network models has been that they are so complicated that even if they reproduce recorded electrophysiological data, they do not provide much understanding of the underlying principles for how networks operate. However, a quantitatively accurate simulation model can be viewed as an 'ideal' test animal, in that neurons and connections can be tuned and explored at will. Such a model would thus be an excellent starting point for exploration of principles, and also for derivation and validation

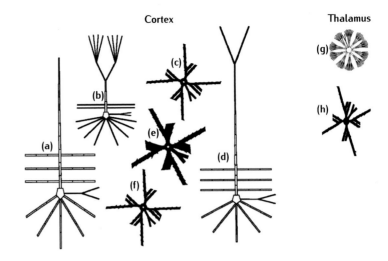

Fig 9.12 Morphological structure of the cell types in the Traub model: **(a)** layer 6 non-tufted pyramidal cells (regular spiking); **(b)** layer 2/3 pyramidal cells (regular spiking, fast rhythmic bursting); **(c)** superficial interneurons (basket cells and axoaxonic cells, low-threshold spiking); **(d)** layer 5 tufted pyramidal cells (intrinsically bursting, regular spiking); **(e)** layer 4 spiny stellate cells; **(f)** deep interneurons; **(g)** nucleus reticularis thalamic cells; and **(h)** thalamocortical relay cells. Figure adapted from Traub et al. (2005), with permission from The American Physiological Society.

of simpler network models of the same system in terms of integrate-and-fire or firing rate models (Koch and Buice, 2015; Einevoll et al., 2019). To allow for such exploration, it will be essential that models are made to be rerun for other parameters and to be made available to the research community.

9.5.1 Traub Thalamocortical Model

The Traub model effectively encompasses a single cortical column and its connectivity with the thalamus (Traub et al., 2005). Space is defined only in the direction of cortical depth, reflecting the layered structure of the cortex. The model network contains 3560 neurons of seven different types (Figure 9.12).

Each cell type is defined by a compartmental model with a particular stylised anatomical structure that captures the principal dendrites with between 50 and 137 compartments. All model cells of a particular type have exactly the same anatomy. The number of compartments was deemed sufficient, on the basis of numerous computer simulations, to reproduce detailed neuronal firing behaviours, including: differences in electrical properties between soma, axon and dendrites, action potential initiation in the axon and back-propagation into the dendrites and dendritic calcium spikes and bursts.

The model network was configured to approximate a variety of experimental slice preparations via scaling of synaptic conductances and application of bias currents to mimic the effects of the bath application of drugs that were ion channels or neuromodulators. Simulations of the model allowed the recording of electrical activity in identified cells, as well as the determination of population activity, presented in the form of a calculated extracellular potential at different cortical depths.

The model makes predictions about the physiology underlying network oscillatory states, including persistent gamma oscillations, sleep spindles and different epileptic states. A key conclusion is that electrical connections between defined neural types make significant contributions to these network states. Another important factor is the presence of strong recurrent interactions between layer four spiny stellate cells. Known network behaviours that are not reproducible by the model are also noted in the paper (Traub et al., 2005).

9.5.2 Blue Brain Neocortex Model

In this project, a model of a hexagonal block of the somatosensory cortex of young rats was constructed based on extensive neurobiological measurements, in particular patch-clamp experiments (Markram et al., 2015). The model comprised 31 000 neurons of 207 different types of biophysically detailed neuron models distinguished by their morphological shapes and electrical properties. The model neurons had different morphologies which were generated from a number of prototypes. In the assembly of the network from the neuronal building blocks, various automatic algorithms motivated by biology were used for positioning and connecting the neurons.

The model was validated against a variety of measurements, and several predictions were provided in Markram et al. (2015). This study also provided some qualitative insights into network behaviour. For example, the model included synapses with short-term plasticity where a key parameter is affected by the extracellular concentration of calcium $[Ca^{2+}]_e$ (Section 7.3). A key finding was that when modelling a virtual brain slice, made by putting seven hexagonal microcircuits next to each other, the spontaneous neural activity was observed to depend sensitively on the value of $[Ca^{2+}]_e$. For large values of $[Ca^{2+}]_e$, slow, oscillatory synchronised bursts of neural activity were observed in the model, mimicking the typical situation in slice experiments. For smaller values of $[Ca^{2+}]_e$, more similar to typical in vivo values, the firing was observed to be more asynchronous and like the in vivo case.

9.5.3 Allen Model for Mouse Visual Cortex

The mouse has become a preferred animal for brain studies in the last decades, partially because of the host of genetic techniques available for manipulation and detailed measurements. Billeh et al. (2020) reported two detailed models for the mouse primary visual cortex (V1), one based on biophysically detailed compartmental neuron models and the other based on integrate-and-fire models. Both models comprise 230 000 neurons generated from a number of exemplars forming a number of layers with a common network structure, and have an outer diameter of 1.7 mm, covering most of V1 in the mouse (Figure 9.13). The biophysically detailed version of the model consists of an inner core (diameter 0.8 mm), with compartmental neurons surrounded by an annulus of integrate-and-fire neurons (Figure 9.13a). The network comprises 17 types of neurons, spread across five layers (Figure 9.13b). Neuron models were obtained from standardised experimental and modelling pipelines.

The models were developed to simulate responses of V1 cells to arbitrary visual stimuli conveyed by thalamic neurons providing inputs to the cortical network. Synaptic connectivity parameters were set by prior experimental findings, as well as requiring the model to satisfy particular criteria, for example, that spontaneous background firing rates and peak firing rates in response to drifting gratings should be in accordance with experimental recordings. In the process of testing and building the model, several new

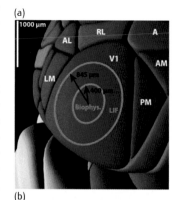

(a)

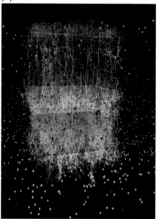

(b)

Fig 9.13 The mouse visual cortex models of Billeh et al. (2020). **(a)** Region covered by the models (400 μm radius for the core; 845 μm radius with surrounding annulus). **(b)** Visualisation of the biophysically detailed network, with 1% of neurons shown (colours show cortical layers; surrounding dots are integrate-and-fire neurons in the annulus). Adapted with permission from Elsevier Inc.: *Neuron* 106, 388–403, © 2020.

specific predictions regarding synaptic connections between various types of neurons were made. The biophysically detailed and the integrate-and-fire model versions performed similarly well for the questions addressed.

9.6 | Summary

Constructing a model network of neurons requires decisions at every stage of model design, starting with the level of description for the individual units of the network, whether they represent single neurons or groups of neurons. Here we have outlined the issues involved and provided guidelines for making appropriate choices. As illustrations of how such choices have been made, we have described existing network models of rate-based neurons and integrate-and-fire neurons, as well as neural mass and neural field models of populations of neurons.

The complexities of constructing biologically detailed models of any size have been illustrated by describing three models in which the aim is to develop faithful representations of parts of the neocortex.

Whilst the ultimate aim of network models may be to help us to understand the cognitive functions of neural subsystems, a realistic first step is to use models to shed light on the complex neural firing patterns recorded from biological neural networks. We have considered network models that explore the mechanisms behind asynchronous and synchronous firing activity, particularly the balance between excitation and inhibition.

Cognitive function covers a large range of topics, to which various modelling strategies have been applied. Generally we do not concentrate on specific cognitive functions, but in Chapter 11, we do consider neurobiological models of learning and memory which have been implemented in model networks.

CHAPTER 10

Brain Tissue

The nervous system consists of not only neurons, but also of other cell types such as glial cells. They can be modelled using the same principles as for neurons. The extracellular space (ECS), that is, the space between cells, contains ions and molecules that affect the activity of both neurons and glial cells. The Nernst potentials of ions are affected by their ECS concentration (Chapter 2), as are slower membrane processes such as ion pumps (Chapter 6). Further, the transport of, in particular, signalling molecules, oxygen and cell nutrients in the irregular landscape of ECS is required for brains to function. This chapter shows how to model such diffusive influences for uncharged molecules, and electrodiffusive influences for ions, involving both diffusion and electrical drift. This electrodiffusive formalism also explains the formation of dense nanometre-thick ion layers around membranes (Debye layers). When ion transport in the ECS stems from electrical drift only, this formalism reduces to the volume conductor theory which is commonly used to model electrical potentials around cells in the ECS. Finally, the chapter outlines how to model ionic and molecular dynamics not only in the ECS, but in the entire brain tissue also comprising neurons, glial cells and blood vessels.

The nervous system is made of not only neurons, but also **glial cells** and **blood vessels**. Glial cells, which in human brains are roughly equal in number to neurons, have in common that they do not fire action potentials, but their functions vary. In the central nervous system, four types of glial cells are found. **Microglia** act as immune defence. **Oligodendrocytes** create myelin sheaths around axons to facilitate propagation of action potentials (Chapter 5). In the peripheral nervous system, this function is instead performed by **Schwann cells**. **Ependymal cells**, amongst other things, produce cerebrospinal fluid (CSF). Finally, **astrocytes** provide nutrients to brain tissue and regulate extracellular ion concentrations, amongst other functions (Kettenmann and Ransom, 2012).

From a biophysical modelling point of view, glial cells are similar to neurons in that they are electrically excitable, a main difference being the

Fig 10.1 Illustration of the extracellular space (ECS) at the nanometre **(a)** and micrometre **(b)** scales. The extracellular matrix consists of various large molecules such as collagen and proteoglycans.

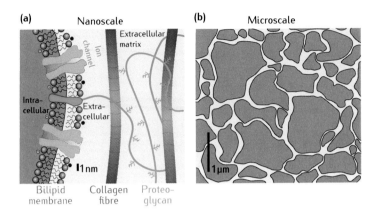

absence of the set of active sodium and potassium ion channels needed to generate action potentials. Thus the principles for modelling electrical glial cell dynamics are essentially the same as for neurons.

Blood vessels, made from **endothelial cells**, constitute the **vascular system** which transports blood rich in oxygen and nutrients to the brain and picks up carbon dioxide and other waste molecules for transportation out.

The space between the cells, the so-called **extracellular space** (**ECS**) (Figure 10.1), is also key for brain function. Even where the brain cell membranes are densely packed, there is a small space between the cells (Figure 10.1b). The continuous ECS structure, which has been likened to the mesh of water surrounding the bubbles in foam, comprises about 20% of the brain volume (Nicholson, 2001). The ECS is important for transporting nutrients and other key molecules over the last stretch from the vascular system to the cells. Furthermore, chemical signalling by molecules moving through the ECS is an important means of communication between cells.

The ECS has a carefully controlled ionic and molecular composition. In particular, the extracellular concentration of potassium, sodium, chloride and calcium must be controlled to give appropriate Nernst potentials so that, for example, neurons can generate action potentials as described in Chapter 3. The ECS also contains long-chain macromolecules that form the **extracellular matrix**. Many of these macromolecules are attached to the membranes; others float around between the cell membranes (Figure 10.1a). So-called **perineuronal nets**, for example, are made of matrix molecules wrapping tightly around the soma and proximal dendrites of a subset of neurons, and they seem to have a key role in long-term storage of memories (Tsien, 2013; Thompson et al., 2018)

Importantly, the same physical laws govern the dynamics of ions and molecules in the ECS as inside cells, that is, diffusion and electrical drift. In this chapter, we describe how the electrical potential and ion concentrations of ions of interest in the ECS can be modelled. In general, the dynamics of these variables are coupled and must be described by a combination of two processes: diffusion (Section 10.1) and electrical drift (Section 10.2). In combination, these are referred to as electrodiffusion, which is described for transport in the ECS in Section 10.3 and in brain tissue as a whole in Section 10.4. We also describe how to model liquid flow in blood vessels or through the ECS, as well as how ion dynamics and liquid flow are connected through the phenomenon of osmosis (Section 10.5).

10.1 Diffusion in Extracellular Space

In Chapter 6, the modelling of diffusion of calcium ions inside cells was described. The diffusion of ions and molecules in the ECS obeys the same physical principles, the main difference being that diffusion now typically occurs in the intricate, heavily convoluted space between the cells. However, careful analysis has shown that this transport process can be accurately described by means of diffusion in **porous media** where one phase is permeant (ECS) and the other is largely impermeant (cells). Here, the diffusion process through the convoluted ECS is represented by a diffusion process in a smooth 3D space with a reduced value for the diffusion coefficient.

The tortuosity factor λ_t can be estimated from measurements of diffusion coefficients for a specific molecule in the ECS (D) and in a free solution (D_0) via the relationship $\lambda_t = \sqrt{D/D_0}$. The tortuosity factor can also be estimated from computational models of diffusion in environments with obstacles mimicking the ECS (Nicholson, 2001).

10.1.1 Diffusion Equation for Extracellular Space

In Chapter 2, Fick's first law in one spatial dimension was used to describe diffusive transport down a concentration gradient across a membrane. Using vector notation (Box 10.1) in the three dimensions of ECS, this 1D Fick's law from Equation 2.2 becomes:

$$\vec{J}_{X,\text{diff}} = -D_X \nabla[X]. \tag{10.1}$$

The ∇ symbol refers to the nabla (or del) operator from vector calculus (Box 10.1). Further, $D_X = D_{X0}/\lambda_t^2$ is the diffusion coefficient for molecule X when diffusing in the ECS. D_{X0} is the diffusion coefficient that would be measured for the molecule in a container filled with the interstitial fluid found between the cells, whilst λ_t is the **tortuosity** of the ECS (see sidebox). This tortuosity factor λ_t is larger than one, and takes into account that diffusion for molecules in the ECS is slowed down by the collision of the diffusing molecules with cell membranes and large molecules. In cortical tissue, a typical value for λ_t is 1.6, which implies that the diffusion coefficient is reduced by about a factor $(1.6)^2 \sim 2.5$ by the convoluted structure of the ECS (Nicholson, 2001, Table 5). Note that [X] refers to the concentration of molecule X per volume of the ECS (not per total volume of brain tissue comprising both the ECS and the volume occupied by the cells). Likewise, $\vec{J}_{X,\text{diff}}$ refers to the particle flux per area of the ECS part only.

To obtain the diffusion equation, a **continuity equation** is needed to account for the fact that the overall number of molecule X is conserved:

$$\frac{\partial[X]}{\partial t} = -\nabla \cdot \vec{J}_{X,\text{diff}} + f_X. \tag{10.2}$$

This equation says that the concentration of X in a small volume around a point may change for two reasons, either by molecules diffusing in from neighbouring volumes in the ECS (first term on the right-hand side) or by molecules entering the ECS from cells (second term on the right-hand side) (Figure 10.3). Here, f_X is the **source density**, the volume density of molecules of type X entering the ECS per unit time (units M s^{-1} = mole L^{-1}s^{-1})).

(a) $\nabla \cdot \vec{G} > 0$

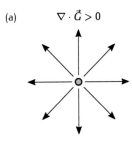

(b) $\nabla \cdot \vec{G} < 0$

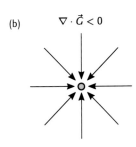

(c) $\nabla \cdot \vec{G} = 0$

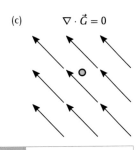

Fig 10.2 Illustration of effects of vector fields around points (blue dots) with positive (a), negative (b) and zero (c) divergence.

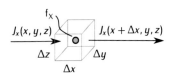

f_x

$J_x(x, y, z)$ $J_x(x + \Delta x, y, z)$

Δz Δy

Δx

Fig 10.3 Illustration of continuity equation in Equation 10.2. The symbols for the particle fluxes in the y- and z-directions, that is, J_y and J_z, are not shown for figure clarity.

Box 10.1 | Nabla and Laplace operators

In Equations 10.1–10.3, the use of the **nabla** (∇) and **Laplace operators** (∇^2) is useful as it allows us to formulate the equations without specifying a particular coordinate system. In Cartesian coordinates, the application of these operators on a scalar function F such as, for example, the molecular concentrations [X] gives:

$$\nabla F = \frac{\partial F}{\partial x}\vec{e}_x + \frac{\partial F}{\partial y}\vec{e}_y + \frac{\partial F}{\partial z}\vec{e}_z,$$

$$\nabla^2 F = \frac{\partial^2 F}{\partial x^2} + \frac{\partial^2 F}{\partial y^2} + \frac{\partial^2 F}{\partial z^2},$$

where $\vec{e}_x$ is a unit vector in the x-direction (1,0,0) and so on.

When the nabla operator is used in a vector dot product acting on a vector function $\vec{G}$, given by:

$$\vec{G} = G_x\vec{e}_x + G_y\vec{e}_y + G_z\vec{e}_z,$$

it computes what is known as the **divergence**:

$$\nabla \cdot \vec{G} = \frac{\partial G_x}{\partial x} + \frac{\partial G_y}{\partial y} + \frac{\partial G_z}{\partial z}.$$

The interpretation of the divergence is illustrated in Figure 10.2. When the divergence $\nabla \cdot \vec{G}$ is positive at a point (displayed as a blue dot in each panel), the vector field $\vec{G}$ is directed out of the point, and the point is referred to as a **source** (Figure 10.2a). When the divergence is negative, the point is referred to as a **sink** (Figure 10.2b). When the divergence is zero, the outward flux of the vector field in a small volume around the point will equal the inward flux (Figure 10.2c).

In situations with spherical symmetry, spherical coordinates are more useful. Here we have:

$$\nabla F = \frac{\partial F}{\partial r}\vec{e}_r,$$

$$\nabla^2 F = \frac{\partial^2 F}{\partial r^2} + \frac{2}{r}\frac{\partial F}{\partial r},$$

$$\nabla \cdot \vec{G} = \frac{\partial G_r}{\partial r},$$

where r is the radial distance, $\vec{e}_r$ is a unit vector in the radial direction and G_r is the radial component of $\vec{G}$.

Insertion of Equation 10.1 into Equation 10.2 to eliminate $\vec{J}_{X,\text{diff}}$ now gives the diffusion equation describing the dynamics of the concentration [X]:

$$\frac{\partial [X]}{\partial t} = D_X \nabla^2 [X] + f_X. \tag{10.3}$$

Here, ∇^2 refers to the Laplace operator from vector calculus (Box 10.1).

10.1.2 Pulse Solution of Diffusion Equation

The physical implications of the diffusion equation for molecular transport are well illustrated by considering how a pulse (see sidebox) of n moles of injected molecule X will be spread out in the ECS (Figure 10.4). The solution to the diffusion equation in this situation can be found in standard textbooks on diffusion (e.g. Crank, 1975):

$$[X](\vec{r}, t) = n \frac{1}{(4\pi D_X t)^{3/2}} e^{-r^2/4D_X t}. \tag{10.4}$$

Here r is the radial distance from the position where the molecules are injected.

The solution to Equation 10.4 is illustrated in Figure 10.5, where the diffusion of a pulse of molecules of type X injected into the ECS is shown. Whilst the concentration of X is always maximal at the point of injection ($r = 0$), the concentration profile gradually smears out. Figure 10.5a shows that, after 1 second, a typical X molecule has diffused some tens of micrometres away from the starting point and the spatial molecule distribution gets broader as time progresses.

This diffusional spread can be quantified further by considering the time t_{max} at which the concentration of X is at its maximum, as a function of the distance from the injection point. This is found by identifying the time for which $\partial[X](\vec{r}, t)/\partial t = 0$. At the injection point ($r = 0$), t_{max} is obviously zero, and for other distances t_{max} is found from Equation 10.4 to be given by $t_{max} = r^2/6D_X$ (Figure 10.5b). A key point here is that t_{max} is proportional to the square of the distance r. For the example in Figure 10.5, t_{max} for $r = 20\,\mu m$ is $\sim$0.3 seconds, whilst it is four times as long ($\sim$1.2 seconds) for $r = 40\,\mu m$.

10.1.3 Extracellular Signalling by Diffusion

Compared to electrical signalling inside and between neurons, extracellular diffusion, as described by Equation 10.4 is a slow process, in particular for long-distance transport. From the expression $t_{max} = r^2/6D$, we can obtain an estimate of the distance r which a typical diffusing molecule moves in a given time t, that is, $r \sim \sqrt{6Dt}$. For a calcium ion where $D \sim 2.3 \times 10^{-6}\,cm^2\,s^{-1}$ (as for the example in Figure 10.5), the distance diffused in 1 second is $r = \sqrt{6 \times 2.3 \times 10^{-6} \times 1.0}\,cm = 37\,\mu m$.

This distance grows only as the square root of time. After 100 seconds, it has increased to only $370\,\mu m$, and diffusion by a millimetre takes more than 10 minutes. Thus, even for a small ion like Ca^{2+}, transport by diffusion is a slow process compared to the millisecond timescale typical for the electrical processes in neurons. For larger molecules, diffusive transport is even slower as the diffusion coefficient decreases with increasing size. For example, for spheroidal molecules, like proteins curled up into a ball-like shape, the diffusion coefficient in water is predicted to be proportional to the inverse of the ball radius (Box 10.2). For diffusion in the ECS, it is likely that the obstacles to diffusion from the convoluted membrane structure will modify this simple relationship, but the qualitative trend of decreasing diffusion coefficients with size will be maintained. A detailed discussion can be found in Nicholson (2001).

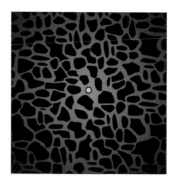

Fig 10.4 Illustration of the diffusion of molecules (blue) in the ECS. The molecules have been injected into the ECS as a pulse from the centre dot. The concentration profile illustrated by the brightness of the blue corresponds to a snapshot at a particular point in time, that is, the concentration profile given by Equation 10.4 for a particular value of t.

Mathematically, the pulse of injected molecules can be represented by means of Dirac δ-functions (Box 7.1). For a pulse of n moles of X injected into the ECS at the origin ($x = y = z = 0$) at time $t = 0$:

$$f(x, y, z, t) = n\delta(x)\delta(y)\delta(z)\delta(t).$$

The δ-function property $\int_{-\infty}^{\infty} \delta(x)dx = 1$ assures that the total number of injected molecules

$$\iiiint_{\text{volume, time}} f(x, y, z, t)dxdydzdt$$

equals n.

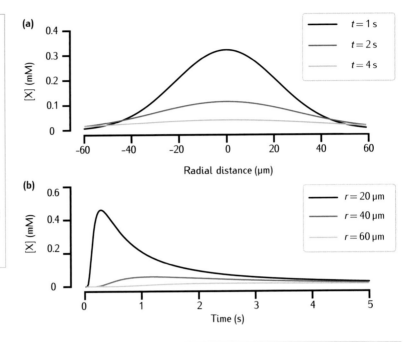

Fig 10.5 Diffusion from point source in the ECS. **(a)** Spatial spread of concentration [X] following injection of pulse of $n = 5 \times 10^{-14}$ moles of X molecules into the ECS as described by Equation 10.4. Three different times following pulse injection at $t = 0$ are considered. **(b)** Development in time of [X] for three different radial distances. The diffusion constant of X has been set to be similar to the value for Ca^{2+}, that is, $D_{X0} = 6.0 \times 10^{-6}$ cm^2 s^{-1}, $\lambda_t = 1.6$. This implies that $D_X = D_{X0}/\lambda_t^2 = 2.3 \times 10^{-6}$ cm^2 s^{-1}.

Box 10.2 | Stokes law

Stokes derived a relationship for the frictional force F_f acting on a small sphere with radius r_s dragged slowly through a viscous fluid with a constant velocity v:

$$F_f = 6\pi\eta r_s v.$$

Here η is the **viscosity** of the fluid. From the definition of **mobility**, $\mu = v/F$, one identifies the particle mobility $\mu = 1/(6\pi\eta r_s)$. The so-called **Einstein relation** relating the mobility and diffusion coefficient D is:

$$\mu = N_A D/RT,$$

where N_A is Avogadro's number. Combining these relationships gives:

$$D = \frac{RT}{6\pi N_A \eta r_s}.$$

The key result is that the diffusion coefficient is inversely proportional to the radius of the sphere. For a more detailed discussion, see Nelson (2008, Chapter 4).

10.2 | Electrical Drift in Extracellular Space

10.2.1 Volume Conductors and Electrical Conductivity

The electrical drift of ions in the ECS is described in a similar way to electrical drift across membranes (Section 2.2.3) or in the longitudinal direction inside thin dendrites or axons. However, as for diffusion, electrical drift in the ECS occurs in three spatial dimensions, and so Equation 2.3 describing the drift of ion type X in one spatial dimension generalises to:

$$\vec{J}_{X,\text{drift}} = -\frac{D_X F}{RT} z_X [X] \nabla V_e. \tag{10.5}$$

Here, the effective diffusion coefficient D_X from Section 10.1.3 has been introduced. The subscript 'e' has also been added to the ECS electric potential V_e to distinguish it from the membrane potential V considered in previous chapters.

The electric current density $\vec{I}_X$ (with units $C\,cm^{-2}\,s^{-1}$ or $A\,cm^{-2}$) associated with this drift is found by multiplying the particle flux with the ion valency z_X and Faraday's constant F:

$$\vec{I}_{X,\text{drift}} = F z_X \vec{J}_{X,\text{drift}}. \tag{10.6}$$

All mobile ions in the ECS will contribute to the net electrical current density, so the net current density is found by summing the contributions from all ion types X:

$$\vec{I} = \sum_X \vec{I}_{X,\text{drift}} = -\frac{F^2}{RT} \left(\sum_X D_X z_X^2 [X] \right) \nabla V_e. \tag{10.7}$$

The subscript 'drift' has been dropped from the net current density $\vec{I}$ to make the notation simpler in the following derivations.

For electrically conducting media like brain tissue, the electrical currents are proportional to the extracellular electrical field $\vec{E}_e$, and the proportionality constant σ_e is referred to as the **electrical conductivity**:

$$\vec{I}_e = \sigma_e \vec{E}_e, \tag{10.8}$$

where $\vec{E}_e = -\nabla V_e$, and the subscript 'e' has been added also to the current density $\vec{I}_e$ to distinguish it from previously encountered intracellular and membrane currents. This relationship is analogous to Ohm's law $I = GV$ for electrical circuits. From Equation 10.7 we can identify σ_e as:

$$\sigma_e = \frac{F^2}{RT} \sum_X D_X z_X^2 [X]. \tag{10.9}$$

Thus the contribution to electrical conductivity from each mobile ion type is proportional to both its concentration and its effective diffusion coefficient.

As an example, we now consider the case where the only mobile ions of relevance are sodium, potassium and chloride. Then the electrical current density is given by:

$$\vec{I}_e = \vec{I}_{\text{Na,drift}} + \vec{I}_{\text{K,drift}} + \vec{I}_{\text{Cl,drift}}$$
$$= -\frac{F^2}{RT} \left(D_{\text{Na}} [\text{Na}^+] + D_K [\text{K}^+] + D_{\text{Cl}} [\text{Cl}^-] \right) \nabla V_e. \tag{10.10}$$

The cations (sodium and potassium) and anions (chloride) will drift in opposite directions in an electric field due to their opposite charges. However, their contributions to the electrical conductivity:

$$\sigma_e = \frac{F^2}{RT} \left(D_{\text{Na}} [\text{Na}^+] + D_K [\text{K}^+] + D_{\text{Cl}} [\text{Cl}^-] \right), \tag{10.11}$$

are seen in Equation 10.9 to be proportional to the square of the valency of the different ion types, so that all ion types give positive contributions to σ_e.

In practice, the value of σ_e is measured in experiments rather than computed from Equation 10.9. Here σ_e is inferred from voltage responses

An estimate of the extracellular conductivity σ_e can be found from the simple formula in Equation 10.11, assuming that only three ion types (sodium, potassium, chloride) contribute to the electrical current. The following parameter values, taken from Halnes et al. (2013), are assumed. Diffusion coefficients:
$D_{\text{Na}0} = 1.33 \times 10^{-5}\,cm^2\,s^{-1}$,
$D_{\text{K}0} = 1.96 \times 10^{-5}\,cm^2\,s^{-1}$,
$D_{\text{Cl}0} = 2.03 \times 10^{-5}\,cm^2\,s^{-1}$;
tortuosity factor: $\lambda_t = 1.6$; ion concentrations: $[\text{K}^+] = 3.0\,\text{mM}$, $[\text{Na}^+] = 145\,\text{mM}$, $[\text{Cl}^-] = 134\,\text{mM}$; temperature: $T = 300$ K. Using the relationship $D_X = D_{X0}/\lambda_t^2$ gives the value $\sigma_e = 0.68\,\text{S}\,\text{m}^{-1}$.

(a)

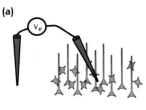

(b)

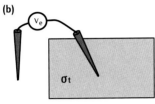

Fig 10.6 Illustration of a key assumption in the volume conductor theory where the electrical potential generated by a current injected into the ECS **(a)** can be modelled as the potential resulting from injecting the same current into a smooth 3D space **(b)** described by a single parameter, the electrical tissue conductivity σ_t. The electrode positioned outside the brain tissue illustrates the **reference electrode**, often positioned far away from the recording electrode to function as an electrical ground.

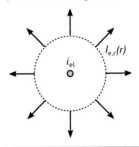

Fig 10.7 Illustration of the derivation of the point-source formula in Equation 10.15. The current i_{el} coming from a current source (blue dot) will spread equally in all directions. Current conservation and spherical symmetry imply that the current density a distance r from the current source will be given by Equation 10.13.

following injection of electrical test currents between electrodes inserted into the ECS. In such measurements, the convention has been to measure the current density per unit area of brain tissue rather than per unit area of the ECS part of brain tissue. This **tissue conductivity** σ_t is related to the ECS conductivity σ_e via:

$$\sigma_t = \alpha \sigma_e, \tag{10.12}$$

where α is the **ECS volume fraction** with a typical value of 0.2.

In cortical grey matter, a value of σ_t between 0.2 and 0.5 S m^{-1} has been found for frequencies lower than a few tens of hertz (Miceli et al., 2017). In comparison, an estimate based on Equation 10.11 and measured values of the ECS diffusion coefficients D_{Na}, D_K and D_{Cl}, as well as typical values for the ECS concentrations of $[Na^+], [K^+]$ and $[Cl^-]$, gives $\sigma_e \sim 0.68$ S m^{-1} (see sidebox). With a volume fraction $\alpha = 0.2$, we thus find $\sigma_t = 0.2 \times 0.68$ S m^{-1} ~ 0.14 S m^{-1}. This is somewhat lower than the experimentally observed values. The discrepancy may at least partially be explained by contributions from other ions than sodium, potassium and chloride. Further, currents may also go through cells, not only around them, thus giving an additional contribution to the measured tissue conductivity; see Halnes et al. (2016) for a discussion.

The main sources of electrical potentials in the ECS are the electrical currents entering the ECS from cells through their membranes. The link between these current sources and the potentials is provided by the **volume conductor theory**, where the starting point is Equation 10.8. In this theory, the ECS is modelled as a smooth 3D space described only by the electrical conductivity σ_t (Figure 10.6), by analogy with the porous media approximation for ECS diffusion in Section 10.1. Thus the individual ion concentrations are not modelled, and the conductivity σ_t is assumed to be a constant over time and not depend dynamically on the ion concentrations as implied by Equation 10.9.

10.2.2 Electric Potential from Current Point Source

If one injects an electrical current via an electrode into a volume conductor, the current will spread radially from the electrode tip (Figure 10.7). Since the medium has an electrical resistance (essentially inversely proportional to the conductivity σ_t), Ohm's law implies that a current forced through it will result in a change in the ECS potential. For the simplest situation when the conductivity σ_t is the same everywhere and in all directions (infinite homogeneous isotropic volume conductor), an analytical formula for the ECS potential generated by a point-like **current source**, like the electrode current in Figure 10.7, can be derived. From current conservation, it follows that with a point-like current source i_{el}, the radial current density $I_{e,r}(r)$ a distance r from the injection point obeys:

$$i_{el} = 4\pi r^2 \alpha I_{e,r}(r), \tag{10.13}$$

because the surface area of a sphere with radius r is $4\pi r^2$. Further, the ECS volume fraction α is included on the right-hand side since the current density $I_{e,r}(r)$ is defined per area of ECS, not per area of brain tissue as a whole.

From Equation 10.8 and the general relationship $\vec{E}_e = -\nabla V_e$, $I_{e,r} = \sigma_e E_{e,r} = -\sigma_e(dV_e/dr)$, cf. radial derivatives in Box 10.1, we then have:

$$\frac{dV_e}{dr} = -E_{e,r} = \frac{-i_{el}}{4\pi\alpha\sigma_e r^2} = \frac{-i_{el}}{4\pi\sigma_t r^2}, \tag{10.14}$$

where we have introduced the tissue conductivity $\sigma_t = \alpha\sigma_e$ in the final step. Integration of this relationship then gives:

$$V_e(r) = -\frac{i_{el}}{4\pi\sigma_t}\int_r^\infty \frac{1}{r'^2}dr' = \frac{i_{el}}{4\pi\sigma_t r}. \tag{10.15}$$

Here the potential $V_e(r)$ has been set to zero infinitely far away from the electrode tip, that is, $V_e(r \to \infty) = 0$. This would correspond to the reference electrode depicted in Figure 10.6 being positioned infinitely far away.

The ECS potential in Equation 10.15 decays as the inverse of the distance from the current source. This formula has the same mathematical form as Coulomb's law for the potential around a point charge in vacuum or a dielectric material (Figure 10.8). However, the underlying physical assumptions are different; Equation 10.15 applies for a current source in a conductor, not a charge in a vacuum or dielectric. The volume conductor theory is a linear theory, meaning that, as can be seen from Equation 10.15, when the electrode current i_{el} is doubled, the generated extracellular potential V_e is doubled as well.

In the above example, we have considered the case where current is injected into the ECS by an electrode. However, the same formula also applies when a point-like current is 'injected' into the ECS by a cellular membrane current. In this case, the contribution from each small patch of membrane can be described by Equation 10.15 with i_{el} replaced by the membrane current i_m (including both ionic currents and capacitive currents; see Chapter 2). Since the volume conductor theory is a linear theory, the total ECS potential from a set of current point sources, whether they are from electrode or membrane currents, can be found by summing the contributions from each point. Likewise, the distribution of membrane currents can be estimated from measurements of V_e, and this is used in so-called current-source density (CSD) analysis of electrical recordings from the ECS (Box 10.3).

10.2.3 Varying Electrical Conductivities

The simple formula in Equation 10.15 assumes an infinite homogeneous (same everywhere) and isotropic (same in all directions) volume conductor. However, in general, the electrical conductivity is different in different types of brain tissue. For example, it is larger in the **grey matter**, where the cell bodies reside, than in the more fatty **white matter**, consisting of neuronal axons and glial cells. The conductivity can also depend on direction as, for example, in the cortical grey matter, it appears somewhat easier for the ions to move in the depth direction (along the apical dendrites of pyramidal neurons) than in the sideways directions (Goto et al., 2010). The volume conductor theory can be modified to incorporate such effects.

A **complex variable** is a variable that takes on complex values. A complex number z has two parts, a real part x and an imaginary part iy, that is, $z = x + iy$. Complex variables are often used in the study of electrical circuits, in particular of their frequency dependence, as they may simplify the mathematical analysis.

Box 10.3 | Current-source density

Measurements of the extracellular electrical potential V_e inside brain tissue have historically been the most important technique for learning about the activity of neurons. The high-frequency part of the signal, that is, frequencies larger than some hundred hertz, measures **spikes**, the extracellular signatures of neuronal action potentials (see Chapter 13). The low-frequency part, commonly called the **local field potential (LFP)**, is more difficult to interpret directly in terms of the underlying neural activity. A common approach has been to use the measured V_e to estimate the **current-source density (CSD)** which quantifies the net volume density of membrane currents that enters the ECS for different positions.

The equation relating the CSD to V_e is found by taking the divergence (Box 10.1) on each side of Equation 10.8:

$$\nabla \cdot \vec{I}_e = \sigma_e \nabla \cdot \vec{E}_e. \tag{a}$$

The CSD c_e can then be defined as:

$$c_e \equiv \nabla \cdot \vec{I}_e. \tag{b}$$

This is a **volume current** density with units A cm^{-3} measuring the current injected into the ECS through cellular membranes.

Equation (a) can then be rewritten as:

$$\nabla \cdot \vec{E}_e = \frac{c_e}{\sigma_e}, \tag{c}$$

and use of the general relationship $\vec{E}_e = -\nabla V_e$ then gives:

$$c_e = -\sigma_e \nabla^2 V_e. \tag{d}$$

In practical use, the standard has been to estimate the CSD per tissue volume c_t rather than per ECS volume c_e. These are related via $c_t = \alpha c_e$, where α is the volume fraction of the ECS in the brain tissue. We then have:

$$c_t = -\sigma_t \nabla^2 V_e. \tag{e}$$

This equation is the starting point for CSD analysis where the goal is to estimate c_t from measured values of V_e. Since the CSD is reflected in the spatial variation of the extracellular potential, simultaneous recordings of V_e at several different positions with **multielectrodes**, electrodes with many contacts, are in practice needed for reliable estimation of c_t (Nicholson and Freeman, 1975; Pettersen et al., 2006; Potworowski et al., 2012).

Anisotropic Conductivity

For the situation where the conductivity varies in the three directions, the formula in Equation 10.15 generalises to (Nicholson and Freeman, 1975):

$$V_e(x, y, z) = \frac{i_{el}}{4\pi \sqrt{\sigma_{ty}\sigma_{tz}x^2 + \sigma_{tz}\sigma_{tx}y^2 + \sigma_{tx}\sigma_{ty}z^2}}. \tag{10.16}$$

To model a larger conductivity in the depth direction (z) of the cortex than in the two sideways directions (x, y), one should set $\sigma_{tz} > \sigma_{tx} = \sigma_{ty}$ in this equation (Figure 10.8b). If electrical isotropy is assumed, $\sigma_{tx} = \sigma_{ty} = \sigma_{tz} = \sigma_t$, Equation 10.16 reduces to Equation 10.15.

Spatially Varying Conductivity

If the electrical conductivity σ_t itself is a function of spatial position, the mathematical formulation of the volume conductor theory becomes more complicated. However, simple analytical solutions can be found for particular situations of interest. For example, in the thin layer of the cortical 'grey matter' where all the cortical neurons are found, the electrical conductivity is lower than in the **cerebrospinal fluid (CSF)** covering the cortical surface. On the other hand, the grey-matter conductivity is larger than in the 'white matter' below, which is more fatty. For such planar stepwise discontinuities in the extracellular conductivity, formulae analogous to Equation 10.15 can be derived by use of the Method of Images from electrostatics (Box 10.4).

Furthermore, the potential V_e generated by any set of current sources can always be found numerically by use of schemes such as **finite element modelling** (FEM) (McIntyre and Grill, 2001; Ness et al., 2015). In these schemes, both spatially varying and anisotropic conductivities can be included. The variable σ_t then becomes a tensor (3×3 matrix) with spatially varying elements (Nicholson and Freeman, 1975).

Frequency-Dependent Conductivity

In the volume conductor models described above, the extracellular conductivity was assumed to be the same for all frequencies, that is, all frequency components of the signal propagate equally easily through the extracellular medium. If the medium has a capacitive component, σ_t will be frequency-dependent. By treating the conductivity as a complex variable, the volume conductor theory can also describe such media (Grimnes and Martinsen, 2015, Chapter 3). Here, the imaginary component represents the capacitive properties of the medium. A fully ohmic medium thus has a frequency-independent and real-valued σ_t (see sidebox on complex variables).

For the frequencies of most relevance to neural systems, less than a few thousand hertz, experiments seem to justify the assumption of an ohmic medium. Only a small variation of σ_t, with modest effects on the potentials, is generally measured in this frequency range (Miceli et al., 2017, Figure 5). However, a frequency-dependent σ_t can, if necessary, be included in the volume-conductor theory by use of Fourier analysis, by individually considering each frequency component of the current source and the resulting ECS potential; for an example, see Miceli et al. (2017).

10.2.4   Ephaptic Interactions

The activity in a neuron can affect the activity in neighbouring neurons as they 'share' their extracellular space (Box 10.5). Such interactions are called **ephaptic** ('touch' or 'junction' in ancient Greek) and provide an alternative way for neurons to interact, in addition to synaptic connections (Holt and Koch, 1999). This interaction is not taken into account in the modelling of the neuronal membrane potential by means of the cable equation or compartmental modelling in the previous chapters. Here it was assumed that the extracellular potential V_e is the same everywhere outside the neuron, and

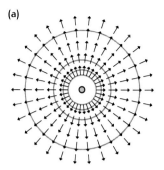

(a)

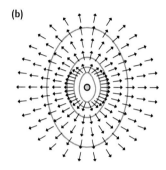

(b)

Fig 10.8 Extracellular electrical potential and electrical currents around current point source (blue dot). **(a)** Circles illustrate equipotential lines with isotropic electrical conductivity, that is, lines along which V_e is constant (Equation 10.15). Arrows illustrate directions of corresponding current densities $I_{e,r} = \sigma_e E_{e,r}$, where $E_{e,r}$ is the radial electric field described in Equation 10.14. **(b)** Similar plot to **(a)** with an anisotropic electrical tissue conductivity where the conductivity in the vertical (z) direction is larger than in the lateral directions, and the potential is given by Equation 10.16, with $\sigma_{tz} > \sigma_{tx} = \sigma_{ty}$.

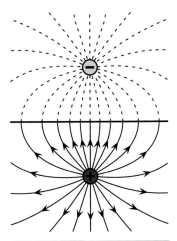

Fig 10.9 Illustration of the use of Method of Images (MoI). The electrical field (lines with arrows) around a positive current source with a highly conductive medium above can be simply modelled by instead assuming a homogeneous electrical conductivity with a negative current source symmetrically placed above the interface to the highly conductive medium (Box 10.4).

Box 10.4 | Method of Images

The formulae for the extracellular potentials generated by current point sources in Equation 10.15 and Equation 10.16 assume that the current source is surrounded by an infinite volume conductor, that is, the electrical conductivity is assumed to be the same everywhere. For planar discontinuous steps in values for the extracellular conductivity, such as at the interfaces between the cortical 'grey matter' and the CSF above or the 'white matter' below, the **Method of Images (MoI)** can be used to generalise the simple expressions in Equations 10.15 and 10.16 (Nicholson and Llinas, 1971; Jackson, 1998; Gold et al., 2006; Ness et al., 2015).

The principle behind MoI is illustrated in Figure 10.9 where a positive current source inside the cortex is positioned below an interface with a highly conductive medium such as the CSF.

For simplicity, we first assume that the medium above the interface has an extremely large conductivity σ_0, like in a metal. In this case, it follows from electrostatics that the electrical field at the interface must be perpendicular to the interface. Otherwise electrical current would constantly flow in the horizontal directions. This boundary condition on the electric field at the interface can be mimicked by a trick, that is, by placing a negative 'image' current source of the same magnitude in the symmetric position in the half plane above the interface (Figure 10.9). The extracellular potential in the region below the interface is then found by summing the contributions from the real and image current sources:

$$V_e(\vec{r}) = \frac{i_s}{4\pi\sigma_t|\vec{r} - \vec{r}_+|} - \frac{i_s}{4\pi\sigma_t|\vec{r} - \vec{r}_-|}. \tag{a}$$

Note that here the current source i_s can be an electrode current i_{el} or a membrane current i_m. For the case where the source is placed in a medium with conductivity σ_t in the vicinity of an interface with a medium with conductivity σ_0, this formula can be generalised to (Ness et al., 2015):

$$V_e(\vec{r}) = \frac{i_s}{4\pi\sigma_t|\vec{r} - \vec{r}_+|} + \left(\frac{1 - \sigma_0/\sigma_t}{1 + \sigma_0/\sigma_t}\right) \frac{i_s}{4\pi\sigma_t|\vec{r} - \vec{r}_-|}. \tag{b}$$

When the conductivity of the medium at the top σ_0 is larger than σ_t (like for cortical current sources below the interface to the CSF at the top of the cortex), we see from the formula that the extracellular potential around the real current source is reduced by the interface as the second term is negative. If the conductivity of the medium at the top σ_0 instead is smaller than σ_t, the extracellular potentials are boosted by the interface. For an illustration of this, see Pettersen et al. (2006, Figure 4).

From the formula, we also see that the effect of the interface on the extracellular potential close to the current source will be reduced as the distance from the current source to the interface increases. This follows since the contribution from the image source will decay as $1/|\vec{r} - \vec{r}_-|$, that is, roughly as the inverse of twice the distance from the real current source to the interface.

Box 10.5 | Ephaptic coupling between MSO neurons

Neurons in the medial superior olive (MSO) perform coincidence detection with high temporal precision to encode the location of sound sources. These neurons each have a bipolar morphology, which can be represented as two sticks protruding in opposite directions out from the soma and are arranged in an orderly spatial pattern. The neurons also tend to receive synchronised excitatory inputs to sound stimuli (Goldwyn et al., 2014). These features together suggest that the dynamics of MSO neurons can be affected by mutual ephaptic interactions (Goldwyn and Rinzel, 2016).

To explore this phenomenon in a model, Goldwyn and Rinzel (2016) considered a bundle of tightly packed cable-stick neurons, as illustrated in Figure 10.10a.

The dynamics of each cable stick is described by the same cable equation as in Chapter 5, the only difference being that the extracellular potential V_e can no longer be assumed to be zero. Thus, in Equation 5.10, $\partial^2 V/\partial x^2$ must be replaced by $V_i/\partial x^2$ where V_i is the intracellular potential. Use of $V = V_i - V_e$ then gives the modified cable equation:

$$C_m \frac{\partial V}{\partial t} = \frac{E_m - V}{R_m} + \frac{d}{4R_a}\frac{\partial^2 V}{\partial x^2} + \frac{d}{4R_a}\frac{\partial^2 V_e}{\partial x^2} + \frac{i_{el}}{\pi d}.$$

(a)

The new term $d/(4R_a)\partial^2 V_e/\partial x^2$ is a fictitious distributed current referred to as the **ephaptic current** (Holt and Koch, 1999). To study ephaptic coupling effects mediated by this current, a population of neurons in the bundle (white neurons in Figure 10.10a) received simultaneous current input at the same positions on the cables (positions marked in Figure 10.10b), The neurons share the same extracellular potential V_e and experience the same ephaptic effect.

Figure 10.10b shows results for three different positions of a stationary excitatory current input on a 2λ-long cable-stick neuron where λ is the cable length constant (Equation 5.12). The resulting depolarisation of the membrane is shown both with (solid lines) and without (dashed lines) ephaptic coupling. As observed, the ephaptic coupling increases the membrane depolarisation at the current injection points and also provides a more rapid decay of depolarisation away from the injection points; the apparent length constant of the cable is reduced. Figure 10.10c shows the corresponding depolarisation of a test neuron (grey neuron in top figure) which does not receive current input like the other neurons. Here the response is only due to ephaptic coupling with the other neurons. The effect can be both depolarising and hyperpolarising, depending on the position along the cable.

The ephaptically induced changes in the membrane potential observed in Figure 10.10c are relatively small, only a couple of millivolts or less. However, in a model where active conductances were added to a 'soma region' of the cable to mimic a spiking MSO neuron, these modest changes were found to affect spike initiation and coincidence detection of synaptic inputs (Goldwyn and Rinzel, 2016).

(a)

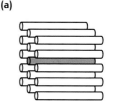

(b)

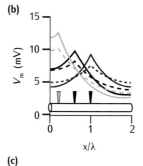

(c)

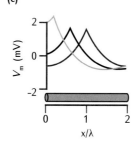

Fig 10.10 Illustration of the model of ephaptic coupling discussed in Box 10.5. **(a)** Model structure with a bundle of neurons (white neurons) receiving constant current inputs at one of three positions on the neurons (marked in **(b)**). The resulting membrane potential in the white neurons for the three different situations are shown in **(b)**, whilst the resulting membrane potential of the grey neuron in the middle of the bundle **(a)**, receiving no current input, is shown in **(c)**. Adapted from Goldwyn and Rinzel (2016) with permission from The American Physiological Society. For more details, see Box 10.5.

also does not change with time. This is an approximation as the membrane currents will affect the extracellular potential, as described by the volume conductor theory above. Specifically, variations in V_e will affect the membrane potential V (since $V = V_i - V_e$) which, in turn, will affect the neural dynamics and membrane currents. Thus a detailed description of the dynamics requires, in principle, the simultaneous solution of the cable equation and the relevant volume conductor equation.

For idealised network models with high symmetry, such as identical simplified neuron models arranged in a sphere or a plane, these coupled equations can be solved with equations that are not much more complicated than the cable equation itself (Rall and Shepherd, 1968; Goldwyn and Rinzel, 2016). However, in most cases, comprehensive numerical simulation schemes are needed to compute solutions to the set of coupled equations (Agudelo-Toro and Neef, 2013; Tveito et al., 2017).

However, variations in the extracellular potentials V_e in brain tissue are typically smaller than a millivolt or so, which is much smaller than the magnitude of the typical membrane potential ($V = V_i - V_e$) of ~60–70 mV. Thus, in relative terms, the membrane potential V will be only slightly modified by the variation in V_e. Since the ion currents across neuronal membranes are governed by the membrane potential, not the extracellular or intracellular potentials by themselves, it is not clear if, or by how much, ephaptic interactions will affect neuron dynamics in brain tissue (Anastassiou and Koch, 2015). They are expected to be more important in tight bundles of axons (Arvanitaki, 1942) or specialised tight-packed neurons, as in the auditory brainstem (Goldwyn and Rinzel, 2016) (Box 10.5) where the neighbouring neurites are highly organised and very closely spaced. Further, ephaptic interactions are expected to be larger when the neural activity is synchronised across the bundle (Holt and Koch, 1999), and can also contribute to such synchronisation (Bokil et al., 2001).

10.3 | Electrodiffusion in Extracellular Space

As described in Chapter 3, every action potential in a neuron is accompanied by a change of ion concentrations both inside and outside the membrane of the soma. During the upstroke of the action potential, sodium ions move into the neuron from the ECS, whilst potassium ions move the opposite way during the later repolarisation phase.

In healthy brains, ion pumps working in the background eventually pump these ions back in the opposite directions to maintain suitable intracellular and extracellular concentrations of sodium and potassium. However, if many neurons fire many action potentials rapidly in succession, the ion pumps may struggle to keep up. As a result, both the potassium and sodium concentrations in the ECS will change. The relative effect will be largest for the ECS potassium concentration which has a low background level of only a few millimolars. During strong firing activity, this concentration may increase by several millimolars, making the Nernst potential of potassium E_K less negative (Section 2.3). This will, in turn, affect the dynamics of the surrounding neurons, and sometimes even lead to epileptic seizures (Somjen, 2004).

Potassium can move away from high-firing regions in the ECS both by diffusion and electrical drift, which together are called electrodiffusion. To model this transport process, it is not sufficient to only keep track of the net movement of charges; it is necessary also to model explicitly the concentrations of potassium and other ions in the ECS solution.

10.3.1 Electrodiffusion Modelling Formalisms

Electrodiffusive processes are challenging to model because the relevant dynamical variables in the equations, the ion concentrations $[X]$ and the electric potential V_e are intimately connected. The ion concentrations determine the electrical potential through the net charge density they set up, and the electrical potential determines how the ions move by electrical drift. The equations for the ion concentrations and the electrical potential must be solved simultaneously.

One element is the Nernst–Planck equation previously encountered in Chapter 2 in the context of 1D ion transport across cell membranes. For 3D ion transport in the ECS, this equation generalises to:

$$\vec{J}_X = -D_X \nabla[X] - \frac{D_X F}{RT} z_X [X] \nabla V_e,$$ (10.17)

which combines the expressions for diffusion and drift fluxes in Equations 10.1 and 10.5, respectively. The dynamics of the ion concentrations is governed by the continuity equation:

$$\frac{\partial [X]}{\partial t} = -\nabla \cdot \vec{J}_X + f_X,$$ (10.18)

which reduces to Equation 10.2 when there is no transport by electrical drift.

Insertion of the Nernst–Planck flux equation into Equation 10.18 then gives:

$$\frac{\partial [X]}{\partial t} = \nabla \cdot \left(D_X \nabla[X] + \frac{D_X F}{RT} z_X [X] \nabla V_e \right) + f_X$$

$$= D_X \nabla^2 [X] + \frac{D_X z_X F}{RT} \left(\nabla[X] \cdot \nabla V_e + [X] \nabla^2 V_e \right) + f_X.$$ (10.19)

Poisson–Nernst–Planck Scheme

In the **Poisson–Nernst–Planck (PNP)** scheme, Equation 10.19 for ion fluxes is combined with the **Poisson equation** from electrostatics describing the link between charge densities and electrical potentials:

$$\nabla^2 V_e = -\frac{\rho}{\epsilon},$$ (10.20)

where the charge density ρ (with units C m^{-3}) is found from summing over the contributions to the charge from all ion types:

$$\rho = F \sum_X z_X [X].$$ (10.21)

Further, ϵ is the electrical permittivity of the solvent, essentially that of water. In Equation 10.21, the sum is over all ion types present in the ECS. For the special case where the only ions present are sodium, potassium and chloride, we have $\rho = F([Na^+] + [K^+] - [Cl^-])$.

Box 10.6 | How fast is charge relaxation in ECS solutions?

For an ohmic medium obeying $\vec{I}_e = \sigma_e \vec{E}_e$, where $\vec{E}_e$ is the electrical field in the ECS, it follows from the charge continuity equation (analogous to the particle continuity equation in Equation 10.2) that:

$$\frac{\partial \rho_e}{\partial t} + \nabla \cdot \vec{I}_e = 0, \tag{a}$$

so that:

$$\frac{\partial \rho_e}{\partial t} + \sigma_e \nabla \cdot \vec{E}_e = 0. \tag{b}$$

With the use of one of Maxwell's equations, $\nabla \cdot \vec{E} = \rho/\epsilon$, a simple first-order differential equation results:

$$\frac{\partial \rho_e}{\partial t} + \frac{\sigma_e}{\epsilon} \rho_e = 0, \tag{c}$$

whose solution is:

$$\rho_e(t) = \rho_e(0) e^{-t/\tau_{relax}}, \tag{d}$$

where $\tau_{relax} \equiv \epsilon/\sigma_e$ is a charge relaxation constant. For 100 mM NaCl, which electrically is similar to the ECS, σ_e is about $0.5\,\mathrm{S\,m^{-1}}$. With a permittivity of $\epsilon \sim 80\epsilon_0 = 80 \times 8.85 \times 10^{-12}$ F/m (corresponding to the value for water), $\tau_{relax} = 1.4$ ns. Consequently, a deviation from charge neutrality in the bulk of the ECS vanishes exponentially over a timescale of nanoseconds.

Solutions for the ion concentrations and associated potential can be found numerically by simultaneous numerical solution of Equations 10.19–10.21. However, this is computationally very demanding since the typical timescale for the relevant charge dynamics is nanoseconds (Box 10.6) and short time-steps are required in the numerical simulations. In practice, it has been difficult to study the dynamics of systems with sizes larger than a few micrometres for longer than a few milliseconds, such as the ECS potential close to an axon during an action potential (Pods et al., 2013) (Figure 10.11).

Electroneutral Schemes

In brain tissue, almost all of the net charge from mobile ions is located in nanometre-thick layers around the cellular membranes (Section 10.3.2). Thus, the ionic solutions both outside and inside the cells are typically **electroneutral**, that is, at all positions, the positive and negative ionic charges cancel out. To a good approximation, the ECS charge density can be set to zero everywhere. This fact can be used to simplify the modelling of electrodiffusive processes and speed up numerical simulations dramatically, compared to the computationally very demanding PNP formalism. The idea is to enforce electroneutrality at each time-step in the numerical simulation.

One such scheme is the **Kirchhoff–Nernst–Planck (KNP)** formalism (Halnes et al., 2016; Solbrå et al., 2018). Here, all net charge density is assumed to be concentrated at the membranes, so that the ECS is electroneutral at all times, that is, ρ in Equation 10.21 is fixed at zero. The mathematical details of the KNP scheme are described in Box 10.7.

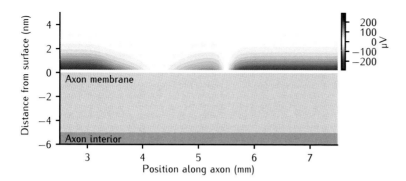

Fig 10.11 Illustration of use of the Poisson–Nernst–Planck (PNP) scheme for simultaneous computation of ECS potentials and ion concentration. Figure shows a snapshot of ECS potential V_e through the thin layer of unbalanced ion charges, the so-called Debye layer (Section 10.3.2), immediately outside an axon membrane during the firing of an action potential. The axon is oriented in the x-direction, and the potential changes sharply over the first nanometres in the y-direction, across the Debye layer. The figure is made based on data from Pods et al. (2013), which also gives details of how the simulation was done.

The KNP scheme is computationally much less demanding than the PNP scheme. This allows for the study of processes in brain tissue occurring on timescales of milliseconds and longer and on spatial scales of millimetres and larger. An example of such a study is given in Section 10.3.3.

However, the PNP scheme is required for studying, for example, how ions form nanometre-thick charged layers around the membranes, so-called Debye layers (Section 10.3.2). Such Debye layers take only nanoseconds to form (Box 10.6). The approximation in KNP of assuming that electroneutrality in the ECS is fulfilled at all times is thus well justified on the millisecond timescale of neural processing.

An overview of the different electrodiffusion modelling formalisms is given in Table 10.1.

10.3.2 Debye Layers

From electrostatics, we know that the electrical field around a point charge in vacuum will decay as the square of the distance from the charge, that is, $E_r \propto 1/(\epsilon_0 r^2)$. This is also the case in dielectrics, such as water, even though the electrical field will be reduced by the dipolar water molecules aligning themselves with the electric field set up by the charge. The net effect is a reduction in the electrical field by a factor $\epsilon_r \sim 80$, corresponding to the value of the **relative permittivity** of water compared to vacuum. Note that $\epsilon = \epsilon_r \epsilon_0$ is the (absolute) permittivity of water.

With small ions present in the water, as in the electrolytic extracellular or intracellular solutions in the brain, the situation changes. The ions will gather around an added charge and form an ion cloud that effectively attenuates the electrical field to zero over a distance of only about a nanometre or so. Likewise, in brain tissue, any net charge in the solutions will reside in thin layers around the cellular membranes called **Debye layers**. The thicknesses of these layers will depend on the ion concentrations, as specified by the **Debye length** λ_D (Box 10.8). For an electrically plausible biological solution, say, a 100 mM concentration of monovalent cations such as Na$^+$, combined with a 100 mM concentration of Cl$^-$, λ_D will be about a nanometre (Figure 10.12).

The important conclusion is that a few nanometres away from an added point charge or a charged membrane, the charges will be completely shielded and thus be electrically invisible to the other molecules. Furthermore, outside these layers, the ionic solution will be electroneutral, the charge carried

Box 10.7 | Kirchhoff–Nernst–Planck (KNP) equations

In the KNP framework (Solbrå et al., 2018), the extracellular potential V_e is required to be such that:

$$\frac{\partial \rho}{\partial t} = -c_{e,cap}, \tag{a}$$

at all points in the system. In KNP, $c_{e,cap}$ is a CSD term (Box 10.3) stemming exclusively from capacitive current across a neuronal membrane, being zero everywhere else. Since $c_{e,cap} \neq 0$ only at the positions of cellular membranes, Equation (a) implies that the charge density can change only there, and that electroneutrality is maintained throughout the rest of the medium where $c_{e,cap} = 0$. To turn this condition into an equation which can be solved for V_e, we substitute the expression for ρ in Equation 10.21, which gives:

$$F \sum_X z_X \frac{\partial [X]}{\partial t} = -c_{e,cap}. \tag{b}$$

Substitution of the expression for the ion concentration dynamics in Equation 10.19 into Equation (b) then gives the final expression needed to solve for V_e when the ion concentrations $[X]$ are known:

$$\nabla \cdot (\sigma_e \nabla V_e + \nabla b) + F \sum_X z_X f_X = -c_{e,cap}. \tag{c}$$

Here σ_e is the conductivity of the extracellular medium given by Equation 10.9, and b is defined as:

$$b = F \sum_X z_X D_X [X]. \tag{d}$$

Equation (c), together with Equation 10.19 forms the KNP equations, which can be solved numerically.

The electrical potential V_e can be separated into the contribution from diffusion in the ECS and the contribution from the membrane currents. These contributions can be analysed separately by replacing Equation (c) with the following equivalent set of equations:

$$V_e = V_{e,drift} + V_{e,diff}, \tag{e}$$

where:

$$\nabla \cdot (\sigma_e \nabla V_{e,drift}) + F \sum_X z_X f_X = -c_{e,cap}, \tag{f}$$

and:

$$\nabla \cdot (\sigma_e \nabla V_{e,diff} + \nabla b) = 0. \tag{g}$$

In this equation set, the ECS is used as reference volume. If instead the tissue volume is used as reference, σ_e must be replaced by σ_t and $c_{e,cap}$ must be replaced by $c_{t,cap} = \alpha c_{e,cap}$ where α is the ECS volume fraction.

by the positive ions being counterbalanced by the charge of the negative ions. This rearrangement of charges happens fast: a charge imbalance in the ECS will typically vanish within nanoseconds (Box 10.6).

Table 10.1 | Schemes for modelling particle transport and electric potentials in ECS

Particle type	Process	Formalism	Typical use
Uncharged molecule	Diffusion	Diffusion equation (Section. 10.1)	Signalling molecules (and low-concentration ions)
Charge (all ions)	Electrical drift	Volume conductor (Section 10.2)	Electrical potentials around cells
Ions	Electrodiffusion	PNP (Section 10.3.1) KNP (Section 10.3.1) Poisson-Boltzmann (Box 10.8)	Microscopic scales (μm, μs) Macroscopic scales (mm, s) Stationary ion distributions on nanometre scale

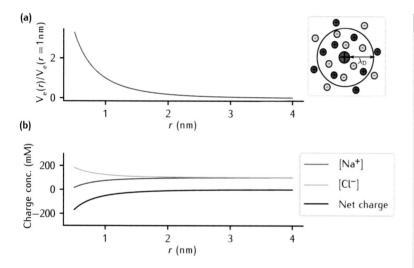

(a)

(b)

Fig 10.12 Illustration of Debye 'cloud' of ions around an added elementary point charge ($e = 1.6 \times 10^{-19}$ C) in a 100 mM NaCl solution. **(a)** The electrical potential V_e (Equation (f) in Box 10.8) decays rapidly and is almost zero, a few Debye lengths λ_D away (Equation (e) in Box 10.8). In this example, $\lambda_D = 1$nm. The corresponding ion concentrations can be found from Equation (a) in Box 10.8, with the use of the Taylor expansion $\exp(-z_X F V_e / RT) \approx 1 - z_X F V_e / RT$. The inset illustrates the electrical shielding by the **Debye cloud** extending a distance of about λ_D from the added positive charge. In panel **(b)**, the concentrations of Na$^+$ and Cl$^-$ are both close to their bulk value of 100 mM, a few nanometres away from the test charge. However, close to the positive test charge, where the screening is incomplete, the concentrations of Na$^+$ and Cl$^-$ are reduced and increased, respectively. The net electrical charge is essentially given by the difference between the Na$^+$ and Cl$^-$ concentrations. Here this net charge, measured as the concentration of elementary charge e in millimolars, is negative, close to the positive test charge.

10.3.3 Electrodiffusion of Ions around Active Neurons

An example application of the KNP formalism for studying electrodiffusion in brain tissue is illustrated in Figure 10.13. Here, the electrodiffusion of ions around a neocortical pyramidal cell, the same compartment model as used in Section 5.6, receiving synaptic inputs across the membrane is depicted. The neuron model is tuned to give a firing rate of about five action potentials per second. This firing activity is accompanied by changes in ion concentrations in the ECS. Specifically, during each action potential, some potassium ions will cross the membrane from the inside moving outwards and increase the ECS concentration of potassium in the soma region (Figure 10.13b). Likewise, some sodium ions will cross the membrane the other way so that the ECS concentration of sodium is decreased (Figure 10.13c). Whilst the concentration changes in this single-neuron example are small, in the micromolar range, concerted activity from many neurons can lead to much larger ECS concentration changes which can affect the ionic reversal potentials and thus the dynamics of neurons (Halnes et al., 2016).

The ECS potential changes accompanying these ion concentration changes are also illustrated in Figure 10.13d. In this example, the spatial pattern of the potential is largely determined by the membrane currents as

Box 10.8 | Poisson–Boltzmann equation and Debye length

The stationary shape of the ionic screening 'clouds' around point charges or ionic screening layers outside membranes can be modelled by the **Poisson–Boltzmann equation**. Here the ion concentrations are assumed to be in thermal equilibrium, so that the ion concentration profiles can be described as Boltzmann distributions known from thermal physics:

$$[X](\vec{r}) = [X]_0 e^{-z_X F V_e(\vec{r})/RT}. \tag{a}$$

Here, $z_X F V_e(\vec{r})$ is the electrostatic potential energy for a mole of ion X. It is convenient to define the electrical potential V_e to be zero far away from the added point charge or charged membrane $V_e(|\vec{r}| \to \infty)=0$. From Equation (a), it follows that $[X]_0$ is the concentration in the bulk solution, far away from the added charge or charged membranes.

As in the PNP scheme, the equation for the ion concentration is combined with the Poisson equation in Equation 10.20. The self-consistent solution for the case where an additional charge density ρ_{add} is added to the ionic solution is thus found by the following Poisson–Boltzmann equation:

$$\nabla^2 V_e = -\frac{F}{\epsilon} \sum_X z_X [X]_0 e^{-z_X F V_e(\vec{r})/RT} - \frac{\rho_{add}(\vec{r})}{\epsilon}. \tag{b}$$

Unlike in the PNP scheme, the ion concentrations enter the model only through their bulk concentrations $[X]_0$ which can be considered constant.

This second-order ordinary differential equation is difficult to solve due to the presence of the potential V_e in the exponential on the right-hand side. However, if the electrical potential energy of the ions is small compared to the thermal energy $|z_X F V_e| \ll RT$, we can do a Taylor expansion $\exp(-z_X F V_e/RT) \approx 1 - z_X F V_e/RT$, so that the equation simplifies to:

$$\nabla^2 V_e = -\frac{F}{\epsilon} \sum_X z_X [X]_0 + \frac{F^2}{\epsilon RT} \left(\sum_X z_X^2 [X]_0 \right) V_e - \frac{\rho_{add}(\vec{r})}{\epsilon}. \tag{c}$$

For electroneutral solutions ($\sum_X z_X [X]_0 = 0$), the first term on the right-hand side is zero, and the equation becomes:

$$\nabla^2 V_e = \frac{1}{\lambda_D^2} V_e - \frac{\rho_{add}(\vec{r})}{\epsilon}, \tag{d}$$

where the Debye length λ_D is introduced as:

$$\lambda_D = \left(\frac{\epsilon RT}{F^2 \sum_X z_X^2 [X]_0} \right)^{1/2}. \tag{e}$$

The mathematical solution to Equation (d), when a point charge Q_{add} is added, is given by:

$$V_e(r) = \frac{Q_{add}}{4\pi\epsilon r} e^{-r/\lambda_D}, \tag{f}$$

where r is the distance from the point charge (see Robinson and Stokes (2002, Chapter 4) for mathematical details).

Box 10.8 (cont.)

Without the ions in the solution, the potential from the added point charge would follow Coulomb's law and would decay as $1/r$ with distance. In ionic solutions, the electrical potential will instead be rapidly attenuated by the spontaneous organisation of a cloud of ions around the charge, with a size roughly corresponding to the Debye length (Figure 10.12a). Likewise, the size of the region around the added charge where the ion concentrations deviate from the bulk values, and the net charge is non-zero, will also be a Debye length or so (Figure 10.12b). For a 0.1 M monovalent salt like NaCl, insertion of $[X]_0 = 0.1$ M, $T = 300$ K and $\epsilon = 80\epsilon_0$ (where ϵ_0 is the vacuum permittivity) into Equation (e) gives $\lambda_D \sim 1$ nm at room temperature.

described by the volume conductor theory (Figure 10.13e). However, the spatial gradients of the ECS ion concentrations also contribute to the potential, in particular in the soma region (Figure 10.13f).

10.3.4 Diffusion Potentials

Potential gradients in the ECS can also be generated without membrane currents. When two NaCl salt solutions with different concentrations are in contact, ions in the more concentrated solution will diffuse into the low-concentration solution. The chloride ions have a diffusion coefficient about 50% larger than that of the sodium ions and will thus diffuse faster into the low-concentration part. As a consequence, a charge imbalance will arise at the liquid interface, and an electrical field will be set up that stops the rapid chloride ions from venturing too far away from their sodium partners. An electrical potential, often referred to as the **liquid junction potential** or **diffusion potential**, will arise at the interface between the two solutions. This potential is analogous to the membrane potential set up by concentration differences across cell membranes, as described in Section 2.3.

The magnitude of this potential will depend on the difference between the diffusion coefficients of the ions involved, as well as the magnitude of the difference in ion concentrations of the two electrolyte solutions involved. Liquid junction potentials arise, for example, in patch–electrode recordings (Box 14.3) where such a potential arises due to the difference between the electrolyte inside the patched neurons and inside the glass electrode. This liquid junction potential must be taken into account when estimating the membrane potential from patch–electrode recordings (Barry and Lynch, 1991).

In general, the ECS potential will be set up by both electrical drift and ion diffusion. An example is provided in Figure 10.13, for which the potential setup by the electrical drift dominates, but ion diffusion nevertheless makes a sizeable contribution.

In a region where neurons fire action potentials rapidly, the local ECS potassium concentration around the somas (or somata) may increase, and concentration gradients may arise compared to the baseline potassium concentration further away. Local increases from 3 mM to up to ~10 mM concentrations may occur also in non-pathological situations. In brain tissue ECS, the local increase in potassium concentration resulting from action potential firing is accompanied by other concentration changes, most notably

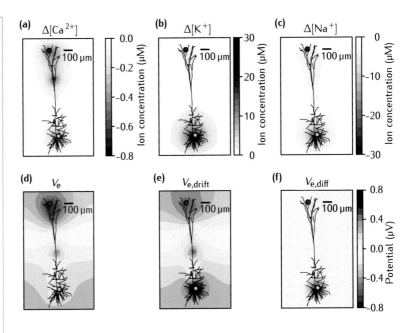

Fig 10.13 Changes in ion concentrations and extracellular potential around a compartmental neuron model computed using the KNP scheme. In the simulation, the neuron is embedded in a 3D cylinder with 0.5 mm radius and 1.5 mm height. The images depict ion concentrations **(a)**–**(c)** and potentials **(d)**–**(f)** in a 2D cross-section after 80 s of random synaptic bombardment across the neuronal membrane. In addition to sodium, potassium and calcium ions, non-specific anions were also included in the simulation. The extracellular potential is low-pass filtered and corresponds to the potential averaged over a 0.1 s interval. **(a)** Changes in calcium concentration $[Ca^{2+}]$. **(b)** Changes in potassium concentration $[K^+]$. **(c)** Changes in sodium concentration $[Na^+]$. **(d)** Total potential V_e (Equation (e), Box 10.7). **(e)** Volume conductor contribution $V_{e,drift}$ (Equation (f), Box 10.7). **(f)** Diffusion contribution $V_{e,diff}$ (Equation (g), Box 10.7). The depicted concentrations are per tissue volume. Figure is made based on data from Solbrå et al. (2018).

the reduction in sodium concentration. In a study using the KNP formalism, simulated neuronal activity that caused local increases from 3 mM to 10 mM of the ECS potassium concentration in cortical tissue led to an ECS diffusion potential of about 0.2 mV (Halnes et al., 2016). In pathological conditions associated with larger concentration changes, larger diffusion potentials can be expected.

10.4 | Electrodiffusion in Brain Tissue

Typically, the intracellular currents are modelled assuming that they are driven exclusively by electrical drift, as described by the cable equation in Chapter 5. In reality, there may be diffusive currents as well, but in the soma and 'thick' dendrites, the ion concentration gradients are commonly thought to be so small that diffusive currents safely can be neglected. However, on spatial scales smaller than a micrometre or so, such as in dendritic spines or thin dendrites, a synaptic input can change substantially the local concentrations of, for example, calcium or chloride ions. If so, electrodiffusive modelling may be required to get accurate results (Qian and Sejnowski, 1989) (Box 6.3).

The repertoire of possible ion dynamics in brain tissue is richer than for neurons operating in isolation. The fast-spiking dynamics of neurons may, for example, interact through the ECS with the slower dynamics of glial cells to produce exotic dynamic phenomena such as spontaneous discharging of spikes and depolarisation block (Somjen et al., 2008; Øyehaug et al., 2012). Some aspects of this interaction may be captured by models including only ion transport across membranes, that is, neglecting the transport of ions within the intracellular and extracellular domains (Somjen et al., 2008;

Øyehaug et al., 2012). However, a more detailed treatment requires inclusion of such intradomain transport which, in general, is electrodiffusive.

Such brain-tissue modelling requires the simultaneous solution of equations describing the intracellular and extracellular electrodiffusive ion dynamics. This set of equations must be self-consistent, and the total number of ions for each ion type must be conserved when summing over both the intracellular and extracellular spaces. Such self-consistent PNP schemes have been developed, but so far their large computational cost has limited their application to the study of individual action potentials (Lopreore et al., 2008; Pods et al., 2013) (Figure 10.11). As discussed in Sections 10.3.1 and 10.3.3, the KNP scheme is much more suitable for studies of slower processes involving larger spatial scales. However, the KNP application in Section 10.3.3 was not fully self-consistent in that the neuronal dynamics was assumed not to be affected by the ion concentration and potential dynamics in the ECS. Self-consistent variations of the KNP scheme where such feedback is included have also been developed (Halnes et al., 2013; Ellingsrud et al., 2020; Sætra et al., 2020). An example of an application of such a self-consistent scheme is given in the next section.

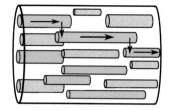

Fig 10.14 Illustration of model used to explore spatial buffering of potassium in the ECS by a KNP scheme (Halnes et al., 2013). The blue cables represent glial cells that contribute to the transport of potassium from a region of high-firing neurons on the left. The grey cables represent cells that do not contribute. The intracellular domains of glial cells are connected via small tunnels (**gap junctions**), and the glial cells thus form a connected **glial syncytium** (Gardner-Medwin, 1983; Chen and Nicholson, 2000).

10.4.1 Spatial Buffering of Extracellular Potassium

One slow process of interest is the problem of spatial **potassium buffering**, that is, how potassium in the ECS can be transported away from high-firing regions to avoid ECS potassium concentrations at unhealthily high levels. Electrodiffusion through the ECS itself provides one such path. Excess potassium can also be channelled out by local glial uptake of potassium and transport inside glial cells to regions with low potassium concentrations.

This process has been studied using the KNP scheme where two spatial domains, the ECS and the intracellular space (ICS), of glial cells were considered (Halnes et al., 2013). The transport was assumed to be essentially 1D where ions move inside and outside astrocytes, modelled using the cable equation, surrounded by the ECS (Figure 10.14). In the model, a steady flux of potassium ions away from the input zone (where the potassium ions are released into the ECS) is established some tens of seconds after the onset of fast neuronal firing. In this steady state, the spatial potassium buffering is mainly driven by local depolarisation of the glial cell membrane. In turn, this increases the local uptake of potassium into glial cells, induces axial transport of potassium inside glia away from the high-firing region and facilitates the release of potassium from glial cells into the ECS in regions where the ECS potassium concentration is low. This glial buffering pathway turned out to be more efficient than electrodiffusion through the ECS alone; in glial cells, the diffusive and electrical drift contributions to potassium transport act in the same direction, whilst in the ECS they instead act in opposite directions (Figure 10.15).

10.4.2 Electrodiffusion at Nanometre Scales

In the PNP scheme, only ion concentrations are modelled. However, both ions and water molecules are discrete entities, and the approximation of using a continuous concentration variable will necessarily break down for scales approaching the atomic nanometre scale. Here, electrodiffusive

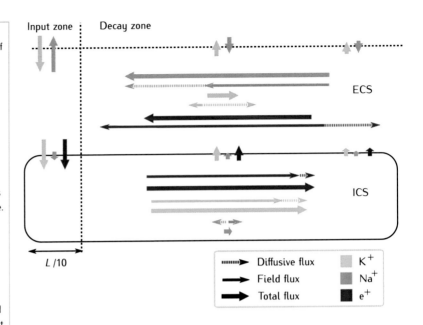

Fig 10.15 Results from exploration of spatial buffering of potassium in the ECS using the KNP scheme (Halnes et al., 2013). Model geometry is depicted in Figure 10.14. An $L = 300\,\mu m$ section of a glial cable (ICS) is considered. The high-firing region where K^+ enters the ECS from neurons (downward pointing blue arrow in upper left corner) corresponds to the left one-tenth of the cable. The model includes three ion types (K^+, Na^+, Cl^-) and a variety of ion channels, as well as a Na–K ion pump embedded in the glial membrane. The flow chart summarises the main transport routes of K^+, Na^+ and net positive electric charge e^+ at steady state, which is reached after some tens of seconds after the onset of high neuronal firing. K^+ enters the ECS in the input zone and leaves the ECS/glia model system some distance away. The transport route of K^+ from entering to leaving the system is predominantly inside the glial cells (ICS), demonstrating their efficiency as a spatial buffer. In contrast, the transport of Na^+, which enters in the decay zone and leaves from the input zone, predominantly takes place in the ECS. Longer arrows mean higher flux densities, but the illustration is not accurate quantitatively. Figure made based on data from Halnes et al. (2013).

processes must be modelled by Monte Carlo simulation techniques for individual ions, analogous to the stochastic model schemes described in Section 6.9, with an electrical drift term added to the equations (Corry et al., 2000a).

Detailed comparisons of such stochastic schemes, with predictions from the PNP scheme (and also the Poisson–Boltzmann equation; cf. Box 10.8), have been done in the context of transport through ion channels (Corry et al., 2000b; Moy et al., 2000). The comparisons found that these continuous schemes give accurate predictions only when the radii of the ion channels were larger than two Debye lengths, that is, a few nanometres (Corry et al., 2000a). Thus the PNP scheme is unsuitable for modelling the detailed transport of ions through narrow ion-selective channels where the 'tunnel' diameter is less than a nanometre (Hille, 2001, Chapter 14).

10.5 | Liquid Flow

When liquid flows, small particles suspended in it will flow with it. Such transport of ions and molecules is called **advective** to distinguish it from the previously encountered diffusive and drift transport. Advective transport through blood flow is particularly important for providing oxygen and nutrients to the brain, as well as for the removal of carbon dioxide and other waste products. There is also a slow extracellular flow of the CSF through brain tissue that regulates the protective fluid buffer surrounding the brain and may also have a role in clearing waste material from brain tissue.

Blood flow and CSF flow are both driven by **hydrostatic pressure**. For water flow across cellular membranes, there is an additional mechanism at play, namely **osmotic flow**, stemming from different net ion concentrations inside and outside cells and the **semipermeable** properties of the membrane (Box 10.9).

The modelling of liquid flow is also important for a quantitative description of several neuroimaging techniques based on measuring blood flow or oxygen content in blood such as **functional magnetic resonance imaging (fMRI)** (see Section 13.11).

10.5.1 Flow across Membranes

With a hydrostatic pressure difference Δp across a membrane with pure water on each side, water molecules will be pushed through it according to the relationship:

$$J_w = -L_p \Delta p. \tag{10.22}$$

Here, J_w is the volume flux of water (volume of water transported across the membrane per area per second), and L_p is the **filtration coefficient** reflecting the water permeability properties of the membrane. The minus sign reflects that the water flows from the high pressure side to the low pressure side of the membrane.

This water flow relationship is analogous to the expression for electrical current flow I across membranes with ohmic conductances encountered in Chapter 2. The pressure difference Δp is analogous to the membrane potential V, and L_p is analogous to the electrical conductance g. In the same way as the electrical conductance g can be increased by adding ion channels to the membrane, the 'water conductance' L_p can be increased by adding water channels, so-called **aquaporins** (Agre et al., 1995). In the nervous system, there is, for example, a high concentration of aquaporins in the part of glial cell membranes close to blood capillaries, suggesting that water flow across the membrane might be particularly important there (Amiry-Moghaddam et al., 2004).

Hydrostatic pressure is not the only force that can push water molecules across membranes. Cellular membranes are semipermeable, so that, whilst water molecules can move across, ions and larger molecules are largely blocked. As a result, water molecules will tend to move to the side of the membrane with the highest concentration of suspended ions and molecules by a mechanism called **osmosis** (Box 10.9). Osmotic flow of water can thus make cells swell or shrink, and thus also regulate the volume of the ECS.

When osmotic effects are included and the concentration of solutes is not too high, Equation 10.22 generalises to:

$$J_w = -L_p(\Delta p - \Delta p_{sol}), \tag{10.23}$$

where the **solute potential** p_{sol} is given by:

$$p_{sol} = \left(\sum_X [X]\right) RT. \tag{10.24}$$

Here R is the gas constant, and T the temperature in kelvins. The sum is over all solutes X in the water, both ions and uncharged molecules, and the contributions from each type to the solute potential are determined by the value of their concentration alone.

In equilibrium, where the net water flow J_w across the membrane is zero, there will thus be an **osmotic pressure** corresponding to the difference in solute potential Δp_{sol} on the two sides of the membrane. For the special case when there is a single solute X on one side of the membrane and pure water on the other, the osmotic pressure is given by:

Box 10.9 | Osmotic flow and the van't Hoff relation

Osmotic flow and the van't Hoff relation in Equation 10.25 can be demonstrated by the simple apparatus depicted in Figure 10.17 where a U-shaped tube, with a semipermeable separating membrane at the bottom, is filled initially with pure water. When, for example, the sugar glucose $C_6H_{12}O_6$, to which the membrane is impermeable, is dissolved in the left part of the tube, water will flow in the leftward direction until a new equilibrium is reached.

In this new equilibrium, the left water column rises a height h above the right column where h is determined by the relation:

$$\rho_m g h = [C_6H_{12}O_6]RT. \tag{a}$$

Here, ρ_m is the mass density of water, g is the acceleration of gravity and $[C_6H_{12}O_6]$ is the molar concentration of glucose on the left side of the membrane.

In this equilibrium situation, the osmotic pressure $\Delta p = [C_6H_{12}O_6]RT$, as described by the van't Hoff relation in Equation 10.25, is equal to the hydrostatic water pressure from the unbalanced water column on the left ($\rho_m g h$). Thus the van't Hoff relation does not describe the pressure across the membrane when water is flowing immediately after adding the sugar, only when the new equilibrium is reached.

This osmotic effect stems from the force acting on the sugar molecules by the membrane when incoming molecules hit the membrane (from the left in Figure 10.17) and bounce back into the sugary solution. A qualitative explanation of the effect (provided in Nelson (2008, Chapter 7)) is that the sugar molecules move in the high-friction regime where this membrane force acting on the sugar molecules (in the leftward direction in the figure) will be balanced by an opposing friction force (in the rightward direction) on the sugar molecules from the surrounding water molecules. This again implies that the sugar molecules will act on the surrounding water molecules with a force in the leftward direction and pull water over to the sugary side of the membrane.

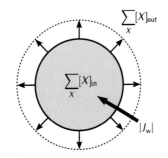

Fig 10.16 Illustration of modelling of swelling of cells according to Equation 10.26. The volume of the cell increases with the influx of water $|J_w|$ due to the higher solute potential $(\sum_X [X])RT$ inside, compared to outside, the cell.

$$p_{osm} = [X]RT. \tag{10.25}$$

This relationship, known as the **van't Hoff relation**, tells us that, when a solute is added on one side of a semipermeable membrane separating two chambers of pure water, water will flow across the membrane to the solute side until a hydrostatic pressure difference $[X]RT$ is established, and the flow stops (Box 10.9).

The process of swelling or shrinking of a cell can be modelled by the following equation describing changes in cell volume v due to water influx into the cell (Figure 10.16):

$$\frac{dv}{dt} = -AJ_w = AL_p(\Delta p_{sol} - \Delta p)$$
$$= AL_p \left(\sum_X [X]_{in} - \sum_X [X]_{out} - \Delta p \right) RT. \tag{10.26}$$

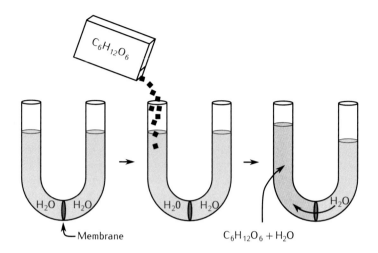

$C_6H_{12}O_6$

H_2O H_2O

H_2O H_2O

H_2O

Membrane

$C_6H_{12}O_6 + H_2O$

Fig 10.17 Illustration of osmosis where addition of sugar molecules on one side of a U-tube filled with pure water with a semipermeable membrane at the bottom, will lead to the water levels on the two sides being different (Box 10.9).

Here, $\Delta p = p_{in} - p_{out}$ is the hydrostatic pressure difference between the inside and outside of the cell and A is the surface area of the cell. If, for example, the concentration of particles inside the cell ($\sum_X [X]_{in}$) suddenly increases, water will flow into the cell and a new equilibrium will be reached with a larger hydrostatic pressure difference Δp, accompanied by a larger cell volume. The larger hydrostatic pressure must be provided ultimately by the surface tension of the cell membrane. With too high concentration differences between the interior and exterior of the cell, the membrane will rupture. In fact, this is a mechanism used by viruses to destroy cells.

Equation 10.26 provides a link between liquid flow and ion dynamics, and a means for modelling the combined regulation of cell volume and ion concentrations. For example, it has been used in combination with the equations describing the ion dynamics outlined in previous chapters, to explore the experimentally observed shrinkage of the ECS following neural stimulation, up to ~30% (Østby et al., 2009). In this model, it was found that the phenomenon could be accounted for by swelling of glial cells associated with the uptake of ECS potassium following neuronal firing. However, the action of specific transport proteins, called **cotransporters**, transporting Na^+, K^+, Cl^- and HCO_3^- ions across the glial cell membrane, was found to be necessary to account quantitatively for the experimentally observed ECS shrinkage.

10.5.2 Capillary Blood Flow

Oxygen and glucose are the key ingredients in brain metabolism and are transported in the blood via a hierarchical vessel structure. For example, **arterioles** at the cortical surface 'dive' into the cortex where they branch out to form an intricate network of **capillaries** (Figure 10.18). From inside the capillaries, the molecules first have to pass through the capillary walls, constituting a key part of the **blood–brain barrier**, either by passive diffusion or by active transport. The final stretch from the outside of the capillaries to the brain cells must be made by diffusion, and this puts an upper limit of about 100 micrometres on the distance from each brain cell to the nearest capillary (Gagnon et al., 2016; Grimes et al., 2016). Due to the intricate

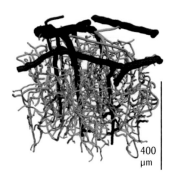

Fig 10.18 Microvasculature in the cortex. Red: arterioles. Blue: veins. Green: capillaries. Adapted from Sweeney et al. (2018) under the Creative Commons Attribution 4.0 International License.

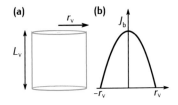

Fig 10.19 Illustration of Hagen–Poiseuille flow through a vessel. For a cylindrical vessel with radius r_v and length L_v (a), a hydrostatic pressure difference Δp between the top and bottom of the tube will give a parabolic steady-state flow profile (b) described by Equation 10.27. The flow velocity will thus be maximum in the centre of the vessel and zero at the vessel wall. The total flow through the vessel is correspondingly given by Equation 10.28.

vessel structure, the requirement that the oxygen molecules must be able to reach brain cells from capillaries is met even when this **vascular system** has been estimated to occupy only 3% of the brain volume (Nicholson, 2001; Blinder et al., 2013).

To understand and model how blood flows in this vascular system, it is useful to consider flow through a cylindrical vessel mimicking flow through a section of an arteriole or a capillary (Figure 10.19). In the context of modelling brain blood flow, the blood is considered to be a so-called **Newtonian fluid**, described only by its mass density and fluid viscosity η. Furthermore, when considering stationary flow, inertial effects are absent and the flux profile of blood flow J_b through the vessel is given by (Nelson, 2008, Chapter 5):

$$J_b(r) = \frac{r_v^2 - r^2}{4 L_v \eta} \Delta p, \tag{10.27}$$

where r_v and L_v are the radius and length of the vessel, respectively, and the flow is driven by the pressure difference Δp between the two vessel ends (Figure 10.19). The total blood flow Q in the vessel is found by integrating this flux over the vessel's cross-sectional area:

$$Q = \int_0^{r_v} 2\pi r J_b \, dr = \frac{\pi r_v^4}{8 L_v \eta} \Delta p = \frac{\Delta p}{Z}, \tag{10.28}$$

where:

$$Z = \frac{8 L_v \eta}{\pi r_v^4}. \tag{10.29}$$

In Equation 10.28, commonly known as the **Hagen–Poiseuille** relation, Z corresponds to a **hydrodynamic resistance**, in analogy to the electric resistance R in Ohm's law $I = \Delta V / R$.

A notable feature of this formula for resistance is the rapid increase of resistance with decreasing radius: halving the radius r_v increases the resistivity by a factor 16. This effect reduces the blood flow velocity in capillaries down to a few millimetres per second (compared to blood velocities of tens of centimetres per second in the aorta), allowing enough time for oxygen molecules to escape from the blood through the capillary walls.

To get a quantitative account of the flow of blood through brain vasculature of the type depicted in Figure 10.18, a vascular network model comprising a hierarchy of interconnected vessels of different sizes must be considered. Examples of such modelling are found in Boas et al. (2008) and Sweeney et al. (2018).

10.5.3 Flow of Cerebrospinal Fluid in ECS

Unlike other body organs, the brain does not have a separate lymph system for removal of waste products, and it has been proposed that waste molecules are removed advectively by pressure-driven flow of the CSF through the ECS (Iliff et al., 2012). Such pressure-driven flow is modelled by **Darcy's law** for flow through porous media:

$$\vec{J}_{CSF} = -\frac{\varkappa}{\eta} \nabla p, \tag{10.30}$$

where x is the intrinsic **hydrodynamic permeability**. The CSF flow, and thus expectedly also the pressure gradients ∇p, has been observed to follow the time course set by the pulsation of the heart (Sakka et al., 2011).

Results from a comprehensive modelling study using FEM has questioned the putative role of pressure-driven advective clearance of waste material from brain tissue (Holter et al., 2017). By solving Darcy's law for a $60 \, \mu m^3$ block of brain tissue, anatomically reconstructed by using **electron microscopy** with nanometre resolution, it was concluded that such advective flow would be too slow to contribute substantially to the clearance. The modelling results suggested that even large molecules would be more effectively removed by diffusion.

10.6 Summary

Neurons are not the only cell type in the brain; they are accompanied by glial cells. These glial cells do not fire action potentials, but are nevertheless electrically excitable cells and can be modelled using the same principles as for neurons. The space between the cells, the ECS, contains ions and molecules that affect the electrical activity of both neurons and glial cells. For one, the ion concentrations in the ECS affect the Nernst potentials, and thus the membrane currents of neurons and glia (Chapter 2). Furthermore, signalling molecules, oxygen and cell nutrients must be transported in the irregular landscape of brain tissue for brains to function.

When modelling the dynamics of neurons, it is often sufficient to model only the flow of electrical currents and how they affect the membrane potentials. When modelling the dynamics of the ECS and its effects on neurons and glial cells, one must generally consider individual ion concentrations explicitly and consider transport by both diffusion and electrical drift. In this chapter, we have described the principles used when modelling such electrodiffusive processes.

When ion transport in the ECS stems from electrical drift only, the electrodiffusive formalism reduces to the volume conductor theory. This theory is commonly used to model electrical potentials around cells in the ECS and can also be used to study ephaptic interactions between cells.

A general approach for studying electrodiffusive processes is to use the PNP equations, which explain, for example, the formation of dense nanometre-thick ion layers, so-called Debye layers, around membranes on a nanosecond timescale. This formalism is computationally very demanding, however. To account for slower diffusion processes on timescales of seconds, for example, the spatial buffering of potassium by glial cells, other formalisms, such as the KNP formalism, must be used.

Nutrients and oxygen are transported to the brain cells by blood vessels. Inside brain tissue, these key molecules are distributed by a complex network of tiny capillaries. Modelling of the liquid flow in this network is required for a quantitative understanding of this transport process. Together with the modelling of ion dynamics in neurons, glial cells and the ECS, this is another element required for a comprehensive physics-based modelling of the dynamics of brain tissue.

CHAPTER 11

Plasticity

Plasticity in the nervous system describes its ability to adapt to change, such as in response to exposure to new information, to fluctuations in the internal environment or to external injury. In each of these cases, questions can be asked of the nervous system at several different levels, requiring computational models at different levels of detail. For example, it is accepted that memory traces are stored in the strengths of individual synapses. It can be asked how the information impinging on the synapses causes appropriate changes in strength, which requires a model of the synapse. It can also be asked how a network of neurons with modifiable synapses can be used for the storage and retrieval of information. Here the modelling is directed at the processing ability of the network as a whole, given a particular mechanism for synaptic modification, which tends to be modelled in less detail. Neurons also exhibit homeostatic plasticity, which is the ability to restore their firing activity to some target level or pattern in response to a fluctuating environment. This can involve modulation of intrinsic membrane currents as well as synaptic plasticity. It must work in concert with synaptic plasticity for learning and memory to enable neural networks to retain and recall stored information whilst still being responsive to new information.

The nervous system is highly plastic. During development, neural precursor cells are born, grow and move into position to form the complex neural networks of the adult brain. Computational modelling is shedding light on the mechanisms underpinning development, as we describe in Chapter 12. Here we concentrate on the forms of plasticity that continue on into adulthood, where changes take place to neurons and their networks in response to sensory input, requirements for motor output and fluctuations in the internal environment such as damage to neural tissue.

Plasticity can be divided into two large classes: (1) functional changes that implement learning and memory; and (2) homeostatic changes that maintain brain activity in appropriate states and compensate for damage. Theoretical consideration and computational modelling are making highly valuable contributions to understanding the mechanisms and functional consequences of plasticity in the brain.

We look firstly at **synaptic plasticity** (Section 11.1). It is widely believed that changes in individual synapses underlie learning and memory, and we start by considering models of how the strengths of individual synapses change in response to the signals impinging on them. Hebb (1949) was the first to propose that synapses are strengthened according to the coincidence of pre- and postsynaptic activity (Section 11.1.1). We explore approaches which aim to find a functional relationship between pre- and postsynaptic activity and the subsequent changes in synaptic strength (Sections 11.2–11.4).

To demonstrate how synaptic plasticity can underpin cognitive function in neural networks, we examine, at three different levels of detail, model networks of neurons containing modifiable synapses which can be used for associative storage and retrieval of information (Section 11.5). We start with networks of two-state (binary) neurons, then move onto spiking integrate-and-fire neurons, before finally looking at networks of compartmental model neurons. All of these examples have a degree of biological plausibility. The binary neural networks, in particular, are related to the field of artificial neural networks (ANNs), in which networks of simple summing neurons, in combination with synaptic learning rules, are used for technological applications. Whilst not covering them in detail, we give a brief overview of ANNs in Box 11.2 and the relationship between ANNs and neuroscience in Box 11.3.

In Section 11.6, we describe models for **homeostasis**: how nerve cells can maintain their level of activity in response to a changing environment. Plasticity to achieve homeostasis may involve changes in intrinsic membrane conductances in a neuron, as well as synaptic plasticity, particularly of inhibitory connections. It may be driven by purely local indicators of activity in a neuron, such as membrane voltage and calcium concentration, or by network-wide indicators, such as via a diffusive signal or the level of network inhibition. We look at a range of **homeostatic plasticity** models that examine the effects of these various mechanisms and possible outcomes for homeostasis.

11.1 | Synaptic Plasticity

Learning and memory in the brain are hypothesised to be at least partly mediated by long-lasting changes in synaptic strength. In his well-known book *The Organization of Behavior*, Hebb (1949) postulated: 'When an axon of cell A is near enough to excite cell B or repeatedly or consistently take part in firing it, some growth or metabolic change takes place in one or both cells such that A's efficiency, as one of the cells firing B, is increased.' From this has arisen the notion of **Hebbian plasticity**: that the activity patterns in the pre- and postsynaptic neurons jointly determine the amount by which the synapse is modified (Figure 11.1). The mantra 'cells that fire together wire together' is often heard. Seung et al. (2000) provide a retrospective view of Hebb's book and the work leading from it.

The concept of Hebbian plasticity was given weight by the discovery of **long-term potentiation** (**LTP**) of synaptic strength (Bliss and Lømo, 1973). **Long-term depression** (**LTD**) (Lynch et al., 1977; Levy and Steward, 1979)

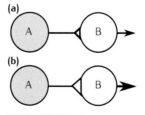

Fig 11.1 Hebbian plasticity. **(a)** Activity in cell A causes activity in cell B. **(b)** Consequently, the synaptic connection from A to B is strengthened.

The first experimental evidence for long-lasting, activity-dependent changes in synaptic strength were obtained by Bliss and Lømo (1973). By inducing strong firing, or **tetanus**, in the granule cells of the dentate fascia of the hippocampus, Bliss and Lømo found that the amplitude of the signal due to granule cell synaptic currents remained larger after the tetanus, for hours or even days thereafter, leading to the term **long-term potentiation**. It soon became apparent that long-lasting decreases in synaptic strength could also occur in certain circumstances (Lynch et al., 1977; Levy and Steward, 1979). The disparate forms of decreasing strength are known as **long-term depression**.

is also possible. These changes correspond to increases or decreases in the magnitude of the postsynaptic response to released neurotransmitter.

The Hebbian postulate and the results on LTP and LTD have been turned into simple mathematical rules for how to change synaptic strengths ('weights') in neural networks in response to pre- and postsynaptic activity levels. We consider a number of such rules in Section 11.1.1. Later, in Section 11.5, we show how Hebbian plasticity can be used to store and retrieve patterns in a neural network. In Sections 11.2 to 11.4, we look at models that are specifically aimed at capturing the processes of LTP and LTD in spiking neurons, matching a variety of experimental data.

11.1.1 Synaptic Learning Rules

Hebbian plasticity embodies the concept of associating the activity of an input neuron with that of an output neuron with which it is synaptically connected. A number of mathematical formulations of this concept have been developed that can be used to allow associated patterns of activity to be stored and retrieved in neural network models (Section 11.5).

Hebbian Rule

A very simple form of a **Hebbian learning rule** is that a synapse is strengthened according to the product of the input (presynaptic) and output (postsynaptic) activities of the appropriate neuron, here interpreted as spiking rates. For an input activity x_i and an output activity x_j, with α as a constant of proportionality, the change in weight w_{ij} is:

$$\Delta w_{ij} = \alpha x_i x_j.$$

Synapses are never weakened, and so ultimately any neural network model using this learning rule will explode with activity as all synapses become large and positive.

BCM Rule

Based on experimental work on the development of ocular dominance in the cortex (Section 12.4), Bienenstock et al. (1982) developed the **Bienenstock–Cooper–Munro (BCM) rule**. They supplemented the basic Hebbian equation with a threshold on the postsynaptic activity to allow both strengthening and weakening of synapses:

$$\Delta w_{ij} = \alpha x_i x_j (x_j - \theta_j).$$

With a fixed threshold θ_j, the rule would cause instabilities. Varying it according to the postsynaptic firing rate can implement competition between synapses on the same postsynaptic cell. Strengthening a synapse on a particular postsynaptic neuron leads to an enhanced postsynaptic firing rate, and therefore a higher threshold, which then decreases the chance that other synapses on the same neuron will be strengthened.

Covariance Rule

An earlier rule proposed by Sejnowski (1977) provides a threshold on both pre- and postsynaptic activity. Synapses are strengthened when the difference of the input activity from its mean value has the same sign as the deviation of the output activity from its mean, and weakened otherwise. If the respective mean values of input and output signals are $\langle x_i \rangle$ and $\langle x_j \rangle$, the change in weight w_{ij} is:

$$\Delta w_{ij} = \alpha(x_i - \langle x_i \rangle)(x_j - \langle x_j \rangle). \tag{11.1}$$

In the special case when input and output activities are binary-valued (neurons are either silent ('0') or firing ('1')) and there are fixed probabilities that the input and output neurons are active during training, Dayan and Willshaw (1991) showed that the **covariance rule** is the optimal linear rule. When both input and output neurons are equally likely to be in the '0' or '1' state, this reduces to the familiar Hopfield rule (Hopfield, 1984).

11.2 Phenomenological Models of LTP/LTD

Models of LTP/LTD seek to map a functional relationship between pre- and postsynaptic activity and changes in the maximum conductance produced by the postsynaptic receptor pool, typically α-amino-3-hydroxy-5-methyl-4-isoxalone propionic acid (AMPA) and N-methyl-D-aspartate (NMDA) receptors. The biology is revealing complex processes that lead to changes in the state and number of receptor molecules (Ajay and Bhalla, 2005). These processes involve calcium-mediated intracellular signalling pathways. Detailed models of such pathways are being developed (Section 11.3).

However, for use in network models of learning and memory, it is necessary to create computationally simpler models of LTP/LTD that capture the essence of these processes, whilst leaving out the detail. Basic models measure pre- and postsynaptic activity in terms of spike times and relate these times directly to changes in synaptic strength. To account for more of the biological data on LTP/LTD, other models describe postsynaptic activity in terms of a filtered membrane potential, capturing some of the spiking history, or by the calcium concentration profile.

11.2.1 Spike Timing-Dependent Plasticity

One approach is to find simple models that account for experimental data on **spike timing-dependent plasticity** (STDP). These data indicate that the precise timing of pre- and postsynaptic spikes determines the magnitude and direction of change of the synaptic conductance (Levy and Steward, 1983; Markram et al., 1997; Bi and Poo, 1998).

Basic STDP Rules
There are many mathematical formulations of an STDP rule. The most basic are two-factor rules that rely only on the times of the pre- and postsynaptic spikes. The simplest one accounts for the change in synaptic strength resulting from a single pre- and postsynaptic spike pair. Suppose that a spike occurs in postsynaptic neuron j at time t^{post}, and one occurs in presynaptic neuron i at time t^{pre}. Defining the time between these spikes as $\Delta t = t^{\text{post}} - t^{\text{pre}}$, an expression for the change in synaptic strength, or weight w_{ij} is (Song et al., 2000; van Rossum et al., 2000):

$$\begin{aligned} \Delta w_{ij} &= A^{\text{LTP}} \exp(-\Delta t / \tau^{\text{LTP}}) && \text{if } \Delta t \geq 0, \\ \Delta w_{ij} &= -A^{\text{LTD}} \exp(\Delta t / \tau^{\text{LTD}}) && \text{if } \Delta t < 0. \end{aligned} \tag{11.2}$$

In addition to tetanic stimulation, it has been found that the relative timing of the pre- and postsynaptic spikes affects the direction of synaptic plasticity. A synapse is more likely to be strengthened if the presynaptic neuron fires within approximately 50 ms before the postsynaptic neuron, conversely, if it fires shortly after the postsynaptic neuron, then the synapse is weakened. This phenomenon is referred to as **spike timing-dependent plasticity** and is found in both developing (Zhang et al., 1998) and adult synapses (Markram et al., 1997; Bi and Poo, 1998).

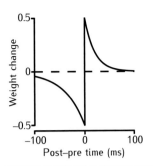

Fig 11.2 Example STDP weight change curves. The weight is increased if the postsynaptic spike occurs at the time of, or later than, the presynaptic spike; otherwise the weight is decreased. The magnitude of the weight change decreases with the time interval between the pre- and postsynaptic spikes. No change occurs if the spikes are too far apart in time.

These weight change curves are illustrated in Figure 11.2. The parameters A^{LTP}, A^{LTD}, τ^{LTP} and τ^{LTD} can be determined experimentally. The data of Bi and Poo (1998) are well fit with $\tau^{\mathrm{LTP}} = 17\,\mathrm{ms}$ and $\tau^{\mathrm{LTD}} = 34\,\mathrm{ms}$ (van Rossum et al., 2000). The magnitude of LTP, A^{LTP}, is often greater than that of LTD, but it is small for synapses that are already strong.

In a synaptic model, the weight w_{ij} could be used as a scaling factor for the maximum postsynaptic receptor conductance $\overline{g}_{\mathrm{syn}}$ (Section 7.2). To calculate weight changes for trains of pre- and postsynaptic spikes, the changes due to individual spike pairs can be summed (Badoual et al., 2006); this could involve all possible spike pairs or be limited to pairing only the most recent postsynaptic spike following a presynaptic spike for LTP and the next presynaptic spike following a postsynaptic spike for LTD (van Rossum et al., 2000). When applied in this way to long trains of random synaptic inputs that generate output spiking in a neuron, the weight change rules in Equation 11.1 are unstable: synapses that begin to be potentiated will more likely contribute to output spiking, leading to further potentiation and runaway growth in synaptic strength. This can be prevented by placing hard maximum and minimum limits on weights, so that they are constrained to $w_{\min} \leq w_{ij} \leq w_{\max}$. However, populations of synapses will tend to an unphysiological bimodal distribution of synaptic weights, with peaks clustered towards $w_{\min}$ and $w_{\max}$ (van Rossum et al., 2000). A unimodal distribution of weights, as measured experimentally, can be achieved if the amount of depression grows with the strength of a synapse to give the modified rules:

$$
\begin{aligned}
\Delta w_{ij} &= A^{\mathrm{LTP}} \exp(-\Delta t/\tau^{\mathrm{LTP}}) && \text{if } \Delta t \geq 0, \\
\Delta w_{ij} &= -A^{\mathrm{LTD}} w_{ij} \exp(\Delta t/\tau^{\mathrm{LTD}}) && \text{if } \Delta t < 0.
\end{aligned}
\tag{11.3}
$$

This increase in the magnitude of depression with synaptic strength counteracts runaway potentiation as the magnitude of potentiation falls relatively with synaptic strength.

More Complex STDP Rules

More complex models attempt to account for data from stimulation protocols that include triplets or more physiological patterns of spikes (Abarbanel et al., 2003; Castellani et al., 2005; Rubin et al., 2005; Badoual et al., 2006). Weight changes to general pre- and postsynaptic spike trains are not simple linear sums of the changes expected from individual spike pairs (Froemke and Dan, 2002). Previous spiking history seems to suppress the magnitude of change for the current spike pair (Froemke and Dan, 2002). Further components can be added to basic STDP models to incorporate effects such as weight saturation and spike triplet interactions (e.g. Badoual et al., 2006).

However, spike timing alone is insufficient to account for many of the experimental protocols that induce synaptic plasticity. To try to account for a wider range of experimental data on LTP/LTD, other synaptic plasticity models are not based purely on spike timing, but also include measures of pre- and postsynaptic activity that provide a longer-term average of activity at a synapse. These could include a trace of neurotransmitter binding to indicate average presynaptic spike rates and a trace of postsynaptic membrane potential that indicates both sub-threshold and spiking activity.

In the model of Clopath et al. (2010), the weight update rules are not direct functions of spike times, but instead use the instantaneous postsynaptic membrane potential x_j, and low-pass filtered versions of this membrane potential $\bar{x}_j$, and the presynaptic spiking $\bar{x}_i$, integrated over spike times t_s^{pre}:

$$\tau \frac{d\bar{x}_j}{dt} = -\bar{x}_j(t) + x_j(t), \tag{11.4}$$

$$\frac{d\bar{x}_i}{dt} = -\frac{\bar{x}_i(t)}{\tau^{\text{pre}}} + \sum_s \delta(t - t_s^{\text{pre}}). \tag{11.5}$$

The weight update rule is (Clopath et al., 2010):

$$\frac{dw_{ij}}{dt} = A^{\text{LTP}} \bar{x}_i(t)[x_j(t) - \Theta^+]_+ [\bar{x}_j^+(t) - \Theta^-]_+$$
$$- \sum_s A^{\text{LTD}} [\bar{x}_j^-(t) - \Theta^-]_+ \delta(t - t_s^{\text{pre}}), \tag{11.6}$$

with hard bounds being applied to the weight update so that $w_{\min} \leq w_{ij} \leq w_{\max}$. The notation $[.]_+$ indicates rectification at zero of the enclosed values.

Here, LTP occurs with a magnitude proportional to the filtered presynaptic activity $\bar{x}_i$, the amount the instantaneous membrane potential x_j is above a high threshold Θ^+ and the amount the filtered potential $\bar{x}_j$ is above a lower threshold Θ^-. These thresholds are typically set to be the firing threshold and the resting membrane potential, respectively. The membrane potential is filtered with the time constant $\tau \equiv \tau^+$ to give $\bar{x}_j^+$ via Equation 11.3. The weight is decreased (LTD) on occurrence of a presynaptic spike when the filtered membrane potential is above the Θ^- threshold, but not otherwise. The membrane potential is now filtered with the time constant $\tau \equiv \tau^-$, which is typically larger than $\tau \equiv \tau^+$, to give $\bar{x}_j^-$.

This model has been fitted to a variety of data with filtering time constants on the order of: $\tau^{\text{pre}} \approx 20$ ms; $\tau^- \approx 10$ ms; and $\tau^+ \approx 5$ ms. Though not indicated here, in neural network simulations, the model modifies the magnitude of LTD, A^{LTD}, as a slowly changing homeostatic function of the mean postsynaptic depolarisation, averaged over a longer time period, one second or more (Clopath et al., 2010). This function increases the magnitude of LTD when a neuron is depolarised for an extended time. This helps to maintain average network spiking activity and such homeostasis is detailed in Section 11.6.

11.2.2 Calcium-Based Rules

The major and most direct determinant of long-term synaptic changes, rather than spike timing, is the postsynaptic calcium level in dendritic spine heads resulting from pre- and postsynaptic activity via calcium entry through NMDA- and voltage-gated calcium channels (Ajay and Bhalla, 2005). Both the magnitude and time course of calcium transients determine the sign and magnitude of synaptic weight changes. Computational models of the complex signalling pathways linking calcium level to the changes in AMPA receptors that result in synaptic weight change are being developed. Examples are presented in Section 11.3. Other models try to capture the essence of this process in computationally simpler ways. Examples include Rackham et al. (2010) and Graupner and Brunel (2012).

In Graupner and Brunel's (2012) model, the change in efficacy ρ of a synapse as a function of calcium concentration at the synapse is captured in a single ordinary differential equation:

$$\tau \frac{d\rho}{dt} = -\rho(1-\rho)(\rho_* - \rho) + \gamma_p(1-\rho)\Theta[c(t)-\theta_p]$$
$$-\gamma_d\rho\Theta[c(t)-\theta_d] + \text{Noise(t)}. \qquad (11.7)$$

Here, the efficacy ρ is bounded between 0 and 1 and so can be used to scale the weight of a particular synapse, w_{ij}, to be between minimum and maximum values. When there is little or no pre- or postsynaptic activity and the calcium concentration $c(t)$ is low, the efficacy slowly moves, with a time constant τ, to either an UP state (LTP with $\rho = 1$) or a DOWN state (LTD with $\rho = 0$) and remains there. Calcium concentration that breaches a threshold (θ_p for LTP and θ_d for LTD; note that Θ is the Heaviside function giving $\Theta[c-\theta] = 0$ if $c < \theta$ and $= 1$ if $c \geq \theta$) will cause a movement in ρ either towards 1 at rate γ_p for LTP or towards 0 at rate γ_d for LTD. The efficacy has an unstable fixed point at $\rho = \rho_*$ and ρ must cross this value to go from LTD to LTP, and vice versa. The noise term adds stochasticity to this transition, as could be expected in the biological synapse.

The model of Rackham et al. (2010) takes a similar approach, but using empirical functions that relate weight changes to local peaks in calcium concentration.

These models seek to capture the essence of the dynamics of the complex signalling pathways that determine LTP and LTD. Though the basic weight change models are simple, they require modelling of the calcium concentration at the synapse, in addition to the spike times and membrane potential used in the LTP/LTD models described above. A basic approach is to model calcium as a simple decaying exponential function of pre- and postsynaptic spike times (Graupner and Brunel, 2012), capturing the increment in calcium at a synapse through activation of NMDA and voltage-gated calcium channels (VGCCs), followed by its extrusion and buffering. More realistic, but computationally expensive, models of calcium concentration, such as described in Chapter 6, could be used. Rackham et al. (2010) used explicit models of NMDA current and VGCCs in a spine head as sources of calcium influx to drive their synaptic plasticity model. Saudargiene et al. (2015) combined a similar calcium influx model with the Graupner and Brunel (2012) plasticity model to explore the effects of phasic patterns of spatially localised excitatory and inhibitory activity on plasticity at a particular synapse in a detailed compartmental model of a hippocampal CA1 pyramidal cell.

11.3 | Biophysical Models of LTP/LTD

The synaptic plasticity models considered so far predict plasticity outcomes but without capturing the underlying biophysical mechanisms with their associated dynamics and constraints. To address this, a model needs to include, in some detail, the biophysical pathways that mediate the translation of synaptic stimulation into changes in synaptic strength. Such models can help elucidate the mechanistic balance between LTP and LTD and how different stimulation protocols may elicit one or the other.

To this end, a large number of modelling studies, at varying levels of detail, have aimed to relate calcium transients at glutamatergic synapses with changes in synaptic strength through phosphorylation of AMPA receptors (AMPARs) or insertion of new receptors (Ajay and Bhalla, 2005).

In one detailed study, Kuroda et al. (2001) modelled the signalling pathways that drive phosphorylation of AMPARs in Purkinje cell spines. This process leads to a decrease in synaptic strength (LTD) through the internalisation of the AMPARs. This model contains 28 molecular species involved in 30 protein–protein interactions and 25 enzymatic reactions.

Other models, such as those of Bhalla and Iyengar (1999) and Castellani et al. (2005), consider the pathways leading to phosphorylation of particular AMPAR subunits found in hippocampal pyramidal cell spines, which results in an increase in synaptic strength (LTP) through an increase in receptor channel conductance. This increase is counteracted by competing pathways that dephosphorylate the subunits, resulting in LTD.

Hayer and Bhalla (2005) model the process of AMPAR recycling by which long-lasting changes in synaptic strength are implemented and maintained. They consider the stability of this process, given the stochasticity due to the actual small numbers of molecules involved.

A particular challenge is to reconcile these models with the wealth of data on LTP/LTD induction under different stimulation protocols, ranging from pairs of pre- and postsynaptic spikes to trains of stimuli at different frequencies repeated over extensive periods of time. The very detailed model of Mäki-Marttunen et al. (2020), described in Box 11.1, attempts to do this.

Detailed synaptic plasticity modelling can be used to explore the physiological induction of LTP/LTD in complex scenarios that are difficult to study experimentally. Such a model has been used to investigate the impacts of dendritic location and inhibitory inputs on plasticity at an excitatory synapse on a CA1 hippocampal pyramidal cell (Saudargiene and Graham, 2015). To achieve this, in the same manner as Mäki-Marttunen et al. (2020), an explicit model of a spine is added to a particular dendritic location of the compartmental cell model and the plasticity model is only included in this one spine head compartment. However, the cell may receive synaptic and electrode stimulation at other cellular locations in parallel with stimulation of the synapse on the spine. The plasticity model used is a little simpler than Mäki-Marttunen et al.'s (2020), but it does combine and retune models of the signalling pathways subserving LTP (Graupner and Brunel, 2007) and LTD (Pi and Lisman, 2008).

It should be noted that calmodulin, a molecule we have previously described as a calcium buffer (Section 6.6), plays an active role in these signalling pathways, particularly in its calcium-bound form. Thus it acts not simply as a buffer that shapes the free calcium transient, but as a calcium sensor that detects the magnitude and time course of a calcium transient and stimulates reactions accordingly (Burgoyne, 2007).

11.4 Other Factors Affecting Synaptic Plasticity

The models of synaptic plasticity we have presented so far are based on data demonstrating that pre- and postsynaptic activity at an excitatory synapse can lead to changes in the strength of that synapse, as expressed through changes in the conductance and number of postsynaptic AMPARs. However, many other factors play a role in the induction and expression of long-term changes in synaptic strength in the nervous system. We now summarise

Box 11.1 | Biophysical LTP/LTD model example

The model by Mäki-Marttunen et al. (2020) extends earlier efforts at modelling the postsynaptic signalling pathways underpinning LTP and LTD to maximise the range of data that can be explained, including some aspects of the effects of neuromodulation on synaptic plasticity. The signalling pathways incorporated into this model are shown in Figure 11.3. These pathways are divided into three main groups: subsuming reactions involving protein kinase A (PKA), protein kinase C (PKC) and calcium/calmodulin-dependent kinase II (CaMKII), respectively. LTP results from PKA- and CaMKII-dependent phosphorylation of AMPAR subunit 1 (GluR1). PKC-dependent phosphorylation and endocytosis of AMPAR subunit 2 (GluR2) lead to synaptic depression. All pathways depend on calcium as a driving signal. PKC also is activated via metabotropic glutamate receptors (mGluRs) and acetylcholine M1 receptors, thus including the effects of a neuromodulator. The model contains around 140 reaction pathways.

Mäki-Marttunen et al. (2020) embedded this plasticity model in a single well-mixed compartment representing a postsynaptic spine head, with mass action laws applying to the signalling pathways. They tested the model in isolation, using directly simulated calcium transients in the spine head. They then added the spine to dendritic locations in a detailed compartmental model of a neocortical layer 2/3 pyramidal cell, with the synapse itself receiving spiking input which generated postsynaptic calcium through activation of NMDA receptors, concurrently with the postsynaptic cell receiving somatic stimulation, causing it to generate action potentials.

Parameter values in the plasticity model were able to be tuned to match experimental plasticity outcomes for a variety of stimulation protocols. Reaction rates in the model were determined from the literature, where either values were available directly or particular pathway activation data were available against which reaction rates could be tuned. Plasticity outcomes were then achieved by optimising the initial concentrations of the different molecular species in the model, whilst keeping reaction rates fixed. This was done because reaction rates are likely to be conserved between different neuronal types and animal ages, whilst concentrations may differ.

what some of these factors are and the modelling work that is helping our understanding of them.

11.4.1 Spine Dynamics

Most excitatory synapses in the mammalian brain are formed on small dendritic protrusions called spines (Kasai et al., 2021). Spines consist of a head, with a volume around $1\,\mu m^3$ connected to the dendrite by a thin neck (Section 5.4.3). They are supported by an actin-based cytoskeleton which is highly labile: spines can change their shape and even appear and disappear on timescales as short as minutes (Kasai et al., 2021). Spine volume is strongly correlated with synaptic strength, with experimental protocols that induce LTP/LTD also resulting in corresponding changes in spine size (Kasai et al., 2021). Spines are also highly dynamic, even in the absence of synaptic activity (Kasai et al., 2021; Shimizu et al., 2021). Small spines, in particular, can ap-

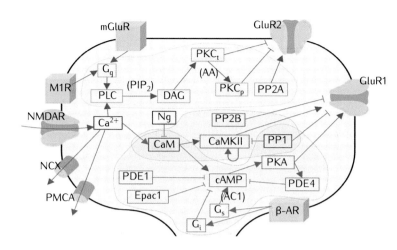

pear and disappear spontaneously, whilst larger spines are more stable. Spine volumes consistently show a long-tailed unimodal distribution on cortical pyramidal cells, with a proliferation of small spines (Humble et al., 2019).

O'Donnell et al. (2011) used a computational model to explore the potential impact of changes in spine volume on calcium-mediated LTP/LTD. An increase in spine volume will result in a decrease in spine head calcium concentrations unless there is some process to counteract this, such as an increase in the number of NMDA receptors. However, even an increase in NMDA receptor density in proportion to the spine surface area is insufficient to provide full compensation (O'Donnell et al., 2011). Thus it is likely that the amplitudes of calcium transients decrease with LTP and an increase in spine volume. O'Donnell et al. (2011) show that this can have the beneficial effect of stabilising large spines which are already strong, whilst promoting LTP/LTD at smaller spines. This results in the unimodal distribution of spine sizes seen experimentally.

How intrinsic spine dynamics interact with the expression of synaptic learning is an open topic to which computational modelling is making a valuable contribution (Shimizu et al., 2021). For example, Humble et al. (2019) and Deger et al. (2018) studied the impact of spontaneous spine dynamics on the distribution of spine sizes in recurrent neural network models in which STDP rules are operating to enable the storage and recall of memory patterns. Both studies demonstrate that intrinsic spine dynamics can help maintain network memory patterns whilst retaining the unimodal distribution of spine sizes. We look at the storage and recall of memory patterns through Hebbian learning in neural networks in Section 11.5.

11.4.2 Consolidation and Maintenance

Most models of synaptic plasticity either ignore the dynamics of its instantiation or only consider the time course of its initial expression. Consolidation of synaptic learning takes place in timescales of minutes to hours and involves protein synthesis operating on a dendritic scale. Early and late phases for the expression of LTP and LTD have been identified. Experimental protocols using brief bursts of high-frequency (HFS) or low-frequency (LFS) stimulation induce an early transient phase of LTP or LTD, respectively, which can decay back to baseline over a few hours. Additional bursts of HFS or LFS can convert early LTP/LTD into late LTP/LTD, which is much more stable and may potentially last indefinitely. Induction of late LTP/LTD requires extra-synaptic protein synthesis, with plasticity-related proteins (PRPs) being delivered to synapses through what is hypothesised to be a **synaptic tagging and capture (STC)** process (Frey and Morris, 1997; Reymann and Frey, 2007; Barrett et al., 2009). Early-phase LTP/LTD at a synapse includes setting a *synaptic tag* at the synapse, which is then able to capture PRPs, when available, and stabilise to late-phase LTP/LTD. Such PRPs can result from sufficient stimulation of the synapse in question or of nearby synapses, allowing heterosynaptic plasticity in which an early-phase LTP/LTD synapse is converted to late-phase plasticity by stimulation only to other synapses and not to itself.

Clopath et al. (2008) and Barrett et al. (2009) developed state-based synaptic models to explore the implications of STC for plasticity at synapses across a neuron, including heterosynaptic transitions from early- to late-phase LTP/LTD. To illustrate, the Barrett et al. (2009) model describes a synapse as existing in six possible states, with time- and activity-dependent transitions between states. The synapse is assumed to be binary, being either weak or strong. Its possible states are weak basal, strong basal, early-LTP, late-LTP, early-LTD and late-LTD. Transitions from basal to early-phase (de)potentiated states require homosynaptic activity. Early-phase states could convert to their equivalent late-phase if there was sufficient activity at this or nearby synapses to open this transition pathway, corresponding to the generation of PRPs. This stochastic Markov model is of the same form as used for modelling ion channels (Section 4.7). These models are able to replicate a variety of experimental data indicating homosynaptic and heterosynaptic plasticity in populations of synapses and the time course of LTP/LTD induction and stabilisation over time periods of hours. This consolidation process can protect established synaptic weight changes from modification during ongoing synaptic plasticity (Clopath et al., 2008).

11.4.3 Presynaptic Plasticity

Changes in synaptic strength also can be expressed presynaptically, through changes in neurotransmitter release. This can arise through changes in action potential-mediated calcium influx or availability of readily releasable vesicles. Presynaptic plasticity can happen independently of, or in concert with, postsynaptic changes (Costa et al., 2017). It may require some mechanism for passing a measure of postsynaptic activity back to the presynaptic terminal. This can be achieved through a diffusible signal, such as postsynaptically released endocannabinoids or nitric oxide (NO) (Sjöström et al., 2007) and may be mediated directly or through glial cells (Manninen et al., 2020).

Manninen et al. (2020) present a complex multi-cell signalling model of presynaptic LTD induction in the developing cortex involving postsynaptic endocannabinoid release that reaches receptors on a glial cell and stimulates calcium-mediated glutamate release from this cell. This extra-synaptic glutamate, along with glutamate spilt over from the synaptic cleft, reaches presynaptic NMDA receptors and so induces calcium influx into the presynaptic terminal. This activates calcineurin, which inhibits vesicle exocytosis, and thus effects LTD.

Costa et al. (2015) propose a phenomenological STDP plasticity rule that includes both pre- and postsynaptic components of LTP and LTD expression. LTP of the postsynaptic strength occurs when a postsynaptic spike follows both a preceding presynaptic and a postsynaptic spike, with a magnitude that is a decaying function of the time of the preceding postsynaptic spike. LTP and LTD are also expressed presynaptically through changes in vesicle release probability, which occur on the advent of a presynaptic spike. The presynaptic term is a function of a filtered version of presynaptic activity plus two filtered versions of postsynaptic activity, representing retrograde signals due to diffusing endocannabinoids and NO. Only the NO component contributes to LTP, with both contributing to LTD. This model matches experimental plasticity data from monosynaptic connections between neocortical layer 5 pyramidal cells.

A presynaptic change in neurotransmitter release has different functional consequences than a postsynaptic change in neurotransmitter response strength. It changes the short-term plasticity of the synapse that determines the magnitude of transmitter release to a train of action potentials over tens to hundreds of milliseconds (Section 7.3). In so doing, it more impacts the dynamics of information storage and retrieval in a neural network, rather than determining what is stored, as mediated by postsynaptic LTP/LTD (Costa et al., 2017). It can act as a homeostatic mechanism for neural gain control. We look at other forms of homeostatic plasticity in Section 11.6.

11.5 Network Models of Learning and Memory

Synaptic plasticity can endow neural networks with learning and memory function. Model neural networks were first developed within the community interested in network models of brain function as well as the development of trainable artificial neural networks (ANNs) for technological applications. To introduce this vast topic, we look at how Hebbian learning rules (Section 11.1.1) can be used in neural networks to enable them to store and recall patterns of information. To give a solid introduction to how different levels of abstraction in both synaptic plasticity and neural network models enable insight into learning and memory, we present a number of approaches to modelling **associative memory** in feedforward and recurrent networks. We give a brief overview of the general field of ANNs in Box 11.2, and the relationship between ANNs and neuroscience in Box 11.3.

11.5.1 Associative Memory Networks

To illustrate the possible function of synaptic plasticity in a neural network, in the following sections, we consider models that embody associative memory. An **associative memory network** can exist in feedforward and recurrent forms, but common to both is the idea of storing and recalling patterns of activity, each of which might represent a stimulus or an event. In the case of the feedforward network, the task is **heteroassociation**, that is, to associate two different patterns of activity with each other. An example of an

Box 11.2 | Artificial neural networks

The Associative Net, described in Section 11.5.2, is an early example of an **artificial neural network (ANN)**. Such networks are inspired by brain architecture and the phenomenon of synaptic plasticity. Instead of trying to replicate neurobiological results, ANNs have largely been developed from an engineering point of view to provide computationally simple, practical solutions to problems. The advent of sophisticated training algorithms has enabled in silico networks of computing units interconnected with variable strength connections to be programmed to learn from experience. These have become the basis of a vast technology, which harnesses the immense computing power now available, using ANNs for pattern recognition, classification and prediction. They are the primary methodology used in artificial intelligence (AI) applications. The development of ANN architectures and algorithms is a vast topic that we cannot cover in any detail here.

Many ANNs are trained by **supervised learning**, under the guidance of a training signal. This differs from Hebbian learning in that training proceeds by adjusting the values of the weights each time an input pattern is presented for training, according to how the network output compares with the desired output for that pattern. The learning rule is now a function of the error between the network and desired outputs. On presentation of a previously learnt input pattern, the network responds with the associated output pattern if the correct mapping has been learnt; in response to a novel input, it generalises to produce an appropriate output. Supervised learning is most often used in feedforward networks with one or more hidden layers between the input and output layers. Errors are back-propagated by calculating them recursively from the errors made by the neurons to which the hidden units project. Rumelhart et al. (1986a) first demonstrated the use of this **back-propagation algorithm** for training neural networks.

One issue that has emerged is that networks with more hidden layers tend to generalise better. The informal argument is that in these networks, more and more salient features of the input can be computed in the hidden layers. With the immense increase in computing power available, networks with a large number of hidden layers, known as **deep neural networks**, and using **deep learning**, can now be used. A landmark was the development of a deep **convolutional neural net**, AlexNet (Krizhevsky et al., 2012), with 60 million parameters and 650 000 neurons, which won the annual **ImageNet Large Scale Visual Recognition Challenge** contest in 2010 and 2012. All subsequent winners have been developments of AlexNet or other convolutional neural nets. Deep neural networks now are used for the extraction of information from every conceivable type of data set. We consider aspects of the relationship between deep learning and neuroscience in Box 11.3.

There are other methods of training ANNs. In **unsupervised learning** (Hinton et al., 1999), the training signal is generated by the network itself. The aim is to detect and extract interesting patterns or categories in the training data. With **reinforcement learning**, the training signal is based on expectations of future reward (Sutton and Barto, 1998, 2018). Such networks discover sequences of activity that move the net from a starting state to a rewarding goal state.

Box 11.2 (cont.)

For more information on the development of neural networks, see the books and papers by Minsky and Papert (1969), Hinton and Anderson (1981), Mc-Clelland et al. (1986), Rumelhart et al. (1986b) and Bishop (2006). Cowan and Sharp (1988) provide a historical overview of the development of neural networks, and Hertz et al. (1991) give a thorough introduction to ANNs from the physics angle. Goodfellow et al. (2016) give a comprehensive introduction to deep learning. Sejnowski (2020) and Poggio et al. (2020) discuss attempts to develop a theoretical basis for deep learning.

association (Marr, 1970) might be a monkey observing a rotten branch (input) and the finding that the branch breaks upon the monkey swinging on it (output). This association may be learnt by experience, and afterwards the monkey will associate rotten branches with falling. In contrast, the recurrent network's task is to store patterns so that each stored pattern can be recalled upon presentation of a fragment of it. For example, seeing part of a familiar face might be sufficient to evoke a memory of the entire face. Effectively, this is associating a pattern with itself, so the task is called **autoassociation**.

The architecture of the network is similar to various neuroanatomical structures; associative networks similar to the ones described here form part of Marr's theories of the cerebellar cortex (Marr, 1969), the neocortex (Marr, 1970) and the hippocampus (Marr, 1971; Willshaw et al., 2015). The structure of the feedforward network is similar to the arrangement of connections in the perforant path from the entorhinal cortex to the granule cells of the dentate gyrus in the hippocampus, and the structure of the recurrent network is similar to hippocampal area CA3 (McNaughton and Morris, 1987).

Firstly, we describe feedforward and recurrent associative networks with binary-valued synapses and binary-valued input and output patterns. Whilst these networks contain very simple elements, they are a good starting point for understanding how a network of neurons can act as memories. Then we demonstrate that the function carried out by these simple networks can be performed by spiking-neuron network models whose connections have been set up using similar principles.

11.5.2 The Feedforward Associative Network

The feedforward associative network model, called the **Associative Net**, introduced by Willshaw et al. (1969), comprises two layers of neurons: an input layer containing N_A neurons, and an output layer containing N_B neurons. Every neuron in the input layer is connected to every neuron in the output layer and so the connections can be visualised in matrix form (Figure 11.4). Each neuron can be either active (1) or inactive (0), and an association is represented by a pair of patterns of 0s and 1s across the input and output layers. In what follows, a fixed number M_A neurons, randomly chosen from the total of N_A neurons, is active in any one input pattern, together with a fixed number M_B in any output pattern, chosen similarly.

The task of the network is to store associations through selected synapses, which can exist in one of two states. During the training phase of the network, it is assumed that input patterns are presented at the same

Box 11.3 | Artificial neural networks and neuroscience

Whilst ANNs are biologically inspired, there is little evidence from neurobiology that the methods of deep learning are used in the nervous system. Nonetheless, attempts are ongoing to reconcile deep learning with brain architecture. It has been demonstrated that the back-propagation of an error signal for learning does not require the precise feedback connections used in ANNs and so is somewhat compatible with the less precise feedback circuitry found in the cortex (Lillicrap et al., 2016). Further, it is argued that the targeting of feedback connections to apical dendrites in cortical pyramidal cells allows the possibility of suitable dendritic processing of error signals for learning (Richards and Lillicrap, 2019).

However, ANNs are providing useful insights into how brain networks may function, particularly for visual processing. One early example of how an ANN was applied in neuroscience is to understand how neurons learn coordinate transformations in analysing the responses of cells in the posterior parietal cortex, which have receptive fields that depend on the position of the eyes. Zipser and Andersen (1988) showed how a network with a hidden layer with inputs that represent the location of an object on the retina and the direction of a head coordinate transformation could learn to transform the position of the object to be relative to the head, rather than the retina. The back-propagation algorithm was used to set up the connections between neurons. Deep neural networks for image recognition have revealed similar feature representations in intermediate layers to those found in the mammalian visual system, thus raising hopes that ANNs can provide valuable insights into brain function (Kriegeskorte, 2015).

Apart from being interpreted as models of brain function, the pattern recognition capabilities of ANNs mean they are a valuable tool in analysing brain signals and extracting features from them. For example, extensive work using deep learning and other pattern recognition and classification techniques from machine learning is being devoted to epileptic seizure prediction and forecasting, using EEG, MRI and other data from humans (Kuhlmann et al., 2018; Shoeibi et al., 2021; Stirling et al., 2021).

time as strong extrinsic inputs coerce the output neurons into firing in the corresponding output pattern. This means that some synapses are potentiated through the firing of both the pre- and postsynaptic neurons. According to the Hebbian prescription, these are the conditions under which synapses are strengthened, and so the strength of each of these potentiated synapses is set to 1, being indicated in the matrix (Figure 11.4) by a filled synapse. This is repeated for the storage of further associations.

To retrieve the output pattern that was stored alongside an input pattern, the input pattern is presented to the network. For each output neuron, the dendritic sum is calculated, which is the number of active input neurons which are presynaptic to potentiated synapses on the output neuron. The dendritic sum is then compared to the threshold, which is set to be equal to M_A. If the sum is equal to the threshold, then the output neuron is considered to be active. It can be seen from the example in Figure 11.5a that associations can be retrieved successfully.

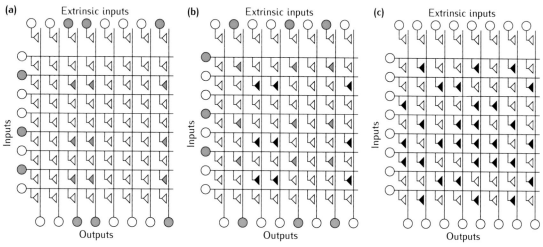

Fig 11.4 The Associative Net, an associative memory model with binary weights. There is an input layer, an output layer and strong extrinsic inputs to each output neuron. Synapses are denoted by triangles. They can be either unpotentiated (unfilled) or potentiated (filled).
(a) Storage of the first association in the network. The input and output patterns are shown as patterns of empty and filled circles. The synapses which are potentiated by this association are indicated by filled synapses; all the other synapses are unpotentiated.
(b) Storage of a second association. The original set of potentiated synapses remain (black synapses), but more synapses corresponding to the input and output patterns are potentiated (blue synapses).
(c) After several associations have been stored in the network and more synapses have been potentiated.

As more memories are stored in the matrix, the number of potentiated synapses increases. It is therefore possible that output cells may be activated when they should not be, because, by chance, M_A synapses from active input neurons have been potentiated (Figure 11.5b). This spurious activity might correspond to faulty recall of a memory. Because of the simplicity of the network, it is possible to compute how the amount of spurious activity will vary as a function of the number of input and output cells and the number of neurons that are active in the input and output patterns. This leads to an estimate of the ultimate capacity of the system. Under optimal conditions, the network is used with very high efficiency (Box 11.4).

By setting a criterion for the frequency of errors that can be tolerated, it is possible to determine how many patterns can be stored in a network. The calculations in Box 11.4 show that the capacity of the network (the number of associations stored reliably) increases as the proportion of neurons that are active in a pattern decreases. This is because the more synapses are potentiated, the more errors are likely to be made; learning a pattern with a low proportion of active neurons leads to a smaller fraction of the synapses in the network being potentiated than learning a pattern with a higher proportion of activated neurons.

11.5.3   The Recurrent Associative Network

Recurrent networks can be used to carry out the task of autoassociation, which is storing a pattern so that upon presentation of part of a previously stored memory pattern, the remainder of the same pattern can be retrieved. As with heteroassociative networks, patterns in these autoassociative networks are stored by setting the synaptic weights using a learning rule which depends on the activity of pre- and postsynaptic neurons (Little, 1974; Hopfield, 1982, 1984; Tsodyks and Feigel'man, 1988).

Rather than using the differential equation for changing the firing rate of cells according to the firing rates of their inputs, given in Equation 9.2, the activity of recurrent networks with binary-valued units is often updated in discrete time-steps. This may be a **synchronous update**, in which at each time-step the activity of all the units is updated on the basis of the activity

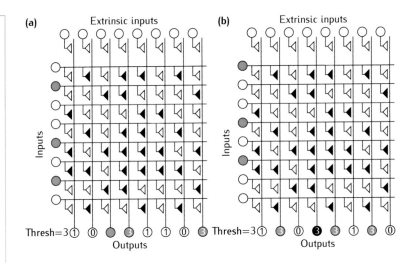

Fig 11.5 Recall in the Associative Net. **(a)** Recall of the first association stored in Figure 11.4. The depolarisation is shown in each output neuron. If the depolarisation reaches the threshold of 3, the neuron is active (blue shading). All the correct neurons are activated. **(b)** Recall of the second pattern stored. All the correct output neurons are activated, but there is also one output neuron (indicated by black shading) that is activated erroneously. This is because the synapses from the active inputs to this neuron have been potentiated by other associations.

of the previous time-step. Alternatively, **asynchronous update** may be used, in which at each time-step a randomly chosen unit is updated. Hertz et al. (1991) provide a fuller discussion of the advantages and disadvantages of each method.

We now consider a network similar to the one studied by Gardner-Medwin (1976), in which the connections from a layer of associative input neurons are replaced by recurrent collaterals from the output neurons (Figure 11.6). Extrinsic inputs are still responsible for activating the output cells during training (pattern storage), but the pre- and postsynaptic patterns are now identical and so the resulting weight matrix is symmetric. The process of recall is now progressive, occurring over multiple time-steps. This allows the network to perform **pattern completion**, that is, recall of an entire pattern when presented with a fragment of it.

The demonstration in Figure 11.6 makes it clear that in order for pattern completion to work, the threshold has to be lower than the number of neurons that are active in the fragment of the pattern which is presented. The lower the threshold, the more powerful the network is as a pattern completion device. However, with a lower threshold, the amount of spurious activity increases. There is therefore the risk that the spurious activity may lead to more spurious activity. In turn, this may lead to a greater number of spurious activations, with the result that the network ends up in an uninformative state in which all neurons are active. This is demonstrated in Figure 11.7a, in which the threshold (nine units) is just below the number of neurons activated initially (10 active neurons selected from patterns containing 100 active neurons). All the correct neurons are active after the first update, but the number of spuriously active neurons in the network increases to its maximum within two updates.

To counteract this problem, inhibitory neurons can be added to the network. The level of inhibition in the network is assumed to be proportional to the number of output neurons which become activated. When only part of a pattern is presented, the activity, and thus the level of inhibition, is low. It is therefore straightforward to activate the low-threshold neurons. However, on the second pass, when more neurons are active, the inhibition is proportionally larger, making it harder to recruit extra neurons to fire.

Box 11.4 | Capacity of the Associative Net

Simple analysis (Willshaw et al., 1969) reveals the conditions under which the network can be used with high efficiency compared with a random access store with no associative capability. Under these conditions, patterns are coded sparsely, that is, each pattern is represented by activity in a relatively small number of neurons.

If M_A of the N_A input neurons and M_B of the N_B extrinsic input neurons are active in the storage of an association (both sets of active neurons being chosen randomly), then the probability of a synapse having the associated input and extrinsic neurons active is $f_A f_B$ where $f_A = M_A/N_A$ and $f_B = M_B/N_B$ are the fractions of neurons that are active in input and extrinsic input patterns, respectively.

We determine the proportion of synapses p that have been potentiated in the storage of R associations by calculating the probability that a synapse has never been potentiated in the storage of any association:

$$1 - p = (1 - f_A f_B)^R.$$

Assuming $f_A f_B \ll 1$ and after rearrangement, this can be rewritten as:

$$R = -\log_e(1 - p)/(f_A f_B). \tag{a}$$

During retrieval, an input pattern that activates M_A input neurons is presented. Some of the $N_B - M_B$ output neurons, which should be silent in recall, may be activated because of erroneous activation of synapses in the storage of other associations. This occurs with probability p^{M_A} and so the mean number of erroneous responses per pattern retrieved is:

$$\epsilon = (N_B - M_B)p^{M_A}.$$

A limit of good recall is at $\epsilon = 1$. A safer limit (Willshaw et al., 1969) is:

$$N_B p^{M_A} = 1,$$

from which it follows that:

$$M_A = -\log_2 N_B / \log_2 p. \tag{b}$$

The efficiency of retrieval E is the ratio of the number of bits in the R patterns retrieved to the number of binary storage registers available. Assuming perfect retrieval, this is:

$$E = R \log_2(C_{M_B}^{N_B})/(N_A N_B),$$

where $C_{M_B}^{N_B}$ is the number of possible combinations of M_B out of N_B elements.

Approximating $\log_2(C_{M_B}^{N_B})$ as $M_B \log_2 N_B$ leads to:

$$E = R M_B \log_2 N_B /(N_A N_B).$$

Substituting for f_A, f_B, R and $\log_2 N_B$ using Equations (a) and (b):

$$E = \log_2 p \log_e(1 - p).$$

E has a maximum value of $\log_e 2$ (69%) when $p = 0.5$. Under these conditions:

$$R = \log_e 2/(f_A f_B), \qquad M_A = \log_2 N_B.$$

This analysis demonstrates that when working under optimal conditions, the network is extremely efficient, sparse coding is required and the number of associations stored reliably scales in proportion to the ratio $N_A N_B/M_A M_B$.

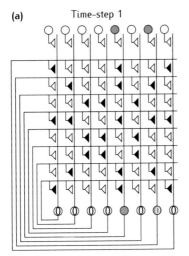
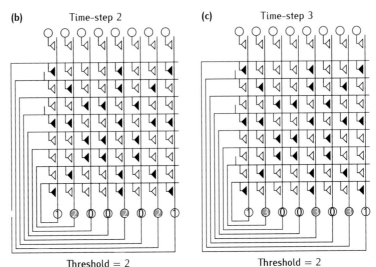

(a) Time-step 1

Threshold = 2

(b) Time-step 2

Threshold = 2

(c) Time-step 3

Threshold = 2

Fig 11.6 Demonstration of pattern completion in the recurrent associative network. **(a)** At the first time-step, two out of three neurons which were active in a previously stored pattern are active in the pattern presented as cue to the extrinsic inputs. This causes the corresponding output neurons to fire. **(b)** At the second time-step, the activity is fed round to the recurrent inputs (indicated by spikes on the input lines). This gives two units of activation at all the neurons in the pattern. Given that the threshold is set to 2, this causes all neurons in the pattern to fire. **(c)** At the third time-step, the activity from all three neurons is fed back to the recurrent inputs. This causes three units of depolarisation on each neuron and, in this case, no spurious neurons are activated.

The step function $\Theta(x)$ has a value of 1 if the value of x is greater than 1, and 0 otherwise (Figure 8.19c).

This principle is demonstrated by the set of simulations summarised in Figure 11.7b. The equation governing the network is:

$$x_j(t+1) = \Theta\left(\sum_i w_{ij} x_i(t) - \gamma \sum_i x_i(t) - \theta\right), \tag{11.8}$$

where $x_j(t)$ is the activity (0 or 1) of neuron j at time-step t, θ is the threshold and γ is the global inhibition parameter. The function $\Theta(\cdot)$ is the step function. The threshold is set to $\theta = 0.5$, and the inhibition $\gamma = 0.9$. Thus, if the network were in a full recall state, with 100 active neurons, it would receive 90 units of inhibition. As with the network in which there is no inhibition and the threshold is just below the number of neurons activated ($\gamma = 0$, $\theta = 9$; Figure 11.7a), recall is tested by presenting patterns in which 10 of the original 100 active neurons are active. With these settings, the network can reach a recall state in which all the correct neurons are active and there are no or few spurious firings.

The recall states of a recurrent network are referred to as attractors because, given a starting configuration of the network sufficiently close to a recall state, the configuration will be attracted towards the recall state. This is made particularly explicit by Hopfield's innovation of the **energy function** (Hopfield, 1982). Any individual configuration of networks of this type can be assigned an energy. Recall states (or attractors) of the network are minima within this energy landscape, and the process by which the state moves towards the attractor is called **attractor dynamics**.

11.5.4 Variations on Associative Networks

Incompletely Connected Networks

Despite the simplicity of the associative network model, examination of its properties demonstrates that network models with very simple elements can give important insights into how networks comprising more complex elements might behave. The simple associative network model has demonstrated how a network might recall a pattern and the importance of the

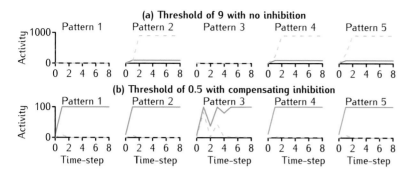

Fig 11.7 The effects of threshold and inhibition on pattern completion in the recurrent associative network. **(a)** One hundred patterns were learnt in a network of 1000 neurons, with 100 neurons active per pattern. The threshold was set at nine units, and then recall was tested in the network by presenting 10% of the active neurons of five different patterns. The graphs show the number of correctly active neurons (solid blue lines) and the number of spuriously active units (dashed grey lines) at each point in time of each of the five simulations of pattern recall. For patterns 2, 4 and 5, whilst the number of correctly active units is 100, the number of spuriously active units rises up to the maximum value of 900. For patterns 1 and 3, there is no recall at all – all units are silent. **(b)** Recall was tested in exactly the same way on the same network, but with a threshold $\theta = 0.5$ and an inhibition parameter $\gamma = 0.9$. After, at most, five time-steps, the network settles to a state where there is the full complement of correct units and no or very few spurious ones. Note the difference in scales between the y-axes of **(a)** and **(b)**.

sparseness of memory patterns and of how the threshold and level of inhibition are set.

The original analysis has been extended in several directions to allow for the examination of more realistic cases. Buckingham (1991) examined threshold setting strategies for heteroassociative recall of noisy cues. Gardner-Medwin (1976) was amongst the first to investigate the biologically more plausible situation where the network is incompletely connected, that is, a synapse between two neurons exists with only a certain probability. This means that the number of synapses activated for a given input varies and requires a lowering of the threshold in order to maintain firing. Buckingham and Willshaw (1993) and Graham and Willshaw (1995) examined efficient threshold setting strategies for heteroassociative recall of noisy cues in incompletely connected heteroassociative networks. Motivated by properties of real synapses in the hippocampal area CA3, the effects of stochastic transmission and variations to the update dynamics have been investigated (Bennett et al., 1994; Graham and Willshaw, 1999).

Linear Associative Networks

Much research has been devoted to analysing networks in which the weights have continuous, rather than binary, values, though the activity states of the model neurons are still binary. The issue of the best synaptic learning rule has been discussed (Section 11.1.1). Typically, synapses can be depressed as well as potentiated. In this case, the optimal learning rule can be calculated (Dayan and Willshaw, 1991), which is the covariance rule (Sejnowski, 1977). A key finding is that, regardless of whether the update is synchronous (Little, 1974) or asynchronous (Hopfield, 1984; Amit et al., 1985; Amit, 1989), the capacity of the network can scale with the size of the network. This scaling occurs only when the learning rule is tuned to the fraction of neurons that are active in any one pattern being present, so that the *average* change in weights caused by learning a pattern is zero (Sejnowski, 1977; Palm, 1988; Dayan and Willshaw, 1991). This principle can be summarised as 'what goes up must come down' and suggests that one role of LTD is to optimise storage capacity (Dayan and Willshaw, 1991). Networks in which the synapses have several discrete conductance levels have also been studied (Amit and Fusi, 1994).

Palimpsests

All the networks described so far are only able to learn a finite number of memories before recall becomes impossible. It is possible to construct

Networks in which old memories are overwritten by new ones are often referred to as **palimpsests**, by analogy to the ancient and medieval practice of scraping away an existing text from vellum or papyrus in order to make way for new text. A faint impression remains of the original text, which can be deciphered using modern archaeological techniques.

networks which can always learn new memories by virtue of forgetting old ones. This can be achieved in networks with binary weights by random depotentiation of synapses (Willshaw, 1971; Amit and Fusi, 1994), and is achieved in networks with continuous weights by various methods (Nadal et al., 1986; Parisi, 1986; Sterratt and Willshaw, 2008). In networks with binary weights, a histogram of the time for which the synapse stays in the potentiated or unpotentiated state, similar to the histogram of channel open times (Section 4.7), can be plotted, and it has an exponential form. In the context of binary synapses, synaptic states with different levels of stability have been considered (Fusi et al., 2005). Transitions between these states are stochastic and on multiple timescales, leading to histograms exhibiting a long-tailed power law dependence ($1/t^{\mu}$ with $\mu > 1$) rather than an exponential one. This mirrors the distribution of ages of exhibited memories measured in psychological experiments (Rubin and Wenzel, 1996).

Associative Networks of Graded Neurons

Associative networks can also be implemented as networks of graded firing rate neurons (Amit and Tsodyks, 1991b). In this case, the neuronal dynamics follow a first-order differential equation (Section 9.2.1) and the firing rate is based on the f–I curve of noisy integrate-and-fire neurons. The cells in this simulation can show a wide range of firing rates.

The approach is also related to that of Treves (1990), who investigated networks comprising rate-based neurons with piecewise-linear f–I curves (Figure 8.19), rather than the binary neurons used in previous models. This type of network can store patterns in which the activity in the neurons is graded, rather than being binary. Roudi and Treves (2006) extended the analysis by comparing associative networks with threshold-linear, binary or smoothly saturating rate-based neurons.

11.5.5 Associative Networks of Simplified Spiking Neurons

The network introduced in Section 11.5.3 demonstrates how memories might be encoded in synaptic strengths, and shows – amongst other things – the importance of setting the level of inhibition so as to prevent activity exploding. However, the neuron model underlying the network was extremely simple, and we might question whether more complicated neurons could support memory storage.

This section presents an approach taken to address this question by Amit and Brunel (1997b), who implemented associative memory in a recurrent network of excitatory and inhibitory integrate-and-fire neurons (Section 9.3.1). The network is intended to represent a highly simplified version of a cortical column, the motivation being recordings from cells in the monkey temporal cortex (Miyashita, 1988). These cells fire at elevated rates for a number of seconds after a familiar visual stimulus has been presented briefly – suggestive of the attractor dynamics exhibited by the recurrent memory networks described earlier. In addition, in the absence of any stimulus, cells fire at a low background rate, and the cells in a column are thought to receive a constant background barrage of spikes from other cortical columns. In Section 9.3.1, we detailed how this network initially was

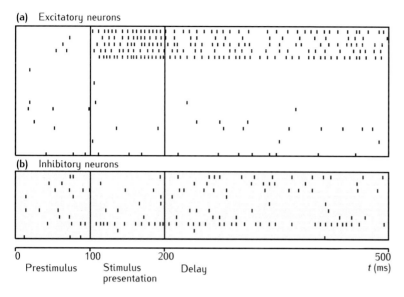

(a) Excitatory neurons

(b) Inhibitory neurons

0   100   200   500

Prestimulus   Stimulus   Delay t (ms)
             presentation

Fig 11.8 Recalling a memory embedded in a network of integrate-and-fire neurons. **(a)** Spike rasters from 20 excitatory neurons, including five active in a recalled pattern. **(b)** Spike rasters of 10 inhibitory neurons. From Amit and Brunel (1997b). Reprinted by permission of the publisher (Taylor & Francis Group, www.informaworld.com).

tuned to reproduce roughly the experimentally recorded firing statistics in the absence of a stimulus.

To embed memories in the network, Amit and Brunel (1997b) used a stochastic learning rule based on Hebbian learning (Section 11.1.1). They assumed that a sequence of binary-valued patterns was presented to each of the excitatory neurons of the network, and that only the connections between pairs of excitatory neurons were modifiable. Whenever the pre- and postsynaptic neurons on either side of a connection were both active, the connection between them was set to a potentiated value with a fixed probability. Whenever only one of the pre- or postsynaptic neurons was active, the connection was set to a depressed level. Connections which had been potentiated or depressed were not eligible for further modification. The result is a synaptic matrix which was then imposed on the network of integrate-and-fire neurons.

Figure 11.8 shows spike rasters of excitatory and inhibitory neurons in the model network. Initially, in the prestimulus period, both the inhibitory and excitatory neurons are spontaneously active. In the stimulus period, current is fed to excitatory neurons that were active in a previously stored pattern. Excitatory neurons which were active in this pattern, a sample of which is shown in the five top neurons in the raster in Figure 11.8a, fire at rates that are higher than their spontaneous rates. In the delay period, after the stimulus is removed, the excitatory neurons in the pattern continue firing, albeit at a lower rate than during the stimulus period. This delay period can persist for a number of seconds, though it is vulnerable to particularly large oscillations in the global activity of the network. This demonstrates that the associative network described in the previous section can be implemented in a network of integrate-and-fire neurons.

11.5.6 Associative Memory in Oscillating Networks

Above, we saw that patterns in an associative memory of spiking neurons can be defined as a group of neurons firing at an elevated rate, compared to the background firing rate of the network (Figure 11.8). An alternative for an oscillating network (Section 9.4) is that a pattern corresponds to those neurons that spike together on a particular oscillation cycle (Singer, 1993; Lisman and Idiart, 1995). This is an example of a temporal code, rather than a rate code. Gamma frequency oscillations, in particular, have been hypothesised to act as a 'clock cycle' for memory patterns, with a new pattern being able to be recalled potentially every 25 ms or so (Lisman and Idiart, 1995). Different patterns being recalled on each gamma cycle may constitute a meaningful pattern sequence, for example, the route ahead through an environment, where each pattern represents a location.

To demonstrate network rhythms and their role in associative memory, we consider a recurrent network of excitatory cells that also includes feedback inhibition. Based on the model of Sommer and Wennekers (2000, 2001), our model network contains 100 excitatory cells that are modelled using the Pinsky–Rinzel two-compartment model of a hippocampal CA3 pyramidal cell (Section 8.1.2). In such a small network, we allow these cells to be connected in an all-to-all manner. In addition, each cell forms an inhibitory connection with all other cells. This provides a level of inhibition that is proportional to the level of excitatory activity in the network. This could be, and would be in a biological network, mediated by an explicit population of inhibitory cells, but for computational simplicity these are omitted. This network model is considerably smaller in size and simpler in structure than the model of Amit and Brunel (1997b). This is allowable precisely because we are modelling an oscillating network in which the principal neurons are firing more or less in synchrony. This is easily achieved with small groups of neurons, whereas the asynchrony required by Amit and Brunel (1997b) requires a large degree of heterogeneity, which is lost in small networks. The simplification of not modelling the inhibitory interneurons introduces a very specific assumption about the connectivity and subsequent activity levels within a feedback inhibitory loop. Models that include explicit interneurons demonstrate that memory recall is rather robust to the precise form of this feedback inhibition (Hunter et al., 2009).

The final ingredient of our model is a structured weight matrix that defines the autoassociative storage of binary patterns, as described earlier for the schematic autoassociative memory network (Section 11.5.3). Here, each pattern consists of 10 active cells out of the population of 100. Fifty patterns are generated by random selection of the 10 cells in each pattern. They are stored in the weight matrix by the binary Hebbian learning scheme described earlier. This binary matrix is used to define the final connectivity between the excitatory cells – an entry of 1 in the matrix means the excitatory connection between these cells is retained, whereas an entry of 0 means the connection is removed. Note that the matrix is symmetric, so if cell i is connected to cell j, then cell j is also connected to cell i.

We test recall of a stored pattern by providing a recall cue in the form of external stimulation to a subset of the cells in a particular pattern, so that they become active. Network activity is then monitored to see if the remaining cells of the pattern subsequently become active and whether any non-pattern (spurious) cells also become active. Fifty patterns is a lot to store for this size of memory network, and errors in the form of spurious activity can be expected during pattern recall. The quality of recall is measured

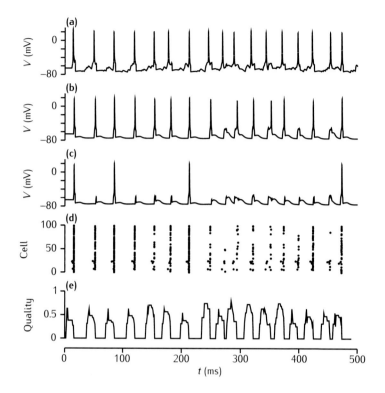

Fig 11.9 Network simulation based on the model of Sommer and Wennekers (2001). The top three traces are the time courses of the membrane potential of three excitatory neurons. **(a)** A cue cell. **(b)** A pattern cell. **(c)** A non-pattern cell. **(d)** A spike raster of 100 excitatory neurons. **(e)** The recall quality over time, with a sliding 10 ms time window. Excitatory synapses have an instantaneous conductance rise time, a decay time of 2 ms and a reversal potential of 0 mV, equivalent to the characteristics of AMPAR-mediated synapses. Inhibitory synapses have an instantaneous conductance rise time, but a decay time of 7 ms and a reversal potential of -75 mV, equivalent to GABA$_A$ receptor-mediated synapses. The excitatory connection weight (maximum AMPA conductance) is 6 nS and the inhibitory connection weight (maximum GABA$_A$ conductance) is 2 nS, with a 2 ms connection delay for all connections. An external stimulus consisting of a continuous 500 Hz Poisson-distributed spike train (representing the convergence of many presynaptic cells) is applied to four cells from one of the stored patterns, to act as a recall cue.

continuously by forming a binary vector defined by all cells that are active (fire action potentials) within the given time window (10 ms) and then calculating the scalar product of this vector against the stored pattern vector, normalised to the length of the recalled vector. A value of 1 results if the pattern is recalled perfectly; otherwise the quality is less than 1 (some pattern cells do not become active or some spurious cells are active).

In the example shown in Figure 11.9, the cued activity results in barrages of network activity roughly every 25 ms (within the gamma band range), with each barrage consisting of the cued cells, many of the remaining pattern cells and some spurious (non-pattern) cells. Close examination of each barrage reveals that the cued cells fire first, followed by the pattern cells, then the spurious cells. Hence, during a barrage, the recall quality rises in stages to a peak before falling back. Thus not only which cells fire on a gamma cycle, but also their phase of firing contains information about the stored pattern. If the strength of inhibition is reduced (not shown), many more spurious cells fire, but late in a gamma cycle, and in addition the cue and pattern cells start to fire in bursts, providing another distinguishing feature of the pattern. This bursting is an intrinsic characteristic of CA3 pyramidal cells, captured by the Pinsky–Rinzel model.

11.6 Plasticity for Homeostasis

Plasticity in the nervous system extends well beyond the synaptic plasticity underpinning learning and memory (Section 11.1). Structural and biochemical changes are needed for neurons and their networks to firstly develop and then to maintain their signal generation and processing characteristics. At the cellular level, this includes the development of morphology (soma, dendrites and the axon), along with the heterogeneous distribution of ion channels that gives a neuron its electrical excitability properties. Developing neurons are incorporated into complex neural networks involving the formation of synaptic connections between different populations of excitatory and inhibitory neurons. We look in detail at computational models of the structural development of neurons and their networks in Chapter 12.

We now consider models of synaptic plasticity and changes in cell physiology that implement **homeostatic plasticity**: the development and maintenance of appropriate neural electrical activity characteristics in the face of a changing environment (Davis, 2006; Marder and Goaillard, 2006; Turrigiano, 2011). Neural networks must be robust in their response to changes in an animal's external environment, to changes in brain state reflected in widespread neuromodulation, to changes in extracellular ion concentrations during ongoing electrical activity and turnover of the proteins that constitute synaptic and membrane-bound ion channels.

Homeostatic plasticity to compensate for such changes can come in a variety of forms (Davis, 2006; Marder and Goaillard, 2006; Turrigiano, 2011; Keck et al., 2017; Kamaleddin, 2022). **Synaptic homeostasis** can involve the scaling of excitatory synaptic strengths to a neuron to prevent saturation in the face of synaptic plasticity for learning and memory and also changes in inhibitory pathways through plasticity of excitatory synapses onto inhibitory neurons or plasticity of inhibitory synapses. **Intrinsic homeostasis** involves changes in the intrinsic excitability of individual neurons through membrane-bound ion channel expression.

Different homeostatic plasticity mechanisms are employed in different brain regions and across developmental stages, from prenatal through to adulthood (Spitzer et al., 2002; Turrigiano, 2011). At the cellular level, the mechanisms are mediated via calcium-driven signalling pathways (Spitzer et al., 2002; Davis, 2006; Marder and Goaillard, 2006), which regulate membrane-bound ion channel expression and synaptic strengths.

Driving signals for homeostatic plasticity must provide information about neuronal activity at the cellular and network levels. As with synaptic plasticity for learning and memory (Section 11.1), at the cellular level, membrane voltage, including spiking, or intracellular calcium concentration are suitable signals for indicating the activity profile of that cell, leading to cell-autonomous homeostasis. At the network level, average activity levels across at least a region of a network can be indicated by diffusing neuromodulators such as nitric oxide (NO). Homeostasis then regulates a neuron's activity on the basis of what is happening across the network, not just at the cell itself. This can also be achieved by cell-autonomous homeostasis working on inhibitory interneurons, rather than on the principal excitatory cells. These interneurons 'sample' network activity in surrounding neurons through their synaptic inputs. Below we present a number of example models of these different forms of homeostatic plasticity.

Homeostatic plasticity may have different outcomes, depending on brain region. Neural circuits, such as **central pattern generators** (CPGs), must generate stable rhythmic activity patterns in the face of perturbations, or even reconfigure to produce a different pattern, if signalled to do so (Marder and Taylor, 2011). Homeostatic plasticity must act to robustly maintain the required firing characteristics of these neural networks and their constituent neurons. In contrast, in the cortex, neurons and their networks must maintain responsiveness to the external signals they receive, to enable suitable processing of such signals and their potential storage and retrieval (Turrigiano, 2011). Thus homeostatic plasticity must now act to maintain the **dynamic range** of neural activity, rather than constrain it to particular levels.

The success of these different forms of homeostatic plasticity relies on the considerable **degeneracy** to be found in the nervous system. At the level of individual neurons, the rich tapestry of membrane-bound ion channels results in many solutions being available to the problem of generating particular patterns of electrical activity (Marder and Goaillard, 2006; Marder and Taylor, 2011; Kamaleddin, 2022). Considerable experimental evidence points to the nervous system making use of such potential variability, and models of homeostatic plasticity demonstrate that such mechanisms may find different solutions depending on the perturbation being compensated for. Degeneracy can be regarded as both a problem and a feature when trying to optimise a neural model against particular data: there is unlikely to be a unique set of parameter values, including ion channel densities and synaptic strengths, that result in the model matching the data. We consider issues of model parameter uniqueness and sensitivity in Chapter 14. Thus degeneracy, along with homeostatic plasticity, also suggests an alternative strategy for computational neuroscientists studying neurons and neural networks. Instead of searching for model parameters for synaptic weights and neuronal ion channel distributions that assure the desired activity profiles, one may introduce homeostatic plasticity rules that allow the neuron or network to evolve into the desired state by itself. Then the challenge for the modeller has shifted to specifying appropriate plasticity rules and tuning their parameters.

11.6.1 Homeostasis at the Cellular Level

Operating at the level of a single neuron, homeostatic plasticity acts to develop and maintain desirable neuronal electrical activity characteristics in a cell-autonomous fashion (Figure 11.10). The effect of homeostatic plasticity may be either to restore a neuron's firing pattern or, alternatively, to match a neuron's electrical responsiveness to the range of inputs it receives. We consider several modelling examples that explore these possibilities via homeostatic changes in intrinsic cell excitability, as determined by ion channel densities, or changes in synaptic strengths.

Developing Faithful Signal Transmission

Arguably a function of the active membrane properties of a neuron is to carry out faithful signal transmission along the dendrites and axon. Incoming synaptic signals need to be propagated to the cell body and then a subsequent action potential needs to travel along an extensive branching axon to reach all of a neuron's synaptic targets.

The early model of Bell (1992) specified an '**anti-Hebbian' learning rule** to generate such faithful signal transmission through changing the maximum conductances of ion channels to minimise voltage fluctuations between local

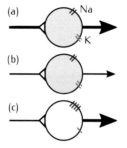

Fig 11.10 Intrinsic plasticity to restore firing rate. **(a)** Strong input results in strong output. **(b)** Weak input leads to weak output. **(c)** Homeostatic plasticity increases cell excitability, in this case by increasing the number of sodium channels and decreasing the number of potassium channels. This restores a strong output firing rate for a sustained weakening of the input.

patches of membrane. The rule is derived from the membrane voltage equation along an active cable, which can be written as:

$$G_a \frac{\partial^2 V}{\partial x^2} = C_m \frac{\partial V}{\partial t} + \sum_k \bar{g}_k g_k (V - E_k) + \sum_s \bar{g}_s g_s (V - E_s), \qquad (11.9)$$

where G_a is the axial conductance and the currents on the right-hand side are divided into the capacitive current, intrinsic ion channel currents (indexed by k) and synaptic currents (indexed by s). The channel and synaptic currents are determined by a conductance that is divided into a fixed maximum $\bar{g}$, multiplied by a normalised time-varying component g. This equation corresponds to Kirchhoff's current law (Section 2.6) in that it describes the current flowing out of a segment of cable as being equal to the sum of the capacitive, ionic and synaptic currents into that segment. So we can write the cable equation in the simpler form of:

$$I_a = \sum_k \bar{g}_k i_k + \sum_s \bar{g}_s i_s, \qquad (11.10)$$

where $I_a = G_a \partial^2 V / \partial x^2$ and we include the capacitive current in the summed intrinsic currents. Written in this form, the cable equation leads to a basic correlational learning rule for maximum conductances:

$$\Delta \bar{g}_k = \epsilon i_k I_a, \qquad (11.11)$$

where a change in maximum conductance for ion channel species k is the product of the channel current i_k, the axial current I_a in the segment of cable and a learning rate ϵ. If the learning rate is negative, then this specifies a rectifying 'anti-Hebbian' learning rule which will try to flatten voltage disparities.

Amongst other examples, this model was used to demonstrate the development of propagation of the classic electrical signal, the action potential, along an axon (Bell, 1992). A compartmental model of an axon containing leak, calcium and potassium conductances was simulated, with conductances set to produce a calcium-mediated action potential at one end of the axon. This homeostatic learning rule was then used to adjust the active conductances, resulting in the ion channel densities in the neighbouring membrane changing to also generate an action potential and so minimise the difference in voltage between adjacent compartments. Final calcium and potassium conductances along the axon eventually matched those in the initial compartment.

Maintaining Firing Characteristics

We now consider examples of homeostasis that act to return a neuron to its normal intrinsic firing state in the face of perturbations to its environment. This state can be complex, ranging from regular spike firing to burst firing of particular frequency and duration.

Multi-sensor Model

A suitable and flexible homeostatic plasticity rule for this is of the form (Liu et al., 1998; Abbott et al., 2003):

$$\tau \frac{d\bar{g}_i}{dt} = \sum_a B_{ia} (\bar{S}_a - S_a) \bar{g}_i, \qquad (11.12)$$

where the maximum conductance $\bar{g}_i$ of a particular ion channel species i is adjusted with time constant τ in proportion B_{ia} to the offset of a number of signal sensors S_a from their setpoint value $\bar{S}_a$. To maintain electrical excitability, if the signal sensors indicate that activity is high, then ionic conductances should be adjusted to reduce excitability, and vice versa if activity is low. Thus the magnitude and sign of each B_{ia} will depend on the type of its ion channel.

The sensors employed in this rule can be functions of calcium concentration, with multiple sensors having different dynamics to characterise calcium transients on different timescales (Liu et al., 1998). One simple sensor may provide an average calcium concentration over a reasonably long time interval, in spiking activity terms, of 1 second or more. This will give an indication of the average cell firing rate, but not of the precise pattern of this firing (Liu et al., 1998). More detail about, say, burst firing, can be obtained with sensors with dynamics in the millisecond to tens of milliseconds range that can detect fluctuations in calcium concentration, and not just its mean value. Hodgkin–Huxley-style (in)activation functions are suitable for these sensors (Liu et al., 1998), but with the time constants and amplitudes of sensor (in)activation as functions of the local calcium concentration, rather than membrane voltage.

Homeostasis operates over longer timescales (hours or days) than the typical neuronal electrical response. Thus the time constant τ in the homeostatic plasticity rule is much larger than the sensor time constants.

Liu et al. (1998) used this rule in a model of how a stomatogastric ganglion neuron can maintain particular firing rate patterns in the face of environmental perturbations. In particular, cultured neurons isolated from this ganglion develop intrinsic burst firing capabilities. They modelled this behaviour using a single compartment neuron model containing eight Hodgkin–Huxley-style ion channel types: fast sodium (Naf), delayed-rectifier potassium (KDR), fast calcium (CaT), slow calcium (CaS), calcium-activated potassium (KCa), A-current, hyperpolarisation-activated h-current and a leak current. All but the leak current were subject to homeostatic plasticity.

They found that three sensors that reacted to fast (milliseconds), slow (tens of milliseconds) and near steady-state (seconds) changes in submembrane calcium concentration were sufficient to detect changes in the average firing rate and spiking patterns, from regular spiking to bursting.

Suitable choices of sensor setpoint $\bar{S}_a$ and conductance alteration per sensor B_{ia} resulted in neurons that were highly robust in their firing patterns, returning to stable bursting behaviour in the face of perturbations. In summary, if the fast sensor was below setpoint, Naf and KDR were increased; if the slow sensor was below setpoint, CaS, CaT and the h-current were increased whilst KDR, KCa and the A-current were decreased; and for a low steady state, KCa and the A-current were decreased whilst the h-current was increased. Due to degeneracy, variations in final channel densities were seen across trials from different starting points, though the same bursting characteristics were restored.

The signals driving homeostasis are detected within seconds, but plasticity outcomes can take hours. Simulation times can be minimised by making relatively rapid changes in ion channel densities in response to a signal averaged over a short period of activity. This assumes that the averaging is sufficient to give an accurate description of the state of the neuron and that it is not necessary to model the precise time course of the adaptation. The major constraint is that adaptation should be slower than signal detection. Liu et al. (1998) were able to make conductance updates with a time constant of $\tau = 5$ seconds.

Multi-expression Model

O'Leary et al. (2014) derived an alternative homeostatic rule for maintaining firing characteristics that has a direct biological interpretation, being based on a simple model of gene regulation. Their rule adjusts ion channel conductances to maintain just a single target, which is the average calcium concentration. Now, to achieve specific dynamic membrane characteristics, different channel conductances are adjusted at different rates. This is based on the known variations in protein expression rates in neurons. The learning rule for changing the maximum conductance $\bar{g}_i$ of ionic species i can be written as:

$$\tau_g \frac{d\bar{g}_i}{dt} = m_i - \bar{g}_i, \tag{11.13}$$

$$\tau_i \frac{dm_i}{dt} = [\text{Ca}^{2+}] - \text{Ca}_\text{T}, \tag{11.14}$$

where Ca_T is the target intracellular calcium concentration. The variable m_i represents the expression rate of the protein for channel type i as a function of the variation of the calcium concentration from its target value. Time constant τ_g is a characteristic time constant for channel expression, which is assumed to be the same for all channel types. Time constant τ_i specifies the coupling rate between expression and calcium concentration, and is particular for each channel type. Note that values for τ_i can be either positive or negative, depending on whether a channel type should be up- or downregulated for calcium concentrations above the target value.

From this rule, it can be deduced that the maximum conductances of different types of ion channel will be correlated according to the ratio of their regulation time constants as $\bar{g}_i/\bar{g}_j \approx |\tau_j/\tau_i|$. Such correlations are seen experimentally (O'Leary et al., 2014) and these ratios determine the neuronal dynamics.

This model was tested by O'Leary et al. (2014) on the same set of conductances as Liu et al. (1998). The regulation time constants τ_i were chosen by initially hand-tuning a model cell to produce the desired behaviour, in this case burst firing. Then the time constant values were set according to the conductance ratios of the seven active channel types in the model. Starting with a quiescent cell and applying this homeostatic rule, the model robustly learnt stable bursting behaviour. Figure 11.11 shows this for our implementation of O'Leary et al.'s (2014) homeostasis rule using the *NEURON* simulator, which is comparable with Figure 3 in O'Leary et al. (2014). Starting with the same small value for all the ionic conductances, the cell initially remains near rest potential (Figure 11.11a), whilst the ionic conductances rapidly separate

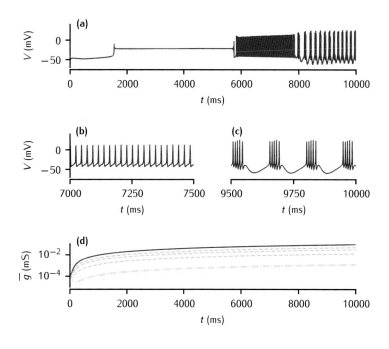

Fig 11.11 Development of cell bursting behaviour with the homeostasis model of O'Leary et al. (2014). **(a)** Membrane potential during 10 s of homeostasis. **(b)** Expanded view of single spiking mode. **(c)** Expanded view of bursting mode. **(d)** Maximum conductance values subject to homeostasis: top line (solid) – Na; next three lines (dashed) – K channels; next two lines (dot-dashed) – Ca channels; bottom (dotted) – h-current.

in value (Figure 11.11d). The conductances then start to grow in fixed proportion, moving the membrane potential firstly to a new steady state near the h-current reversal potential before entering the single spiking mode (Figure 11.11b), which then migrates to the final bursting mode (Figure 11.11c). O'Leary et al. (2014) also demonstrate that suitable choices for the regulation time constants enable the cell to learn a variety of active behaviours, including Type I and Type II spiking (Section 4.4.2), rebound bursting and regular spiking.

As we will discuss below, this type of cell-level intrinsic homeostasis model can be used in a network setting to determine activity across multiple, connected neurons.

Regulating Firing Activity

Firing activity at the cell level can also be regulated through synaptic plasticity. Plasticity for learning and memory, such as implementing Hebbian learning, can lead to an average increase or decrease in the strengths of excitatory synaptic input to a neuron, thus distorting its average activity level. This can be compensated for through scaling of all excitatory synaptic strengths to maintain a constant average strength across the synaptic population (Turrigiano, 2011).

As described in Section 8.5.3, cortical neurons typically exhibit a strong variability in their firing pattern. This variability can be reproduced in models where the neurons receive, on average, the same amount of excitatory and inhibitory inputs, that is, they receive balanced inputs. It has been demonstrated that homeostatic adjustment of inhibitory synapses can automatically tune a neuron to achieve a balance between its excitatory and inhibitory inputs (Vogels et al., 2011; Luz and Shamir, 2012). This homeostasis can be achieved through an STDP rule that operates on inhibitory, rather than excitatory, synapses. The rule proposed by Vogels et al. (2011) maintains an activity trace x in both pre- and postsynaptic neurons, in which x is incremented at each neuronal spike time t_s and then decays with time constant τ^{STDP}:

$$\tau^{\mathrm{STDP}} \frac{dx}{dt} = -x + \delta(t - t_s). \tag{11.15}$$

The weight between a presynaptic inhibitory neuron and its postsynaptic target is then updated on each pre- and postsynaptic spike according to:

$$\begin{aligned} \Delta w(t_s^{\mathrm{pre}}) &= \eta(x^{\mathrm{pre}} - \alpha), \\ \Delta w(t_s^{\mathrm{post}}) &= \eta x^{\mathrm{post}}, \end{aligned} \tag{11.16}$$

with learning rate η and depression factor α. The depression factor determines the preferred firing rate of the postsynaptic neuron.

In trials with a model spiking neuron receiving a large number of excitatory and inhibitory inputs that were grouped into different signal channels, this STDP rule created a detailed balance between excitation and inhibition, in which the neuron exhibited sparse, irregular firing, but with transient elevated rates at the onset of preferred signals (Vogels et al., 2011). The detailed balance relied on suitably fast correlations between excitatory and inhibitory inputs in each signal channel. Such a detailed balance was also created in a recurrent network of neurons in which patterns had been embedded via Hebbian learning at connections between excitatory neurons. Transient recall of an embedded pattern could be elicited by elevated input to a subset of the excitatory neurons in such a pattern.

Maintaining Dynamic Range

Another possibility is that homeostatic plasticity rules drive ion channel distributions to match a cell's electrical excitability to the range of synaptic input it receives, so that its output firing rate encodes an input strength. This contrasts with homeostatic principles which operate to maintain a firing rate in the face of changing inputs.

This can be formalised in a plasticity rule designed to maximise information transfer between strengths of synaptic input and the cell output firing rate (Stemmler and Koch, 1999). If such a rule acts on the conductance magnitudes of ion channels, then the aim is to increase the conductance of those channel types contributing to electrical excitability and whose voltage activation range most closely matches the membrane depolarisation range provided by the most common synaptic inputs. This will result in the neuron being most responsive over the range of synaptic strengths it is receiving, thus providing the best firing rate coding of its inputs. Such a rule is (Stemmler and Koch, 1999):

$$\Delta \bar{g}_i = \frac{\eta}{\bar{g}_i} \left\langle \frac{\delta I_i}{\delta V(t)} + c(\langle V \rangle) I_i \right\rangle, \tag{11.17}$$

where the maximum conductance $\bar{g}_i$ of a particular ion channel species i is adjusted by a fractional amount $\Delta \bar{g}_i$ at time t, with a learning rate η. The amount of change is determined by a function of the variational derivative of the ionic current I_i with respect to the instantaneous local membrane voltage V plus a constraining function c of the average voltage.

Stemmler and Koch (1999) demonstrated the efficacy of this rule using a two-compartment model neuron containing a variety of sodium, potassium and calcium currents. Applying this plasticity rule to the potassium and calcium currents in the dendritic compartment, where membrane voltage is a combination of synaptic inputs and back-propagating somatic action potentials, enabled the neuron to self-tune to the amplitude range of the inputs. Note that the rule is applied to each ion channel type independently.

This ability to maintain the dynamic range of a neuron is particularly important in neural networks where synaptic plasticity for learning affects the range of synaptic input strengths seen by different neurons in the network. We consider computational modelling work addressing this scenario below.

11.6.2 Homeostasis at the Network Level

Neurons are not isolated entities but operate within networks, so homeostatic plasticity, even if operating locally at a cell level, will impact activity across a network of neurons. Considering homeostatic plasticity at the neural network level also opens up the range of possible mechanisms through which such plasticity may be implemented. The models we have considered so far have explored homeostasis acting autonomously on individual excitatory neurons. Such homeostasis can still form part of network plasticity and we look below at models that explore its effects. However, instead of being driven by activity levels in an individual neuron, homeostatic plasticity can also be driven by activity levels measured across a network. This can work through the agency of an activity-dependent diffusive signal, such as NO, or through plasticity in inhibitory neurons whose activity represents a sample of excitatory activity in the surrounding network. We also look below at models that explore these possibilities.

An important feature of homeostasis at the network level is its interaction with synaptic plasticity for learning and memory (Box 11.5). Homeostasis must both regulate network activity whilst maintaining the memory traces imprinted through synaptic plasticity. This has implications for the form of homeostatic learning rules as well as the dynamics of homeostatic plasticity (Keck et al., 2017; Zenke et al., 2017).

Network Homeostasis and Firing Patterns

Intrinsic homeostasis of ion channel densities can also produce appropriate activity patterns at the network level. In the network context of the crustacean stomatogastric ganglion, intrinsic homeostasis can restore rhythmic activity across the network following perturbation (Golowasch et al., 1999; O'Leary et al., 2014). The individual neuronal active properties that result in the network are not the same as those created through homeostasis in the isolated neurons (Golowasch et al., 1999), since calcium-based activity sensors for homeostasis in the network include activity measured from synaptically connected cells.

Alternatively, it has been demonstrated that, for fixed individual neuronal properties, plasticity of the inhibitory connections between the

Box 11.5 | Homeostasis versus learning and memory

A key issue to which computational modelling is playing a major role in helping our understanding is how synaptic plasticity, underpinning learning and memory, and homeostatic plasticity, underpinning neural activity regulation, interact (Keck et al., 2017; Zenke et al., 2017). Hebbian learning, as underpins associative memory (Section 11.5), is inherently unstable: when pre- and postsynaptic neurons are co-active, their synaptic connection is strengthened, leading to a positive feedback loop of increased activity. Some stabilising influences are needed to prevent runaway activity in these neural networks. Homeostatic plasticity in principle provides suitable mechanisms for stabilisation, but there are issues with the apparent slowness of homeostasis compared with synaptic plasticity: the negative feedback provided by homeostasis acting on timescales of hours to days cannot prevent runaway activity due to changes in synaptic strength happening at millisecond to second timescales. Models demonstrate that suitably fast stabilising mechanisms can maintain network activity levels whilst allowing imprinting of memory patterns through synaptic plasticity. More experimental and modelling work is required to identify and understand the properties of such rapid compensatory processes (Zenke et al., 2017) and the full variety and roles of homeostasis in the nervous system.

neurons in the stomatogastric ganglion can also implement homeostasis of burst firing patterns (Soto-Treviño et al., 2001; Abbott et al., 2003). Similarly to Vogels et al. (2011), this plasticity rule depends on both pre- and postsynaptic activity levels, but sensed through presynaptic voltage and postsynaptic calcium current and concentration, rather than through spike times, as in Vogels et al. (2011).

Network Homeostasis and Balanced Networks

Section 9.3.1 describes a network of integrate-and-fire neurons that exhibits a balanced state with irregular and asynchronous firing, obtained by appropriately tuning the magnitudes of synaptic weight parameters. Vogels et al. (2011) demonstrated that cell-autonomous homeostatic plasticity of inhibitory synapses can achieve such a globally balanced network state. Now we consider models that demonstrate the use of network-wide activity signals to drive homeostatic plasticity for maintaining a detailed balance in a neural network. This means that the average activity level across a network is regulated in such a way that a range of activity levels for individual neurons is maintained, rather than all neurons being driven to the same average firing rates. This is important for the network responding to diverse input signals and for imprinting and recalling memory patterns.

Network Homeostasis via Nitric Oxide

A diffusive signal, produced by neurons in proportion to their activity, should be a suitable driver of homeostatic plasticity across a network, providing each neuron with a measure of average activity across their neighbouring network (Figure 11.12). Sweeney et al. (2015) studied computationally the diffusive neuromodulator NO as a network-wide driving signal for homeostatic plasticity. Neurons may produce NO as a function of their activity level and so act as point sources of NO, which may diffuse on the order of 100 μm. NO acts to modulate the intrinsic excitability of neurons, principally through the modulation of potassium conductances.

Sweeney et al. (2015) modelled NO synthesis and diffusion across a network of integrate-and-fire model neurons (80% excitatory; 20% inhibitory) laid out on a 2D toroidal surface. The firing threshold of individual neurons was adjusted by an NO-dependent plasticity rule, to mimic the effect of changes in intrinsic conductances. Each neuron produced nitric oxide synthase (nNOS) as a function of the intracellular calcium level induced by spiking activity, described by the equations:

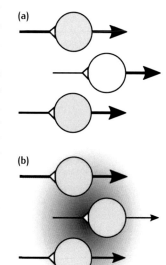

$$\frac{d[Ca^{2+}]}{dt} = -\frac{[Ca^{2+}]}{\tau_{Ca}} + [Ca^{2+}]_{spike}\,\delta(t_{spike}), \tag{11.18}$$

$$\frac{d[nNOS]}{dt} = -\frac{1}{\tau_{nNOS}} \frac{[Ca^{2+}]^n}{[Ca^{2+}]^n + K^n}, \tag{11.19}$$

where, for this study: $[Ca^{2+}]_{spike} = 1$ (unitless); $\tau_{Ca} = 10\,\text{ms}$; $\tau_{nNOS} = 100\,\text{ms}$; $n = 3$; $K = 1$. Each neuron then acted as a point source of NO, with a concentration proportional to their nNOS level. NO concentration across the network surface was determined by free diffusion:

$$\frac{d[NO]}{dt} - D\nabla^2[NO] = [nNOS] - \lambda[NO], \tag{11.20}$$

with $D = 1000\,\mu\text{m}^2\,\text{s}^{-1}$ and $\lambda = 0.1\,\text{s}^{-1}$.

A neuron adjusted its firing threshold θ_i as a function of its local NO concentration via:

$$\frac{d\theta_i}{dt} = \frac{1}{\tau_{hom}} \frac{[NO] - [NO]_0}{[NO]}, \tag{11.21}$$

where $\tau_{hom} = 2500\,\text{ms}$ was the time constant for homeostasis. Thus the firing threshold increases for [NO] above a target value $[NO]_0$.

Sweeney et al. (2015) demonstrated that this mechanism was capable of maintaining stable average firing rates across the network, whilst at the same time allowing considerable heterogeneity in the firing rates of individual neurons. This was needed for the establishment and maintenance of memory patterns stored via Hebbian synaptic plasticity. If it was assumed that NO did not diffuse, then the resultant purely local (individual neuron level) homeostasis, on the other hand, produced regulated and uniform firing rates across the network.

Fig 11.12 **(a)** Non-diffusive intrinsic homeostasis restores firing rates through individual changes in cell excitability. **(b)** Diffusive homeostasis establishes similar cell excitability in a local area, thus maintaining diverse, input-dependent firing rates.

Network Homeostasis via Inhibitory Plasticity

In a similar study, Kaleb et al. (2021) considered input-dependent inhibitory plasticity (IDIP) as a network-wide homeostatic mechanism in networks of integrate-and-fire neurons. Inhibitory neurons scaled their outgoing synaptic weights as a function of the summed input activity they received (Figure 11.13). If the input activity is above a target value, then the inhibitory weights were increased in an attempt to decrease network activity levels. A

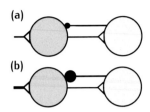

Fig 11.13 (a) Excitatory cell receiving recurrent inhibition. **(b)** Inhibitory homeostasis increases the inhibitory connection weight to restore local network excitatory firing levels if they were increased due to stronger stimuli.

filtered trace x_j of the synaptic input received by inhibitory neuron j was maintained via:

$$\tau_{\text{IDIP}}\frac{\mathrm{d}x_j}{\mathrm{d}t} = -x_j + \sum_i g^{\text{E}}_{ij}(t), \tag{11.22}$$

where g^{E}_{ij} was the conductance time course of the synaptic input from excitatory neuron i and $\tau_{\text{IDIP}} = 160$ ms determined the length of time over which inputs could influence homeostasis. As a result, every time the inhibitory neuron j spiked, the strength of each of its output synapses was changed via:

$$\Delta w_{jk} = \eta_{\text{IDIP}}(x_j(t) - \theta_{\text{IDIP}})\delta(t_{\text{spike}}), \tag{11.23}$$

where w_{jk} was the weight of the inhibitory synapse onto excitatory neuron k; η_{IDIP} determined the learning rate, and θ_{IDIP} the direction of change.

Amongst a number of variants of this rule, they also considered changing the inhibitory neuron firing threshold θ_i, rather than the synaptic weights:

$$\Delta\theta_i = \eta_{\text{IDIP}}(\theta_{\text{IDIP}} - x_j(t))\delta(t_{\text{spike}}). \tag{11.24}$$

Now if input activity was above the target, the firing threshold was decreased to increase inhibitory neuron firing and decrease overall network activity.

As with Sweeney et al. (2015), this network-level homeostasis was able to regulate network activity levels whilst at the same time allowing heterogeneity in individual neuron firing rates due to locality of inhibitory feedback.

11.7 | Summary

Computational modelling is making very valuable contributions to our understanding of the mechanisms and dynamics of plasticity, in all its forms. In this chapter, we have concentrated on plasticity in the adult nervous system. In Chapter 12, we consider the structural changes in cell morphology and neural network architecture taking place during development.

Synaptic plasticity is thought to be the major mechanism underpinning learning and memory in the brain. Modelling is helping to understand the relationship between pre- and postsynaptic activity patterns and associated changes in the strength of the connection between two neurons, known as long-term potentiation (LTP) and long-term depression (LTD). Much experimental data can be captured by synaptic learning rules which map the timing of pre- and postsynaptic spikes to changes in synaptic strength. This is known as spike timing-dependent plasticity (STDP). Other models try to capture more of the biophysical mechanisms underpinning weight change by modelling the calcium-driven postsynaptic signalling pathways that lead to alteration in AMPAR expression and conductance.

Synaptic learning rules also can be reduced to simple formulae that can be used in neural network models to help elucidate the mechanisms of learning and memory in the brain. To give an introduction to this, we have presented models of associative memory, in which Hebbian learning is used to enable the storage and retrieval of patterns of activity. The field of artificial neural networks (ANNs) is inspired by brain architecture and plasticity, but has largely taken an engineering approach to building networks and learning rules to solve difficult problems in pattern recognition and classification and other applications. We have provided a brief overview of ANNs and their relationship to neuroscience.

In parallel with synaptic plasticity for learning and memory, other forms of plasticity implement homeostasis that regulates firing patterns in neural networks in the face of environmental changes. Modelling is needed to understand what homeostatic plasticity learning rules may apply and what their effects may be. We have considered a number of examples here, involving neuronal intrinsic plasticity as well as synaptic plasticity. Modelling is also helping to understand how synaptic plasticity for learning and memory and plasticity for homeostasis can work together.

CHAPTER 12

Development of the Nervous System

So far we have been discussing how to model accurately the electrical and chemical properties of neurons and how these cells interact within the networks of cells forming the nervous system. The existence of a correct structure is essential for the proper functioning of the nervous system, and we now discuss modelling of the development of the nervous system. Most existing models of developmental processes are not as widely accepted as, for example, the Hodgkin–Huxley model of nerve impulse propagation. They are designed on the basis of usually unverified assumptions to test a particular theory for neural development. Our aim is to cast light on the different types of issues that arise when constructing a model of development through discussing several case examples of models applied to particular neural developmental phenomena. We look at models constructed at the levels of individual neurons and of ensembles of nerve cells.

A variety of different processes are involved in the generation of the nervous system, from embryonic development to adulthood. This chapter concentrates on the principles used in modelling the development of specific neural structures. Some of the same mechanisms of plasticity that we discuss may apply to adult systems, described in Chapter 11.

Modelling in developmental computational neuroscience is usually restricted to models at the levels of single neurons or networks of neurons, other levels being investigated within other disciplines. For example, computational theories of neural precursors, or stem cells, are not considered, although they could be; the elegant mathematical treatment of morphogenetic fields, providing possible mechanisms by which continuous gradients of molecules across the brain called **morphogens** can be read out to specify regions of the brain in early development (Turing, 1952; Meinhardt, 1983; Murray, 1993), is regarded as the province of theoretical biology rather than of computational neuroscience.

Many of the subcellular mechanisms underlying the understanding of neural development are not understood and so strong assumptions have to be made. Different modellers make different assumptions, leading to very few widely accepted theories for how the machinery of development operates at the level of detail needed to construct a useful computational simulator, as has been done for compartmental modelling with simulators such as *NEURON*. The developmental mechanisms hypothesised may be driven by neural activity, or by molecular signalling, or by a combination of the two.

In this chapter, we describe how computational modelling is applied to cases where there has been a long period of model development. This will illustrate how modellers develop their ideas in response to new data or to other models. One important question is whether a developmental model is intended to model the developmental process, as in most of the examples described here, or merely constructs the adult configuration. Other issues are concerned with the interpretation of available data, making plausible assumptions about the underlying processes and model comparison.

The development of the nervous system occurs after a complex series of developmental steps, many of which are common to the development of very different multicellular organisms.

The fertilised egg goes through a series of cell divisions and rearrangements, ultimately forming several layers of cells. The layers give rise to the various organs of the body, including the nervous system. Amongst the stages of development are:

Cell division. A large collection of cells is generated from the fertilised egg.

Gastrulation. These cells are rearranged into three layers. The inner layer (endoderm) forms the gut and associated organs; the middle layer (mesoderm) forms muscle, cartilage and bone, as well as the notochord, the precursor of the vertebral column; and the outer layer (ectoderm) forms the epidermis, the outer layer of the body, and the neural plate from which the nervous system develops.

Neurulation. Lying along the dorsal surface of the embryo, the edges of the neural plate fuse together to form the neural tube. At the same time, so-called neural crest cells migrate from the epidermis to the mesoderm to form the peripheral nervous system.

Development of the nervous system. In vertebrates, the rostral end of the neural tube enlarges to form the three primary vesicles of the brain, the remainder of the neural tube giving rise to the spinal cord. The retina also develops from the neural tube and the other sensory organs are formed from thickenings of the ectoderm. Formation of the individual structures of the brain involves a combination of accumulation of a large population of neurons through cell division, the migration of these cells, a significant amount of cell death and the formation and rearrangement of nerve connections.

There are many standard texts describing development, amongst them being those by Gilbert (1997) and Wolpert et al. (2002). There are very few texts specifically on neural development, exceptions being those by Sanes et al. (2000), Price and Willshaw (2000) and Price et al. (2011).

A representative collection of topics covered in modelling neural development can be found in van Ooyen (2003). A later review is by Goodhill (2018).

In chronological order of development, the research problems discussed here are:

The development of neuronal morphology. The development of specific shapes of neurons, with respect specifically to the complex dendrites (Section 12.1).

The development of pattern within a set of neurons. How individual ganglion cells come to be positioned within the retina (Section 12.2).

The development of patterns of nerve connections. Three examples are discussed:

- Development of the characteristic pattern of neuromuscular innervation in which each muscle fibre develops contact with a single axonal branch (Section 12.3).
- How individual neurons in the visual cortex come to respond to specific types of visual stimuli. Here we examine modelling of the development of patterns of **ocular dominance** in the visual cortex (Section 12.4). Another application frequently addressed by modellers is the development of maps of orientation selectivity. This is not discussed here and relevant references are Ferster and Miller (2000), Priebe (2016) and Goodhill (2018).
- The development of retinotopically ordered maps of connections in vertebrates (Section 12.5).

12.1 Neuronal Morphology

The first example to be considered is the development of the characteristic shape of a neuron, where there are good morphological data available. From birth to adulthood, a neuron changes from being a typically round cell into a complex branched structure of dendrites and an axon, which have distinct properties and functions. The cell's morphological development can be characterised into a number of stages (Figure 12.1):

(1) Neurite initiation;
(2) Neurite differentiation;

Fig 12.1 Three stages of neurite development: (1) initiation; (2) differentiation into an axon and dendrites; and (3) elaboration, including elongation and branching.

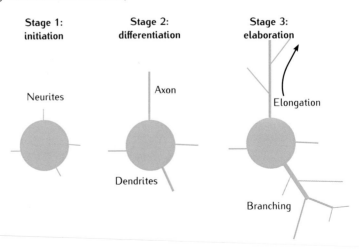

Stage 1: initiation

Stage 2: differentiation

Stage 3: elaboration

Neurites

Axon

Dendrites

Elongation

Branching

(3) Neurite (axon and dendrites) elaboration:
- Elongation and branching;
- Axon pathfinding;
- Dendrite space filling.

As an example of the use of mathematical modelling and computer simulation to study neuron development, we consider models of the morphological development of neurites in stage 3. A comprehensive overview of modelling the various stages of neuron development can be found in van Ooyen (2003). A summary of different modelling approaches and their technical considerations is provided in Kiddie et al. (2005) and Graham and van Ooyen (2006).

A model of neurite development will seek to capture aspects of the neurite's morphological development, including elongation, branching and changes in the diameter of neurite segments over time. As introduced in Section 5.5, such characteristics were formalised by Hillman (1979) who specified seven fundamental parameters that define the morphology of neurites. These are listed in Box 12.1 and were illustrated in Figure 5.21. A model of neurite development may seek to reproduce all or only some of these quantities, depending on the purpose of the model. For example, the emphasis might be on the topology of the growing neurite, captured by the length of segments and their branching structure, but without concern for their orientation in 3D space.

In contrast to the construction algorithms considered in Section 5.5, which construct a neurite model for a particular point in time, models of neurite development seek to capture how neuritic morphology evolves with time. Such **growth algorithms** (van Pelt and Uylings, 1999) may still be statistical, like the construction algorithms, but now need to specify how the distributions of the fundamental parameters change over time. Two approaches have been followed. The first tries to formulate the simplest possible description of the elongation and branching rates that generates trees with realistic morphology (Berry and Bradley, 1976; van Pelt and Verwer, 1983; Ireland et al., 1985; Horsfield et al., 1987; Nowakowski et al., 1992; van Pelt and Uylings, 1999). The parameters of such models may indicate dependencies of these rates on developmental time or tree outgrowth, without identifying particular biophysical causes. The second approach tries to describe branching and elongation rates as functions of identifiable biophysical parameters (Hely et al., 2001; Kiddie et al., 2005). We look at examples of both approaches.

Box 12.1 | Fundamental shape parameters of the neurite

(1) Stem branch diameters, P: parent; C: child; S: sibling;

(2) Terminal branch diameters;

(3) Branch lengths;

(4) Branch taper: ratio of branch diameter at its root to the diameter at its distal end;

(5) Branch ratio between diameters of daughter branches, S:C;

(6) Branch power e, relating the diameter of the parent to its daughters, $P^e = C^e + S^e$;

(7) Branch angle between sibling branches.

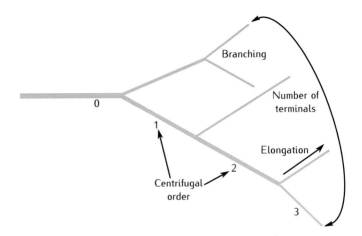

Fig 12.2 The BESTL growth algorithm. A terminal segment's branching probability depends on its centrifugal order and the number of terminals in the tree. Terminal segment elongation rates may differ between the initial branching phase and a final, elongation-only phase.

12.1.1 Statistical Modelling of Neurite Growth

An example of the first approach that captures successfully much of the growth dynamics of a wide variety of dendrite types is the **BESTL algorithm** (van Pelt and Uylings, 1999; van Pelt et al., 2001, 2003). This algorithm aims to reproduce the branching structure and segment lengths of dendrites (Figure 12.2). It does not consider diameters, though a Rall-type power rule (Section 5.4.4) can be used to add diameters to segments. Only terminal segments are assumed to lengthen and branch, with branching events resulting in bifurcations. Segment branching and elongation are handled as independent processes specified by a branching rate and an elongation rate.

The branching rate of each terminal segment j is (van Pelt et al., 2003):

$$p_j(t) = D(t)C(t)2^{-S\gamma_j}n(t)^{-E}, \tag{12.1}$$

where

$$C(t) = n(t)/\sum_{j=1}^{n(t)} 2^{-S\gamma_j}, \tag{12.2}$$

$D(t)$ is the basic branching rate at time t;
$C(t)$ is a normalising factor;
$n(t)$ is the number of terminal segments;
γ_j is the centrifugal order of terminal segment j (Figure 12.2);
S is a constant determining the dependence of branching on centrifugal order;
E is a constant determining the dependence of branching on the number of terminals.

Elongation is handled independently of branching. Following a branching event, the algorithm proceeds by giving each daughter branch an initial, short length and an elongation rate, both drawn from gamma distributions (van Pelt et al., 2003). Elongation may continue after branching has ceased, with developmental time being divided into an initial branching phase followed by an elongation-only phase. Elongation rates in the latter phase may be different from the branching phase (van Pelt and Uylings, 1999).

A discrete-time version of this algorithm is summarised in Box 12.2. The basic branching rate is calculated as $D(t) = B/T$ for a constant branching parameter B over developmental time T. Example dendrograms taken at

Box 12.2 | The BESTL algorithm

Divide developmental time T into N time bins. For each time bin Δt_i ($i = 1$ to N):

(1) During the initial branching phase:
    (a) For each terminal j, calculate the probability of branching within the time interval t_i to $t_i + \Delta t_i$ from Equation 12.1:

$$p_j(t_i)\Delta t_i.$$

    (b) If branching occurs, add new daughter branches, with given initial lengths and elongation rates drawn from gamma distributions.
    (c) Lengthen all terminal branches according to their branching phase elongation rate l_b: $\Delta l = l_b\Delta t_i$.
(2) During the final elongation phase, lengthen all terminal branches according to their elongation-only phase elongation rate l_e: $\Delta l = l_e\Delta t_i$.

increasing time points during a single run of the algorithm are shown in Figure 12.3.

The small number of parameters in the model enables parameter values to be optimised over experimental data sets. This is done largely using data from adult dendrites. The algorithm is run many times to produce distributions of the number of terminals, their centrifugal order and intermediate and terminal segment lengths. Model parameters B, S and E and terminal elongation rates are then adjusted using, say, the method of maximum likelihood estimation (Box 14.7) to optimise these distributions against equivalent experimental data, usually to match the mean and standard deviation (van Pelt et al., 1997; van Pelt and Uylings, 1999; van Pelt et al., 2001, 2003). In this case, the model predicts the temporal evolution of tree development. Where data from immature dendrites are available, this can also be used in the optimisation process (van Pelt et al., 2003). Since the model tracks development over time, it is possible to use the model to make predictions about the effect of interventions during the growth process, such as pruning of particular branches (van Pelt, 1997).

Diameters must be added to the resultant dendrograms to get neurite models that may be used in compartmental modelling. This can be done by, for example, using a power law rule that relates a parent segment diameter to its two daughter diameters via branch power e (Box 12.1). A new terminal segment is given a diameter drawn from a suitable distribution, and the branch power e at each bifurcation also is chosen from a distribution of values (van Pelt and Uylings, 1999). Then all segment diameters are updated recursively following a branching event, starting with intermediate segments whose daughters are both terminals. The diameter of a parent segment is then $P = (C^e + S^e)^{1/e}$, for diameters C and S for child and sibling. For a suitable distribution of e, a reasonable fit may be obtained to real diameter distributions for a number of types of dendrite (van Pelt and Uylings, 1999).

The relation between time bins and real developmental time may not be linear, that is, the duration of each time bin might correspond to a different period of real time (van Pelt and Uylings, 2002; van Pelt et al., 2003).

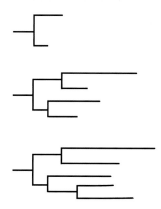

Fig 12.3 Dendrograms for a developing neurite, produced at increasing time points during a single run of the BESTL algorithm.

12.1.2 Biophysical Models of Neurite Growth

An alternative approach to a statistical growth algorithm for modelling neurite development is one based on the known biophysics of neurite outgrowth. Such an algorithm may be more complex than a statistical algorithm like BESTL, and consequently it may be harder to optimise its parameters against known data. However, it gives a more direct handle on the key biochemical processes that underpin neuronal development. Biophysically based models have been used to study all of the different developmental phases outlined above. The common style is to model the production and transport of key growth-determining molecules through both intracellular and extracellular space, with molecular interactions taking place at specific spatial locations, such as in a growth cone. The modelling techniques are similar to those we have considered for compartmental modelling of membrane voltage and for modelling intracellular signalling pathways (Chapters 5 and 6). However, in this case, the physical structures being modelled change shape over time. This introduces extra numerical complexity into setting up and solving these models.

Intracellular Models

The simplest approach involves modelling the production and transport (by diffusion and active transport) of molecules within a 1D intracellular space, as illustrated in Figure 12.4. The multidimensional extracellular space is assumed to be homogeneous and therefore can be ignored as its effects are represented intrinsically within elongation and branching rates and other model parameters. The single intracellular dimension is the longitudinal axis of a neurite, with all molecular concentrations assumed to be uniform in the radial direction. The only complication arises from the assumption that neurite outgrowth – both elongation and branching – is dependent on molecular concentration, which will change as the cellular spaces change in volume. Care must be taken to ensure conservation of mass in all numerical calculations with such a model.

A basic example of this approach considers the production and transport of the protein **tubulin** along a growing neurite (Figure 12.4). Free tubulin has concentration $c(x, t)$ at a point x along the neurite at time t. Tubulin molecules move by active transport (a) and diffusion (D), and degrade at rate g (van Veen and van Pelt, 1994; Miller and Samuels, 1997; McLean and Graham, 2004; Graham et al., 2006). In the cell body ($x = 0$), synthesis of tubulin occurs at a fixed rate $\epsilon_0 c_0$. At the distal end of the neurite ($x = l$), assembly of tubulin onto **microtubules** occurs at rate $\epsilon_l c$, and spontaneous disassembly at rate ζ_l. These processes are summarised by the following equations for how the tubulin concentration $c(x, t)$ changes over time and space:

$$\frac{\partial c}{\partial t} + a\frac{\partial c}{\partial x} = D\frac{\partial^2 c}{\partial x^2} - gc, \qquad (12.3)$$

$$-\frac{\partial c}{\partial x} = \epsilon_0 c_0 \quad \text{at } x = 0, \qquad (12.4)$$

$$-\frac{\partial c}{\partial x} = \epsilon_l c - \zeta_l \quad \text{at } x = l. \qquad (12.5)$$

Tubulin is a protein that polymerises into long filaments to form the cytoskeleton of a neurite.

Fig 12.4 Development of the intracellular cytoskeleton of a neurite, determined by transport of tubulin and assembly of microtubules.

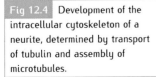

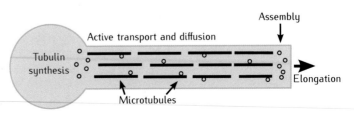

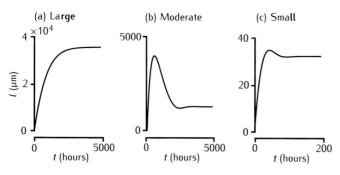

Fig 12.5 Elongation of a model neurite over time in three growth modes (McLean and Graham, 2004; Graham et al., 2006). Achieved steady-state lengths are stable in each case, but there is a prominent overshoot and retraction before reaching steady state in the moderate growth mode.

Equations 12.4 and 12.5 describe the boundary conditions for the neurite in terms of the flux of tubulin into the neurite due to synthesis in the cell body (Equation 12.4) and its flux into microtubules at the growing end of the neurite (Equation 12.5).

This model has been used to investigate the dynamics of neurite outgrowth by specifying the elongation rate of the neurite as a function of the net microtubule assembly rate:

$$\frac{\mathrm{d}l}{\mathrm{d}t} = k\epsilon_l c(l,t) - k\zeta_l, \tag{12.6}$$

where the scaling factor k converts microtubule assembly and disassembly into a change in the length of the neurite (McLean and Graham, 2004; Graham et al., 2006). Three growth modes are evident in this model (Figure 12.5). Growth to long lengths is determined by active transport of tubulin to the growing tip. In contrast, growth to short lengths is dominated by diffusion. Moderate lengths are achieved by a balance of active transport and diffusion. An updated version of this model has been derived by Diehl et al. (2014), in which they introduce an explicit growth cone compartment of fixed volume at the growing end of the neurite and also take into account the growth velocity when calculating the flux of free tubulin from the neurite into the growth cone. The dynamics of outgrowth and tubulin concentration profiles along a growing neurite show some differences from the model presented here.

Similar growth models have been used to demonstrate that competition for resources, such as tubulin, may underlie differential growth rates seen between different branches in real growing neurites (van Ooyen et al., 2001; Hjorth et al., 2014). In these models, multiple branches are modelled as above, but with each attached to a single source of the growth-promoting substance. Branches may actually retract as a consequence of a lack of resource.

An alternative model assumes that, rather than determining the elongation rate, the tubulin concentration at a neurite tip determines the branching rate $p(t)$ of the terminal:

$$p(t) = D(t)c(l,t). \tag{12.7}$$

This is on the basis that branching is more likely, the higher the rate at which microtubules are being assembled to provide the scaffold for new branches (van Pelt et al., 2003; Graham and van Ooyen, 2004). As with the elongation model, branching rates are sensitive to tubulin synthesis and transport rates. A close match to the BESTL model can be achieved with biophysical transport rates.

A natural extension to these models is to describe both elongation and branching as simultaneous functions of tubulin concentration and other factors affecting microtubule assembly. One approach introduces microtubule stability as a function of the phosphorylation state of **microtubule-associated proteins**, or **MAPs** (Hely et al., 2001; Kiddie et al., 2005). The premise is that phosphorylated MAP loses its microtubule cross-linking ability, destabilising microtubule bundles and promoting branching. This necessitates modelling the production and transport of the MAPs and the influx and diffusion of calcium (Figure 12.6). The local calcium concentration determines the phosphorylation state of the MAPs. This phosphorylation process could be modelled in more or less detail using the techniques for modelling intracellular signalling pathways described in Chapter 6. Model results indicate that neurite tree topology is a strong function of the binding rate of the MAPs to microtubules and the phosphorylation rate of the MAPs (Hely et al., 2001; Kiddie et al., 2005).

Numerical Methods

The numerical simulation of these 1D intracellular growth models is very similar in principle to the compartmental modelling approach for the membrane potential (Chapter 5), but now the compartmental quantities are chemical concentrations that are assumed to be well mixed within each compartment. Movement of molecules between compartments is by diffusion (Section 6.5.4 for longitudinal diffusion of calcium) and active transport, for which appropriate compartmental coupling terms must be derived (Graham and van Ooyen, 2001, 2004; Kiddie et al., 2005). A simple compartmentalisation of the partial differential equation (PDE) in Equation 12.3 is:

$$\frac{dc_j}{dt} = \hat{D}_{j,j-1}(c_{j-1} - c_j) + \hat{D}_{j,j+1}(c_{j+1} - c_j) + \hat{a}_{j,j-1}c_{j-1} - \hat{a}_{j,j+1}c_j - gc_j, \quad (12.8)$$

where $\hat{D}_{j,j-1} = D(a_{j,j-1}/v_j \Delta_{j,j-1})$ is the effective diffusion rate for substance c when $a_{j,j-1}$ is the cross-sectional area between compartments j and $j-1$, v_j is the volume of compartment j and $\Delta_{j,j-1}$ is the distance between the midpoints of compartments j and $j-1$. Active transport is unidirectional towards the distal end of the neurite at an effective rate

Fig 12.6 Development of the intracellular microtubule cytoskeleton of a neurite. Phosphorylated MAP results in weak cross-linking of microtubule bundles and a consequent higher likelihood of branching. Strong cross-linking, when MAP is dephosphorylated, promotes elongation.

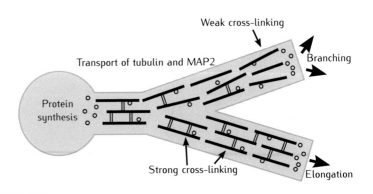

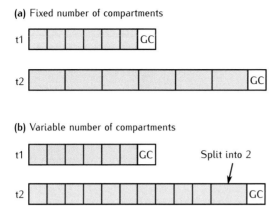

(a) Fixed number of compartments

t1

t2

(b) Variable number of compartments

t1   Split into 2

t2

Fig 12.7 Alternative methods for compartmental modelling of neurite growth. **(a)** The number of compartments is fixed and all compartments are elongated equally; numerically, the compartmental equations are solved over a normalised unit length. **(b)** The number of compartments increases with neurite length; only the compartment proximal to the fixed-size growth cone (GC) is elongated; this compartment splits into two when its length reaches twice that of the static compartments.

$\hat{a}_{j,j-1} = a(a_{j,j-1}/v_j)$ into compartment j from compartment $j-1$. More sophisticated spatial and temporal discretisations may be necessary for computational efficiency and accuracy (Graham et al., 2006; Diehl et al., 2016).

The main difference compared with compartmental modelling of voltage in neurons with a static morphology is that now the lengths and diameters of neurite segments may change over time. Any such changes result in a change in the volume of a compartment and this must be accounted for by the appropriate changes to the concentrations being modelled. It may also be necessary to alter the compartmental structure of the model to account sensibly for changes in segment length. Typically only terminal branches are growing and so only the compartmental structure of these branches needs to be dynamically adjusted. Two main approaches have been taken for this (Kiddie et al., 2005).

One approach is to keep the number of compartments constant in the growing segment and change the length of all these compartments uniformly at each time-step according to the growth rate (Figure 12.7a). This becomes implicit by using suitable numeric schemes in which spatial discretisation is normalised against segment length, so that all calculations are done against a fixed pseudo-length of one unit (McLean and Graham, 2004; Kiddie et al., 2005; Graham et al., 2006; Diehl et al., 2016). Absolute lengths and concentrations are obtained by the inverse transformation, when required. This approach requires choosing the number of compartments for a segment in advance so that, as the neurite grows, concentration calculations remain accurate over multiple length scales, from the very short to the very long.

Alternatively, as many compartments as possible could be kept at a fixed length (Graham and van Ooyen, 2001; Kiddie et al., 2005). It is assumed that only the section of neurite immediately proximal to the growth cone is elongating and so only the length of this compartment needs to change (Figure 12.7b). Once this compartment reaches twice the length of the static compartments, it is subdivided into two compartments. A number of studies have used this method (Graham and van Ooyen, 2001; Hely et al., 2001; Kiddie et al., 2005; Hjorth et al., 2014; Mironov et al., 2014). Note that branching events also require the addition of new compartments and these must be of some initial non-zero length for numerical stability. Retraction of branches requires the removal of compartments when they become too short.

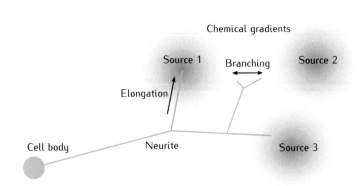

Fig 12.8 Neurite outgrowth in an external environment containing three sources of a diffusible attractive chemical.

Extracellular Models

Neurite growth is influenced by extracellular chemical signals too. The intracellular models considered above assume such signals are uniform throughout the growing environment and thus have a constant effect on elongation and branching rates. However, these signals may emanate from discrete sources located at one or more points in the environment. When the extracellular space is not homogeneous, it must be modelled explicitly. Often, this is done in terms of a 2D flat plane, rather than in 3D (Li et al., 1992, 1995; van Veen and van Pelt, 1992; Hentschel and van Ooyen, 1999; Aeschlimann and Tettoni, 2001; Maskery et al., 2004; Feng et al., 2005; Krottje and van Ooyen, 2007; Mironov et al., 2014). Physical and chemical cues may be located at specific locations in this environment, with chemical cues diffusing freely from their source, influencing directional guidance and branching of neurite growth cones (Figure 12.8). The spatial location of each growth cone needs to be specified and tracked over time. The neurite itself is assumed not to occupy any volume so that only growth cone locations need to be calculated. The visco-elastic properties of the growth cone and trailing neurite may be included (van Veen and van Pelt, 1992; Li et al., 1994, 1995; Aeschlimann and Tettoni, 2001). Extracellular chemical gradients are calculated on a spatial grid, given source locations and rates of diffusion (Hentschel and van Ooyen, 1999; Krottje and van Ooyen, 2007; Mironov et al., 2014). The numerical solution of these models may use finite element techniques, which are the extension to 2D or 3D space of the 1D spatial discretisations used in compartmental modelling (Section 5.7).

12.2 | Pattern within a Set of Neurons

12.2.1 Development of Pattern within Morphogenesis

We approach the notion of pattern in a population of cells by discussing the emergence of pattern in morphogenesis, at stages long before the nervous system is formed. Although early development is not usually addressed in computational neuroscience, a body of mathematics developed to analyse morphogenesis is also applicable to some of the neurobiological questions we are considering.

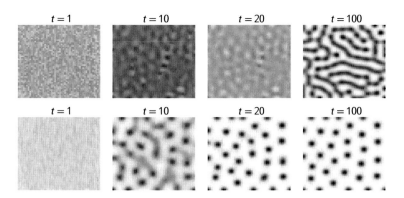

$t = 1$ $t = 10$ $t = 20$ $t = 100$

$t = 1$ $t = 10$ $t = 20$ $t = 100$

Fig 12.9 Two numerical simulations of the Gierer–Meinhardt model. Each row of plots shows the development over time of patterns generated using the Gierer–Meinhardt equations (a) and (b) (Box 12.3), which have been discretised on a 50 × 50 pixel grid. The starting conditions are randomised, and the patterns are allowed to develop over time until a steady state is reached. In the top row, stripe-like patterns emerge. Parameter values are: $\rho = 1.0$, $\rho_1 = 1 \times 10^{-5}$, $\rho_2 = 1 \times 10^{-4}$, $\mu_1 = 1$, $\mu_2 = 1$, $D_1 = 0.0$ and $D_2 = 0.4$. In the bottom row, employing different parameter values leads to a pattern of spots emerging. Parameter values are as for the top row, except that the diffusion constants are $D_1 = 0.12$ and $D_2 = 1.6$. Simulations generated using a modified version of code written by Soetaert et al. (2010).

A simple example of a biological pattern is where there is a systematic variation of some identifiable feature over a 2D biological structure, or field. Well-known biological examples are the patterns on seashells or butterfly wings and the distribution of melanocytes over animal skins, such as the pattern of zebra stripes (Meinhardt, 1983; Murray, 1993) or the individual digits in the vertebrate limb or body segmentation in the insects.

Two different types of model for the development of biological pattern have been explored, both based on the idea that there are hypothetical molecules, or morphogens, distributed across a field of cells (Green and Sharpe, 2015). The suggestion that pattern is specified by gradients of morphogen dates back over 100 years (Lawrence, 2001). Its modern incarnation is the concept of **Positional Information** (Wolpert, 1969; Wolpert et al., 2002) and applied widely, such as to the formation of digits in a developing limb bud. The term reaction–diffusion is often used to describe the alternative notion, that several morphogens diffuse over the substrate, or field, usually assumed to be 2D, and react with each other. Turing (1952) was probably the first to compute the types of pattern that can be formed in a reaction–diffusion system. The underlying mathematical principles (Edelstein–Keshet, 1988; Murray, 1993) are used widely in modelling the emergence of pattern in neural development (Green and Sharpe, 2015).

The basic requirements for the generation of spatial pattern in this way are that there must be (1) at least two different morphogens; (2) different rates of diffusion for the different morphogens; (3) specific interactions between the morphogens. One case that is often considered is that of two morphogens which interact according to non-linear dynamics (Box 12.3).

The mathematics of this type of scheme to develop periodic patterns have been explored widely and the results applied to the generation of many different types of periodic pattern, such as the pattern of cilia on the surface of a frog embryo, the bristles on the cuticle of the bug *Rhodnius* or of the spacing of leaves (Meinhardt, 1983). One crucial finding from this analysis is that the values of the fixed parameters, compared to the size of the field, determine the periodicity of the patterns found. Murray (1993) observed that in a developing system, **morphogenetic fields** will change size and so the ratios of parameter values to field size change over time. He suggested that to generate many of the different types of patterns of animal coat marking seen, reaction–diffusion could be switched on at different stages of development, allowing patterns of different periodicity to be formed.

Miura and Maini (2004) give an introduction to Turing's model (1952) of pattern formation, with the aim of motivating non-mathematicians to carry out their own simulations of the model.

Box 12.3 | Reaction–diffusion: the Gierer–Meinhardt model

Without any spatial effects, the interaction between the two morphogens U and V can be described by the following two coupled equations for how the morphogens vary over time:

$$dU/dt = R_1(U, V),$$
$$dV/dt = R_2(U, V),$$

where R_1 and R_2 are non-linear functions which control the rates of production of U and V.

When the morphogens vary over 2D space as well as time, the two relevant equations are:

$$\partial U/\partial t = R_1(U, V) + D_1 \partial^2 U/\partial x^2 + D_1 \partial^2 U/\partial y^2,$$
$$\partial V/\partial t = R_2(U, V) + D_2 \partial^2 V/\partial x^2 + D_2 \partial^2 V/\partial y^2.$$

D_1 and D_2 are the diffusion coefficients controlling the rate of spread of the morphogens over the surface.

One specific choice for the functions R_1 and R_2, which control the production of the two morphogens U and V, is given in the **Gierer–Meinhardt model** (Gierer and Meinhardt, 1972; Meinhardt, 1983):

$$\partial U/\partial t = \rho U^2/V - \mu_1 U + \rho_1 + D_1 \partial^2 U/\partial x^2 + D_1 \partial^2 U/\partial y^2, \tag{a}$$
$$\partial V/\partial t = \rho U^2 - \mu_2 V + \rho_2 + D_2 \partial^2 V/\partial x^2 + D_2 \partial^2 V/\partial y^2. \tag{b}$$

Morphogen U is an activator as it promotes the synthesis of itself and V; morphogen V is an inhibitor as it inhibits its own growth and ultimately limits the growth of U (Equations a and b). Numerical simulations of this model with two sets of parameters are shown in Figure 12.9. Provided the parameter values in these equations satisfy certain conditions, stable spatial patterns of morphogen will emerge from almost uniform initial distributions of morphogen.

A simple way of modelling the production of a mosaic uses a random number generator to place cells on a 2D surface, one by one. Before each new cell is added in, a check is made as to whether its position is within the prespecified exclusion zone of any cells already present. If it is, new random positions are tried until a position is found which is sufficiently distant from all existing cells. This is a **pairwise interaction point process (PIPP) model** (Diggle, 2002).

12.2.2 Development of Retinal Mosaics

Neurons are arranged in many different types of pattern. In the mammalian system, perhaps the most remarkable is the 3D arrangements of neurons of six different types and their processes in the mammalian cerebellar cortex (Eccles et al., 1967). One particular type of cell pattern that has been analysed quantitatively is the pattern of neurons of the same type over the retina (Cook and Chalupa, 2000). For example, ganglion cells are distributed over the surface of the vertebrate retina to form regular patterns, called **retinal mosaics**. From analysis of the spatial autocorrelation of these types of pattern, it has been suggested that the way these retinal mosaics are formed is consistent with each new cell being placed at a random position subject to minimum spacing between any two cells (Galli-Resta et al., 1997). Production of mosaics in this way can be modelled mathematically.

In order to test the applicability of this phenomenological model, attributes of the mathematical distributions can be measured and compared with what is found experimentally. The minimum spacing rule has been found to reproduce successfully the distribution of several retinal mosaics,

(a) (b)

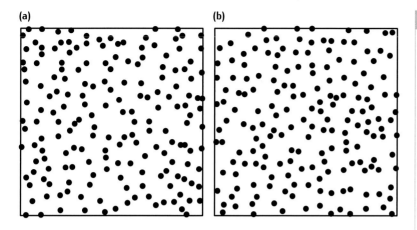

Fig 12.10 Real and simulated distributions of retinal neurons. **(a)** Positions of rat cholinergic amacrine bodies (Resta et al., 2005). Each cell is drawn with a 10 μm diameter and the field of view is 400 μm × 400 μm. **(b)** Simulation of the observed field in panel **(a)**. Here, the minimal distance exclusion zone was drawn from a normal distribution, with mean = 22 μm and standard deviation = 6 μm. Images kindly supplied by Stephen Eglen.

including rat cholinergic amacrine cells (Galli-Resta et al., 1997; Figure 12.10) and nicotinamide adenine dinucleotide phosphate-diaphorase (NADPH-d)-active retinal ganglion cells in chick (Cellerino et al., 2000).

Construction of a model at this level does not shed any light as to the mechanism. In order to understand how in the retina an irregular spatial distribution of undifferentiated cells is transformed into a regular distribution of differentiated cells, the idea of **lateral inhibition** has been invoked. In one approach, the **Delta** and **Notch molecules**, two types of signalling molecule, are used in a type of reaction–diffusion scheme to impart **primary fate** or secondary fate to an ensemble of differentiating cells. Those cells with primary fate develop into mature cells of a particular type, for example, retinal ganglion cells of a particular class. Cells with secondary fate can be switched to different types. Various authors have modelled this phenomenon and we adopt the formulation due to Eglen and Willshaw (2002), who applied the model of Delta–Notch signalling developed by Collier et al. (1996). This is a discrete, rather than continuous, problem, which is formulated in terms of the levels D_i and N_i of the two molecules in a typical cell i. The amounts of these two quantities in a cell are used to express whether it is acquiring primary (D) or secondary (N) fate. The reaction equations are:

$$\mathrm{d}D_i/\mathrm{d}t = -D_i + g(N_i), \tag{12.9}$$

$$\mathrm{d}N_i/\mathrm{d}t = -N_i + f(\overline{D_i}), \tag{12.10}$$

with $f(x) = x^2/(A + x^2)$ and $g(x) = 1/(1 + Bx^2)$.

A and B are constants and $\overline{D_i}$ is the average value of D computed over the neighbours of cell i. The neighbours of a cell are found using the Voronoi tessellation (Okabe et al., 1992).

In this form of Equations 12.9 and 12.10, the quantities N_i and D_i are dimensionless. The functions f and g are chosen to be monotonically increasing and monotonically decreasing, respectively. This means that higher levels of D in a cell lead to higher levels of N in its neighbours, which leads to lower levels of D in these cells (Equation 12.10). Consequently, the values of D and N in neighbouring cells tend to go to opposite extremes. The result is that if a particular cell has primary fate (high D), the neighbouring

The term lateral inhibition is used widely in neurobiology to describe situations where the activity in one nerve cell can diminish the activity in its neighbours. Usually 'activity' means the electrical activity of a neuron, but in other contexts, this can refer to the amount of a particular signalling molecule present in a cell.

cells will have secondary fate (high N). In this way, a mechanism of lateral inhibition is introduced, ensuring that cells are regularly spaced out over the retinal surface.

The purpose of this work was to investigate whether this mechanism of lateral inhibition would be sufficient to generate mosaic patterns with the required regularity or whether an additional mechanism, in this case cell death, was needed (Eglen and Willshaw, 2002). Computer simulations were carried out to assess whether the regularity in an initially irregular distribution of cells could be improved by this method of assigning primary and secondary fate to cells.

The set of equations defined by Equations 12.9 and 12.10 was iterated until the values of D and N had stabilised in all cells. A cell i acquired primary fate if D_i exceeded 90% of its maximum value of 1; otherwise it was said to have acquired secondary fate.

It was found that lateral inhibition makes the distribution of primary fate cells more regular than in the initial irregular distribution, but is insufficient to produce mosaics to the required degree of regularity. A combination of lateral inhibition and cell death working together would suffice, but the resulting pattern will depend on the precise type of cell death assumed in the model.

There has been interest in the relation between the distributions of the class of β retinal ganglion cells which is subdivided according to whether they are on-centre or off-centre in response to light. The combination of responses to the two types of cell shapes the visual responses. Eglen et al. (2005) had developed quantitative methods to compare different mosaic patterns and to simulate these patterns. To generate a mosaic, they used a pairwise interaction point process (**PIPP**-type) method (Diggle, 2002) and concluded that the on-centre and off-centre mosaics are most likely to be generated independently, except for the physical constraint that no two cell bodies can occupy the same space. In contrast, Paik and Ringach (2011) suggested that the different mosaic patterns of on-centre and off-centre cells form Moiré interference patterns, the points of interference forming the framework for the quasi-periodic maps of orientation selectivity seen in the visual cortex. Hore et al. (2012) argued that this implies that the on-centre and off-centre mosaics each have a hexagonal structure, which is not seen in the data, whereas their PIPP method of generating mosaics gives excellent fits to the data and the Moiré lattice model gives the incorrect periodicity seen in the orientation maps. Jang and Paik (2017) criticised this method of generating mosaics and used an energy measure to argue for the existence of interactions between on-centre and off-centre mosaics. Roy et al. (2021) took a different approach. They trained a set of 100 noisy filters by maximising the mutual information between their output and a set of natural images presented to them. They found that the filters became either on-centre or off-centre cells, thereby defining two separate mosaics. They then generated mosaics where the mean distance in the visual field between all pairs of points from different mosaics is: (1) the same as ('independent'); (2) less than ('aligned'); or (3) greater than ('anti-aligned') what is expected if the two mosaics are generated at random. According to this comparison, the mosaics they computed by optimising mutual information are anti-aligned, which is in accordance

A Moiré pattern is a visual effect formed by superposing two similar regular patterns which differ in size, orientation or spacing.

with measurements from rats and primates and supports the idea of heterotypic repulsion. However, these authors did not model directly the mosaics formed in the retina and made inferences from the mosaics generated in the visual field.

Resolution of these issues will be helped by new and higher-quality data, as well as by consideration of whether the models used to generate mosaics could feasibly be used as computational models to generate cellular patterns, rather than being a mathematically convenient way to produce the required structure. Work (Kling et al., 2020) using large-scale multielectrode recording from macaques and humans may help to provide such data.

12.3 Neuromuscular Connections

This is the first of three examples of modelling the development of patterns of connections. Here we look at the elimination of connections between nerve and muscle in early development to leave each muscle fibre with contact from a single motor neuron.

How this pattern of connections develops is a classic problem in developmental neurobiology which has been addressed through computational modelling. For reviews of the neurobiology see Jansen and Fladby (1990) and Sanes and Lichtman (1999), and for modelling approaches, see van Ooyen (2001) and Goodhill (2018). Compared with the development of many other neural systems, in this case basic neuroanatomical and physiological information is known, in terms of how many neurons contact how many muscle fibres and how the pattern of neuromuscular connections develops. However, some very basic information is lacking and so the typical modelling approach has been to focus on working out what patterns of connection are generated in a model based on as yet untested assumptions.

A set of closely related models is now described to show how modelling studies can build on previous work. We illustrate how models are constructed at different levels of detail and how different assumptions are necessary and made use of in the models.

In vertebrate skeletal muscle, each muscle fibre is innervated along its length at a single region, the endplate, where the acetylcholine receptors are concentrated. Here the binding of acetylcholine from the activated nerve terminals causes muscle contraction. In adult muscle, each endplate receives innervation from a single motor neuron. In most cases, there are more muscle fibres than motor neurons, and so the axon of each motor neuron branches profusely. In contrast, in neonatal muscle, each endplate is innervated by as many as 5–10 different motor neurons (Jansen and Fladby, 1990; Sanes and Lichtman, 1999; Figure 12.11). One basic question is how the transformation from superinnervation to single innervation of individual muscle fibres takes place. Since there is very little, if any, motor neuron death and little evidence for the making of new connections during this stage of development, it is generally held that axons withdraw contacts from individual muscle fibres until the state of single innervation is reached. This phenomenon is also seen in the development of connections between neurons such as in the

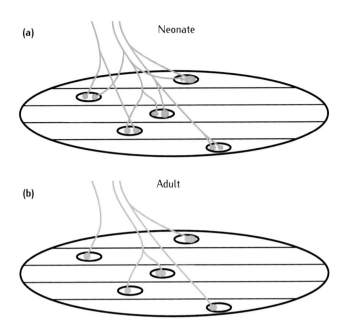

Fig 12.11 The transformation between the state of **(a)** superinnervation of individual fibres in neonatal skeletal muscle and **(b)** that of single innervation in adult muscle.

superior cervical ganglion (Purves and Lichtman, 1980) and the cerebellar cortex (Crepel et al., 1980).

Most modelling approaches are concerned with how synaptic contacts are withdrawn from the initial configuration of superinnervation to attain a pattern of single innervation. In all models, several crucial untested assumptions have to be made. The principal assumption is that there is some physical property possessed by each synapse, the amount of which determines whether a synapse will be withdrawn or will become stabilised. The amount of this property is assumed to vary between synapses and change during development. As will be described, another key assumption made is whether the physical property is in limited supply, as this will affect the type of interactions possible within the model.

This physical property is usually assumed to be some measure of synaptic efficacy, such as the area or volume of the synaptic contact or the number of receptors or the degree of depolarisation of the synaptic membrane by a given amount of transmitter. In the absence of any definitive answer, in most models, the physical nature of this property is left unspecified; it is assumed that a scalar quantity, often called the **synaptic strength**, is assigned to each synapse. Synaptic strengths vary over time in a manner prescribed in the model. Synapses reaching a constant positive value are deemed to have been stabilised and those reaching zero strength to have been withdrawn. The problem for the computational modeller is to design a set of biologically plausible equations for how synaptic strengths change over time that fit the known facts and lead to useful predictions.

12.3.1 Competition for a Single Resource

Willshaw (1981) proposed two simple equations to express how axons compete with each other whilst maintaining total synaptic strength, the interactions being mediated through competition between the terminals at each

---

Box 12.4 Competition for a presynaptic or a
                 postsynaptic resource

In the model due to Willshaw (1981), the rate of change of the strength S_{nm} of the synapse of neuron n at endplate m, with mean synaptic strength M_m, is:

$$\frac{dS_{nm}}{dt} = -\alpha M_m + \beta_n S_{nm}. \tag{a}$$

α can be thought of as a constant, with β_n set to maintain the total strength of the synapses made by each neuron n at a constant level:

$$\beta_n = \alpha \frac{\Sigma_k M_k}{\Sigma_{nk}}, \tag{b}$$

where the sum is over only those endplates with which axon n has a terminal.

In the model due to Gouzé et al. (1983), the rate equation for an individual terminal is:

$$\frac{dS_{nm}}{dt} = K I_n \mu_m S_{nm}^\alpha, \tag{c}$$

where K and α are constants and I_n represents the electrical impulse activity in axon n. μ_m is the amount of resource available at muscle fibre m, and α is assumed to be greater than 1 so that the stronger a synapse is, the faster resource is taken up by it, until all the resource is exhausted.

---

endplate. Following O'Brien et al. (1978), he assumed that each synapse emits into its endplate region an amount of a substance which degrades all synapses at that endplate. In addition, all synapses are strengthened continuously, thereby counterbalancing the degradation so as to keep the total synaptic strength of all the synapses of each motor neuron constant (Box 12.4).

It is straightforward to show that under the action of these two counterbalancing influences, any initial pattern of superinnervation is converted into a pattern in which just one synapse at each endplate has a positive synaptic strength and all the other synapses have zero strength, thus specifying a pattern of single innervation. Willshaw (1981) applied this model to some of the experimental results known at the time. For example, in the model, initially large **motor units** lose more connections than smaller ones, in agreement with experimental observations (Brown et al., 1976).

This is a high-level model, relying on the assumption of competition at an endplate through conservation of resource in terms of the total synaptic strength associated with a motor neuron. This assumption is intuitively attractive, but there is little direct evidence for it. It is a type of **sum rule**, often used in artificial neural networks (Box 11.2) and introduced for computational necessity rather than biological realism. A problem for all models with presynaptic sum rules is how information about synaptic strengths on widely separated distal parts of the axon could be exchanged to enable the total strength of the synapses of each motor neuron to be kept constant.

A complementary model of competition was introduced by Gouzé et al. (1983). In this model, axons compete at the endplate directly for the finite amount of resource assigned to each muscle fibre (Box 12.4). Here a

The **motor unit** is made up of a motor neuron and all the muscle fibres that it innervates. The size of the motor unit is the number of muscle fibres in it.

postsynaptic sum rule is imposed as it is assumed that the synapses are built out of the limited resource available at each muscle fibre. The synapses, having different initial strengths, will take up resources at different rates. Once all resources have been used up at an endplate, those synapses that have reached a prespecified strength are regarded as stable and all others as having been withdrawn. This threshold value is set so that only one synapse survives at each endplate.

In this model, many activity-dependent effects can be modelled through the term I_n, representing the rate of electrical activity. This is an innovation that was not present in Willshaw's 1981 model. The sum rule acts locally, at each endplate, and so does not require exchange of information over distance between synapses from the same axon. However, the emergence of single innervation after reducing the number of motor neurons by partial denervation of the motor nerve (Brown et al., 1976; Betz et al., 1980) requires the number of motor axons to be communicated to each endplate. This is in order to set the threshold value of synaptic strength so that only one synapse survives at each endplate.

12.3.2 Competition for Presynaptic and Postsynaptic Resources

Bennett and Robinson (1989) took a more biophysical approach using a formulation which makes more explicit the way in which synaptic strength is conserved. They suggested that there is reversible binding between presynaptic factor A, originating in the motor neuron, and postsynaptic factor B, originating in the muscle fibre, to form stabilising factor C, playing the same role as synaptic strength. The supply of both A and B is assumed to be limited; as more and more C is formed, A and B are used up. The model described here is the extension by Rasmussen and Willshaw (1993) of Bennett and Robinson's (1989) formulation. It is specified by a set of differential equations for the production of stabilising factor C from locally available A and B (Box 12.5). One molecule of A and one molecule of B make one molecule of C (Figure 12.12).

Binding reactions of this type are described in Section 6.1.1.

- The forward reaction in the rate equation for the production of C at a synapse is proportional to the product of the locally available amounts of A, B and C, the involvement of C being a simple way of favouring the growth of synapses; the backward reaction is proportional to the amount of C.
- The amount of A available at each neuron is the initial amount minus the amount made available to each synapse and that converted into C. Rasmussen and Willshaw (1993) assume an equal rate of uptake of A into each of the synapses of the neuron.
- The amount of B available at each muscle fibre is the initial amount minus the amount converted into C.

Since there are two constraints on this system, concerning the amounts of A and B available, Rasmussen and Willshaw (1993) called this the **Dual Constraint Model** (DCM). Results of a simulation of DCM with $N = 6$ axons and $M = 240$ muscle fibres, typical values for the mouse lumbrical muscle, are shown in Figure 12.13.

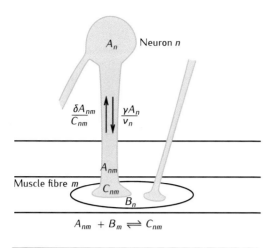

Fig 12.12 Schematic diagram of the Dual Constraint Model.

In the original model (Bennett and Robinson, 1989), the total amount A_0 of factor A initially available was set equal to the total amount B_0 of factor B initially available. Rasmussen and Willshaw (1993) looked at the more general case when these quantities are not matched. By carrying out stability analysis (Box 8.3) of C_{nm}, they showed that in the stable state, all those synapses with a strength C_{nm} of less than $B_0/2$ will not survive.

This result is relevant to the debate over the significance of the phenomenon of intrinsic withdrawal (Fladby and Jansen, 1987). There is a lot of evidence that synapses compete at the endplate for sole occupancy of the endplate. However, if competition for postsynaptic resources is the sole mechanism, endplates should never be denervated, as once single innervation is reached, there will be nothing left for the surviving synapse to compete against and so it will not be eliminated. Since there are cases of denervated endplates, or intrinsic withdrawal (Fladby and Jansen, 1987), it was argued that competition for sole occupancy of the endplate does not operate. In the DCM, only synapses with at least a minimal amount $B_0/2$ of stabilising factor will survive. In the model, since stabilising factor C is produced from one molecule of A and one of B, the maximum number of synapses that any given axon can make will be limited by the number of times that $B_0/2$ divides into A_0. If the initial number of connections made by an axon is greater than this ratio, $2A_0/B_0$, some of the synapses will be below critical strength and will then withdraw, with or without competition.

This model is a more biologically grounded model than the previous ones described, being expressed in terms of biological rate equations for the putative factors and not requiring information exchange over long distances to normalise synapses. It accounts for a wider range of phenomena as it addresses the finding of intrinsic withdrawal, but it does not address any activity-dependent effects.

In a further attempt to link the proposed molecular factors to biology, Van Ooyen and Willshaw (1999a, b, 2000) adopted a formalism used widely in population biology, such as in models of consumer–resource systems (Yodzis, 1989). The probably unrealistic constraint of a fixed amount of resource used in Rasmussen and Willshaw (1993) was replaced by the generation of

Box 12.5 | Competition for two resources

A_n is the amount of A in neuron n, and B_m the amount of B in muscle fibre m. A_{nm} and C_{nm} are the amounts of A and C in synapse nm.

Dual Constraint Model (Rasmussen and Willshaw, 1993).

Rates of uptake of C and A at synapse nm are:

$$\frac{dC_{nm}}{dt} = \alpha A_{nm} B_m C_{nm} - \beta C_{nm}, \tag{a}$$

$$\frac{dA_{nm}}{dt} = \gamma A_n / v_n - \delta A_{nm} / C_{nm}, \tag{b}$$

where α, β, γ and δ are constants and v_n is the number of axonal branches in neuron n. Conservation equations for A and B are:

$$A_0 = A_n + \Sigma_k A_{nk} + \Sigma_k C_{nk}, \tag{c}$$

$$B_0 = B_m + \Sigma_j C_{jm}. \tag{d}$$

Neurotrophic Model (Van Ooyen and Willshaw, 1999a).

Rates of uptake of bound receptor C, free receptor A, produced at variable rate ϕ, and neurotrophin B, produced at fixed rate σ:

$$\frac{dC_n}{dt} = (\kappa_a A_n B - \kappa_d C_n) - \rho C_n, \tag{e}$$

$$\frac{dA_n}{dt} = \phi_n - (\kappa_a A_n B - \kappa_d C_n) - \gamma A_n, \tag{f}$$

$$\frac{dB}{dt} = \sigma - \delta B - \Sigma_k (\kappa_a A_k B - \kappa_d C_n), \tag{g}$$

where κ_a, κ_d, γ, ρ and δ are all constants.

The rate of production of receptor ϕ_n is assumed to be a function F of the density of bound receptor C_n. Since axonal growth takes place on a relatively slow timescale, with the time constant τ being of the order of days, the dependency is expressed by:

$$\tau \frac{d\phi_n}{dt} = F(C_n) - \phi_n. \tag{h}$$

At **steady state**, the functions F and ϕ_n are identical.

Since the precise form of F is unknown, Van Ooyen and Willshaw (1999a) investigated what pattern of innervation of a target cell would evolve for different types of upregulation function F. They examined the general class of function $F(C_n)$, defined as:

$$F(C_n) = \alpha_n C_n^p / (K_n^p + C_n^p), \tag{i}$$

where K_n and α_n are constants for each neuron.

presynaptic and postsynaptic resources at specific rates balanced out by continual unspecific loss of resource. By exploring different forms of the rate function for the presynaptic resource, Van Ooyen and Willshaw (1999a) examined the situations under which patterns of innervation other than single innervation could result.

One novel feature of this model is the inclusion of neurotrophic signalling, which guides the formation and stabilisation of synaptic contacts; neurotrophic factors may have positive feedback effects on synaptic growth

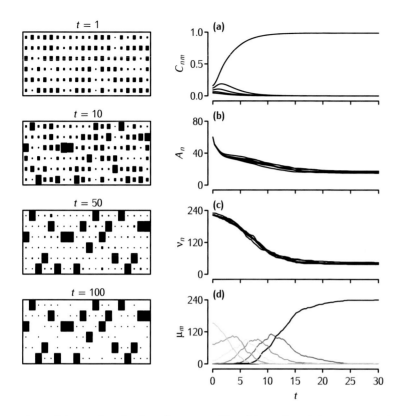

Fig 12.13 The DCM model applied to the mouse lumbrical muscle. The left-hand column shows the development through time of the pattern of connectivity as displayed in a connection matrix C_{nm}. Each motor neuron is associated with a row and each muscle fibre with a column. The area of each square indicates the strength of that element of C_{nm}. For clarity, only 20 of the 240 endplates are shown. The right-hand column shows the development through time of four different quantities. **(a)** The value of C_{n1}, that is, the strength of connections from each of the six axons competing for the first muscle fibre (Equation a). **(b)** The value of the amount of available presynaptic resource A_n at each of the six axons (Figure (a)). **(c)** The motor unit size v_n of each of the six axons. **(d)** The number of endplates μ which are singly or multiply innervated. The colour of the line indicates the level of multiple innervation, ranging from 1 (black) to 6 (blue). Parameters: $N = 6$, $M = 240$, $A_0 = 80$, $B_0 = 1$, $\alpha = 45$, $\beta = 0.4$, $\gamma = 3$, $\delta = 2$. Only terminals larger than $\theta = 0.01$ were considered as being viable. Terminals that were smaller than this value were regarded as not contributing to v_n and therefore did not receive a supply of A.

through increasing the size of synapses (Garofalo et al., 1992) or upregulating receptor density (Holtzmann et al., 1992).

In this model, the development of contacts from several different motor neurons to a single muscle fibre is analysed (Figure 12.14). To aid comparison, we explain the model with the terminology used for the DCM. All reaction equations are shown in Box 12.5.

- At each synapse, free receptor A and neurotrophin B bind to form bound receptor C. The rate of production of C follows standard reaction dynamics. The forward rate is proportional to the product of receptor A and locally available neurotrophin B. The backward rate is proportional to C. In addition, there is a non-specific loss term proportional to C.
- The free receptor A in a neuron is generated at a variable rate. Some free receptor is lost on being converted to bound receptor and there are also non-specific losses.
- Neurotrophin B is generated from muscle fibres at a constant rate. Neurotrophin is lost on conversion to bound receptor C and there are also non-specific losses.
- To represent the finding that the density of free receptor A in a neuron can be upregulated by the amount of bound receptor C (Holtzmann et al., 1992), it was assumed that the rate of production ϕ is a function of the density of C.

When the **steady state** is reached, ϕ_n changes no more and so $\frac{d\phi_n}{dt}$ equals 0. From Equation (h) in Box 12.5, it follows directly that
$$\phi_n = F(C_n).$$

Van Ooyen and Willshaw (1999a) examined four qualitatively different cases which are motivated from biology (Figure 12.15):

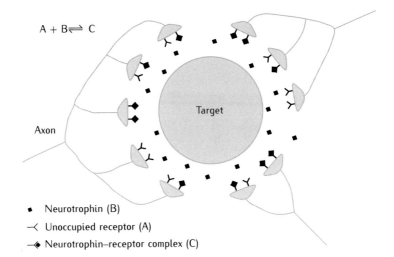

Fig 12.14 Competitive model due to Van Ooyen and Willshaw (1999a). Single target with three axons. Neurotrophin emitted by the target binds to the receptors on the axon terminals. Reproduced with permission from The Royal Society of London.

$A + B \rightleftharpoons C$

Axon

Target

• Neurotrophin (B)

⊰ Unoccupied receptor (A)

⊸ Neurotrophin–receptor complex (C)

(1) The rate of upregulation of receptor is independent of the amount of bound receptor. There is no elimination of contacts and all the connections made initially survive (Figure 12.15a).

(2) $F(C_n)$ depends linearly on C_n over a large range. Elimination of axons occurs until single innervation is reached (Figure 12.15b).

(3) F is a Michaelis–Menten function (Section 6.1.2). Elimination of contacts occurs and either single or multiple innervation results, depending on the precise values of the growth function (Figure 12.15c, d).

(4) $F(C_n)$ is a Hill function, a special case of the Michaelis–Menten function (Section 6.1.2). Unlike the other cases, where there is just one stable equilibrium pattern of innervation, here there are multiple possible stable states. Which equilibrium will be reached in any situation depends on the fixed parameter values and the initial values of ϕ_n (Figure 12.15e, f).

Examination of this model for the development of nerve connections by competition shows that using different assumptions for the nature of the competitive process leads to a model with different behaviours. This model, involving the generation and consumption of neurotrophins, provides a set of predictions which, in principle, are testable. Prominent amongst these is that different receptor upregulation functions lead to different patterns of connections; that is, if the upregulation function is known, then the pattern of connections can be predicted, and vice versa.

12.3.3 Detailed Modelling at the Endplate

A later model is based on work on transgenic mice in which individual axons express fluorescent proteins, making individual axons identifiable (Turney and Lichtman, 2012). Here, a laser was used to remove axonal terminals at an endplate during the initial period of superinnervation. The subsequent reorganisation of terminals was then followed. Previous work had shown that during the period of superinnervation, the axonal terminal occupying the largest fraction of the endplate tends to survive at the expense of the other terminals there. One question is whether the presence of a terminal drives

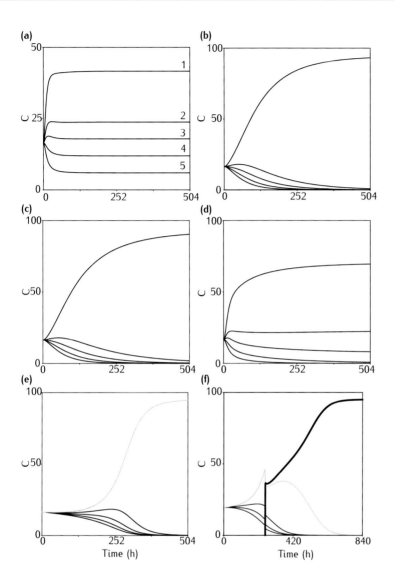

Fig 12.15 Results from simulations of the model due to Van Ooyen and Willshaw (1999a) for four different forms of the receptor upregulation function
$F(C_n) = \alpha_n C_n^p / (K_n^p + C_n^p)$, with K_n and α_n constants for each neuron. The figures show how the concentration of bound receptor in different axonal terminals, taken as a measure of the terminal's ability to survive, varies over time. **(a)** $p = 0$ so that $F(C)$ is constant: all the initial contacts survive. **(b)** $p = 1$ and K_n is large, much greater than C_n, so that effectively $F(C)$ is linear over a large range of C: contacts are eliminated until just one survives. **(c, d)** $p = 1$ and K_n is smaller. $F(C)$ is a Michaelis–Menten function. Either single innervation **(c)** or multiple innervation **(d)** results. **(e, f)** $p = 2$, so that $F(C)$ is a Hill function. **(e)** Single innervation (blue line) results. **(f)** Initial conditions influence the outcome. The simulation shown in **(e)** was stopped after 250 time-steps and an additional axon was introduced. The value of C for the axon that had survived in **(e)** (blue line) gradually went down to 0 and the survivor was the newly introduced axon. Figures from Van Ooyen and Willshaw (1999a), where the parameter values are noted. Our **(a)** is Figure 3 of Van Ooyen and Willshaw (1999a); **(b)** is 4a; **(c)** is 6c; **(d)** is 5b; **(e)** is 6a; **(f)** is 7c. Reproduced with permission from The Royal Society of London.

off other terminals or terminals grow in to occupy empty sites. This latter idea is supported by experiments in which, when endplates were innervated by two axons initially, the removal of the dominant axon by laser led to the surviving axon taking over the territory. The question arises of whether this also occurs in normal development. Previous work (Walsh and Lichtman, 2003), which had shown an unoccupied site being occupied a day later, had insufficient time resolution to show whether withdrawal is a stimulation for reinnervation. However, using a new technique to visualise terminals abutting on the endplate, in one serendipitous case, an axon terminal was seen withdrawing, with immediate occupancy by a second terminal. Turney and Lichtman (2012) concluded that the elimination of superinnervation is a dynamic process, involving receptor sites being vacated, causing reinnervation. They formulated a simple mathematical model for this idea in which terminals are withdrawn and innervated at random, with the probability of reinnervating a site by a particular axon depending on the number of its terminals occupying sites nearby.

These authors concentrated on the details of withdrawal of superinnervation at an individual endplate, made possible by advances in visualisation of the neural structure, and enabling modelling at a level of detail not attempted before. In their model, at each endplate, axon terminals compete for a fixed resource, the receptors at the endplate, which are in limited supply. This is a postsynaptic sum rule model where withdrawal of superinnervation at each endplate acts independently. It would be interesting to see how far this model accounts for other phenomena associated with the system as a whole rather than at a single endplate, as follows: (1) it should be possible to calculate the change in the distribution of motor unit sizes over the population of neurons during development and relate this to that found (Brown et al., 1976); (2) this model does not seem to suffer from the problem of the postsynaptic sum rule inherent in the model by Gouzé et al. (1983) to explain the reestablishment of single innervation after partial denervation (Brown et al., 1976; Betz et al., 1980); (3) it is unclear whether intrinsic withdrawal (Fladby and Jansen, 1987) of contacts from an endplate is predictable from this model. The connections between motor neurons and muscle fibres are dramatically reorganised in early postnatal life.

Another hypothesis which is being explored by modern experimental methods, and hitherto not investigated by modellers, is that elimination of superinnervation is activity-based (Harris, 1981) and related to the **Size Principle** (Henneman, 1957) of motor action. According to this principle, during muscular activity the motor neuron generating the smallest amount of force (assumed to activate the smallest number of muscle fibres) is activated first, followed by the neurons generating larger amounts of force, in order, until the required force is generated. From experiments showing that electrical stimulation inhibits synaptic elimination (Buffelli et al., 2002), it was investigated whether competition between neurons that are recruited sequentially with similar patterns of neural activity is less likely to lead to synaptic elimination than competition between neurons with more asynchronous activity. Meirovitch et al. (2021) used a **connectome** approach to make a map of the connections between motor neurons and muscle fibres in a small mouse muscle, in both neonates and adults. They used serial section scanning electron microscopy and **Brainbow technology** (Livet et al., 2007), enabling different neurons to be labelled by differently coloured fluorescent proteins. This yielded a database of over 6000 neuromuscular junctions and several complete connectomes at different ages. They found that already at birth there was a correlation between the size (cross-sectional area) of the muscle fibres and the number of motor neurons that provided innervation to them, similar to that found in the adult, which could underlie the Size Principle. They also found that axons of similar motor unit size tended to share muscle fibres more often than expected in a random model. These conclusions and other evidence suggested that one interpretation of the data is that there is an activity-based linear ranking of motor neuron axons where neurons with similar activity innervate the same muscle fibres. This work offers possibilities to modellers, such as examining the consequences of whether activity in the contacts at one endplate can affect the contacts at nearby endplates on the same muscle fibre.

12.4 | Patterns of Ocular Dominance

In many species, information from the two eyes is combined in central brain structures to form binocular regions. Hubel and Wiesel (1962) discovered cells in cat visual cortex area V1 that are selectively responsive to stimulation through one eye. Similar ocularity preferences are maintained down through the cortical layers, from which the concept of **ocular dominance columns** was established. The anatomical basis for ocularity columns was confirmed later (LeVay et al., 1975), with the distribution of the ocularity preferences over the surface of the binocular cortex resembling a pattern of zebra stripes (Hubel et al., 1977; Hubel and Wiesel, 1977; LeVay et al., 1980). Similar patterns of ocular dominance are found in the retinotectal projection in **three-eyed frogs** where there is a projection from two eyes onto the same target (Constantine-Paton and Law, 1978) and in *Xenopus* where a compound eye made from two halves of matching retinal origin makes a superposed projection on the optic tectum (Fawcett and Willshaw, 1982).

Most work on ocular dominance columns has been in cat and monkey. Before birth, innervation from the two eyes forms rudimentary ocular dominance columns, which segregate over a period of about six weeks (Blakemore and Van Sluyters, 1975; Rakic, 1976; LeVay et al., 1978). There has been much research on investigating the plasticity of ocular dominance columns during the critical period in development; for reviews, see Katz and Crowley (2002) and Adams and Horton (2009). There has been extensive application of models of activity-dependent development based on Hebbian principles and the plasticity of ocular dominance in response to altered visual input, the first of which are due to von der Malsburg and Willshaw (1976) and Swindale (1980).

Swindale (1980) assumed that at each point in layer IVc, treated as a 2D surface, the growth of the synapses, from the axons originating from the right and left eyes via the lateral geniculate nucleus, was influenced by other synapses from other positions on the 2D surface. Synapses from the same eye which terminated on the cortex a short distance away exerted a positive effect, and those further away a negative effect. The influences from synapses from the opposite eye were reversed.

These assumptions enabled him to formulate a set of equations for the development of the synaptic densities $n_L(x, y, t)$ and $n_R(x, y, t)$ at any position (x, y) on the cortical surface. The function $w_{RL}(x - x', y - y')$ specifies the influence on a synapse at position (x, y) from an axon originating from the right eye by a synapse at position (x', y') from an axon originating from the left eye. The functions w_{RL}, w_{RR} and w_{LL} are similar functions for the other types of possible between-eye and within-eye interactions. The rates of change of n_L and n_R are:

$$\frac{dn_L}{dt} = (n_L * w_{LL} + n_R * w_{LR})h(n_L), \tag{12.11}$$

$$\frac{dn_R}{dt} = (n_R * w_{RR} + n_L * w_{RL})h(n_R). \tag{12.12}$$

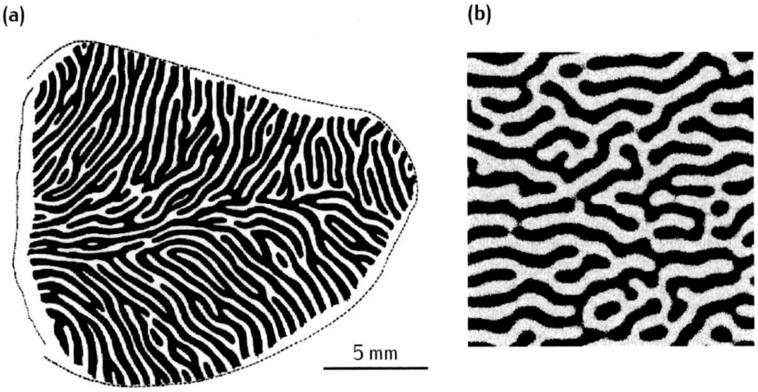

Fig 12.16 Comparison of anatomical and simulation patterns of ocular dominance. **(a)** Ocular dominance pattern in the monkey visual cortex (Hubel and Wiesel, 1977). **(b)** Simulated pattern. Figure in **(a)** reproduced with permission from The Royal Society of London: figure in **(b)** kindly supplied by Nick Swindale.

(a)

(b)

5 mm

The **convolution** of two functions $A(x, y)$ and $B(x, y)$ is:

$$\iint A(x', y')B(x-x', y-y')dx'dy'.$$

Convolution can be thought of as the function that results when spatial filter B is applied to function A.

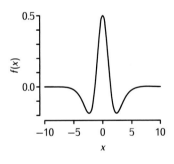

Fig 12.17 A Mexican hat function. This is described by a function $f(x)$ that is formed by subtracting one Gaussian function from another with different parameter values:
$f(x) = \exp(-x^2/2)$
$- 0.5\exp(-(x/2)^2/2).$

The asterisk denotes spatial convolution, the 2D version of convolution introduced in Section 8.6, and the terms in the equations involving convolution provide for the effect of interactions between synapses depending on distance.

A function $h(n)$ is used to keep the values of n_L and n_R within bounds. A suitable form for $h(n)$ which keeps n between 0 and a maximum value N is:

$$h(n_L) = n_L(N - n_L), \tag{12.13}$$
$$h(n_R) = n_R(N - n_R). \tag{12.14}$$

One special case that was considered is where, for all positions (x, y) on the surface, the sums of the left-eye and right-eye synapses are kept constant:

$$n_L + n_R = N. \tag{12.15}$$

One case that Swindale explored analytically was where the functions w_{LL} and w_{RR} describing the within-eye interactions are chosen to be excitatory at short range and inhibitory at long range (the **Mexican hat function**) and the between-eye interactions w_{LR} and w_{RL} are inverted Mexican hat functions (Figure 12.17). He examined the special case when the sum rule (Equation 12.15) is applied to the total synaptic density. In this case, the two identical within-eye functions are the exact negatives of the two identical between-eye functions. By looking at the Fourier components of the patterns of left- and right-eye synapses that emerge, he proved that, from an initial state in which the left- and right-eye synapses are randomly distributed, the cortical surface becomes partitioned amongst the left- and right-eye synapses to form a branching pattern of ocular dominance stripes (Figure 12.16a), which has the features of the patterns seen experimentally (Figure 12.16b). The width of the stripes is controlled by the spread of the interaction functions. He was then able to account for several findings in the experimental literature. For example, the fact that monocular deprivation leads to the stripe width of the synapses from the deprived eye shrinking in favour of the width of the undeprived eye synapses is explicable if it is assumed that the monocular deprivation causes reductions in the short-range interactions in both within-eye and between-eye densities (Hubel et al., 1977). Whilst Swindale could provide a mathematical description of the formation of ocular dominance stripes together with a number of specific predictions, the physical basis of his model's components are not specified. In particular, his notion of a synaptic density is not tied to any measurable quantity. In addition, in discussing the monocular deprivation experiments, he does not propose a precise mechanism for the synaptic interaction function. In this case, he cites the explanation given by Hubel et al. (1977) that 'the deprived eye is presumed to lose its effectiveness locally'.

An earlier paper (von der Malsburg and Willshaw, 1976) had given a more specific model for the formation of ocular dominance stripes, which is held to be influenced by electrical neural activity. Synapses from the same eye are assumed to fire with correlated activity, and the activity from different eyes is anticorrelated; in addition, there are short-range excitatory connections and long-range inhibitory connections between cortical cells. Combining the effects of these separate mechanisms enables Swindale's four synaptic interaction functions to be implemented. In addition, a Hebbian synaptic modification rule (Section 11.1.1) enables individual cells to acquire a specific ocularity, leading to regular patterns of ocularity over the 2D cortical surface. In this model more assumptions are made about the properties of neurons and synapses than Swindale's, and therefore the model is less general. On the other hand, its specificity makes it easier to test experimentally.

These two early models demonstrate the advantages and disadvantages of devising models at different levels of detail. Since then, a vast range of models have been developed at differing levels of complexity, all of which contain mechanisms for synaptic modification based on Hebbian principles, reviewed in Swindale (1996) and Goodhill (2018). In models where synaptic modification is unlimited, one problem is how cell activity can be kept with bounds. One solution is through a mechanism of competition for limited synaptic resources, which is an important topic in Section 12.3. Another solution is to use a combination of Hebbian modification with a mechanism of homeostatic plasticity, introduced in Section 11.6, which maintains average cell activity at a constant level. There is evidence for this. For example, Mrsic-Flogel et al. (2007) used **two-photon imaging** to show that after extensive monocular deprivation, the cortical response to the deprived eye was depressed initially and then recovered to the normal level. In modelling work, Toyoizumi et al. (2013) showed how changes in inhibition in the cortex during development can initiate the critical period for ocular dominance plasticity. Instead of viewing both homeostatic and Hebbian modification as acting on each synapse, they then showed that when the slow-acting homeostasis is controlled by postsynaptic activity alone, the development of ocular dominance with the correct time course results (Toyoizumi et al., 2014).

12.5 Retinotopic Maps

Many **topographically ordered maps** of connections exist in the nervous system. Mechanisms involving both electrical and chemical signalling are believed to determine map formation. Interest in this subject as a topic for modelling started over 80 years ago (Sperry, 1943, 1963) and has increased dramatically over the last 10 years, owing to improvements in visualisation techniques and new experimental methods for investigating the properties of topographic maps.

In a topographically ordered map of connections between two neural structures, the axons from one set of cell bodies make connections with the cells of another so that there is a geographical map of one structure on the other.

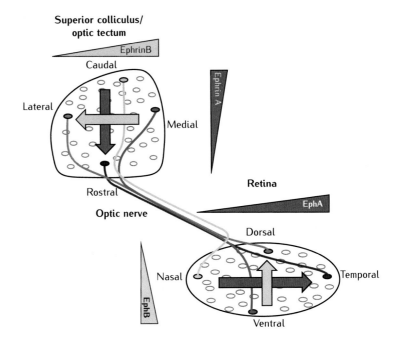

Fig 12.18 Retinotopic maps in the vertebrate visual system. Axons of the retinal ganglion cells (RGCs) form the optic nerve and travel to innervate the contralateral superior colliculus, or optic tectum. As the axons pass along the optic pathway to their targets, the topographic order initially present in the retina is largely destroyed, but it is recreated in the target region. The large arrows indicate the principal axes of the retina and the target region and their correspondence in the mature map. Each RGC contains EphA and EphB receptors. EphA receptor density increases smoothly from the nasal to temporal retina, and EphB density increases from the dorsal to ventral retina. In the colliculus, the density of the ligand ephrinA increases from rostral to caudal, and that of ephrinB from lateral to medial.

Langley (1895) was the first to propose a theory for the formation of specific nerve connections based on his experiments on the regeneration of nerve fibres in the sympathetic nervous system. At that time, only information about the regeneration of connections was available, which he used to argue about development. This situation lasted until the 1970s, when information about development started to become available.

In all vertebrates, the axons from the retinal ganglion cells grow out to form the optic nerve. In non-mammalian vertebrates, the main target structure of the retinal ganglion cell axons in the optic nerve is the **optic tectum**; in mammals, the homologous target is the **superior colliculus**. The projection of axonal terminals on their target is in the form of a 2D map of the retina onto the optic tectum or superior colliculus, in a fixed orientation (Gaze, 1970; Figure 12.18). Schmidt (2019) reviewed in depth what is known about the formation of the retinotectal projection.

How does each axon find its target site? The general modelling approach has been to develop a model to account for the production of ordered maps in normal animals and then to challenge the model to reproduce maps obtained in animals which have suffered disruption by surgical or genetic intervention.

Discussion of the wide variety of approaches to modelling the development of topographic maps will give us insights into many issues that arise in computational modelling of developing systems.

12.5.1 Experimental Data

The data come from neuroanatomical and electrophysiological investigations. Methods tracing the destination of axonal fibres that develop from one structure to another can yield the relationship between individual points in the two structures to high precision. However, since only a few mapping points can be derived from a single animal, the data have to be collected from a large number of animals and then replotted onto a standard brain, thereby removing any variation between animals. More information from a single animal can be derived from extracellular recordings from the tectum or colliculus in response to visual stimulation (Gaze, 1970). Using Fourier-based intrinsic optical imaging (Kalatsky and Stryker, 2003) enables higher-

Box 12.6 | Measuring maps using electrophysiological recording methods

In extracellular recording, the visual field position (or positions) evoking a response at each of a series of electrode positions in the colliculus or the tectum is found by exploring the entire visual field. Intrinsic imaging takes advantage of the spectral properties of haemoglobin which has different absorption properties when oxygenated or deoxygenated. The location of neuronal activity in the exposed brain evoked by a point stimulus in the visual field is measured as changes in light reflectance of the blood to show where the oxygenated blood, and therefore neuronal activity, is located. **Fourier-based** intrinsic imaging involves scanning the visual field repeatedly with a bar stimulus in two orthogonal directions, usually azimuthal (nasal to temporal) and elevational (dorsal to ventral). By extracting the component of the recorded signal at the scanning frequency, many artefacts encountered with the standard method can be removed. The position along one axis of the visual field that corresponds to each position on the exposed part of the brain is given by the phase of the signal. Scanning in two orthogonal directions reveals the 2D position in the visual field which is most strongly linked to the given position in the brain.

These methods are suitable for revealing a one-to-one map and a one-to-many map, where one point in the visual field maps to several points on the brain. However, problems arise in the case of a many-to-one map (e.g. two distinct areas of the visual field maps to one area on the target). In extracellular recording, electrical spikes recorded from different visual positions are often distinguishable, being of different heights when visualised on an oscilloscope. In Fourier-based intrinsic imaging, since only a single position in visual space associated with each point on the target structure can be calculated, in areas where there is a double projection, only an average position will be signalled, which prevents the entire multiple map from being reconstructed accurately. In standard intrinsic imaging, the visual system is stimulated by presenting single spots of light and so this problem does not arise.

resolution maps than from extracellular recording or from standard intrinsic imaging (Grinvald et al., 1986). All methods are ideal for reconstructing one-to-one maps but can be less suited for more complex maps, such as double representations of the retina on the target structure (Box 12.6; Willshaw and Gale, 2022).

The basic requirement of a model is that it must account for the development of an ordered 2D retinotopic map in a prespecified orientation. It must then also account for the different types of experimental maps found where retinal axons make connections with cells other than the ones that they would have contacted if part of a normal map – *connectivity plasticity* (Figure 12.19). Four types of connectivity plasticity have been observed.

In the adult. In adult fish, after removal of part of the retina or the tectum to form a size **mismatch**, an expanded or contracted ordered projection regenerates (Figure 12.19b, c). For example, a surgically

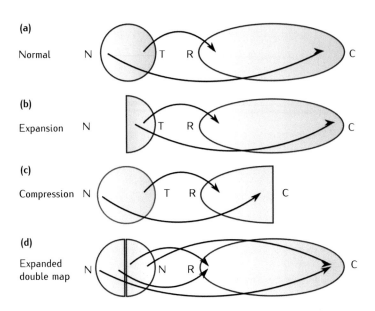

Fig 12.19 Schematic showing the results of various experiments on the retinotectal system of non-mammalian vertebrates where the retina or tectum has been altered surgically. (a) Normal projection showing that the temporal retina (T) projects to the rostral tectum (R), and the nasal retina (T) projects to the caudal tectum (C). (b) A half-retina expands its projection over the entire tectum. (c) A whole retina compresses its projection into a half-tectum. (d) A double nasal compound eye in *Xenopus* makes a double projection, with each of the two originally nasal poles of the retina projecting to the caudal tectum and the vertical midline projecting rostrally.

constructed half-retina eventually regenerates a map over the entire tectum, although normally it would be restricted to one-half (Gaze and Sharma, 1970; Gaze and Keating, 1972; Sharma, 1972; Schmidt et al., 1978). This result is referred to as **systems-matching** (Gaze and Keating, 1972).

During normal development. In the amphibians *Xenopus laevis* and *Rana*, as more and more retina and tectum develop, retinotectal connections are continually adjusted (Gaze et al., 1974, 1979; Reh and Constantine-Paton, 1983) so that the entire retina existing at each stage is mapped over the entire colliculus.

In compound eye maps. In embryonic *Xenopus laevis*, before the optic nerve has formed, a **compound eye** is constructed by replacing a half-eye rudiment by another half-eye rudiment of different embryonic origin. In the adult, compound eyes are of normal size and appearance, and a single optic nerve develops and innervates the contralateral tectum in the normal fashion. However, the projection made by a compound eye, as assessed by extracellular recording, is grossly abnormal. Each half-eye corresponding to the two half-eye rudiments brought together to make the compound eye projects in order across the entire optic tectum (Gaze et al., 1963; Gaze and Straznicky, 1980), instead of being localised to just one-half of the tectum. When the two half-eye rudiments are of matching origin, a double map results (Figure 12.19d). Later fibre tracing experiments showed that in these cases, ocular dominance columns are formed so that the two projections interdigitate (Fawcett and Willshaw, 1982).

In genetically modified animals. These results from mouse were not available to the earlier modellers. Here the retina and colliculus do not change significantly in size, whilst retinocollicular connections are being made, and so systems-matching might not be as important as in other species. Connection plasticity was seen in genetically modified

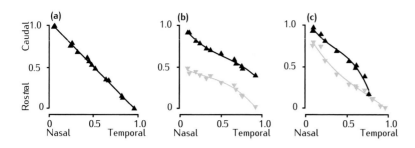

Fig 12.20 The projection of the nasotemporal retina onto the rostrocaudal superior colliculus in EphA3 knockin mice compared with wild types obtained by fibre tracing. **(a)** Wild-type projection. **(b)** Homozygous knockin (EphA3^{++}). **(c)** Heterozygous knockin (EphA3^{+-}). Black-filled triangles show projections from ganglion cells containing no EphA3; blue inverted triangles relate to EphA3-positive RGCs. From Willshaw (2006), redrawn from the results of retrograde tracing experiments (Brown et al., 2000). Original data kindly provided by Greg Lemke.

animals where use was made of the discovery of the existence of the **Eph receptor** and its ligand **ephrin**, in graded form across the axes of the retina and tectum/colliculus, respectively (Flanagan and Vanderhaeghen, 1998). These molecules exist in two forms, A and B, which could in theory specify a retinotopic order independently along the two axes of the map, in line with Sperry's ideas of chemospecificity (Sperry, 1943, 1944, 1945), described in Section 12.5.4 (Figure 12.21).

> **Knockout of ephrinA.** Knockout of ephrinA molecules thought to label the rostrocaudal axis of the mouse colliculus results in axons terminating in abnormal positions. In some cases, a small region of the retina projects to more than one position on the colliculus (Feldheim et al., 2000).
>
> **Knockin of EphA.** Introducing an extra type of EphA receptor, EphA3, into the Islet2 cells, distributed across the entire retina (Brown et al., 2000) and making up 50% of the entire population of retinal ganglion cells, defines two populations of ganglion cells. In homozygous knockins, each population makes an ordered map in the normal orientation on different parts of the colliculus (Figure 12.20b). In heterozygous knockins, the two maps coincide in the rostral colliculus (Figure 12.20). Where the map collapsed has been characterised (Reber et al., 2004). Similar findings resulted when EphA3 knockin was combined with the knockout of an engenic EphA type normally present in the retina (Bevins et al., 2011).

12.5.2 Mechanisms for Map Formation

Five types of possible mechanism have been discussed (Prestige and Willshaw, 1975; Price and Willshaw, 2000), with the latter three featuring in most contemporary models.

(1) **Fibre ordering.** Axons are guided to their target due to the ordering of nerve fibres within the pathway. In some species, there is order within the mature pathway, such as in the *Cichlid* fish (Scholes, 1979), but this is not maintained over the entire extent of the pathway.

(2) **Timing.** During neurogenesis, the earliest fibres reaching their target make contact with the earliest differentiating cells. This assumes a mechanism that converts positional information into temporal information. In invertebrates, Macagno (1978) showed that, in the *Daphnia* visual system, which connections are made depends on the time of arrival of the innervating retinal fibres. For 2D map-making, variation is

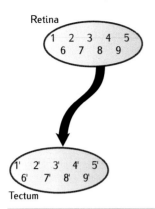

Retina

Tectum

Fig 12.21 The simplest form of the mechanism of chemoaffinity (Sperry, 1943, 1944, 1945) for the formation of specific retinotectal connections. Each retinal cell carries a molecular label which identifies it uniquely. Each tectal cell carries a similar identifying label. Each axon must be able to find and connect with the tectal cell carrying the corresponding label (1 with 1', 2 with 2', etc.).

Sperry (1943, 1944) observed that, after cutting the optic nerve in adult newt, the nerve regenerated and normal function was restored. When he did the same experiment after rotating the eye, the response to visual stimulation did not adapt but followed the degree of rotation. The inference drawn was that each regenerating axon had found its original tectal partner. By extension, the same mechanism could act to form the initial map. Later experiments in goldfish (Levine and Jacobson, 1974; Hope et al., 1976; Gaze and Hope, 1983) suggested that, after regeneration of the optic nerve following interchange of pieces of the optic tectum, regenerating retinal nerve fibres could locate their original partners in their new location.

required along two independent dimensions, whereas timing provides for variation along one dimension.

(3) **Chemoaffinity.** Sperry (1943, 1944, 1945) proposed that there are pre-existing sets of biochemical **markers** which label both retinal and tectal cells; and that the ordered pattern of connections observed during development is generated by the interconnection of each retinal cell with the tectal cell with the matching marker or label (Figure 12.21). Although various molecules had been proposed as molecular labels, evidence was lacking until the discovery in a variety of species of gradients of Eph receptors and of the matching ligands, the ephrins, in both the retina and colliculus (Section 12.5.1; Flanagan and Vanderhaeghen, 1998).

(4) **Neighbour–neighbour interactions through neural activity.** The patterns of electrical activity in neurons encode information enabling the appropriate contacts to be made. Lettvin (cited in Chung, 1974) suggested that 'electrical activities in the optic nerve may be utilised by the nervous system in maintaining spatial contiguity between fibres'. From this general idea, a set of models has arisen based on the concept that activity-based nearest-neighbour interactions amongst retinal and target cells lead to neighbouring retinal cells developing contacts preferentially with neighbouring target cells. This is an extension of the mantra 'cells that fire together wire together' used in the discussion of Hebbian plasticity (Section 11.1).

There is a lot of evidence about the effects of electrical activity in the retinocortical pathway, but comparatively little about the effects on retinotectal or retinocollicular map development. Retinal axons regenerating to the adult goldfish tectum either in the absence of neural activity through blockage of the sodium channels by tetrodotoxin (TTX) (Meyer, 1983; Schmidt and Edwards, 1983) or with the animal kept under stroboscopic illumination (Cook and Rankin, 1986) make less precise connections than controls. In mutant mice in which the $\beta 2$ subunit of the acetylcholine receptor has been deleted, leading to changes in neural activity, the receptive fields of collicular neurons become elongated along the nasotemporal axis and the topography of the retinocollicular map is degraded slightly (Feldheim et al., 1998; Mrsic-Flogel et al., 2005; Pfeiffenberger et al., 2006; Lyngholm et al., 2019).

For neural activity mechanisms to operate, several conditions must be satisfied. At the time when connections are formed, there must already be activity in the retinal cells. In addition, cells that are neighbours in the retina or in the tectum/colliculus must fire more strongly than non-neighbours. Usually it is assumed that this is through short-range excitatory connections between retinal cells and tectal cells. There is evidence that the retina is spontaneously active and that the degree of correlation between the firing patterns at two points on the retina decreases with increasing distance between them. Strongly correlated spontaneous activity has been demonstrated amongst ganglion cells in the adult at least (Rodieck, 1967; Arnett, 1978). Finally, synapses between the more active retinal and tectal cells must be strengthened preferentially. Usually a correlative mechanism is assumed of the type standardly invoked for models of synaptic plasticity in adults, such as the simple Hebbian form (Section 11.1.1) or spike timing-dependent plasticity (STDP; Section 11.2).

(5) Competition. There are different types of constraint imposed on the formation of a synapse arising from information gained from other synapses. Some of the mechanisms of competition are expressed at a high level.

> **Sum rule.** If the total strength of all the contacts made by a given retinal cell is kept constant, there will be competition amongst its synapses because the larger a particular synapse is, the smaller the other synapses made by that cell will be. Normally a sum rule is represented by a mathematical equation, rather than through a specific mechanism. One example of a sum rule already encountered is in the model for the elimination of superinnervation in developing muscle described in Section 12.3.1. Sum rules are needed on computational grounds to prevent instability through runaway of synaptic strengths and are justified in general and intuitive terms rather than through the existence of a specific mechanism.

> **Constant innervation density.** Another possible form of competition is through a mechanism keeping the density of contacts over the target structure at a constant level. Some authors have proposed this as a mechanism that overrides the assumed sets of fixed chemoaffinities so as to spread the connections from a surgically reduced retina over an intact tectum.

> **Homeostatic plasticity.** There is evidence that homeostatic plasticity mechanisms adjust synaptic strengths to maintain stability of the system in terms of the mean levels of neural activity. Such mechanisms can promote competitive effects during activity-dependent development, and these are reviewed by Turrigiano and Nelson (2004). Homeostasis is described in Section 11.6.

Although spike timing-dependent plasticity is usually invoked in models of learning and memory, it also occurs in a developing system, as shown in the retinotectal system in *Xenopus laevis* (Zhang et al., 1998).

12.5.3 Scope of the Models and Principal Assumptions

Most models are intended to apply once retinal axons have reached their target region, the optic tectum or the superior colliculus, where they make specific connections to form the ordered map. Therefore, any possible guidance cues in the 3D world which axons have to navigate to find their targets are usually not represented in the model. Here is a list of the principal assumptions:

(1) Both the retina and optic tectum/superior colliculus are usually treated as 2D planar structures. In some cases, the mapping between two 1D structures is assumed.

(2) The numbers of neurons in the two structures which are to interconnect are chosen to be large enough that a mapping of some precision can develop, but are usually much smaller than in the real neural system.

(3) The strength of connection between each retinal ganglion cell and each target neuron is expressed in terms of some known or assumed physical attribute of the contact between them, often referred to as the synaptic strength.

Fig 12.22 Sets of affinities between eight retinal cells and eight tectal cells displayed in matrix form for the two different types of chemoaffinity scheme identified by Prestige and Willshaw (1975). Each retinal cell is identified with a row and each tectal cell with a column, and the individual entries represent affinities. **(a)** In a **direct matching** scheme, each retinal/tectal cell has highest affinity with one particular tectal/retinal cell and less affinity with others. **(b)** In a **graded matching** scheme, each retinal/tectal cell has a graded affinity with all tectal/retinal cells. In their model of graded matching, Prestige and Willshaw (1975) interpreted affinities as contact lifetimes and connections were made at random. Once made, a connection would remain for the specified lifetime, which would be extended if, during this time, another contact was made between the same retinal and tectal cells.

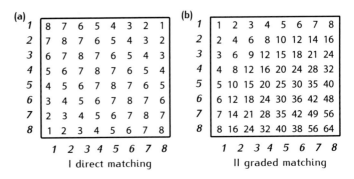

(4) The more detailed models are made up of a set of rules, usually expressed as a set of differential equations, for how synaptic strengths change over time.

(5) Most models involve the mechanisms of chemoaffinity, neighbour–neighbour interactions, mediated by neural activity, and competition.

12.5.4 Computational Models

To illustrate the types of mechanism that have been proposed, we describe how they are embedded in a representative sample of the large range of computational models, classified into three groups. In the first two groups, the driving mechanism is either chemoaffinity or neighbour–neighbour interactions, usually, but not exclusively, mediated by neural activity. The third group involves both types of mechanism.

Chemoaffinity

Sperry (1943, 1944, 1945) assumed that retinal and tectal labels match together like locks and keys. In a later paper, well before the involvement of Eph receptors and their associated ligands, the ephrins, had been recognised, the labels were assumed to be molecules that exist in graded form across the retina or tectum, with each molecule labelling a separate axis (Sperry, 1963). Each retinal cell would connect to the tectal cell with the matching label.

To account for systems-matching results, it was suggested that the surgical removal of cells had caused a reorganisation of labels in the surgically affected structure (Meyer and Sperry, 1973). For example, retinal hemiablation would cause the set of labels initially deployed over the remaining half-retina to become rescaled, such that the set of labels possessed by a normal retina would now be spread across the half-retina. This would allow the half-retina to come to project in order across the entire tectum (Figure 12.19). This subsidiary mechanism was called **regulation** by analogy with similar findings in the field of morphogenesis, where a complete structure can regenerate from a partial structure (Weiss, 1939).

An alternative explanation was that the labels do not change, and the flexibility in connectivity is possible through a form of competition (Gaze and Keating, 1972), where axons have graded affinity for a particular tectal cell. Their argument was that axons from a normal retina innervating a surgically reduced tectum will be subject to high competitive pressure from other axons and squeeze in to innervate the tectum available; conversely, the terminals from axons from a half-retina innervating a normal tectum will have less competitive pressure and would therefore be spread out over the entire tectum. How a combination of chemoaffinity and this informal notion of competition could work requires a computational model.

Prestige and Willshaw (1975) distinguished between two types of chemoaffinity schemes (Figure 12.22):

Type I (direct matching). Each retinal cell has the greatest affinity for a small group of tectal cells and less for all other cells, as in the original concept of chemoaffinity (Sperry, 1963). According to this scheme, there is no scope for flexibility in the connections made.

Type II (graded matching). There are graded affinities between the two sets of cells. All axons have high affinity for making connections at one end of the tectum and progressively less for tectal cells elsewhere. Conversely, tectal cells have high affinity for axons from one pole of the retina and less from others.

Prestige and Willshaw (1975) explored computational models of graded matching (Figure 12.22). Simulations of a 1D retina connecting to a 1D tectum showed that ordered maps can be formed only when competition is introduced by limiting the number of contacts that each cell can make. This ensured an ordered mapping. Without imposing a limit, the majority of the connections would be made between the retinal and tectal cells of highest affinity. In order to produce systems-matching when the two systems are of different sizes, the additional assumption had to be made that the number of connections made by each cell can be altered. The conclusion was that graded matching is not sufficient on its own and requires to be complemented by an additional mechanism.

Chemoaffinity – Energy-Based Models

Another approach to chemoaffinity that was originated by Fraser (1981) is to assume that signals from both axons and target cells generate an energy field for each axon, which gives rise to forces that guide it to its position of minimum energy.

Gierer (1983) pointed out that this formulation can be interpreted in different ways, giving fundamentally different implementations. If axons search randomly until hitting the position of minimum energy, at which point they make contact, then this is effectively the proposal due to Sperry (1963). If they are assumed to be directed by the forces generated in the field to move down the direction of the maximal slope of the field, then their resting position is where opposing forces are in equilibrium. This latter idea is reflected exactly in the notion of gradients and countergradients, which is favoured by a number of contemporary neuroscientists, for example, Rashid et al. (2005) and Marler et al. (2008).

A simple example given by Gierer (1983) of a 1D countergradient model is one where a molecule distributed as a single gradient over the retina is inhibited by a molecule from a similar gradient produced in the tectum. The tectal molecule is inhibited by the retinal molecule in a complementary fashion. We now use a discussion of this model to illustrate the issues arising in implementing models developed by others (Box 12.7).

Sperry's theory was formulated as a reaction to the **retrograde modulation** hypothesis of his PhD supervisor (Weiss, 1937a, b), another example of an abstract model. According to retrograde modulation, growing nerve fibres make contact with their target at random. Different retinal cells send out electrical signals of different types, with each tectal cell tuned to respond to the signal which is characteristic of a different retinal location. In this way, specificity between individual retinal and tectal cells is accomplished. How a tectal cell is tuned to a particular retinal signal was not addressed and therefore, as a mechanism for map-making, this hypothesis is incomplete.

Box 12.7 | Implementation of the Gierer (1983) model

The ability to reproduce a model is just as important as the ability to reproduce an experiment. It has not always been the case that authors publish the code which produced the published results. In order to reproduce model results, often a close reading of the original paper is required.

Throughout this book, we have endeavoured to reimplement simulations as much as possible. Here we describe in more detail than usual the exercise of implementing a model from the published details. We chose the **Gierer model** (Gierer, 1983) of development of connections from the retina to the tectum specified by Equations 12.18 and 12.19. He showed that, after removal of 12 of the original 24 retinal axons, the surviving 12 retinal axons form an expanded map over 24 tectal positions. According to Gierer (1983: p. 85):

> To allow for graded density distributions, each fibre unit is subdivided into 16 equal density units, and the position of each density unit is calculated separately. Density $\rho \ldots$ is the number of density units per unit area of tectum. If from [Equation 12.18] $\Delta p/\Delta x$ is negative, the density unit is shifted one position to the right, if $\Delta p/\Delta x$ is positive, the unit is shifted to the left. The process is carried out for all positions and density units, and reiterated, the time of development being given as the number of iterations. In the computer simulation, retinal and tectal areas extended from $x = u = 0.15$ (position 1) to $x = u = 3.6$ (position 24) with $\alpha = 1 \ldots$ and $\epsilon \ldots$ was taken as 0.005.

To describe the locations in the tectum and retina, we created two arrays, x and u, with components $x_j = 0.15i$, $i = 1, \ldots, 24$ and $u_i = 0.015j$, $j = 1, \ldots, 12$. The 12×16 matrix $\mathbf{L}$ records the locations of each axon's 16 terminals ('density units'). Each element of $\mathbf{L}$ contains the index of the tectal location to which the terminal projects. Each element of the density array ρ_j, $j = 1, \ldots, 24$, is computed by counting how many elements in $\mathbf{L}$ are equal to j.

It is difficult to understand what is calculated at each update step. Is just one or all of the 192 density units updated? Is the value of r updated every time a unit is shifted to the left or right? Exactly how is the gradient computed? For example, when comparing the values of the p to the left and right of a unit, what happens if *both* are lower? After some experimentation, we settled on the following. A time-step $\Delta t = 1/(16 \times 12)$ is chosen. At each time-step, a density unit (i, k) is chosen at random. Its position j is read out from L_{ik}. The gradient is estimated as $p_{i,j+1} - p_{i,j-1}$, a slightly different expression being used at the edges. The shift of i from position j to $j + 1$ or $j - 1$ is made as described by Gierer (1983), and the values of L_{ik} and ρ_j and ρ_{j-1} or ρ_{j+1} are updated. r is then updated: $r_j(t + \Delta t) = r_j(t) + \epsilon\rho_j\Delta t$.

Our simulations are almost certainly different from Gierer's. However, our results agree reasonably well with his, though in our simulations, development occurs about twice as quickly (Figure 12.23).

Consider an axon from retinal position u, where there is an amount $e^{-\alpha u}$ of the retinal molecule. At time t, the axon is at tectal position x, where there is an amount $e^{-\alpha x}$ of the tectal molecule, with α being a constant. The energy p at position x and the consequent force $\frac{dp}{dx}$ acting on the axon are:

$$p(x) = e^{-\alpha u}/e^{-\alpha x} + e^{-\alpha x}/e^{-\alpha u}, \tag{12.16}$$

$$\frac{dp}{dx} = \alpha(e^{-\alpha u}/e^{-\alpha x} - e^{-\alpha x}/e^{-\alpha u}). \tag{12.17}$$

The axon moves under the influence of the force so as to reduce the value of p until it reaches its unique minimum value, which is at $x = u$. This means that each axon, with a different value of u, will terminate at a different tectal coordinate x. Since the molecules are distributed in smooth gradients across the retina and tectum, respectively, an ordered map will result.

This mechanism is of the direct matching type, and so cannot account for the systems-matching behaviour following removal of retinal or tectal tissue from adult fish. To remedy this, Gierer (1983) added a competitive mechanism to equalise the density of innervation by including in the energy function (Equation 12.16) a term $r(x,t)$, which increases at a rate proportional to the local density of terminals $\rho(x,t)$. The effect of this is to maintain a constant density of terminals over the entire tectum. With the constant ϵ specifying the rate of increase of p, Equation 12.16 for energy p now becomes:

$$p(x,t) = e^{-\alpha u}/e^{-\alpha x} + e^{-\alpha x}/e^{-\alpha u} + r(x,t), \tag{12.18}$$

$$\frac{\partial r}{\partial t} = \epsilon \rho(x,t). \tag{12.19}$$

Consider the case of a surgically reduced half-retina reinnervating the goldfish optic tectum, where initially there is a map in the normal half of the tectum before expansion (Schmidt et al., 1978). According to the augmented model (Gierer, 1983), an axon would be subject to a chemoaffinity force directing it to return to its normal position, together with one that directs it into uninnervated tectum. The compromise between these two forces allows the half retinal projection to be distributed over the entire tectum. The reason for this is that, with the axons returning initially to the positions they would occupy if part of a normal map, the distribution of accumulated synaptic density will always provide a force directing axons into the uninnervated territory. Figure 12.23 gives representative results from our implementation of Gierer's (1983) model. Once synaptic density over the tectum has become uniform, at each time-step a constant value is added to energy p (Equation 12.18) and so the two opposing forces will always remain in equilibrium. The simulations presented by Gierer (1983) were for maps in 1D. Hjorth et al. (2015) extended this model to the formation of 2D maps.

The **Servomechanism model** (Honda, 1998, 2003) bears similarities to the Gierer (1983) model. It was the first to incorporate information about EphA receptors and ephrinAs in a computational model. Following experimental results (Cheng et al., 1995; Flanagan and Vanderhaeghen, 1998), it was assumed that EphA receptors and ephrinA ligands are arranged in gradients so that axons from the high end of the EphA gradient in the retina project to the cells at the low end of the ephrinA gradient in the colliculus,

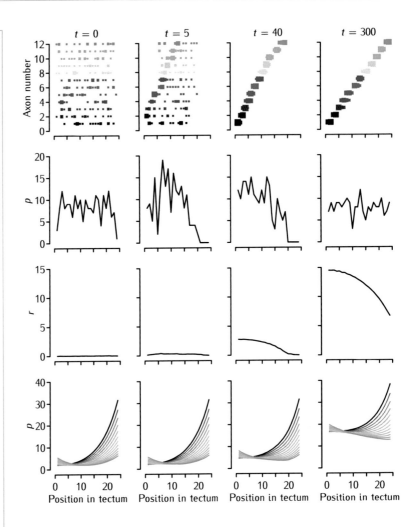

Fig 12.23 Simulation of the establishment of connections in the mismatch experiments, after Gierer (1983). Each column shows the system at a different time. **Top row.** The projection of each of the 16 density units from each of 12 axons onto the 24 tectal positions. Each axon is colour-coded, from axon 1 (black) to axon 12 (blue). The area of each square is proportional to the number of density units contributed by the axon to that tectal position. **Row 2.** The density of connections ρ at each tectal position. **Row 3.** The distribution of the accumulated density r over the tectum. **Row 4.** The value of the potential p at each tectal location for each of the 12 axons. The colour of the curve corresponds to the colours in the top row. At $t = 0$, the initial random state of the mapping with uniform density ρ over the tectum and r is uniformly zero. The minimum value of p for each axon is at the tectal position appropriate to a normal map. By $t = 5$, most of the axons fill the left-hand part of the tectum in a topographic fashion, with the preferred positions of the axons corresponding to the positions of the minima of potential p. By $t = 40$, prolonged occupancy of the axons in the left half of the tectum has caused the accumulated density r to increase in the left part of the tectum, decreasing towards the right. This causes the minima of the p curves to be shifted to the right, so that the optimal positions of axons are shifted to the right (top row). By $t = 300$, the process is essentially complete, with an ordered and expanded mapping from the retina to the tectum.

and vice versa, by means of the interactions between EphA receptors and ephrinA being repulsive.

Only the projection from the nasotemporal retina to the rostrocaudal colliculus was modelled. Axons are assumed to grow into the colliculus at the rostral pole. They are subject to a constant signal directing them towards the posterior colliculus. A second signal acts in the opposite direction, with its strength determined by the number of receptors bound by ligands. For an axon with amount R of receptor which binds to an amount L of ligand, the strength of the variable signal is the product RL. In the final stable state, when the mapping has been set up, each axon terminal on the colliculus is located where the constant force on the axon is counterbalanced by the variable force.

As is the case for the Gierer (1983) model, Honda's original model (Honda, 1998) is of the direct matching type, with no flexibility of connection. To remedy this, Honda (2003) allowed axon terminals at crowded target sites to move to adjacent sites in order to make the terminal density over the extent of the target more uniform. This competitive mechanism requires a method of measuring and comparing target density and is less effective than the one used by Gierer (1983) because, once the terminal density becomes uniform, axons tend to return to innervate the targets that they would do when part of a normal map. The difference may be because in the Gierer (1983) model, the extra mechanism to ensure uniform contact density involves changes in the energy function rather than in the individual forces.

Neighbour–Neighbour Matching

An alternative approach to accounting for systems-matching is based on the observation that in retinotopic maps, neighbours in the retina project to neighbours on the optic tectum.

Willshaw and von der Malsburg (1976) proposed the **Neural Activity Model** where the establishment of retinotopic maps is by means of a mechanism that connects neighbouring retinal cells to neighbouring tectal cells and showed how this can be implemented in electrical terms. They argued that the action of lateral inhibition, within both the retina and the tectum, coupled with synaptic modification through a Hebbian mechanism (Section 11.1.1), would achieve short-range order. For long-range order, a separate mechanism is needed to ensure the desired polarity of the map. With these two mechanisms acting together, systems-matching can be achieved.

Hope et al. (1976) proposed the **Arrow Model**, according to which two axons terminating at neighbouring target locations exchange positions if they are in the wrong relative position. A simple 1D analogy is that soldiers are required to line up in order of height by swapping positions with their neighbours. To do this, they need to distinguish the 'tall' end of the line from the 'short' end. This is a high-level model and how this knowledge is communicated to the axons to enable sorting to be carried out is not specified. There is a discrete number of target sites and contacts, which seems unbiological, as is also the process of pairwise swapping over of connections which has to be done without axons being tangled up. To account for the expansion and contraction of projections in the mismatch experiments, in the model, the swapping-over process is alternated with one of random exploration, so that all the available target structure is covered.

The Arrow Model is an example of an abstract model that is testable (Hope et al., 1976). It predicts that the map produced by allowing optic nerve fibres to reinnervate the adult tectum after a portion of the tectum has been removed, rotated and then replaced will be normal, except that the small portion of the map identified with the rotated part of the tectum will be rotated by a corresponding amount. However, since there is no information about absolute position on the tectum, if two parts of the tectum are interchanged without rotation, the model predicts a normal map. Hope et al. (1976) reported that the maps obtained after translocation do contain matching translocated portions, which falsifies this model.

Chemoaffinity with Neighbour–Neighbour Interactions

The **Marker Induction Model** (von der Malsburg and Willshaw, 1977; Willshaw and von der Malsburg, 1979) is a chemoaffinity model of direct matching, where the labels in the tectum change throughout development through neighbour–neighbour interactions (Box 12.8).

Box 12.8 | The Marker Induction Model

It is assumed that over the retina, there are concentration gradients of a number of transportable substances, at least one for each spatial dimension. Each retinal location is identified uniquely by the blend of concentrations of the various substances, which thereby act as labels or markers. These substances are transported down the axons to the axonal terminals, where they are injected onto the tectal surface at a rate proportional both to their concentration and the strength of the synapse. They then spread out to label tectal cells. Each synapse between a retinal fibre and a tectal cell has a **strength**, specifying the rate of injection of markers from axon to cell. In addition, it has a **fitness**, which expresses the similarity between the marker blends carried by the retinal and tectal cells.

Synaptic strengths are governed by three rules, which are implemented in the underlying dynamics:

(1) The strength of a synapse is set according to its fitness.
(2) Competition is introduced by imposing the condition that the sum of the strengths of all of the synapses from each retinal cell is fixed; if some synapses are strengthened, then others are weakened. This assures stability as no synapse can grow without bound.
(3) Each axon forms new terminal branches nearby existing ones, and branches with synapses whose strengths fall below a threshold value are removed.

Each tectal cell acquires, by induction through the synapses, the markers that are characteristic of the retinal cells that innervate it, at rates determined by the current strengths of these synapses. The synaptic strengths are changed continuously according to the similarity between the respective retinal and tectal configurations of markers, with lateral transport of markers in the tectum ensuring that neighbouring tectal cells carry similar blends of markers. As a result, each tectal cell acquires the markers that are similar to those in these retinal cells. The result is a map of synaptic strengths that is locally continuous, but without map orientation being specified. This can be done either by biasing the initial pattern of innervation or by providing weak polarising gradients of markers in the tectum. With the addition of this second mechanism, a continuous map in the required orientation results, with a copy of the retinal markers becoming induced onto the tectum.

In this model, the problem of how a set of markers, or labels, in one structure can be reproduced in a second structure in a way that is resistant to variations in the developmental programme for the individual structures is solved. The model is able to account for the systems-matching sets of results (Gaze and Keating, 1972), as well as those on the reinnervation of the optic tectum following graft translocation and rotation that suggest that in some, but not all, cases, retinal fibres return to innervate the same tectal cells (Yoon, 1971, 1980; Levine and Jacobson, 1974; Jacobson and Levine, 1975; Hope et al., 1976; Gaze and Hope, 1983). This behaviour is the crucial difference between implementing the nearest-neighbour idea in molecular terms rather than electrical terms as in the Neural Activity Model (Willshaw and von der Malsburg, 1976).

Support for the idea of marker induction came from Schmidt et al. (1978). They carried out a set of ingenious experiments in adult goldfish, in which half of one retina was removed. Subsequently, either the normal or the half-retina was forced to innervate the ipsilateral tectum that already carried a retinal projection. The nature of the tectal labels was inferred from knowledge of the retinal projection that it already carried. From the map made by the ipsilateral retina, they concluded that in these experiments, the surgically constructed half-retina always retained its original labels and the changes in labels were in the tectum, that is, the changes were induced from the retina (Schmidt et al., 1978; Schmidt, 2019).

Subsequently Willshaw (2006) explored the idea that the molecular labels required for the Marker Induction Model (Willshaw and von der Malsburg, 1979) are supplied by the EphA and EphB receptors for the two axes of the retina, and the ephrinA and ephrinB ligands for the colliculus. Based on fibre tracing experiments (Brown et al., 2000; Figure 12.20), Willshaw (2006) showed that the double maps formed in mutant mice in which half the retinal ganglion cells have an extra EphA receptor type are predicted from this model.

Several authors developed variants of previous models. Whitelaw and Cowan (1981) explored mappings in 1D where chemoaffinity (direct matching) and neural activity-based mechanisms act together to continually increase synaptic strengths, with a procedure of synaptic normalisation used to keep synaptic strengths within bounds. Overton and Arbib (1982) added a second influence to the Arrow Model which directed axons to their appropriate tectal positions, thereby adding a chemoaffinity mechanism which accounts for these tectal graft translocation experiments.

Chemoaffinity – Energy-Based – with Neighbour–Neighbour Interactions

We now examine in detail a series of energy models (Koulakov and Tsigankov, 2004; Tsigankov and Koulakov, 2006, 2010; Triplett et al., 2011), which were applied to the development of the retinocollicular map in the mouse.

The energy measure is made of three terms: (1) chemoaffinity; (2) nearest-neighbour interactions mediated through electrical activity; and (3) competition which limits the number of contacts. The chemoaffinity mechanism is of the graded matching type (Figure 12.22). In paradigms involving disturbance of the chemoaffinity gradients, this allows axons to connect with cells to which they would not normally connect.

In an earlier model, called here TK2006 (Tsigankov and Koulakov, 2006, 2010), there are as many retinal cells as collicular cells. In this case, the energy measure has just two terms, with competition being present implicitly through imposing one-to-one connectivity. The energy minimisation procedure involves a stochastic procedure whereby two randomly chosen retinal axons swap their connections probabilistically if this would lead to a decrease in energy.

In the Tsigankov/Koulakov models, axonal contacts are rearranged probabilistically. For example, in Tsigankov and Koulakov (2006), in the case of the mapping of the nasotemporal retina onto the rostrocaudal colliculus, if two axons with EphA levels R_1 and R_2 terminate at neighbouring positions on the colliculus where the ephrinA levels are L_1 and L_2, the probability that these two axons exchange their positions is:

$$P = 1/2 + \alpha(R_1 - R_2)(L_1 - L_2),$$

where α is a positive constant.

This means that exchange of axonal positions is likely when the difference between R_1 and R_2 has the same sign as the difference between L_1 and L_2. This will force axons with high EphA to terminate in regions of low ephrinA, and vice versa.

The converse mapping of the dorsoventral retinal axis to the mediolateral axis in the colliculus where high EphB should match to high ephrinB, and vice versa, can be achieved by reversing the plus sign in the equation.

In a later version of the model, called here TK2011 (Triplett et al., 2011), axons and cells were each allowed to have more than one contact. The energy measure now has an explicit competition term, penalising large or small numbers of contacts per cell. In this model, the optimisation procedure is the repeated application of two steps: a new axonal contact is selected at random for inclusion, and an existing contact chosen at random is selected for deletion. If the effect of each change is to decrease the overall energy, the change is accepted probabilistically.

Both of these models have been applied to various experimental paradigms involving the genetic manipulation of the distributions of EphA and ephrinA gradients over the retina and colliculus. Tsigankov and Koulakov (2006) showed that when ephrinAs are removed from the model colliculus, the model TK2006 shows the same qualitative features found in a variety of experiments involving the knockout of certain classes of ephrinA found in the normal mouse colliculus (Frisén et al., 1998; Feldheim et al., 2000).

Most attention has been paid to modelling the double maps produced in the EphA3 knockin experiments (Brown et al., 2000; Figure 12.20). Tsigankov and Koulakov (2010) showed that in the case of the homozygous knockin of EphA3, two separated maps were formed by TK2006 (as in Figure 12.20b). For the heterozygous case, in the model these two maps fuse in the rostral colliculus, as found experimentally (Figure 12.20c). This finding was explained on the basis that the signal-to-noise ratio of the probabilistic fibre switching is lower in the rostral colliculus because of lower EphA3 levels. The effect depends on the precise shapes of the Eph and ephrin profiles in the retina and colliculus.

Owens et al. (2015) analysed data from their Fourier-based intrinsic imaging studies of heterogenous EphA3 knockin maps. They analysed data from the azimuthal scan and characterised each map by two numbers calculated from sampling the rostrocaudal profile at three different mediolateral positions. They used these numbers to propose that the maps in heterozygous EphA3 knockins represent a phase transition between wild-type maps, where there is a single representation of the visual field on the colliculus, to maps in homozygous knockins where there is a double projection. This transition was proposed to be mediated by neural activity. They also reported that their own simulations of TK2006 (Tsigankov and Koulakov, 2006) confirmed their proposal.

Willshaw and Gale (2022) analysed the same data. They used the information from both elevational and azimuthal scans, enabling them to reconstruct the maps directly in 2D using their Lattice Method (Willshaw et al., 2014). They found no evidence that the heterozygous maps represent a transient state and instead found a variety of double heterozygous maps. They suggested that the variable potency from animal to animal of the extra EphA3 in the heterozygous retina is a cause of the variety of maps seen. They simulated this variability using the later model TK2011. Using the earlier model TK2006 was less satisfactory. The constraint of one-to-one connectivity imposed in the model forces the two populations of retinal cells to project to mutually exclusive regions, in patterns resembling zebra stripes, which makes the double maps more difficult to interpret.

Model Comparison

It is difficult to compare models for the topographic mapping problem. Different models may apply to different data sets. In addition, models tend to be formulated in different ways so that to enable comparison, they have to be reformulated in a common framework. Notwithstanding, two attempts have been made.

Triplett et al. (2011) compared three different models within the energy framework used for both TK2006 and TK2011 against the properties of the retinocollicular map in the mouse seen in wild types and in the **Math5 mutant** (Triplett et al., 2011). Here the number of retinal ganglion cells is reduced to 5–10% of the normal figure, and the map on the colliculus is restricted to the anteromedial colliculus. The models compared were TK2006 with single gradients and with dual gradients in both the retina and the colliculus and the Servomechanism model (Section 12.5.4). In all three models, the activity and competition components of the energy function were the same and the chemoaffinity term differed. The models produced equally good wild-type maps and TK2006 was adjudged to model the nonlinearities seen in the Math5 mutants best.

Hjorth et al. (2015) reimplemented four models in a common computational framework, so that differences in implementation were eliminated. They then selected a set of experimental data against which model performance could be assessed. The focus was on the mouse retinocollicular system where the plasticity of connections found in the amphibian system after surgical manipulation is not evident. The models were required to account for:

- The existence of a normal topographic map;
- The double maps reported in mutants with the homozygous knockin of EphA3;
- The partially double maps reported in mutants with the heterozygous knockin of EphA3;
- The compressed maps seen in mutants with homozygous knockout of Math5 leading to a reduced population of retinal cells.

In each model, the same numbers of cells were simulated and the same set of gradients used. The models are described in Section 12.5.4 and are:

(1) The Gierer model;
(2) The Whitelaw and Cowan model;
(3) The Marker Induction Model;
(4) The model called here TK2011.

All models contain mechanisms of chemoaffinity and competition with all but the Gierer model having a neighbour–neighbour mechanism, in the Whitelaw and Cowan model and TK2011 being mediated by neural activity.

Hjorth et al. (2015) found that to reproduce wild-type and the EphA3 knockin maps required a graded matching mechanism of chemoaffinity (Prestige and Willshaw, 1975). The only way in which these could be generated using a direct affinity mechanism was when it was coupled with a mechanism for inducing the tectal labels, as in Marker Induction. Only TK2011 reproduced a collapse point in heterozygous EphA3 knockin mice, on account of the strong activity-dependent mechanism. However, analysis (Willshaw and Gale, 2022) has shown that the existence and location of the collapse point depend on the relative amount of EphA3 knocked in. No model could account for the consistent, residual global order along the rostrocaudal axis in maps when all ephrinA ligands were removed (Cang et al., 2008; Willshaw et al., 2014). However, both TK2011 and the Marker Induction Model produced some order along the anterior–posterior axis. Only TK2011 and the Gierer model reproduced the Math5 phenotype (Triplett et al., 2011) where the projection is restricted to one portion of the colliculus.

12.5.5 General Pointers about Modelling the Development of Nerve Connections

We have given an extensive, but still not exhaustive, account of modelling attempts in order to illustrate the types of issues that arise and decisions that have been made in model construction and application. Here we summarise some key issues:

Availability and selection of data. Although the purpose is to model the development of a neural system, in most models, there is no comparison between experiment and model of the intermediate states of development. This is probably due to the fact that little data are available about the time course of the developmental process. Within this constraint, modellers tend to focus on one experimental paradigm. For example, the early models of topographic maps (Section 12.5) were of the retinotectal map in non-mammalian vertebrates where there is plasticity during normal development. At the time of writing, the focus is on the homologous retinocollicular projection in the mouse, where there is little plasticity, but manipulation of the system is possible by genetic means. The question here is how tailor-made a model should be to the data on which it is based. It is quite possible that in future, the focus will switch to the zebrafish retinotectal projection (Kita et al., 2015; Bollmann, 2019) where the modelling of the retinotectal projection may return to prominence. This system is accessible, displays plasticity during its entire life and is manipulable by surgical, genetic and chemical means. A variety of advanced tools are available to visualise the connectivity patterns at a detailed level.

Broadening the question. In some cases, the availability of new visualisation methods has expanded the focus of interest. For example, linked with the ability to visualise individual axons innervating a muscle, computational models for how individual axons in developing muscle make or retract connections have been developed (Section 12.3). Such models have yet to be linked to the older models of the withdrawal of superinnervation by a population of motor neurons.

Interpretation of experimental data. It is important for the modeller to be aware of the source of the data and how the data should be interpreted. For example, the estimates of the number of motor neurons remaining after partial denervation (discussed in Section 12.3.1) were derived indirectly from physiological measurements, whereas in the subsequent study of axonal movement at the endplate (Section 12.3.3), axons were visualised directly. As pointed out earlier (Box 12.6), using any of the known methods to reconstruct topographic maps has its limitations.

Realistic modelling. In most developmental situations, to use Marr's terminology (Marr, 1982), whilst usually it is clear as to what computation should be performed (e.g. the formation of ocular dominance columns or the elimination of superinnervation in developing muscle), there are different types of algorithm possible and different ways of implementing them. This means that it is possible to design different types of model for the same developmental phenomenon, with each model embodying different types of assumed mechanism. How far is it intended that these implementations can be realised directly in neural tissue, given that, in many cases, they are built on unverified assumptions? Using mechanisms in the model that are biologically plausible can be beneficial in choosing between alternatives. For example, different mechanisms of competition yield different results. In some cases, the mechanisms involve local operations which could be feasibly implemented by local signalling. A particularly difficult case to justify is the implementation of a synaptic conservation rule, which occurs in all models of the development of nerve connections discussed. For example, in the case of development of the elimination of superinnervation in developing muscle (Section 12.3), to maintain the total synaptic strength from all the contacts made by a motor neuron, constant information about the strengths of all the synapses has to be relayed back to the motor neuron. In this application, a possible solution is offered in the DCM (Section 12.3.2). Another case in point is the development of a topographic mapping according to the Koulakov series of models (Section 12.5.4) through changing connections to gradually lower an energy function which also requires a global calculation to be made.

Marr (1982) divided a computational task into (1) the computation to be performed; (2) the algorithm or method adopted; (3) the implementation of the algorithm in the available hardware.

12.6 Summary

The modelling of neural development has been a minority occupation amongst computational neuroscientists. One reason for this could be that, for many computational neuroscientists, the best-studied systems in developmental neuroscience, such as the development of single neuron morphology in the cerebellar cortex (McKay and Turner, 2005), of topographically ordered connections in the vertebrate retinotectal or retinocollicular system (Section 12.5) or of olfactory connections (Mombaerts, 2006), are straightforward examples of the instructions in the genome playing out during development. Therefore, it might be thought that to understand these systems does not present such a challenge to computational neuroscientists compared with trying to understand the role played by the mammalian neocortex in perception or learning. Another reason may be that, for many computational neuroscientists, the field of computational neuroscience studies how the brain computes (Churchland and Sejnowski, 1992), where the neurons are the computational elements and the currency of neural computation is the nerve impulse.

In this chapter, we have reviewed examples of modelling work in the exciting field of developmental neuroscience. Of the modelling methods reviewed in this book, this topic is the only one where both molecular and electrical mechanisms play pivotal roles. The future for developmental computational neuroscience is bright, with the advent of new animal models and new methods, such as genetic manipulation, to investigate the core mechanisms at work and to challenge existing models, together with new methods of visualisation of the development of the nervous system.

CHAPTER 13

Modelling Measurements and Stimulation

Candidate models for how neurons or networks operate must be validated against experimental data. For this, it is necessary to have a good model for the measurement itself. For example, to compare model predictions from cortical networks with electrical signals recorded by electrodes placed on the cortical surface or the head scalp, the so-called volume conductor theory is required to make a proper quantitative link between the network activity and the measured signals. Here we describe the physics and modelling of electric, magnetic and other measurements of brain activity. The physical principles behind electric and magnetic stimulation of brain tissue are the same as those covering electric and magnetic measurements, and are also outlined.

In previous chapters, the main focus has been on how to build computational models of brain activity. In order to test such models, their predictions must ultimately be compared with experiments. Action potentials can be measured with intracellular electrodes measuring the membrane potential directly. Action potentials can also be seen with sharp extracellular electrodes positioned next to the soma of the neuron where each action potential event gives a standardised signature in the recorded extracellular potentials (EPs), a spike. Here the link between the measurement and the modelled times of action potential events of a neuron in, say, a network model is clear.

For other experimental measures (Figure 13.1), the link between the model and what is measured is much less clear. For example, the **local field potential (LFP)**, that is, the low-frequency part of the **extracellular potential (EP)** measured inside the brain with extracellular electrodes, reflects the electrical activity of thousands of surrounding neurons. The same applies for **electrocorticography (ECoG)** where the recording electrode is placed on the cortical surface. Systems-level non-invasive measurements of brain activity, such as **electroencephalography (EEG)**, **magnetoencephalography (MEG)**, **functional magnetic resonance imaging (fMRI)** and **positron emission tomography (PET)**, measure the activity of hundreds of thousands or millions of neurons.

In optical measurements of neural activity, light of chosen wavelengths is shone on brain tissue and the resulting returning light is measured and

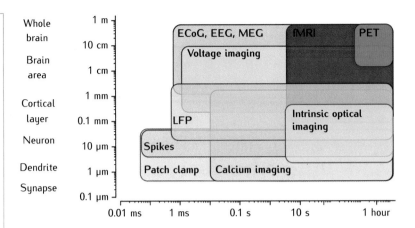

Fig 13.1 Overview of techniques for measuring neural activity with approximate bounds of their spatial and temporal ranges of applicability. Electrical (Sections 13.3–13.6) and magnetic techniques (Section 13.9) are in blue, whilst optical techniques (Section 13.10) are in yellow. fMRI (red) instead measures haemodynamic activity (Section 13.11), whilst PET (green) measures metabolic activity (Section 13.11).

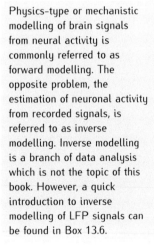

Physics-type or mechanistic modelling of brain signals from neural activity is commonly referred to as forward modelling. The opposite problem, the estimation of neuronal activity from recorded signals, is referred to as inverse modelling. Inverse modelling is a branch of data analysis which is not the topic of this book. However, a quick introduction to inverse modelling of LFP signals can be found in Box 13.6.

analysed. One type of such imaging was discussed in Box 12.6. Often reporter molecules are added to the brain tissue, such as in **voltage imaging**. In **voltage-sensitive dye imaging (VSDI)**, dyes are injected into the brain tissue, or genetically encoded to be produced in the cell, and the recorded signal measures the membrane potential in populations of neurons. In **calcium imaging**, specific fluorescent dyes binding calcium inside cells are added or genetically encoded to be present, and the fluorescence intensity gives a measure of the intracellular calcium concentration.

The traditional way of analysing such physiological data has been to look for statistical correlations between recorded signals and what the animal is doing or what stimulus is presented to it. For example, the receptive fields of neurons in the visual system described in Section 9.2 are found by correlating the recorded spiking of neurons with visual stimuli shown to the animal. This is an example of **inverse modelling** (see sidebox) where properties of neurons are inferred directly from measurements.

Instead this chapter is about **forward modelling** (see sidebox) of the link between models for brain activity, in particular neural and neural network models, and the various types of experimental measurements that are used to probe brain activity. Such modelling gives numbers for what the models predict for the various measurement modalities, and thus allows for a quantitative comparison of the models with experiments. It also provides intuition about what the various types of measurement can and cannot tell about the underlying neural activity, and thus also aids the qualitative interpretation of experiments. Compared to the modelling of brain activity itself, these **measurement physics** problems have received relatively little attention in the computational neuroscience community. For a thorough exposition of this subject, see Brette and Destexhe (2012).

There is also a growing interest in electric, magnetic and optical **stimulation** of neurons, both as a research tool and for use in neuroprosthetic devices. As the biophysics governing measurements and stimulation is very similar, we will also discuss the modelling of neural stimulation.

This chapter focuses on techniques for measuring brain activity. Anatomical techniques for measuring brain structure are not covered. We first describe the principles for modelling EPs (Section 13.2) before describing applications for modelling spikes (Section 13.3), LFP (Sections 13.4–13.5), ECoG and EEG signals (Section 13.6); see Figure 13.2. Electrical stimulation

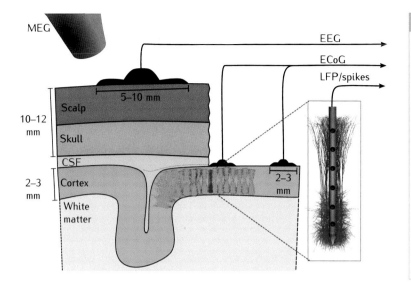

MEG

EEG

ECoG

LFP/spikes

5–10 mm

10–12 mm

Scalp

Skull

CSF

2–3 mm

Cortex

White matter

2–3 mm

Fig 13.2 Illustration of commonly used devices for electric and magnetic measurement of brain activity measurements at the network and whole-brain levels. EEG electrodes are placed on the scalp and ECoG electrodes on the cortical surface, whilst electrodes placed inside the cortex record LFP signals and spikes. In MEG, the tiny magnetic fields stemming from brain activity are measured by SQUIDs placed outside the head. The numbers refer to typical spatial dimensions in the human head.

is described next (Sections 13.7–13.8) before the principles for magnetic (Section 13.9) and optical (Section 13.10) measurements and stimulation are briefly outlined. Other techniques are mentioned in the final Section 13.11.

13.1 Membrane Potentials

The variable receiving most interest in neural modelling is the membrane potential. The principle for measuring the membrane potential at a particular position is straightforward: measure the potential difference between two sharp electrodes placed on each side of the cellular membrane. In practice, there are many technical challenges related to doing such measurements, including how to model the electrodes themselves so that accurate and reliable estimates of the membrane potentials can be obtained. One issue is the liquid junction potential that arises in a glass electrode due to different electrolyte solutions inside the neuron and inside the electrode (Section 10.3.4). For more discussion, see Box 14.3.

13.2 Modelling Extracellular Potentials

With electrodes placed inside the cortex, the action potential firing of nearby neurons is reflected in the high-frequency part of the recorded EP, frequencies larger than a few hundred hertz (Gold et al., 2006; Pettersen and Einevoll, 2008). In contrast, the low-frequency part of this EP, the so-called LFP, largely reflects how currents flow in dendrites of the neurons surrounding the recording contact following synaptic inputs (Einevoll et al., 2013a). Measurements of EPs are also used to study how action potentials propagate in axons (McColgan et al., 2017; Emmenegger et al., 2019) and activate synapses (Swadlow et al., 2002; Hagen et al., 2017).

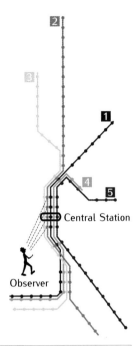

Fig 13.3 Illustration of the difference between membrane potentials and extracellular potentials by analogy with passenger traffic at the Oslo subway system; see Box 13.1 for explanation.

Box 13.1 | Membrane versus extracellular potentials

Modellers have sometimes confused membrane potentials and extracellular potentials, and incorrectly assumed them to be essentially the same when comparing with experiments. However, the signals are certainly not the same, and the connection between them is not trivial.

To illustrate this, we consider in Figure 13.3 an old map of the Oslo subway system where the structure of different lines ('dendrites') stretching out from Oslo Central Station ('soma') makes it resemble a neuron. The subway stations (marked as dots) could then correspond to 'neuronal compartments' and the net number of passengers entering the subway system at each station to the net 'membrane current' at this 'compartment'. If more passengers enter than leave a subway station at a point in time, the number of people, that is, the 'membrane potential', at this subway station increases. The soma membrane potential, crucial for predicting the generation of neuronal action potentials, would then correspond to the number of passengers at the subway station at Oslo Central Station. On the other hand, the extracellular potential would be more similar to what could be measured by an (eccentric) external observer counting passengers flowing in and out of a few neighbouring subway stations. Whilst the analogy is not perfect, it illustrates that membrane potentials and extracellular potentials really measure two different aspects of the same underlying neuronal activity (Pettersen et al., 2012).

The main source of EPs is electrical currents entering or leaving the extracellular space (ECS) through cell membranes. According to the volume conductor theory outlined in Section 10.2, these current sources will result in changes in the ECS potentials as the currents are driven through the ohmic extracellular medium. Computational modelling of these EPs thus follows the same principles: membrane currents computed from neuron models are combined with the volume conductor theory to produce signal predictions. In the simplest case where the potential is assumed to be recorded by an ideal point electrode and the brain tissue is assumed to be an infinite and isotropic volume conductor, Equation 13.4 below provides the key formula linking the EP and the membrane current. Whilst membrane potentials and EPs both are electrical potentials, their physical interpretations are quite different. A light-hearted illustration of this difference by use of an analogy with observations of passenger traffic in the Oslo subway system is given in Box 13.1; see also Figure 13.3.

As described in Section 10.3, diffusion from ion concentration gradients in the ECS may also set up electrical potential differences in the ECS, but, if present, these are expected to affect only the lowest frequencies of the recorded signals, that is, lower than a few hertz, since diffusion is a slow process.

13.2.1 Extracellular Potentials from Two-Compartment Neurons

The simplest neuron model generating EPs is the two-compartment neuron with a soma compartment and a single dendrite compartment (as for the

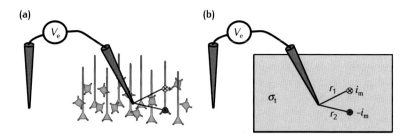

(a)

(b)

Fig 13.4 Illustration of the principles for modelling the EP generated by neural activity. **(a)** A single two-compartment neuron receives a single excitatory synaptic input onto the top (dendrite) compartment (red triangle). A sharp electrode is inserted into a piece of brain tissue, and the potential difference between the electrode tip and a reference electrode positioned (infinitely) far away is measured. **(b)** In the volume conductor theory, the conductive properties of the ECS is described by the extracellular conductivity σ_t. Assuming an infinite volume conductor with a constant conductivity, the potential measured by the electrode is given by Equation 13.1 where $i_m(t)$ is the net current entering the top (dendrite; blue dot) compartment and leaving the bottom (soma; red dot) compartment. Further, r_1 and r_2 are the distances from the electrode tip to the top and bottom compartments, respectively.

Pinsky–Rinzel model in Section 8.1). The principles of how to compute the EPs generated in this model are illustrated in Figure 13.4. Here a piece of brain tissue is considered, and an extracellular recording electrode measures the potential difference between the electrode tip and a reference electrode far away. For simplicity, we assume that only a single neuron is active, that is, only this cell has non-zero membrane currents.

This two-compartment neuron receives an excitatory synaptic input from a single presynaptic action potential at its dendrite. As described in Chapter 7, such a synaptic input will result in a (positive) current entering the neuron, corresponding to negative values for the current i_m depicted in Figure 13.4. This negative current is called a **current sink**. From a two-compartment description of neuron dynamics, it follows that a current sink in the dendrite compartment must be accompanied by a membrane current of identical magnitude leaving the soma compartment, a current source (Box 13.2). From the point of view of an electrode positioned outside the neuron, as in Figure 13.4, these will be the two electrical currents originating in the active neuron. According to the volume conductor theory (Section 10.2), each current will give a contribution to the potential in the ECS.

For the case where the extracellular conductivity σ_t is isotropic, the contribution from each membrane current to the recorded potential will be inversely proportional to the distance between the position of the membrane current and the electrode tip (Equation 10.15). Thus the EP recorded by the electrode is predicted to be:

$$V_e(\vec{r}_{el}, t) = \frac{i_m(t)}{4\pi\sigma_t|\vec{r}_{el} - \vec{r}_d|} - \frac{i_m(t)}{4\pi\sigma_t|\vec{r}_{el} - \vec{r}_s|}. \tag{13.1}$$

Here i_m is the membrane current that includes both the ionic membrane currents and the capacitive membrane current (see sidebox). Further, $\vec{r}_d$ and $\vec{r}_s$ are, respectively, the centre positions of the dendrite and soma compartments (where the membrane currents are assumed to enter the ECS), and $\vec{r}_{el}$ is the position of the electrode tip.

This formula illustrates why a single-compartment neuron model does not generate EPs: if the two-compartment neuron is 'shrunken' so that the two compartments are located at the same position ($\vec{r}_d = \vec{r}_s$), it effectively becomes a single-compartment neuron. For this situation, the formula in Equation 13.1 predicts zero EP.

The EP that would be measured around a two-compartment neuron model is illustrated in Figure 13.6a. With the electrode tip positioned around the dendrite compartment, a negative potential deflection will be observed

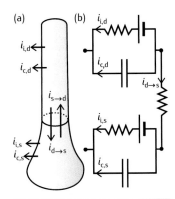

Fig 13.5 Source and sink currents in a two-compartment neuron with a dendrite and soma compartment (a) and its associated circuit diagram (b).

Box 13.2 | Source and sink currents are balanced

The source and sink membrane currents across the surface of a neuron always sum to zero. This can be shown by considering a two-compartment neuron with a dendrite and soma compartment (Figure 13.5a) and its associated circuit diagram (Figure 13.5b).

For this simple neuron, Kirchhoff's current law introduced in Chapter 2 gives:

$$i_{c,d} + i_{i,d} + i_{d \to s} = 0, \tag{a}$$

$$i_{c,s} + i_{i,s} + i_{s \to d} = 0. \tag{b}$$

Equation (a) is for the dendrite compartment; Equation (b) is for the soma compartment, and i_c and i_i represent the capacitive and ionic currents across the membrane, respectively. Furthermore, $i_{d \to s}$ is the axial current from the dendrite to the soma compartment, and $i_{s \to d}$ is the axial current the other way. These axial currents are obviously the same but have opposite signs, so that $i_{s \to d} = -i_{d \to s}$.

Since the total membrane current i_m is the sum of the capacitive and ionic currents, we have:

$$i_{m,d} = -i_{d \to s}, \tag{c}$$

$$i_{m,s} = -i_{s \to d}. \tag{d}$$

By summing each side of this equation set, we find:

$$\sum_{j=d,s} i_{m,j} = i_{m,d} + i_{m,s} = -(i_{d \to s} + i_{s \to d}) = -i_{d \to s} + i_{d \to s} = 0, \tag{e}$$

demonstrating that the total membrane current across the neuronal surface sums to zero.

This argument demonstrates that current monopoles cannot exist for two-compartment neuron models. The same reasoning can be used to show that it also applies for multicompartmental neuron models with any number N compartments, that is, $\sum_{j=1}^{N} i_{m,j} = 0$ (Equation 13.5).

because the sink contribution (first term in Equation 13.1) dominates. Around the soma compartment, there is the exact opposite pattern of potential because the soma term (second term in Equation 13.1) dominates. At the horizontal midline between the soma and dendrite compartments, the EP is exactly zero as the contributions from the soma and dendrite membrane currents cancel out.

Figure 13.6b shows that with the model parameters chosen for this particular two-compartment model, the net membrane current i_m in the dendrite compartment is much smaller than the synaptic current i_{syn}. This is because a large fraction of the input current returns as passive leak currents i_{return} in the same compartment.

Whilst the membrane current entering the dendrite compartment is the same as the membrane current leaving the soma compartment, the membrane depolarisation is initially much larger in the dendrite compartment than in the soma compartment (Figure 13.6c). However, after 20 ms or so, the membrane potential has equilibrated within the neuron and decays exponentially with the membrane time constant. The same phenomenon

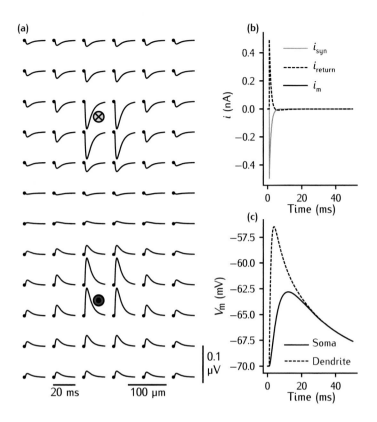

Fig 13.6 **(a)** EP from a two-compartment model, with passive membranes receiving an excitatory synaptic current in the dendrite compartment (blue dot). Black dots at the start of EP traces denote the spatial positions of the EP. **(b)** Currents in dendrite compartment following synaptic activation: synaptic input current $i_{\text{syn}}(t)$, passive return current $i_{\text{return}}(t)$, net membrane current (synaptic + return) $i_{\text{m}}(t)$. **(c)** Membrane potentials V_{m} in soma and dendrite compartments following synaptic activation. The resting potential is $-70\,\text{mV}$. Analogous to Figure 4.2 in Pettersen et al. (2012).

with initial spatial variation of the membrane potential followed by spatial equilibration and slow relaxation to the resting potential was observed for a semi-infinite cable receiving a transient current input in Figure 5.13.

In the so-called far-field limit where the recording electrode is far away from the neuron, the EP around a two-compartment neuron gives rise to a **dipolar** pattern where the expression in Equation 13.1 simplifies to (Box 13.3):

$$V_{\text{e}}(r,t) = \frac{i_{\text{m}}(t)l_d}{4\pi\sigma_{\text{t}}r^2}\cos\theta. \tag{13.2}$$

Here l_d is the distance between the current source and sink, that is, the distance between centre points of the two compartments. Further, θ is the angle of the recording electrode with the line between the source and sink, and r is the distance between the midpoint of this line and the recording electrode. The product of the current i_{m} and the source–sink separation l_d corresponds to the **current dipole moment** p. The far-field potential, that is, the potential at distances much larger than the source–sink separation, set up by a **current dipole** can thus be written as:

$$V_{\text{e}}(r,t) = \frac{p(t)}{4\pi\sigma_{\text{t}}r^2}\cos\theta. \tag{13.3}$$

A notable feature, in addition to the angular variation of the potential, is that the potential around a dipole decays faster than around a single source or sink, that is, as $1/r^2$ rather than as $1/r$. From the dipolar formula in Equation 13.2, it follows that when the spatial distance d becomes small, the generated EP also becomes small. In fact, in the limit when $d \to 0$, the EP

The membrane current i_{m} that enters EP formulae like in Equation 13.1 includes both the ionic membrane currents and the capacitive membrane current. According to the volume conductor theory (Section 10.2), the EP arises from currents entering or leaving the ECS at the outside of the cellular membranes, and the solely ionic currents immediately outside the membrane are the sums of the ionic and capacitive currents through the membranes.

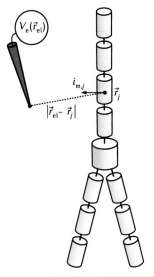

Fig 13.7 Illustration of the multicompartmental forward–modelling formula (Equation 13.4). The EP $V_e(\vec{r}_{el}, t)$ measured by an electrode with the tip at the point $\vec{r}_{el}$ is proportional to the sum of the membrane currents $i_{m,j}$ from each compartment j centred at $\vec{r}_j$, weighted by the inverse of the distance between the compartment j and the electrode.

vanishes. This again demonstrates that a spatial separation between a source and a sink is required to generate an EP and that single-compartment neuron models do not generate such signals.

13.2.2 Extracellular Potentials from Multicompartmental Neurons

The formula for the EP around a two-compartment neuron in Equation 13.1 generalises in a straightforward manner to the case when the neuron model has many compartments. For a neuron model with N compartments, the EP is given by (Figure 13.7):

$$V_e(\vec{r}_{el}, t) = \frac{1}{4\pi\sigma_t} \sum_{j=1}^{N} \frac{i_{m,j}(t)}{|\vec{r}_{el} - \vec{r}_j|}. \tag{13.4}$$

The requirement of a balance between current sources and sinks (Box 13.2) necessitates that:

$$\sum_{j=1}^{N} i_{m,j}(t) = 0. \tag{13.5}$$

For $N = 1$, Equation 13.5 gives that $i_m(t) = 0$. This current balance requirement implies, as discussed in Box 13.2, that neurons do not give rise to **current monopoles**. However, note that a current injected into the interior of a neuron by an electrode will give an EP with a monopolar component. In this situation, the return current leaving the neuron into the ECS will not be balanced by a membrane current going the opposite way, and the net membrane current will no longer be zero.

The formula in Equation 13.4 gives the recipe for computing EPs generated by arbitrarily complex multicompartmental neurons. Since the EPs generated by several neurons add up linearly, the formula can be used to compute the EPs generated by populations of neurons as well.

13.2.3 Effects from Spatially Varying Conductivities

The formula in Equation 13.4 assumes that the extracellular conductivity σ_t has the same value everywhere. Whilst σ_t has been measured to be quite constant within the cortical grey matter (Goto et al., 2010), it is, for example, lower in the white matter immediately below it due to the insulating glial cell wrapping of the axons there. Moreover, the cortical surface is enveloped by a poorly conducting skull with a low value of σ, with a thin layer of a highly conductive fluid (the cerebrospinal fluid (CSF)) sandwiched between the skull and grey matter. These discontinuities in electrical conductivity do not only affect the values of the EPs above or below the cortex, they also affect the values of the EPs close to the interfaces inside the cortex: inside the cortex close to an interface with the lower-conducting white matter, the EP will, for example, be larger compared to the situation with no discontinuity in σ_t.

For positions close to a planar step-like discontinuity in the extracellular conductivity, the EPs can be computed by a simple extension of the formula in Equation 13.4 valid for infinite homogeneous conductors, that is, the same value of σ_t everywhere. This is done by means of the Method of Images (MoI) (Box 10.4). Here the discontinuity in σ_t is accommodated in the modelling by adding appropriately chosen virtual current sources to

Box 13.3 | Dipolar EP around the two-compartment neuron

Far away from the cell, the EP set up by the two-compartment neuron gives a characteristic dipolar pattern. In this regime, the EP pattern can be described by a current dipole moment $\vec{p}$. This is illustrated in Figure 13.8, where the computation of the EP around such a neuron is depicted.

As described in Equation 13.1, the EP generated by two such balanced currents is given by (Figure 13.8a):

$$V_e(\vec{r}) = \frac{i_m}{4\pi\sigma_t r_+} - \frac{i_m}{4\pi\sigma_t r_-} = \frac{i_m(r_- - r_+)}{4\pi\sigma_t r_+ r_-}. \tag{a}$$

From trigonometry, it follows that: $r_-^2 = r_+^2 + l_d^2 + 2r_+ l_d \cos\theta'$ or $r_- - r_+ = l_d(l_d + 2r_+ \cos\theta')/(r_- + r_+)$ so that:

$$V_e(\vec{r}) = \frac{i_m l_d(l_d + 2r_+ \cos\theta')}{4\pi\sigma_t r_+ r_-(r_- + r_+)} \approx \frac{i_m l_d \cos\theta}{4\pi\sigma_t r^2}. \tag{b}$$

In the final step, the **far-field limit** $l_d \ll r$ has been assumed so that $r_- \to r, r_+ \to r$, and $\theta' \to \theta$.

The current dipole moment is defined as $\vec{p} = i_m l_d \vec{e}_p$, where $\vec{e}_p$ is a unit vector pointing from the sink current $(-i_m)$ to the source current $(+i_m)$. Then the dipole EP is given by:

$$V_e(\vec{r}) = \frac{\vec{p} \cdot \vec{e}_r}{4\pi\sigma_t r^2}, \tag{c}$$

where $\vec{e}_r$ is a unit vector pointing in the radial direction towards the electrode tip.

This dipole formula is only applicable far away from the neuron. This is illustrated in Figure 13.8b comparing the EPs predicted by the dipolar formula (Equation c) with the two-monopole formula (Equation a). In the present example, a dipole length $l_d = 0.8$ mm is assumed and the two formulae are seen to give very similar results for distances larger than 1–2 mm. The upper pair of curves corresponds to $\theta = 0$, which is in the vertical direction where the EP is maximal, whilst the lower pair of curves corresponds to $\theta = 60°$. A current $i_m = 0.05$ nA is used, roughly the maximum magnitude of the net membrane current in Figure 13.6.

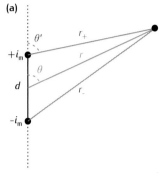

(a)

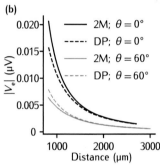

(b)

Fig 13.8 Dipole approximation for the two-compartment neuron model. **(a)** Mathematical variables used in derivation of the dipole formula in Box 13.3. **(b)** Comparison of magnitudes of potentials around the two-compartment neuron computed using the two-monopole formula (2M, Equation 13.1) and dipole approximation the (DP, Equation 13.2). See Box 13.3 for further details.

impose the correct boundary condition on the electrical potential at the planar interface. For examples of the use of this method, see Pettersen et al. (2006, Figure 4).

The MoI can be applied only to highly symmetric situations with one or more step-like planar discontinuities in the extracellular conductivities; Box 13.5 gives an example. However, solutions for the EP V_e can always be found numerically for any spatial variation of σ_t by use of mathematical schemes such as finite element modelling (FEM) (McIntyre and Grill, 2001; Ness et al., 2015).

13.2.4 Modelling Extracellular Recording Electrodes

The formulae for the EPs V_e in Equations 13.1, 13.3 and 13.4 all assume that the recording electrodes are point-like and ideal in the sense that the electrode is assumed to measure the potential at a single position and also not affect the recording itself. In practice, such ideal point electrodes do not exist. The electrode device occupies physical space, and the contacts themselves are never completely point-like. For metal electrodes, the metallic contacts, which pick up the potentials, vary in size from a few micrometres in diameter up to many millimetres. This will lead to a spatial smearing effect where the measured potential reflects a spatial average of the EP that would have been measured by an ideal point electrode. This effect can be modelled by means of the **disc-electrode approximation** (Moulin et al., 2008; Lindén et al., 2014) where the potential is computed as the average of point-electrode potentials computed across the contact surface of the electrode:

$$V_{e,\text{disc}} = \frac{1}{A_{\text{disc}}} \iint_{A_{\text{disc}}} V_e(\vec{r}') \, dA'. \tag{13.6}$$

Here the integral is over the surface of the electrode contact with area A_{disc}.

In addition to this spatial averaging effect, metal electrodes also affect the electric field, and thus the electrical potential around them. Metals are highly conductive, and from the electrostatic theory, it is known that at a metallic surface, the electric field is always oriented perpendicular to the surface. This effect can be included in the model by considering a spatially varying electrical conductivity $\sigma_t(\vec{r})$ with a very large value for σ_t for the positions corresponding to the metal electrode. Furthermore, the non-conducting electrode shanks, for which $\sigma \approx 0$, can be quite sizeable and also affect the recorded potential.

For the cases of spatially varying electrical conductivity in the brain considered above, solutions for the EP V_e can always be found numerically for any shape or size of the electrode by FEM (McIntyre and Grill, 2001; Mechler and Victor, 2012; Lempka and McIntyre, 2013; Ness et al., 2015; Buccino et al., 2019). This numerical scheme is more cumbersome to use than the disc-electrode approximation where the potential is given by the formula in Equation 13.6. Thus the disc-electrode approximation is preferred, given that it provides sufficiently accurate results. Comparison of results using the disc-electrode approximation and FEM calculations suggests that this approximation is accurate within 1% or so when the current source is positioned further away from the contact than the diameter of the electrode contact (Ness et al., 2015, Figure 5).

In some experiments, slices of brain tissue are excised from animal brains and studied in vitro in a dish of saline. The measurement dish and saline will have different electrical conductivities than the brain slice, and this must be taken into account for proper modelling of the measured electrical signals; see Box 13.5 for an example.

13.3 | Spikes

When a sharp electrode is placed close to the soma of a neuron, an action potential will be seen extracellularly as a sharp 'spike' in the EP (Figure 13.9). The magnitude of this signal is quite different from an intracellularly recorded action potential. Whilst action potentials have amplitudes of about 100 mV, the amplitudes of extracellular spikes are typically less than 1 mV (Figure 13.9c). Nevertheless, the detection of these spiking events from

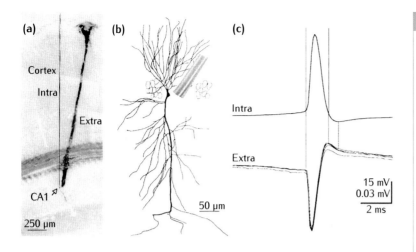

Fig 13.9 Simultaneous recording of intracellular and extracellular action potentials ('spikes') from a pyramidal cell. **(a)** Arrangement of intracellular and extracellular electrodes in the brain. **(b)** Tracing of neuron recorded from extracellular electrode is indicated. **(c)** Intracellular (top) and extracellular (bottom) recordings of a spike. Depicted potentials are the average over 849 spikes. The four lines shown for the EPs correspond to recordings from the four contacts constituting the multichannel tetrode electrode. The scale bar indicates the different scales for the two recordings. Adapted from Henze et al. (2000). Reproduced with permission from the American Physiological Society.

the recorded signal is relatively easy with modern electrodes and electronics, and faithful recordings of the firing activity of neurons can be obtained. Such measurements have been instrumental in learning about **neural representations** and receptive fields, that is, how neurons in different parts of the brain represent or encode information, for example, about visual and other sensory input. An example of modelling of a receptive field for a neuron in the visual system is given in Section 9.2.2.

An illustration of the difference in the shape and magnitude of the action potential and the corresponding extracellular signal (spike) for a ball-and-stick neuron is given in Figure 13.10. Whilst the signal characteristics are very different, nevertheless the spike provides a clear signature of an action potential in the adjacent neuron. This allows the times of occurrences of action potentials to be reliably measured in experiments, provided the spike amplitudes are large compared to the ambient noise level. A sharp recording electrode placed in brain tissue will pick up spikes typically from several neurons positioned in the vicinity of the electrode tip. Often one would like to know the train of spikes coming from individual neurons, and if so, the recorded spike data must be sorted in a process referred to as **spike sorting** (Box 13.4).

13.3.1 Spikes from Two-Compartment Neuron Models

Intracellular potentials can be modelled with a single-compartment neuron model. However, as described in the previous section, a two-compartment neuron model is the simplest model that can produce an EP such as a spike. An example of this is provided in Figure 13.11d where active sodium and potassium conductances have been added to the soma compartment of an otherwise passive two-compartment neuron model, and the EP is computed from Equation 13.1. Around the soma, a characteristic EP spike with a sharp negative peak followed by a slower positive hump is seen, in accordance with typical experimental spike recordings as exemplified by Figure 13.9c. Around the dendrite compartment, inverted spikes of the same sizes are seen, but such large inverted spikes are rarely, if ever, seen in experiments. This suggests that modelling dendrites as a single compartment is inadequate

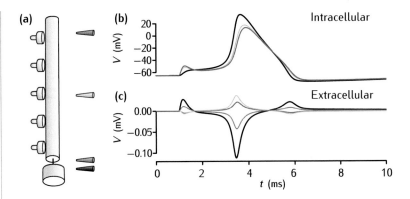

Fig 13.10 Simulation of EPs during an action potential ('spike'). **(a)** Extracellular electrodes (black, grey, blue and dark blue) are placed close to a ball-and-stick model neuron with an active soma and synapses on its dendrite. The soma is 40 μm long and 40 μm in diameter. The single dendritic cable is 200 μm long and 4 μm in diameter. **(b)** Intracellular recordings when the synapses are activated enough to cause an action potential to be fired. Traces are from the soma (black), halfway down the dendrite (blue) and in the distal dendrite (dark blue). The initial synaptic stimulation can be seen in the dendritic traces. **(c)** EPs computed using Equation 13.4. Colours refer to colours of extracellular electrodes depicted in (a). During the synaptic stimulation, the dendrites act as a sink of extracellular current and the soma acts as a source. This can be seen in the negative deflection of the EP in medial and distal dendrites and in the positive deflection of the EP close to the soma. During the action potential, the soma is a sink of current and the dendrites are current sources; this is reflected in the large negative deflection of the EP close to the soma and the smaller deflections of the EP near the dendrites. As the neuron repolarises, the roles of the soma and dendrites are again reversed.

when one is interested in predicting the detailed spatial pattern of spike shapes that would be recorded by an electrode at different positions around a neuron.

13.3.2 Spikes from Multicompartment Neuron Models

A more detailed picture of the spike shapes that can be expected around real neurons is obtained by considering a detailed multicompartmental neuron model with a comprehensive branching structure typical for real neurons, as in Figure 13.11b. With this dendritic morphology, the membrane currents through the dendrites are spread over a larger membrane area. As a result, Equation 13.4 predicts that the largest spikes will be seen around the soma for the example pyramidal neuron. Around the apical dendrites, the spikes will still have an inverted shape, compared to spikes close to the soma. However, their amplitudes will be small, so they will not be seen in most experiments.

For the multicompartment spike model, the spike width increases with distance. The sharpest spikes are seen for positions close to the soma (bottom panel in Figure 13.11b). This corresponds to a low-pass filtering in the sense that distant spikes lose some of their high-frequency components, compared to the spikes close to the soma. This filtering effect is absent for the spike generated by the two-compartment neuron model. The effect stems from the cable properties of the neuron, and has been referred to as 'intrinsic dendritic filtering' (Lindén et al., 2010), as opposed to, say, possible filtering by the extracellular medium itself.

13.3.3 Dependence of Spike Size and Shape on Neuronal Properties

Extracellular measurement of spikes from a neuron in a living brain is blind in the sense that it is not known what type of neuron is recorded from when an electrode is lowered into the brain. Some neuron types produce spikes with larger amplitudes and/or broader shapes than others, and as seen in Figure 13.11, both the shape and amplitude depend critically on recording positions. Large spike amplitudes imply that they will be more dominant in electrical recordings, and ideally this bias should be considered in the analysis of joint recordings of spikes from many neurons.

To understand the link between the morphology of neurons and their spike amplitudes and shapes, it is convenient to consider **ball-and-stick** neurons where a passive dendrite cable 'stick' is connected to a point-like

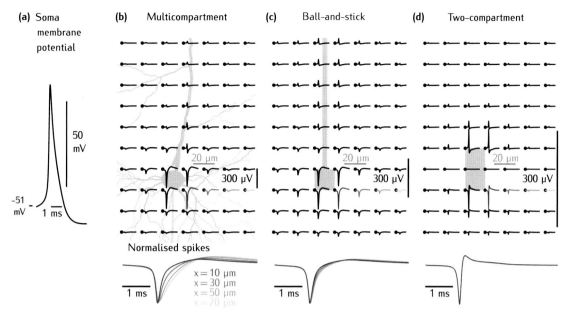

Fig 13.11 Comparison of extracellular spikes from an action potential for different neuron models. **(a)** Membrane potential in the soma during a spiking event, generated by injection of a synaptic current into the soma of a pyramidal neuron model (Hay et al., 2011), the same as used in Section 5.6. Top panels in **(b–d)** show spikes computed at different positions around the neuron model. Black dots at the start of spike traces denote the spatial positions of the spikes. **(b)** Same pyramidal neuron model as in **(a)**. **(c)** Ball-and-stick model with soma membrane potential from **(a)** imposed on the soma of the ball-and-stick model. **(d)** Two-compartment neuron model with soma membrane potential in **(a)** similarly imposed. Bottom panels show normalised spikes for several lateral distances from the soma. Here spikes are normalised to have the same magnitude of the negative peak.

soma. Despite its simplicity, the ball-and-stick neuron model exhibits several of the key qualitative features observed for the pyramidal neurons with extensive dendritic branch patterns, including increased spike width as the distance from the soma increases (Figure 13.11b).

Pettersen and Einevoll (2008) took advantage of the mathematical tractability of the ball-and-stick model to derive analytical expressions for how the amplitude of the recorded spike depends on distance from the neuron as well as the electric properties of the neuron. A qualitative rule derived from this work was that close to a spiking neuron, the spike amplitude is roughly proportional to the sum of the diameters of the dendritic branches attached to the soma raised to the power 3/2, that is, $\sum d^{3/2}$. For spikes recorded far away, the amplitude is instead proportional to $\sum d^2$, that is, to the sum of the cross-sectional areas of the same dendritic branches. Neurons with many thick dendritic branches attached to the soma will thus generate the largest spikes. Their analytical expressions also explained why spikes recorded far away from the neuron will be blunter and have larger spike widths than those recorded close by. See Pettersen and Einevoll (2008) and Pettersen et al. (2012) for more discussion.

13.3.4 Spike Effects from Active Dendritic Conductances

In the models described in the previous subsections, the active sodium and potassium conductances generating the action potentials have been assumed to be present in the soma only; the dendrites have been passive. However, real neurons have active conductances also in the dendrites. These dendritic active conductances typically are slower-acting than the rapid conductances making the spike, and are expected to affect the low-frequency part of the EPs, the LFP, more than the shape of the spike. The above models have also neglected the part of the axon closest to the soma in generating the spike shapes, even though the density of active channels are known to be particularly high there. For modelling studies investigating these issues, see Gold et al. (2006) and Telenczuk et al. (2018).

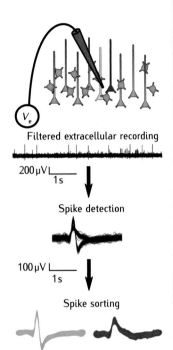

Filtered extracellular recording

200 μV L___
       1s

Spike detection

100 μV L___
       1s

Spike sorting

Fig 13.12 Overview of spike sorting. An extracellular electrode records the time course of the EP near to neurons, for example, the blue and red neurons. The spike detection algorithm identifies sections of the filtered extracellular trace that correspond to spikes from neurons near to the electrode. The spike-sorting algorithm then classifies the spikes according to characteristics of their shape. Since neurons at different positions produce different characteristic extracellular signatures, it can be inferred that the sorted spikes originated in different neurons. Box 13.4 has more explanation. Scale bars are indicative only. Figure adapted from original in Rey et al. (2015) released under Creative Commons BY 4.0 licence.

Box 13.4 | Spike sorting

A sharp electrode placed in brain tissue will pick up spiking signals from several neurons (Figure 13.12). However, the shapes of the spikes will be different for the different neurons, and this can be used to sort the spikes according to their neurons of origin. This is referred to as spike sorting, and is a problem of great practical importance both for neuroscience research and development of neuroprosthetic devices.

In the present day, with electrodes with hundreds of electrode contacts, fast and accurate automatic spike-sorting methods are needed to replace time-consuming manual ones (Quiroga, 2007). To develop and test such automatic methods, one needs benchmarking spiking data where the 'ground truth', the actual spiking times for the contributing neurons, is known (Einevoll et al., 2012). One use of the EP modelling scheme for spikes has been to generate such benchmarking data (Camuñas-Mesa and Quiroga, 2013; Hagen et al., 2015; Mondragón-González and Burguière, 2017; Buccino and Einevoll, 2021).

A modern electrode has numerous recording contacts, often placed only some micrometres apart (Jun et al., 2017). Thus a spike can be measured at several contacts simultaneously, with each contact recording a slightly different shape reflecting the different positions of the contacts relative to the spiking neuron. This allows for not only accurate spike sorting, but also estimation of the spatial position of the neuron. Likewise, the spatial variation of the spike shape around the neuronal soma (Figure 13.11) depends on the details of the action potential and dendritic morphology, thus also allowing for the identification of neuron type (Buccino et al., 2018).

13.3.5 Soma Spikes Initiated in the Axon

In most neurons, the action potential is initiated some distance away from the soma along the axon initial segment (Bender and Trussell, 2012; Goethals and Brette, 2020), rather than in the soma itself. It may then be expected that prior to the ignition of the full soma action potential, there will be a briefly lasting current dipole where current enters at the axon initial segment and returns through the soma. This would be observed as a tiny and narrow positive peak in the spike prior to the negative sodium peak for recording close to the soma (Telenczuk et al., 2018). Experimental techniques for exploration of this and other details of spike patterns are now becoming possible by means of high-density microelectrode arrays (HD-MEAs) (Emmenegger et al., 2019).

13.4 | Local Field Potentials

13.4.1 Local Field Potentials from Single Neurons

Synaptic inputs onto neurons also give rise to EPs. An example is given in Figure 13.14 showing the computed EP around a pyramidal neuron receiving a single synaptic input current. This neuron model has no active conductances, so that generation of action potentials is not possible. Compared to spikes, the EPs generated by synaptic inputs vary less rapidly with time, and

Box 13.5 | Spikes in microelectrode arrays

As an alternative to measuring spikes in living brains, small slices of brain tissue can be put in dishes where electrophysiological cellular properties can be studied for a few hours. Such **in vitro recordings** allows for more detailed and better-controlled investigations than what can be achieved in living brains. In one type of such recordings, the brain tissue is placed on a **microelectrode array** (MEA). The bottom of the device contains a grid of electrode contacts picking up electrical potentials generated by the neural activity of the slice of brain tissue placed on top of it. The slice and MEA are further covered with a liquid, typically saline, to keep the cells in the brain tissue functioning for the duration of the experiment (Figure 13.13a).

The experimental setup itself has an effect on the spikes recorded. The microelectrode contacts are embedded in an insulating glass plate with very low electrical conductivity, whilst the saline has a higher electrical conductivity than the brain slice it covers. For such planar stepwise discontinuities in the conductivity (assuming for the moment the MEA substrate, slice and saline all are infinite planes), formulae analogous to Equation 10.15 can be derived by use of the Method of Images from electrostatics (Box 10.4).

Whilst the largest effect on the recorded spike shape comes from the insulating glass substrate, which roughly doubles the size of the recorded spikes, Figure 13.13b illustrates the effect from the saline covering the brain slice. As seen in the example results, the highly conductive saline cover reduces the size of the recorded spike compared to the hypothetical situation where the saline had the same value of the electrical conductivity as the brain slice.

---

thus the signal has more low-frequency components. For historical reasons, such low-frequency EPs are referred to as LFPs (Section 9.4) (see sidebox).

For the example in Figure 13.14a, the shape and amplitude of the LFP depend strongly on position. Close to the position of the synaptic input at the apical dendritic branch, the LFP is always negative. This is because the sink provided by the excitatory synaptic current itself dominates the sum in the formula in Equation 13.4. At positions around the soma, the LFP is instead always positive, with the return currents leaving the neuron in the soma region now dominating the sum. This qualitative dipolar feature of the LFP pattern is in accordance with the prediction from the much simpler two-compartment model (Figure 13.6). At other positions, for example, above the synaptic input, a more complicated biphasic LFP signal is observed.

Inspection of the membrane current curves in Figure 13.14b reveals that the soma current not only is oppositely directed to the membrane current in the apical compartment, but also is blunter and lasts longer. This reflects the low-pass filtering properties of the neuronal cables, which implies that high-frequency components of membrane currents are dampened en route from the apical synapse to the soma (Box 5.7). This phenomenon has implications for the frequency filtering of the LFP signal, as discussed in Section 13.4.2. Likewise, and for the same physical reason, the soma membrane potential trace is much blunter than the membrane potential trace in the apical compartment where the active synapse is positioned.

The term local field potential (LFP) refers to the low-frequency part of the electrical potential recorded between cells in brain tissue. In typical electrophysiological experiments, spikes are extracted from the part of the EP above 300–500 hertz or so; the LFP is the part of the EP below this frequency. Historically, the term 'local' was used to signify that the sources of the LFP signal is more local than EEG signals measured from the head scalp. The term 'field' is a bit misleading since what is measured is the electrical potentials, not the electrical fields. Since LFP is a well-established term, it is practical to keep it, but maybe a better interpretation of the acronym LFP would be 'low-frequency potential'.

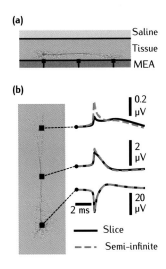

Fig 13.13 **(a)** Sketch of spike recording where a slice of brain tissue containing the neuron is placed on a MEA and covered with saline. **(b)** Spikes computed by use of Method of Images (Box 10.4) for the sandwich structure in **(a)** ('slice') and the hypothetical situation where the saline has the same value of the electrical conductivity as the brain slice ('semi-infinite'). For further details, see Ness et al. (2015).

LFP Pattern Depends on Position of Synaptic Inputs

A neuronal action potential always originates in the same place, typically in the soma or in a part of the axon close by, and the resulting extracellular spike pattern is quite standardised. The spike pattern is essentially independent of the position of synaptic inputs that excited the cell to fire, whereas the extracellular LFP pattern depends critically on the synaptic position. For example, for an excitatory synaptic current positioned on the soma, the dipolar pattern will have a roughly inverted structure compared to when it is positioned at a top dendritic branch. This is illustrated by the colour plots in Figure 13.15 illustrating the maximum positive or negative deviation of the LFP from baseline following stimulation by an excitatory synaptic current. For spatial positions above the apical synaptic input in Figure 13.15a, the largest signal amplitudes typically correspond to negative LFPs. For positions below the synaptic input, the largest LFP amplitudes instead typically correspond to positive LFPs.

An excitatory input onto the soma will give an 'inverted' dipolar pattern, as seen in Figure 13.15b. Here negative LFPs are seen around and below the soma, and positive LFPs along the apical dendrite. This illustrates the critical dependence of the generated LFP on the spatial position of the synaptic input.

Open-Field versus Closed-Field

The pyramidal neuron is a characteristic **'open-field'** dendritic structure where the synaptic input currents and a large part of the return currents may be substantially separated in space. This implies a large current dipole and also a larger contribution to the experimentally recorded LFPs. However, with numerous simultaneous synaptic inputs spread evenly across the dendrites, there will be cancellations of the LFPs generated by the different synaptic input currents. With such a **'closed-field'** structure, the generated LFP will be thus much reduced (Lindén et al., 2010, 2011).

Spatial Decay of LFP Signals around Neurons

The contour plots in Figure 13.15 illustrates that the LFP from a single neuron decays fairly rapidly with distance, although not as rapidly as for a spike. This decay with distance was further investigated in Lindén et al. (2010, 2011): decay close to the neuron, that is, within the bush of dendritic branches, is moderate. However, outside the dendritic bushes, the LFP decays much more sharply. At distances larger than a millimetre or so from the neuron, the LFP amplitude decays roughly as the square of the distance. This is in the far-field regime where the dipolar approximation is applicable, so that the dipole formula in Equation 13.3 can be used to compute the LFP; see also Box 13.3.

13.4.2 Frequency Filtering of LFPs

Close inspection of the pattern of LFPs generated by a single synaptic input in Figure 13.14a reveals that the negative LFP close to the synapse is sharper than the corresponding positive LFP close to the soma. This reflects that some of the high-frequency components of the LFP in the soma region have been filtered out due to the capacitive properties of the membrane.

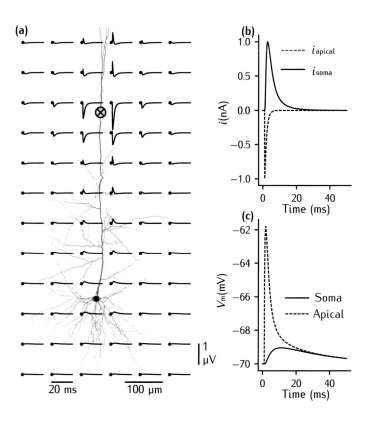

(a)

(b)

(c)

20 ms   100 μm

Fig 13.14 **(a)** LFP from multicompartment model computed using the formula in Equation 13.4, assuming an infinite volume conductor. The pyramidal neuron model is from Hay et al. (2011), the same as used in Section 5.6, with active conductances set to zero. The neuron receives a single excitatory synaptic current $i_{syn}(t)$ at a synapse on one of the apical dendritic branches (blue dot). Black dots at the start of LFP traces denote the spatial positions of the LFP.
(b) Normalised membrane currents at the soma and at the apical compartment where the synapse is positioned.
(c) Membrane potential at the soma and at the same apical compartment. Analogous to Figure 4.2 in Pettersen et al. (2012).

This frequency filtering of the LFP signal is illustrated more clearly in Figure 13.16a, showing LFP patterns generated when oscillatory currents are injected at the position of the synapse. For the case when the injected current is oscillating at the slow frequency of 1 Hz, the resulting LFP has a distinct dipolar pattern, with the two poles positioned at the synapse and in the soma region, respectively. With a high frequency of 100 Hz, the pattern is very different. Here the dipolar pattern has shrunk and the LFP is now very small near the soma. Thus high-frequency LFP components are attenuated near the soma relative to the low-frequency components, in accordance with the observation of blunter LFP signals close to the soma in Figure 13.14. Note that the origin of this phenomenon, referred to as 'intrinsic dendritic filtering' (Lindén et al., 2010), is the capacitive property of the neuronal membrane. A qualitatively similar attenuation of high-frequency EP components, compared to low-frequency components, is thus also observed for spikes (Pettersen and Einevoll, 2008; Pettersen et al., 2012), as seen in the bottom panel of Figure 13.11b.

Figure 13.16b illustrates the core biophysical effect underlying this low-pass filtering. Here the depth distributions of the total return current along the depth axis of the neuron are shown. For 1 Hz, the return membrane current following synaptic current injection is substantial across the full depth profile of the neuron, with a large contribution also from the soma region. This implies that there will be sizeable LFPs across the whole depth. For 100 Hz, almost all the return membrane current instead goes through

Fig 13.15 LFP from multicompartment model computed using the formula in Equation 13.4 assuming an infinite volume conductor. The pyramidal neuron model is the same as in Figure 13.14. The neuron receives a single excitatory synaptic current on **(a)** one of the apical dendritic branches or **(b)** the soma, marked as blue dots. LFPs at selected spatial positions (marked with black dots on trace starting points) are shown. Colour plots show the maximum-magnitude LFPs during the time course of the signal, that is, maximum positive or negative deviation of the LFP from baseline following the synaptic input. Note the logarithmic colour scale.

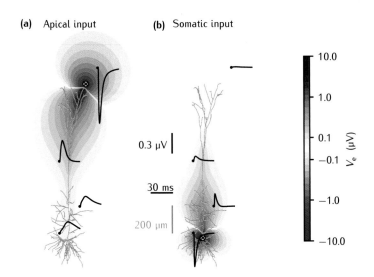

the apical dendrites, and there will be only sizeable LFPs around the apical branches (Figure 13.16a).

Both the 1 Hz and the 100 Hz situations depicted in Figure 13.16 correspond to 'open-field' situations where the averaged return currents are spatially displaced from the current injection point at the synapse. However, the current dipole length, the distance between the current injection point and the mean position of the return current, is much larger for the 1 Hz case than for the 100 Hz case. The magnitude of the current dipole moment is proportional to this distance (Box 13.3), and will thus be larger for the 1 Hz situation than for the 100 Hz situation. As described by Equation 13.3, the LFP can be expected to be approximately proportional to this current dipole moment some distance away from the neuron. Consequently, even if the amplitude of the 100 Hz oscillatory currents and that of the 1 Hz oscillatory currents setting up the LFP are identical, the 1 Hz component of the LFP will be larger than the 100 Hz component. The LFP measurement is thus biased towards measuring low-frequency activity of the neural activity.

13.4.3 LFPs from Many Neurons

As we saw in Section 13.3, spikes, the extracellular signatures of action potentials, decay quite rapidly as the recording electrode is moved some tens of micrometres away from the soma (Figure 13.11). A spike is thus quite 'local' in space. The spike can also be said to be 'local' in time as it only lasts a few milliseconds. This locality both in space and time allows for the identification of individual spikes from individual neurons under in vivo conditions, so-called spike sorting (Box 13.4).

For the LFPs, the situation is different. As seen above, the LFP contribution from individual neurons extends much further from the neuron, and the individual LFP contributions from synaptic inputs easily last 10 ms or more. Thus, in practice, the LFP signal recorded in living brains will typically come from a collection of thousands of neurons located around the recording electrode. This complicates the interpretation of the LFP in terms of the underlying activity in neurons. However, due to the linearity of

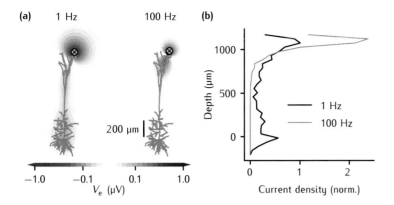

Fig 13.16 (a) Time snapshots
of LFP patterns set up by
sinusoidal current injections,
1 Hz and 100 Hz, respectively, at
a synapse (blue dot) on one of
the apical branches. The neuron
model is the same as in
Figure 13.14. Note the
logarithmic colour scale.
(b) Depth distributions of
membrane return currents for
1 Hz and 100 Hz current
injections. The depicted linear
current density, that is, linear
density of membrane currents
summed across all dendritic
branches at depth in question, is
shown as a function of depth,
that is, the vertical position
relative to the soma in (a). The
current density is normalised to
the maximum value for the 1 Hz
case.

the volume conductor theory, the LFP contributions from several neurons simply add up. The modelling of multi-neuron LFPs is thus, in principle, not more complicated than the modelling of single-neuron LFPs described above.

An example of the modelling of such a multi-neuron LFP is given in Figure 13.17. Here we consider a cylindrical population of 100 pyramidal neurons surrounding a recording electrode measuring the electrical potentials at eight positions along the cylindrical axis (Figure 13.17b). The neurons in the populations receive numerous excitatory synaptic inputs from neurons outside the populations (Figure 13.17a) which generates an LFP signal that is recorded by the electrode (Figure 13.17c).

Over the time course of the simulation, three scenarios are considered (Figure 13.17a):

(1) An epoch where all synaptic inputs are positioned on basal dendrites during which a negative LFP signal is observed at the lower channels, whilst a positive LFP signal is observed for the upper channels. For the channel immediately above the synaptic inputs (channel 5), the LFP signal is small.

(2) An epoch where all synaptic inputs are positioned on the apical dendrites, during which the opposite dipolar pattern with negative LFPs at the top and positive LFPs at the bottom is observed. Again the LFP in channel 5 is observed to be small, and an overall depth variation qualitatively similar to that observed for the simple two-compartment model receiving a single excitatory input in Figure 13.6 is seen.

(3) A final epoch where synaptic inputs are spread uniformly across the entire dendritic structure of the neurons, during which a very small LFP signal is observed in all recording channels. This can be explained by the different sign of the dipolar contributions from apical and basal synaptic inputs, leading to almost perfect cancellation of the LFP in all channels.

Both (1) and (2) correspond to what is called 'open-field' configurations and qualitatively agree with the LFP patterns observed from single neurons receiving single synaptic inputs (Figure 13.15). Scenario (3) is called a 'closed-field' configuration.

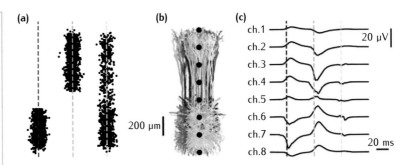

Fig 13.17 LFP signals **(c)** as recorded by a linear multielectrode with eight contacts positioned in the centre of a population of pyramidal neurons of the same type as in Figure 13.14. The population contains 100 neurons placed in a uniform distribution on a disc of radius 200 μm **(b)**. Three epochs with different spatial distributions of excitatory inputs are considered: basal only (black dashed line), apical only (dark cyan dashed line) and uniform (light cyan dashed line) **(a)**.

This illustrates that the LFP signal cannot necessarily be used as a measure of spiking activity in networks. When synaptic inputs are placed either all on apical dendrites or all on basal dendrites, the magnitude of the measured LFP signals reflects the spiking of the presynaptic neurons. But if the synaptic inputs from the same presynaptic neuron are uniformly spread out across the dendrites, the LFP signal is almost absent. This happens even if the net excitation of the somata of the receiving neurons can be just as large, or larger, than in configurations giving large LFPs.

Two key take-home messages are that, unlike spikes, a large LFP signal is not only observed around the positions of the synaptic inputs driving the LFP. Furthermore, an LFP signal set up by a neuron may be virtually zero around its soma (cf. Figure 13.17c). Also, an LFP signal recorded in a particular layer of the cortex, say, does not necessarily stem from neurons with somata in the same layer (cf. LFPs recorded in apical channels in Figure 13.17a, b).

As illustrated above, the LFP signal is more difficult to analyse than spikes. Whilst a spike can be identified directly as an action potential from a neuron, computational modelling is generally required for a proper interpretation of the LFP signal (Einevoll et al., 2013a).

The modelling of LFP signals described in this chapter is commonly referred to as forward modelling. Inverse modelling, the estimation of neuronal activity from recorded LFP signals, is not the focus here. However, a quick introduction to inverse modelling of LFP signals can be found in Box 13.6.

13.5 Case Study: The Spatial Extent of the Local Field Potential

The recording of LFPs has a long history in neuroscience. Therefore, maybe it is surprising that the question 'how local is the local field potential?' is still debated. How many neurons contribute to the LFP recorded by an electric contact? Alternatively, what is the typical radius of the volume surrounding the contact containing the neurons generating (almost) all of the recorded LFP, referred to as the **spatial reach**? Experimental answers to this question have been conflicting, giving estimates for the spatial reach spanning from a few hundred micrometres (Katzner et al., 2009; Xing et al., 2009) to

---

Box 13.6 | Inverse modelling of LFPs

In this chapter, the focus is on forward modelling of LFPs, that is, biophysics-based modelling of LFPs stemming from electrical activity in neurons. The opposite problem, estimation of the neural activity from measured LFPs, is referred to as the inverse problem. Unlike the forward problem, the inverse problem is ill-posed in that there is no unique solution: any set of measured LFP signals can, in principle, be explained by an infinite number of combinations of neural sources. Thus, to provide a unique solution, additional a priori assumptions must be made.

A standard way to analyse LFPs recorded in laminarly organised brain structures like the cortex or hippocampus is to do a current-source density (CSD) analysis. The CSD is a measure of the net volume density of electrical currents entering or leaving the ECS (Box 10.3). Whilst the CSD does not directly reveal which neurons are active, it provides a more localised measure of neural activity than the LFP signals due to unavoidable volume conduction of the latter (Pettersen et al., 2012). However, CSD estimation is also ill-posed, and additional assumptions must be made to allow for unique estimates. For example, in the standard CSD analysis, the CSD is estimated by assuming the neural activity to be constant in the horizontal directions in the layered cortical or hippocampal structures (Nicholson and Freeman, 1975). Other more recent approaches like the **inverse CSD** (**iCSD**) (Pettersen et al., 2006) and the **kernel CSD** (**kCSD**) methods (Potworowski et al., 2012) make different assumptions about the underlying structure of the CSD.

An alternative to CSD analysis is to decompose the measured LFPs into contributions from individual populations of neurons, for example, by application of general mathematical decomposition techniques like **principal components analysis** (**PCA**) (Barth and Di, 1991) or **independent components analysis** (**ICA**) (Leski et al., 2010; Makarov et al., 2010). These involve assumptions like orthogonality or statistical independence of LFP contributions from the various populations, assumptions that a priori cannot be expected to be obeyed by real neuronal populations. Another alternative is to impose more physiological constraints and, when available, take advantage of simultaneous recordings of action potentials, as in **laminar population analysis** (**LPA**) (Einevoll et al., 2007). For reviews of these and other methods for analysing LFPs, see Einevoll et al. (2013a) and Pesaran et al. (2018).

---

several millimetres (Kreiman et al., 2006) or more (Kajikawa and Schroeder, 2011). Modelling has revealed that these experimental observations can be reconciled, the key finding being that the spatial reach depends strongly on how synchronous the synaptic inputs driving the LFP-generating neurons are (Lindén et al., 2011).

In Lindén et al. (2011), this question was addressed by calculating the spatial size of the pool of neurons contributing to the LFP recorded by an electrode. The model used to explore this question is shown in Figure 13.18a. Here a population of identical pyramidal neurons receiving synaptic inputs is positioned with circular symmetry in a disc around the recording electrode.

Fig 13.18 Illustration of modelling approach used to explore the question about the 'locality' of the LFP. **(a)** Sketch of model setup where neurons are evenly distributed on a disc of radius R around the electrode tip. **(b)** The population LFP $V_e(t)$ is given as a sum over contributions $V_{e,i}(t)$ from individual cells at distances r_i. **(c)** Single-cell LFP shape function $F(r)$, assuming dipolar sources. **(d)** Number $N(r)$ of cells on a ring of radius r. **(e)** Illustration of dependence of compound amplitude, defined as standard deviation of signal $V_{e,\sigma}$ (cf. **(a)**), on population radius R for the cases where the single-neuron LFP contributions are either uncorrelated or fully correlated. The dependence of $V_{e,\sigma}$ on the population radius R defines the spatial reach R_{reach} of the electrode (see text). Redrawn based on information from Lindén et al. (2011).

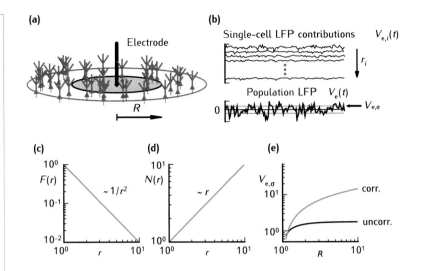

The LFP contributions from the individual neurons $V_{e,i}(t)$ sum up to give the total population LFP, $V_e(t)$ (Figure 13.18b).

This total population LFP signal can be computed by using Equation 13.4 to compute the signal contribution from each neuron and then sum up all the single neuron contributions to get the compound LFP from the entire population of neurons (Figure 13.18b). With increasing radius R of the population, more and more neurons will contribute to the compound LFP, and the amplitude $V_e(R)$ is thus expected to increase with R. On the other hand, the contribution to the compound signal from a single neuron decreases with distance from the electrode, as seen in Section 13.4.1.

With many neurons, each receiving numerous synaptic inputs contributing to the LFP, the compound LFP will vary over time. To get a single number quantifying the magnitude of the population LFP, Lindén et al. (2011) used the standard deviation of the signal $V_{e,\sigma}$ (Figure 13.18b). Thus intuitively one could expect that $V_{e,\sigma}(R)$ approaches a constant value $V_{e,\sigma}^{\infty}$ as the population size R increases. The spatial reach R_{reach} is defined as the radius for which the compound LFP has reached a certain fraction α (e.g. $\alpha = 0.95$ as in Lindén et al. (2011)) of this limiting value $V_{e,\sigma}^{\infty}$.

However, it is not clear at the outset that as R increases, $V_{e,\sigma}(R)$ converges towards a fixed value and that a spatial reach defined in this way exists. It turns out that a finite spatial reach is obtained under some conditions, but not all. We now consider a simple qualitative model that illuminates what factors determine $V_{e,\sigma}(R)$ and the conditions under which it converges to a finite value as R increases.

Two factors are a priori expected to be key in determining how the compound population LFP $V_{e,\sigma}(R)$ increases with R: (1) how sharply the single-neuron contribution decays with distance from the neurons; and (2) the number density $N(r)$ of neurons positioned in a ring of radius r around the electrode. The decay of the single-neuron contribution can be captured by a **shape function** $F(r)$ (Lindén et al., 2011; Łęski et al., 2013). As we are interested in how the compound LFP increases with R for large values of R, only the behaviour of the shape function $F(r)$ far away from the neuron is of interest. For such large distances, the far-field dipole approximation can be assumed, $F(r) \propto 1/r^2$ (Figure 13.18c). The number of neurons within each ring will be proportional to the circumference $2\pi r$ of the ring, and thus $N(r)$ will be proportional to r (Figure 13.18d).

A third key factor is the level of **correlations** between the single-neuron contributions to the compound LFP.

13.5.1 Uncorrelated Single-Neuron LFPs

In a situation where the neurons in the population each receive synaptic inputs at completely random times, the contributions will be uncorrelated and the individual LFP contributions will tend to cancel out. In this case, the amplitude of the compound signal $V_{e,\sigma}(R)$ will not be proportional to the number of individual sources. However, the variance of the compound LFP, that is, the square of $V_{e,\sigma}(R)$, will be given as the sum of the variances of the individual single-neuron contributions. This is analogous to the situation of a random walk where the squared distance from the starting point after N random kicks is proportional to N; see, for example, Nelson (2008, Chapter 4). In this case (Lindén et al., 2011):

$$V_{e,\sigma}(R)^2 \propto \sum_i V_{e,i}^2 \propto \sum_i F(r_i)^2. \tag{13.7}$$

Here the sum over i goes over all neurons with their somata positioned within a distance R from the recording electrode (Figure 13.18a), and $V_{e,i}$ represents the amplitude of the LFP contribution from neuron i (Figure 13.18b). This contribution is assumed to be proportional to the shape function $F(r_i)$ where r_i is the distance from neuron i to the recording electrode.

With many neurons contributing to the compound LFP, we can approximate the sum in Equation 13.7 with the integral:

$$V_{e,\sigma}(R)^2 \propto \int_0^R N(r)F(r)^2 dr. \tag{13.8}$$

The current focus is on the contribution from the distant neurons where the far-field limit of $F(r)$ applies, and $F(r) \propto 1/r^2$. For the contributions from these distant neurons, we can approximate the integral in Equation 13.8 with:

$$V_{e,\sigma}(R)^2 \sim \int_{R_x}^R N(r)\left(\frac{1}{r^2}\right)^2 dr \propto \int_{R_x}^R \frac{1}{r^3} dr \propto \frac{1}{R_x^2} - \frac{1}{R^2}. \tag{13.9}$$

For convenience, we have introduced a lower cut-off radius R_x in the integral to remove the unphysical divergence that would appear by assuming the far-field relationship $F(r) \sim 1/r^2$ for distances r approaching zero.

The key observation from Equation 13.9 is that the amplitude $V_{e,\sigma}(R)$, that is, the square root of the variance in Equation 13.9, will approach a fixed finite value $V_{e,\sigma}^\infty$ as $R \to \infty$ (since $1/R^2 \to 0$ in this limit) (Figure 13.18e). This, in turn, implies that a spatial reach corresponding to the value of R_{reach} for which $V_{e,\sigma}(R_{\text{reach}}) = \alpha V_{e,\sigma}^\infty$ can be found.

In Lindén et al. (2011), this finding of a finite spatial reach R_{reach} was confirmed in comprehensive numerical simulations summing over uncorrelated single-neuron LFP contributions, as illustrated in Figure 13.18a,b. Moreover, for LFP recordings at the soma level, R_{reach} was found to be between 100 and

Box 13.7 | Olbers' paradox and population LFPs

Two centuries ago, the German astronomer Heinrick Olbers wondered why the night sky is dark, and not light as it should be according to the following reasoning (Figure 13.19): (1) the light intensity from a single star falls off as $1/r^2$; (2) in a homogeneous universe, the number of stars in a spherical shell of a certain thickness around Earth should increase as r^2; (3) the contribution from each such shell of stars to the illumination on Earth should thus be independent of the distance r; and (4) with an infinite universe, there will be an infinite number of shells, and thus an infinite light intensity on Earth! The resolution of this paradox, known as Olbers' paradox (Harrison, 1989), has some analogies to the question of summation of LFP signals from many neurons.

The paradox is resolved if one takes into account our modern knowledge that the universe is not infinitely old and the speed of light is finite, so that there can be no contributions from stars in shells at distances of 14 billion light years or more away. This summation of light from numerous stars is analogous to the summation of LFPs recorded at a particular position from numerous neurons in the surrounding brain (Einevoll et al., 2013b).

Whilst a finite brain is not required for a finite LFP, a finite universe with a finite age seems to be required to resolve Olbers' paradox.

Fig 13.19 Olbers' paradox (Box 13.7): why is the night sky dark?

200 μm for the cortical neuron populations considered where a value of α of 0.95 was chosen. These values roughly corresponded to lateral extension of the dendritic branches around the soma.

With explicit formulae for the shape function $F(r)$, quantitative models providing explicit formulae for the amplitude of the compound signal $V_{e,\sigma}(R)$ can be obtained. In Łęski et al. (2013) and Einevoll et al. (2013a), this approach was used with analytical shape function $F(r)$ estimated from numerical simulation for different types of neurons. This allowed for, for example, derivation of explicit formulae for the spatial reach for the population LFP in terms of parameters describing the spatial shape of the LFP around single neurons.

13.5.2 Correlated Single-Neuron LFPs

For fully correlated synaptic inputs onto the neurons, the single-neuron LFP contributions to the compound population LFP will overlap in time. Further, if the different synaptic inputs have similar spatial positions on the receiving neurons, for example, all placed on the apical dendrites, there will be little cancellation of their contributions to the single-neuron LFP contributions. This is akin to constructive interference in wave physics where water waves from several sources add up to give interference patterns, and a summation of wave crests from the individual wave sources gives a large compound wave.

In this case, the compound LFP can be approximated by simply summing the individual amplitude contributions. Formulated as an integral, this gives:

$$V_{e,\sigma}(R) \propto \int_0^R N(r)f(r)\mathrm{d}r. \tag{13.10}$$

When we now insert $N(r) \sim r$ and $F(r) \sim 1/r^2$, the contributions from the neurons positioned outside a radial distance R_x from the electrode is given by:

$$V_{e,\sigma}(R) \propto \int_{R_x}^{R} r \frac{1}{r^2} dr = \int_{R_x}^{R} \frac{1}{r} dr = \ln \frac{R}{R_x}. \tag{13.11}$$

The key observation here is that the amplitude $V_{e,\sigma}(R) \to \infty$ when $R \to \infty$. Thus, unlike for the uncorrelated case, the LFP amplitude increases without bound as the population size R increases, and the compound LFP signal thus has an 'infinite' spatial reach.

Whilst this argument was based on assuming the single-neuron LFP contributions to be fully correlated, the principled conclusion about the divergence of the LFP population amplitude as the population size $R \to \infty$ also holds for partially correlated single-neuron LFPs (Lindén et al., 2011; Łęski et al., 2013).

In reality, no neuron population in the brain can be infinitely large. However, this reasoning illustrates why correlations plays a key role in determining features of the population LFP. In addition to the effect on the spatial reach, correlations can strongly boost the overall amplitude of the compound population LFP, and experimentally observed LFPs are expected to be dominated by such contributions (Lindén et al., 2011; Łęski et al., 2013; see also reviews in Einevoll et al. (2013a, b)).

The question of whether the spatial reach of the LFP signal is finite or infinite has some resemblance to the ancient question of why the night sky is dark given the numerous stars in the universe. Olbers' paradox and the link to the spatial reach question is described in Box 13.7.

Another way to phrase the question 'how local is the local field potential?', is to ask how far outside an LFP-generating population the LFP extends. Whilst not discussed here, this question can also be addressed by means of similar techniques to the ones used to investigate the spatial reach (Lindén et al., 2011; Łęski et al., 2013; Hagen et al., 2017).

13.6 Extracellular Potentials at Cortical Surfaces (ECoG) and on the Head Scalp (EEG)

In the previous sections, we focused on modelling spikes and LFPs recorded inside the grey matter, such as the cortex, where the neuronal somata and dendrites are located. Extracellular potential EPs are also recorded elsewhere, such as at cortical surfaces (ECoG) or on the scalp (EEG) (Figure 13.2).

ECoG electrodes are placed on the cortical surface at the interface between the cortex and the cerebrospinal fluid (CSF), which has a higher electrical conductivity σ than the underlying cortical matter. If an infinite planar, step-like discontinuity in electrical conductivity is assumed at the cortex–CSF interface, the MoI (Box 10.4) can be used to generalise Equation 13.4 to give the potential recorded by an ECoG electrode from cortical neural activity. In this method, the effects of the discontinuity in σ at

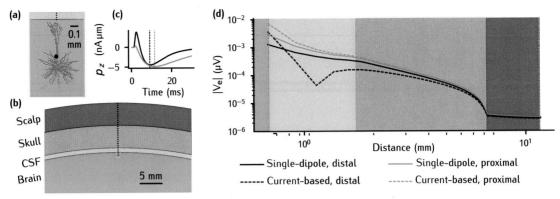

Fig 13.20 Illustration of dampening of electrical signal from a single neuron from cortex to scalp. **(a)** Human layer 2/3 neuron (Eyal et al., 2016) receiving single transient excitatory synaptic input currents. **(b)** Four-sphere head model used in the modelling where the different colours represent the brain, CSF, skull and scalp, respectively. The tiny box at the cortical surface corresponds to **(a)**. Parameters in Table 13.1. **(c)** Current dipole moment in z-direction p_z for two simulations corresponding to synaptic inputs at either a proximal (black dot) or a distal (cyan dot) location, see **(a)**. **(d)** Magnitude of EP V_e as a function of distance from the top of the neuron for either the proximal or the distal synaptic input seen in **(a)**. Maximal magnitudes during the time course in **(c)** are shown. Dashed lines show results computed based on the membrane currents in all neural compartments. Technically this is done by considering the potential generated by a large set of small current dipoles, effectively giving the same result (Næss et al., 2021, Figure 2). Solid lines show results using a single dipole representing the neural activity, (cf. Næss et al., 2021). Note that the curve for current-based potential for the distal input has a kink in the CSF, reflecting a sign change of the potential.

the interface are accounted for by adding appropriately chosen virtual current sources above the cortex–CSF interface to impose the correct boundary condition on the electrical potential at the interface (Box 10.4). However, note that this application of MoI assumes the planar layer of the CSF to be infinitely thick in the vertical direction. The CSF layer in humans has a thickness of only around 1 mm and is covered by the poorly conducting skull.

EEG recordings are non-invasive since the recording electrodes are placed directly on the head scalp, so that no surgery is required for their use. They have thus been of key importance for studying human brain activity for a century. To properly interpret EEG recordings in terms of the underlying activity in neurons, numerous models for the propagation of electrical signals from neurons to EEG electrodes have been developed. In one type of model, the head is considered to be spherical, and different parts of the head (e.g., brain, CSF, **skull**, **scalp**) are modelled as spheres or spherical shells with different electrical conductivities (Nunez and Srinivasan, 2006). The advantage of these spherical models is that they can provide analytical formulae for the EEG potentials generated by neural activity. Another approach is comprehensive numerical modelling, including detailed geometries reconstructed from detailed brain imaging (Bangera et al., 2010; Vorwerk et al., 2014).

Here we consider one of the spherical head models, the **four-sphere head model**, which assumes a four-layered spherical head where the layers represent the brain tissue, CSF, skull and scalp, respectively (Figure 13.20b). The CSF layer is quite thin (~1 mm) but has a larger conductivity than the spherical brain it encapsulates. The skull is thicker (~5 mm) but has a much smaller conductivity than brain tissue. The scalp has about the same thickness as the skull, but a similar conductivity to brain tissue (Næss et al., 2017). Analytical formulae for the potential setup by a current dipole (Box 13.3) positioned in the spherical brain in the innermost part of the model can be derived. The formulae are cumbersome and not listed here, but can be found in Næss et al. (2017).

This combination of layer thicknesses and electrical conductivities gives a characteristic profile of the decay of the potential generated by a single neuronal current dipole as measured across the different layers (Figure 13.20d). In particular, the potential decays by about a factor of 100–1000 from the cortical surface, where ECoG is recorded, to the scalp, where EEG is recorded.

Table 13.1 Parameters used for four-sphere head model in numerical examples in Figures 13.20–13.22

	Brain	CSF	Skull	Scalp
Outer radius (cm)	8.9	9.0	9.5	10.0
Conductivity σ (S/m)	0.276	1.65	0.01	0.465

As expected, the potential decay is strongest in the highly resistive skull: from Ohm's law it follows that a current pushed through a large resistance will give a large potential drop.

For the two examples considered in Figure 13.20, the single-dipole approximation gives accurate results for computation of potentials at the scalp, that is, EEG signals. For ECoG signals recorded at the brain surface, this is not so. Here a membrane-current based formalism must generally be used, as for the LFP (Næss et al., 2021).

In general, the EEG signal will stem from multiple current dipoles and their geometrical alignment will determine the compound EEG. This is illustrated in Figure 13.21. Four vertically oriented current dipoles will give a large, radially symmetric EEG signal (Figure 13.21a). When the same dipoles are staggered, there will be large cancellations and consequently a small EEG signal (Figure 13.21b). With four dipoles aligned in the lateral direction, like in a cortical sulcus (Figure 13.21c), a large EEG signal with a qualitatively different spatial pattern than that in Figure 13.21a will be produced. With the same dipoles grouped into two pairs pointing at each other across a cortical sulcus, there will instead be essentially no EEG signal (Figure 13.21d).

Not all neurons contribute equally to recorded EPs. This is illustrated for the single-neuron EEG signal in Figure 13.22. Here the largest EEG signals arise from the human pyramidal neuron, whilst the more stellate human interneuron barely produce an EEG signal at all. However, as for the LFP signal in Figure 13.17, the pyramidal neuron only produces a sizeable signal when the synaptic inputs onto these neurons are positioned unevenly on the dendritic tree, that is, either on the basal dendrites including the soma or on the apical dendrites. Another thing to note is that the EEG signal effectively is only set up by the z-component, that is, the vertical component, of the current-dipole moment (Figure 13.22c).

The results in Figures 13.20–13.22 assumed the electrode to be ideal and point-like, that is, the electrode is so small that it does not affect the electrical field around the electrode and also measures the potential at a single spatial point. For the ECoG and EEG electrodes which typically measure several millimetres across, there will generally be a spatial smearing effect. This effect can be modelled by the disc-electrode approximation where the recorded potential is assumed to be given by the spatial average of the computed point-electrode potential across the contact surface of the electrode (Equation 13.6). An example of the computation of ECoG signals with the combined use of MoI and the disc-electrode approximation can be found in Hagen et al. (2018, Figure 5D).

Fig 13.21 Spatial pattern of EEG potentials across the scalp computed with the four-sphere head model for different configurations of current dipoles (Næss et al., 2017). Head model parameters given in Table 13.1. **(a)** Set of four adjacent identical current dipoles oriented in the same direction. **(b)** Same current dipoles with staggered orientations. **(c)** Same current dipoles oriented in the same lateral direction, mimicking a situation with four parallel dipoles in a cortical sulcus (illustrated with dotted lines). **(d)** Alternative dipole arrangement compared to the situation around the cortical sulcus in **(c)**, with two pairs of dipoles positioned on opposite sides of a sulcus with current dipoles pointing in opposite directions. The magnitude of p is set to 10^7 nA μm.

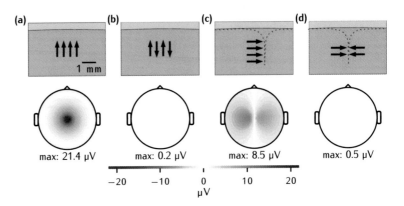

In addition to the spatial smearing effect due to the physical extension of the recording electrode, the highly conductive metallic electrode contacts will also by themselves modify the electrical field in their immediate vicinity. This effect, which also will affect the recorded potential, can be incorporated by use of FEM (Section 13.2.4).

The examples of LFP and EEG modelling presented so far have focused on illustrating the key principles behind signal generation. Box 13.8 instead shows joint modelling of LFP and EEG signals from a comprehensive cortical network with 80 000 neurons representing a 1 mm^2 piece of cat visual cortex (Potjans and Diesmann, 2014).

13.7 | Electrical Stimulation of Brain Cells

So far in this chapter, we have described how electrical potentials generated by neural activity can be modelled. Recordings of such electrical signals have been key for exploration of how the brain works and to monitor brain activity. The same mathematical formalism can also be used to model **electric stimulation** of brain cells by means of extracellular electrodes. Such electrical stimulation is used, for example, in cochlear implants to restore hearing for deaf people (Clark et al., 1977; Zeng et al., 2008) or to alleviate movement disorders in patients with Parkinson's disease by **deep brain stimulation** (DBS) (Benabid et al., 2009). Several clinical applications of such stimulation paradigms have been explored (Joucla and Yvert, 2012).

The principle behind all these applications is the same: the electrical current is imposed on brain tissue by means of electrodes placed on the scalp or cortical surface or inside brain tissue. In **transcranial electrical stimulation** (TES), the electrodes are placed on the scalp and both direct (DC) and alternating currents (AC) are used (Herrmann et al., 2013). With this technique, the size of the electric field imposed in the brain is typically on the order of 0.1–0.5 V/m (Miranda et al., 2018). This electric field is too low to induce an action potential by itself (Fröhlich and McCormick, 2010), and TES is thus thought to only have a modulatory effect on action potential generation.

In TES, the imposed electrical fields are quite diffuse. In DBS, the electrical current is delivered by electrodes positioned deep inside the brain and

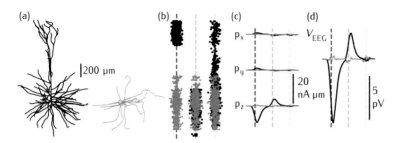

Fig 13.22 Dependence of single-neuron EEG signals on spatial distribution of synaptic inputs. **(a)** Two human neurons with passive conductances only are considered: a layer-3 pyramidal cell (black) and a layer-3 interneuron (grey) downloaded from Allen Institute for Brain Sciences (celltypes.brain-map.org). **(b)** Three epochs with different spatial distributions of excitatory inputs on the pyramidal neuron: basal only (black dashed line), apical only (dark cyan dashed line) and uniform (light cyan dashed line). For the interneuron, the synapses are uniformly distributed in all epochs. **(c)** Components of current-dipole moments, with z being the vertical direction. **(d)** EEG signals computed from four-sphere head model for an electrode placed on the scalp in the direction of the apical dendrite of the excitatory cell. Head model parameters are listed in Table 13.1.

much more focused electrical stimulation can be imposed (Butson and McIntyre, 2008).

13.7.1 Electrode Arrangements and Stimulation

The physical picture of electrical brain stimulation is that the current delivered between two electrodes generates gradients in the EP V_e as it is forced through the ohmic brain tissue. The relationship between V_e and the imposed current density in the ECS is given by Equation 10.8:

$$\vec{I}_e = -\sigma \nabla V_e. \tag{13.12}$$

There are no current sources in the ECS itself, that is, $\nabla \cdot \vec{I}_e = 0$; see Box 10.1 for a description of the ∇ operator. All currents are provided either by the electrodes or by membrane current sources. Thus, inside the ECS, the EP obeys:

$$\nabla^2 V_e = \frac{1}{\sigma} \nabla \cdot \vec{I}_e = 0. \tag{13.13}$$

The effect of the electrode stimulation can now be included by imposing the boundary conditions $V_e = V_{e0}$ at one of the electrodes, and $V_e = 0$ at the other (Joucla and Yvert, 2012). For DC stimulation, V_{e0} is a fixed constant, whilst for AC stimulation, V_{e0} is time-dependent. The membrane potential which governs the dynamics of neurons is the difference between the intracellular potential V_i and the EP V_e. The change in V_e imposed by the electrical stimulation will therefore change the membrane potential and possibly the output spike pattern (Rattay, 1999; Fröhlich and McCormick, 2010).

Analytical solutions for the EP imposed by stimulating electrodes can be found for some special cases. With current injected from a spherical electrode, with the return electrode placed infinitely far away, the potential applied will decay radially with distance from the centre of the electrode, that is, $V_e \sim 1/r$ (Figure 13.24a). For two parallel plate-like electrodes, the electric field will be directed perpendicularly to the plates and the gradient of the potential dV_e/dx will be constant (Figure 13.24b). For other electrode configurations with more complicated geometries, such as those used in DBS where multiple stimulation electrodes are arranged on a single electrode shank, Equation 13.13 must be solved numerically by means of FEM to find imposed EPs (Butson and McIntyre, 2005, 2008).

In transcranial stimulation, the imposed current has to penetrate the scalp, skull and CSF before it can enter the brain tissue where the neurons are located. Here the connection between the current delivered by the stimulation electrodes and the induced potential is less transparent. However,

---

Box 13.8 EEG and LFP signals from the cortical circuit

Whilst single-compartment neurons, such as the integrate-and-fire model, are popular when studying network dynamics (Section 9.3), they do not produce EPs. To compute EEGs and LFPs stemming from such network models, a hybrid strategy has been used (Hagen et al., 2016, 2022; Næss et al., 2021). Here the occurrence of action potentials in the network of integrate-and-fire neurons is computed and stored in a first computational step, and then replayed in a second step where each neuron has been replaced by a multicompartmental neuron. This approach was used by Næss et al. (2021) to compute LFP and EEG signals stemming from the **Potjans–Diesmann** cortical microcircuit model encountered in Section 9.3.3.

Figure 13.23 shows neural activity around a time when the network model receives stimulus-evoked spiking input from the visual thalamus (time = 900 ms). **(a)** Results from the first computational step where dots indicate spike times of individual neurons and populations are represented with different colours (I = inhibitory; E = excitatory). **(b)** Multicompartment model neurons (one example morphology per population) used in the second computational step to compute LFP and EEG signals with colours corresponding to **(a)**. Depths of layer boundaries are shown relative to the cortical surface, $z = 0$. A linear multielectrode with 16 contacts separated by 100 μm (black dots) is positioned through the centre of the populations. **(c)** LFPs calculated at the positions of these contacts. **(d)** Components of the current-dipole moment $\vec{p}$ for the two layer-5 populations (L5I and L5E), both for individual cells (grey) and for the population as a whole (black). The LFP signals are computed assuming an infinite volume conductor with constant extracellular conductivity σ_t (Equation 13.4). **(e)** Four-sphere head model used when computing the EEG, with the red column depicting the position of the population and the blue disc the position of the EEG electrode. **(f)** EEG signal from each population found by summing the single-cell EEG contribution of all individual cells within each population (different colours, same colour scheme as in **(a, b)**), together with the total summed EEG signal (black). The dashed red curve shows the EEG computed by first summing the z-component of the current-dipole moments for all pyramidal cells, that is, L2/3E, L5E and L6E, and then calculating the EEG from these population dipoles. The close agreement between the two computed EEG signals demonstrates that the vertical component (z-component) of the population dipoles in practice determines the EEG signal.

---

numerical studies have indicated that stimulation currents on the order of a few milliamperes may give intracranial electric fields of 0.1–0.5 V m^{-1} (Miranda et al., 2018) or so, large enough to affect neural spiking (Fröhlich and McCormick, 2010).

For a current-carrying metal electrode embedded in an electrolyte like brain tissue, there will be a drop in potential across the electrode–tissue interface, and the magnitude of this drop will depend on the amount of current delivered by the stimulation electrode. This potential drop can be explicitly incorporated in the modelling formalism either by including an

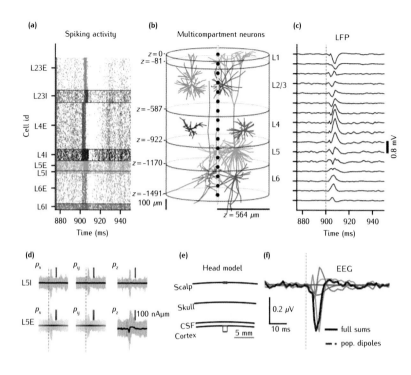

(a) Spiking activity

Cell id

L23E
L23I
L4E
L4I
L5E
L5I
L6E
L6I

880 900 920 940
Time (ms)

(b) Multicompartment neurons

$z = 0$
$z = -81$

$z = -587$

$z = -922$

$z = -1170$

$z = -1491$

100 μm

$\bar{z} = 564$ μm

L1
L2/3
L4
L5
L6

(c) LFP

880 900 920 940
Time (ms)

0.8 mV

(d)

p_x p_y p_z

L5I

p_x p_y p_z

L5E

100 nAμm

(e) Head model

Scalp
Skull
CSF
Cortex

5 mm

(f) EEG

0.2 μV
10 ms

—— full sums
—·— pop. dipoles

Fig 13.23 LFP and EEG signals computed from a comprehensive cortical network with 80 000 neurons representing a 1 mm^2 piece of the cat visual cortex (Figure 13.8). The structure of the network is illustrated in Figure 9.11, and the LFP and EEG signals are computed with the software tool **LFPy**. See Box 13.8 for further explanation. Figure made based on data from Næss et al. (2021).

interface impedance or by modifying the boundary condition for V_e at the electrode surfaces (Joucla and Yvert, 2012).

13.7.2 Electrical Excitation of Neural Cables

In the modelling of neural dynamics in Chapters 2, 3 and 5, the EP V_e was assumed to be fixed at zero in the ECS outside the neurons. In Chapter 10, it was shown how neural activity will modify V_e and possibly lead to ephaptic interactions between neurons (Section 10.2.4). The electrical stimulation of neurons similarly works through the modification of V_e around the neurons by imposing extracellular currents with stimulation electrodes. Thus the effect on the dynamics of neurons can be modelled by allowing V_e to vary with position across the neuron. In general, such a scheme must be pursued by similar numerical schemes as those used to model ephaptic interactions.

However, for a passive dendritic stick described by the cable equation, effects of stimulation by extracellular electrodes can be investigated also analytically. The cable equation for such a stick in the absence of any DC input through the membrane is given by:

$$C_m \frac{\partial V}{\partial t} = \frac{E_m - V}{R_m} + \frac{d}{4R_a} \frac{\partial^2 V_i}{\partial x^2}.$$

(13.14)

This is a generalisation of Equation 5.10 in that the membrane potential V has been replaced by the intracellular potential V_i in the second term on the right. This term represents the effects from the axial currents inside the dendrite, which depends on V_i, not V (unless $V_e = 0$). Since the membrane potential $V = V_i - V_e$, we can reformulate this equation to:

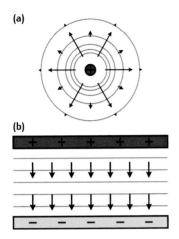

$$C_m \frac{\partial V}{\partial t} = \frac{E_m - V}{R_m} + \frac{d}{4R_a} \frac{\partial^2 V}{\partial x^2} + \frac{d}{4R_a} \frac{\partial^2 V_e}{\partial x^2}, \tag{13.15}$$

or:

$$C_m \frac{\partial V}{\partial t} = \frac{E_m - V}{R_m} + \frac{d}{4R_a} \frac{\partial^2 V}{\partial x^2} + I_{ex}, \tag{13.16}$$

where a new quantity I_{ex} has been introduced:

$$I_{ex} = \frac{d}{4R_a} \frac{\partial^2 V_e}{\partial x^2}. \tag{13.17}$$

Thus the effect of the imposed extracellular field set up by the stimulation electrode can be accounted for by a fictive current I_{ex} proportional to $\partial^2 V_e / \partial x^2$.

A key insight from this analysis is that an imposed constant electrical field $\vec{E} = E_x \vec{e}_x = -dV/dx\, \vec{e}_x$ will not have any effect on an infinitely long dendritic stick, since a constant electric field implies $dE_x/dx = -d^2V/dx^2 = 0$. Thus such an infinitely long dendritic stick would not be affected by a constant electric field. Note, however, that real dendrites have finite lengths. Thus a straight dendritic stick placed between two parallel plates, like in Figure 13.24b, will still be affected by the constant electrical field due to end effects. When positioned close to a spherical electrode, as in Figure 13.24a, the electrical field decays with distance from the electrode. Then I_{ex} will be non-zero for a dendritic stick oriented, for example, in the radial direction (Equation 13.17).

Fig 13.24 Schematic illustration of EPs V_e (grey contour lines) and extracellular electrical fields $\vec{E}_e$ (black arrows) (or alternatively extracellular current $\vec{I}_e$) imposed by stimulation electrodes.
(a) Spherical electrode where currents go from the depicted electrode towards a return electrode infinitely far away.
(b) Two parallel plate-like electrodes, with current leaving the upper plate and entering the lower plate. Length of arrows reflects the strength of the electric field.

13.8 | Case Study: Deep Brain Stimulation

DBS has become an important clinical tool for the treatment of medical conditions, such as **Parkinson's disease**, essential tremor, dystonia and epilepsy, that are believed to be caused by dysfunction of neural circuitry (Lozano and Lipsman, 2013). Through DBS, abnormal brain activity can be targeted and suppressed through electrical stimulation.

Parkinson's disease, first described by James Parkinson in 1817, is a neurodegenerative disease with symptoms of slowness of movement, tremor, rigidity and postural instability (Jankovic, 2008). It is associated with loss of dopamine cells from the **substantia nigra**, leading to significant structural and functional changes in the **striatum** (Day et al., 2006). These are two nuclei of the **basal ganglia**, which also include the external and internal segments of the **globus pallidus** and the **subthalamic nucleus** (STN). The resulting perturbations propagate through the basal ganglia and are likely to play a primary role in the symptoms of Parkinson's disease (Bevan et al., 2002). The so-called **beta band oscillation**, in the range of 15–30 Hz, recorded in the STN is correlated with Parkinsonian symptoms and is suppressed by dopamine-enhancing drugs, with the extent of the suppression being correlated with the degree of relief of motor symptoms (Levy et al., 2002; Kühn et al., 2008).

The use of DBS arises from the discovery that high-frequency stimulation of the **thalamus** relieves tremor (Benabid et al., 1991) and currently it is the primary surgical treatment for Parkinson's disease. A stimulating electrode is placed chronically in the basal ganglia and a subcutaneously located signal generator unit provides current to the electrode. The most common anatomical target for the electrode is the STN, although other basal ganglia nuclei can be targeted. Continuous high-frequency (> 100 Hz) electrical stimulation of the STN improves significantly motor symptoms in patients. A study of 51 DBS Parkinsonian patients assessed 17 years after operation at the same clinic showed that the time spent in typical Parkinsonian movements was reduced by 60–70% and the need for dopamine-enhancing drugs by 50% compared to baseline (Bove et al., 2021).

Despite its success and widespread use, the mechanisms by which DBS leads to clinical benefits for Parkinsonian patients are still debated. Since ablation of the same target nuclei has similar clinical benefits to that of DBS, this led to the initial hypothesis that DBS leads to a similar suppression of neural activity (Filho et al., 2001). Neurophysiological evidence for this is divided. In DBS, the glutamatergic STN projection neurons can easily discharge well above the stimulation frequencies of 100–150 Hz used clinically (Kita et al., 1983). In addition, elevated levels of extracellular glutamate in the substantia nigra and other nuclei targeted by the STN projection neurons have been measured during high-frequency stimulation (Windels et al., 2000). Directly recording the physiological effects of DBS within the STN is experimentally difficult due to interference from the stimulating electrode.

13.8.1 Modelling Deep Brain Stimulation of the Basal Ganglia

It is vital to understand the mechanisms by which DBS achieves its clinical benefits if the procedure is to be effectively optimised and further developed. DBS operates at the network level, affecting large numbers of neural processes within range of the stimulating contacts on the electrode. However, measuring the effects of the stimulation close to the stimulating electrode is difficult. The development of a model combining accurate physiology of the neurons and processes, together with a physical model of DBS, can provide unique insight into its underlying mechanisms. In this section, we review one of the first detailed network-level models of DBS within the basal ganglia, which was developed in conjunction with electrical recordings from Parkinsonian macaque monkeys (Miocinovic et al., 2006). We then discuss how this type of approach has been expanded in contemporary modelling work.

A model of the action of DBS needs to encompass all of the key neural systems influenced by the stimulating electrode. Which parts of the basal ganglia are affected, and the effectiveness of the therapy, depends critically on the placement of the stimulating electrode. An accurate model of the electrode itself, its location and the electric field it generates within neural tissue during stimulation must be defined. Electrophysiological models of the neural elements must then be made. These include populations of entire neurons, axon tracks/bundles and terminating axons and synaptic boutons targeting cells within the vicinity of the electrode.

Primate models of Parkinson's disease have provided an invaluable research tool in understanding the underlying pathophysiology. Single injec-

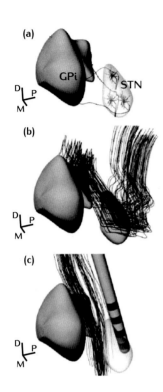

The internal capsule is a compact subcortical structure containing all the nerve fibres coming from and going to the neocortex.

tions of 1-methyl-4-phenyl-1,2,3,6-tetrahydropyridine (MPTP) through the internal carotid artery leads to reversible effects of contralateral rigidity and **bradykinesia** (slowness of movement), two cardinal symptoms of Parkinson's disease. This primate model provides a framework in which to examine the key parameters in DBS (Hashimoto et al., 2003). Use of scaled clinical DBS electrodes implanted in Parkinsonian macaques has provided a valuable tool for investigating its underlying mechanisms.

Miocinovic et al. (2006) brought together models of the DBS electrode and its field effect on axons (McIntyre et al., 2004) with models of the STN projection neurons (Gillies and Willshaw, 2006), together with data from macaque models of Parkinson's disease and high-frequency stimulation (Hashimoto et al., 2003). Their goal was to replicate in their model the effects of STN DBS that improved bradykinesia and rigidity in two macaque monkeys. They modelled the electrode and the electric field it generates by representing the ECS as a network of resistors of equal value (Butson and McIntyre, 2005).

Anatomical Model

As the position of the stimulating electrode plays a critical role in the effectiveness of the treatment, determining the location of neural processes with respect to the stimulating electrode is a critical part of the computational model. Three-dimensional reconstruction from histological slices of the brain of the Parkinsonian macaque provided the trajectory of the electrode in relation to key basal ganglia nuclei. Figure 13.25 illustrates a 3D reconstruction of a single macaque brain and the relative position of the clinically effective stimulating electrode.

A significant proportion of STN projection neurons is likely to be influenced by the stimulating electrode. The positions of the neurons and their axons must be specified accurately to estimate the extent of this influence. Three-dimensional reconstruction of macaque STN projection neurons that target the pallidum showed that their axons course dorsally along the ventral border of the thalamus or ventrally along the lateral border of the STN (Sato et al., 2000). Anatomical models of a single reconstructed STN neuron morphology were adapted to create two additional populations whose pallidal projections follow these paths (Figure 13.25a). This provided three anatomical types of STN neurons that may be influenced differentially by the electrode.

The axons of the globus pallidus internal segment (GPi) cross the **internal capsule** dorsal to the STN. This tract is called the lenticular fasciculus and was also modelled due to its proximity to the stimulating electrode. These fibres course along the dorsal border of the STN before targeting the thalamus (Figure 13.25b).

Since clinical DBS can lead to evoked motor responses resulting from activation of corticospinal fibres in the internal capsule, which lies directly along the lateral border of the STN, corticospinal fibres were also modelled running lateral to the STN (Figure 13.25c).

In the model, each of the five reconstructed pathways (three populations of subthalamic–pallidal axon paths, the lenticular fasciculus and the corticospinal fibres of the internal capsule) was replicated by placing nerve fibres

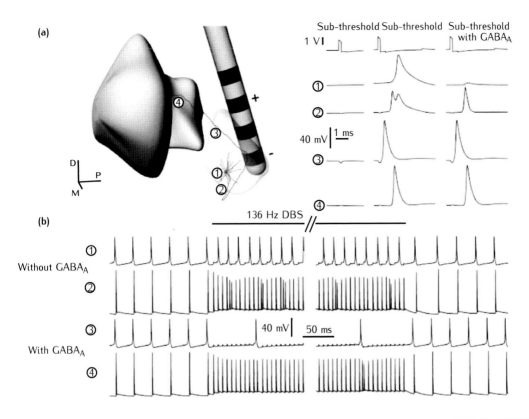

Fig 13.26 Three-dimensional simulation of DBS. **(a)** A DBS electrode is shown next to a model of an STN neuron. The DBS electrode induces an extracellular electric field (not shown) which stimulates the STN neuron, leading it to fire when the DBS electrode pulse is super-threshold. The traces at the right indicate the simulated membrane potential measured from the corresponding positions of the STN neuron shown in the figure, when the DBS pulse is either sub-threshold or super-threshold, and when there is GABAergic input to the soma. **(b)** Simulated STN activity in response to high-frequency DBS stimulation with and without GABAergic input. Adapted from Miocinovic et al. (2006), with permission from The American Physiological Society.

at randomly chosen positions within their respective anatomical boundaries. The model STN neuron morphologies were duplicated randomly within the STN to create a population of projection neurons. This created an anatomical model of neural processes within the vicinity of the DBS electrode. Any processes or neurons and dendritic fields that intersected significantly with the electrode itself were removed from the model (effectively eliminating 'damaged' cells or axons). In total, 150 STN projection neurons, 80 axons of the lenticular fasciculus and 70 corticospinal fibres of the internal capsule were included in the anatomical model.

Network Compartmental Models

A compartmental modelling approach was used to specify the physiological properties of the neurons in the network. Compartmental models have the advantage of being able to describe accurately the wide range of neural physiology observed in STN projection neurons (Gillies and Willshaw, 2006). However, the disadvantages include the enormous numbers of parameters involved and the computational resources required. This can significantly restrict the numbers of neurons that can be simulated. The most common approach to dealing with the large numbers of parameters is to use published models of single neurons, the code for which are available in databases such as ModelDB. These models are then replicated to create network populations.

The compartmental model of the STN projection neuron (Figure 13.26) is based on that due to Gillies and Willshaw (2006), which used anatomical and physiological data from the rat STN projection neuron. Adaptation of the model to the reconstructed macaque STN morphology was made via adjustments to the maximum channel conductances, leaving the underlying channel kinetic properties unchanged. In particular, the model conductances were retuned to reproduce the physiological characteristics observed in macaque STN neurons in the Parkinsonian state.

The myelinated axon compartmental models of McIntyre et al. (2002) were adapted to model the axons of the STN neurons, the lenticular fasciculus and the corticospinal fibres of the internal capsule. As only the pallidal axons of the lenticular fasciculus were modelled, and not their originating neural body, tonic firing patterns characteristic of neurons of the GPi were induced in these fibres using simulated current injection at the origin of the axon.

The influence of high-frequency stimulation-induced trans-synaptic conductances on STN neuron activity was also investigated in the simulations. STN neurons receive a GABAergic input from the pallidum that primarily terminates proximally (Smith et al., 1990). $GABA_A$ synaptic conductances were added to the somatic compartment of the STN model. The effects of high-frequency stimulation-induced trans-synaptic conductances were modelled via synchronous activation of these conductances in each STN neuron in response to stimulation pulse onset.

Modelling the DBS Electrode

To model the DBS electrode and the surrounding ECS, Miocinovic et al. (2006) used a 3D finite element model. The extracellular conductivity was assumed to be homogeneous, although a later study (Miocinovic et al., 2009) used anisotropic conductivities inferred from diffusion tensor imaging studies. The simulated dimensions were taken from those for a human, scaled down for a monkey. The electrode shaft was modelled as an insulator, and each of the four contacts as a conductor (Figure 13.26). For bipolar stimulation, the potential at two of the four contacts had to be specified. The stimulus waveforms were biphasic square voltage pulses, which were modified to take account of the effects of electrode capacitance.

13.8.2 Results

In the model, the neurons were stimulated, with the electrodes positioned as for the two Parkinsonian macaques used in this study. This enabled direct comparisons to be made between the model findings and the single unit recordings made in the macaques.

The performance of the model was assessed by calculating the levels of tonic activation during clinically effective and ineffective STN DBS for these three populations:

(1) Corticospinal fibres of passage;
(2) STN projection neurons;
(3) Fibres of passage of neurons from the GPi.

Activation levels of simulated corticospinal fibres were low, as found experimentally by measuring muscle contraction. This indicated that the voltage spread in the tissue around the electrode was predicted well by the finite element model.

Activity in both STN projection neurons (Figure 13.26) and GPi fibres of passage was found to be important in the context of STN DBS. Under simulated STN DBS conditions, both STN neurons and GPi fibres were activated and there were statistically significant differences between clinically effective and clinically ineffective levels of activity. There was little effect of frequency, although it is known that DBS at frequencies greater than 100 Hz is more beneficial, and at frequencies less than 50 Hz sometimes less beneficial (Rizzone et al., 2001).

The results obtained from simulation studies did not change markedly when different variants of the STN model neuron were used. This seemed to be due to the fact that action potential propagation is initiated in the myelinated axon rather than in the cell body (McIntyre et al., 2004).

Electrophysiological Recordings

The responses of both GPi neurons and STN neurons to STN DBS were measured using single unit recordings. By looking at the effects of moving the electrode 0.25 mm from its original position in different directions, it was found that electrode placement was crucial. For example, for clinically effective stimulation, the STN activity varied by between 18% and 42% compared to the levels with the electrode at its original position, and between 6% and 16% for GPi. There were differences between the two macaques in the rates of firing of GPi neurons, and these differences were predicted from the model.

Whereas in the simulations STN activity was assessed at the distal end of the axon, microelectrode recordings in the STN picked up primarily somatic output. Therefore, recordings were also made in the STN axons. Activation was heavily dependent on the positioning of the electrode. With the stimulating electrode placed far away from the STN cell, the soma could become inhibited through the stimulation of the inhibitory inputs to the soma; with the stimulating electrode very close, so that the axon was excited directly, axonal firing was dominated by this stimulation.

13.8.3 Comments about the Success of This Study

The main finding of this joint modelling/experimental study was that the model predicted that during STN DBS, there can be activation of both STN projection neurons and GPi fibres of passage, and this was borne out by the experimental results. The relative proportion of activity in these two neuron types was highly dependent on electrode position. Indications from the experiments on the macaques are that for clinical benefit, approximately one-half of the STN should be active. Activation of GPi fibres is also beneficial, but large-scale activation may not be necessary.

The model of STN used was as comprehensive a model as possible of the prime actors in DBS of the basal ganglia, but substantial simplifications had to be made. Many of the parameter values describing biophysical properties came from rat STN neurons (Gillies and Willshaw, 2006) and so they do not necessarily apply to macaques. In addition, the simulated time courses of activation onset were very rapid, whereas the effects of DBS evolve over a much longer timescale. Two other limitations were that it was difficult to estimate the tissue conductivity in 3D, which was needed in modelling the effects of the electrode; and only the most likely candidates for involvement in STN DBS – STN projection neurons and GPi fibres – were considered, although there are other, much less likely, candidates.

One major simplification was that the influence of afferent activity through cortical axons was not considered. There is recent evidence that direct stimulation of these axons can cause therapeutic effects (Dejean et al., 2009; Gradinaru et al., 2009).

Despite these qualifications, this is the first demonstration of an accurate simulation model for STN DBS, with important conclusions being made about the action of STN DBS. In subsequent work, several of the avenues highlighted for improvement have been followed, which are now described.

13.8.4 Pathway Activation Modelling

The fact that there are several axonal pathways associated with the STN requires more detailed models of activation through the application of DBS. Gunalan et al. (2018) reviewed a range of patient-specific models for action potential initiation in response to activation. In a Field Cable model, the electrical field, axonal trajectories and ion channels are modelled directly. The two other approaches involve approximations to this model, requiring less computing resource than the computationally intensive Field Cable approach. In a Driving Force model, ion channels are represented indirectly, and in a Volume of Tissue Activated model, the calculation of axonal response in an electric field is estimated by a predictive algorithm. Neither model represents good approximations to a full Field Cable model, which points towards the need for developing this type of model despite its severe computational demands. It should be emphasised that although the Field Cable approach is the current modelling gold standard, current experimental techniques do not allow comparison of the outputs of this model with experiment.

13.8.5 Electrode Design and Stimulation Protocols

Modelling work has contributed both to the design of the optimal electrode for DBS and to finding the optimal patient-specific stimulus regime, reviewed by Yu et al. (2020). Several approaches have attempted to improve on the traditional electrode arrangement with four cylindrical contacts spaced along the electrode (Figure 13.25). Computational modelling is being used to simulate the electrical field and the volume of tissue activated. The aim is to maximise the coverage of the area to be stimulated whilst minimising current spread elsewhere. Using a larger number of contacts enables the direction and range of stimulation to be controlled.

In DBS, the electrodes are implanted permanently and usually a fixed pattern of stimulation is applied continually. However, one area of increasing interest is the online modulation of the electrical stimulation (Rosin et al., 2011; Beuter et al., 2014). This technique has been used in epileptic patients to suppress synchronised cortical activity in the gamma frequency band (35–100 Hz) (Sohal and Sun, 2011). In Parkinsonian patients, two frequency bands are associated with motor dysfunction: the alpha band (8–12 Hz) and the more significant beta band (13–30 Hz). The usual computational approaches focus on reducing activity in the beta band. For example,

Little et al. (2016) used a simple procedure to turn on the stimulus when the power in the beta band exceeded a fixed threshold and turn it off when it was lower. In clinical trials, they found significant improvement.

These approaches rely on a simplified model of the target structure. Since it has been impossible to obtain comparative recordings from Parkinsonian patients for comparison, most work shows potential benefits in theory rather than in methods which have been validated in clinical applications. One example of modelling work is by Grado et al. (2018). They used a Bayesian adaptive control system to reduce the beta band power in a mean-field model of the basal ganglia–thalamocortical system subject to stimulation by a continuous series of pulses. In this model, the mean firing rates and voltages of nine interlinked basal ganglia brain structures are simulated for the normal state and also for a Parkinsonian case where there are strong oscillations in the beta band. Operating over a short timescale, the beta band oscillations are assessed and the stimulator emits a pulse, provided that the power and phase each exceed an adjustable threshold. Over a longer timescale, how the power varies as a function of stimulus amplitude and these two threshold values is used to set these three parameters using a Bayesian method (Section 14.7). One simplification made by Grado et al. (2018) is that the stimulating electrode supplies a single pulse, whereas multiple contacts are commonly used. How to design controllers to closed-loop DBS across the more realistic types of an electrode is a subject of ongoing theoretical interest (Weerasinghe et al., 2021).

13.8.6 Alternative Targets for DBS

Use of **optogenetic technology** in a freely moving Parkinsonian rat (Gradinaru et al., 2009) offers a promising avenue for the direct monitoring of the individual components of basal ganglia circuitry. Valverde et al. (2020) found that in a Parkinsonian rat, stimulation of the STN ameliorated the hyperactivity in the pyramidal cells of the motor cortex. In addition, two classes of inhibitory interneuron were affected: somatostatin neurons were activated and parvalbumin neurons were inhibited. Optogenetic activation of somatostatin interneurons also alleviated Parkinsonian symptoms. To explore further the interactions within the motor cortex, these authors constructed a simple model of integrate-and-fire neurons (Section 2.7) comprising three populations of cells interconnected with both excitatory and inhibitory synapses. The properties of the model neurons and their interconnectivity were based on experimental findings. DBS was modelled by applying instantaneous excitation to all cells and the Parkinsonian condition represented by lowering the threshold of firing in the pyramidal cells. The model was used to test the information processing capabilities of the motor cortex by showing that, for example, Parkinsonian cortical hyperactivity can be counteracted either by DBS or by somatostatin activation without compromising the underlying dynamics. Whilst several aspects of the model can be criticised, such as that the authors have been selective in the pieces of the complex thalamocortical network that they have modelled, this is a good example of how modelling can aid in the treatment of Parkinson's disease, in this case through helping to identify more selective and less invasive targets of DBS.

Optogenetics is a technology of introducing by genetic means light-sensitive channels into membranes which are often used to activate selectively a specific class of nerve cell (Section 13.10.2).

13.9 Modelling Magnetic Signals and Magnetic Stimulation

13.9.1 Magnetic Field Recordings

Of the techniques for measuring electrical potentials already described, EEG is the only one that is non-invasive so that it can be used on humans without surgical intervention. Another important non-invasive technique is MEG where the tiny **magnetic fields** generated by neural activity are measured outside the head (Figure 13.2). These magnetic fields are tiny, on the order of 10^{-14} T which is about a billionth of the Earth's magnetic field. They must thus be recorded with extremely sensitive measurement devices, and the standard method has been to use so-called **superconducting quantum interference devices** (SQUIDs) (Hämäläinen et al., 1993; Ilmoniemi and Sarvas, 2019). Recently other techniques have been developed, for example, **optically pumped magnetometers** (OPMs) where the effect of the magnetic field on optically polarised electron spin is used (Kominis et al., 2003; Iivanainen et al., 2017). For an overview of such recent developments, see Ilmoniemi and Sarvas (2019).

Whilst EEG signals arise from volume currents in the ECS, the MEG signal stems from both intracellular and extracellular currents. However, similar to EEG signals, the MEG signals generated by neurons can, in many situations of practical interest, assuming, for example, a spherically symmetric electric head model, be computed from knowledge of neuronal current dipoles alone (Ilmoniemi and Sarvas, 2019). For the simplest case, assuming an infinite volume conductor as a head model, the magnetic field $\vec{B}_{\mathrm{d}}$ set up by a current dipole $\vec{p}$ can be computed using a variant of the **Biot–Savart law** from magnetostatics (Hämäläinen et al., 1993):

$$\vec{B}_{\mathrm{d}} = \frac{\mu_0}{4\pi} \frac{\vec{p} \times \vec{R}}{R^3} . \tag{13.18}$$

Here $\vec{R} = \vec{r} - \vec{r}'$ is the displacement between the position of the dipole $\vec{r}'$ and the position of the recording SQUID $\vec{r}$, and $R = |\vec{R}|$.

When modelling EEG signals, the key material parameter is the electrical conductivity σ which varies across the different parts of the head (brain, CSF, skull, scalp), and the signal has to propagate through on the way to the recording electrode. When modelling magnetic fields recorded by MEG, the analogous material parameter is the **magnetic permeability** μ. In biological materials, μ is very close to the magnetic constant, the magnetic permeability of vacuum μ_0. Thus μ_0 is used in the forward modelling formula in Equation 13.18 (Hämäläinen et al., 1993).

Most studies with magnetic brain signals have been based on MEG recordings where the neuronal sources are so far away from the magnetic field sensors that the far-field dipole approximation assumed in Equation 13.18 can be applied. However, measurement devices for measuring magnetic fields close to the neurons have also been developed (Barbieri et al., 2016; Caruso et al., 2017). To compute the magnetic fields close to neurons, another version of the Biot–Savart law must be used (Blagoev et al., 2007):

$$\vec{B}(\vec{r}) = \frac{\mu_0}{4\pi} \sum_{m=1}^{m_a} i_m^a \frac{\vec{d}_m \times (\vec{r} - \vec{r}_m)}{|\vec{r} - \vec{r}_m|^3} . \tag{13.19}$$

This formula provides the magnetic field for m_a axial currents i_m^a where $\vec{d}_m$ are axial line element vectors and $\vec{r}_m$ are the midpoint positions of each axial current. When used in combination with multicompartmental modelling, the sum goes over the intracellular axial currents between all adjacent compartments (Hagen et al., 2018). Note that this formula assumes that the contributions to the magnetic field from volume currents in the ECS are negligible (Hämäläinen et al., 1993, p. 427).

For examples of computation of magnetic fields stemming from neural activity, see Box 13.9 and Figure 13.27c and f.

13.9.2 Magnetic Field Stimulation

Analogous to electrical brain stimulation as described above, neurons can also be activated by application of magnetic fields. A key technique used in both basic research and clinical applications is **transcranial magnetic stimulation (TMS)** (Barker et al., 1985) where a magnetic coil placed immediately outside the head delivers brief pulses of strong magnetic fields into the brain. These brief magnetic pulses, with amplitudes of up to 2 T lasting about 0.1 ms, induce a substantial electrical field and, in turn, a substantial electrical current in the underlying cortex (Hallett, 2007). Unlike for TES, these induced currents are large enough to not only modulate the firing of action potentials, but also induce action potentials by themselves (Ilmoniemi et al., 2016).

13.10 Optical Measurements and Stimulation

13.10.1 Optical Measurements

The use of light and other types of electromagnetic radiation to observe and probe nature is ubiquitous in science. Indeed, our understanding of the universe is based on such measurements. Since brain activity stems from electrochemical processes which are difficult to observe optically, historically the use of such techniques to measure activity in living brains has played a secondary role. However, over the last decades, a host of new optical techniques for probing brain activity have been developed, which are being used increasingly (Frostig, 2009; Roe, 2009). Many different techniques have been developed, but the principle is essentially the same in all: send light into brain tissue and measure the light being reflected or emitted (Figure 13.28).

Intrinsic Optical Imaging

It has been known for a long time that blood changes colour depending on oxygen content. Arterial oxygen-rich blood is redder than venous blood containing less oxygen. This oxygen dependence of light reflectance of blood is used in **intrinsic optical imaging** (Frostig and Chen-Bee, 2012) where light in the visible or near-infrared spectrum is applied. The signal recorded is not easy to interpret in terms of the underlying activity in neurons, however.

Box 13.9 | Multimodal modelling

The term **multimodal modelling** refers to the joint modelling of data from several different types of measurements, that is, several measurement modalities. An example of multimodal modelling of electric and magnetic signals generated by a single multicompartment model neuron is given in Figure 13.27.

Figure 13.27a shows the EP V_e around a multicompartment neuron model computed using the membrane-current based formula in Equation 13.4 The excitatory synaptic input current $i_{syn}(t)$ (inset I), at position of the red triangle, results in positive deflections of the membrane potential throughout the neuron, including at the soma ($V_{soma}(t)$, inset II), and also for the EP $V_e(t)$ for the depicted example position. The image plot shows the EP at the time when the synapse current is the largest ($t = 2.25$ ms).

Figure 13.27b shows an expanded view of the EP in Figure 13.27a, computed with membrane-current based formula (V_e; Equation 13.4) inside the dotted white circle with radius 500 µm and the dipole formula ($V_{e,p}$; Equation 13.2) outside the white circle. The computed current dipole moment $\vec{p}$ used in the dipole calculation is shown as an arrow. Time course of $V_{e,p}$ at positions of the coloured dots is shown in the inset. As observed, the prediction from the simpler dipole formula agrees excellently with the predictions from the membrane currents at the dotted–circle interface, implying that the far-field limit is reached at this distance from the soma.

Figure 13.27c correspondingly shows the y-component of the magnetic field associated with the same neural activity, computed with the Biot–Savart formula (B_y; Equation 13.19) inside the dotted circle and with the dipole formula ($B_{y,p}$; Equation 13.18) outside. Further, $B_{y,p}(t)$ at positions of the coloured dots are shown in the inset. An excellent agreement between the two modelling approaches, Biot–Savart versus dipole formula, is seen at the dotted–circle interface.

Corresponding EEG signals computed with the four-sphere head model for nine positions at the head (Figure 13.27d) are shown in Figure 13.27e. The numbered sites 1–9 mark the EEG recording locations, each offset by an angle $\pi/16$ in the xz-plane. The parameters used in the four-sphere model are given in Table 13.1, and the current dipole is positioned under recording location 5, with a radial distance of 8.75 cm from the centre of the head model.

The tangential component of the scalp magnetic field B_ϕ at the same positions are shown in Figure 13.27. Here, rather than assuming an infinite volume conductor formula in Equation 13.18), we assume the more realistic situation of a spherically symmetric head conductor and use Equation 34 in Hämäläinen et al. (1993). B_ϕ is the component in the angular direction in the xy-plane, with the z-axis going from the origin through recording position 5. At position 5, the magnetic field in the positive y-direction is depicted.

The main point here is not the detailed results for this particular neuron, rather the illustration of physics-based multimodal modelling of neural activity: the whole set of electric and magnetic signals can be computed from a multicompartment neuron model or a network of neuron models.

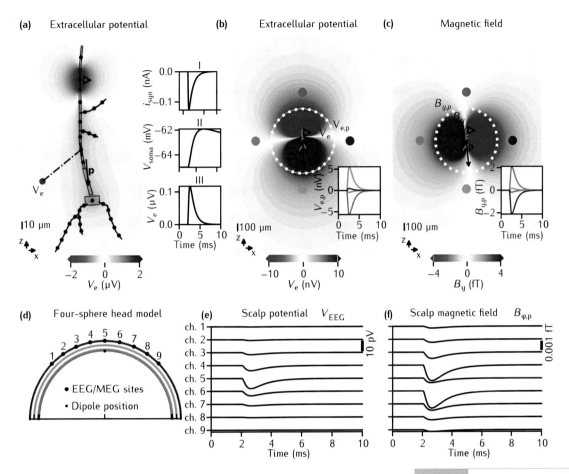

Fig 13.27 Illustration of multimodal modelling of electric and magnetic signals generated by a single multicompartment model neuron (Box 13.9). The signals are computed using the simulation tool LFPy (see Hagen et al., 2018). Figure made based on data from Hagen et al. (2018).

One reason is that light is strongly scattered by brain tissue so that the reflected light is diffuse and unfocused, and imaging based on these techniques is thus often termed **diffusive optical imaging**. Further, the reflected signal does not directly measure the electrical activity of neurons. Rather, the recorded signal is a measure of blood volume and blood oxygen content. Since the amount of oxygen is actively regulated according to the needs of neurons, this is correlated with neural activity, but the timescale of the signal is given by the timescale of blood dynamics which is seconds, rather than milliseconds. As an example of its use, some of the modelling of neural maps in Section 12.5 relies on this type of optical data.

Voltage-Sensitive Dye Imaging

Optical signals directly reflecting the electrical dynamics of neurons can be recorded by voltage-sensitive dye imaging (VSDI) (Chemla and Chavane, 2012). Here particular marker molecules are added to the brain tissue prior to optical imaging or, in a more recent technique, genetically encoded (Bando et al., 2019). These marker molecules are embedded in the neuronal membranes and are designed such that the amount of reflected light depends on the local membrane potential.

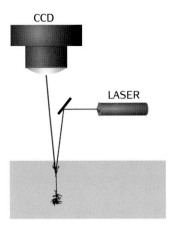

CCD

LASER

Fig 13.28 Illustration of the principle behind optical measurements of brain activity. Laser light with a suitably chosen wavelength is shone onto brain tissue, and the resulting emitted light is measured by a charge-coupled device (CCD) camera. Note: in two-photon calcium imaging, two separate laser beams are used instead to excite the brain tissue.

Biophysical modelling of the recorded VSDI signals will thus involve modelling of the membrane potentials averaged over the neuronal membrane areas with embedded marker molecules which are reached by light stimulation (Chemla and Chavane, 2012). The propagation of light from the stimulating laser to the neuron and back to the detector can also be modelled (Tian et al., 2011).

Calcium Imaging

As outlined in Chapter 6, calcium is involved in many neuronal processes. **Calcium imaging** thus offers a unique window into the dynamics of cells (Grienberger and Konnerth, 2012; Helmchen, 2012). Here specific fluorescent dyes inside the cell bind to free calcium, and the fluorescence intensity gives a measure of the intracellular calcium concentration. A model relating the free calcium concentration to this intensity is described in Section 6.6.3.

In **two-photon calcium imaging**, very high spatial resolution, in the micrometre range, is obtained by utilising the quantum nature of light. Here two photons, each with too little energy to excite the fluorescent calcium indicator molecule, impinge on the region of interest. Fluorescence is only possible in the region where the two beams overlap, a region which can be as small as a micrometre.

When neurons fire action potentials, calcium flows into the cell, and two-photon calcium thus offers the possibility to monitor action potential firing in individual cells. However, the timescale of calcium dynamics is typically on the order of 100 ms, much longer than the millisecond timescale of action potential dynamics (Helmchen, 2012). Thus, whilst two-photon calcium imaging offers the opportunity to measure activity in many individual neurons at the same time, at present, electrical recordings are required to monitor the generation of individual action potentials in detail. Nevertheless, two-photon imaging provides a good qualitative measure of neural activity, where increased neural firing gives a larger signal.

13.10.2 Optical Stimulation

Some bacteria can register light by means of light-sensitive ion channels embedded in their membranes. In **optogenetics**, such ion channels are implanted into neuronal membranes so that cells can be electrically stimulated by application of light (Thompson et al., 2014; Appasani, 2017). This technology has many potential clinical applications, for example, in providing retinal prostheses to restore vision for blind people (Busskamp et al., 2010; Nirenberg and Pandarinath, 2012). It also offers a unique research tool for probing neural circuits as, for example, subgroups of neurons can be selectively stimulated by light to fire action potentials so that the network response to this particular activity can be investigated more easily.

Two types of light-sensitive ion channels are of particular interest. **Channelrhodopsin** responds to blue light by opening up a channel, allowing sodium and other cations to flow into the cell, thus depolarising it. **Halorhodopsin** instead responds to green/yellow light by moving chloride ions into the cell, thus hyperpolarising it. The action of these light-gated ion channels can be modelled using the same principles as those used to model voltage- and ligand-gated ion channels (Grossman et al., 2013; Williams et al., 2013). Effects of such stimulation on neuronal dynamics can thus be modelled using the principles outlined in this book; for example, see Foutz et al. (2012).

13.11 Other Measurement Techniques

Many techniques are now available for measuring physiological activity in the brain at different spatial and temporal scales (Figure 13.1). In this chapter, the focus has been on experiments directly measuring electrical activity in neurons and networks and where biophysical modelling is key to making a proper link between neural activity and what is measured. Electrical and magnetic measurements stand out here, and to some extent optical measurements. Other measurement modalities have also provided important insights into brain activity.

Magnetic resonance imaging (MRI) is a common systems-level imaging technique. It is non-invasive and is thus frequently used in human studies. The technique comes in two main variants: static MRI, measuring the anatomical structure of the brain, and fMRI, measuring brain activity. The measurement is based on the magnetic properties of atomic nuclei, as these have been found to depend sensitively on the surrounding tissue. The **BOLD signal** from fMRI measures blood flow and oxygen content, physical variables that are only indirectly related to the electrical activity of neurons (Bartels et al., 2012). Unlike for other systems-level measures like EEG and MEG, a clear forward model linking neural activity to the BOLD signal is not yet established, but see Lundengård et al. (2016) and Sten et al. (2017). Also, the spatial and temporal resolutions of the signal are limited and typically measured in millimetres and seconds, respectively.

PET is another commonly used systems-level technique (Heurling et al., 2017). Here injected radioactive substances emit positrons, essentially positively charged electrons, which annihilate with electrons in the brain, resulting in the emission of two gamma-frequency photons being ejected in opposite directions. These two photons are then picked up by detectors placed around the head, allowing for estimation of the point of positron emission. The positron-emitting radioactive nuclei, for example, radioactive isotopes of carbon, oxygen and fluorine, are incorporated into molecules involved in the brain process of interest, and then inserted into the subject's blood. The intensity of gamma photons stemming from a particular position in the brain then gives a measure of the neural activity at this position. As the emitted positron often travels a millimetre or more prior to annihilation, the PET technique, like fMRI, also has a limited spatial resolution.

13.12 Summary

In this chapter, we focused on modelling of what is measured in various types of physiological brain experiments. The goal of such modelling is twofold:

(1) to make quantitative experimental predictions from neuron and neural network models; and (2) to develop an intuition about what the measured signal can tell you about the underlying brain activity.

The emphasis was on electrical measurements (spikes, LFP, ECoG, EEG) where the membrane currents of the cells provide the physical link between the neuron activity and recorded signals and the volume conductor theory is used to provide the quantitative link. The forward modelling of magnetic signals, such as MEG, also stemming from membrane currents, was also outlined. Furthermore, stimulation by electric currents and magnetic fields by methods such as TES and TMS are governed by essentially the same physical principles and can be modelled similarly.

In optical measurements of neural activity, light of selected wavelengths are shone on brain tissue and the resulting returned light is measured and analysed. Intrinsic optical imaging, that is, optical imaging without any signalling molecules added to the brain tissue, is the conceptually and technically simplest version. The signal largely reflects blood dynamics, however, and has limited spatial and temporal resolution. In VSDI, dye molecules are added to the brain tissue to make the returned light signal be proportional to the membrane potentials of the target neurons, giving a more direct measure of neural activity. In calcium imaging, specific reporter molecules in the brain tissue instead make the return signal be a measure of the intracellular calcium concentration, and thus also of neuronal firing activity. However, the temporal resolution is limited to 0.1 seconds or so. Combined with two-photon light stimulation, this technique offers a very high spatial resolution. Furthermore, with the advent of optogenetic techniques, it is now also possible to stimulate neurons optically, not only record their activity.

Whilst harder to interpret in terms of neural activity, the non-invasive measurements EEG and MEG are particularly important for assessing activity in human brains. Other important non-invasive measurement techniques are fMRI and PET. Here the physical link between the activity of neurons and what is measured is less established than for EEG and MEG.

CHAPTER 14

Model Selection and Optimisation

Modelling a neural system involves both the selection of the mathematical form of the model's components, such as neurons, synapses and ion channels, plus assigning values to all of the model's parameters. This may be based on different criteria, including matching the known biophysics, fitting a suitable empirical function to data, or computational simplicity. Thus **model selection** is a non-trivial problem that needs careful consideration. Once the form of the model has been established, it is likely that at best only a few parameter values will be available through existing experimental measurements or computational models. Instead it will be necessary to obtain values either by **parameter estimation** from experimental data or through optimisation of model output against required criteria. Many mathematical techniques are available for these purposes and we outline them here. We focus on how to specify suitable criteria against which a model can be optimised. For most complex models, it is likely that ranges of parameter values can provide equally good outcomes against performance criteria. Exploring the parameter space can lead to valuable insights into how particular model components, such as membrane-bound ion channels, contribute to particular patterns of neuronal activity. It is important to establish the sensitivity of the model to particular parameter values.

Building a model of any system requires two major stages: firstly, choosing the components that make up the model, known as **model selection**; and then setting values for all of the parameters in the model so that it reproduces data about the system to a given level of accuracy, known as **model optimisation** or **model fitting**. These stages are not mutually exclusive and may well run in an iterative manner before the final model, or set of models, is settled upon. For instance, optimising the model may highlight that certain parameters in the selected model are not important for reproducing the data of interest, in which case the components corresponding to those parameters potentially can be discarded from the model without impairing model performance. Both model selection and model optimisation can be pursued ad hoc and in formal ways. In this chapter, we outline the emerging formal approaches to these modelling stages for neural systems.

Throughout this book, we have presented different ways of modelling the same components of neural systems. For example, a range of approaches to modelling neurons is covered in Chapters 5 and 8, and for neural networks in Chapter 9. We have tried to emphasise the different motivations that underpin the choices to be made in selecting models for neurons and their networks and other subcomponents, such as synapses. These choices generally concern the complexity of the model. Often we want to choose the simplest model that can help us answer the scientific questions under study. Alternatively, we might want to build as detailed and biophysically realistic a model as possible to help us explore the features of such a model and carry out experiments on our model that are not possible on real neural tissue. In Section 14.1, we revisit the model selection problem and introduce some more formal ways of choosing between different possible models for the same system.

Once we have a model structure, we will be faced with the issue of assigning values to the parameters of the model so that we can run numerical simulations or analyse model properties. This can be a complex and time-consuming process that may proceed through many stages. We look at this problem in detail in Section 14.2. Issues of directly fitting to experimental data or feature-based fitting are considered, including the error measures that need to be applied to judge model output against experiment. Numerical optimisation procedures for adjusting model parameter values to minimise the measured errors are outlined (Section 14.3). Two case studies involving setting the parameter values of the passive (Section 14.4) and active (Section 14.5) membrane properties of a neuronal compartmental model are given to illustrate the procedures.

Two important issues should be explored when carrying out any model optimisation. Firstly, we need to ascertain whether the parameter values obtained are unique. It is highly likely that the data we have available against which we are optimising are insufficient to constrain all the model's parameters (Marder and Taylor, 2011; O'Leary et al., 2015), in which case, many sets of parameter values may provide an optimal fit to the data. This is an important outcome that ties in with what we know about the robustness of neurobiology (Marder and Taylor, 2011; Migliore et al., 2018; Kamaleddin, 2022). Associated with this is that we need to assess the sensitivity of model output to particular parameter values (Section 14.6). The model may be sensitive to the values of certain parameters, such that uncertainty in these parameters will lead to uncertainty in the model output. Ideally we can obtain probability distributions for parameter values, rather than single values (Section 14.7).

14.1 Model Selection

When starting the development of a model of part of the nervous system, we are faced with considerable choice in the mathematical descriptions we could use for the neural components involved. This is the problem of **model selection**. The reasons for our final choices can be quite varied, but hopefully start with deciding on the level of detail needed to answer the scientific questions under consideration (O'Leary et al., 2015). The aim might be to use our model to provide support for an explanation for experimental data. In this case, the components of the model should have readily identifiable biophysical counterparts, such as ion channels. Alternatively, we may be more interested in the computational capabilities of a system that exhibits certain characteristics of the nervous system. For example, we might want to explore the capabilities of large networks of spiking neurons. So we might forgo some of the detailed biophysics in favour of computational simplicity and the ability to analyse the resultant model, leading us to choose an integrate-and-fire neuron model as the fundamental component of our neural networks.

These rationales for choosing different models do not address one significant question: what is the best model for the system under study, given that it needs to reproduce certain nervous system behaviour that is characterised by available experimental data?

Most of the models in this book are based on the known laws of neuronal biophysics, constituting **law-driven modelling**. As discussed in Saltelli et al. (2008, Chapter 1), such models are typically over-parameterised so that available experimental data are not sufficient to uniquely allocate values to all of the parameters. When optimising such a model against the experimental data, often there is no single optimal solution and we can find many different sets of parameter values for our model that perform equally well. This might not be an issue for the scientific questions we are trying to address. In our consideration of parameter sensitivity (Section 14.6), **ensemble modelling** looks to explore the total behavioural space of such models, seeking to understand the contribution of different model components to producing the same or different model outcomes (Section 14.6.1). Such outcomes may correspond to known behaviour in the system under study, or may be aberrant, perhaps relating to disease states.

However, we may want to reduce our law-driven model to the simplest possible model that reproduces the desired behaviour captured in our available experimental data. Parameter sensitivity analysis can help with this too. **Uncertainty quantification** (Section 14.6.2) will provide information on which parameters strongly determine the desired behaviour and which do not. If the behaviour is only weakly or not at all sensitive to the value of a particular parameter, then the model component associated with that parameter is a candidate for removal from the model. For example, a particular type of ion channel in particular spatial locations in a compartmental neuron model may not influence the specific neuron-firing characteristics of interest.

This contrasts with **data-driven modelling** (Saltelli et al., 2008, Chapter 1) where typically the aim is to find the simplest model with the fewest parameters that adequately captures the statistics of the available data. Such models are often abstract mathematical constructs without direct links to underlying physical laws. Data-driven modelling is at the heart of **machine learning** (Bishop, 2006; Goodfellow et al., 2016) in artificial intelligence (Box 11.2).

14.1.1 Model Reduction and Comparison

How much reduction in model complexity is possible is likely to be a matter of some judgement on how far the model should fit the available experimental data, compared with how well the model might generalise to new data, which may encompass new behavioural states. This is the classic problem in machine learning, often called the bias versus variance dilemma (Bishop, 2006; Goodfellow et al., 2016). For any set of data, a sufficiently complex mathematical function (model) can be found to fit the data exactly. But such a model is overfitting as it is also capturing any noise in the data. Consequently, our apparently perfect model may not match well any new set of data that was not used in setting the model parameter values. A simpler model, with fewer parameters, might perform worse on the original set of data but actually do better on the new set. That is, the simple model captures more of the underlying process, minus the noise, and so generalises better. There is no absolute way to choose the best model that has the fewest parameters and generalises well, as this depends on how much weight is given to closely matching data versus the cost of having many parameters. However, a number of criteria have been developed to allow models at least to be compared with each other. Following on from a discussion of probabilistic model parameter fitting (Section 14.7), we outline two such model comparison criteria that make use of similar probabilistic techniques (Section 14.7.2).

14.2 Estimating Model Parameter Values

Regardless of the type of model being developed, the mathematical equations that define the model's components all contain parameters that need to be assigned values before the model can be solved or simulated. So we now assume that we have decided on a particular model structure, through a model selection process, but we need to determine values for all the parameters. In what follows, we detail the parameter value estimation process as applied to compartmental models of neurons, but the steps of the process are applicable to any form of model.

In principle, the values of many of the parameters in a biophysically based model can be determined from experiments on real neurons. Hodgkin and Huxley's development of the action potential model (Chapter 3) is the classic example of designing experiments to enable the determination of parameter values for specific model components, namely the dynamics of the sodium and potassium ion channels, directly from data obtained from the system being modelled, in this case the squid giant axon. However, in wider practice, experimental data for the parameters of interest may not be available because the required experiments have not been performed or even may not be possible. Where data are available, it has often been obtained from neurons in different species, and even neurons of a different type, from that being modelled. Thus we need principled ways of adjusting parameter values obtained from experiment and determining values for those parameters for which no data are available at all.

14.2.1 The Parameter Value Estimation Process

Parameter estimation is the process of finding parameter values that best fit our mathematical model to a prescribed set of experimental data. There

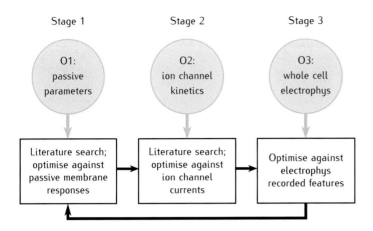

Fig 14.1 Typical stages involved in parameter estimation for a compartmental neuron model with a given morphology. Stages are likely to be iterated, resulting in adjustments to passive membrane and ion channel kinetic parameters following their initial estimation.

are two crucial issues in the parameter estimation problem: (1) determining which parameters need their values to be estimated; and (2) specifying which performance criteria the model should be optimised against to obtain estimated values. Choosing which parameters to estimate will likely result in a **multi-stage optimisation** (MSO) process. For a complex model such as a compartmental model of a neuron, before arriving at our final model which can reproduce certain behaviour of the entire neuron, we will need to carry out parameter estimation on subcomponents of the model, based on data pertinent to the subcomponent (Figure 14.1).

In Section 14.4, we consider the problem of estimating the passive electrical characteristics of a neuron (membrane resistance, membrane capacitance and axial resistance) and the form of experimental data required for this. Active membrane-bound ion channel characteristics can be obtained by fitting a model to data taken from specialised experiments yielding voltage and current traces that give information about the specific ion channel dynamics, when other channel types have been blocked pharmacologically. This is the approach taken by Hodgkin and Huxley when studying the action potential (Chapter 3).

Obtaining such passive and active characteristics constitutes two parameter estimation tasks that need to be carried out before we can begin to optimise our complete neuron model. The compartmental modeller is likely to obtain at least first-pass estimates of these parameters from the literature, relying on others to have done the required experiments and analysis. However, these parameters may or may not then be adjusted when optimising the full model neuron's performance against reference data. As we shall see in Section 14.5, in a complex compartmental model, it is not possible to estimate the values of all parameters simultaneously, due to a combination of a lack of constraining data and the computer resources required. Thus we must resort to fixing certain parameter values, whilst estimating others, in a multi-stage optimisation process (Figure 14.1; see also Figure 14.7b).

This leads to the second issue in parameter estimation, namely choosing what performance criteria to optimise a model against. The more parameters whose values we have to estimate, the greater the number of **degrees of freedom** in the model, that is, parameter values that can be changed in the model to match the desired behaviour. Simple behaviours can be used as criteria to

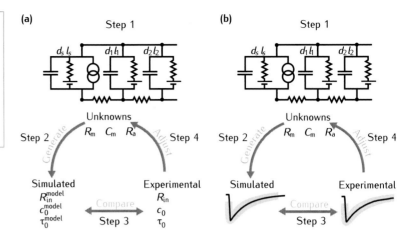

Fig 14.2 Steps involved in parameter estimation for an example passive model. **(a)** Example steps using derived physiological data (input resistance R_{in}, time constant τ_0 and coefficient c_0). **(b)** The same steps with direct comparison of transients.

optimise against, such as the firing rate of a neuron as a function of injected current. However, such criteria may be readily satisfied by wide-ranging sets of parameter values. In this situation, the optimisation problem is under-constrained. It may be possible, and desirable, to add further criteria, creating a **multi-objective optimisation** (MOO) problem (Section 14.3.3). This will add constraints and likely narrow the range of possible solutions.

Step-by-Step Parameter Estimation

Whether for a model subcomponent or for the entire model, a parameter estimation stage proceeds in a step-by-step manner to achieve reasonable values for unknown parameters:

Step 1 Fix the known parameters, such as those specifying the cell morphology, and make educated guesses for the remaining unknown parameter values.

Step 2 Use the model to simulate experiments, such as the neuronal voltage response to current stimulation, producing model data.

Step 3 Compare the model data with experimental data by calculating the value of an **error measure**.

Step 4 Adjust one or more unknown parameter values and repeat from Step 2 until the simulated data sufficiently matches the experimental data, that is, minimise the error measure.

Step 5 Use the model to simulate new experiments not used in the steps above and compare the resulting model data with the new experimental data to verify that the chosen parameter values are robust.

This approach is illustrated in Figure 14.2, where the task is to estimate values for the neuronal passive parameters C_m, R_m and R_a on the basis of small-amplitude voltage responses due to current injection (Section 14.4). Figure 14.2a shows **feature-based fitting**, where features are extracted from the recorded voltage responses from both the model and the experiments, in this case: cell input resistance R_{in}, the largest time constant of the voltage response τ_0 and the associated response amplitude c_0. It is the difference between these simulated and the experimentally recorded feature values that is to be reduced by re-estimating values for our parameters C_m, R_m and R_a.

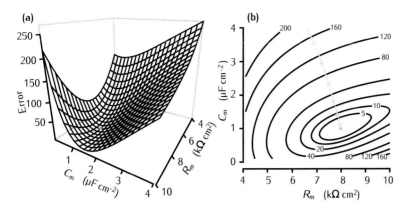

Fig 14.3 **(a)** An error surface generated from direct comparison of transients in a single-compartment RC circuit model. For convenience of visualisation, the error surface is plotted over only two unknown variables, R_m and C_m. The minimum is indicated by a blue point ($C_m = 1\,\mu\mathrm{F\,cm}^{-2}$, $R_m = 8\,\mathrm{k\Omega\,cm}^2$). **(b)** A contour plot of the same surface demonstrating a single minimum. The arrows illustrate a gradient descent approach to finding the minimum.

A suitable **error measure** is the sum of the squared differences between these model and experimental values:

$$\mathrm{Error} = w_1(R_{\mathrm{in}} - R_{\mathrm{in}}^{\mathrm{model}})^2 + w_2(\tau_0 - \tau_0^{\mathrm{model}})^2 + w_3(c_0 - c_0^{\mathrm{model}})^2, \qquad (14.1)$$

where we are using the superscript 'model' to indicate simulated observables; w_i are weights that can be used to scale the components of the error measure, but are all set to 1 in this example. Squaring the errors ensures that it is the magnitudes of the errors that are counted, and not their sign (positive or negative).

Figure 14.2b shows **direct fitting**, where the error measure is the sum-of-squared difference between the recorded V_t and the simulated V_t^{model} membrane potential transient at a possibly large number n of discrete time points t:

$$\mathrm{Error} = \sum_{t=1}^{n}\left(V_t - V_t^{\mathrm{model}}\right)^2. \qquad (14.2)$$

As discussed below, such direct fitting may be very sensitive to noise in the experimental recordings, so the use of extracted features may be more robust for parameter estimation.

14.2.2 Error Measures

In this example, for every set of estimated values for R_m, C_m and R_a, we can produce a simulated output and then calculate the error. Plotting the error against different sets of parameter values defines a surface in a space with one dimension for each unknown parameter and one dimension for the value of the error. An example with two unknowns is shown in Figure 14.3. Although visualising high dimensions is difficult, an **error surface** provides a convenient concept in parameter estimation. The task of finding parameter values that lead to a minimum error translates to moving around this surface in order to find the minimum.

One problem is that an error surface may contain multiple hills and valleys, leading to the possibility that algorithms that follow the gradient become trapped in local minima rather than finding the global or overall minimum (Figure 14.4). Parameter spaces and error surfaces for compartmental models are generally complex, with multiple local minima likely (Vanier and Bower, 1999). Searching for parameter values therefore requires approaches

An important aspect of defining a good error measure is that all values that contribute to the measure are of similar magnitude. This can usually be achieved by using suitable units for each quantity. If it is decided that one component of the error measure is more important, then this can be reflected by weighting the components individually.

Instead of minimising a model error, we could maximise some **fitness measure** of the model. The error or **fitness** surfaces produced by the corresponding measures are often referred to as the **error function** or the **fitness function**, respectively.

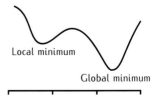

Local minimum

Global minimum

Fig 14.4 One-dimensional plot of an error surface showing one local and one global minimum.

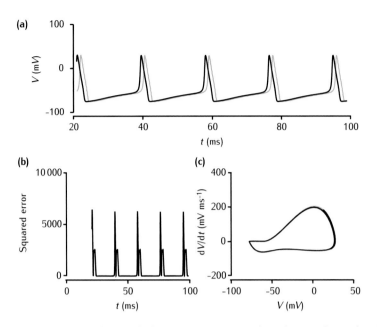

Fig 14.5 **(a)** Two action potential trains, slightly shifted in time. **(b)** The squared error between these two voltage traces can be large at specific time points. **(c)** The phase plot is identical in both cases.

that reduce the possibility of choosing parameter values that result in a local minimum error rather than the global minimum.

Continuing in the context of compartmental modelling of neurons, we now consider specific issues with direct fitting to voltage traces and illustrate the range of features that could be used in feature-based fitting.

Direct Fitting

A potential difficulty with directly fitting to data is that the measured error can be very sensitive to noise in the experimental data. For example, when trying to match a recorded action potential, any small time-base mismatch (phase shift) between the model and experimental traces can lead to very large errors (Figure 14.5a, b), even when the shape of the model action potential is a very close match to the experimental potential (van Geit et al., 2008). This sensitivity can be removed by basing the error measure on a phase plot of V versus dV/dt instead (LeMasson and Maex, 2001; Achard and De Schutter, 2006; van Geit et al., 2008). The phase plot captures details of the shape of the action potential, independently of its exact occurrence in time (Figure 14.5c).

Feature-Based Fitting

Rather than trying to match a neuronal model directly to electrophysiological traces, as considered above, it can be more robust and informative to optimise model parameters against features derived from these recordings. Neuronal firing characteristics can be complex and typical features that may be considered for use in model optimisation include (Druckmann et al., 2007; Hay et al., 2011):

- Spike (action potential) rate;
- Spike rate accommodation;
- Latency to first spike;
- Spike shape: amplitude, half-width, afterhyperpolarisation (AHP);

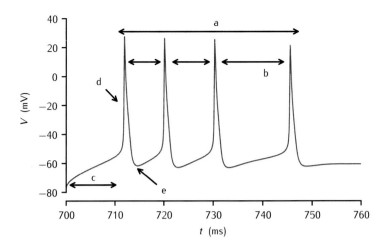

Fig 14.6 Features derived from the somatic spiking response to an injected step current in a simulation of a layer 5 pyramidal cell: a, spike rate; b, spike rate accommodation (from ISIs); c, latency to first spike; d, spike half-width; e, spike AHP amplitude.

- Spike bursting: burst interspike interval (ISI), number of spikes per burst, interburst interval;
- Dendritic back-propagating spike amplitude with distance from soma;
- Dendritic calcium spike shape: amplitude and width.

Some of these features are illustrated in Figure 14.6. They may have been obtained by averaging over many sets of experimental data and so are much more robust to noise than parameter estimation by direct fitting to single recordings. It is important here also to measure the variation in feature values across experiments, by calculating the standard deviation, in addition to the mean response (Druckmann et al., 2007). A feature that does not vary much should be reproduced accurately by the model, whereas more leeway can be accepted for a feature that varies widely between experimental recordings. The error between the model response and the mean experimental value can be measured in units of the experimental standard deviation to account for this (Druckmann et al., 2007).

Particularly with feature-based fitting, it may be attempted to meet more than one criterion at the same time, with different criteria potentially being in conflict (Druckmann et al., 2007). Multi-objective error measures usually weigh each criterion, or feature, both to normalise all numerical values and to either equalise or bias the contribution of particular features to the error measure (e.g. Equation 14.1).

14.3 Optimisation Algorithms

Finding a good parameter set involves searching for parameter values that minimise the error measure. A complex error measure can have many troughs and peaks. Troughs form local minima, which may fall well short of the best minimisation that can be achieved (Figure 14.4). We require an algorithm that can find the bottom of a trough that is at least close to the deepest possible, that is, results in the smallest possible error. This algorithm must be able to search the error surface and eventually settle on the bottom of a deep trough without getting prematurely stuck in an unnecessarily shallow trough. Suitable algorithms typically include a stochastic component in their search strategy that may result in an occasional increase in the error measure during searching, with the hope that a small error point will be found eventually, but not too slowly. Such methods include **simulated annealing** and **evolutionary algorithms (EAs)**. Different algorithms can be combined effectively to provide an initial widespread search followed by an efficient descent to the bottom of a deep trough in the error surface. A useful review of suitable optimisation algorithms is to be found in van Geit et al. (2008). Software packages aimed at providing user-friendly access to algorithms for optimising neural models are being developed (Friedrich et al., 2014; van Geit et al., 2016).

14.3.1 Deterministic Algorithms

Algorithms that use available information to compute a new set of parameter values that represent a better solution than the current set can often find a minimum very quickly. Many algorithms make use of the local gradient of the error measure to determine the direction in which to explore (Figure 14.3). They then find the minimum in that direction before determining a new direction to explore. If all directions result in an increase in the error, then a minimum has been found. Algorithms differ in the choice of the direction in which to search, given the local gradient. A significant issue is that such algorithms may find a solution that is only a local minimum and not the best possible solution.

The **steepest gradient descent** algorithm uses the intuitively appealing notion of searching for a minimum in the direction of a maximally descending gradient (Figure 14.3b). Successive search directions must be perpendicular to each other by virtue of the fact that the minimum in the previous direction has already been found. This can result in lots of small steps being taken to descend a long, narrow valley (Press et al., 1987). A better alternative is **conjugate gradient descent** (Press et al., 1987) which uses the direction of descent, calculated on the basis of local gradient information, that is as independent as possible from the previous step directions, so that large steps can be taken in each new direction towards the minimum. Conjugate gradient descent is useful in parameter fitting of compartmental models with relatively few parameters (Bhalla and Bower, 1993; Vanier and Bower, 1999).

Typical error measures do not readily yield an explicit gradient expression involving the parameters being optimised. Consequently, at each iteration, the local gradient must be calculated numerically using many error measure evaluations at points around the current search point. To minimise the computational cost of gradient calculations, **stochastic gradient descent** is used in which, at each iteration, the error gradient is only calculated at a small random subset of the data points. This is the default standard algorithm in machine learning for training artificial neural networks (Bishop, 2006; Goodfellow et al., 2016).

To ensure that gradient descent is likely to find a useful minimum, a highly exploratory search, such as **brute force search**, can be used initially to narrow down the region of parameter space to be searched by gradient descent (Bhalla and Bower, 1993).

Another deterministic algorithm that does not require gradient calculations is the **downhill simplex** method (Press et al., 1987; LeMasson and

Maex, 2001). This involves selecting an initial set of N points in parameter space, called the 'simplex'. The fitness of each point in the simplex is then evaluated and the worst-performing point is moved towards the other points. Different algorithms vary in exactly how points are moved. In any case, the simplex contracts in space as a minimum is approached.

14.3.2 Stochastic Algorithms

Stochastic algorithms combine exploiting local information with exploration of the search space. This exploration may involve moving into areas of worse fitness, but ultimately can result in a global minimum being found.

Simulated Annealing

A popular algorithm that combines the exploitation of gradient descent and simplex methods with exploration is **simulated annealing** (Kirkpatrick et al., 1983). At each time-step, the current solution is replaced probabilistically with a new solution The new candidate solution is usually chosen on the basis of some local information, which can be as simple as choosing a point randomly within a certain distance of the current solution. The probability that the new candidate y replaces the current solution x is calculated as a function of the *temperature* parameter T using a Boltzmann–Gibbs distribution (van Geit et al., 2008):

$$P_{\text{repl}} = \begin{cases} 1 & \text{if } f(y) < f(x), \\ \exp[(f(x) - f(y))/cT] & \text{otherwise} \end{cases} \tag{14.3}$$

where f is the error measure and c is a positive scaling constant. If the new candidate yields a lower error than the current solution, it replaces this solution. If it has a higher error, the probability of replacement depends on temperature T. Initially, T is high, allowing exploration of the search space. As the algorithm proceeds, T is decreased, or 'annealed', leading to an increasingly deterministic choice of new candidates that reduce the error. Different annealing schedules can be used, such as exponential cooling, where T is updated on each iteration k via:

$$T_k = \rho T_{k-1}, \text{with } 0 < \rho < 1. \tag{14.4}$$

Simulated annealing often works well for optimising parameters in compartmental neural models (Vanier and Bower, 1999; Nowotny et al., 2008).

Evolutionary Algorithms

Complex optimisation problems can be tackled using the family of **EAs**. These include **genetic algorithms** (GAs) (Holland, 1975) and **evolution strategies** (ES) (Achard and De Schutter, 2006; van Geit et al., 2008). The essence of these algorithms is the evolution of a population of candidate solutions towards ever fitter solutions. This evolution includes reproduction, mutation and selection. Such algorithms proceed using the following steps:

(1) Randomly generate an initial population of candidate solutions.
(2) Evaluate the fitness of each candidate solution.
(3) Generate (breed) new solutions from a set of the best candidate solutions by using reproduction via crossover operations between selected parents, which combine part of one parent with part of the other to form a new solution, and mutation of current solutions.

EAs in this family differ in how candidate solutions are represented and precisely how evolution takes place. Both GA (Vanier and Bower, 1999) and ES (Achard and De Schutter, 2006) approaches are effective for neural model parameter optimisation. One advantage of ES is that it allows direct representation of real-valued parameters, whereas GA requires binary encoding of parameter values.

(4) Evaluate the fitness of the new solutions.

(5) Replace the least fit candidate solutions with better new solutions.

(6) Repeat from Step 3 until a suitable level of fitness is reached or specified number of repeats is exceeded.

An important feature of EAs is the fact that they always maintain a population of candidate solutions, many of which may have equally high fitness values. The set of equally good solutions is known as the pareto front. Search with an EA will be terminated when the quality (fitness) of the pareto front reaches a predefined level, or at least is no longer increasing, with the outcome being multiple sets of parameter values that all provide good solutions. For a multi-objective problem, such solutions may differ in their quality across the different objectives, thus representing trade-offs between these objectives (Druckmann et al., 2007). This leads to the concept of domination: instead of one solution simply being better than another in terms of fitness, a solution may dominate another if it is better in at least one objective and equally good in all of the others. This gives a fine-grained view of fitness across multiple objectives. Two solutions may have the same fitness and not dominate each other by providing best performance for different objectives, whilst being no worse for all the others.

14.3.3 Multi-objective Optimisation

The more parameters whose values we have to estimate, the greater the number of degrees of freedom in the model. Simple behaviours to be used as criteria to optimise against, such as the firing rate of a neuron as a function of injected current, may be readily satisfied by wide-ranging sets of parameter values. In this situation, the optimisation problem is under-constrained. It may be possible, and desirable, to add further criteria, creating a MOO problem. This will add constraints and likely narrow the range of possible solutions. So it might seem that all we need to do is to specify a suitably rich set of criteria and then optimise all the parameter values in our model to arrive at a single optimal solution. In practice, this is an unlikely situation to achieve. Even if it is possible in theory, for large numbers of parameters and large numbers of criteria, it may not be practical to solve the optimisation problem in a reasonable amount of time, even on today's most powerful computers. One approach is to break the problem down into subproblems:

(1) Accept estimates from the literature (or other resources, such as model databases) for as many parameters as possible, and then optimise the remaining parameters against a limited set of criteria;

(2) Use the set of solutions obtained as the starting point for optimising against the remaining criteria; and/or

(3) Reoptimise over the same criteria, but for the parameters whose values were initially taken from elsewhere, whilst holding the already optimised parameters at one of the optimal solutions.

This process may be repeated iteratively until good solutions for all the desired criteria are found. Thus our parameter estimation process will likely involve both MSO and MOO (Figure 14.7).

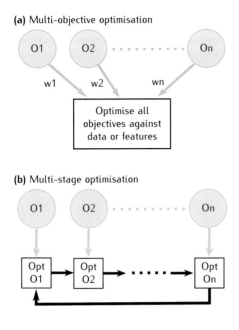

(a) Multi-objective optimisation

(b) Multi-stage optimisation

Fig 14.7 Comparison of **(a)** MOO and **(b)** MSO. In MOO, a group of objectives, Ox, with perhaps different significance weightings, wx, are optimised simultaneously. In MSO, subsets of parameters are optimised against specific objectives at each stage. These stages are likely to be iterated, resulting in readjustments to parameter subsets.

14.4 Case Study: Determining Passive Neuronal Properties

To set the scene of the complex problem of choosing values for the many parameters in a compartmental model, we consider the problem of obtaining values for the key passive membrane parameters: membrane resistance (R_m) and capacitance (C_m), and the axial resistance (R_a), given that we have an initially passive compartmental model embodying a fixed neuronal morphology and we assume these properties are uniform throughout the morphology. This is far from straightforward, as these passive parameters generally cannot be directly measured for a neuron by experiment. Instead, they must be estimated from the time course of voltage transients by either matching passive compartmental model responses directly to the recorded transients (direct fitting) or by matching the model with derived features of the transient responses (feature-based fitting). We will consider both approaches here, highlighting the issues with appropriate treatment of the experimental data and its implications for model fitting. An excellent overview, particularly of the issues involved, is to be found in Rall et al. (1992). The assumptions underlying these approaches are listed in Box 14.1.

Building a compartmental model from both morphological and physiological data from the same cell and the general procedure of selecting passive properties for the model that reproduces recorded physiology have proved useful in estimating the capacitance and passive resistive properties of a variety of cells (Lux et al., 1970; Clements and Redman, 1989; Rall et al., 1992; Major et al., 1994; Stuart and Spruston, 1998; Thurbon et al., 1998).

In the next case study (Section 14.5), we consider an example of choosing values for all of the parameters in a compartmental model, including active

> **Box 14.1** | Assumptions of the RC circuit representation
>
> (1) The membrane is passive, its properties being described by fixed resist-
>   ance and capacitance only.
> (2) Specific membrane resistance, capacitance and cytoplasmic resistivity
>   are uniform throughout the morphology.
> (3) There is no depletion or accumulation of the ion pools on either side
>   of the membrane.
> (4) The cytoplasm provides only ohmic resistance.
> (5) Voltage dependence within the cytoplasm is 1D, with no radial or an-
>   gular dependence.
> (6) The extracellular medium is isopotential.
>
> See Rall (1977) for a review of the assumptions for a passive RC circuit
> compartmental model, and some of their justifications.

Injected current pulses can be used to assess whether the cell has a passive membrane resistance R_m (Thurbon et al., 1998). Averaged voltage transients produced with, for example, -0.5 nA current pulses can be compared to transients produced with -1.0 nA pulses. If the voltage transient arising from the larger-amplitude pulse is twice that of the first, then the response is linear and the membrane resistance can be treated as being passive.

ion channel densities distributed across the different compartments, so that the model reproduces electrophysiological recordings of neuronal spiking activity.

14.4.1   Data for Passive Parameter Estimation

In this process, it is important to choose appropriate passive physiological properties that we wish the model to match or reproduce. Primarily, we need to ensure that they are effectively passive, that is, that a model with a constant R_m can reproduce the phenomena. Not only does this restrict experiments to those involving small voltage changes, but also there should be experimental verification that the recorded responses are indeed passive. Similarly, we need to ensure that the passive responses we are modelling require C_m, R_a and R_m for their generation.

We can only expect to estimate values of these parameters from observed physiology if these parameters are relevant to that physiology. For example, we cannot expect to determine realistic values of specific membrane capacitance by only comparing maximum stable voltage offsets induced by small current injection. Solutions to the membrane equation when the voltage is not changing do not depend on membrane capacitance (Box 5.4). Therefore, responses from periods during which the membrane potential is changing, for example, transient responses, can give more information about electrotonic architecture than steady-state responses, as transient behaviour involves charging (or discharging) membrane capacitance.

We now consider different approaches to estimating these passive parameters. These include recording and comparing the membrane time constants and their coefficients. Directly comparing experimental and simulated membrane transients that arise from small and brief current injections is also a common approach. Such transients are dependent on all three passive parameters.

14.4.2   Feature-Based Passive Parameter Estimation

Suitable data from which the passive parameters can be estimated through fitting of our passive compartmental model to match experimentally measured values include:

- Time constants of the passive change in membrane voltage following a current step;
- Time constants of the current transient during a voltage clamp;
- Measurement of the total cell input resistance from steady-state voltage steps;
- The full time course of voltage transients to short current pulses.

The first three constitute features derived from recorded voltage responses. We will consider each of these in turn.

Voltage Time Constants

Using a stimulating electrode to inject a current pulse into a neuron will result in a transient change (depolarisation or hyperpolarisation) in the membrane voltage at the site of stimulation and then throughout the cell. If the membrane has purely passive properties over the range of this voltage change, then the time course of the passive transient in the membrane potential can be described as a sum of exponentials (Rall, 1969), each with a time constant and an amplitude coefficient (Equation 5.14 and Box 5.5):

$$V(t) = c_0 e^{-t/\tau_0} + c_1 e^{-t/\tau_1} + c_2 e^{-t/\tau_2} + \ldots \tag{14.5}$$

By convention, the time constant τ_0 and associated amplitude coefficient c_0 account for the slowest response component, that is, τ_0 is the largest time constant.

If the neuron is isopotential throughout, then only the first time constant has a non-zero amplitude and it will equal the passive membrane time constant $\tau_0 = R_m C_m$. The more usual situation in spatially extensive neurons is that in addition to the membrane current at the site of stimulation, there will be a number of faster equalising time constants due to rapid flow of current intracellularly along the dendrites and capacitively (Rall, 1969). These time constants and their amplitude coefficients derived from experimental data can be used to estimate the passive properties (C_m, R_a, R_m) by demonstrating that a model with these properties and equivalent morphology generates the same transients with the same time constants.

Multiple time constants (τ_0, τ_1, etc.) and amplitude coefficients (c_0, c_1, etc.) can be derived from experimental data. Essentially, the voltage decay waveform is plotted with a logarithmic voltage axis, which enables straight line regression to be applied to the final decay phase (Figure 14.8). This obtains an estimate of the apparent time constant τ_0 and amplitude constant c_0 of the slowest response component. The response due to this exponential component is subtracted from the original response and the process is repeated for successively smaller time constants. For a full description of the approach, see Rall (1969). This method is a good way of estimating the slowest time constant τ_0. However, it is less reliable for estimating the remaining, faster time constants (Nitzan et al., 1990; Holmes et al., 1992). This is unfortunate, as they are largely determined by C_m and R_a and even a measurement of τ_1 would help constrain these parameters, along with R_m. If only τ_0 and c_0 are available, then further data, such as the cell's input resistance, or time constants from current transients during voltage clamp (Holmes et al., 1992), are needed to help with uniquely determining the passive parameter values.

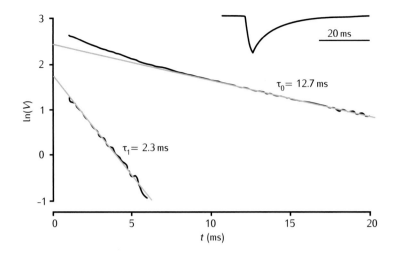

Fig 14.8 A plot of the natural logarithm of voltage following brief hyperpolarisation of a guinea pig motor neuron (black lines). The transient shown in the insert is an average of 50 responses in the same cell. A straight line fit of the later portion of the transient is used to calculate τ_0 (upper blue line). Subtracting this first exponential allows a second time constant τ_1 (lower blue line) to be estimated from the straight line fit to the remainder of the earlier portion of the transient. Adapted from Nitzan et al. (1990), with permission from The American Physiological Society.

(a) Sealed end

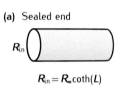

$$R_{in} = R_\infty \coth(L)$$

(b) Leaky end

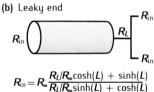

$$R_{in} = R_\infty \frac{R_L/R_\infty \cosh(L) + \sinh(L)}{R_L/R_\infty \sinh(L) + \cosh(L)}$$

Fig 14.9 **(a)** Input resistance to a sealed-end cable. Recall from Chapter 5 that $L = l/\lambda$ and $R_\infty = R_m/\pi d\lambda$. **(b)** The input resistance to the first branch of a tree can be calculated from a leaky-end cable, where the leak resistance at the end is calculated from the combined input resistances of child branches.

Voltage Clamp Time Constants

Voltage clamp protocols applied through a stimulating electrode will result in injected current transients that implement and maintain the specified voltage profile, such as a step to a new steady-state voltage level. Similarly to voltage transients under current clamp, voltage clamp current transient time constants can be recorded and compared with a model under simulated voltage clamp conditions (Rall, 1969). Voltage clamp techniques can ameliorate certain types of noise; for example, control of the voltage in the dendrites can limit the role of active channels in random synaptic events.

Input Resistance

The steady-state input resistance can be measured from the steady-state voltage response to small current steps. An experimentally determined value can be compared to the value determined from the model through simulations of the same experimental protocol and subsequent analysis, or the theoretical input resistance can be calculated directly from the model.

In a single sealed-end cable, the input resistance is related to both the membrane and axial (cytoplasmic) resistances (Figure 14.9a; Box 5.4). In a more realistic branching dendritic tree, the relationship between these resistances is more complex. The input resistance to any non-terminal cylindrical branch can be calculated by treating that branch as a cable with a leaky end (Figure 14.9b). The resistance R_L at the leaky end is calculated from the input resistances of connected branches. For example, if the distal end of a branch connects to two child branches, then the resistance R_L for this branch can be calculated from the input resistances of the child branches R_{in1} and R_{in2} by:

$$\frac{1}{R_L} = \frac{1}{R_{in1}} + \frac{1}{R_{in2}}. \tag{14.6}$$

Assuming sealed ends at all terminal branches, we can use an algorithm to calculate the input resistance to any tree and consequently for the complete model cell itself (Box 14.2).

Note that R_{in} is significantly less sensitive to changes in intracellular resistance compared to membrane resistance (Nitzan et al., 1990). This

Box 14.2 | Input resistance in a branching tree

We can use an algorithmic approach to calculate the input resistance at the base of any complex branching tree by using the cable equation and different boundary conditions depending on the type of branch. Terminal branches are assumed to have a sealed end, and internal branches have leaky ends (with the leak calculated from the input resistances of connected branches):

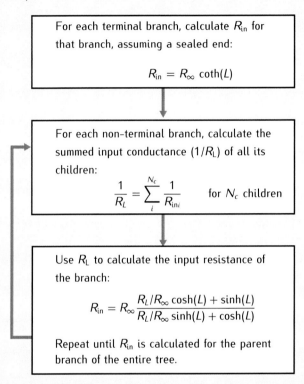

For each terminal branch, calculate R_{in} for that branch, assuming a sealed end:

$$R_{in} = R_\infty \coth(L)$$

For each non-terminal branch, calculate the summed input conductance $(1/R_L)$ of all its children:

$$\frac{1}{R_L} = \sum_i^{N_c} \frac{1}{R_{in_i}} \quad \text{for } N_c \text{ children}$$

Use R_L to calculate the input resistance of the branch:

$$R_{in} = R_\infty \frac{R_L/R_\infty \cosh(L) + \sinh(L)}{R_L/R_\infty \sinh(L) + \cosh(L)}$$

Repeat until R_{in} is calculated for the parent branch of the entire tree.

The input resistance of an entire model cell with N_t branches emerging from the soma is then given by:

$$\frac{1}{R_{in}} = \frac{1}{R_{soma}} + \sum_{i=1}^{N_t} \frac{1}{R_{in_i}}.$$

R_{soma} is the input resistance of the soma compartment, generally given by R_m/a_{soma} where a_{soma} is the surface area of the soma compartment.

is important to consider when comparing the computed and physiologically recorded value of R_{in}. The same value of R_{in} can be calculated from models with very different intracellular resistances and only small differences in their membrane resistances.

14.4.3 Voltage Transients for Direct Model Fitting

Transients recorded from the cell, for example, the voltage decay after a brief current injection, can be compared directly to simulated transients generated from the model under the same simulated experimental conditions

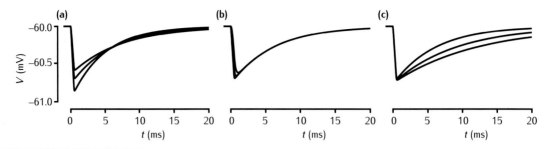

Fig 14.10 Changes in RC circuit passive properties affect the shape of a simulated transient generated from stimulating a single cable with a brief 480 μs current pulse of −0.5 nA. **(a)** C_m varied, with values of 0.8, 1.0 and 1.2 μF cm^{-2}. **(b)** R_a varied, with values of 32.4, 97.2 and 194.4 Ω cm. **(c)** R_m varied, with values of 6, 9 and 12 kΩ cm^2. The standard passive circuit parameters are C_m 1.0 μF cm^{-2}, R_a 32.4 Ω cm and R_m 6 kΩ cm^2.

(Clements and Redman, 1989). Short current pulses used in the generation of transients have an advantage over longer ones as they reduce the time during which the membrane potential deviates from its resting level, thereby lessening the risk of activating slow voltage-dependent conductances.

This direct approach avoids the errors that may be introduced when determining electrophysiological properties such as R_{in}, τ_0 and c_0. These errors are passed into the estimation of R_m, C_m and R_a. Direct fitting has only one parameter estimation process, which is selecting values of R_m, C_m and R_a that minimise the differences between simulated and experimental voltage traces.

To ensure that transients do not engage active channels, it is important to test that their form does not change over time when repeatedly producing transients. Large numbers of transients generated by the same stimulation protocol can be pooled and averaged to give a prototype target transient for the model to reproduce. Not only does this reduce noise but also, with sufficiently large numbers of recorded samples, it is possible to place approximate bounds around the averaged target transient within which an individual transient will fall, say, 95% of the time. We can have confidence in a fit if it remains within these bands; otherwise the fit can be rejected (Major et al., 1994).

The interval of time over which the comparison between model and recorded transients is made is critical. Immediately at the end of the stimulation, the recorded experimental transient is likely to include electrode artefacts (Section 14.4.4). Starting the comparison too soon so as to include these may lead to the selection of passive parameter values which are tainted by electrical properties of the electrode. Starting too late will reduce the information contained within the transient. If we wish to estimate the three passive properties of the RC circuit using the comparison, we need to ensure there are sufficient data in the time interval to constrain each parameter. Figure 14.10 illustrates areas of voltage transients affected by the passive properties. The early rise of the transient is affected significantly by both C_m and R_a (Figure 14.10a, b). R_m has a dominant influence on the decay phase (Figure 14.10c). Estimating the three passive parameters by only examining the late decay phase would not constrain the minimisation sufficiently nor generate useful estimates of R_a and C_m.

14.4.4 Sources of Error

Ensuring the passive RC model assumptions (Box 14.1) are reasonable is the first step in gaining confidence in the final passive parameter values attained.

Box 14.3 | Modelling intracellular electrodes

The voltage and current recordings needed to determine passive cell physiology will be obtained using an intracellular **sharp electrode** (Purves, 1981) or a whole-cell patch electrode (Hamill et al., 1981). High-quality recordings require the removal of the contributions of the electrical properties of the electrode and the rest of the recording apparatus from the recorded signals. This is done as a recording is made by appropriate circuitry in the recording apparatus. This circuitry subtracts signals from the recorded signal, based on a simple model of the electrode as an RC circuit (Figure 14.11).

Passing current through the electrode results in a large voltage drop across the electrode tip due to the tip resistance, which is in series with the cell's membrane resistance. The voltage measured is the sum of the tip voltage and membrane voltage. The membrane voltage is obtained at the recording apparatus by subtracting a signal proportional to the injected current (which models the tip voltage) from the recorded voltage, a process known as **bridge balance**. This works well, provided that the tip resistance is constant and the electrode time constant is much faster than the membrane time constant. These assumptions may be more or less reasonable in different recording setups.

The wall of the glass electrode contributes a capacitance that needs to be allowed for as well. Similarly to the electrode resistance, **capacitance compensation** is obtained by circuitry that feeds the output of the preamplifier through a capacitor and adds it to the preamplifier input, thus effectively adding in a capacitive current which is the inverse of the current lost through the capacitance of the electrode.

High-quality recordings rely on accurate calibration of the bridge balance and capacitance compensation (Park et al., 1983; Wilson and Park, 1989), and the validity of the underlying assumptions forming the electrode 'model'. Even so, care still needs to be taken when analysing the slow measurements used for determining passive membrane properties (Section 14.4.1). The situation is worse when recording faster synaptic signals and action potentials. Particularly demanding are **dynamic clamp** experiments (Figure 14.12) in which rapidly changing currents, mimicking, for example, synaptic signals, are injected into the cell and the resulting voltage transients are recorded. Serious artefacts can be introduced into the recording due to flaws in this simple electrode model, but techniques based on better electrode models are being developed (Brette et al., 2008). For a thorough presentation of the modelling of electrodes, see Brette and Destexhe (2012).

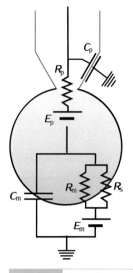

Fig 14.11 A simple electrical circuit representing either an intracellular sharp or whole-cell pipette electrode attached to a soma. The electrode is modelled as tip potential E_p and resistance R_p in parallel with pipette wall capacitance C_p. For small-diameter sharp electrodes, R_p is on the order of 10–100s of megaohms, whereas for larger-bore whole-cell electrodes, it is around $10\,M\Omega$ (Major et al., 1994). C_p is typically on the order of $10\,pF$. An intracellular sharp electrode may introduce a shunt resistance R_s on the order of $100\,M\Omega$. With a tight-seal whole-cell electrode, R_s is on the order of $10\,G\Omega$ (Li et al., 2004).

For some of the assumptions, such as that the resistance and capacitive membrane properties are passive, this is possible experimentally. For example, it is important to verify that the experimental protocol does not activate or inactivate voltage-dependent currents. Other assumptions, such as the uniformity of specific passive properties over the cell morphology, are harder to verify.

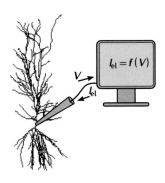

Fig 14.12 The dynamic clamp experimental setup (Sharp et al., 1993; Economo et al., 2010). In real time, the recorded membrane voltage V is input to a computational model, which generates the injected current I_{el}. This current may represent a synaptic or an ion channel current. Its calculation could involve voltage-gated channel kinetics or the response of an entire compartmental model cell. This is an important use for the neural computational models described in this book.

Electrode Artefacts

The presence of recording/stimulus electrodes can introduce artefacts (Box 14.3). In particular, sharp electrodes can lead to a shunt, or a change in membrane resistance, at the electrode site (Figure 14.11). For example, the entry of calcium ions through a membrane puncture can activate calcium-dependent channels which are selective for potassium ions. **Whole-cell patch electrodes** with seal resistances above 5 GΩ can minimise the risk of shunts. Li et al. (2004) have carried out a detailed comparison of the artefacts introduced by sharp and whole-cell electrodes in the same experimental preparation.

The model can be extended to include a membrane shunt in the soma that represents the change in membrane resistance in this compartment (Figure 14.11). This extension to the model allows the possibility of changes in membrane resistance at the soma to be captured at the expense of introducing an additional free parameter, the **soma shunt conductance** g_{shunt}. This additional parameter can be included in the general parameter estimation process, optimising now over four parameters.

It is possible to assess if an obvious soma shunt conductance is exhibited in the recorded transients. In this approach, a reference model is constructed, with realistic values of R_m, R_a and C_m. Perturbations of interest, such as a soma shunt, can be introduced into the reference model. Parameters of a model with no shunt (i.e. the three passive parameters) are then optimised to match the transients generated by the reference model. In other words, a simulated scenario is constructed where we optimise three parameters of a model to make it mimic, as best it can, the data generated from a model using four parameters. Examining the residual left after the reference transient is subtracted from the optimised model transient can prove useful in identifying cells with specific non-uniformities (Major and Evans, 1994; Thurbon et al., 1998).

The electrode itself can also be responsible for sources of error in the final parameters. The combined capacitive properties of a patch pipette and amplifier can alter the early phase of transients. Subtracting extracellular controls from the response does not eliminate the introduced pipette aberrations (Major and Evans, 1994). Building a model of the pipette is one approach to quantifying its effect (Major et al., 1994). In turn, this introduces additional electrode-specific parameters into the model, with values that can often be difficult to obtain. Alternatively, carefully selecting the times over which model and experimental transients are matched can avoid these errors. In particular, deciding on an appropriate post-stimulus start time for the comparison is crucial (Major et al., 1994). Optimising parameters to fit transients too soon after the stimulus may yield values that are contaminated by the electrode artefacts. Starting the comparison too late will fail to sufficiently constrain the passive properties (Figure 14.10).

Uniqueness of Estimates

A further issue to consider is whether the set of optimal parameter values obtained is unique. In the passive model, if any of the parameters defining the model morphology, such as soma diameter, are also unknown, an infinite number of combinations of the parameter values R_m, R_a and C_m and

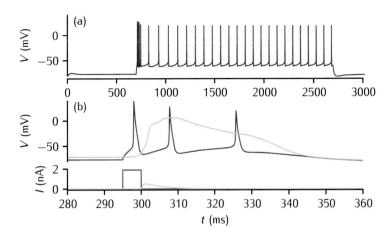

Fig 14.13 Current injection responses of layer 5 pyramidal cell model that accurately match experimental data. (a) Somatic spiking response to a 2 s-long step current in the soma. (b) Membrane voltage (top) at somatic (black) and dendritic (blue) recording points to (bottom) injected current at these two sites.

the unknown soma diameter d_s can generate identical simulated data. There is no unique set of parameter values that can be selected to underlie the passive physiology. In general, with unknowns in the model morphology, it is very difficult to calculate unique parameter values from experimental data recorded with electrodes.

There may be multiple sets of parameter values which generate identical model transients in response to one stimulation protocol, but generate different responses under other protocols. This type of non-uniqueness is particularly problematic for an unknown soma shunt (Major et al., 1994). Consequently, it is important to keep aside, or have access to, experimental data not used in the parameter estimation procedure. These data can then be used to test the model and assess whether the chosen parameter values act appropriately in different situations.

The axial resistance in a passive model consisting of a single cylindrical compartment, diameter d_s and length l_s, is:

$$\frac{4R_a l_s}{\pi d_s^2}.$$

With both R_a and d_s as unknown parameters, there is an infinite number of combinations of R_a and d_s that yield the same axial resistance.

14.5 Case Study: Neocortical Pyramidal Cell Model

A neuronal compartmental model is tuned typically to reproduce firing characteristics that depend on heterogeneous active membrane properties, and not just the passive characteristics. To give an illustrative example of such a parameter estimation problem, the Hay layer 5 pyramidal cell model (Hay et al., 2011), introduced in Section 5.6, is the outcome of a complex, multi-stage parameter estimation procedure. The aim of this model was to replicate two major characteristics of the electrophysiological responsiveness of neocortical pyramidal cells (Figure 14.13): (1) output spiking activity due to somatic current injection, and (2) dendritic back-propagating action potential-activated calcium (BAC) spike firing.

Twenty features were extracted from experimental data: 10 for somatic firing alone and 10 for somatic spikes and dendritic calcium spikes during BAC firing (Table 14.1). Some of these are illustrated in Figure 14.6. To enable parameter estimation to be carried out, these features were combined to give eight objectives that had to be satisfied for the output spiking characteristics and five objectives for BAC firing. Model performance was assessed

Table 14.1 Features for optimisation of the Hay layer 5 pyramidal cell model

Step current firing	BAC firing
Spike frequency (Hz)	Ca^{2+} spike peak (mV)
Spike rate accommodation	Ca^{2+} spike width (ms)
Coefficient of variation of spike ISI	Somatic spike count during BAC firing
Initial burst ISI (ms)	Mean somatic spike ISI (ms)
First spike latency (ms)	Somatic AHP depth (mV)
Spike peak (mV)	Somatic spike peak (mV)
Fast AHP depth (mV)	Somatic spike half-width (ms)
Slow AHP depth (mV)	Spike count with somatic stimulation only
Slow AHP time	BAP amplitude at 620 μm (mV)
Spike half-width (ms)	BAP amplitude at 800 μm (mV)

Summarised from Table 1 of Hay et al. (2011). Left-hand column contains features of somatic action potentials (spikes) with somatic step current injection only. Right-hand column contains features for combined somatic and dendritic stimulation leading to BAC firing. ISI, interspike interval; AHP, afterhyperpolarisation potential; BAP, back-propagating potential.

by quantifying model output in terms of its distance from each objective in units of the experimental standard deviation from the experimental mean. Good model performance was deemed to be an output that was within 2–3 standard deviations of the mean for all the objectives. This somewhat broad range recognises the variability in the experimental data itself, as well as the likely difficulties of a single optimised model being able to more tightly fit all of the objectives.

Around 200 compartments were used to approximate the cell morphology obtained from the digitisation of a biological neuron. Each compartment contained up to 10 different ion channel types, plus a model for calcium influx and diffusion. As is typically done, passive characteristics and ion channel dynamics largely were taken from existing models found in the literature and remained fixed. An exception was the specific membrane resistance R_m, which was included in the optimisation process in the form of the maximum leak conductance. Initially, optimisation was restricted to ion channel densities (maximum conductances) and calcium dynamics (two parameters) in each compartment. Even so, this would require optimising around 2000 parameters if all compartments were treated individually. This is both impractical and certainly not sufficiently constrained by the available electrophysiological features. Consequently, it was assumed that most ion channels were uniformly distributed in particular neuronal segments, such as the apical dendrites or the perisomatic region. This reduced the number of free parameters down to 22, which is still a large number to deal with.

Optimisation was carried out using an elitist non-dominated sorting EA (Box 14.4) using particular methods for genetic selection, crossover and

Box 14.4 | Elitist non-dominated sorting EA

The aim of an elitist non-dominated sorting evolutionary algorithm is to maintain a diverse population of non-dominated solutions during the optimisation process. To achieve this, at each iteration, the population of newly generated solutions (of size N) is considered together with the current set of solutions (also of size N). These $2N$ solutions are sorted into sets of non-dominated solutions of different rank: the set of solutions of rank 1 consists of those solutions, from the full population of $2N$, of equal best fit that do not dominate each other; the set of rank 2 is drawn in the same way from the reduced population obtained by withdrawing solutions of rank 1, and so on. Thus solutions are sorted into order with decreasing performance as the rank increases and equal performance within each rank. The new population of size N is taken by adding non-dominated solutions of rank 1, then rank 2 and so on until N solutions are obtained. If all solutions of rank k cannot be taken (as it would increase the solution population beyond N), then those solutions that are most different from other solutions of this rank (according to some distance measure) are chosen to ensure diversity in the new population. This approach is known as *elitist* because it ensures that the best members of the current population are carried over to the new population at each iteration.

mutation (Deb et al., 2002; Druckmann et al., 2007). The EA was run for 500 generations with a population size of 1000 sets of parameter values (possible solutions). On each generation, a new set of 1000 solutions was produced from the previous generation. Computer runtime to carry out an optimisation took from two to five days, even using from 240 to 1024 compute cores to parallelise the computations.

Optimising the model against both sets of objectives at once proved to be impractical, as good solutions to all criteria were not found with the population size and generation number specified above. Given the very long computation times with this specification, increasing population size or number of generations would only increase this time. Instead, the model was optimised against one or other set of features individually, in a stepwise fitting approach. Good solutions were then obtained for the characteristic being optimised against, with from tens to hundreds of the models in the final EA population having acceptable performance over all objectives, but these solutions did not produce satisfactory performance for other characteristics not optimised against. For example, optimising against spike firing characteristics for somatic current injection resulted in around 50 acceptable models from the final EA population. However, these models could not produce BAC firing as their apical dendrites were largely passive. This makes sense since dendritic activity does not contribute significantly to output spiking in response to a somatic current injection alone. Based on this logic, Hay et al. (2011) successfully tuned their model on both sets of features by doing the optimisation in the following stages:

(1) Optimise for BAC firing to produce a family of possible solutions.
(2) Select the solution that gave the best output spiking behaviour as well.

(3) Fix the dendritic ion channel conductance parameters of that model.
(4) Optimise the remaining perisomatic ion channel conductances to produce suitable output spiking behaviour.

The resulting model still produced good BAC firing when now tuned to give good output spiking as well (Figure 14.13). This exemplifies that a degree of prior knowledge may be needed to construct workable optimisation procedures for complex neuronal models. A similar staged approach is detailed in Mäki-Marttunen et al. (2018).

14.6 | Parameter Sensitivity Analysis

The ability of widely varying sets of parameter values to satisfy performance criteria has been recognised in models of many biological systems and is often referred to as **'sloppy' parameter sensitivity** (Gutenkunst et al., 2007) or 'sloppy' modelling (O'Leary et al., 2015).

It is important to note that here an ensemble of models refers to a single structural model, such as a compartmental model of a neuron, with the ensemble being different sets of parameter values for that model, such as ion channel densities in the different compartments. This should be distinguished from having a set of structurally different models, with different associated parameters, that we might want to compare in performance as models of the system under study. This is the field of model selection, which we introduced in Section 14.1.

An important adjunct to model optimisation is **parameter sensitivity analysis**, which involves gaining an understanding of how sensitive the model output is to variation in individual, or groups of, parameter values. The family of good solutions obtained using an evolutionary algorithm should give some useful insights, such as how different combinations of ion channels may produce very similar electrophysiological behaviour in a neuron. However, systematic approaches can be taken to quantify how model output varies with changes in parameter values. A thorough introduction to this important topic for modelling in general can be found in Saltelli et al. (2008).

14.6.1 Ensemble Modelling

Ensemble modelling approaches build databases of model performance across different sets of parameter values (Prinz et al., 2003; Marder and Taylor, 2011; Sekulić et al., 2014; Migliore et al., 2018; Marín et al., 2021). Examination of the variation in parameter values across an ensemble of models should give insight into the sensitivity of good model performance to particular parameters. For example, the spatial distribution and maximum conductance of a particular ion channel type may be within tight bounds in all ensemble models, indicating this ion channel must be tightly regulated in real neurons to provide observed properties. Other ion channel types may vary widely in their spatial density across the ensemble, indicating neuronal output is not particularly sensitive to these channels. Alternatively, this variation may correlate with changes in relative performance across different criteria in multi-objective optimisation, whilst overall performance is similar, indicating particular functional roles for the channels. An example of such a study is given in Box 14.5.

Obtaining an interesting ensemble can be done by using an EA for model optimisation: the resulting pareto front is an ensemble of models that all perform well for the given optimisation criteria. Modelling workflows built around using the pareto front to create such ensembles have been developed and used to investigate possible parameter sensitivity in cerebellar granule cells (Marín et al., 2021) and hippocampal neurons (Migliore et al., 2018).

Wider exploration of parameter space, beyond the optimal solutions provided by the pareto front of an EA, can give considerable insight into the significance of different parameters for model output. A simple approach to generating a large ensemble is brute-force searching of the parameter space over a coarse grid of parameter values (Prinz et al., 2003; Sekulić et al., 2014). When model output for each chosen set of parameter values is ranked against performance criteria, this approach will provide both high-performing parameter sets (i.e. optimisation) as well as probably rather poorly performing models. High-performing parameter sets could provide the starting point for further optimisation, using optimisation algorithms, as outlined above. However, exploring performance variation across the full parameter space should reveal specific parameter sensitivities. For example, it might be of particular interest to determine which ion channel densities and their distributions most contribute to specific electrophysiological responses. Another feature that may be seen is the co-regulation of ion channel densities (Box 14.5), such that an increase in conductance of one ion channel type is accompanied by either an increase or a decrease in conductance of another ion channel type, to maintain the same neuronal behaviour (Sekulić et al., 2014; Migliore et al., 2018).

Given the large number of parameters in a typical compartmental neuron model, brute-force searching will be expensive computationally. However, it is naturally parallelisable, since each model run with a particular parameter set is independent of any other run, and so full use can be made of multi-core computers.

14.6.2 Uncertainty Quantification of Model Output

Another way to view parameter sensitivity is to consider how uncertainty in a parameter's value affects uncertainty in model output. If a model is highly sensitive to a particular parameter and we do not know very precisely that parameter's value, then we must be uncertain about the model output. Formal methodologies and neurally relevant examples of such **uncertainty quantification** are provided in Eriksson et al. (2018) and Tennøe et al. (2018). Eriksson et al. (2018) used their approach to assess the sensitivity of model outcomes to chemical reaction rates for a complex signalling pathway model of postsynaptic plasticity. Tennøe et al. (2018) describe several examples and provide a general software package for carrying out this analysis.

To quantify the uncertainty in model output, firstly we need methods for determining the expected distribution of parameter values, given knowledge of the system under study and the data we have available for fitting parameter values to. In Section 14.7, we consider Bayesian methods for identifying probability distributions for parameter values, rather than the point estimates we have so far considered. Eriksson et al. (2018) carried out such distribution estimation as the first stage in their uncertainty quantification workflow. However, in the absence of strong experimental data for the distribution of parameter values, also it is possible to assume simple probability distributions, such as uniform over a range of values or Gaussian, using mean values found from model optimisation, for example, one parameter set found from an EA. Variance of the distributions should be set to give a range for each parameter whilst keeping possible values within sensible physiological bounds (Tennøe et al., 2018).

Given this quantification of parameter value uncertainty in the form of probability distributions, we then need to assess how this uncertainty transfers to uncertainty in the model output. To carry out such an uncertainty quantification, the distribution of model output, given the specified parameter distributions, needs to be determined. This is done using Monte Carlo (MC) sampling methods (Box 14.6) to draw a large number of sets of parameter values from their joint probability distribution and determine the model

Box 14.5 | Ensemble modelling example

Sekulić et al. (2014) used an ensemble modelling approach to examine the potential characteristics of oriens-lacunosum molecular (O-LM) interneurons in the mammalian hippocampus. The key question under study was the likelihood of the presence of the h current I_h in the dendrites of these cells. They examined the densities (maximum conductance) of nine ion channel types distributed uniformly in the soma, dendrites or axon. Each conductance was given from three to five possible maximum values, covering a physiologically reasonable range from low to high, with a search grid containing 233 280 possible parameter sets. As with optimisation, the model outputs for each parameter set were ranked according to a performance value calculated from the fit to various electrophysiological data from real O-LM neurons. The challenge was to categorise performance across these sets. This was achieved by considering the rate of change in performance as a function of model rank order. Parameter sensitivity was assessed by how rapidly model performance changed with changes in parameter values: the compartmental model was highly sensitive to a particular parameter if changes in that parameter were correlated with large changes in model rank.

The models were most sensitive to the densities of dendritic fast sodium (Nad), A-type potassium (KA), h and fast and slow delayed-rectifier potassium (KDRf, KDRs) channels. They were least sensitive to somatic fast sodium (Nas), M-type potassium (KM), AHP and, L- and T-type calcium (CaL, CaT) channels.

Of particular interest is co-regulation between ion channel types, in which a change in one channel density is accompanied by a change in the other, in high-performing models. As might be expected, the ensemble clearly showed co-regulation between fast sodium and fast delayed-rectifier potassium channels where an increase in density in one was accompanied by an increase in density in the other. Crucial to the key question of the study, it was also seen that an increase in dendritic h channel density was accompanied by a decrease in KDRs or KA. No such co-regulation was seen when h channels were present only on the soma. Thus the ensemble models reveal that the presence of h channels in the dendrites of these neurons acts as a signature that could be used to direct experimental confirmation and analysis.

output for each obtained parameter set. This will gradually build up sample data for the distribution of the model output, given the distribution of parameter values. The major difficulty with this method is that it may require thousands to tens of thousands of model evaluations and so be computationally impractical if each model evaluation takes a significant amount of time. This could well be the case for a detailed compartmental neuron model.

To make this approach computationally tractable, it is necessary to both minimise the number of model evaluations required and speed up each evaluation as much as possible. Minimising the number of evaluations involves using MC sampling methods that cover the parameter distribution space with as few samples as possible (Box 14.6). Speeding up model evaluations

Box 14.6 | Monte Carlo sampling methods

Determining a model's parameter sensitivity and determining probability distributions for parameter values both involve calculating functions of random variables to produce posterior probability distributions. In very many cases, particularly when our functions are complex models that must be simulated, it is necessary to generate samples for the posterior distribution, as it cannot be calculated directly. This involves drawing random samples from the prior parameter distributions to act as inputs to our model. Such use of random numbers is known as a **Monte Carlo (MC) sampling method**. A good introduction to the theory of these methods can be found in MacKay (2003), Chapter 29.

For a complex function (model), simply drawing prior samples at random may require a very large number of such samples, and hence function evaluations, to provide a good estimate of the posterior distribution. This can be very computationally challenging, and thus a range of methods have been developed to minimise the number of prior samples, and hence function evaluations, required. Different approaches include Markov-Chain Monte Carlo methods and quasi-Monte Carlo methods.

Markov-Chain Monte Carlo (MCMC) methods bias parameter value selection towards areas that correspond with high density in the posterior distribution, on the basis that parameter values close to a value that gives a high posterior probability will also likely result in a high probability. Thus each new sample of parameter values is chosen to be within a metric distance of the currently tested sample: this is the Markov Chain aspect of these methods. Different MCMC methods select or reject new samples using different metrics. Popular methods are the Metropolis–Hastings algorithm and Gibbs sampling.

Quasi-Monte Carlo methods replace the random selection of parameter values with quasi-random sequences of values that, hopefully, adequately cover the prior parameter space with fewer samples than simple random selection.

requires suitable model reduction techniques that attempt to simplify the model, at least around the model states of interest. For example, in the synaptic reaction pathway model considered in Eriksson et al. (2018), certain reaction equations can be replaced by steady-state values if their dynamics do not significantly affect model behaviour. Alternatively, the model can be approximated by a simpler model, with the approximation being accurate around the behaviour of interest. Tennøe et al. (2018) explores the use of polynomial chaos expansions to achieve this.

Given sufficient sample data of the model output, we can assess the characteristics of the model output distribution to give us information about model output uncertainty. Basic statistics are the expected mean and variance of the output, for possibly one or more features, such as spike rate. Further statistics might include the bounds within which, say, 90% of model outputs fall. The range between such bounds gives an indication of the general sensitivity of the model to parameter values, with a larger range indicating heightened sensitivity. What is also needed is some way to quantify

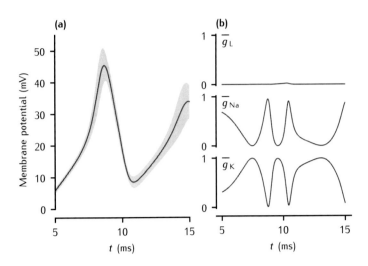

Uncertainty quantification in the Hodgkin–Huxley model of action potential generation. The maximum sodium, potassium and leak conductances are varied by ±10% around their original values, with all other parameters being fixed at their original values. **(a)** Membrane potential due to a current step (solid line: mean; shading: 90% prediction interval). **(b)** First-order Sobol indices for the maximum leak, sodium and potassium conductances, respectively.

Uncertainpy is a freely available Python toolbox for carrying out uncertainty quantification and sensitivity analysis in neural (and other) systems. It makes use of the *Chaospy* and *SALib* Python packages.

the contribution of individual parameters to model uncertainty. Sobol sensitivity indices provide one suitable set of measures (Tennøe et al., 2018). The first-order Sobol sensitivity index measures the variance in the expected value of model output, given that the value of a particular parameter is known, normalised against the total variance in model output. The higher this index is for a particular parameter, the more sensitive the model output is to this parameter's value.

Tennøe et al. (2018) show several illustrative examples of uncertainty quantification, calculated using their software package *Uncertainpy* (see sidebox). One example is the examination of the Hodgkin–Huxley model of action potential generation, showing which particular parameters most affect action potential shape during its time course, as first studied by Torres Valderrama et al. (2015). Analysing all 11 parameters in the model shows that the action potential is most sensitive to two parameters: the maximum conductance of the delayed-rectifier potassium conductance and the reversal potential of the fast sodium conductance. A simpler analysis using only three parameters, the maximum conductances of the sodium, potassium and leak currents, is illustrated in Figure 14.14. Clearly here the sodium and potassium conductances most influence the action potential characteristics when all other parameters have fixed values.

14.7 | Probabilistic Model Optimisation

In the previous section, we considered some implications of uncertainty in model parameter values on model outcomes. This is a part of the wider issue of the statistical estimation of parameter values. The model parameter fitting procedures we have described so far are designed to produce a point-valued estimate of the parameters, that is, a single value for each model parameter. This set of parameter values will be the best, as measured by some error function, for fitting the model to the experimental data we have to hand. However, we have not considered the fact that any experimental data will be subject to measurement errors and noise, or that the system we are modelling is itself stochastic, such as neurotransmitter release at chemical synapses and ion channel opening and closing. If we recorded a new set of experimental data and reoptimised our model against this, it is likely we would end up with a different point-valued estimate for the optimal parameter values.

If we accept there is noise in the system under study and any data recorded from it, then we must accept that there is uncertainty in any estimate of the parameter values based on those data and we should quantify this uncertainty by estimating a probability distribution for the value of each parameter, rather than producing a point-valued estimate. This will require that our model itself includes stochastic terms that account for any randomness in the data against which we are fitting our model.

We have considered specific examples of such stochastic models in this book: single ion channel models (Section 4.7); stochastic action potential generation due to ion channel noise (Section 4.7.3); and probabilistic release of neurotransmitter at synapses (Section 7.3). Other models introduce noise that is not specific to a single stochastic system component. For example, noise in the membrane potential of a neuron can be accounted for in integrate-and-fire neurons by adding Gaussian noise with a zero mean and suitable variance and amplitude to the membrane potential at each simulation time-step (Section 8.5), or such noise can be added to the firing threshold of the integrate-and-fire neuron.

Above we have considered compartmental neuron models as being deterministic, that is, producing the exact same output for the same inputs, but membrane potential noise can be added in one or many compartments, as for integrate-and-fire neurons. This is often only done for the somatic compartment, to introduce variability in action potential firing.

Once we have a stochastic model, even for a fixed set of parameter values, each simulation of the model will produce different results. So rather than matching a single, consistent model output to our data, we are now faced with determining the probability distribution of model outputs to compare against the distribution of our experimental data. **Bayesian inference** provides a way of tackling this problem that allows us to determine the likely probability distributions of the parameter values, given our data set. It works by refining prior (initial) beliefs about parameter values using the evidence provided by the data, compared against model simulations, to produce updated posterior (inferred) beliefs of the probability distributions for the parameter values. An excellent introduction to Bayesian parameter estimation is Chapter 9 of the online book by Franke (2021).

14.7.1 Bayesian Inference for Parameter Estimation

Our aim now is to find reasonable probability distributions for each of the model's parameter values, given that the distribution of our stochastic model output should match a given set of experimental data with some degree of accuracy. Let's say we have m parameters, which we denote by the m-dimensional vector $\vec{\theta}$, and a set of n data points (observations), denoted by the vector $\vec{y}^{\text{expt}}$. We might also have some idea of the range of sensible parameter values, allowing us to define initial (prior) probability distributions $p(\vec{\theta})$, which are our prior beliefs in the likely parameter values before observing the experimental data set. Then Bayesian inference gives us a way of refining these prior beliefs on the basis of the set of data observations to provide an updated set of probability distributions referred to as the posterior distributions. By Bayes' rule, the posterior distribution of the parameter values given the data, $p(\vec{\theta}|\vec{y}^{\text{expt}})$, is proportional to the product of their prior probability $p(\vec{\theta})$ and the likelihood $p(\vec{y}^{\text{expt}}|\vec{\theta})$, which is the probability of our model generating the data $\vec{y}^{\text{expt}}$, given a specified set of parameter values $\vec{\theta}$:

$$p(\vec{\theta}|\vec{y}^{\text{expt}}) = \frac{p(\vec{y}^{\text{expt}}|\vec{\theta})p(\vec{\theta})}{p(\vec{y}^{\text{expt}})}. \tag{14.7}$$

Calculating the posterior probability distribution for the parameters using this formula is often challenging, especially if the model we are fitting is complex. This is the case with the neuroscientific models we are considering, such as compartmental models of neurons. However, a variety of numerical approaches are available that enable estimates of the posterior to be developed through running model simulations many times for different candidate parameter values.

The first step is to choose suitable prior distributions $p(\vec{\theta})$ from which sample parameter values can be readily obtained. We might have some idea of the expected value of a parameter and its likely variance around that expectation, leading us to define a normal (Gaussian) distribution as its prior. At the least we should have an idea of the limits on the range of a parameter, due to known physical constraints or estimates from experiments. This enables us to define a uniform prior over this range, that is, we assign an equal probability to every value in this range. This form of prior actually leads to a direct connection between Bayesian inference and obtaining a point-estimate of parameter values by minimising the error between the data and model output, via maximum likelihood estimation (MLE) (Box 14.7).

Calculating the likelihood term $p(\vec{y}^{\text{expt}}|\vec{\theta})$ is likely to be much more challenging. The probability of the data $\vec{y}^{\text{expt}}$, given a set of parameter values $\vec{\theta}$, depends on our model, which is unlikely to allow tractable calculation of the probability. Instead, we assume some functional form for the likelihood, known as an auxiliary likelihood, and then run large numbers of model simulations to build up estimates of this likelihood on the basis of model outputs, for the currently selected parameter values $\vec{\theta}$.

This is the approach taken in methods known as **Bayesian synthetic likelihood** (BSL) (Price et al., 2018) and **approximate Bayesian computation** (ABC) (Marin et al., 2012). These methods typically work with summary statistics $\vec{s}^{\text{expt}}$ of the data $\vec{y}^{\text{expt}}$ being fitted to, rather than the raw data itself, to reduce the dimensionality of the problem. For example, we might match the number of action potentials emitted by a neuron in a given time interval, rather than the time-varying membrane potential.

The methods proceed by generating k independent sets of N model outcomes for the current parameter values $\vec{\theta}$, where N must be large enough to allow for reasonable estimates of the summary statistics, for example, mean and variance of the model outcomes. The summary statistics $\vec{s}^{\text{model}}$ are then calculated for each of these k sets of simulated outcomes and then compared

Box 14.7 | Minimising MSE versus MLE

When optimising a deterministic model, the usual error function when determining an optimum point-valued estimate of the model's parameter values is the mean squared error (MSE) between the data and the model output (Section 14.2). Under certain assumptions, this is equivalent to finding the maximum likelihood estimate (MLE) for the parameter values from a stochastic model, as we now show.

If our model is deterministic, such as the compartmental models of neurons considered in Section 14.2, we can use Bayesian inference to determine parameter value distributions if we add a suitable noise term to the model. A common assumption is that each recorded data point is subject to independent Gaussian noise with a constant variance σ^2, and so we add such noise to our model. The likelihood $p(\vec{y}^{\mathrm{expt}}|\vec{\theta})$ of our data arising from the current set of parameters is then given by the joint Gaussian probability density function:

$$p(\vec{y}^{\mathrm{expt}}|\vec{\theta}) = \prod_{i=1}^{n} \frac{1}{\sqrt{2\pi\sigma^2}} \exp\left(-\frac{(y_i^{\mathrm{expt}} - y_i^{\mathrm{model}}(\vec{\theta}))^2}{2\sigma^2}\right). \tag{a}$$

Note that the same squared error used in deterministic model optimisation has now appeared in our likelihood function.

To be able to generate a posterior probability density for our parameter values using Bayesian inference (Equation 14.7), we need to specify a prior probability density for our parameters. If we do not know anything about our parameters beyond their expected range of values, we could assume they each have a uniform probability distribution over their expected range. That is, any particular value is equally likely. This is known as a **flat** or an **uninformative prior**. Then the Bayes equation reduces to:

$$p(\vec{\theta}|\vec{y}^{\mathrm{expt}}) \propto p(\vec{y}^{\mathrm{expt}}|\vec{\theta}). \tag{b}$$

Suppose we want to get the most likely point-valued estimate for the parameter values. This means we want to maximise $p(\vec{\theta}|\vec{y}^{\mathrm{expt}})$, which requires finding the maximum of $p(\vec{y}^{\mathrm{expt}}|\vec{\theta})$ with respect to $\vec{\theta}$, as specified by the function in Equation (a). This is maximum likelihood estimation (MSE). It is convenient to take the log of the likelihood, since the maximum will still be the same, given that the log is a monotonic function and also this converts the product in the likelihood into a sum:

$$\ln p(\vec{y}^{\mathrm{expt}}|\vec{\theta}) = -\sum_{i=1}^{n} \frac{(y_i^{\mathrm{expt}} - y_i^{\mathrm{model}}(\vec{\theta}))^2}{2\sigma^2} - \frac{n}{2}\ln 2\pi\sigma^2 \tag{c}$$

The first term of this equation, which contains $\vec{\theta}$, is proportional to the negative mean squared error. Thus finding the maximum a posteriori (MAP) estimate of the parameters in a posterior distribution with a flat prior is equivalent to finding the minimum squared error of the deterministic model.

with the experimental data set's summary statistics to obtain an estimate of the likelihood of the data. This is done by assuming the likelihood can be approximated by some auxiliary distribution which can be calculated.

In BSL, it is assumed that the summary statistics are normally (Gaussian) distributed. The mean $\mu(\vec{\theta})$ and variance $\Sigma(\vec{\theta})$ of this auxiliary distribution are estimated from the k sets of summary statistics $\vec{s}^{\text{model}}$, generated from the simulations. The likelihood of the experimental data summary statistics can then be calculated from this normal distribution $\mathcal{N}(.)$ (Price et al., 2018):

$$p(\vec{s}^{\text{expt}}|\vec{\theta}) = \mathcal{N}(\vec{s}^{\text{expt}}; \mu(\vec{\theta}), \Sigma(\vec{\theta})). \tag{14.8}$$

ABC works in a similar way but uses a non-parametric auxiliary likelihood (Price et al., 2018):

$$p(\vec{s}^{\text{expt}}|\vec{\theta}) = \frac{1}{k}\sum_{i=1}^{k} K(\rho(\vec{s}^{\text{expt}}; \vec{s}_i^{\text{model}}(\vec{\theta}))), \tag{14.9}$$

where $\rho(\vec{s}^{\text{expt}}; \vec{s}_i^{\text{model}})$ measures the distance between the experimental and simulated statistics and K is a suitable kernel function that controls the tolerance of the distance, giving larger weight to smaller values of ρ.

The overall procedure for building up the posterior distribution is:

(1) Draw one sample of parameter values $\vec{\theta}$ from the prior distribution $p(\vec{\theta})$.
(2) Using Equation 14.8 or 14.9, estimate the likelihood of the summary statistics $p(\vec{s}^{\text{expt}}|\vec{\theta})$, given this set of parameter values $\vec{\theta}$.
(3) Compute and save the unnormalised posterior $p(\vec{\theta}|\vec{s}^{\text{expt}}) \propto p(\vec{s}^{\text{expt}}|\vec{\theta})p(\vec{\theta})$.
(4) Repeat from Step 1 until the parameter space has been sufficiently covered by samples.
(5) Finally, normalise the posterior to give a valid distribution (note that this circumvents explicit calculation of the denominator $p(\vec{s}^{\text{expt}})$ in Bayes' rule, which only acts as a normalising factor).

This whole procedure also could be repeated a number of times, using the obtained posterior as the prior for the next iteration, to increase accuracy (Eriksson et al., 2018). It can also be repeated as more experimental data become available. This will result in a shift from the prior strongly influencing the posterior distribution to the posterior being more strongly determined by the data.

Use of BSL and ABC in Bayesian inference involves many statistical and computational subtleties that are beyond the scope of this book. In addition to accuracy, a major concern is computational tractability, that is, that the methods can achieve a good result in a reasonable computation time. The two limiting factors here are how long a single model simulation takes and how many simulations are required. Ideally the full model will be used for simulations, but it may be necessary and possible to use a simpler approximate model that is cheaper to calculate and accurate enough for matching the data under study. This approach was taken by Tennøe et al. (2018) in their work on parameter sensitivity analysis (Section 14.6). A key aspect of limiting the number of required simulations is to explore the parameter space as efficiently as possible, that is, covering the likely range of values with as few samples as possible. This is the aim of the many variants of MC sampling methods, which are outlined in Box 14.6.

Once a sample set for the posterior distribution has been obtained, then we further want to characterise this distribution to gain information about the likely values for, and uncertainty in, our parameter values. We could fit a functional form to the distribution, such as a multivariate Gaussian, exponential or uniform distribution. Such functional forms could then be used from which to draw sample parameter values when running future simulations of our model. The multivariate distribution may also provide information about any correlations between parameters (Eriksson et al., 2018). In addition, it is likely that we want to know the most probable range of parameter values. Point-valued estimates can be obtained as either the mean of the posterior distribution or as its maximum value, known as the MAP estimate. These may be similar or quite different, depending on the shape of the distribution. The mean, at least, does depend on the entire distribution, unlike the MAP estimate (Franke, 2021, Chapter 9). The likely range of values also can be measured in different ways, such as credible intervals and inner-quartile intervals. These two measures are similar for unimodal, symmetric distributions, but may differ significantly in multimodal or highly skewed distributions (Franke, 2021, Chapter 9). The credible interval has the advantage that it will always include the MAP point estimate, whereas the inner-quartile may not.

14.7.2 Probabilistic Model Comparison for Model Selection

We have now seen that we can use probabilistic techniques to determine probability distributions for the likely values of model parameters. We have treated the situation where we have a defined model with a fixed set of parameters. More generally, we might have a family of models which we want to choose between. Similar probabilistic approaches enable the specification of criteria against which models can be ranked to aid in the selection of the best-performing model. **Model comparison** criteria enable the calculation of the relative performance of different models, where performance incorporates both the accuracy of the model output and the complexity of the model. In general, an accurate model of low complexity is to be preferred. A clear introduction to formal model comparison using such criteria can be found in Franke (2021, Chapter 10). We outline two popular criteria here.

Akaike Information Criterion

One of the simplest model comparison criteria to calculate is the **Akaike Information Criterion** (AIC), which is based on information-theoretic concepts. Suppose now we have a set of models $\mathcal{M}$ for our data, of which each model M_i has its own set of k_i parameters $\vec{\theta}_i$. Following the approach of Bayesian inference used above to find parameter values, we consider the likelihood of the data, given the model along with its parameters $p(\vec{y}^{\text{expt}}|\vec{\theta}_i, M_i)$. We optimise each model's parameter values just using this likelihood, to get the point-valued estimate:

$$\vec{\theta}_i^{\text{opt}} = \text{argmax}_{\vec{\theta}_i}\, p(\vec{y}^{\text{expt}}|\vec{\theta}_i, M_i). \tag{14.10}$$

Then the AIC is given by:

$$\text{AIC}(M_i, \vec{y}^{\text{expt}}) = 2k_i - 2\log p(\vec{y}^{\text{expt}}|\vec{\theta}_i^{\text{opt}}, M_i). \tag{14.11}$$

Given our set of models $\mathcal{M}$, the best model in the AIC sense is the one with the lowest AIC score, given that all models' parameters are optimised against the same set of data $\vec{y}^{\text{expt}}$. From the criterion, it can be seen clearly that the number of parameters k_i increases the AIC (first term on the right-hand side of Equation 14.11), whilst a closer fit to the data, in the maximum likelihood sense, lowers the AIC (second term of Equation 14.11).

Bayes Factors

Full Bayesian inference can also be used to carry out such a model comparison, by calculating what are called **Bayes factors**. Naively, we might like to know the posterior distribution of models, given the available data, which by Bayes' rule is given by:

$$p(M_i|\vec{y}^{\text{expt}}) = \frac{p(\vec{y}^{\text{expt}}|M_i)p(M_i)}{p(\vec{y}^{\text{expt}})}. \tag{14.12}$$

Given that there are infinitely many models that may be defined, it is not possible to calculate this posterior distribution over all possible models. However, given two well-defined models, M_1 and M_2, we can compare them by calculating the posterior odds:

$$\frac{p(M_1|\vec{y}^{\text{expt}})}{p(M_2|\vec{y}^{\text{expt}})} = \frac{p(\vec{y}^{\text{expt}}|M_1)}{p(\vec{y}^{\text{expt}}|M_2)} \frac{p(M_1)}{p(M_2)}. \tag{14.13}$$

Even without knowing the prior odds $p(M_1)/p(M_2)$, we can get a measure of how much evidence the data provide in favour of one model over another by calculating the ratio of the likelihoods, known as the Bayes factor (BF):

$$BF_{12} = \frac{p(\vec{y}^{\text{expt}}|M_1)}{p(\vec{y}^{\text{expt}}|M_2)}. \tag{14.14}$$

Remembering that the models are defined over sets of parameters, then the BF_{12} is the ratio of the marginal likelihoods:

$$\frac{p(\vec{y}^{\text{expt}}|M_1)}{p(\vec{y}^{\text{expt}}|M_2)} = \frac{\int p(\vec{\theta}_1|M_1)p(\vec{y}^{\text{expt}}|\vec{\theta}_1,M_1)d\vec{\theta}_1}{\int p(\vec{\theta}_2|M_2)p(\vec{y}^{\text{expt}}|\vec{\theta}_2,M_2)d\vec{\theta}_2}. \tag{14.15}$$

As with calculating the posterior distributions of parameter values (Section 14.7.1), the main difficulty here lies in calculating the likelihoods $p(\vec{y}^{\text{expt}}|\vec{\theta}_i,M_i)$. However, a number of numerical approaches, including MC methods, can be used.

The greater the value of BF_{12} is above 1, the stronger the evidence is for model M_1 compared to M_2. Though this is qualitative, once BF_{12} is on the order of tens, then the evidence for M_1 is very strong, rising to essentially conclusive for $BF_{12} \geq 100$ (Franke, 2021, Chapter 10). In its integration over all possible parameter values, the BF punishes model complexity, since models with large numbers of parameters whose values are not well known (diffuse priors) will likely score poorly.

14.8 | Summary

Developing a computational model of a system is a process that involves many choices and decisions. Is my model going to be based on what is known about the system (law-driven), such as neural biophysics? Or am I building an empirical model fitted to data drawn from experiments on the system (data-driven) such as ion channel currents? We have considered both modelling approaches in this book and summarise the issues of this **model selection** problem in this chapter. We also introduced concepts from machine learning that can enable a quantitative comparison between probabilistic models for the same system, leading to a choice of the best model in terms of a trade-off between model accuracy and its simplicity and ability to generalise.

Having decided on a particular model form and structure, the remaining major challenge is to assign values to all of the model's parameters, the **parameter estimation** problem. This involves adjusting parameter values to optimise the accuracy of model output against certain data about the system. Many numerical optimisation methods are available to automate this model-fitting process, but the fitting process is often subject to difficulties due to a lack of constraining data and potentially competing fitness objectives. Models are often under-determined by the data: more than one set of parameter values may provide equally good fits to the data (Gutenkunst et al., 2007; Marder and Taylor, 2011; O'Leary et al., 2015). Multiple objectives, such as matching distinct dendritic and somatic voltage responses in morphologically complex compartmental neuron models, may be difficult to satisfy in a single parameter optimisation procedure. Some parameters may be more important for one objective than another. Thus a multi-stage optimisation process involving at each stage fixing some parameter values, whilst optimising others against a particular objective, may be needed.

An important outcome of parameter value estimation (model fitting) is to gain an appreciation of the range of parameter values that may provide good model performance and further, how sensitive model output is to particular parameters (O'Leary et al., 2015). This has parallels in biology where there is evidence for neurons being robust in their electrophysiological responses to precise distributions of membrane-bound ion channels (Marder and Taylor, 2011; Kamaleddin, 2022).

CHAPTER 15

Farewell

In this book, we have aimed to explain the principles of computational neuroscience by showing how the underlying mechanisms are being modelled, together with presenting critical accounts of examples of their use. In some chapters, we have placed the modelling work described in its historical context where we felt this would be interesting and useful. We now make some brief comments about where the field of computational neuroscience came from and where it might be going.

15.1 | The Development of Computational Modelling in Neuroscience

The field of computational modelling in neuroscience has been in existence for almost 100 years. During that time, it has gone through several stages of development. From the 1930s, researchers in the field of **mathematical biophysics** conceived of mathematical and physical models of biophysical phenomena. Relating to neuroscience, there was work on how networks of nerve cells could store and retrieve information through the adaptive tuning of the thresholds of selected nerve cells (Shimbel, 1950), on linking psychophysical judgements with the underlying neural mechanisms (Landahl, 1939) and on mathematical accounts of how nerve cells could act as logical elements and what the computational capabilities of a network of such elements were (McCulloch and Pitts, 1943). From the 1950s onwards, there was interest, particularly in the physics community, in treating the brain as a uniform medium and calculating its modes of behaviour, such as the conditions under which it would support the propagation of waves of activity across it (Beurle, 1956). Much of this work laid the foundation for the emergence of the field of artificial neural networks in the 1950s and 1960s (Box 11.2). This field enjoyed a renaissance in the 1980s, and once more in the 2010s, and is now regarded as an essential machine learning technique.

From the 1950s, computational models which addressed much more targeted questions in neuroscience were being developed and different models were constructed at different levels of detail. One of the most famous of these is the model of the propagation of the nerve impulse (Hodgkin and Huxley, 1952d), the subject of Chapter 3 of this book. Together with the advent of compartmental modelling of neurons (Rall, 1962, 1964), the methodology was available to make detailed models of neurons. In contrast, the model of how the cerebellar cortex could function as a memory device (Marr, 1969) was formulated at a higher level, with much less detail about the individual components of the model. For example, the modifiable synapses in the model are represented as two-state devices. Notwithstanding, this model has been a source of inspiration for physiologists investigating how the cerebellar cortex can be used to learn to associate motor commands with actions.

Since the 1970s, there has been an ever-increasing involvement of computational modellers in the study of the nervous system. Most computational models are aimed at answering specific, rather than general, questions and they rely heavily on experimental data to constrain the model. Whereas previous generations of modellers were content to represent the properties of the nerve cell at a somewhat abstract level, many models incorporate detail at the subcellular level. In many cases, computational models can provide answers at different spatial scales (Einevoll et al., 2019). Computational modelling now includes modelling of the complex extracellular influences on nerve cells (Chapter 10). Many experimental neuroscientists now regard computational methods as a valid approach for neuroscience and have collaborations with computational neuroscientists; all premier neuroscience journals accept papers describing computational neuroscience modelling work that links to experimental results. This heightened interest in modelling in neuroscience could be due to the increased sophistication of models, the availability of data and the better analytical methods now available. More generally, modelling has been successful in other spheres such as in weather forecasting where multiscale modelling is providing ever more accurate predictions (Bauer et al., 2015). The immense activity from 2020 in modelling the spread of SARS-CoV-2 virus, mentioned in Chapter 1, has captured the attention of both decision-makers and the general public.

15.2   The Future of Computational Neuroscience

What advances in computational modelling in neuroscience might be expected to take place over the next few decades? As reflected in this book, with the growing acceptance of the utility of the modelling paradigm, the initial concentration on modelling at the levels of single neurons and of networks of neurons has broadened to encompass models at more detailed levels, as well as incorporating extracellular influences. Further progress will be driven by the increased availability of data and code, advances in computer technology and the development of new artificial intelligence (AI) algorithms, coupled with advances in new measurement techniques in neuroscience.

15.2.1   Sharing Data and Models

Modellers require experimental data to constrain and validate their models. There are now vast quantities of different types of data available, ranging from whole brain imaging to the use of genetic markers to identify specific classes of nerve cells and synapses. The worldwide trend to make data freely available is of great benefit to modellers. Many funding bodies now actively promote the sharing of high-quality data. Most journals require their authors to share their code and data, either directly on request or by placing the data in a publicly accessible database. It is more difficult for modellers to obtain unpublished data, but then the question of data quality arises. With an increased quantity and complexity of neuroscience data being available, there will be an increased demand for new methods of data analysis and presentation. Resources of data and accompanying analytical tools are now being assembled for specific areas of neuroscience. The EU-funded **Human Brain Project** has established the *EBRAINS* initiative to bring together a wide variety of neuroscience data and tools for analysis and model specification and simulation. Whilst increasingly public funding bodies and scientific journals encourage the sharing of data, it is paradoxical that the Allen Institute, the organisation that has pushed most actively the sharing of data and other resources, is funded privately.

With the diversity of different types of data now available, it will be increasingly important to model the measurement process itself, a topic which is treated extensively in Chapter 13. Modellers also may have to adapt their ideas as new ways of thinking about the basic concepts in neuroscience emerge. For example, the possibility that the extracellular perineuronal nets have a role in learning and memory may have to be incorporated into models of plasticity (Tsien, 2013; Thompson et al., 2018).

In addition to the availability of code due to the requirements of funders and journals, many computational neuroscientists have developed their own archives of computational models and made them freely available. Communities exist to support users of simulation packages, such as *NEURON*, and developers can contribute their models to freely accessible repositories. One established archive is *ModelDB* containing over 1000 computational models. The *OpenSourceBrain* (*OSB*) repository has been set up to promote and archive simulator-independent models specified using *NeuroML* (Cannon et al., 2014) or *PyNN* (Davison et al., 2009). This enables a model to be run on the user's simulator of choice. *OpenSourceBrain* itself provides facilities for a user to run models and visualise results online.

One familiar problem is to ensure that shared code is error-free. Ideally journals could require this for published code. The *Codecheck* initiative is developing tools to enable the code published in a scientific paper to be given a stamp of authenticity. Under this initiative, the influential code used in the UK to model the spread of the SARS-CoV-2 virus was, in March 2020, made available on *Github* and was found to be reproducible (Chawla, 2020).

15.2.2 Advances in Computer Technology

As the power of computers increases, it is possible to simulate even more detailed neurons and larger and larger numbers of networks of neurons. The human nervous system contains around 10^{11} neurons and 10^{14} synapses. At the time of writing, it is possible to simulate networks of tens of thousands of compartmental model neurons or a larger number of integrate-and-fire neurons.

The Allen Institute actively promotes Open Science in the study of the biosciences, including the neurosciences. A large variety of neuroscience data and tools from mouse to human is made freely available, including the *Allen Brain Map*.

It is difficult to know what the ultimate limit will be, depending on how the speed and storage capacity of computers increase. According to **Moore's law** (Moore, 1965), speed and storage capacity were held to double every year, but this is now not the case.

In addition to the increase in power of familiar digital computers, specialised neuromorphic hardware-based computers, such as *SpiNNaker* and *BrainScaleS*, now provide the ability to simulate large networks of simple spiking neurons in real time or even faster than real time. These facilities open up the possibility of simulating brain-scale neural networks in reasonable computation time. The *EBRAINS* initiative provides researchers with remote access to these neuromorphic computers.

15.2.3 The Need for Abstraction

As highlighted in the related field of **systems biology**, there is a growing interest in constructing simulation models of the many complex molecular interactions within a single cell. This has highlighted the more general issue that, to make progress in the computational modelling of biological systems, abstraction of the elements of the system will be required. Without abstraction, the resulting models will be too complex to understand and it will not be feasible to carry out the desired computations, even using the fastest computers that might be developed over the next 20 years. As discussed in earlier chapters in this book, the level at which the abstraction is carried out will depend on the question being asked. For example, to calculate how much information could be stored in an associative memory model, the precise dynamics of the synaptic modification rule may not be important. But to understand the time course of memory acquisition, it probably will be. A related challenge is to connect models at different levels of abstraction. For example, under what conditions, if any, can a biophysically detailed multi-compartmental model be replaced by a simplified single-compartment model of the integrate-and-fire type? Likewise, when can the dynamics of a network of integrate-and-fire-type neurons be captured sufficiently accurately by a firing rate model?

15.2.4 The Impact of New AI Techniques and Advances in New Methods of Measurement

As already described, computational modelling in neuroscience and neural networks, now the primary technology underlying AI, share a common lineage (Box 11.3) and so there will continue to be a symbiosis between the two. In general, AI methods will have a great influence on model selection and optimisation (Chapter 14) and on methods for data analysis. The pattern recognition capabilities of AI technologies are already being put to use to identify relevant patterns of brain activity in diverse spatial and temporal data recorded using MRI, EEG and other techniques. This is useful both for theorising about, and modelling, cognition in the brain and for clinical diagnosis and prediction, such as for epileptic seizures (Box 11.3).

15.2.5 Applications of Computational Modelling

In this book, we have concentrated on using computational modelling towards gaining an understanding of the function and development of the

The original formulation of **Moore's law** is that the number of components that can be placed on a computer chip, and hence the power of the computer, doubles every year. There have been various predictions about when the fundamental physical barriers will cause this law to break down. So far, this has not happened, although it is now accepted that this law, and other similar laws, should be modified. Moore (1975) modified his statement to give a doubling every 18 months or two years. How this slowing down is offset by parallelisation is not clear.

The term systems biology was coined to describe attempts to understand biological systems at many different levels, a dynamic interaction being envisaged between the processes of modelling and the acquisition of experimental data. The prime example was to understand heart physiology (Noble, 2006). Advances in the availability of data and models have made it possible to study the function and dysfunction of the heart at many levels, from genes to the entire system. Noble (2006) pointed out that computational neuroscience and systems biology share a common line of descent from Hodgkin and Huxley's work (Chapter 3). Systems biology is now concerned with understanding subcellular pathways.

nervous system. We anticipate that the number of application areas will increase, and here we mention just three of them.

Close integration between experiment and theory. There are ways in which experiment and computational modelling can be used interactively. Computational models could be run so as to provide instant feedback during the course of experiments as to how the experimental conditions should be changed. The model simulation would need to take seconds or minutes, at most. Hence the complexity of the model will be constrained by the power of the computer on which it is running. At a more specific level, the technique of dynamic clamp (Figure 14.12) allows the electrophysiological properties of a neuron to be investigated by injecting current into it through electrodes, with the current being calculated as the output of a model which may take measurements, such as membrane voltage, as inputs (Destexhe and Bal, 2009). This approach can be used to simulate the action of a single synapse on the neuron or the insertion of a new ionic channel current (Milescu et al., 2008), for example.

The modelling of clinical interventions. Neural compartmental models are ideal for use in drug discovery studies to understand, at the cellular and network levels, the effects of the application of drugs to, say, block specific ion channels (Aradi and Érdi, 2006). Systemic effects would be more difficult to predict as models at this level remain to be established. Another type of intervention is seen in the use of deep brain stimulation (DBS) for treatment of neurological conditions. An example of computational modelling that holds out great promise is modelling the effects of continuous electrical stimulation of the subthalamic nucleus of the basal ganglia described in Section 13.8, which is used for the relief of Parkinson's disease. Our understanding of why DBS works and how to configure it optimally will be greatly enhanced once satisfactory computational models have been developed. As described in Section 13.8, the initial modelling of the basal ganglia has been extended to include modelling of targets other than the basal ganglia, as well as modelling the stimulating electrode to explore different modes of stimulation.

Brain–machine interaction and neural prostheses. Already there are clear demonstrations that brain-derived signals can be used to drive external devices, such as computer screen cursors and robotic arms (Lebedev and Nicolelis, 2006). Signals include EEGs and spiking from small ensembles of neurons. Models that interpret these signals are typically artificial neural networks or other machine learning algorithms. In principle, spiking neural network models could be used, provided the simulations run in real time. This is still a significant barrier. However, such models could accept spiking inputs directly from biological neurons and produce a spiking output that could be used to stimulate downstream neurons. Such technology will open up the promise of neural prostheses, in which computer models replace damaged brain nuclei or augment brain function. To date, cochlea implants (Zeng et al., 2008) that turn sound into electrical impulses that are sent directly to the auditory nerve represent the most successful neural prosthesis, with more than three-quarters of a million devices having been implanted world wide (as of December 2019; NIH Publication No. 00-4798). Computational modelling of the cochlea has an important role to play in optimising electrode design, placement and stimulation protocols (Seeber and Bruce, 2016).

15.3   And Finally ...

In comparison with the enormous leap forward in computer power that has been seen over the past 50 years, the fundamental principles of modelling in computational neuroscience have remained largely unchanged. The increase in computing power enabling more powerful modelling is a double-edged sword. Whilst powerful computers make possible new types of modelling and analysis which were unimaginable in the past, they also carry the risk of distracting us from these fundamental principles. However, we feel confident that the principles of computational neuroscience that we have described in this book will be valid in 50 years' time. We hope that new generations of researchers working in the neurosciences will use our book to formulate their theories precisely, express their theories as computational models and validate them through computer simulation, analysis and experiment.

References

Abarbanel H.D.L., Gibb L., Huerta R. and Rabinovich M.I. (2003). Biophysical model of synaptic plasticity dynamics. *Biol. Cybern.* **89**, 214–226.

Abbott L. and Marder E. (1998). Modeling small networks. In *Methods in Neuronal Modeling: From Ions to Networks*, eds. C. Koch and I. Segev (MIT Press, Cambridge, MA), chapter 10, pp. 361–410. Second edition.

Abbott L.F., Thoroughman K.A., Prinz A.A., Thirumalai V. and Marder E. (2003). Activity-dependent modification of intrinsic and synaptic conductances in neurons and rhythmic networks. In *Modeling Neural Development*, ed. A. van Ooyen (MIT Press, Cambridge, MA), chapter 8, pp. 151–166.

Abraham M.J., Murtola T., Schulz R., Páll S., Smith J.C., Hess B. and Lindahl E. (2015). GROMACS: High performance molecular simulations through multi-level parallelism from laptops to supercomputers. *SoftwareX* **1–2**, 19–25.

Achard P. and De Schutter E. (2006). Complex parameter landscape for a complex neuron model. *PLoS Comput. Biol.* **2**, e94.

Adams D.L. and Horton J.C. (2009). Ocular dominance columns: enigmas and challenges. *The Neuroscientist* **15**, 62–77.

Adrian E.D. (1928). *The Basis of Sensation: The Action of the Sense Organs* (W W Norton and Co., London).

Aeschlimann M. and Tettoni L. (2001). Biophysical model of axonal pathfinding. *Neurocomputing* **38–40**, 87–92.

Agmon-Snir H., Carr C.E. and Rinzel J. (1998). The role of dendrites in auditory coincidence detection. *Nature* **393**, 268–272.

Agre P., Brown D. and Nielsen S. (1995). Aquaporin water channels: unanswered questions and unresolved controversies. *Curr. Opin. Cell Biol.* **7**, 472–483.

Agudelo-Toro A. and Neef A. (2013). Computationally efficient simulation of electrical activity at cell membranes interacting with self-generated and externally imposed electric fields. *J. Neural Eng.* **10**, 026019.

Ajay S.M. and Bhalla U.S. (2005). Synaptic plasticity in vitro and in silico: insights into an intracellular signaling maze. *Physiology (Bethesda)* **21**, 289–296.

Alexander S.P., Kelly E., Mathie A., Peters J.A., Veale E.L., Armstrong J.F., Faccenda E., Harding S.D., Pawson A.J., Southan C. et al. (2021). The concise guide to pharmacology 2021/22: Introduction and other protein targets. *Br. J. Pharmacol.* **178**, S1–S26.

Alexander S.P.H., Mathie A., Peters J.A., Veale E.L., Striessnig J., Kelly E., Armstrong J.F., Faccenda E., Harding S.D., Pawson A.J., Sharman J.L., Southan C., Davies J.A. and CGTP Collaborators. (2019). The concise guide to pharmacology 2019/20: Ion channels. *Br. J. Pharmacol.* **176**, S142–S228.

Amari S. (1977). Dynamics of pattern formation in lateral-inhibition type neural fields. *Biol. Cybern.* **27**, 77–87.

Amiry-Moghaddam M., Frydenlund D. and Ottersen O. (2004). Anchoring of aquaporin-4 in brain: molecular mechanisms and implications for the physiology and pathophysiology of water transport. *Neuroscience* **129**, 997–1008.

Amit D.J. (1989). *Modeling Brain Function: The World of Attractor Networks* (Cambridge University Press, Cambridge).

Amit D.J. and Brunel N. (1997a). Model of global spontaneous activity and local structured activity during delay periods in the cerebral cortex. *Cereb. Cortex* **7**, 237–252.

Amit D.J. and Brunel N. (1997b). Dynamics of a recurrent network of spiking neurons before and following learning. *Network Comp. Neural Syst.* **8**, 373–404.

Amit D.J. and Fusi S. (1994). Learning in neural networks with material synapses. *Neural Comput.* **6**, 957–982.

Amit D.J., Gutfreund H. and Sompolinsky H. (1985). Spin-glass models of neural networks. *Phys. Rev. A* **32**, 1007–1018.

Amit D.J. and Tsodyks M.V. (1991a). Quantitative study of attractor neural network retrieving at low spike rates: I. substrate – spikes, rates and neuronal gain. *Network Comp. Neural Syst.* **2**, 259–273.

Amit D.J. and Tsodyks M.V. (1991b). Quantitative study of attractor neural network retrieving at low spike rates: II. low-rate retrieval in symmetric networks. *Network Comp. Neural Syst.* **2**, 275–294.

Anastassiou C.A. and Koch C. (2015). Ephaptic coupling to endogenous electric field activity: why bother? *Curr. Opin. Neurobiol.* **31**, 95–103.

Anderson D.J., Rose J.E., Hind J.E. and Brugge J.F. (1971). Temporal position of discharges in single auditory nerve fibers within the cycle of a sine-wave stimulus: frequency and intensity effects. *J. Acoust. Soc. Am.* **49**, 1131–1139.

Andrade R., Foehring R.C. and Tzingounis A.V. (2012). The calcium-activated slow AHP: cutting through the Gordian knot. *Front. Cell. Neurosci.* **6**, 47.

Andrews S.S. (2017). Smoldyn: particle-based simulation with rule-based modeling, improved molecular interaction and a library interface. *Bioinformatics* **33**, 710–717.

Andrews S.S. (2018). Particle-based stochastic simulators. In *Encyclopedia of Computational Neuroscience*, eds. D. Jaeger and R. Jung (Springer-Verlag, New York, NY), pp. 1–5.

Andrews S.S., Addy N.J., Brent R. and Arkin A.P. (2010). Detailed simulations of cell biology with Smoldyn 2.1. *PLoS Comput. Biol.* **6**, e1000705.

Appasani K. (2017). *Optogenetics: From Neuronal Function to Mapping and Disease Biology* (Cambridge University Press, Cambridge).

Aradi I. and Érdi P. (2006). Computational neuropharmacology: dynamical approaches in drug discovery. *Trends Pharmacol. Sci.* **27**, 240–243.

Aradi I., Santhakumar V. and Soltesz I. (2004). Impact of heterogeneous perisomatic IPSC populations on pyramidal cell firing rates. *J. Neurophysiol.* **91**, 2849–2858.

Aradi I. and Soltesz I. (2002). Modulation of network behaviour by changes in variance in interneuronal properties. *J. Physiol.* **538**, 227–251.

Arbas E.A. and Calabrese R.L. (1987). Slow oscillations of membrane potential in interneurons that control heartbeat in the medicinal leech. *J. Neurosci.* **7**, 3953–3960.

Arisi I., Cattaneo A. and Rosato V. (2006). Parameter estimate of signal transduction pathways. *BMC Neurosci.* **7 (Suppl 1)**, S6.

Armstrong C.M. and Bezanilla F. (1973). Currents related to movement of the gating particles of the sodium channels. *Nature* **242**, 459–461.

Arnett D.W. (1978). Statistical dependence between neighbouring retinal ganglion cells in goldfish. *Exp. Brain Res.* **32**, 49–53.

Arrhenius S. (1889). Über die Reaktionsgeschwindigkeit bei der Inversion von Rohrzucker in Säuren. *Z. Phys. Chem.* **4**, 226–248.

Arvanitaki A. (1942). Effects evoked in an axon by the activity of a contiguous one. *J. Neurophysiol.* **5**, 89–108.

Ascher P. and Nowak L. (1988). The role of divalent cations in the N-methyl-D-aspartate responses of mouse central neurones in culture. *J. Physiol.* **399**, 247–266.

Ascoli G.A. (2002). Neuroanatomical algorithms for dendritic modelling. *Network Comp. Neural Syst.* **13**, 247–260.

Ascoli G.A. (2006). Mobilizing the base of neuroscience data: the case of neuronal morphologies. *Nat. Rev. Neurosci.* **7**, 318–324.

Ascoli G.A., Krichmar J.L., Nasuto S.J. and Senft S.L. (2001). Generation, description and storage of dendritic morphology data. *Philos. Trans. R. Soc. Lond. B Biol. Sci.* **356(1412)**, 1131–1145.

Attwell D. and Laughlin S.B. (2001). An energy budget for signaling in the grey matter of the brain. *J. Cereb. Blood Flow Metab.* **21**, 1133–1145.

Auerbach A.A. and Bennett M.V.L. (1969). A rectifying electronic synapse in the central nervous system of a vertebrate. *J. Gen. Physiol.* **53**, 211–237.

Badoual M., Zou Q., Davison A.P., Rudolph M., Bal T., Frégnac Y. and Destexhe A. (2006). Biophysical and phenomenological models of multiple spike interactions in spike-timing dependent plasticity. *Int. J. Neural Sys.* **16**, 79–97.

Baek M., DiMaio F., Anishchenko I., Dauparas J., Ovchinnikov S., Lee G.R., Wang J., Cong Q., Kinch L.N., Schaeffer R.D. et al. (2021). Accurate prediction of protein structures and interactions using a three-track neural network. *Science* **373**, 871–876.

Baker P.F., Hodgkin A.L. and Ridgway E.B. (1971). Depolarization and calcium entry in squid giant axons. *J. Physiol.* **218**, 709–755.

Bando Y., Grimm C., Cornejo V.H. and Yuste R. (2019). Genetic voltage indicators. *BMC Biol.* **17**, 71.

Bangera N.B., Schomer D.L., Dehghani N., Ulbert I., Cash S., Papavasiliou S., Eisenberg S.R., Dale A.M. and Halgren E. (2010). Experimental validation of the influence of white matter anisotropy on the intracranial EEG forward solution. *J. Comput. Neurosci.* **29**, 371–387.

Barbieri F., Trauchessec V., Caruso L., Trejo-Rosillo J., Telenczuk B., Paul E., Bal T., Destexhe A., Fermon C., Pannetier-Lecoeur M. et al. (2016). Local recording of biological magnetic fields using giant magneto resistance-based micro-probes. *Sci. Rep.* **6**, 39330.

Barbour B. and Häusser M. (1997). Intersynaptic diffusion of neurotransmitter. *Trends Neurosci.* **20**, 377–384.

Barker A.T., Jalinous R. and Freeston I.L. (1985). Non-invasive magnetic stimulation of human motor cortex. *Lancet* **325**, 1106–1107.

Barrett A.B., Billings G.O., Morris R.G. and van Rossum M.C. (2009). State based model of long-term potentiation and synaptic tagging and capture. *PLoS Comput. Biol.* **5**, e1000259.

Barry P.H. and Lynch J.W. (1991). Liquid junction potentials and small cell effects in patch-clamp analysis. *J. Membrane Biol.* **121**, 101–117.

Bartels A., Goense J. and Logothetis N. (2012). Functional magnetic resonance imaging. In *Handbook of Neural Activity Measurement*, eds. R. Brette and A. Destexhe (Cambridge University Press, Cambridge), pp. 410–469.

Barth D.S. and Di S. (1991). Laminar excitability cycles in neocortex. *J. Neurophysiol.* **65**, 891–898.

Bartos M., Vida I. and Jonas P. (2007). Synaptic mechanisms of synchronized gamma oscillations in inhibitory interneuron networks. *Nat. Rev. Neurosci.* **8**, 45–56.

Bauer P., Thorpe A. and Brunet G. (2015). The quiet revolution of numerical weather prediction. *Nature* **525**, 47–55.

Bean B.P. (2007). The action potential in mammalian central neurons. *Nat. Rev. Neurosci.* **8**, 451–465.

Beining M., Mongiat L.A., Schwarzacher S.W., Cuntz H. and Jedlicka P. (2017). T2n as a new tool for robust electrophysiological modeling demonstrated for mature and adult-born dentate granule cells. *Elife* **6**, e26517.

Bell A.J. (1992). Self-organisation in real neurons: anti-Hebb in 'channel space'? In *Neural Information Processing Systems 4*, eds. J.E. Moody, S.J. Hanson and R.P. Lippmann (Morgan Kaufmann, San Mateo, CA), pp. 35–42.

Ben-Yishai R., Bar-Or R.L. and Sompolinsky H. (1995). Theory of orientation tuning in visual cortex. *Proc. Natl. Acad. Sci. USA* **92**, 3844–3848.

Benabid A.L., Chabardes S., Mitrofanis J. and Pollak P. (2009). Deep brain stimulation of the subthalamic nucleus for the treatment of Parkinson's disease. *Lancet Neurol.* **8**, 67–81.

Benabid A.L., Pollak P., Hoffmann D., Gervason C., Hommel M., Perret J., De Rougemont J. and Gao D. (1991). Long-term suppression of tremor by chronic stimulation of the ventral intermediate thalamic nucleus. *Lancet* **337**, 403–406.

Bender K.J. and Trussell L.O. (2012). The physiology of the axon initial segment. *Annu. Rev. Neurosci.* **35**, 249–265.

Bennett M.R., Farnell L. and Gibson W.G. (2000a). The probability of quantal secretion near a single calcium channel of an active zone. *Biophys. J.* **78**, 2201–2221.

Bennett M.R., Farnell L. and Gibson W.G. (2000b). The probability of quantal secretion within an array of calcium channels of an active zone. *Biophys. J.* **78**, 2222–2240.

Bennett M.R., Gibson W.G. and Robinson J. (1994). Dynamics of the CA3 pyramidal neuron autoassociative memory network in the hippocampus. *Philos. Trans. R. Soc. Lond. B Biol. Sci.* **343**, 167–187.

Bennett M.R. and Robinson J. (1989). Growth and elimination of nerve terminals at synaptic sites during polyneuronal innervation of muscle cells: a trophic hypothesis. *Proc. R. Soc. Lond. B Biol. Sci.* **235**, 299–320.

Bennett M.V.L. and Zukin R.S. (2004). Electrical coupling and neuronal synchronization in the mammalian brain. *Neuron* **41**, 495–511.

Benzi R., Sutera A. and Vulpiani A. (1981). The mechanism of stochastic resonance. *J. Phys. A Math. Gen.* **14**, L453–L457.

Berridge M.J. (1998). Neuronal calcium signalling. *Neuron* **21**, 13–26.

Berridge M.J., Bootman M.D. and Roderick H.L. (2003). Calcium signalling: dynamics, homeostasis and remodelling. *Nat. Rev. Mol. Cell Biol.* **4**, 517–529.

Berry M. and Bradley P.M. (1976). The application of network analysis to the study of branching patterns of large dendritic trees. *Brain Res.* **109**, 111–132.

Bertozzi A.L., Franco E., Mohler G., Short M.B. and Sledge D. (2020). The challenges of modeling and forecasting the spread of COVID-19. *Proc. Natl. Acad. Sci. USA* **117**, 16732–16738.

Bertram R., Sherman A. and Stanley E.F. (1996). Single-domain/bound calcium hypothesis of transmitter release and facilitation. *J. Neurophysiol.* **75**, 1919–1931.

Betz W.J. (1970). Depression of transmitter release at the neuromuscular junction of the frog. *J. Physiol.* **206**, 620–644.

Betz W.J., Caldwell J.H. and Ribchester R.R. (1980). The effects of partial denervation at birth on the development of muscle fibres and motor units in rat lumbrical muscle. *J. Physiol.* **303**, 265–279.

Beurle R.L. (1956). Properties of a mass of cells capable of regenerating pulses. *Philos. Trans. R. Soc. Lond. B Biol. Sci.* **240**, 55–94.

Beuter A., Lefaucheur J.P. and Modolo J. (2014). Closed-loop cortical neuromodulation in Parkinson's disease: An alternative to deep brain stimulation? *Clin. Neurophysiol.* **125**, 874–885.

Bevan M.D., Magill P.J., Terman D., Bolam J.P. and Wilson C.J. (2002). Move to the rhythm: oscillations in the subthalamic nucleus–external globus pallidus network. *Trends Neurosci.* **25**, 525–531.

Bevins N., Lemke G. and Reber M. (2011). Genetic dissection of EphA receptor signaling dynamics during retinotopic mapping. *J. Neurosci.* **31**, 10302–10310.

Bezaire M.J., Raikov I., Burk K., Vyas D. and Soltesz I. (2016). Interneuronal mechanisms of hippocampal theta oscillations in a full-scale model of the rodent CA1 circuit. *Elife* **5**, e18566.

Bezanilla F. and Armstrong C.M. (1977). Inactivation of the sodium channel. I. Sodium current experiments. *J. Gen. Physiol.* **70**, 549–566.

Bhalla U.S. (1998). The network within: signaling pathways. In *The Book of GENESIS: Exploring Realistic Neural Models with the General Neural Simulation System*, eds. J.M. Bower and D. Beeman (Springer-Verlag, New York, NY), chapter 10, pp. 169–192. Second edition.

Bhalla U.S. (2001). Modeling networks of signaling pathways. In *Computational Neuroscience: Realistic Modeling for Experimentalists*, ed. E. De Schutter (CRC Press, Boca Raton, FL), chapter 2, pp. 25–48.

Bhalla U.S. (2004a). Models of cell signaling pathways. *Curr. Opin. Genet. Dev.* **14**, 375–381.

Bhalla U.S. (2004b). Signaling in small subcellular volumes. I. Stochastic and diffusion effects on individual pathways. *Biophys. J.* **87**, 733–744.

Bhalla U.S. (2004c). Signaling in small subcellular volumes. II. Stochastic and diffusion effects on synaptic network properties. *Biophys. J.* **87**, 745–753.

Bhalla U.S. (2014). Molecular computation in neurons: a modeling perspective. *Curr. Opin. Neurobiol.* **25**, 31–37.

Bhalla U.S. and Bower J.M. (1993). Exploring parameter space in detailed single neuron models: simulations of the mitral and granule cells of the olfactory bulb. *J. Neurophysiol.* **60**, 1948–1965.

Bhalla U.S. and Iyengar R. (1999). Emergent properties of networks of biological signaling pathways. *Science* **283**, 381–387.

Bi G.Q. and Poo M.M. (1998). Synaptic modifications in cultured hippocampal neurons: dependence on spike timing, synaptic strength, and postsynaptic cell type. *J. Neurosci.* **18**, 10464–10472.

Bienenstock E.L., Cooper L.N. and Munro P.W. (1982). Theory for the development of neuron selectivity: orientation specificity and binocular interaction in visual cortex. *J. Neurosci.* **2**, 32–48.

Billeh Y.N., Cai B., Gratiy S.L., Dai K., Iyer R., Gouwens N.W., Abbasi-Asl R., Jia X., Siegle J.H., Olsen S.R., Koch C., Mihalas S. and Arkhipov A. (2020). Systematic integration of structural and functional data into multi-scale models of mouse primary visual cortex. *Neuron* **106**, 388–403.

Billups B., Graham B.P., Wong A.Y.C. and Forsythe I.D. (2005). Unmasking group III metabotropic glutamate autoreceptor function at excitatory synapses in the rat CNS. *J. Physiol.* **565**, 885–896.

Bishop C.M. (2006). *Pattern Recognition and Machine Learning* (Springer, New York, NY).

Blackwell K.T. (2005). Modeling calcium concentration and biochemical reactions. *Brains Minds Media* **1**, bmm224.

Blackwell K.T. (2006). An efficient stochastic diffusion algorithm for modeling second messengers in dendrites and spines. *J. Neurosci. Methods* **157**, 142–153.

Blackwell K.T. and Hellgren Kotaleski J. (2002). Modeling the dynamics of second messenger pathways. In *Neuroscience Databases: A Practical Guide*, ed. R. Kötter (Kluwer, Norwell, MA), chapter 5, pp. 63–80.

Blackwell K.T. and Koh W. (2019). Stochastic simulators. In *Encyclopedia of Computational Neuroscience*, eds. D. Jaeger and R. Jung (Springer-Verlag, New York, NY), pp. 1–10.

Blagoev K., Mihaila B., Travis B., Alexandrov L., Bishop A., Ranken D., Posse S., Gasparovic C., Mayer A., Aine C. et al. (2007). Modelling the magnetic signature of neuronal tissue. *NeuroImage* **37**, 137–148.

Blakemore C. and Van Sluyters R.C. (1975). Innate and environmental factors in the development of the kitten's visual cortex. *J. Physiol.* **248**, 663–716.

Blaustein A.P. and Hodgkin A.L. (1969). The effect of cyanide on the efflux of calcium from squid axons. *J. Physiol.* **200**, 497–527.

Blinder P., Tsai P.S., Kaufhold J.P., Knutsen P.M., Suhl H. and Kleinfeld D. (2013). The cortical angiome: an interconnected vascular network with noncolumnar patterns of blood flow. *Nat. Neurosci.* **16**, 889–897.

Bliss T.V. and Lømo T. (1973). Long-lasting potentiation of synaptic transmission in the dentate area of the anaesthetized rabbit following stimulation of the perforant path. *J. Physiol.* **232**, 331–356.

Blomquist P., Devor A., Indahl U.G., Ulbert I., Einevoll G.T. and Dale A.M. (2009). Estimation of thalamocortical and intracortical network models from joint thalamic single-electrode and cortical laminar-electrode recordings in the rat barrel system. *PLoS Comput. Biol.* **5**, e1000328.

Blomquist P., Wyller J. and Einevoll G.T. (2005). Localized activity patterns in two-population neuronal networks. *Physica D: Nonlinear Phenomena* **206**, 180–212.

Bloomfield S.A., Hamos J.E. and Sherman S.M. (1987). Passive cable properties and morphological correlates in neurons of the lateral geniculate nucleus of the cat. *J. Physiol.* **383**, 653–668.

Boas D.A., Jones S.R., Devor A., Huppert T.J. and Dale A.M. (2008). A vascular anatomical network model of the spatio-temporal response to brain activation. *NeuroImage* **40**, 1116–1129.

Bokil H., Laaris N., Blinder K., Ennis M. and Keller A. (2001). Ephaptic interactions in the mammalian olfactory system. *J. Neurosci.* **21**, RC173.

Bollmann J.H. (2019). The zebrafish visual system: from circuits to behavior. *Annu. Rev. Vis. Sci.* **5**, 269–293.

Booth V. and Rinzel J. (1995). A minimal, compartmental model for a dendritic origin of bistability of motoneuron firing patterns. *J. Comput. Neurosci.* **2**, 299–312.

Borg-Graham L.J. (1989). Modelling the somatic electrical response of hippocampal pyramidal neurons. Technical Report AITR-1161, MIT AI Laboratory, Cambridge, MA.

Borg-Graham L.J. (1999). Interpretations of data and mechanisms for hippocampal pyramidal cell models. In *Cerebral Cortex, Volume 13: Models of Cortical Circuits*, eds. P.S. Ulinski, E.G. Jones and A. Peters (Plenum Publishers, New York, NY), pp. 19–138.

Bormann G., Brosens F. and De Schutter E. (2001). Diffusion. In *Computational Modeling of Genetic and Biochemical Networks*, eds. J.M. Bower and H. Bolouri (MIT Press, Cambridge, MA), chapter 7, pp. 189–224.

Born G., Schneider-Soupiadis F.A., Erisken S., Vaiceliunaite A., Lao C.L., Mobarhan M.H., Spacek M.A., Einevoll G.T. and Busse L. (2021). Corticothalamic feedback sculpts visual spatial integration in mouse thalamus. *Nat. Neurosci.* **24**, 1711–1720.

Bos H., Diesmann M. and Helias M. (2016). Identifying anatomical origins of coexisting oscillations in the cortical microcircuit. *PLoS Comput. Biol.* **12**, e1005132.

Boutillier P., Maasha M., Li X., Medina-Abarca H.F., Krivine J., Feret J., Cristescu I., Forbes A.G. and Fontana W. (2018). The Kappa platform for rule-based modeling. *Bioinformatics* **34**, i583–i592.

Bove F., Mulas D., Cavallieri F., Castrioto A., Chabardès S., Meoni S., Schmitt E., Bichon A., Di Stasio E., Kistner A. et al. (2021). Long-term outcomes (15 years) after subthalamic nucleus deep brain stimulation in patients with Parkinson disease. *Neurology* **97**, e254–e262.

Brette R. (2006). Exact simulation of integrate-and-fire models with synaptic conductances. *Neural Comput.* **18**, 2004–2027.

Brette R. and Destexhe A., eds. (2012). *Handbook of Neural Activity Measurement* (Cambridge University Press, Cambridge).

Brette R. and Gerstner W. (2005). Adaptive exponential integrate-and-fire model as an effective description of neuronal activity. *J. Neurophysiol.* **94**, 3637–3642.

Brette R., Piwkowska Z., Monier C., Rudolph-Lilith M., Fournier J., Levy M., Frégnac Y., Bal T. and Destexhe A. (2008). High-resolution intracellular recordings using a real-time computational model of the electrode. *Neuron* **59**, 379–391.

Brette R., Rudolph M., Carnevale T., Hines M., Beeman D., Bower J.M., Diesmann M., Morrison A., Goodman P.H., Harris Jr F.C., Zirpe M., Natschläger T., Pecevski D., Ermentrout B., Djurfeldt M., Lansner A., Rochel O., Vieville T., Muller E., Davison A.P., El Boustani S. and Destexhe A. (2007). Simulation of networks of spiking neurons: A review of tools and strategies. *J. Comput. Neurosci.* **23**, 349–398.

Brown A., Yates P.A., Burrola P., Ortuno D., Vaidya A., Jessell T.M., Pfaff S.L., O'Leary D.D. and Lemke G. (2000). Topographic mapping from the retina to the midbrain is controlled by relative but not absolute levels of EphA receptor signalling. *Cell* **102**, 77–88.

Brown A.M., Schwindt P.C. and Crill W.E. (1993). Voltage dependence and activation kinetics of pharmacologically defined components of the high-threshold calcium current in rat neocortical neurons. *J. Neurophysiol.* **70**, 1530–1543.

Brown D.A. and Griffith W.H. (1983). Calcium-activated outward current in voltage-clamped hippocampal neurones of the guinea-pig. *J. Physiol.* **337**, 287–301.

Brown M.C., Jansen J.K.S. and van Essen D.C. (1976). Polyneuronal innervation of skeletal muscle in new-born rats and its elimination during maturation. *J. Physiol.* **261**, 387–422.

Brunel N. (2000). Dynamics of sparsely connected networks of excitatory and inhibitory spiking neurons. *J. Comput. Neurosci.* **8**, 183–208.

Buccino A.P. and Einevoll G.T. (2021). MEArec: A fast and customizable testbench simulator for ground-truth extracellular spiking activity. *Neuroinformatics* **19**, 185–204.

Buccino A.P., Kordovan M., Ness T.V., Merkt B., Häfliger P.D., Fyhn M., Cauwenberghs G., Rotter S. and Einevoll G.T. (2018). Combining biophysical modeling and deep learning for multielectrode array neuron localization and classification. *J. Neurophysiol.* **120**, 1212–1232.

Buccino A.P., Kuchta M., Jæger K.H., Ness T.V., Berthet P., Mardal K.A., Cauwenberghs G. and Tveito A. (2019). How does the presence of neural probes affect extracellular potentials? *J. Neural Eng.* **16**, 026030.

Buckingham J. and Willshaw D. (1993). On setting unit thresholds in an incompletely connected associative net. *Network Comp. Neural Syst.* **4**, 441–459.

Buckingham J.T. (1991). Delicate nets, faint recollections: a study of partially connected associative network memories. Ph.D. thesis, University of Edinburgh, Edinburgh.

Buffelli M., Busetto G., Cangiano L. and Cangiano A. (2002). Perinatal switch from synchronous to asynchronous activity of motoneurons: link with synapse elimination. *Proc. Natl. Acad. Sci. USA* **99**, 13200–13205.

Buice M.A. and Cowan J.D. (2009). Statistical mechanics of the neocortex. *Prog. Biophys. Mol. Biol.* **99**, 53–86.

Burgoyne R.D. (2007). Neuronal calcium sensor proteins: generating diversity in neuronal Ca^{2+} signalling. *Nat. Rev. Neurosci.* **8**, 182–193.

Burke R.E. and Marks W.B. (2002). Some approaches to quantitative dendritic morphology. In *Computational Neuroanatomy: Principles and Methods*, ed. G.A. Ascoli (The Humana Press Inc., Totawa, NJ), chapter 2, pp. 27–48.

Burke R.E., Marks W.B. and Ulfhake B. (1992). A parsimonious description of motoneuron dendritic morphology using computer simulation. *J. Neurosci.* **12**, 2403–2416.

Busskamp V., Duebel J., Balya D., Fradot M., Viney T.J., Siegert S., Groner A.C., Cabuy E., Forster V., Seeliger M., Biel M., Humphries P., Paques M., Mohand-Said S., Trono D., Deisseroth K., Sahel J.A., Picaud S. and Roska B. (2010). Genetic reactivation of cone photoreceptors restores visual responses in retinitis pigmentosa. *Science* **329**, 413–417.

Butson C.R. and McIntyre C.C. (2005). Tissue and electrode capacitance reduce neural activation volumes during deep brain stimulation. *Clin. Neurophysiol.* **116**, 2490–2500.

Butson C.R. and McIntyre C.C. (2008). Current steering to control the volume of tissue activated during deep brain stimulation. *Brain Stimul.* **1**, 7–15.

Byrne A., O'Dea R.D., Forrester M., Ross J. and Coombes S. (2020). Next-generation neural mass and field modeling. *J. Neurophysiol.* **123**, 726–742.

Cain N., Iyer R., Koch C. and Mihalas S. (2016). The computational properties of a simplified cortical column model. *PLoS Comput. Biol.* **12**, e1005045.

Camuñas-Mesa L.A. and Quiroga R.Q. (2013). A detailed and fast model of extracellular recordings. *Neural Comput.* **25**, 1191–1212.

Cang J., Wang L., Stryker M.P. and Feldheim D.A. (2008). Roles of ephrin-As and structured activity in the development of functional maps in the superior colliculus. *J. Neurosci.* **28**, 11015–11023.

Cannon R.C., O'Donnell C. and Nolan M.F. (2010). Stochastic ion channel gating in dendritic neurons: morphology dependence and probabilistic synaptic activation of dendritic spikes. *PLoS Comput. Biol.* **6**, e1000886.

Cannon R.C. and D'Alessandro G. (2006). The ion channel inverse problem: neuroinformatics meets biophysics. *PLoS Comput. Biol.* **2**, 862–868.

Cannon R.C., Gleeson P., Crook S., Ganapathy G., Marin B., Piasini E. and Silver R.A. (2014). LEMS: a language for expressing complex biological models in concise and hierarchical form and its use in underpinning NeuroML 2. *Front. Neuroinform.* **8**, 79.

Cannon R.C., Turner D.A., Pyapali G.K. and Wheal H.V. (1998). An on-line archive of reconstructed hippocampal neurons. *J. Neurosci. Methods* **84**, 49–54.

Carnevale N.T. and Hines M.L. (2006). *The NEURON Book* (Cambridge University Press, Cambridge).

Carriquiry A.L., Ireland W.P., Kliemann W. and Uemura E. (1991). Statistical evaluation of dendritic growth models. *Bull. Math. Biol.* **53**, 579–589.

Caruso L., Wunderle T., Lewis C.M., Valadeiro J., Trauchessec V., Rosillo J.T., Amaral J.P., Ni J., Jendritza P., Fermon C., Cardoso S., Freitas P.P., Fries P. and Pannetier-Lecoeur M. (2017). In vivo magnetic recording of neuronal activity. *Neuron* **95**, 1283–1291.e4.

Castellani G.C., Quinlan E.M., Bersani F., Cooper L.N. and Shouval H.Z. (2005). A model of bidirectional synaptic plasticity: from signaling network to channel conductance. *Learn. Mem.* **12**, 423–432.

Catterall W.A., Dib-Hajj S., Meisler M.H. and Pietrobon D. (2008). Inherited neuronal ion channelopathies: new windows on complex neurological diseases. *J. Neurosci.* **28**, 11768–11777.

Catterall W.A., Goldin A.L. and Waxman S.G. (2005a). International Union of Pharmacology. XLVII. Nomenclature and structure–function relationships of voltage-gated sodium channels. *Pharmacol. Rev.* **57**, 397–409.

Catterall W.A., Lenaeus M.J. and Gamal El-Din T.M. (2020). Structure and pharmacology of voltage-gated sodium and calcium channels. *Annu. Rev. Pharmacol. Toxicol.* **60**, 133–154.

Catterall W.A., Perez-Reyes E., Snutch T.P. and Striessnig J. (2005b). International Union of Pharmacology. XLVIII. Nomenclature and structure–function relationships of voltage-gated calcium channels. *Pharmacol. Rev.* **57**, 411–425.

Cellerino A., Novelli E. and Galli-Resta L. (2000). Retinal ganglion cells with NADPH-diaphorase activity in the chick form a regular mosaic with a strong dorsoventral asymmetry that can be modeled by a minimal spacing rule. *Eur. J. Neurosci.* **12**, 613–620.

Chang C.C. (1999). Looking back on the discovery of α-bungarotoxin. *J. Biomed. Sci.* **6**, 368–375.

Chawla D.S. (2020). Critiqued coronavirus simulation gets thumbs up from code-checking efforts. *Nature* **582**, 323–325.

Chemla S. and Chavane F. (2012). Voltage-sensitive dye imaging. In *Handbook of Neural Activity Measurement*, eds. R. Brette and A. Destexhe (Cambridge University Press, Cambridge), pp. 327–361.

Chen K.C. and Nicholson C. (2000). Spatial buffering of potassium ions in brain extracellular space. *Biophys. J.* **78**, 2776–2797.

Cheng H.J., Nakamoto M., Bergemann A.D. and Flanagan J.G. (1995). Complementary gradients in expression and binding of ELF-1 and Mek4 in development of the topographic retinotectal projection map. *Cell* **82**, 371–381.

Cherniak C., Changizi M. and Kang D.W. (1999). Large-scale optimization of neuronal arbors. *Phys. Rev. E* **59**, 6001–6009.

Cherniak C., Mokhtarzada Z. and Nodelman U. (2002). Optimal-wiring models of neuroanatomy. In *Computational Neuroanatomy: Principles and Methods*, ed. G.A. Ascoli (The Humana Press Inc, Totawa, NJ), chapter 4, pp. 71–82.

Chow C.C. and White J.A. (1996). Spontaneous action potentials due to channel fluctuations. *Biophys. J.* **71**, 3013–3021.

Chung S.H. (1974). In search of the rules for nerve connections. *Cell* **3**, 201–205.

Churchland P.S. and Sejnowski T.J. (1992). *The Computational Brain* (MIT Press, Cambridge, MA).

Chylek L.A., Harris L.A., Faeder J.R. and Hlavacek W.S. (2015). Modeling for (physical) biologists: an introduction to the rule-based approach. *Phys. Biol.* **12**, 045007.

Clapham D.E., Julius D., Montell C. and Schultz G. (2005). International Union of Pharmacology. XLIX. Nomenclature and structure–function relationships of transient receptor potential channels. *Pharmacol. Rev.* **57**, 427–450.

Clark G.M., Black R., Dewhurst D.J., Forster I.C., Patrick J.F. and Tong Y.C. (1977). A multiple-electrode hearing prosthesis for cochlea implantation in deaf patients. *Med. Prog. Technol.* **5**, 127–140.

Clements J.D., Lester R.A., Tong G., Jahr C.E. and Westbrook G.L. (1992). The time course of glutamate in the synaptic cleft. *Science* **258**, 1498–1501.

Clements J.D. and Redman S.J. (1989). Cable properties of cat spinal motorneurons measured by combining voltage clamp, current clamp and intracellular staining. *J. Physiol.* **409**, 63–87.

Clerx M., Beattie K.A., Gavaghan D.J. and Mirams G.R. (2019). Four ways to fit an ion channel model. *Biophys. J.* **117**, 2420–2437.

Clopath C., Büsing L., Vasilaki E. and Gerstner W. (2010). Connectivity reflects coding: a model of voltage-based STDP with homeostasis. *Nat. Neurosci.* **13**, 344–352.

Clopath C., Ziegler L., Vasilaki E., Büsing L. and Gerstner W. (2008). Tag-trigger-consolidation: A model of early and late long-term-potentiation and depression. *PLoS Comput. Biol.* **4**, e10000248.

Coggan J.S., Bartol T.M., Esquenazi E., Stiles J.R., Lamont S., Martone M.E., Berg D.K., Ellisman M.H. and Sejnowski T.J. (2005). Evidence for ectopic neurotransmission at a neuronal synapse. *Science* **309**, 446–451.

Cole K.S. (1968). *Membranes, Ions, and Impulses: A Chapter of Classical Biophysics* (University of California Press, Berkeley, CA).

Collier J.E., Monk N.A.M., Maini P.K. and Lewis J.H. (1996). Pattern formation with lateral feedback: a mathematical model of Delta-Notch intracellular signalling. *J. Theor. Biol.* **183**, 429–446.

Colquhoun D., Hatton C.J. and Hawkes A.G. (2003). The quality of maximum likelihood estimates of ion channel rate constants. *J. Physiol.* **547**, 699–728.

Colquhoun D., Hawkes A.G. and Srodzinski K. (1996). Joint distributions of apparent open times and shut times of single ion channels and the maximum likelihood fitting of mechanisms. *Philos. Trans. R. Soc. Lond. A* **354**, 2555–2590.

Connor J.A. and Stevens C.F. (1971a). Inward and delayed outward membrane currents in isolated neural somata under voltage clamp. *J. Physiol.* **213**, 1–19.

Connor J.A. and Stevens C.F. (1971b). Prediction of repetitive firing behaviour from voltage clamp data on an isolated neurone soma. *J. Physiol.* **213**, 31–53.

Connor J.A. and Stevens C.F. (1971c). Voltage clamp studies of a transient outward membrane current in gastropod neural somata. *J. Physiol.* **213**, 21–30.

Connor J.A., Walter D. and Mckown R. (1977). Neural repetitive firing: modifications of the Hodgkin–Huxley axon suggested by experimental results from crustacean axons. *Biophys. J.* **18**, 81–102.

Connors B.W. and Long M.A. (2004). Electrical synapses in the mammalian brain. *Annu. Rev. Neurosci.* **27**, 393–418.

Constantine-Paton M. and Law M.I. (1978). Eye-specific termination bands in tecta of three-eyed frogs. *Science* **202**, 639–641.

Cook J.E. and Chalupa L.M. (2000). Retinal mosaics: new insights into an old concept. *Trends Neurosci.* **23**, 26–34.

Cook J.E. and Rankin E.C.C. (1986). Impaired refinement of the regenerated retinotectal projection of the goldfish in stroboscopic light: a quantitative WGA-HRP study. *Exp. Brain Res.* **63**, 421–430.

Coombes S. (2005). Waves, bumps, and patterns in neural field theories. *Biol. Cybern.* **93**, 91–108.

Coombes S., beim Graben P., Potthast R. and Wright J. (2014). *Neural Fields: Theory and Applications* (Springer, Berlin).

Coombs J.S., Curtis D.R. and Eccles J.C. (1956). Time courses of motoneuronal responses. *Nature* **178**, 1049–1050.

Corry B., Kuyucak S. and Chung S.H. (2000a). Invalidity of continuum theories of electrolytes in nanopores. *Chem. Phys. Lett.* **320**, 35–41.

Corry B., Kuyucak S. and Chung S.H. (2000b). Tests of continuum theories as models of ion channels. II. Poisson–Nernst–Planck theory versus Brownian dynamics. *Biophys. J.* **78**, 2364–2381.

Costa R.P., Froemke R.C., Sjöström P.J. and van Rossum M.C.W. (2015). Unified pre- and postsynaptic long-term plasticity enables reliable and flexible learning. *Elife* **4**, e09457.

Costa R.P., Mizusaki B.E., Sjöström P.J. and van Rossum M.C. (2017). Functional consequences of pre- and postsynaptic expression of synaptic plasticity. *Philos. Trans. R. Soc. Lond. B Biol. Sci.* **372**, 20160153.

Cowan J.D. and Sharp D.H. (1988). Neural nets. *Q. Rev. Biophys.* **21**, 365–427.

Crank J. (1975). *The Mathematics of Diffusion* (Oxford University Press, Oxford).

Crank J. and Nicholson P. (1947). A practical method for numerical evaluation of solutions of partial differential equations of the heat-conduction type. *P. Camb. Philos. Soc.* **43**, 50–67.

Crepel F., Delhaye-Bouchaud N., Guastavino J.M. and Sampaio I. (1980). Multiple innervation of cerebellar Purkinje cells by climbing fibres in staggerer mutant mouse. *Nature* **283**, 483–484.

Cui J., Cox D.H. and Aldrich R.W. (1997). Intrinsic voltage dependence and Ca^{2+} regulation of *mslo* large conductance Ca-activated K^+ channels. *J. Gen. Physiol.* **109**, 647–673.

Cuntz H., Forstner F., Borst A. and Hausser M. (2010). One rule to grow them all: A general theory of neuronal branching and its practical application. *PLoS Comput. Biol.* **6**, e1000877.

D'Angelo E. (2016). Challenging Marr's theory of the cerebellum. In *Computational Theories and Their Implementation in the Brain: The Legacy of David Marr*, eds. L.M. Vaina and R.E. Passingham (Oxford University Press, Oxford), pp. 62–78.

Danos V., Feret J., Fontana W. and Krivine J. (2007). Scalable simulation of cellular signaling networks. In *Programming Languages and Systems*, ed. Z. Shao (Springer, Berlin, Heidelberg), volume 4807 of *Lecture Notes in Computer Science*, chapter 10, pp. 139–157.

Davis G.W. (2006). Homeostatic control of neural activity: from phenomenology to molecular design. *Annu. Rev. Neurosci.* **29**, 307–323.

Davison A.P., Brüderle D., Eppler J.M., Kremkow J., Muller E., Pecevski D., Perrinet L. and Yger P. (2009). PyNN: a common interface for neuronal network simulators. *Front. Neuroinform.* **2**, 11.

Day M., Wang Z., Ding J., An X., Ingham C.A., Shering A.F., Wokosin D., Ilijic E., Sun Z., Sampson A.R., Mugnaini E., Deutch A.Y., Sesack S.R., Arbuthnott G.W. and Surmeier D.J. (2006). Selective elimination of glutamatergic synapses on striatopallidal neurons in Parkinson disease models. *Nat. Neurosci.* **9**, 251–259.

Dayan P. and Abbott L.F. (2001). *Theoretical Neuroscience: Computational and Mathematical Modeling of Neural Systems* (MIT Press, Cambridge, MA).

Dayan P. and Willshaw D.J. (1991). Optimising synaptic learning rules in linear associative memories. *Biol. Cybern.* **65**, 253–265.

De Schutter E. and Bower J.M. (1994a). An active membrane model of the cerebellar Purkinje cell: I. Simulation of current clamps in slice. *J. Neurophysiol.* **71**, 375–400.

De Schutter E. and Bower J.M. (1994b). An active membrane model of the cerebellar Purkinje cell: II. Simulation of synaptic responses. *J. Neurophysiol.* **71**, 401–419.

De Schutter E. and Smolen P. (1998). Calcium dynamics in large neuronal models. In *Methods in Neuronal Modeling: From Ions to Networks*, eds. C. Koch and I. Segev (MIT Press, Cambridge, MA), chapter 6, pp. 211–250. Second edition.

De Young G.W. and Keizer J. (1992). A single-pool inositol 1,4,5-trisphosphate-receptor-based model for agonist-stimulated oscillations in Ca^{2+} concentration. *Proc. Natl. Acad. Sci. USA* **89**, 9895–9899.

Deb K., Pratap A., Agarwal S. and Meyarivan T. (2002). A fast and elitist multiobjective genetic algorithm: NSGA-II. *IEEE Trans. Evolut. Comput.* **6**, 182–197.

Deger M., Seeholzer A. and Gerstner W. (2018). Multicontact co-operativity in spike-timing-dependent structural plasticity stabilizes networks. *Cereb. Cortex* **28**, 1396–1415.

Dejean C., Hyland B. and Arbuthnott G. (2009). Cortical effects of subthalamic stimulation correlate with behavioral recovery from dopamine antagonist induced akinesia. *Cereb. Cortex* **19**, 1055–1063.

Del Castillo J. and Katz B. (1954a). Quantal components of the end-plate potential. *J. Physiol.* **124**, 560–573.

Del Castillo J. and Katz B. (1954b). Statistical factors involved in neuromuscular facilitation and depression. *J. Physiol.* **124**, 574–585.

DeMarco K.R., Bekker S. and Vorobyov I. (2019). Challenges and advances in atomistic simulations of potassium and sodium ion channel gating and permeation. *J. Physiol.* **597**, 679–698.

Destexhe A. and Bal T., eds. (2009). *Dynamic-Clamp from Principles to Applications* (Springer-Verlag, New York, NY).

Destexhe A. and Huguenard J.R. (2000). Nonlinear thermodynamic models of voltage-dependent currents. *J. Comput. Neurosci.* **9**, 259–270.

Destexhe A., Mainen Z.F. and Sejnowski T.J. (1994a). An efficient method for computing synaptic conductances based on a kinetic model of receptor binding. *Neural Comput.* **6**, 14–18.

Destexhe A., Mainen Z.F. and Sejnowski T.J. (1994b). Synthesis of models for excitable membranes, synaptic transmission and neuromodulation using a common kinetic formalism. *J. Comput. Neurosci.* **1**, 195–230.

Destexhe A., Mainen Z.F. and Sejnowski T.J. (1998). Kinetic models of synaptic transmission. In *Methods in Neuronal Modeling: From Ions to Networks*, eds. C. Koch and I. Segev (MIT Press, Cambridge, MA), chapter 1, pp. 1–25. Second edition.

Destexhe A. and Sejnowski T.J. (1995). G-protein activation kinetics and spill-over of GABA may account for differences between inhibitory synapses in the hippocampus and thalamus. *Proc. Natl. Acad. Sci. USA* **92**, 9515–9519.

Diehl S., Henningsson E. and Heyden A. (2016). Efficient simulations of tubulin-driven axonal growth. *J. Comput. Neurosci.* **41**, 45–63.

Diehl S., Henningsson E., Heyden A. and Perna S. (2014). A one-dimensional moving-boundary model for tubulin-driven axonal growth. *J. Theor. Biol.* **358**, 194–207.

Diggle P. (2002). *Statistical Analysis of Spatial Point Patterns* (second edition). (Edward Arnold, London).

Dittman J.S., Kreitzer A.C. and Regehr W.G. (2000). Interplay between facilitation, depression, and residual calcium at three presynaptic terminals. *J. Neurosci.* **20**, 1374–1385.

Dittman J.S. and Regehr W.G. (1998). Calcium dependence and recovery kinetics of presynaptic depression at the climbing fiber to Purkinje cell synapse. *J. Neurosci.* **18**, 6147–6162.

Dittrich M., Pattillo J.M., King J.D., Cho S., Stiles J.R. and Meriney S.D. (2013). An excess-calcium-binding-site model predicts neurotransmitter release at the neuromuscular junction. *Biophys. J.* **104**, 2751–2763.

Doi T., Kuroda S., Michikawa T. and Kawato M. (2005). Inositol 1,4,5-triphosphate-dependent Ca^{2+} threshold dynamics detect spike timing in cerebellar Purkinje cells. *J. Neurosci.* **25**, 950–961.

Donohue D.E. and Ascoli G.A. (2005). Local diameter fully constrains dendritic size in basal but not apical trees of CA1 pyramidal neurons. *J. Comput. Neurosci.* **19**, 223–238.

Donohue D.E., Scorcioni R. and Ascoli G.A. (2002). Generation and description of neuronal morphology using L-neuron: a case study. In *Computational Neuroanatomy: Principles and Methods*, ed. G.A. Ascoli (The Humana Press Inc., Totawa, NJ), chapter 3, pp. 49–69.

Douglas J.K., Wilkens L., Pantazelou E. and Moss F. (1993). Noise enhancement of the information transfer in crayfish mechanoreceptors by stochastic resonance. *Nature* **365**, 337–340.

Doyle D.A., Morais Cabral J., Pfuetzner R.A., Kuo A., Gulbis J.M., Cohen S.L., Chait B.T. and MacKinnon R. (1998). The structure of the potassium channel: molecular basis of K^+ conduction and selectivity. *Science* **280**, 69–77.

Druckmann S., Banitt Y., Gidon A., Schürmann F., Markram H. and Segev I. (2007). A novel multiple objective optimization framework for constraining conductance-based neuron models by experimental data. *Front. Neurosci.* **1**, 7–18.

Duncan A.L., Song W. and Sansom M.S.P. (2020). Lipid-dependent regulation of ion channels and G protein-coupled receptors: Insights from structures and simulations. *Annu. Rev. Pharmacol. Toxicol.* **60**, 31–50.

Dyhrfjeld-Johnsen J., Santhakumar V., Morgan R.J., Huerta R., Tsimring L. and Soltesz I. (2007). Topological determinants of epileptogenesis in large-scale structural and functional models of the dentate gyrus derived from experimental data. *J. Neurophysiol.* **97**, 1566–1587.

Eccles J.C., Ito M. and Szentagothai J. (1967). *The Cerebellum as a Neuronal Machine* (Springer-Verlag, Berlin).

Economo M.N., Fernandez F.R. and White J.A. (2010). Dynamic clamp: alteration of response properties and creation of virtual realities in neurophysiology. *J. Neurosci.* **30**, 2407–2413.

Edelstein–Keshet L. (1988). *Mathematical Models in Biology*. Birkhäuser Mathematics Series (McGraw-Hill, Inc., New York, NY).

Eglen S.J., Diggle P.J. and Troy J.B. (2005). Homotypic constraints dominate positioning of on- and off-center beta retinal ganglion cells. *Vis. Neurosci.* **22**, 859–871.

Eglen S.J. and Willshaw D.J. (2002). Influence on cell fate mechanisms upon retinal mosaic formation: a modelling study. *Development* **129**, 5399–5408.

Einevoll G.T., Destexhe A., Diesmann M., Grün S., Jirsa V., de Kamps M., Migliore M., Ness T.V., Plesser H.E. and Schürmann F. (2019). The scientific case for brain simulations. *Neuron* **102**, 735–744.

Einevoll G.T., Franke F., Hagen E., Pouzat C. and Harris K.D. (2012). Towards reliable spike-train recordings from thousands of neurons with multielectrodes. *Curr. Opin. Neurobiol.* **22**, 11–17.

Einevoll G.T. and Heggelund P. (2000). Mathematical models for the spatial receptive-field organization of nonlagged X-cells in dorsal lateral geniculate nucleus of cat. *Vis. Neurosci.* **17**, 871–886.

Einevoll G.T., Kayser C., Logothetis N.K. and Panzeri S. (2013a). Modelling and analysis of local field potentials for studying the function of cortical circuits. *Nat. Rev. Neurosci.* **14**, 770–785.

Einevoll G.T., Lindén H., Tetzlaff T., Łęski S. and Pettersen K.H. (2013b). Local field potential: biophysical origin and analysis. In *Principles of Neural Coding*, eds. R.Q. Quiroga and S. Panzeri (CRC Press, Boca Raton, FL), pp. 37–59.

Einevoll G.T., Pettersen K.H., Devor A., Ulbert I., Halgren E. and Dale A.M. (2007). Laminar population analysis: estimating firing rates and evoked synaptic activity from multielectrode recordings in rat barrel cortex. *J. Neurophysiol.* **97**, 2174–2190.

Einevoll G.T. and Plesser H.E. (2012). Extended difference-of-Gaussians model incorporating cortical feedback for relay cells in the lateral geniculate nucleus of cat. *Cognitive Neurodynamics* **6**, 307–324.

Einstein A. (1905). Über die von der molekulartheoretischen Theorie der Wärme geforderte Bewegung von in ruhenden Flüssigkeiten suspendierten Teilchen. *Ann. Phys. Leipzig* **17**, 549.

Ellingsrud A.J., Solbrå A., Einevoll G.T., Halnes G. and Rognes M.E. (2020). Finite element simulation of ionic electrodiffusion in cellular geometries. *Front. Neuroinform.* **14**, 11.

Elmqvist D. and Quastel D.M.J. (1965). A quantitative study of end-plate potentials in isolated human muscle. *J. Physiol.* **178**, 505–529.

Emmenegger V., Obien M.E.J., Franke F. and Hierlemann A. (2019). Technologies to study action potential propagation with a focus on HD-MEAs. *Front. Cell. Neurosci.* **13**, 159.

Epstein M., Calderhead B., Girolami M.A. and Sivilotti L.G. (2016). Bayesian statistical inference in ion-channel models with exact missed event correction. *Biophys. J.* **111**, 333–348.

Eriksson O., Jauhiainen A., Maad Sasane S., Kramer A., Nair A.G., Sartorius C. and Hellgren Kotaleski J. (2018). Uncertainty quantification, propagation and characterization by Bayesian analysis combined with global sensitivity analysis applied to dynamical intracellular pathway models. *Bioinformatics* **35**, 284–292.

Ermentrout G.B. and Kopell N. (1998). Fine structure of neural spiking and synchronization in the presence of conduction delays. *Proc. Natl. Acad. Sci. USA* **95**, 1259–1264.

Ermentrout G.B. and Terman D.H. (2010). *Mathematical Foundations of Neuroscience*, volume 35 (Springer Science & Business Media, New York, NY).

Eyal G., Verhoog M.B., Testa-Silva G., Deitcher Y., Lodder J.C., Benavides-Piccione R., Morales J., DeFelipe J., de Kock C.P., Mansvelder H.D. and Segev I. (2016). Unique membrane properties and enhanced signal processing in human neocortical neurons. *Elife* **5**, e16553.

Eyring H. (1935). The activated complex in chemical reactions. *J. Chem. Phys.* **3**, 107–115.

Faeder J.R., Blinov M.L. and Hlavacek W.S. (2009). Rule-based modeling of biochemical systems with BioNetGen. In *Systems Biology*, ed. I.V. Maly (Humana Press, Totowa, NJ), volume 500 of *Methods in Molecular Biology*, pp. 113–167.

Fawcett J. and Willshaw D. (1982). Compound eyes project stripes on the optic tectum in *Xenopus*. *Nature* **296**, 350–352.

Feldheim D.A., Kim Y.I., Bergemann A.D., Frisén J., Barbacid M. and Flanagan J.G. (2000). Genetic analysis of ephrin-A2 and ephrin-A5 shows their requirement in multiple aspects of retinocollicular mapping. *Neuron* **25**, 563–574.

Feldheim D.A., Vanderhaeghen P., Hansen M.J., Frisén J., Lu Q., Barbacid M. and Flanagan J.G. (1998). Topographic guidance labels in a sensory projection to the forebrain. *Neuron* **21**, 1303–1313.

Feng N., Ning G. and Zheng X. (2005). A framework for simulating axon guidance. *Neurocomputing* **68**, 70–84.

Fenwick E.M., Marty A. and Neher E. (1982). Sodium and calcium channels in bovine chromaffin cells. *J. Physiol.* **331**, 599–635.

Ferster D. and Miller K.D. (2000). Neural mechanisms of orientation selectivity in the visual cortex. *Annu. Rev. Neurosci.* **23**, 441–471.

Fick A. (1855). Über Diffusion. *Ann. Phys. Chem.* **94**, 59–86.

Filho O., Silva D., Souza H., Cavalcante J., Sousa J., Ferraz F., Silva L. and Santos L. (2001). Stereotactic subthalamic nucleus lesioning for the treatment of Parkinson's disease. *Stereotact. Funct. Neurosurg.* **77**, 79–86.

Firth C.A.J.M. and Bray D. (2001). Stochastic simulation of cell signaling pathways. In *Computational Modeling of Genetic and Biochemical Networks*, eds. J.M. Bower and H. Bolouri (MIT Press, Cambridge, MA), chapter 9, pp. 263–286.

FitzHugh R. (1961). Impulses and physiological states in theoretical models of nerve membrane. *Biophys. J.* **1**, 445–466.

Fladby T. and Jansen J.K.S. (1987). Postnatal loss of synaptic terminals in the partially denervated mouse soleus muscle. *Acta Physiol. Scand.* **129**, 239–246.

Flanagan J.G. and Vanderhaeghen P. (1998). The ephrins and Eph receptors in neural development. *Annu. Rev. Neurosci.* **21**, 309–345.

Flood E., Boiteux C., Lev B., Vorobyov I. and Allen T.W. (2019). Atomistic simulations of membrane ion channel conduction, gating, and modulation. *Chem. Rev.* **119**, 7737–7832.

Földy C., Aradi I., Howard A. and Soltesz I. (2003). Diversity beyond variance: modulation of firing rates and network coherence by GABAergic subpopulations. *Eur. J. Neurosci.* **19**, 119–130.

Földy C., Dyhrfjeld-Johnsen J. and Soltesz I. (2005). Structure of cortical microcircuit theory. *J. Physiol.* **562.1**, 47–54.

Forsythe I.D., Tsujimoto T., Barnes-Davies M., Cuttle M.F. and Takahashi T. (1998). Inactivation of presynaptic calcium current contributes to synaptic depression at a fast central synapse. *Neuron* **20**, 797–807.

Fourcaud-Trocmé N., Hansel D., van Vreeswijk C. and Brunel N. (2003). How spike generation mechanisms determine the neuronal response to fluctuating inputs. *J. Neurosci.* **23**, 11628–11640.

Foutz T.J., Arlow R.L. and McIntyre C.C. (2012). Theoretical principles underlying optical stimulation of a channelrhodopsin-2 positive pyramidal neuron. *J. Neurophysiol.* **107**, 3235–3245.

Fox A.P., Nowycky M.C. and Tsien R.W. (1987). Kinetic and pharmacological properties distinguishing three types of calcium currents in chick sensory neurones. *J. Physiol.* **394**, 149–172.

Fraiman D. and Dawson S.P. (2004). A model of the IP3 receptor with a luminal calcium binding site: stochastic simulations and analysis. *Cell Calcium* **35**, 403–413.

Franke M. (2021). *An Introduction to Data Analysis* (Online book at https://michael-franke.github.io/intro-data-analysis/).

Franks K.M., Bartol T.M. and Sejnowski T.J. (2002). A Monte Carlo model reveals independent signaling at a central glutamatergic synapse. *Biophys. J.* **83**, 2333–2348.

Fraser S.E. (1981). A different adhesion approach to the patterning of neural connections. *Dev. Biol.* **79**, 453–464.

Frey U. and Morris R.G. (1997). Synaptic tagging and long-term potentiation. *Nature* **385**, 533–536.

Friedrich P., Vella M., Gulyás A.I., Freund T.F. and Káli S. (2014). A flexible, interactive software tool for fitting the parameters of neuronal models. *Front. Neuroinform.* **8**, 63.

Frisén J., Yates P.A., McLaughlin T., Friedman G.C., O'Leary D.D. and Barbacid M. (1998). Ephrin-a5 (AL-1/RAGS) is essential for proper retinal axon guidance and topographic mapping in the mammalian visual system. *Neuron* **20**, 235–243.

Froemke R.C. and Dan Y. (2002). Spike-timing-dependent synaptic modification induced by natural spike trains. *Nature* **416**, 433–438.

Fröhlich F. and McCormick D.A. (2010). Endogenous electric fields may guide neocortical network activity. *Neuron* **67**, 129–143.

Frostig R.D., ed. (2009). *In Vivo Optical Imaging of Brain Function* (CRC Press, Boca Raton, FL), second edition.

Frostig R.D. and Chen-Bee C.H. (2012). Intrinsic signal optical imaging. In *Handbook of Neural Activity Measurement*, eds. R. Brette and A. Destexhe (Cambridge University Press, Cambridge), pp. 92–135.

Fuhrmann G., Segev I., Markram H. and Tsodyks M. (2002). Coding of temporal information by activity-dependent synapses. *J. Neurophysiol.* **87**, 140–148.

Furshpan E.J. and Potter D.D. (1959). Transmission at the giant motor synapses of the crayfish. *J. Physiol.* **145**, 289–325.

Fusi S., Drew P.J. and Abbott L.F. (2005). Cascade models of synaptically stored memories. *Neuron* **45**, 599–611.

Fuster J.M. and Alexander G.E. (1971). Neuron activity related to short-term memory. *Science* **173**, 652–654.

Gabbiani F. and Koch C. (1998). Principles of spike train analysis. In *Methods in Neuronal Modeling: From Ions to Networks*, eds. C. Koch and I. Segev (MIT Press, Cambridge, MA), chapter 9, pp. 313–360. Second edition.

Gabbiani F., Midtgaard J. and Knöpfel T. (1994). Synaptic integration in a model of cerebellar granule cells. *J. Neurophysiol.* **72**, 999–1009.

Gagnon L., Smith A.F., Boas D.A., Devor A., Secomb T.W. and Sakadžić S. (2016). Modeling of cerebral oxygen transport based on in vivo microscopic imaging of microvascular network structure, blood flow, and oxygenation. *Front. Comput. Neurosci.* **10**, 82.

Galli-Resta L., Resta G., Tan S.S. and Reese B.E. (1997). Mosaics of Islet-1 expressing amacrine cells assembled by short range cellular interactions. *J. Neurosci.* **17**, 7831–7838.

García-Nafría J. and Tate C.G. (2020). Cryo-electron microscopy: Moving beyond X-ray crystal structures for drug receptors and drug development. *Annu. Rev. Pharmacol. Toxicol.* **60**, 51–71.

Gardiner C.W. (1985). *Handbook of Stochastic Methods* (Springer-Verlag, Berlin).

Gardner-Medwin A. (1983). A study of the mechanisms by which potassium moves through brain tissue in the rat. *J. Physiol.* **335**, 353–374.

Gardner-Medwin A.R. (1976). The recall of events through the learning of associations between their parts. *Proc. R. Soc. Lond. B Biol. Sci.* **194**, 375–402.

Garofalo L., da Sikva A.R. and Cuello C. (1992). Nerve growth factor induced synaptogenesis and hypertrophy of cortical cholinergic terminals. *Proc. Natl. Acad. Sci. USA* **89**, 2639–2643.

Gaze R.M. (1970). *The Formation of Nerve Connections: A Consideration of Neural Specificity Modulation and Comparable Phenomena* (Academic Press, Cambridge, MA).

Gaze R.M., Feldman J.D., Cooke J. and Chung S.H. (1979). The orientation of the visuo-tectal map in *Xenopus*: development aspects. *J. Embryol. Exp. Morphol.* **53**, 39–66.

Gaze R.M. and Hope R.A. (1983). The visuotectal projection following translocation of grafts within an optic tectum in the goldfish. *J. Physiol.* **344**, 257.

Gaze R.M., Jacobson M. and Székely G. (1963). The retinotectal projection in *Xenopus* with compound eyes. *J. Physiol.* **165**, 384–499.

Gaze R.M. and Keating M.J. (1972). The visual system and 'neuronal specificity'. *Nature* **237**, 375–378.

Gaze R.M., Keating M.J. and Chung S.H. (1974). The evolution of the retinotectal map during development in *Xenopus*. *Proc. R. Soc. Lond. B Biol. Sci.* **185**, 301–330.

Gaze R.M. and Sharma S.C. (1970). Axial differences in the reinnervation of the goldfish optic tectum by regenerating optic nerve fibres. *Exp. Brain Res.* **10**, 171–181.

Gaze R.M. and Straznicky C. (1980). Stable programming for map orientation in disarranged embryonic eyes in *Xenopus*. *J. Embryol. Exp. Morphol.* **55**, 143–165.

Geisler C.D. and Goldberg J.M. (1966). A stochastic model of the repetitive activity of neurons. *Biophys. J.* **6**, 53–69.

Gentet L.J., Stuart G.J. and Clements J.D. (2000). Direct measurement of specific membrane capacitance in neurons. *Biophys. J.* **79**, 314–320.

Gerstner W. (2000). Population dynamics of spiking neurons: fast transients, asynchronous states, and locking. *Neural Comput.* **12**, 43–89.

Gerstner W., Kistler W.M., Naud R. and Paninski L. (2014). *Neuronal Dynamics: From Single Neurons to Networks and Models of Cognition* (Cambridge University Press, Cambridge).

Gibson M.A. and Bruck J. (2000). Efficient exact simulation of chemical systems with many species and many channels. *J. Phys. Chem. A* **104**, 1876–1889.

Gierer A. (1983). Model for the retino-tectal projection. *Proc. R. Soc. Lond. B Biol. Sci.* **218**, 77–93.

Gierer A. and Meinhardt H. (1972). A theory of biological pattern formation. *Kybernetik* **12**, 30–39.

Gilbert S.F. (1997). *Developmental Biology* (Sinauer Associates, Inc., Sunderland, MA), fifth edition.

Gillespie D. (1977). Exact stochastic simulation of coupled chemical reactions. *J. Phys. Chem.* **81**, 2340–2361.

Gillespie D. (2001). Approximate accelerated stochastic simulation of chemically reacting systems. *J. Chem. Phys.* **115**, 1716–1733.

Gillies A. and Willshaw D. (2006). Membrane channel interactions underlying rat subthalamic projection neuron rhythmic and bursting activity. *J. Neurophysiol.* **95**, 2352–2365.

Gin E., Kirk V. and Sneyd J. (2006). A bifurcation analysis of calcium buffering. *J. Theor. Biol.* **242**, 1–15.

Gingrich K.J. and Byrne J.H. (1985). Simulation of synaptic depression, post-tetanic potentiation, and presynaptic facilitation of synaptic potentials from sensory neurons mediating gill-withdrawal reflex in *Aplysia*. *J. Neurophysiol.* **53**, 652–669.

Gleeson P., Cantarelli M., Marin B., Quintana A., Earnshaw M., Sadeh S., Piasini E., Birgiolas J., Cannon R.C., Cayco-Gajic N.A. et al. (2019). Open Source Brain: a collaborative resource for visualizing, analyzing, simulating, and developing standardized models of neurons and circuits. *Neuron* **103**, 395–411.

Goethals S. and Brette R. (2020). Theoretical relation between axon initial segment geometry and excitability. *Elife* **9**, e53432.

Gold C., Henze D.A., Koch C. and Buzsáki G. (2006). On the origin of the extracellular action potential waveform: A modeling study. *J. Neurophysiol.* **95**, 3113–3128.

Goldbeter A., Dupont G. and Berridge M.J. (1990). Minimal model for signal-induced Ca^{2+} oscillations and for their frequency encoding through protein phosphorylation. *Proc. Natl. Acad. Sci. USA* **87**, 1461–1465.

Golding N.L., Kath W.L. and Spruston N. (2001). Dichotomy of action-potential backpropagation in CA1 pyramidal neuron dendrites. *J. Neurophysiol.* **86**, 2998–3010.

Goldman D.E. (1943). Potential, impedance, and rectification in membranes. *J. Gen. Physiol.* **27**, 37–60.

Goldman L. and Schauf C.L. (1972). Inactivation of the sodium current in *Myxicola* giant axon. *J. Gen. Physiol.* **59**, 659–675.

Goldstein S.A., Bayliss D.A., Kim D., Lesage F., Plant L.D. and Rajan S. (2005). International Union of Pharmacology. LV. Nomenclature and molecular relationships of two-P potassium channels. *Pharmacol. Rev.* **57**, 527–540.

Goldstein S.S. and Rall W. (1974). Changes in action potential shape and velocity for changing core conductor geometry. *Biophys. J.* **14**, 731–757.

Goldwyn J.H., Mc Laughlin M., Verschooten E., Joris P.X. and Rinzel J. (2014). A model of the medial superior olive explains spatiotemporal features of local field potentials. *J. Neurosci.* **34**, 11705–11722.

Goldwyn J.H. and Rinzel J. (2016). Neuronal coupling by endogenous electric fields: cable theory and applications to coincidence detector neurons in the auditory brain stem. *J. Neurophysiol.* **115**, 2033–2051.

Golowasch J., Casey M., Abbott L. and Marder E. (1999). Network stability from activity-dependent regulation of neuronal conductances. *Neural Comput.* **11**, 1079–1096.

Goodfellow I., Bengio Y. and Courville A. (2016). *Deep Learning* (MIT Press, Cambridge, MA).

Goodhill G.J. (2018). Theoretical models of neural development. *iScience* **8**, 183–199.

Goto T., Hatanaka R., Ogawa T., Sumiyoshi A. and Riera Jorge Kawashima R. (2010). An evaluation of the conductivity profile in the somatosensory barrel cortex of Wistar rats. *J. Neurophysiol.* **104**, 3388–3412.

Gouzé J.L., Lasry J.M. and Changeux J.P. (1983). Selective stabilization of muscle innervation during development: a mathematical model. *Biol. Cybern.* **46**, 207–215.

Gradinaru V., Mogri M., Thompson K.R., Henderson J.M. and Deisseroth K. (2009). Optical deconstruction of Parkinsonian neural circuitry. *Science* **324**, 354–359.

Grado L.L., Johnson M.D. and Netoff T.I. (2018). Bayesian adaptive dual control of deep brain stimulation in a computational model of Parkinson's disease. *PLoS Comput. Biol.* **14**, e1006606.

Graham B. and Willshaw D. (1999). Probabilistic synaptic transmission in the associative net. *Neural Comput.* **11**, 117–137.

Graham B.P., Lauchlan K. and McLean D.R. (2006). Dynamics of outgrowth in a continuum model of neurite elongation. *J. Comput. Neurosci.* **20**, 43–60.

Graham B.P. and Redman S.J. (1994). A simulation of action potentials in synaptic boutons during presynaptic inhibition. *J. Neurophysiol.* **71**, 538–549.

Graham B.P. and van Ooyen A. (2001). Compartmental models of growing neurites. *Neurocomputing* **38–40**, 31–36.

Graham B.P. and van Ooyen A. (2004). Transport limited effects in a model of dendritic branching. *J. Theor. Biol.* **230**, 421–432.

Graham B.P. and van Ooyen A. (2006). Mathematical modelling and numerical simulation of the morphological development of neurons. *BMC Neurosci.* **7 (Suppl 1)**, S9.

Graham B.P. and Willshaw D.J. (1995). Improving recall from an associative memory. *Biol. Cybern.* **72**, 337–346.

Graupner M. and Brunel N. (2007). STDP in a bistable synapse model based on CaMKII and associated signaling pathways. *PLoS Comput. Biol.* **3**, e221.

Graupner M. and Brunel N. (2012). Calcium-based plasticity model explains sensitivity of synaptic changes to spike pattern, rate, and dendritic location. *Proc. Natl. Acad. Sci. USA* **109**, 3991–3996.

Green J.B. and Sharpe J. (2015). Positional information and reaction-diffusion: two big ideas in developmental biology combine. *Development* **142**, 1203–1211.

Grienberger C. and Konnerth A. (2012). Imaging calcium in neurons. *Neuron* **73**, 862–885.

Griffith J.S. (1963). A field theory of neural nets: I. Derivation of field equations. *Bull. Math. Biophys.* **25**, 111–120.

Grimes D.R., Kannan P., Warren D.R., Markelc B., Bates R., Muschel R. and Partridge M. (2016). Estimating oxygen distribution from vasculature in three-dimensional tumour tissue. *J. R. Soc. Interface* **13**, 20160070.

Grimnes S. and Martinsen O.G. (2015). *Bioimpedance and Bioelectricity Basics* (Academic Press, Cambridge, MA), third edition.

Grinvald A., Lieke E., Frostig R.D., Gilbert C.D. and Wiesel T.N. (1986). Functional architecture of cortex revealed by optical imaging of intrinsic signals. *Nature* **324**, 361–364.

Grossman N., Simiaki V., Martinet C., Toumazou C., Schultz S.R. and Nikolic K. (2013). The spatial pattern of light determines the kinetics and modulates backpropagation of optogenetic action potentials. *J. Comput. Neurosci.* **34**, 477–488.

Gunalan K., Howell B. and McIntyre C.C. (2018). Quantifying axonal responses in patient-specific models of subthalamic deep brain stimulation. *Neuroimage* **172**, 263–277.

Gutenkunst R.N., Waterfall J.J., Casey F.P., Brown K.S., Myers C.R. and Sethna J.P. (2007). Universally sloppy parameter sensitivities in systems biology models. *PLoS Comput. Biol.* **3**, e189.

Gutman G.A., Chandy K.G., Grissmer S., Lazdunski M., McKinnon D., Pardo L.A., Robertson G.A., Rudy B., Sanguinetti M.C., Stühmer W. and Wang X. (2005). International Union of Pharmacology. LIII. Nomenclature and molecular relationships of voltage-gated potassium channels. *Pharmacol. Rev.* **57**, 473–508.

Hagen E., Dahmen D., Stavrinou M.L., Lindén H., Tetzlaff T., van Albada S.J., Grün S., Diesmann M. and Einevoll G.T. (2016). Hybrid scheme for modeling local field potentials from point-neuron networks. *Cereb. Cortex* **26**, 4461–4496.

Hagen E., Fossum J.C., Pettersen K.H., Alonso J.M., Swadlow H.A. and Einevoll G.T. (2017). Focal local field potential signature of the single-axon monosynaptic thalamocortical connection. *J. Neurosci.* **37**, 5123–5143.

Hagen E., Magnusson S.H., Ness T.V., Halnes G., Babu P.N., Linssen C., Morrison A. and Einevoll G.T. (2022). Brain signal predictions from multi-scale networks using a linearized framework. *PLoS Comput. Biol.* **18**, e1010353.

Hagen E., Næss S., Ness T.V. and Einevoll G.T. (2018). Multimodal modeling of neural network activity: computing LFP, ECoG, EEG and MEG signals with LFPy2.0. *Front. Neuroinform.* **12**, 92.

Hagen E., Ness T.V., Khosrowshahi A., Sørensen C., Fyhn M., Hafting T., Franke F. and Einevoll G.T. (2015). ViSAPy: A Python tool for biophysics-based generation of virtual spiking activity for evaluation of spike-sorting algorithms. *J. Neurosci. Methods* **245**, 182–204.

Hagiwara S. and Saito N. (1959). Voltage-current relations in nerve cell membrane of *Onchidium verruculatum*. *J. Physiol.* **148**, 161–179.

Halavi M., Hamilton K.A., Parekh R. and Ascoli G.A. (2012). Digital reconstructions of neuronal morphology: three decades of research trends. *Front. Neurosci.* **6**, 49.

Halavi M., Polavaram S., Donohue D.E., Hamilton G., Hoyt J., Smith K.P. and Ascoli G.A. (2008). Neuromorpho.Org implementation of digital neuroscience: dense coverage and integration with the NIF. *Neuroinformatics* **6**, 241–252.

Hale J.K. and Koçak H. (1991). *Dynamics and Bifurcations*. Texts in Applied Mathematics (Springer-Verlag, New York, NY).

Hallett M. (2007). Transcranial magnetic stimulation: a primer. *Neuron* **55**, 187–199.

Halliwell J.V. and Adams P.R. (1982). Voltage-clamp analysis of muscarinic excitation in hippocampal neurons. *Brain Res.* **250**, 71–92.

Halnes G., Mäki-Marttunen T., Keller D., Pettersen K.H., Andreassen O.A. and Einevoll G.T. (2016). Effect of ionic diffusion on extracellular potentials in neural tissue. *PLoS Comput. Biol.* **12**, e1005193.

Halnes G., Ostby I., Pettersen K.H., Omholt S.W. and Einevoll G.T. (2013). Electrodiffusive model for astrocytic and neuronal ion concentration dynamics. *PLoS Comput. Biol.* **9**, e1003386.

Hämäläinen M., Hari R., Ilmoniemi R., Knuutila J. and Lounasmaa O.V. (1993). Magnetoencephalography – theory, instrumentation, and applications to noninvasive studies of the working human brain. *Rev. Mod. Phys* **65**, 413–497.

Hamill O.P., Marty A., Neher E., Sakmann B. and Sigworth F.J. (1981). Improved patch-clamp techniques for high-resolution current recording from cells and cell-free membrane patches. *Pflügers Arch.* **391**, 85–100.

Hansel D. and Mato G. (2000). Existence and stability of persistent states in large neuronal networks. *Phys. Rev. Lett.* **86**, 4175–4178.

Hansel D., Mato G., Meunier C. and Neltner L. (1998). On numerical simulations of integrate-and-fire neural networks. *Neural Comput.* **10**, 467–483.

Harris W.A. (1981). Neural activity and development. *Annu. Rev. Physiol.* **43**, 689–710.

Harrison E. (1989). *Darkness at Night: A Riddle of the Universe* (Harvard University Press, Cambridge, MA).

Hashimoto T., Elder C., Okun M., Patrick S. and Vitek J. (2003). Stimulation of the subthalamic nucleus changes the firing pattern of pallidal neurons. *J. Neurosci.* **23**, 1916–1923.

Hay E., Hill S., Schürmann F., Markram H. and Segev I. (2011). Models of neocortical layer 5b pyramidal cells capturing a wide range of dendritic and perisomatic active properties. *PLoS Comput. Biol.* **7**, e1002107.

Hayer A. and Bhalla U.S. (2005). Molecular switches at the synapse emerge from receptor and kinase traffic. *PLoS Comput. Biol.* **1**, e20.

Hebb D.O. (1949). *The Organization of Behavior* (Wiley, New York, NY).

Hedger G. and Sansom M.S.P. (2016). Lipid interaction sites on channels, transporters and receptors: Recent insights from molecular dynamics simulations. *Biochimica et Biophysica Acta* **1858**, 2390–2400.

Heinemann C., von Rüden, Chow R.H. and Neher E. (1993). A two-step model of secretion control in neuroendocrine cells. *Pflügers Arch.* **424**, 105–112.

Heinemann S.H., Rettig J., Graack H.R. and Pongs O. (1996). Functional characterization of Kv channel beta-subunits from rat brain. *J. Physiol.* **493 (Pt 3)**, 625–633.

Helmchen F. (2012). Calcium imaging. In *Handbook of Neural Activity Measurement*, eds. R. Brette and A. Destexhe (Cambridge University Press, Cambridge), pp. 362–409.

Hely T.A., Graham B.P. and van Ooyen A. (2001). A computational model of dendrite elongation and branching based on MAP2 phosphorylation. *J. Theor. Biol.* **210**, 375–384.

Henneman E. (1957). Relation between size of neurons and their susceptibility to discharge. *Science* **126**, 1345–1347.

Hentschel H.G.E. and van Ooyen A. (1999). Models of axon guidance and bundling during development. *Proc. R. Soc. Lond. B Biol. Sci.* **266**, 2231–2238.

Henze D.A., Borhegy Z., Csicsvari J., Mamiya A., Harris K.D. and Buzsaki G. (2000). Intracellular features predicted by extracellular recordings in the hippocampus in vivo. *J. Neurophysiol.* **84**, 390–400.

Hepburn I., Chen W., Wils S. and De Schutter E. (2012). STEPS: efficient simulation of stochastic reaction–diffusion models in realistic morphologies. *BMC Syst. Biol.* **6**, 36.

Herrmann C.S., Rach S., Neuling T. and Strüber D. (2013). Transcranial alternating current stimulation: a review of the underlying mechanisms and modulation of cognitive processes. *Front. Hum. Neurosci.* **7**, 279.

Hertz J.A., Krogh A.S. and Palmer R.G. (1991). *Introduction to the Theory of Neural Computation* (Addison-Wesley, Reading, MA).

Herz A.V., Gollisch T., Machens C.K. and Jaeger D. (2006). Modeling single-neuron dynamics and computations: A balance of detail and abstraction. *Science* **314**, 80–85.

Heurling K., Leuzy A., Jonasson M., Frick A., Zimmer E.R., Nordberg A. and Lubberink M. (2017). Quantitative positron emission tomography in brain research. *Brain Res.* **1670**, 220–234.

Hill A.V. (1936). Excitation and accommodation in nerve. *Proc. R. Soc. Lond. B Biol. Sci.* **119**, 305–355.

Hill T.L. and Chen Y.D. (1972). On the theory of ion transport across the nerve membrane, VI. Free energy and activation free energies of conformational change. *Proc. Natl. Acad. Sci. USA* **69**, 1723–1726.

Hille B. (2001). *Ion Channels of Excitable Membranes* (Sinauer Associates, Sunderland, MA), third edition.

Hillman D.E. (1979). Neuronal shape parameters and substructures as a basis of neuronal form. In *The Neurosciences: Fourth Study Program*, eds. F.O. Schmitt and F.G. Worden (MIT Press, Cambridge, MA), pp. 477–498.

Hines K.E. (2015). A primer on Bayesian inference for biophysical systems. *Biophys. J.* **108**, 2103–2113.

Hines M.L. and Carnevale N.T. (2008). Translating network models to parallel hardware in NEURON. *J. Neurosci. Methods* **169**, 425–455.

Hines M.L., Morse T., Migliore M., Carnevale N.T. and Shepherd G.M. (2004). ModelDB: A database to support computational neuroscience. *J. Comput. Neurosci.* **17**, 7–11.

Hinton G.E. and Anderson J.A., eds. (1981). *Parallel Models of Associative Memory* (Lawrence Erlbaum Associates, Hillsdale, NJ).

Hinton G.E., Sejnowski T.J. et al. (1999). *Unsupervised Learning: Foundations of Neural Computation* (MIT Press, Cambridge, MA).

Hirschberg B., Maylie J., Adelman J.P. and Marrion N.V. (1998). Gating of recombinant small-conductance Ca-activated K^+ channels by calcium. *J. Gen. Physiol.* **111**, 565–581.

Hjorth J.J., Sterratt D.C., Cutts C.S., Willshaw D.J. and Eglen S.J. (2015). Quantitative assessment of computational models for retinotopic map formation. *Dev. Neurobiol.* **75**, 641–666.

Hjorth J.J.J., Kozlov A., Carannante I., Frost Nylén J., Lindroos R., Johansson Y., Tokarska A., Dorst M.C., Suryanarayana S.M., Silberg G., Hellgren Kotaleski J. and Grillner S. (2020). The microcircuits of striatum in silico. *Proc. Natl. Acad. Sci. USA* **117**, 9554–9565.

Hjorth J.J.J., van Pelt J., Mansvelder H.D. and van Ooyen A. (2014). Competitive dynamics during resource-driven neurite outgrowth. *PLoS One* **9**, e86741.

Hodgkin A.L. (1948). The local electric changes associated with repetitive action in a non-medullated axon. *J. Physiol.* **107**, 165–181.

Hodgkin A.L. (1964). *The Conduction of the Nervous Impulse* (Charles C. Thomas, Springfield, IL).

Hodgkin A.L. (1976). Chance and design in electrophysiology: an informal account of certain experiments on nerve carried out between 1934 and 1952. *J. Physiol.* **263**, 1–21.

Hodgkin A.L. and Huxley A.F. (1952a). Currents carried by sodium and potassium ions through the membrane of the giant axon of *Loligo*. *J. Physiol.* **116**, 449–472.

Hodgkin A.L. and Huxley A.F. (1952b). The components of membrane conductance in the giant axon of *Loligo*. *J. Physiol.* **116**, 473–496.

Hodgkin A.L. and Huxley A.F. (1952c). The dual effect of membrane potential on sodium conductance in the giant axon of *Loligo*. *J. Physiol.* **116**, 497–506.

Hodgkin A.L. and Huxley A.F. (1952d). A quantitative description of membrane current and its application to conduction and excitation in nerve. *J. Physiol.* **117**, 500–544.

Hodgkin A.L., Huxley A.F. and Katz B. (1952). Measurement of current–voltage relations in the membrane of the giant axon of *Loligo*. *J. Physiol.* **116**, 424–448.

Hodgkin A.L. and Katz B. (1949). The effect of sodium ions on the electrical activity of the giant axon of the squid. *J. Physiol.* **108**, 37–77.

Hoffman D.A., Magee J.C., Colbert C.M. and Johnston D. (1997). K^+ channel regulation of signal propagation in dendrites of hippocampal pyramidal neurons. *Nature* **387**, 869–875.

Hofmann F., Biel M. and Kaupp B.U. (2005). International Union of Pharmacology. LI. Nomenclature and structure-function relationships of cyclic nucleotide-regulated channels. *Pharmacol. Rev.* **57**, 455–462.

Holland J.H., ed. (1975). *Adaptation in Natural and Artificial Systems: An Introductory Analysis with Applications to Biology, Control, and Artificial Intelligence* (University of Michigan Press, Ann Arbor, MI).

Holmes W.R. and Rall W. (1992a). Electrotonic length estimates in neurons with dendritic tapering or somatic shunt. *J. Neurophysiol.* **68**, 1421–1437.

Holmes W.R. and Rall W. (1992b). Electrotonic models of neuronal dendrites and single neuron computation. In *Single Neuron Computation*, eds. T. McKenna, J. Davis and S.F. Zornetzer (Academic Press, Boston, MA), chapter 1, pp. 7–25.

Holmes W.R. and Rall W. (1992c). Estimating the electrotonic structure of neurons with compartmental models. *J. Neurophysiol.* **68**, 1438–1452.

Holmes W.R., Segev I. and Rall W. (1992). Interpretation of time constant and electrotonic length estimates in multicylinder or branched neuronal structures. *J. Neurophysiol.* **68**, 1401–1420.

Holt G.R. and Koch C. (1999). Electrical interactions via the extracellular potential near cell bodies. *J. Comput. Neurosci.* **6**, 169–184.

Holter K.E., Kehlet B., Devor A., Sejnowski T.J., Dale A.M., Omholt S.W., Ottersen O.P., Nagelhus E.A., Mardal K.A. and Pettersen K.H. (2017). Interstitial solute transport in 3D reconstructed neuropil occurs by diffusion rather than bulk flow. *Proc. Natl. Acad. Sci. USA* **114**, 9894–9899.

Holtzmann D.M., Li Y., Parada L.F., Kinsmann S., Chen C.K., Valletta J.S., Zhou J., Long J.B. and Mobley W.C. (1992). p140[trk] mRNA marks NGF-responsive forebrain neurons: evidence that *trk* gene expression is induced by NGF. *Neuron* **9**, 465–478.

Honda H. (1998). Topographic mapping in the retinotectal projection by means of complementary ligand and receptor gradients: a computer simulation study. *J. Theor. Biol.* **192**, 235–246.

Honda H. (2003). Competition for retinal ganglion axons for targets under the servomechanism model explains abnormal retinocollicular projection of Eph receptor-overexpressing or ephrin-lacking mice. *J. Neurosci.* **23**, 10368–10377.

Hoops S., Sahle S., Gauges R., Lee C., Pahle J., Simus N., Singhal M., Xu L., Mendes P. and Kummer U. (2006). COPASI – COmplex PAthway SImulator. *Bioinformatics* **22**, 3067–3074.

Hope R.A., Hammond B.J. and Gaze R.M. (1976). The arrow model: retinotectal specificity and map formation in the goldfish visual system. *Proc. R. Soc. Lond. B Biol. Sci.* **194**, 447–466.

Hopfield J.J. (1982). Neural networks and physical systems with emergent collective computational abilities. *Proc. Natl. Acad. Sci. USA* **79**, 2554–2558.

Hopfield J.J. (1984). Neurons with graded response have collective computational properties like those of two-state neurons. *Proc. Natl. Acad. Sci. USA* **81**, 3088–3092.

Hore V.R., Troy J.B. and Eglen S.J. (2012). Parasol cell mosaics are unlikely to drive the formation of structured orientation maps in primary visual cortex. *Vis. Neurosci.* **29**, 283–299.

Horsfield K., Woldenberg M.J. and Bowes C.L. (1987). Sequential and synchronous growth models related to vertex analysis and branching ratios. *Bull. Math. Biol.* **49**, 413–430.

Hospedales T.M., van Rossum M.C.W., Graham B.P. and Dutia M.B. (2008). Implications of noise and neural heterogeneity for vestibulo-ocular reflex fidelity. *Neural Comput.* **20**, 756–778.

Howarth C., Gleeson P. and Attwell D. (2012). Updated energy budgets for neural computation in the neocortex and cerebellum. *J. Cereb. Blood Flow Metab.* **32**, 1222–1232.

Hoyle G. (1983). Origins of intracellular microelectrodes. *Trends Neurosci.* **6**, 163.

Hoyt R.C. (1963). The squid giant axon: mathematical models. *Biophys. J.* **3**, 399–431.

Hoyt R.C. (1968). Sodium inactivation in nerve fibres. *Biophys. J.* **8**, 1074–1097.

Hubel D.H. and Wiesel T.N. (1962). Receptive fields, binocular interaction and functional architecture in the cat's visual cortex. *J. Physiol.* **160**, 106–154.

Hubel D.H. and Wiesel T.N. (1977). Functional architecture of macaque monkey visual cortex. *Proc. R. Soc. Lond. B Biol. Sci.* **1130**, 1–59.

Hubel D.H., Wiesel T.N. and LeVay S. (1977). Plasticity of ocular dominance columns in monkey striate cortex. *Philos. Trans. R. Soc. Lond. B Biol. Sci.* **278**, 377–409.

Huguenard J.R. and McCormick D.A. (1992). Simulation of the currents involved in rhythmic oscillations in thalamic relay neurons. *J. Neurophysiol.* **68**, 1373–1383.

Hull M.J., Soffe S.R., Willshaw D.J. and Roberts A. (2015). Modelling the effects of electrical coupling between unmyelinated axons of brainstem neurons controlling rhythmic activity. *PLoS Comput. Biol.* **11**, e1004240.

Humble J., Hiratsuka K., Kasai H. and Toyoizumi T. (2019). Intrinsic spine dynamics are critical for recurrent network learning in models with and without autism spectrum disorder. *Front. Comput. Neurosci.* **13**, 38.

Hunter R., Cobb S. and Graham B.P. (2009). Improving associative memory in a network of spiking neurons. *Neural Network World* **19**, 447–470.

Iivanainen J., Stenroos M. and Parkkonen L. (2017). Measuring MEG closer to the brain: Performance of on-scalp sensor arrays. *NeuroImage* **147**, 542–553.

Iliff J.J., Wang M., Liao Y., Plogg B.A., Peng W., Gundersen G.A., Benveniste H., Vates G.E., Deane R., Goldman S.A., Nagelhus E.A. and Nedergaard M. (2012). A paravascular pathway facilitates CSF flow through the brain parenchyma and the clearance of interstitial solutes, including amyloid β. *Sci. Transl. Med.* **4**, 147ra111.

Ilmoniemi R.J., Mäki H., Saari J., Salvador R. and Miranda P.C. (2016). The frequency-dependent neuronal length constant in transcranial magnetic stimulation. *Front. Cell. Neurosci.* **10**, 194.

Ilmoniemi R.J. and Sarvas J. (2019). *Brain Signals – Physics and Mathematics of MEG and EEG* (MIT Press, Cambridge, MA).

Ireland W., Heidel J. and Uemura E. (1985). A mathematical model for the growth of dendritic trees. *Neurosci. Lett.* **54**, 243–249.

Irvine L.A., Jafri M.S. and Winslow R.L. (1999). Cardiac sodium channel Markov model with temperature dependence and recovery from inactivation. *Biophys. J.* **76**, 1868–1885.

Izhikevich E.M. (2003). Simple model of spiking neurons. *IEEE Trans. Neural Netw.* **14**, 1569–1572.

Izhikevich E.M. (2004). Which model to use for cortical spiking neurons? *IEEE Trans. Neural Netw.* **15**, 1063–1070.

Izhikevich E.M. (2007). *Dynamical Systems in Neuroscience: The Geometry of Excitability and Bursting* (MIT Press, Cambridge, MA).

Izhikevich E.M. and Edelman G.M. (2008). Large-scale model of mammalian thalamocortical systems. *Proc. Natl. Acad. Sci. USA* **105**, 3593–3598.

Jackson J.D. (1998). *Classical Electrodynamics* (John Wiley & Sons, Hoboken, NJ), third edition.

Jacobson M. and Levine R.L. (1975). Plasticity in the adult frog brain: filling in the visual scotoma after excision or translocation of parts of the optic tectum. *Brain Res.* **88**, 339–345.

Jaeger D. and Bower J.M. (1999). Synaptic control of spiking in cerebellar Purkinje cells: dynamic current clamp based on model conductances. *J. Neurosci.* **19**, 6090–6101.

Jaffe D.B., Ross W.N., Lisman J.E., Lasser-Ross N., Miyakawa H. and Johnston D. (1994). A model for dendritic Ca^{2+} accumulation in hippocampal pyramidal neurons based on fluorescence imaging measurements. *J. Neurophysiol.* **71**, 1065–1077.

Jahr C.E. and Stevens C.F. (1990a). A quantitative description of NMDA receptor-channel kinetic behavior. *J. Neurosci.* **10**, 1830–1837.

Jahr C.E. and Stevens C.F. (1990b). Voltage dependence of NMDA-activated macroscopic conductances predicted by single-channel kinetics. *J. Neurosci.* **10**, 3178–3182.

Jang J. and Paik S.B. (2017). Interlayer repulsion of retinal ganglion cell mosaics regulates spatial organization of functional maps in the visual cortex. *J. Neurosci.* **37**, 12141–12152.

Jankovic J. (2008). Parkinson's disease: clinical features and diagnosis. *J. Neurol. Neurosurg. Psychiatr.* **79**, 368–376.

Jansen J.K.S. and Fladby T. (1990). The perinatal reorganization of the innervation of skeletal muscle in mammals. *Prog. Neurobiol.* **34**, 39–90.

Jentsch T.J. (2000). Neuronal KCNQ potassium channels: physiology and role in disease. *Nat. Rev. Neurosci.* **1**, 21–30.

Jentsch T.J., Neagoe I. and Scheel O. (2005). CLC chloride channels and transporters. *Curr. Opin. Neurobiol.* **15**, 319–325.

Jirsa V. and Haken H. (1997). A derivation of a macroscopic field theory of the brain from the quasi-microscopic neural dynamics. *Physica D* **99**, 503–526.

Johnston D. and Wu S.M.S. (1995). *Foundations of Cellular Neurophysiology* (MIT Press, Cambridge, MA).

Jolivet R., Lewis T.J. and Gerstner W. (2004). Generalized integrate-and-fire models of neuronal activity approximate spike trains of a detailed model to a high degree of accuracy. *J. Neurophysiol.* **92**, 959–976.

Jolivet R., Schürmann F., Berger T., Naud R., Gerstner W. and Roth A. (2008). The quantitative single-neuron modeling competition. *Biol. Cybern.* **99**, 417–426.

Joucla S. and Yvert B. (2012). Modeling extracellular electrical neural stimulation: from basic understanding to MEA-based applications. *J. Physiol. Paris* **106**, 146–58.

Jumper J., Evans R., Pritzel A., Green T., Figurnov M., Ronneberger O., Tunyasuvunakool K., Bates R., Žídek A., Potapenko A., Bridgland A., Meyer C., Kohl S.A.A., Ballard A.J., Cowie A., Romera-Paredes B., Nikolov S., Jain R., Adler J., Back T., Petersen S., Reiman D., Clancy E., Zielinski M., Steinegger M., Pacholska M., Berghammer T., Bodenstein S., Silver D., Vinyals O., Senior A.W., Kavukcuoglu K., Kohli P. and Hassabis D. (2021). Highly accurate protein structure prediction with AlphaFold. *Nature* **596**, 583–589.

Jun J.J., Steinmetz N.A., Siegle J.H., Denman D.J., Bauza M., Barbarits B., Lee A.K., Anastassiou C.A., Andrei A., Aydın Ç., Barbic M., Blanche T.J., Bonin V., Couto J., Dutta B., Gratiy S.L., Gutnisky D.A., Häusser M., Karsh B., Ledochowitsch P., Lopez C.M., Mitelut C., Musa S., Okun M., Pachitariu M., Putzeys J., Rich P.D., Rossant C., Lung Sun W., Svoboda K., Carandini M., Harris K.D., Koch C., O'Keefe J. and Harris T.D. (2017). Fully integrated silicon probes for high-density recording of neural activity. *Nature* **551**, 232–236.

Kajikawa Y. and Schroeder C.E. (2011). How local is the local field potential? *Neuron* **72**, 847–858.

Kalatsky V.A. and Stryker M.P. (2003). New paradigm for optical imaging: temporally encoded maps of intrinsic signal. *Neuron* **38**, 529–545.

Kaleb K., Pedrosa V. and Clopath C. (2021). Network-centered homeostasis through inhibition maintains hippocampal spatial map and cortical circuit function. *Cell Reports* **36**, 109577.

Kamaleddin M.A. (2022). Degeneracy in the nervous system: from neuronal excitability to neural coding. *Bioessays* **44**, 2100148.

Kasai H., Ziv N.E., Okazaki H., Yagishita S. and Toyoizumi T. (2021). Spine dynamics in the brain, mental disorders and artificial neural networks. *Nat. Rev. Neurosci.* **22**, 407–422.

Kasimova M.A., Tarek M., Shaytan A.K., Shaitan K.V. and Delemotte L. (2014). Voltage-gated ion channel modulation by lipids: Insights from molecular dynamics simulations. *Biochimica et Biophysica Acta* **1838**, 1322–1331.

Katz B. and Miledi R. (1968). The role of calcium in neuromuscular facilitation. *J. Physiol.* **195**, 481–492.

Katz L.C. and Crowley J.C. (2002). Development of cortical circuits: lessons from ocular dominance columns. *Nat. Rev. Neurosci.* **3**, 34–42.

Katzner S., Nauhaus I., Benucci A., Bonin V., Ringach D.L. and Carandini M. (2009). Local origin of field potentials in visual cortex. *Neuron* **61**, 35–41.

Keck T., Toyoizumi T., Chen L., Doiron B., Feldman D.E., Fox K., Gerstner W., Haydon P.G., Hübener M., Lee H.K., Lisman J.E., Rose T., Sengpiel F., Stellwagen D., Stryker M.P., Turrigiano G.G. and van Rossum M.C. (2017). Integrating Hebbian and homeostatic plasticity: the current state of the field and future research directions. *Philos. Trans. R. Soc. Lond. B Biol. Sci.* **372**, 20160158.

Keizer J. and Levine L. (1996). Ryanodine receptor adaptation and Ca^{2+}-induced Ca^{2+} release-dependent Ca^{2+} oscillations. *Biophys. J.* **71**, 3477–3487.

Kepler T.B., Abbott L.F. and Marder E. (1992). Reduction of conductance-based neuron models. *Biol. Cybern.* **66**, 381–387.

Kerr R.A., Bartol T.M., Kaminsky B., Dittrich M., Chang J.C., Baden S.B., Sejnowski T.J. and Stiles J.R. (2008). Fast Monte Carlo simulation methods for biological reaction–diffusion systems in solution and on surfaces. *SIAM J. Sci. Comput.* **30**, 3126.

Kerti K., Lorincz A. and Nusser Z. (2012). Unique somato-dendritic distribution pattern of Kv4.2 channels on hippocampal CA1 pyramidal cells. *Eur. J. Neurosci.* **35**, 66–75.

Kettenmann H. and Ransom B.R. (2012). *Neuroglia* (Oxford University Press, Oxford), third edition.

Kiddie G., McLean D.R., van Ooyen A. and Graham B.P. (2005). Biologically plausible models of neurite outgrowth. *Prog. Brain Res.* **147**, 67–80.

Kim B., Hawes S.L., Gillani F., Wallace L.J. and Blackwell K.T. (2013). Signaling pathways involved in striatal synaptic plasticity are sensitive to temporal pattern and exhibit spatial specificity. *PLoS Comput. Biol.* **9**, e1002953.

Kim S.S., Rouault H., Druckmann S. and Jayaraman V. (2017). Ring attractor dynamics in the *Drosophila* central brain. *Science* **356**, 849–853.

Kim U., Bal T. and McCormick D.A. (1995). Spindle waves are propagating synchronized oscillations in the ferret LGNd in vitro. *J. Neurophysiol.* **74**, 1301–1323.

Kirkpatrick S., Gelatt C.D. and Vecchi M.P. (1983). Optimization by simulated annealing. *Science* **220**, 671–680.

Kita E.M., Scott E.K. and Goodhill G.J. (2015). Topographic wiring of the retinotectal connection in zebrafish. *Dev. Neurobiol.* **75**, 542–556.

Kita H., Chang H. and Kitai S. (1983). Pallidal inputs to the subthalamus: intracellular analysis. *Brain Res.* **264**, 255–265.

Kliemann W.A. (1987). Stochastic dynamic model for the characterization of the geometric structure of dendritic processes. *Bull. Math. Biol.* **49**, 135–152.

Kling A., Gogliettino A.R., Shah N.P., Wu E.G., Brackbill N., Sher A., Litke A.M., Silva R.A. and Chichilnisky E. (2020). Functional organization of midget and parasol ganglion cells in the human retina. *BioRxiv* 2020.08.07.240762.

Knight B.W. (1972). Dynamics of encoding in a population of neurons. *J. Gen. Physiol.* **59**, 734–766.

Koch C. (1999). *Biophysics of Computation: Information Processing in Single Neurons* (Oxford University Press, New York, NY).

Koch C. and Buice M.A. (2015). A biological imitation game. *Cell* **163**, 277–280.

Köhling R. (2002). Voltage-gated sodium channels in epilepsy. *Epilepsia* **43**, 1278–1295.

Kominis I.K., Kornack T.W., Allred J.C. and Romalis M.V. (2003). A subfemtotesla multichannel atomic magnetometer. *Nature* **422**, 596–599.

Kopell N., Börgers C., Pervouchine D., Malerba P. and Tort A. (2010). Gamma and theta rhythms in biophysical models of hippocampal circuits. In *Hippocampal Microcircuits: A Computational Modeler's Resource Book*, eds. V. Cutsuridis, B. Graham, S. Cobb and I. Vida (Springer, New York, NY), pp. 423–457.

Köpfer D.A., Song C., Gruene T., Sheldrick G.M., Zachariae U. and de Groot B.L. (2014). Ion permeation in K^+ channels occurs by direct coulomb knock-on. *Science* **346**, 352–355.

Koulakov A.A. and Tsigankov D.N. (2004). A stochastic model for retinocollicular map development. *BMC Neurosci.* **5**, 30–46.

Krauskopf B., Osinga H.M. and Galán-Vioque J., eds. (2007). *Numerical Continuation Methods for Dynamical Systems: Path Following and Boundary Value Problems* (Springer, Dordrecht).

Kreiman G., Hung C.P., Kraskov A., Quiroga R.Q., Poggio T. and DiCarlo J.J. (2006). Object selectivity of local field potentials and spikes in the macaque inferior temporal cortex. *Neuron* **49**, 433–445.

Kriegeskorte N. (2015). Deep neural networks: a new framework for modeling biological vision and brain information processing. *Annu. Rev. Vis. Sci.* **1**, 417–446.

Krizhevsky A., Sutskever I. and Hinton G.E. (2012). ImageNet classification with deep convolutional neural networks. *Adv. Neural Inf. Process. Syst.* **25**, 1097–1105.

Krottje J.K. and van Ooyen A. (2007). A mathematical framework for modeling axon guidance. *Bull. Math. Biol.* **69**, 3–31.

Kubo Y., Adelman J.P., Clapham D.E., Jan L.Y., Karschin A., Kurachi Y., Lazdunski M., Nichols C.G., Seino S. and Vandenberg C.A. (2005). International Union of Pharmacology. LIV. Nomenclature and molecular relationships of inwardly rectifying potassium channels. *Pharmacol. Rev.* **57**, 509–526.

Kuhlmann L., Lehnertz K., Richardson M.P., Schelter B. and Zaveri H.P. (2018). Seizure prediction – ready for a new era. *Nat. Rev. Neurol.* **14**, 618–630.

Kühn A.A., Kempf F., Brücke C., Doyle L.G., Martinez-Torres I., Pogosyan A., Trottenberg T., Kupsch A., Schneider G.H., Hariz M.I. et al. (2008). High-frequency stimulation of the subthalamic nucleus suppresses oscillatory β activity in patients with Parkinson's disease in parallel with improvement in motor performance. *J. Neurosci.* **28**, 6165–6173.

Kuroda S., Schweighofer N. and Kawato M. (2001). Exploration of signal transduction pathways in cerebellar long-term depression by kinetic simulation. *J. Neurosci.* **21**, 5693–5702.

Kusano K. and Landau E.M. (1975). Depression and recovery of transmission at the squid giant synapse. *J. Physiol.* **245**, 13–32.

Kutzner C., Köpfer D.A., Machtens J.P., de Groot B.L., Song C. and Zachariae U. (2016). Insights into the function of ion channels by computational electrophysiology simulations. *Biochimica et Biophysica Acta* **1858**, 1741–1752.

Laker D., Tolle F., Stegen M., Heerdegen M., Köhling R., Kirschstein T. and Wolfart J. (2021). Kv7 and Kir6 channels shape the slow AHP in mouse dentate gyrus granule cells and control burst-like firing behavior. *Neuroscience* **467**, 56–72.

Landahl H. (1939). A contribution to the mathematical biophysics of psychophysical discrimination II. *Bull. Math. Biophys.* **1**, 159–176.

Langley J.N. (1895). Note on regeneration of pre-ganglionic fibres of the sympathetic ganglion. *J. Physiol.* **18**, 280–284.

Lapicque L. (1907). Recherches quantitatives sur l'excitation électrique des nerfs traitée comme une polarisation. *J. Physiol. Paris* **9**, 620–635.

Larkman A., Stratford K. and Jack J. (1991). Quantal analysis of excitatory synaptic action and depression in hippocampal slices. *Nature* **350**, 344–347.

Larkum M. (2013). A cellular mechanism for cortical associations: an organizing principle for the cerebral cortex. *Trends Neurosci.* **36**, 141–151.

Latham P.E., Richmond B.J., Nelson P.G. and Nirenberg S. (2000). Intrinsic dynamics in neuronal networks. I. Theory. *J. Neurophysiol.* **83**, 808–827.

Lau C., Hunter M.J., Stewart A., Perozo E. and Vandenberg J.I. (2018). Never at rest: insights into the conformational dynamics of ion channels from cryo-electron microscopy. *J. Physiol.* **596**, 1107–1119.

Laurent G. (1996). Dynamical representation of odors by oscillating and evolving neural assemblies. *Trends Neurosci.* **19**, 489–496.

Lawrence P.A. (2001). Morphogens: how big is the big picture? *Nature Cell Biology* **3**, E151–E154.

Lebedev M.A. and Nicolelis M.A.L. (2006). Brain-machine interfaces: past, present and future. *Trends Neurosci.* **29**, 536–546.

Lei C.L., Clerx M., Whittaker D.G., Gavaghan D.J. and de Boer T.P. (2020). Accounting for variability in ion current recordings using a mathematical model of artefacts in voltage-clamp experiments. *Philos. Trans. R. Soc. Lond. A* **378**, 20190348.

LeMasson G. and Maex R. (2001). Introduction to equation solving and parameter fitting. In *Computational Neuroscience: Realistic Modeling for Experimentalists*, ed. E. De Schutter (CRC Press, Boca Raton, FL), chapter 1, pp. 1–23.

Lempka S.F. and McIntyre C.C. (2013). Theoretical analysis of the local field potential in deep brain stimulation applications. *PLoS One* **8**, e59839.

Leski S., Kublik E., Swiejkowski D.A., Wrobel A. and Wojcik D.K. (2010). Extracting functional components of neural dynamics with independent component analysis and inverse current source density. *J. Comput. Neurosci.* **29**, 459–473.

Łęski S., Lindén H., Tetzlaff T., Pettersen K.H. and Einevoll G.T. (2013). Frequency dependence of signal power and spatial reach of the local field potential. *PLoS Comput. Biol.* **9**, e1003137.

LeVay S., Hubel D.H. and Wiesel T.N. (1975). The pattern of ocular dominance columns in macaque visual cortex revealed by a reduced silver stain. *J. Comp. Neurol.* **159**, 559–575.

LeVay S., Stryker M.P. and Shatz C.J. (1978). Ocular dominance columns and their development in layer IV of the cat's visual cortex: a quantitative study. *J. Comp. Neurol.* **179**, 223–244.

LeVay S., Wiesel T.N. and Hubel D.H. (1980). The development of ocular dominance columns in normal and visually deprived monkeys. *J. Comp. Neurol.* **191**, 1–51.

Levine R.L. and Jacobson M. (1974). Deployment of optic nerve fibers is determined by positional markers in the frog's tectum. *Exp. Neurol.* **43**, 527–538.

Levy R., Ashby P., Hutchison W.D., Lang A.E., Lozano A.M. and Dostrovsky J.O. (2002). Dependence of subthalamic nucleus oscillations on movement and dopamine in Parkinson's disease. *Brain* **125**, 1196–1209.

Levy W.B. and Steward O. (1979). Synapses as associative memory elements in the hippocampal formation. *Brain Res.* **175**, 233–245.

Levy W.B. and Steward O. (1983). Temporal contiguity requirements for long-term associative potentiation/depression in the hippocampus. *Neuroscience* **8**, 791–797.

Li G.H., Qin C.D. and Li M.H. (1994). On the mechanisms of growth cone locomotion: modeling and computer simulation. *J. Theor. Biol.* **169**, 355–362.

Li G.H., Qin C.D. and Wang L.W. (1995). Computer model of growth cone behavior and neuronal morphogenesis. *J. Theor. Biol.* **174**, 381–389.

Li G.H., Qin C.D. and Wang Z.S. (1992). Neurite branching pattern formation: modeling and computer simulation. *J. Theor. Biol.* **157**, 463–486.

Li W.C., Soffe S.R. and Roberts A. (2004). A direct comparison of whole cell patch and sharp electrodes by simultaneous recording from single spinal neurons in frog tadpoles. *J. Neurophysiol.* **92**, 380–386.

Li Y.X. and Rinzel J. (1994). Equations for InsP$_3$ receptor-mediated [Ca^{2+}]$_i$ oscillations derived from a detailed kinetic model: a Hodgkin–Huxley like formalism. *J. Theor. Biol.* **166**, 461–473.

Liley A.W. and North K.A.K. (1953). An electrical investigation of effects of repetitive stimulation on mammalian neuromuscular junction. *J. Neurophysiol.* **16**, 509–527.

Liley D.T.J., Cadusch P.J. and Dafilis M.P. (2002). A spatially continuous mean field theory of electrocortical activity. *Network Comp. Neural Syst.* **13**, 67–113.

Lillicrap T.P., Cownden D., Tweed D.B. and Akerman C.J. (2016). Random synaptic feedback weights support error backpropagation for deep learning. *Nat. Commun.* **7**, 1–10.

Lindén H., Hagen E., Łęski S., Norheim E.S., Pettersen K.H. and Einevoll G.T. (2014). LFPy: A tool for biophysical simulation of extracellular potentials generated by detailed model neurons. *Front. Neuroinform.* **7**, 41.

Lindén H., Pettersen K.H. and Einevoll G.T. (2010). Intrinsic dendritic filtering gives low-pass power spectra of local field potentials. *J. Comput. Neurosci.* **29**, 423–444.

Lindén H., Tetzlaff T., Potjans T.C., Pettersen K.H., Grün S., Diesmann M. and Einevoll G.T. (2011). Modeling the spatial reach of the LFP. *Neuron* **72**, 859–872.

Lindsay K.A., Maxwell D.J., Rosenberg J.R. and Tucker G. (2007). A new approach to reconstruction models of dendritic branching patterns. *Math. Biosci.* **205**, 271–296.

Lisman J., Yasuda R. and Raghavachari S. (2012). Mechanisms of CaMKII action in long-term potentiation. *Nat. Rev. Neurosci.* **13**, 169–182.

Lisman J.E. and Idiart M.A.P. (1995). Storage of 7 ± 2 short-term memories in oscillatory subcycles. *Science* **267**, 1512–1514.

Little S., Beudel M., Zrinzo L., Foltynie T., Limousin P., Hariz M., Neal S., Cheeran B., Cagnan H., Gratwicke J. et al. (2016). Bilateral adaptive deep brain stimulation is effective in Parkinson's disease. *J. Neurol. Neurosurg. Psychiatr.* **87**, 717–721.

Little W.A. (1974). The existence of persistent states in the brain. *Math. Biosci.* **19**, 101–120.

Liu Z., Golowasch J., Marder E. and Abbott L.F. (1998). A model neuron with activity-dependent conductances regulated by multiple calcium sensors. *J. Neurosci.* **18**, 2309–2320.

Livet J., Weissman T.A., Kang H., Draft R.W., Lu J., Bennis R.A., Sanes J.R. and Lichtman J.W. (2007). Transgenic strategies for combinatorial expression of fluorescent proteins in the nervous system. *Nature* **450**, 56–62.

Lopreore C.L., Bartol T.M., Coggan J.S., Keller D.X., Sosinsky G.E., Ellisman M.H. and Sejnowski T.J. (2008). Computational modeling of three-dimensional electrodiffusion in biological systems: application to the node of Ranvier. *Biophys. J.* **95**, 2624–2635.

Lotka A.J. (1925). *Elements of Physical Biology* (Willliams and Wilkins, Baltimore, MD).

Lozano A.M. and Lipsman N. (2013). Probing and regulating dysfunctional circuits using deep brain stimulation. *Neuron* **77**, 406–424.

Lundengård K., Cedersund G., Sten S., Leong F., Smedberg A., Elinder F. and Engström M. (2016). Mechanistic mathematical modeling tests hypotheses of the neurovascular coupling in fMRI. *PLoS Comput. Biol.* **12**, e1004971.

Lux H.D., Schubert P. and Kreutzberg G.W. (1970). Direct matching of morphological and electrophysiological data in the cat spinal motorneurons. In *Excitatory Synaptic Mechanisms*, eds. P. Anderson and J.K. Janson (Universitetsforlaget, Oslo), pp. 189–198.

Luz Y. and Shamir M. (2012). Balancing feed-forward excitation and inhibition via Hebbian inhibitory synaptic plasticity. *PLoS Comput. Biol.* **8**, e1002334.

Lynch G.S., Dunwiddie T. and Gribkoff V. (1977). Heterosynaptic depression: a postsynaptic correlate of long-term potentiation. *Nature* **266**, 737–739.

Lyngholm D., Sterratt D.C., Hjorth J.J., Willshaw D.J., Eglen S.J. and Thompson I.D. (2019). Measuring and modelling the emergence of order in the mouse retinocollicular projection. *bioRxiv* p. 713628.

Macagno E.R. (1978). Mechanism for the formation of synaptic projections in the arthropod visual system. *Nature* **275**, 318–320.

MacKay D.J.C. (2003). *Information Theory, Inference, and Learning Algorithms* (Cambridge University Press, Cambridge).

Magee J.C. and Johnston D. (1995). Characterization of single voltage-gated Na$^+$ and Ca^{2+} channels in apical dendrites of rat CA1 pyramidal neurons. *J. Physiol.* **487**, 67–90.

Magistretti J. and Alonso A. (1999). Biophysical properties and slow voltage-dependent inactivation of a sustained sodium current in entorhinal cortex layer-II principal neurons: a whole-cell and single-channel study. *J. Gen. Physiol.* **114**, 491–509.

Magistretti J., Castelli L., Forti L. and D'Angelo E. (2006). Kinetic and functional analysis of transient, persistent and resurgent sodium currents in rat cerebellar granule cells in situ: an electrophysiological and modelling study. *J. Physiol.* **573**, 83–106.

Magleby K.L. (1987). Short-term changes in synaptic efficacy. In *Synaptic Function*, eds. G.M. Edelman, L.E. Gall, W. Maxwell and W.M. Cowan (Wiley, New York, NY), pp. 21–56.

Major G. and Evans J.D. (1994). Solutions for transients in arbitrarily branching cables: IV. Nonuniform electrical parameters. *Biophys. J.* **66**, 615–638.

Major G., Larkman A.U., Jonas P., Sackmann B. and Jack J.J. (1994). Detailed passive cable models of whole-cell recorded CA3 pyramidal neurons in rat hippocampal slices. *J. Neurosci.* **14**, 4613–4638.

Makarov V.A., Makarova J. and Herreras O. (2010). Disentanglement of local field potential sources by independent component analysis. *J. Comput. Neurosci.* **29**, 445–457.

Mäki-Marttunen T., Halnes G., Devor A., Metzner C., Dale A.M., Andreassen O.A. and Einevoll G.T. (2018). A stepwise neuron model fitting procedure designed for recordings with high spatial resolution: Application to layer 5 pyramidal cells. *J. Neurosci. Methods* **293**, 264–283.

Mäki-Marttunen T., Iannella N., Edwards A.G., Einevoll G.T. and Blackwell K.T. (2020). A unified computational model for cortical post-synaptic plasticity. *Elife* **9**, e55714.

Malthus T.R. (1798). *An Essay on the Principle of Population* (Oxford University Press, Oxford).

Mangold K.E., Wang W., Johnson E.K., Bhagavan D., Moreno J.D., Nerbonne J.M. and Silva J.R. (2021). Identification of structures for ion channel kinetic models. *PLoS Comput. Biol.* **17**, e1008932.

Manninen T., Saudargiene A. and Linne M.L. (2020). Astrocyte-mediated spike-timing-dependent long-term depression modulates synaptic properties in the developing cortex. *PLoS Comput. Biol.* **16**, e1008360.

Manor Y., Gonczarowski J. and Segev I. (1991a). Propogation of action potentials along complex axonal trees: model and implementation. *Biophys. J.* **60**, 1411–1423.

Manor Y., Koch C. and Segev I. (1991b). Effect of geometrical irregularities on propagation delay in axonal trees. *Biophys. J.* **60**, 1424–1437.

Maravall M., Mainen Z.F., Sabatini B.L. and Svoboda K. (2000). Estimating intracellular calcium concentrations and buffering without wavelength ratioing. *Biophys. J.* **78**, 2655–2667.

Marder E. (2009). Electrical synapses: rectification demystified. *Curr. Biol.* **19**, R34–R35.

Marder E. and Goaillard J.M. (2006). Variability, compensation and homeostasis in neuron and network function. *Nat. Rev. Neurosci.* **7**, 153–160.

Marder E. and Taylor A.L. (2011). Multiple models to capture the variability in biological neurons and networks. *Nat. Rev. Neurosci.* **14**, 133–138.

Marin J.M., Pudlo P., Robert C.P. and Ryder R.J. (2012). Approximate Bayesian computational methods. *Statistics and Computing* **22**, 1167–1180.

Marín M., Cruz N.C., Ortigosa E.M., Sáez-Lara M.J., Garrido J.A. and Carrillo R.R. (2021). On the use of a multimodal optimizer for fitting neuron models. application to the cerebellar granule cell. *Front. Neuroinform.* **15**, 17.

Markram H. (2006). The Blue Brain Project. *Nat. Rev. Neurosci.* **7**, 153–160.

Markram H., Gupta A., Uziel A., Wang Y. and Tsodyks M. (1998). Information processing with frequency-dependent synaptic connections. *Neurobiol. Learn. Mem.* **70**, 101–112.

Markram H., Luebke J., Frotscher M. and Sakmann B. (1997). Regulation of synaptic efficacy by coincidence of postsynaptic APs and EPSPs. *Science* **275**, 213–215.

Markram H., Muller E., Ramaswamy S., Reimann M.W., Abdellah M., Sanchez C.A., Ailamaki A., Alonso-Nanclares L., Antille N., Arsever S., Kahou G.A.A., Berger T.K., Bilgili A., Buncic N., Chalimourda A., Chindemi G., Courcol J.D., Delalondre F., Delattre V., Druckmann S., Dumusc R., Dynes J., Eilemann S., Gal E., Gevaert M.E., Ghobril J.P., Gidon A., Graham J.W., Gupta A., Haenel V., Hay E., Heinis T., Hernando J.B., Hines M., Kanari L., Keller D., Kenyon J., Khazen G., Kim Y., King J.G., Kisvarday Z., Kumbhar P., Lasserre S., Le Bé J.V., Magalhães B.R.C., Merchán-Pérez A., Meystre J., Morrice B.R., Muller J., Muñoz-Céspedes A., Muralidhar S., Muthurasa K., Nachbaur D., Newton T.H., Nolte M., Ovcharenko A., Palacios J., Pastor L., Perin R., Ranjan R., Riachi I., Rodríguez J.R., Riquelme J.L., Rössert C., Sfyrakis K., Shi Y., Shillcock J.C., Silberberg G., Silva R., Tauheed F., Telefont M., Toledo-Rodriguez M., Tränkler T., Van Geit W., Díaz J.V., Walker R., Wang Y., Zaninetta S.M., DeFelipe J., Hill S.L., Segev I. and Schürmann F. (2015). Reconstruction and simulation of neocortical microcircuitry. *Cell* **163**, 456–492.

Marler K.J.M., Becker-Barroso E., Martinez A., Llovera M., Wentzel C., Poopalasundaram S., Hindges R., Soriano E., Comella J. and Drescher U. (2008). A TrkB/ephrinA interaction controls retinal axon branching and synaptogenesis. *J. Neurosci.* **28**, 12700–12712.

Marmont M. (1949). Studies on the axon membrane. *J. Cell. Comp. Physiol.* **34**, 351–382.

Marr D. (1969). A theory of cerebellar cortex. *J. Physiol.* **202**, 437–470.

Marr D. (1970). A theory for cerebral neocortex. *Proc. R. Soc. Lond. B Biol. Sci.* **176**, 161–234.

Marr D. (1971). Simple memory: a theory for archicortex. *Philos. Trans. R. Soc. Lond. B Biol. Sci.* **262**, 23–81.

Marr D. (1982). *Vision* (MIT Press, Cambridge, MA).

Mascagni M.V. and Sherman A.S. (1998). Numerical methods for neuronal modeling. In *Methods in Neuronal Modeling: From Ions to Networks*, eds. C. Koch and I. Segev (MIT Press, Cambridge, MA), chapter 14, pp. 569–606. Second edition.

Maskery S., Buettner H.M. and Shinbrot T. (2004). Growth cone pathfinding: a competition between deterministic and stochastic events. *BMC Neurosci.* **5**, 22.

Matveev V. and Wang X.J. (2000). Implications of all-or-none synaptic transmission and short-term depression beyond vesicle depletion: a computational study. *J. Neurosci.* **20**, 1575–1588.

Maurice N., Mercer J., Chan C.S., Hernandez-Lopez S., Held J., Tkatch T. and Surmeier D.J. (2004). D_2 dopamine receptor-mediated modulation of voltage-dependent Na^+ channels reduces autonomous activity in striatal cholinergic interneurons. *J. Neurosci.* **24**, 10289–10301.

McClelland J.L., Rumelhart D.E. and the PDP Research Group, eds. (1986). *Parallel Distributed Processing: Explorations in the Microstructure of Cognition. Volume 2: Psychological and Biological Models* (MIT Press, Cambridge, MA).

McColgan T., Liu J., Kuokkanen P.T., Carr C.E., Wagner H. and Kempter R. (2017). Dipolar extracellular potentials generated by axonal projections. *Elife* **6**, e26106.

McCulloch W.S. and Pitts W. (1943). A logical calculus of the ideas immanent in nervous activity. *Bull. Math. Biophys.* **5**, 115–133.

McIntyre C., Grill W., Sherman D. and Thakor N. (2004). Cellular effects of deep brain stimulation: model-based analysis of activation and inhibition. *J. Neurophysiol.* **91**, 1457–1469.

McIntyre C., Richardson A. and Grill W. (2002). Modelling the excitability of mammalian nerve cells: influence of afterpotentials on the recovery cycle. *J. Neurophysiol.* **87**, 995–1006.

McIntyre C.C. and Grill W.M. (2001). Finite element analysis of the current-density and electric field generated by metal microelectrodes. *Ann. Biomed. Eng.* **29**, 227–235.

McIntyre C.C., Richardson A.G. and Grill W.M. (2002). Modeling the excitability of mammalian nerve fibers: influence of afterpotentials on the recovery cycle. *J. Neurophysiol.* **87**, 995–1006.

McKay B.E. and Turner R.W. (2005). Physiological and morphological development of the rat cerebellar Purkinje cell. *J. Physiol.* **567**, 829–850.

McLean D.R. and Graham B.P. (2004). Mathematical formulation and analysis of a continuum model for tubulin-driven neurite elongation. *Proc. R. Soc. Lond. A* **460**, 2437–2456.

McNaughton B.L. and Morris R.G.M. (1987). Hippocampal synaptic enhancement and information storage within a distributed memory system. *Trends Neurosci.* **10**, 408–415.

Mechler F. and Victor J.D. (2012). Dipole characterization of single neurons from their extracellular action potentials. *J. Comput. Neurosci.* **32**, 73–100.

Meijering E. (2010). Neuron tracing in perspective. *Cytometry A* **77**, 693–704.

Meinhardt H. (1983). *Models of Biological Pattern Formation* (Academic Press, Cambridge, MA), second edition.

Meirovitch Y., Kang K., Draft R.W., Pavarino E.C., Echeverri M.F.H., Yang F., Turney S.G., Berger D.R., Peleg A., Schaleck R.L. et al. (2021). Neuromuscular connectomes across development reveal synaptic ordering rules. *bioRxiv* 2021.09.20.460480.

Mejias J.F., Murray J.D., Kennedy H. and Wang X.J. (2016). Feedforward and feedback frequency-dependent interactions in a large-scale laminar network of the primate cortex. *Sci. Adv.* **2**, e1601335.

Mel B.W. (1993). Synaptic integration in an excitable dendritic tree. *J. Neurophysiol.* **70**, 1086–1101.

Menon V., Spruston N. and Kath W.L. (2009). A state-mutating genetic algorithm to design ion-channel models. *Proc. Natl. Acad. Sci. USA* **106**, 16829–16834.

Metz A.E., Spruston N. and Martina M. (2007). Dendritic D-type potassium currents inhibit the spike afterdepolarization in rat hippocampal CA1 pyramidal neurons. *J. Physiol.* **581**, 175–187.

Meyer R.L. (1983). Tetrodotoxin inhibits the formation of refined retinotopography in goldfish. *Dev. Brain Res.* **6**, 293–298.

Meyer R.L. and Sperry R.W. (1973). Tests for neuroplasticity in the anuran retinotectal system. *Exp. Neurol.* **40**, 525–539.

Miceli S., Ness T.V., Einevoll G.T. and Schubert D. (2017). Impedance spectrum in cortical tissue: Implications for propagation of LFP signals on the microscopic level. *eNeuro* **4**, 0291.

Michaelis L. and Menten M.L. (1913). Die Kinetik der Invertinwirkung. *Biochem. Z.* **49**, 333–369.

Michalski P.J. and Loew L.M. (2012). CaMKII activation and dynamics are independent of the holoenzyme structure: an infinite subunit holoenzyme approximation. *Phys Biol* **9**, 036010.

Michelson S. and Schulman H. (1994). CaM kinase: a model for its activation and dynamics. *J. Theor. Biol.* **171**, 281–290.

Migliore M., Cavarretta F., Marasco A., Tulumello E., Hines M.L. and Shepherd G.M. (2015). Synaptic clusters function as odor operators in the olfactory bulb. *Proc. Natl. Acad. Sci. USA* **112**, 8499–8504.

Migliore M., Cook E.P., Jaffe D.B., Turner D.A. and Johnston D. (1995). Computer simulations of morphologically reconstructed CA3 hippocampal neurons. *J. Neurophysiol.* **73**, 1157–1168.

Migliore M., Ferrante M. and Ascoli G.A. (2005). Signal propagation in oblique dendrites of CA1 pyramidal cells. *J. Neurophysiol.* **94**, 4145–4155.

Migliore M. and Shepherd G.M. (2002). Emerging rules for the distributions of active dendritic conductances. *Nat. Rev. Neurosci.* **3**, 362–370.

Migliore R., Lupascu C.A., Bologna L.L., Romani A., Courcol J.D., Antonel S., van Geit W.A.H., Thomson A.M., Mercer A., Lange S., Falck J., Roössert C.A., Shi Y., Hagens O., Pezzoli M., Freund T.F., Kali S., Muller E.B., Schürmann F., Markram H. and Migliore M. (2018). The physiological variability of channel density in hippocampal CA1 pyramidal cells and interneurons explored using a unified data-driven modeling workflow. *PLoS Comput. Biol.* **14**, e1006423.

Milescu L.S., Akk G. and Sachs F. (2005). Maximum likelihood estimation of ion channel kinetics from macroscopic currents. *Biophys. J.* **88**, 2494–2515.

Milescu L.S., Yamanishi T., Ptak K., Mogri M.Z. and Smith J.C. (2008). Real-time kinetic modeling of voltage-gated ion channels using dynamic clamp. *Biophys. J.* **95**, 66–87.

Miller K.E. and Samuels D.C. (1997). The axon as a metabolic compartment: protein degradation, transport and maximum length of an axon. *J. Theor. Biol.* **186**, 373–379.

Minsky M.L. and Papert S.A. (1969). *Perceptrons* (MIT Press, Cambridge, MA).

Miocinovic S., Lempka S.F., Russo G.S., Maks C.B., Butson C.R., Sakaie K.E., Vitek J.L. and McIntyre C.C. (2009). Experimental and theoretical characterization of the voltage distribution generated by deep brain stimulation. *Exp. Neurol.* **216**, 166–176.

Miocinovic S., Parent M., Buston C., Hahn P., Russo G., Vitek J. and McIntyre C. (2006). Computational analysis of subthalamic nucleus and lenticular fasciculus activation during therapeutic deep brain stimulation. *J. Neurophysiol.* **96**, 1569–1580.

Miranda P.C., Callejón-Leblic M.A., Salvador R. and Ruffini G. (2018). Realistic modeling of transcranial current stimulation: The electric field in the brain. *Curr. Opin. Biomed. Eng.* **8**, 20–27.

Mironov V.I., Romanov A.S., Simonov A.Y., Vedunova M.V. and Kazantsev V.B. (2014). Oscillations in a neurite growth model with extracellular feedback. *Neurosci. Lett.* **570**, 16–20.

Mishina M., Kurosaki T., Tobimatsu T., Morimoto Y., Noda M., Yamamoto T., Terao M., Lindstrom J., Takahashi T., Kuno M. and Numa S. (1984). Expression of functional acetylcholine receptor from cloned cDNAs. *Nature* **307**, 604–608.

Miura T. and Maini P.K. (2004). Periodic pattern formation in reaction-diffusion systems: an introduction for numerical simulation. *Anat. Sci. Int* **79**, 112–113.

Miyashita Y. (1988). Neuronal correlate of visual associative long-term memory in the primate temporal cortex. *Nature* **335**, 817–820.

Mobarhan M.H., Halnes G., Martinez-Canada P., Hafting T., Fyhn M. and Einevoll G.T. (2018). Firing-rate based network modeling of the dLGN circuit: Effects of cortical feedback on spatiotemporal response properties of relay cells. *PLoS Comput. Biol.* **14**, e1006156.

Moczydlowski E. and Latorre R. (1983). Gating kinetics of Ca^{2+}-activated K^+ channels from rat muscle incorporated into planar lipid bilayers. Evidence for two voltage-dependent Ca^{2+} binding reactions. *J. Gen. Physiol.* **82**, 511–542.

Mombaerts P. (2006). Axonal wiring in the mouse olfactory system. *Annu. Rev. Cell Dev. Biol.* **22**, 713–737.

Mondragón-González S.L. and Burguière E. (2017). Bio-inspired benchmark generator for extracellular multi-unit recordings. *Sci. Rep.* **7**, 43253.

Moore G.E. (1965). Cramming more components onto integrated circuits. *Electronics* **8**, 1–4.

Moore G.E. (1975). Progress in digital integrated electronics. *1975 International Electronic Devices Meeting* **1**, 11–13.

Morris C. and Lecar H. (1981). Voltage oscillations in the barnacle giant muscle fiber. *Biophys. J.* **35**, 193–213.

Moulin C., Glière A., Barbier D., Joucla S., Yvert B., Mailley P. and Guillemaud R. (2008). A new 3-D finite-element model based on thin-film approximation for microelectrode array recording of extracellular action potential. *IEEE Trans. Biomed. Eng.* **55**, 683–92.

Moy G., Corry B., Kuyucak S. and Chung S.H. (2000). Tests of continuum theories as models of ion channels. I. Poisson–Boltzmann theory versus Brownian dynamics. *Biophys. J.* **78**, 2349–2363.

Mrsic-Flogel T.D., Hofer S.B., Creutzfeldt C., Cloëz-Tayarani I., Changeux J.P., Bonhoeffer T. and Hübener M. (2005). Altered map of visual space in the superior colliculus of mice lacking early retinal waves. *J. Neurosci.* **25**, 6921–6928.

Mrsic-Flogel T.D., Hofer S.B., Ohki K., Reid R.C., Bonhoeffer T. and Hübener M. (2007). Homeostatic regulation of eye-specific responses in visual cortex during ocular dominance plasticity. *Neuron* **54**, 961–972.

Murray J.D. (1993). *Mathematical Biology* (Springer-Verlag, Berlin, Heidelberg, New York, NY), second edition.

Nadal J.P., Toulouse G., Changeux J.P. and Dehaene S. (1986). Networks of formal neurons and memory palimpsests. *Europhys. Lett.* **1**, 535–542.

Næss S., Chintaluri H.C., Ness T.V., Dale A.M., Einevoll G.T. and Wójcik D.K. (2017). Corrected four-sphere head model for EEG signals. *Front. Hum. Neurosci.* **11**, 490.

Næss S., Halnes G., Hagen E., Hagler D.J., Dale A.M., Einevoll G.T. and Ness T.V. (2021). Biophysically detailed forward modeling of the neural origin of EEG and MEG signals. *NeuroImage* **225**, 117467.

Nagumo J., Arimoto S. and Yoshizawa S. (1962). An active pulse transmission line simulating nerve axon. *Proc. IRE* **50**, 2061–2070.

Narahashi T., Moore J.W. and Scott W.R. (1964). Tetrodotoxin blockage of sodium conductance increase in lobster giant axons. *J. Gen. Physiol.* **47**, 965–974.

Neher E. and Sakmann B. (1976). Single-channel currents recorded from membrane of denervated frog muscle fibres. *Nature* **260**, 779–802.

Nelson P. (2008). *Biological Physics* (WH Freeman, New York, NY), updated first edition.

Nernst W. (1888). Zur Kinetik der Lösung befindlichen Körper: Theorie der Diffusion. *Z. Phys. Chem.* **2**, 613–637.

Ness T.B., Potworowski J., Leski S., Chintaluri H.C., Glabska H., Wójcik D.K. and Einevoll G.T. (2015). Modelling and analysis of electrical potentials recorded in microelectrode arrays (MEAs). *Neuroinformatics* **13**, 403–426.

Netoff T.I., Clewley R., Arno S., Keck T. and White J.A. (2004). Epilepsy in small-world networks. *J. Neurosci.* **24**, 8075–8083.

Newport T.D., Sansom M.S.P. and Stansfeld P.J. (2018). The MemProtMD database: a resource for membrane-embedded protein structures and their lipid interactions. *Nucleic Acids Res.* **47**, D390–D397.

Neymotin S.A., Dura-Bernal S., Lakatos P., Sanger T.D. and Lytton W.W. (2016). Multitarget multiscale simulation for pharmacological treatment of dystonia in motor cortex. *Front. Pharmacol.* **7**, 157.

Nicholson C. (2001). Diffusion and related transport mechanisms in brain tissue. *Rep. Prog. Phys.* **64**, 815.

Nicholson C. and Freeman J.A. (1975). Theory of current source-density analysis and determination of conductivity tensor for anuran cerebellum. *J. Neurophysiol.* **38**, 356–368.

Nicholson C. and Llinas R. (1971). Field potentials in the alligator cerebellum and theory of their relationship to Purkinje cell dendritic spikes. *J. Neurophysiol.* **34**, 509–531.

Nicolai C. and Sachs F. (2013). Solving ion channel kinetics with the QuB software. *Biophys. Rev. Lett.* **8**, 191–211.

Nilius B., Hess P., Lansman J.B. and Tsien R.W. (1985). A novel type of cardiac calcium channel in ventricular cells. *Nature* **316**, 443–446.

Nirenberg S. and Pandarinath C. (2012). Retinal prosthetic strategy with the capacity to restore normal vision. *Proc. Natl. Acad. Sci. USA* **109**, 15012–15017.

Nitzan R., Segev I. and Yarom Y. (1990). Voltage behavior along the irregular dendritic structure of morphologically and physiologically characterized vagal motorneurons in the guinea pig. *J. Neurophysiol.* **63**, 333–346.

Noble D. (2006). Systems biology and the heart. *Biosystems* **83**, 75–80.

Nowakowski R.S., Hayes N.L. and Egger M.D. (1992). Competitive interactions during dendritic growth: a simple stochastic growth algorithm. *Brain Res.* **576**, 152–156.

Nowotny T., Levi R. and Selverston A.I. (2008). Probing the dynamics of identified neurons with a data-driven modeling approach. *PLoS One* **3**, e2627.

Nunez P. (1974). The brain wave equation: A model for the EEG. *Math. Biosci.* **21**, 279–297.

Nunez P.L. and Srinivasan R. (2006). *Electric Fields of the Brain: The Neurophysics of EEG* (Oxford University Press, Oxford), second edition.

O'Brien R.A.D., Østberg A.J.C. and Vrbova G. (1978). Observations on the elimination of polyneural innervation in developing mammalian skeletal muscle. *J. Physiol.* **282**, 571–582.

O'Donnell C., Nolan M.F. and van Rossum M.C. (2011). Dendritic spine dynamics regulate the long-term stability of synaptic plasticity. *J. Neurosci.* **31**, 16142–16156.

Okabe A., Boots B. and Sugihara K. (1992). *Spatial Tessellations: Concepts and Applications of Voronoi Diagrams* (Wiley, New York, NY).

O'Keefe J. and Recce M.L. (1993). Phase relationship between hippocampal place units and the EEG theta rhythm. *Hippocampus* **3**, 317–330.

O'Leary T., Sutton A.C. and Marder E. (2015). Computational models in the age of large datasets. *Curr. Opin. Neurobiol.* **32**, 87–94.

O'Leary T., Williams A.H., Franci A. and Marder E. (2014). Cell types, network homeostasis, and pathological compensation from a biologically plausible ion channel expression model. *Neuron* **82**, 809–821.

Orbán G., Kiss T. and Érdi P. (2006). Intrinsic and synaptic mechanisms determining the timing of neuron population activity during hippocampal theta oscillation. *J. Neurophysiol.* **96**, 2889–2904.

Osborne H., Lai Y.M., Lepperød M.E., Sichau D., Deutz L. and de Kamps M. (2021). MIIND : A model-agnostic simulator of neural populations. *Front. Neuroinform.* **15**.

Østby I., Øyehaug L., Einevoll G.T., Nagelhus E.A., Plahte E., Zeuthen T., Lloyd C.M., Ottersen O.P. and Omholt S.W. (2009). Astrocytic mechanisms explaining neural-activity-induced shrinkage of extraneuronal space. *PLoS Comput. Biol.* **5**, e1000272.

Overton K.J. and Arbib M.A. (1982). The extended branch-arrow model of the formation of retino-tectal connections. *Biol. Cybern.* **45**, 157–175.

Owens M.T., Feldheim D.A., Stryker M.P. and Triplett J.W. (2015). Stochastic interaction between neural activity and molecular cues in the formation of topographic maps. *Neuron* **87**, 1261–1273.

Øyehaug L., Østby I., Lloyd C.M., Omholt S.W. and Einevoll G.T. (2012). Dependence of spontaneous neuronal firing and depolarisation block on astroglial membrane transport mechanisms. *J. Comput. Neurosci.* **32**, 147–165.

Paik S.B. and Ringach D.L. (2011). Retinal origin of orientation maps in visual cortex. *Nat. Neurosci.* **14**, 919–925.

Palm G. (1988). On the asymptotic information storage capacity of neural networks. In *Neural Computers*, eds. R. Eckmiller and C. von der Malsburg (Springer-Verlag, Berlin), volume F41 of *NATO ASI Series*, pp. 271–280.

Parisi G. (1986). A memory which forgets. *J. Phys. A Math. Gen.* **19**, L617–L620.

Park M.R., Kita H., Klee M.R. and Oomura Y. (1983). Bridge balance in intracellular recording; introduction of the phase-sensitive method. *J. Neurosci. Methods* **8**, 105–125.

Parnas I. and Segev I. (1979). A mathematical model for conduction of action potentials along bifurcating axons. *J. Physiol.* **295**, 323–343.

Patlak J. (1991). Molecular kinetics of voltage-dependent Na^+ channels. *Physiol. Rev.* **71**, 1047–1080.

Peng H., Hawrylycz M., Roskams J., Hill S., Spruston N., Meijering E. and Ascoli G.A. (2015). BigNeuron: Large-scale 3D neuron reconstruction from optical microscopy images. *Neuron* **87**, 252–256.

Pesaran B., Vinck M., Einevoll G.T., Sirota A., Fries P., Siegel M., Truccolo W., Schroeder C.E. and Srinivasan R. (2018). Investigating large-scale brain dynamics using field potential recordings: analysis and interpretation. *Nat. Neurosci.* **21**, 903–919.

Pettersen K.H., Devor A., Ulbert I., Dale A.M. and Einevoll G.T. (2006). Current-source density estimation based on inversion of electrostatic forward solution: effects of finite extent of neuronal activity and conductivity discontinuities. *J. Neurosci. Methods* **154**, 116–133.

Pettersen K.H. and Einevoll G.T. (2008). Amplitude variability and extracellular low-pass filtering of neuronal spikes. *Biophys. J.* **94**, 784–802.

Pettersen K.H., Lindén H., Dale A.M. and Einevoll G.T. (2012). Extracellular spikes and CSD. In *Handbook of Neural Activity Measurement*, eds. R. Brette and A. Destexhe (Cambridge University Press, Cambridge), pp. 92–135.

Pfeiffenberger C., Yamada J. and Feldheim D.A. (2006). Ephrin-As and patterned retinal activity act together in the development of topographic maps in the primary visual system. *J. Neurosci.* **26**, 12873–12884.

Phelan P., Goulding L.A., Tam J.L.Y., Allen M.J., Dawber R.J., Davies J.A. and Bacon J.P. (2009). Molecular mechanism of rectification at identified electrical synapses in the *Drosophila* giant fiber system. *Curr. Biol.* **18**, 1955–1960.

Phillips J.C., Hardy D.J., Maia J.D.C., Stone J.E., Ribeiro J.A.V., Bernardi R.C., Buch R., Fiorin G., Hénin J., Jiang W., McGreevy R., Melo M.C.R., Radak B.K., Skeel R.D., Singharoy A., Wang Y., Roux B., Aksimentiev A., Luthey-Schulten Z., Kalé L.V., Schulten K., Chipot C. and Tajkhorshid E. (2020). Scalable molecular dynamics on CPU and GPU architectures with NAMD. *J. Chem. Phys.* **153**, 044130.

Pi H.J. and Lisman J.E. (2008). Coupled phosphatase and kinase switches produce the tristability required for long-term potentiation and long-term depression. *J. Neurosci.* **28**, 13132–13138.

Pinsky P.F. and Rinzel J. (1994). Intrinsic and network rhythmogenesis in a reduced Traub model for CA3 neurons. *J. Comput. Neurosci.* **1**, 39–60.

Pinto D.J., Hartings J.A., Brumberg J.C. and Simons D.J. (2003). Cortical damping: Analysis of thalamocortical response transformations in rodent barrel cortex. *Cereb. Cortex* **13**, 33–44.

Pinto D.J., Patrick S.L., Huang W.C. and Connors B.W. (2005). Initiation, propagation, and termination of epileptiform activity in rodent neocortex in vitro involve distinct mechanisms. *J. Neurosci.* **25**, 8131–8140.

Planck M. (1890). Über die Erregung von Electricität und Wärme in Electrolyten. *Ann. Phys. Chem. Neue Folge* **39**, 161–186.

Podlaski W.F., Seeholzer A., Groschner L.N., Miesenböck G., Ranjan R. and Vogels T.P. (2017). Mapping the function of neuronal ion channels in model and experiment. *Elife* **6**, e22152.

Pods J., Schönke J. and Bastian P. (2013). Electrodiffusion models of neurons and extracellular space using the Poisson–Nernst–Planck equations – numerical simulation of the intra- and extracellular potential for an axon model. *Biophys. J.* **105**, 242–254.

Poggio T., Banburski A. and Liao Q. (2020). Theoretical issues in deep networks. *Proc. Natl. Acad. Sci. USA* **117**, 30039–30045.

Poirazi P., Brannon T. and Mel B.W. (2003). Arithmetic of subthreshold synaptic summation in a model CA1 pyramidal cell. *Neuron* **37**, 977–987.

Pollak E. and Talkner P. (2005). Reaction rate theory: What it was, where is it today, and where is it going? *Chaos* **15**, 026116.

Potjans T.C. and Diesmann M. (2014). The cell-type specific cortical microcircuit: relating structure and activity in a full-scale spiking network model. *Cereb. Cortex* **24**, 785–806.

Potworowski J., Jakuczun W., Łęski S. and Wójcik D. (2012). Kernel current source density method. *Neural Comput.* **24**, 541–575.

Pozzorini C., Mensi S., Hagens O., Naud R., Koch C. and Gerstner W. (2015). Automated high-throughput characterization of single neurons by means of simplified spiking models. *PLoS Comput. Biol.* **11**, e1004275.

Prescott S.A., De Koninck Y. and Sejnowski T.J. (2008). Biophysical basis for three distinct dynamical mechanisms of action potential initiation. *PLoS Comput. Biol.* **4**, e1000198.

Press W.H., Flannery B.P., Teukolsky S.A. and Vetterling W.T., eds. (1987). *Numerical Recipes: The Art of Scientific Computing* (Cambridge University Press, Cambridge).

Prestige M.C. and Willshaw D.J. (1975). On a role for competition in the formation of patterned neural connexions. *Proc. R. Soc. Lond. B Biol. Sci.* **190**, 77–98.

Price D.J., Kind P., Jarman A. and Mason J. (2011). *An Introduction to Neural Mechanisms of Development* (John Wiley, Hoboken, NJ).

Price D.J. and Willshaw D.J. (2000). *Mechanisms of Cortical Development* (Oxford University Press, Oxford).

Price L.F., Drovandi C.C., Lee A. and Nott D.J. (2018). Bayesian synthetic likelihood. *J. Comput. Graph. Stat.* **27**, 1–11.

Priebe N.J. (2016). Mechanisms of orientation selectivity in the primary visual cortex. *Annu. Rev. Vis. Sci.* **2**, 85–107.

Prinz A.A., Billimoria C.P. and Marder E. (2003). Alternative to hand-tuning conductance-based models: construction and analysis of databases of model neurons. *J. Neurophysiol.* **90**, 3998–4015.

Protopapas A.D., Vanier M. and Bower J.M. (1998). Simulating large networks of neurons. In *Methods in Neuronal Modeling: From Ions to Networks*, eds. C. Koch and I. Segev (MIT Press, Cambridge, MA), chapter 12, pp. 461–498. Second edition.

Purves D. (1981). *Microelectrode Methods for Intracellular Recording and Ionophoresis* (Academic Press, London).

Purves D. and Lichtman J.W. (1980). Elimination of the synapses in the developing nervous system. *Science* **210**, 153–157.

Qian N. and Sejnowski T.J. (1989). An electro-diffusion model for computing membrane potentials and ionic concentrations in branching dendrites, spines and axons. *Biol. Cybern.* **62**, 1–15.

Qin F., Auerbach A. and Sachs F. (1996). Estimating single-channel kinetic parameters from idealized patch clamp data containing missed events. *Biophys. J.* **70**, 264–280.

Qin F., Auerbach A. and Sachs F. (1997). Maximum likelihood estimation of aggregated Markov processes. *Proc. R. Soc. Lond. B Biol. Sci.* **264**, 375–383.

Quiroga R.Q. (2007). Spike sorting. *Scholarpedia* **2**, 3583.

Rackham O., Tsaneva-Atanasova K., Ganesh A. and Mellor J. (2010). A Ca^{2+}-based computational model for NMDA receptor-dependent synaptic plasticity at individual post-synaptic spines in the hippocampus. *Front. Synaptic Neurosci.* **2**, 31.

Radojević M. and Meijering E. (2019). Automated neuron reconstruction from 3D fluorescence microscopy images using sequential Monte Carlo estimation. *Neuroinformatics* **17**, 423–442.

Rakic P. (1976). Prenatal genesis of connections subserving ocular dominance in the rhesus monkey. *Nature* **261**, 467–471.

Rall W. (1957). Membrane time constant of motorneurons. *Science* **126**, 454.

Rall W. (1962). Electrophysiology of a dendritic neuron model. *Biophys. J.* **2**, 145–167.

Rall W. (1964). Theoretical significance of dendritic trees for neuronal input–output relations. In *Neural Theory and Modeling*, ed. R.F. Reiss (Stanford University Press, Palo Alto, CA), pp. 73–97.

Rall W. (1967). Distinguishing theoretical synaptic potentials computed for different somadendritic distributions of synaptic inputs. *J. Neurophysiol.* **30**, 1138–1168.

Rall W. (1969). Time constants and electrotonic length of membrane cylinders and neurons. *Biophys. J.* **9**, 1438–1541.

Rall W. (1977). Theoretical significance of dendritic trees for neuronal input–output relations. In *Handbook of Physiology. The Nervous System. Cellular Biology of Neurons* (American Physiological Society, Bethesda, MD), pp. 39–97.

Rall W. (1990). Perspectives on neuron modeling. In *The Segmental Motor System*, eds. M.D. Binder and L.M. Mendell (Oxford University Press, New York, NY), pp. 129–149.

Rall W., Burke R.E., Holmes W.R., Jack J.J., Redman S.J. and Segev I. (1992). Matching dendritic neuron models to experimental data. *Physiol. Rev.* **72**, 159–186.

Rall W. and Shepherd G.M. (1968). Theoretical reconstruction of field potentials and dendrodendritic synaptic interactions in olfactory bulb. *J. Neurophysiol.* **31**, 884–915.

Rall W., Shepherd G.M., Reese T.S. and Brightman M.W. (1966). Dendrodendritic synaptic pathway for inhibition in the olfactory bulb. *Exp. Neurol.* **14**, 44–56.

Ramón y Cajal S. (1911). *Histologie du Système Nerveux de l'Homme et des Vertébrés* (Maloine, Paris). English translation by P. Pasik and T. Pasik (1999). *Texture of the Nervous System of Man and the Vertebrates* (Springer, New York, NY).

Ran I., Quastel D.M.J., Mathers D.A. and Puil E. (2009). Fluctuation analysis of tetanic rundown (short-term depression) at a corticothalamic synapse. *Biophys. J.* **96**, 2505–2531.

Ranjan R., Logette E., Marani M., Herzog M., Tâche V., Scantamburlo E., Buchillier V. and Markram H. (2019). A kinetic map of the homomeric voltage-gated potassium channel (Kv) family. *Front. Cell. Neurosci.* **13**, 358.

Rao-Mirotznik R., Buchsbaum G. and Sterling P. (1998). Transmitter concentration at a three-dimensional synapse. *J. Neurophysiol.* **80**, 3163–3172.

Rashid T., Upton A.L., Blentic A., Ciossek T., Knöll B., Thompson I.D. and Drescher U. (2005). Opposing gradients of ephrin-As and EphA7 in the superior colliculus are essential for topographic mapping in the mammalian visual system. *Neuron* **47**, 57–69.

Rasmussen C.E. and Willshaw D.J. (1993). Presynaptic and postsynaptic competition in models for the development of neuromuscular connections. *Biol. Cybern.* **68**, 409–419.

Rattay F. (1999). The basic mechanism for the electrical stimulation of the nervous system. *Neuroscience* **89**, 335–346.

Reber M., Burrola P. and Lemke G. (2004). A relative signalling model for the formation of a topographic neural map. *Nature* **431**, 847–853.

Redman S.J. (1990). Quantal analysis of synaptic potentials in neurons of the central nervous system. *Physiol. Rev.* **70**, 165–198.

Reh T.A. and Constantine-Paton M. (1983). Retinal ganglion cell terminals change their projection sites during larval development of *Rana pipiens*. *J. Neurosci.* **4**, 442–457.

Reimann M.W., Anastassiou C.A., Perin R., Hill S.L., Markram H. and Koch C. (2013). A biophysically detailed model of neocortical local field potentials predicts the critical role of active membrane currents. *Neuron* **79**, 375–390.

Resta V., Novelli E., Di Virgilio F. and Galli-Resta L. (2005). Neuronal death induced by endogenous extracellular ATP in retinal cholinergic neuron density control. *Development* **132**, 2873–2882.

Rey H.G., Pedreira C. and Quian Quiroga R. (2015). Past, present and future of spike sorting techniques. *Brain Res. Bull.* **119**, 106–117.

Reymann K.G. and Frey J.U. (2007). The late maintenance of hippocampal LTP: requirements, phases, 'synaptic tagging', 'late-associativity' and implications. *Neuropharmacology* **52**, 24–40.

Richards B.A. and Lillicrap T.P. (2019). Dendritic solutions to the credit assignment problem. *Curr. Opin. Neurobiol.* **54**, 28–36.

Rieke F., Warland D., de Ruyter van Steveninck R. and Bialek W. (1997). *Spikes: Exploring the Neural Code* (MIT Press, Cambridge, MA).

Rinzel J. and Ermentrout B. (1998). Analysis of neural excitability and oscillations. In *Methods in Neuronal Modeling: From Ions to Networks*, eds. C. Koch and I. Segev (MIT Press, Cambridge, MA), pp. 251–291. Second edition.

Rizzoli S.O. and Betz W.J. (2005). Synaptic vesicle pools. *Nat. Rev. Neurosci.* **6**, 57–69.

Rizzone M., Lanotte M., Bergamasco B., Tavella A., Torre E., Faccani G., Melcarne A. and Lopiano L. (2001). Deep brain stimulation of the subthalamic nucleus in Parkinson's disease: effects of variation in stimulation parameters. *J. Neurol. Neurosurg. Psychiatr.* **71**, 215–219.

Robinson R.A. and Stokes R.H. (2002). *Electrolyte Solutions* (Dover, New York, NY).

Rodieck R.W. (1965). Quantitative analysis of cat retinal ganglion cell response to visual stimuli. *Vision Res.* **5**, 583–601.

Rodieck R.W. (1967). Maintained activity of cat retinal ganglion cells. *J. Neurophysiol.* **30**, 1043–1071.

Rodriguez B.M., Sigg D. and Bezanilla F. (1998). Voltage gating of *Shaker* K^+ channels: the effect of temperature on ionic and gating currents. *J. Gen. Physiol.* **112**, 223–242.

Roe A., ed. (2009). *Imaging the Brain with Optical Methods* (Springer, Berlin).

Rosenblatt F. (1958). The perceptron: a probabilistic model for information storage and organization in the brain. *Psychol. Rev.* **65**, 386–408.

Roshchin M.V., Ierusalimsky V.N., Balaban P.M. and Nikitin E.S. (2020). Ca^{2+}-activated KCa3.1 potassium channels contribute to the slow afterhyperpolarization in L5 neocortical pyramidal neurons. *Sci. Rep.* **10**, 14484.

Rosin B., Slovik M., Mitelman R., Rivlin-Etzion M., Haber S.N., Israel Z., Vaadia E. and Bergman H. (2011). Closed-loop deep brain stimulation is superior in ameliorating parkinsonism. *Neuron* **72**, 370–384.

Roudi Y. and Treves A. (2006). Localized activity profiles and storage capacity of rate-based autoassociative networks. *Phys. Rev. E* **73**, 061904.

Roux B., Allen T., Bernèche S. and Im W. (2004). Theoretical and computational models of biological ion channels. *Q. Rev. Biophys.* **37**, 15–103.

Roy S., Jun N.Y., Davis E.L., Pearson J. and Field G.D. (2021). Inter-mosaic coordination of retinal receptive fields. *Nature* **592**, 409–413.

Rubin D.C. and Wenzel A.E. (1996). One hundred years of forgetting: a quantitative description of retention. *Psychol. Rev.* **103**, 734–760.

Rubin J.E., Gerkin R.C., Bi G.Q. and Chow C.C. (2005). Calcium time course as a signal for spike-timing-dependent plasticity. *J. Neurophysiol.* **93**, 2600–2613.

Rudolph M. and Destexhe A. (2006). Analytical integrate-and-fire neuron models with conductance-based dynamics for event-driven simulation strategies. *Neural Comput.* **18**, 2146–2210.

Rumelhart D.E., Hinton G.E. and Williams R.J. (1986a). Learning representations by back-propagating errors. *Nature* **323**, 533–536.

Rumelhart D.E., McClelland J.L. and the PDP Research Group, eds. (1986b). *Parallel Distributed Processing: Explorations in the Microstructure of Cognition. Volume 1: Foundations* (MIT Press, Cambridge, MA).

Sabatini B.L., Oertner T.G. and Svoboda K. (2002). The life cycle of Ca^{2+} ions in dendritic spines. *Neuron* **33**, 439–452.

Sætra M.J., Einevoll G.T. and Halnes G. (2020). An electrodiffusive, ion conserving Pinsky-Rinzel model with homeostatic mechanisms. *PLoS Comput. Biol.* **16**, e1007661.

Sah P. and Faber E.S. (2002). Channels underlying neuronal calcium-activated potassium currents. *Prog. Neurobiol.* **66**, 345–353.

Sah P., Gibb A.J. and Gage P.W. (1988). Potassium current activated by depolarization of dissociated neurons from adult guinea pig hippocampus. *J. Gen. Physiol.* **92**, 263–278.

Sakka L., Coll G. and Chazal J. (2011). Anatomy and physiology of cerebrospinal fluid. *Eur. Ann. Otorhinolaryngol. Head Neck Dis.* **128**, 309–316.

Saltelli A., Ratto M., Andres T., Campolongo F., Cariboni J., Gatelli D., Saisan M. and Tarantola S. (2008). *Global Sensitivity Analysis. The Primer.* (John Wiley & Sons, Ltd., Hoboken, NJ).

Samsonovich A.V. and Ascoli G.A. (2003). Statistical morphological analysis of hippocampal principal neurons indicates cell-specific repulsion of dendrites from their own cell. *J. Neurosci. Res.* **71**, 173–187.

Samsonovich A.V. and Ascoli G.A. (2005a). Algorithmic description of hippocampal granule cell dendritic morphology. *Neurocomputing* **65-66**, 253–260.

Samsonovich A.V. and Ascoli G.A. (2005b). Statistical determinants of dendritic morphology in hippocampal pyramidal neurons: a hidden Markov model. *Hippocampus* **15**, 166–183.

Sanes D.H., Reh T.A. and Harris W.A. (2000). *Development of the Nervous System* (Academic Press, San Diego, CA).

Sanes J.R. and Lichtman J.W. (1999). Development of the vertebrate neuromuscular junction. *Annu. Rev. Neurosci.* **22**, 389–342.

Santhakumar V., Aradi I. and Soltetz I. (2005). Role of mossy fiber sprouting and mossy cell loss in hyperexcitability: a network model of the dentate gyrus incorporating cell types axonal typography. *J. Neurophysiol.* **93**, 437–453.

Sanz Leon P., Knock S.A., Woodman M.M., Domide L., Mersmann J., McIntosh A.R. and Jirsa V. (2013). The virtual brain: a simulator of primate brain network dynamics. *Front. Neuroinform.* **7**, 10.

Sato F., Parent M., Lévesque M. and Parent A. (2000). Axonal branching pattern of neurons of the subthalamic nucleus in primates. *J. Comp. Neurol.* **424**, 142–152.

Saudargiene A., Cobb S. and Graham B.P. (2015). A computational study on plasticity during theta cycles at Schaffer collateral synapses on CA1 pyramidal cells in the hippocampus. *Hippocampus* **25**, 208–218.

Saudargiene A. and Graham B.P. (2015). Inhibitory control of site-specific synaptic plasticity in a model CA1 pyramidal neuron. *Biosyst.* **130**, 37–50.

Schiegg A., Gerstner W., Ritz R. and van Hemmen J.L. (1995). Intracellular Ca^{2+} stores can account for the time course of LTP induction: a model of Ca^{2+} dynamics in dendritic spines. *J. Neurophysiol.* **74**, 1046–1055.

Schmidt J.T. (2019). *Self-Organizing Neural Maps: The Retinotectal Map and Mechanisms of Neural Development: From Retina to Tectum* (Academic Press, Cambridge, MA).

Schmidt J.T., Cicerone C.M. and Easter S.S. (1978). Expansion of the half retinal projection to the tectum in goldfish: an electrophysiological and anatomical study. *J. Comp. Neurol.* **177**, 257–278.

Schmidt J.T. and Edwards D.L. (1983). Activity sharpens the map during the regeneration of the retinotectal projection in goldfish. *Brain Res.* **269**, 29–39.

Scholes J. (1979). Nerve fibre topography in the retinal projection to the tectum. *Nature* **278**, 620–624.

Schrempf H., Schmidt O., Kümmerlen R., Hinnah S., Müller D., Betzler M., Steinkamp T. and Wagner R. (1995). A prokaryotic potassium ion channel with two predicted transmembrane segments from *Streptomyces lividans*. *EMBO J.* **14**, 5170–5178.

Schuster S., Marhl M. and Höfer (2002). Modelling of simple and complex calcium oscillations: from single-cell responses to intercellular signalling. *Eur. J. Biochem.* **269**, 1333–1355.

Scorcioni R., Lazarewicz M.T. and Ascoli G.A. (2004). Quantitative morphometry of hippocampal pyramidal cells: differences between anatomical classes and reconstructing laboratories. *J. Comp. Neurol.* **473**, 177–193.

Seeber B.U. and Bruce I.C. (2016). The history and future of neural modeling for cochlear implants. *Network Comp. Neural Syst.* **27**, 53–66.

Segev I. (1990). Computer study of presynaptic inhibition controlling the spread of action potentials into axonal terminals. *J. Neurophysiol.* **63**, 987–998.

Segev I., Rinzel J. and Shepherd G., eds. (1995). *The Theoretical Foundation of Dendritic Function: Selected Papers of Wilfrid Rall with Commentaries* (MIT Press, Cambridge, MA).

Sejnowski T.J. (1977). Statistical constraints on synaptic plasticity. *J. Theor. Biol.* **69**, 385–389.

Sejnowski T.J. (2020). The unreasonable effectiveness of deep learning in artificial intelligence. *Proc. Natl. Acad. Sci. USA* **117**, 30033–30038.

Sekulić V., Lawrence J.J. and Skinner F.K. (2014). Using multi-compartment ensemble modeling as an investigative tool of spatially distributed biophysical balances: Application to hippocampal oriens-lacunosum/moleculare (O-LM) cells. *PLoS One* **9(10)**, e106567.

Seung H.S., Lee D.D., Reis B.Y. and Tank D.W. (2000). Stability of the memory of eye position in a recurrent network of conductance-based neurons. *Neuron* **26**, 259–271.

Shadlen M.N. and Newsome W.T. (1994). Noise, neural codes and cortical organization. *Curr. Opin. Neurobiol.* **4**, 569–579.

Shadlen M.N. and Newsome W.T. (1995). Is there a signal in the noise? *Curr. Opin. Neurobiol.* **5**, 248–250.

Shadlen M.N. and Newsome W.T. (1998). The variable discharge of cortical neurons: implications for connectivity, computation, and information coding. *J. Neurosci.* **18**, 3870–3896.

Shai A.S., Anastassiou C.A., Larkum M.E. and Koch C. (2015). Physiology of layer 5 pyramidal neurons in mouse visual cortex: coincidence detection through bursting. *PLoS Comput. Biol.* **11**, e1004090.

Shannon C.E. (1948). A mathematical theory of communication. *Bell. Syst. Tech. J.* **27**, 379–423.

Sharma S.C. (1972). The retinal projection in adult goldfish: an experimental study. *Brain Res.* **39**, 213–223.

Sharp A.A., O'Neil M.B., Abbott L.F. and Marder E. (1993). The dynamic clamp: artificial conductances in biological neurons. *Trends Neurosci.* **16**, 389–394.

Shepherd G.M. (1996). The dendritic spine: a multifunctional integrative unit. *J. Neurophysiol.* **75**, 2197–2210.

Shimbel A. (1950). Contributions to the mathematical biophysics of the central nervous system with special reference to learning. *Bull. Math. Biophys.* **12**, 241–275.

Shimizu G., Yoshida K., Kasai H. and Toyoizumi T. (2021). Computational roles of intrinsic synaptic dynamics. *Curr. Opin. Neurobiol.* **70**, 34–42.

Shoeibi A., Khodatars M., Ghassemi N., Jafari M., Moridian P., Alizadehsani R., Panahiazar M., Khozeimeh F., Zare A., Hosseini-Nejad H., Khosravi A., Atiya A.F., Aminshahidi D., Hussain S., Rouhani M., Nahavandi S. and Acharya U.R. (2021). Epileptic seizures detection using deep learning techniques: A review. *Int. J. Environ. Res. Public Health* **18**, 5780.

Siekmann I., Sneyd J. and Crampin E.J. (2012). MCMC can detect nonidentifiable models. *Biophys. J.* **103**, 2275–2286.

Singer W. (1993). Synchronization of cortical activity and its putative role in information processing and learning. *Annu. Rev. Physiol.* **55**, 349–374.

Sivilotti L. and Colquhoun D. (2016). In praise of single channel kinetics. *J. Gen. Physiol.* **148**, 79–88.

Sjöström P.J., Turrigiano G.G. and Nelson S.B. (2007). Multiple forms of long-term plasticity at unitary neocortical layer 5 synapses. *Neuropharmacology* **52**, 176–184.

Skaar J.E.W., Stasik A.J., Hagen E., Ness T.V. and Einevoll G.T. (2020). Estimation of neural network model parameters from local field potentials (LFPs). *PLoS Comput. Biol.* **16**, e1007725.

Skinner F.K., Turrigiano G.G. and Marder E. (1993). Frequency and burst duration in oscillating neurons and two-cell networks. *Biol. Cybern.* **69**, 375–383.

Smart J.L. and McCammon J.A. (1998). Analysis of synaptic transmission in the neuromuscular junction using a continuum finite element model. *Biophys. J.* **75**, 1679–1688.

Smith D.J. and Rubel E.W. (1979). Organization and development of brain stem auditory nuclei of the chicken: dendritic gradients in nucleus laminaris. *J. Comp. Neurol.* **186**, 213–239.

Smith G.D. (2001). Modeling local and global calcium signals using reaction–diffusion equations. In *Computational Neuroscience: Realistic Modeling for Experimentalists*, ed. E. De Schutter (CRC Press, Boca Raton, FL), chapter 3, pp. 49–85.

Smith Y., Bolam J.P.P. and Von Krosigk M. (1990). Topographical and synaptic organization of the GABA-containing pallidosubthalamic projection in the rat. *Eur. J. Neurosci.* **2**, 500–511.

Sneddon M.W., Faeder J.R. and Emonet T. (2011). Efficient modeling, simulation and coarse-graining of biological complexity with NFsim. *Nat. Methods* **8**, 177–183.

Soetaert K., Petzoldt T. and Setzer R.W. (2010). Solving differential equations in R: Package deSolve. *J. Stat. Softw.* **33**, 1–25.

Soffe S.R., Roberts A. and Li W.C. (2009). Defining the excitatory neurons that drive the locomotor rhythm in a simple vertebrate: insights into the origin of reticulospinal control. *J. Physiol.* **587**, 4829–4844.

Softky W.R. and Koch C. (1993). The highly irregular firing of cortical cells is inconsistent with temporal integration of random EPSPs. *J. Neurosci.* **13**, 334–350.

Sohal V.S. and Sun F.T. (2011). Responsive neurostimulation suppresses synchronized cortical rhythms in patients with epilepsy. *Neurosurg. Clin. N. Am.* **22**, 481–488.

Solbrå A., Bergersen A.W., van den Brink J., Malthe-Sørenssen A., Einevoll G.T. and Halnes G. (2018). A Kirchhoff–Nernst–Planck framework for modeling large scale extracellular electrodiffusion surrounding morphologically detailed neurons. *PLoS Comput. Biol.* **14**, e1006510.

Somjen G., Kager H. and Wadman W. (2008). Computer simulations of neuron-glia interactions mediated by ion flux. *J. Comput. Neurosci.* **25**, 349–365.

Somjen G.G. (2004). *Ions in the Brain: Normal Function, Seizures, and Stroke* (Oxford University Press, New York, NY).

Sommer F.T. and Wennekers T. (2000). Modelling studies on the computational function of fast temporal structure in cortical circuit activity. *J. Physiol. Paris* **94**, 473–488.

Sommer F.T. and Wennekers T. (2001). Associative memory in networks of spiking neurons. *Neural Netw.* **14**, 825–834.

Somogyi P. and Klausberger T. (2005). Defined types of cortical interneurone structure space and spike timing in the hippocampus. *J. Physiol.* **562.1**, 9–26.

Song S., Miller K.D. and Abbott L.F. (2000). Competitive Hebbian learning through spike-timing-dependent synaptic plasticity. *Nat. Neurosci.* **3**, 919–926.

Sosinsky G.E., Deerinck T.J., Greco R., Buitenhuys C.H., Bartol T.M. and Ellisman M.H. (2005). Development of a model for microphysiological simulations: small nodes of Ranvier from peripheral nerves of mice reconstructed by electron tomography. *Neuroinformatics* **3**, 133–162.

Soto-Treviño C., Thoroughman K.A., Marder E. and Abbott L. (2001). Activity-dependent modification of inhibitory synapses in models of rhythmic neural networks. *Nat. Neurosci.* **4**, 297–303.

Sperry R.W. (1943). Visuomotor co-ordination in the newt (*Triturus viridescens*) after regeneration of the optic nerve. *J. Comp. Neurol.* **79**, 33–55.

Sperry R.W. (1944). Optic nerve regeneration with return of vision in anurans. *J. Neurophysiol.* **7**, 57–69.

Sperry R.W. (1945). Restoration of vision after crossing of optic nerves and after contralateral transplantation of the eye. *J. Neurophysiol.* **8**, 15–28.

Sperry R.W. (1963). Chemoaffinity in the orderly growth of nerve fiber patterns and connections. *Proc. Natl. Acad. Sci. USA* **50**, 703–710.

Spitzer N.C., Kingston P.A., Manning Jr T.J. and Conklin M.W. (2002). Outside and in: development of neuronal excitability. *Curr. Opin. Neurobiol.* **12**, 315–323.

Srinivasan R. and Chiel H.J. (1993). Fast calculation of synaptic conductances. *Neural Comput.* **5**, 200–204.

Starek G., Freites J.A., Bernèche S. and Tobias D.J. (2017). Gating energetics of a voltage-dependent K^+ channel pore domain. *J. Comput. Chem.* **38**, 1472–1478.

Stefan M.I., Bartol T.M., Sejnowski T.J. and Kennedy M.B. (2014). Multi-state modeling of biomolecules. *PLoS Comput. Biol.* **10**, e1003844.

Stefan M.I., Marshall D.P. and Le Novère N. (2012). Structural analysis and stochastic modelling suggest a mechanism for calmodulin trapping by CaMKII. *PLoS One* **7**, e29406.

Stein R.B. (1965). A theoretical analysis of neuronal variability. *Biophys. J.* **5**, 173–194.

Stemmler M. and Koch C. (1999). How voltage-dependent conductances can adapt to maximize the information encoded by neuronal firing rate. *Nat. Neurosci.* **2**, 521–527.

Sten S., Lundengård K., Witt S.T., Cedersund G., Elinder F. and Engström M. (2017). Neural inhibition can explain negative BOLD responses: A mechanistic modelling and fMRI study. *NeuroImage* **158**, 219–231.

Sterratt D.C. and Willshaw D. (2008). Inhomogeneities in heteroassociative memories with linear learning rules. *Neural Comput.* **20**, 311–344.

Stevens C.F. (1978). Interactions between intrinsic membrane protein and electric field. *Biophys. J.* **22**, 295–306.

Stiles J.R. and Bartol T.M. (2001). Monte Carlo methods for simulating realistic synaptic microphysiology using MCell. In *Computational Neuroscience: Realistic Modeling for Experimentalists*, ed. E. De Schutter (CRC Press, Boca Raton, FL), chapter 4, pp. 87–127.

Stirling R.E., Cook M.J., Grayden D.B. and Karoly P.J. (2021). Seizure forecasting and cyclic control of seizures. *Epilepsia* **62**, S2–S14.

Stocker M. (2004). Ca^{2+}-activated K^+ channels: molecular determinants and function of the SK family. *Nat. Rev. Neurosci.* **5**, 758–770.

Storm J.F. (1988). Temporal integration by a slowly inactivating K^+ current in hippocampal neurons. *Nature* **336**, 379–381.

Strassberg A.F. and DeFelice L.J. (1993). Limitations of the Hodgkin–Huxley formalism: effects of single channel kinetics on transmembrane voltage dynamics. *Neural Comput.* **5**, 843–855.

Stratford K., Mason A., Larkman A., Major G. and Jack J.J.B. (1989). The modelling of pyramidal neurons in the visual cortex. In *The Computing Neuron*, eds. R. Durbin, C. Miall and G. Mitchison (Addison-Wesley, Workingham), pp. 296–321.

Stuart G. and Spruston N. (1998). Determinants of voltage attenuation in neocortical pyramidal neuron dendrites. *J. Neurosci.* **18**, 3501–3510.

Stundzia A.B. and Lumsden C.J. (1996). Stochastic simulation of coupled reaction-diffusion processes. *J. Comput. Phys.* **127**, 196–207.

Sumikawa K., Houghton M., Emtage J.S., Richards B.M. and Barnard E.A. (1981). Active multi-subunit ACh receptor assembled by translation of heterologous mRNA in *Xenopus* oocytes. *Nature* **292**, 862–864.

Sutton R.S. and Barto A.G. (1998). *Reinforcement Learning: An Introduction* (MIT Press, Cambridge, MA).

Sutton R.S. and Barto A.G. (2018). *Reinforcement Learning: An Introduction* (MIT Press, Cambridge, MA), second edition.

Swadlow H.A., Gusev A.G. and Bezdudnaya T. (2002). Activation of a cortical column by a thalamocortical impulse. *J. Neurosci.* **22**, 7766–7773.

Sweeney P.W., Walker-Samuel S. and Shipley R.J. (2018). Insights into cerebral haemodynamics and oxygenation utilising in vivo mural cell imaging and mathematical modelling. *Sci. Rep.* **8**, 1373.

Sweeney Y., Hellgren Kotaleski J. and Hennig M.H. (2015). A diffusive homeostatic signal maintains neural heterogeneity and responsiveness in cortical networks. *PLoS Comput. Biol.* **11**, e1004389.

Swindale N. (1996). The development of topography in the visual cortex: a review of models. *Network Comp. Neural Syst.* **7**, 161–247.

Swindale N.V. (1980). A model for the formation of ocular dominance stripes. *Proc. R. Soc. Lond. B Biol. Sci.* **208**, 243–264.

Szilágyi T. and De Schutter E. (2004). Effects of variability in anatomical reconstruction techniques on models of synaptic integration by dendrites: a comparison of three Internet archives. *Eur. J. Neurosci.* **19**, 1257–1266.

Tamori Y. (1993). Theory of dendritic morphology. *Phys. Rev. E* **48**, 3124–3129.

Tang Y. and Othmer H.G. (1994). A model of calcium dynamics in cardiac myocytes based on the kinetics of ryanodine-sensitive calcium channels. *Biophys. J.* **67**, 2223–2235.

Tapia J.J., Saglam A.S., Czech J., Kuczewski R., Bartol T.M., Sejnowski T.J. and Faeder J.R. (2019). MCell-R: A particle-resolution network-free spatial modeling framework. *Methods Mol. Biol.* **1945**, 203–229.

Telenczuk M., Brette R., Destexhe A. and Telenczuk B. (2018). Contribution of the axon initial segment to action potentials recorded extracellularly. *eNeuro* **5**, ENEURO.0068-18.2018.

Tennøe S., Halnes G. and Einevoll G.T. (2018). Uncertainpy: a Python toolbox for uncertainty quantification and sensitivity analysis in computational neuroscience. *Front. Neuroinform.* **12**, 49.

Thompson A.C., Stoddart P.R. and Jansen E.D. (2014). Optical stimulation of neurons. *Curr. Mol. Imaging* **3**, 162–177.

Thompson E.H., Lensjø K.K., Wigestrand M.B., Malthe-Sørenssen A., Hafting T. and Fyhn M. (2018). Removal of perineuronal nets disrupts recall of a remote fear memory. *Proc. Natl. Acad. Sci. USA* **115**, 607–612.

Thomson A.M. (2000a). Facilitation, augmentation and potentiation at central synapses. *Trends Neurosci.* **23**, 305–312.

Thomson A.M. (2000b). Molecular frequency filters at central synapses. *Prog. Neurobiol.* **62**, 159–196.

Thurbon D., Lüscher H., Hofstetter T. and Redman S.J. (1998). Passive electrical properties of ventral horn neurons in rat spinal cord slices. *J. Neurophysiol.* **79**, 2485–2502.

Tian P., Devor A., Sakadzic S., Dale A.M. and Boas D.A. (2011). Monte Carlo simulation of the spatial resolution and depth sensitivity of two-dimensional optical imaging of the brain. *J. Biomed. Opt.* **16**, 016006.

Tombola F., Pathak M.M. and Isacoff E.Y. (2006). How does voltage open an ion channel? *Annu. Rev. Cell Dev. Biol.* **22**, 23–52.

Torben-Nielsen B., Vanderlooy S. and Postma E.O. (2008). Non-parametric algorithmic generation of neuronal morphologies. *Neuroinformatics* **6**, 257–277.

Torres Valderrama A., Witteveen J., Navarro M. and Blom J. (2015). Uncertainty propagation in nerve impulses through the action potential mechanism. *J. Math. Neurosc.* **5**, 3.

Toyoizumi T., Kaneko M., Stryker M.P. and Miller K.D. (2014). Modeling the dynamic interaction of Hebbian and homeostatic plasticity. *Neuron* **84**, 497–510.

Toyoizumi T., Miyamoto H., Yazaki-Sugiyama Y., Atapour N., Hensch T.K. and Miller K.D. (2013). A theory of the transition to critical period plasticity: inhibition selectively suppresses spontaneous activity. *Neuron* **80**, 51–63.

Traub R.D., Bibbig A., LeBeau F.E.N., Buhl E.H. and Whittington M. (2004). Cellular mechanisms of neuronal population oscillations in the hippocampus in vitro. *Annu. Rev. Neurosci.* **27**, 247–278.

Traub R.D., Contreras D., Cunningham M.O., Murray H., LeBeau F.E.N., Roopun A., Bibbig A., Wilent W.B., Higley M.J. and Whittington M. (2005). Single-column thalamocortical network model exhibiting gamma oscillations, sleep spindles, and epileptogenic bursts. *J. Neurophysiol.* **93**, 2194–2232.

Traub R.D., Jefferys J.G., Miles R., Whittington M.A. and Tóth K. (1994). A branching dendritic model of a rodent CA3 pyramidal neurone. *J. Physiol.* **481**, 79–95.

Traub R.D., Jefferys J.G.R. and Whittington M.A. (1999). *Fast Oscillations in Cortical Circuits* (MIT Press, Cambridge, MA).

Traub R.D. and Llinás R. (1977). The spatial distribution of ionic conductances in normal and axotomized motor neurons. *Neuroscience* **2**, 829–850.

Traub R.D., Miles R. and Buzsáki G. (1992). Computer simulation of carbachol-driven rhythmic population oscillations in the CA3 region of the in vitro rat hippocampus. *J. Physiol.* **451**, 653–672.

Traub R.D., Wong R.K., Miles R. and Michelson H. (1991). A model of a CA3 hippocampal pyramidal neuron incorporating voltage-clamp data on intrinsic conductances. *J. Neurophysiol.* **66**, 635–650.

Treves A. (1990). Threshold-linear formal neurons in auto-associative networks. *J. Phys. A Math. Gen.* **23**, 2631–2650.

Triplett J.W., Pfeiffenberger C., Yamada J., Stafford B.K., Sweeney N.T., Litke A.M., Sher A., Koulakov A.A. and Feldheim D.A. (2011). Competition is a driving force in topographic mapping. *Proc. Natl. Acad. Sci. USA* **108**, 19060–19065.

Trommershäuser J., Schneggenburger R., Zippelius A. and Neher E. (2003). Heterogeneous presynaptic release probabilities: functional relevance for short-term plasticity. *Biophys. J.* **84**, 1563–1579.

Tsien R.W. and Noble D. (1969). A transition state theory approach to the kinetics of conductance changes in excitable membranes. *J. Membrane Biol.* **1**, 248–273.

Tsien R.Y. (2013). Very long-term memories may be stored in the pattern of holes in the perineuronal net. *Proc. Natl. Acad. Sci. USA* **110**, 12456–12461.

Tsigankov D. and Koulakov A.A. (2010). Sperry versus Hebb: Topographic mapping in Isl2/EphA3 mutant mice. *BMC Neurosci.* **11**, 1–15.

Tsigankov D.N. and Koulakov A.A. (2006). A unifying model for activity-dependent and activity-independent mechanisms predicts complete structure of topographic maps in ephrin-A deficient mice. *J. Comput. Neurosci.* **1**, 101–114.

Tsodyks M.V. and Feigel'man M.V. (1988). The enhanced storage capacity in neural networks with low activity level. *Europhys. Lett.* **6**, 101–105.

Tsodyks M.V. and Markram H. (1997). The neural code between neocortical pyramidal neurons depends on neurotransmitter release probability. *Proc. Natl. Acad. Sci. USA* **94**, 719–723.

Tsodyks M.V., Pawelzik K. and Markram H. (1998). Neural networks with dynamic synapses. *Neural Comput.* **10**, 821–835.

Tuckwell H.C. (1988). *Introduction to Theoretical Neurobiology: Volume 2, Nonlinear and Stochastic Theories*, volume 8 of *Cambridge Studies in Mathematical Biology* (Cambridge University Press, Cambridge).

Turing A.M. (1952). The chemical basis of morphogenesis. *Philos. Trans. R. Soc. Lond. B Biol. Sci.* **237**, 37–72.

Turner R.W., Asmara H., Engbers J.D.T., Miclat J., Rizwan A.P., Sahu G. and Zamponi G.W. (2016). Assessing the role of IKCa channels in generating the sAHP of CA1 hippocampal pyramidal cells. *Channels* **10**, 313–319.

Turney S.G. and Lichtman J.W. (2012). Reversing the outcome of synapse elimination at developing neuromuscular junctions in vivo: evidence for synaptic competition and its mechanism. *PLoS Biol.* **10**, e1001352.

Turrigiano G. (2011). Too many cooks? Intrinsic and synaptic homeostatic mechanisms in cortical circuit refinement. *Annu. Rev. Neurosci.* **34**, 89–103.

Turrigiano G.G. and Nelson S.B. (2004). Homeostatic plasticity in the developing nervous system. *Nat. Rev. Neurosci.* **5**, 97–107.

Tveito A., Jæger K.H., Lines G.T., Paszkowski Ł., Sundnes J., Edwards A.G., Mäki-Marttunen T., Halnes G. and Einevoll G.T. (2017). An evaluation of the accuracy of classical models for computing the membrane potential and extracellular potential for neurons. *Front. Comput. Neurosci.* **11**, 27.

Tyrrell L.R.T. and Willshaw D.J. (1992). Cerebellar cortex: its simulation and the relevance of Marr's theory. *Philos. Trans. R. Soc. Lond. B Biol. Sci.* **336**, 239–257.

Uemura E., Carriquiry A., Kliemann W. and Goodwin J. (1995). Mathematical modeling of dendritic growth in vitro. *Brain Res.* **671**, 187–194.

Valverde S., Vandecasteele M., Piette C., Derousseaux W., Gangarossa G., Arbelaiz A.A., Touboul J., Degos B. and Venance L. (2020). Deep brain stimulation-guided optogenetic rescue of parkinsonian symptoms. *Nat. Commun.* **11**, 1–17.

Van Elburg R.A.J. and van Ooyen A. (2009). Generalization of the event-based Carnevale-Hines integration scheme for integrate-and-fire models. *Neural Comput.* **21**, 1913–1930.

Van Geit W., De Schutter E. and Achard P. (2008). Automated neuron model optimization techniques: a review. *Biol. Cybern.* **99**, 241–251.

Van Geit W., Gevaert M., Chindemi G., Rössert C., Courcol J.D., Muller E.B., Schörmann F., Segev I. and Markram H. (2016). BluePyOpt: Leveraging open source software and cloud infrastructure to optimise model parameters in neuroscience. *Front. Neuroinform.* **10**, 17.

Van Ooyen A. (2001). Competition in the development of nerve connections: a review of models. *Network Comp. Neural Syst.* **12**, R1–R47.

Van Ooyen A., ed. (2003). *Modeling Neural Development* (MIT Press, Cambridge, MA).

Van Ooyen A., Graham B.P. and Ramakers G. (2001). Competition for tubulin between growing neurites during development. *Neurocomputing* **38–40**, 73–78.

Van Ooyen A. and Willshaw D.J. (1999a). Competition for neurotrophic factor in the development of nerve connections. *Proc. R. Soc. Lond. B Biol. Sci.* **266**, 883–892.

Van Ooyen A. and Willshaw D.J. (1999b). Poly- and mononeural innervation in a model for the development of neuromuscular connections. *J. Theor. Biol.* **196**, 495–511.

Van Ooyen A. and Willshaw D.J. (2000). Development of nerve connections under the control of neurotrophic factors: parallels with consumer-resource models in population biology. *J. Theor. Biol.* **206**, 195–210.

Van Pelt J. (1997). Effect of pruning on dendritic tree topology. *J. Theor. Biol.* **186**, 17–32.

Van Pelt J., Dityatev A.E. and Uylings H.B.M. (1997). Natural variability in the number of dendritic segments: model-based inferences about branching during neurite outgrowth. *J. Comp. Neurol.* **387**, 325–340.

Van Pelt J., Graham B.P. and Uylings H.B.M. (2003). Formation of dendritic branching patterns. In *Modeling Neural Development*, ed. A. van Ooyen (MIT Press, Cambridge, MA), chapter 4, pp. 75–94.

Van Pelt J. and Uylings H.B.M. (1999). Natural variability in the geometry of dendritic branching patterns. In *Modeling in the Neurosciences: From Ionic Channels to Neural Networks*, ed. R.R. Poznanski (CRC Press, Boca Raton, FL), chapter 4, pp. 79–108.

Van Pelt J. and Uylings H.B.M. (2002). Branching rates and growth functions in the outgrowth of dendritic branching patterns. *Network Comp. Neural Syst.* **13**, 261–281.

Van Pelt J., van Ooyen A. and Uylings H.B.M. (2001). Modeling dendritic geometry and the development of nerve connections. In *Computational Neuroscience: Realistic Modeling for Experimentalists*, ed. E. De Schutter (CRC Press, Boca Raton, FL), chapter 7, pp. 179–208.

Van Pelt J. and Verwer R.W.H. (1983). The exact probabilities of branching patterns under segmental and terminal growth hypotheses. *Bull. Math. Biol.* **45**, 269–285.

Van Rossum M.C.W. (2001). The transient precision of integrate and fire neurons: effect of background activity and noise. *J. Comput. Neurosci.* **10**, 303–311.

Van Rossum M.C.W., Bi G.Q. and Turrigiano G.G. (2000). Stable Hebbian learning from spike timing-dependent plasticity. *J. Neurosci.* **20**, 8812–8821.

Van Rossum M.C.W., Turrigiano G.G. and Nelson S.B. (2002). Fast propagation of firing rates through layered networks of noisy neurons. *J. Neurosci.* **22**, 1956–1966.

Van Veen M. and van Pelt J. (1992). A model for outgrowth of branching neurites. *J. Theor. Biol.* **159**, 1–23.

Van Veen M. and van Pelt J. (1994). Neuritic growth rate described by modeling microtubule dynamics. *Bull. Math. Biol.* **56**, 249–273.

Van Vreeswijk C. and Sompolinsky H. (1996). Chaos in neuronal networks with balanced excitatory and inhibitory activity. *Science* **274**, 1724–1726.

Vandenberg C.A. and Bezanilla F. (1991). A sodium channel gating model based on single channel, macroscopic ionic, and gating currents in the squid giant axon. *Biophys. J.* **60**, 1511–1533.

Vanier M. and Bower J. (1999). A comparative survey of automated parameter-search methods for compartmental neural models. *J. Comput. Neurosci.* **7**, 149–171.

Vasudeva K. and Bhalla U.S. (2004). Adaptive stochastic-deterministic chemical kinetic simulations. *Bioinformatics* **20**, 78–84.

Vere-Jones D. (1966). Simple stochastic models for the release of quanta of transmitter from a nerve terminal. *Aust. J. Stat.* **8**, 53–63.

Verhulst P.F. (1845). Recherches mathématiques sur la loi d'accroissement de la population. *Nouv. mém. de l'Academie Royale des Sci. et Belles-Lettres de Bruxelles* **18**, 1–41.

Verma P., Kienle A., Flockerzi D. and Ramkrishna D. (2020). Computational analysis of a 9D model for a small DRG neuron. *J. Comput. Neurosci.* **48**, 429–444.

Vogels T.P., Sprekeler H., Zenke F., Clopath C. and Gerstner W. (2011). Inhibitory plasticity balances excitation and inhibition in sensory pathways and memory networks. *Science* **334**, 1569–1573.

Volterra V. (1926). Fluctuations in the abundance of a species considered mathematically. *Nature* **118**, 558–560.

von der Malsburg C. and Willshaw D.J. (1976). A mechanism for producing continuous neural mappings: ocularity dominance stripes and ordered retino-tectal projections. *Exp. Brain Res.* **Suppl. 1**, 463–469.

von der Malsburg C. and Willshaw D.J. (1977). How to label nerve cells so that they can interconnect in an ordered fashion. *Proc. Natl. Acad. Sci. USA* **74**, 5176–5178.

Vorwerk J., Cho J.H., Rampp S., Hamer H., Knösche T.R. and Wolters C.H. (2014). A guideline for head volume conductor modeling in EEG and MEG. *NeuroImage* **100**, 590–607.

Wadiche J.I. and Jahr C.E. (2001). Multivesicular release at climbing fiber–Purkinje cell synapses. *Neuron* **32**, 301–313.

Wagner J. and Keizer J. (1994). Effects of rapid buffers on Ca^{2+} diffusion and Ca^{2+} oscillations. *Biophys. J.* **67**, 447–456.

Walmsley B., Alvarez F.J. and Fyffe R.E.W. (1998). Diversity of structure and function at mammalian central synapses. *Trends Neurosci.* **21**, 81–88.

Walmsley B., Graham B. and Nicol M. (1995). Serial E-M and simulation study of presynaptic inhibition along a group Ia collateral in the spinal cord. *J. Neurophysiol.* **74**, 616–623.

Walsh M.K. and Lichtman J.W. (2003). In vivo time-lapse imaging of synaptic takeover associated with naturally occurring synapse elimination. *Neuron* **37**, 67–73.

Wang J., Chen S., Nolan M.F. and Siegelbaum S.A. (2002). Activity-dependent regulation of HCN pacemaker channels by cyclic AMP: signaling through dynamic allosteric coupling. *Neuron* **36**, 451–461.

Wang X.J., Tegnér J., Constantinidis C. and Goldman-Rakic P.S. (2004). Division of labor among distinct subtypes of inhibitory neurons in a cortical microcircuit of working memory. *Proc. Natl. Acad. Sci. USA* **101**, 1368–1373.

Watts D.J. and Strogatz S.H. (1998). Collective dynamics of 'small-world' networks. *Nature* **393**, 440–442.

Weerasinghe G., Duchet B., Bick C. and Bogacz R. (2021). Optimal closed-loop deep brain stimulation using multiple independently controlled contacts. *PLoS Comput. Biol.* **17**, e1009281.

Wei A.D., Gutman G.A., Aldrich R., Chandy G.K., Grissmer S. and Wulff H. (2005). International Union of Pharmacology. LII. Nomenclature and molecular relationships of calcium-activated potassium channels. *Pharmacol. Rev.* **57**, 463–472.

Weis S., Schneggenburger R. and Neher E. (1999). Properties of a model of Ca^{++}-dependent vesicle pool dynamics and short term synaptic depression. *Biophys. J.* **77**, 2418–2429.

Weiss J.N. (1997). The Hill equation revisited: uses and misuses. *FASEB J.* **11**, 835–841.

Weiss P. (1937a). Further experimental investigations on the phenomenon of homologous response in transplanted amphibian limbs. I. Functional observations. *J. Comp. Neurol.* **66**, 181–209.

Weiss P. (1937b). Further experimental investigations on the phenomenon of homologous response in transplanted amphibian limbs. II. Nerve regeneration and the innervation of transplanted limbs. *J. Comp. Neurol.* **66**, 481–536.

Weiss P. (1939). *Principles of Development* (Holt, New York, NY).

Weisstein E.W. (2017). Necklace. From MathWorld – A Wolfram Web Resource. (Online at https://Mathworld.wolfram.com. Retrieved 30th August 2017).

Whitelaw V.A. and Cowan J.D. (1981). Specificity and plasticity of retinotectal connections: a computational model. *J. Neurosci.* **1**, 1369–1387.

Whittington M.A., Traub R.D., Kopell N., Ermentrout B. and Buhl E.H. (2000). Inhibition-based rhythms: experimental and mathematical observations on network dynamics. *Int. J. Psychophysiol.* **38**, 315–336.

Wigmore M.A. and Lacey M.G. (2000). A Kv3-like persistent, outwardly rectifying, Cs^+-permeable, K^+ current in rat subthalamic nucleus neurones. *J. Physiol.* **527**, 493–506.

Williams J.C., Xu J., Lu Z., Klimas A., Chen X., Ambrosi C.M., Cohen I.S. and Entcheva E. (2013). Computational optogenetics: empirically-derived voltage- and light-sensitive channelrhodopsin-2 model. *PLoS Comput. Biol.* **9**, e1003220.

Willshaw D. (1971). Models of distributed associative memory. Ph.D. thesis, University of Edinburgh, Edinburgh.

Willshaw D.J. (1981). The establishment and the subsequent elimination of polyneural innervation of developing muscle: theoretical considerations. *Proc. R. Soc. Lond. B Biol. Sci.* **212**, 233–252.

Willshaw D.J. (2006). Analysis of mouse EphA knockins and knockouts suggests that retinal axons programme target cells to form ordered retinotopic maps. *Development* **133**, 2705–2717.

Willshaw D.J., Buneman O.P. and Longuet-Higgins H.C. (1969). Non-holographic associative memory. *Nature* **222**, 960–962.

Willshaw D.J., Dayan P. and Morris R.G.M. (2015). Memory, modelling and Marr: a commentary on Marr (1971) 'Simple memory: a theory of archicortex'. *Philos. Trans. R. Soc. Lond. B Biol. Sci.* **370**, 20140383.

Willshaw D.J. and Gale N.M. (2022). Reanalysis of double EphA3 knockin maps in mouse shows the importance of chemoaffinity in topographic map formation. *bioRxiv* 2022.03.29.486226.

Willshaw D.J., Sterratt D.C. and Teriakidis A. (2014). Analysis of local and global topographic order in mouse retinocollicular maps. *J. Neurosci.* **34**, 1791–1805.

Willshaw D.J. and von der Malsburg C. (1976). How patterned neural connections can be set up by self-organization. *Proc. R. Soc. Lond. B Biol. Sci.* **194**, 431–445.

Willshaw D.J. and von der Malsburg C. (1979). A marker induction mechanism for the establishment of ordered neural mappings: its application to the retinotectal problem. *Philos. Trans. R. Soc. Lond. B Biol. Sci.* **287**, 203–234.

Wilson C.J. and Park M.R. (1989). Capacitance compensation and bridge balance adjustment in intracellular recording from dendritic neurons. *J. Neurosci. Methods* **27**, 51–75.

Wilson E., Vant J., Layton J., Boyd R., Lee H., Turilli M., Hernández B., Wilkinson S., Jha S., Gupta C., Sarkar D. and Singharoy A. (2021). Large-scale molecular dynamics simulations of cellular compartments. In *Structure and Function of Membrane Proteins*, eds. I. Schmidt-Krey and J.C. Gumbart (Humana Press, New York, NY), volume 2302 of *Methods in Molecular Biology*, pp. 335–356.

Wilson H.R. (1999). *Spikes, Decisions and Actions: The Dynamical Foundations of Neuroscience* (Oxford University Press, New York, NY).

Wilson H.R. and Cowan J.D. (1972). Excitatory and inhibitory interactions in localized populations of model neurons. *Biophys. J.* **12**, 1–24.

Wilson H.R. and Cowan J.D. (1973). A mathematical theory of the functional dynamics of cortical and thalamic nervous tissue. *Kybernetik* **13**, 55–80.

Windels F., Bruet N., Poupard N., Urbain N., Chouvet G., Feuerstein C. and Savasta M. (2000). Effects of high frequency stimulation of subthalamic nucleus on extracellular glutamate and GABA in substantia nigra and globus pallidus in the normal rat. *Eur. J. Neurosci.* **12**, 4141–4146.

Winslow J.L., Jou S.F., Wang S. and Wojtovicz J.M. (1999). Signals in stochastically generated neurons. *J. Comput. Neurosci.* **6**, 5–26.

Wisedchaisri G., Tonggu L., McCord E., Gamal El-Din T.M., Wang L., Zheng N. and Catterall W.A. (2019). Resting-state structure and gating mechanism of a voltage-gated sodium channel. *Cell* **178**, 993–1003.

Wolpert L. (1969). Positional information and the spatial pattern of cellular differentiation. *J. Theor. Biol.* **25**, 1–47.

Wolpert L., Beddington R., Jessell T., Lawrence P., Meyerowitz E. and Smith J. (2002). *Principles of Development* (Oxford University Press, Oxford), second edition.

Wong A.Y.C., Graham B. P. Billups B. and Forsythe I.D. (2003). Distinguishing between presynaptic and postsynaptic mechanisms of short term depression during action potential trains. *J. Neurosci.* **23**, 4868–4877.

Woolf T.B., Shepherd G.M. and Greer C.A. (1991). Serial reconstruction of granule cell spines in the mammalian olfactory bulb. *Synapse* **7**, 181–192.

Worden M.K., Bykhovskaia M. and Hackett J.T. (1997). Facilitation at the lobster neuromuscular junction: a stimulus-dependent mobilization model. *J. Neurophysiol.* **78**, 417–428.

Xing D., Yeh C.I. and Shapley R.M. (2009). Spatial spread of the local field potential and its laminar variation in visual cortex. *J. Neurosci.* **29**, 11540–11549.

Yamada W.M., Koch C. and Adams P.R. (1998). Multiple channels and calcium dynamics. In *Methods in Neuronal Modeling: From Ions to Networks*, eds. C. Koch and I. Segev (MIT Press, Cambridge, MA), chapter 4, pp.137–170. Second edition.

Yamada W.M. and Zucker R.S. (1992). Time course of transmitter release calculated from simulations of a calcium diffusion model. *Biophys. J.* **61**, 671–682.

Yodzis P. (1989). *Introduction to Theoretical Ecology* (Harper and Row, New York, NY).

Yoon M. (1971). Reorganization of retinotectal projection following surgical operations on the optic tectum in goldfish. *Exp. Neurol.* **33**, 395–411.

Yoon M.G. (1980). Retention of the topographic addresses by reciprocally translated tectal re-implant in adult goldfish. *J. Physiol.* **308**, 197–215.

Yousaf M., Kriener B., Wyller J. and Einevoll G.T. (2013). Generation and annihilation of localized persistent-activity states in a two-population neural-field model. *Neural Netw.* **46**, 75–90.

Yu F.H., Yarov-Yarovoy V., Gutman G.A. and Catterall W.A. (2005). Overview of molecular relationships in the voltage-gated ion channel superfamily. *Pharmacol. Rev.* **57**, 387–395.

Yu Y., Wang X., Wang Q. and Wang Q. (2020). A review of computational modeling and deep brain stimulation: applications to Parkinson's disease. *Appl. Math. Mech.* **41**, 1747–1768.

Zador A., Agmon-Snir H. and Segev I. (1995). The morphoelectotonic transform: a graphical approach to dendritic function. *J. Neurosci.* **15**, 1669–1682.

Zador A. and Koch C. (1994). Linearized models of calcium dynamics: formal equivalence to the cable equation. *J. Neurosci.* **14**, 4705–4715.

Zador A., Koch C. and Brown T.H. (1990). Biophysical model of a Hebbian synapse. *Proc. Natl. Acad. Sci. USA* **87**, 6718–6722.

Zeng F.G., Rebscher S., Harrison W., Sun X. and Feng H. (2008). Cochlear implants: System design, integration, and evaluation. *IEEE Rev. Biomed. Eng.* **1**, 115–142.

Zenke F., Gerstner W. and Ganguli S. (2017). The temporal paradox of Hebbian learning and homeostatic plasticity. *Curr. Opin. Neurobiol.* **43**, 166–176.

Zhang L.I., Tao H.W., Holt C.E., Harris W.A. and Poo M.M. (1998). A critical window for cooperation and competition among developing retinotectal synapses. *Nature* **395**, 37–44.

Zipser D. and Andersen R.A. (1988). A back-propagation programmed network that simulates response properties of a subset of posterior parietal neurons. *Nature* **331**, 679–684.

Zucker R.S. (1974). Characteristics of crayfish neuromuscular facilitation and their calcium dependence. *J. Physiol.* **241**, 91–110.

Zucker R.S. (1989). Short-term synaptic plasticity. *Annu. Rev. Neurosci.* **12**, 13–31.

Zucker R.S. (1999). Calcium- and activity-dependent synaptic plasticity. *Curr. Opin. Neurobiol.* **9**, 305–313.

Zucker R.S. and Fogelson A.L. (1986). Relationship between transmitter release and presynaptic calcium influx when calcium enters through discrete channels. *Proc. Natl. Acad. Sci. USA* **83**, 3032–3036.

Zucker R.S. and Regehr W.G. (2002). Short-term synaptic plasticity. *Annu. Rev. Physiol.* **64**, 355–405.

Index

A bold font denotes a definition of the term. 'b' refers to a box, 'f' refers to a figure, 's' refers to a sidebox and 't' refers to a table.

Printed in the United States
by Baker & Taylor Publisher Services